Applied Probability and Statistics (Continued)

BARTHOLOMEW · Stochastic Models for Social Processes, *Second Edition*
BENNETT and FRANKLIN · Statistical Analysis in Chemistry and the Chemical Industry
BHAT · Elements of Applied Stochastic Processes
BOX and DRAPER · Evolutionary Operation: A Statistical Method for Process Improvement
BROWNLEE · Statistical Theory and Methodology in Science and Engineering, *Second Edition*
BURY · Statistical Models in Applied Science
CHERNOFF and MOSES · Elementary Decision Theory
CHOW · Analysis and Control of Dynamic Economic Systems
CLELLAND, deCANI, BROWN, BURSK, and MURRAY · Basic Statistics with Business Applications, *Second Edition*
COCHRAN · Sampling Techniques, *Second Edition*
COCHRAN and COX · Experimental Designs, *Second Edition*
COX · Planning of Experiments
COX and MILLER · The Theory of Stochastic Processes, *Second Edition*
DANIEL and WOOD · Fitting Equations to Data
DAVID · Order Statistics
DEMING · Sample Design in Business Research
DODGE and ROMIG · Sampling Inspection Tables, *Second Edition*
DRAPER and SMITH · Applied Regression Analysis
DUNN and CLARK · Applied Statistics: Analysis of Variance and Regression
ELANDT-JOHNSON · Probability Models and Statistical Methods in Genetics
FLEISS · Statistical Methods for Rates and Proportions
GOLDBERGER · Econometric Theory
GROSS and CLARK · Survival Distributions
GROSS and HARRIS · Fundamentals of Queueing Theory
GUTTMAN, WILKS and HUNTER · Introductory Engineering Statistics, *Second Edition*
HAHN and SHAPIRO · Statistical Models in Engineering
HALD · Statistical Tables and Formulas
HALD · Statistical Theory with Engineering Applications
HARTIGAN · Clustering Algorithms
HOEL · Elementary Statistics, *Third Edition*
HOLLANDER and WOLFE · Nonparametric Statistical Methods
HUANG · Regression and Econometric Methods
JAGERS · Branching Processes with Biological Applications
JOHNSON and KOTZ · Distributions in Statistics
Discrete Distributions
Continuous Univariate Distributions-1
Continuous Univariate Distributions-2
Continuous Multivariate Distributions
JOHNSON and LEONE · Statistics and Experimental Design: In Engineering and the Physical Sciences, Volumes I and II
LANCASTER · The Chi Squared Distribution

continued on back

Nonparametic Methods in Multivariate Analysis

A WILEY PUBLICATION IN MATHEMATICAL STATISTICS

Nonparametric Methods in Multivariate Analysis

MADAN LAL PURI

Professor of Mathematics
Indiana University, Bloomington

PRANAB KUMAR SEN

Professor of Biostatistics and Statistics
University of North Carolina, Chapel Hill

John Wiley & Sons, Inc.

New York · London · Sydney · Toronto

Library of Congress Catalogue Card Number: 79-129052

ISBN 0-471-70240-4

Printed in the United States of America

10 9 8 7 6 5 4 3

To Uma and Gauri

Preface

The purpose of this book is to present some aspects of the theory of nonparametric multivariate statistical analysis in a systematic and logically integrated form. It incorporates what we believe to be the most important developments that have taken place in the subject in recent years.

The book is written for specialists, teachers, and advanced graduate students with a background knowledge of parametric multivariate statistical theory. We also assume that the reader knows probability theory at the graduate level.

The book is divided into ten chapters. Chapter 1 provides a brief outline of the material covered in the book. Chapter 2 examines the basic concepts and the tools of probability theory which are used in the statistical theory developed in subsequent chapters. Chapter 3 surveys the univariate nonparametric theory which has relevance to the multivariate theory developed in later chapters. A large number of problems are given at the end of the chapter, and a student who masters this chapter is well equipped to do research in nonparametric statistics. Chapter 4 discusses the multivariate one-sample tests of location, and Chapter 5 is devoted to the study of multivariate several-sample tests of location and scale. Chapter 6 deals with the problems of point as well as interval estimation in some linear models. Chapter 7 provides nonparametric procedures in multifactor designs. Chapter 8 considers the problem of testing independence of different subsets of a stochastic vector. Chapter 9 deals with the problem of testing the identity of dispersion matrices, and Chapter 10 (appendix) with the derivation of some of the results used in the previous chapters.

Throughout the book the emphasis is mostly on Hoeffding's U-statistics and suitable families of rank-order statistics. For lack of space, we have not been able to incorporate the recent works of Hájek (1968) and Pyke and Shorack (1968). The work of these authors and some allied problems in nonparametric multivariate analysis, for example, problems relating to paired comparison models and general linear hypotheses, which we would have liked to have included in this book, will be published in a separate volume.

We have presented substantial parts of the book in courses at the University of North Carolina, Chapel Hill, Courant Institute of Mathematical Sciences, New York University, and at Indiana University, Bloomington. We have also presented different portions of the book at professional meetings and at colloquiums at various universities in the United States and abroad. During these years, we benefited greatly from the comments of our students and colleagues, and we regret that we cannot thank them individually here.

At different stages of the writing, we benefited immensely from the valuable suggestions offered by Professor Wassily Hoeffding. Professor Hoeffding's critical reading of Chapter 3 resulted in many improvements. Our sincere gratitude goes to him for all the time which he gave us. Thanks are also due to Professors H. A. David, Dana Quade, J. S. Rao, and Malay Ghosh for pointing out several errors in the original drafts of the manuscript. Mr. Kalyan Dutta's help in the preparation of the bibliography and Mr. Sujit Basu's help in the preparation of the index are very much appreciated. Mrs. Delores Gold and Mrs. Barbara Fortson deserve our deepest gratitude for the typing of the manuscript.

The writing of the book was supported in part by the National Institute of Health, Grant GM-12868, the Army Research Office, Durham, Grant DA-ARO-D-31-124-G746, the National Science Foundation, Grant GP-12462, and the Aerospace Research Laboratories, Office of Aerospace Research, United States Air Force, Contract number F33615-70-C-1124. The support from these organizations is very much appreciated.

Thanks are due to Professors B. G. Greenberg and J. E. Grizzle for their keen interest and their valuable assistance from the initiation to the completion of this project.

M. L. Puri
P. K. Sen

Bloomington, Indiana
Chapel Hill, North Carolina
September 1970

Contents

CHAPTER

1. A BRIEF OUTLINE OF THE MATERIAL COVERED IN THE BOOK

1.1 Introduction . . . 1
1.2 The organization of the book . . . 2
1.3 A brief review of some allied problems in nonparametric multivariate analysis . . . 4

2. PRELIMINARIES

2.1 Introduction . . . 8
2.2 Probability spaces . . . 8
2.3 Convergences . . . 13
2.4 Sums of independent random variables . . . 15
2.5 Laws of large numbers . . . 17
2.6 Convergence of distribution functions . . . 18
2.7 The central limit theorem . . . 22
2.8 Distribution of quadratic forms . . . 25
2.9 Some results on multivariate normal distribution . . . 27
2.10 Martingales . . . 29
2.11 Weak convergence of some stochastic processes . . . 31
2.12 Some results in measure theory . . . 43

3. A SURVEY OF NONPARAMETRIC INFERENCE

3.1 Introduction . . . 49
3.2 Distribution theory of regular functions of distribution functions . 51
3.3 Theory of permutation tests . . . 66
3.4 Permutational limit theorems . . . 70

3.5 Asymptotic power of nonparametric tests 86
3.6 Distribution theory of rank order statistics 92
3.7 Locally most powerful rank tests 107
3.8 The asymptotic efficiency of tests 112
Exercises 124

4. RANK TESTS FOR THE MULTIVARIATE SINGLE-SAMPLE LOCATION PROBLEMS

4.1 Introduction 144
4.2 A bivariate sign test for location 144
4.3 The hypothesis of sign-invariance and the basic permutation principle 148
4.4 A class of (linear) rank order statistics and the allied tests . . 150
4.5 Tests based on U-statistics. 167
4.6 Efficiency considerations 173
Exercises 177

5. MULTIVARIATE MULTISAMPLE RANK TESTS FOR LOCATION AND SCALE

5.1 Introduction 181
5.2 Formulation of the multivariate multisample location and scale problems 182
5.3 Basic rank permutation principle 182
5.4 Some permutation rank order tests 184
5.5 Asymptotic multinormality of $T_{Ni}^{(k)}$, $k = 1, \ldots, c$, $i = 1, \ldots, p$ for arbitrary $F_1(\mathbf{x}), \ldots, F_c(\mathbf{x})$ 196
5.6 The limiting distribution of rank order tests for sequences of translation and scale alternatives 203
5.7 Multivariate tests based on the U-statistics 206
5.8 Asymptotic relative efficiency of rank order tests 211
5.9 Rank order tests for the analysis of covariance problem . . 215
Exercises 219

6. ESTIMATORS IN LINEAR MODELS (ONE WAY LAYOUTS) BASED ON RANK TESTS

6.1 Introduction 221
6.2 Point estimation of location of a symmetric distribution . . 222
6.3 Point estimation of location in the multivariate two-sample problem 227
6.4 Asymptotic relative efficiency of the estimators 230
6.5 An alternative approach to the two-sample case 234
6.6 Point estimation of contrasts in one-way layout 236
6.7 Interval estimation in univariate one-sample and two-sample location problems 240
6.8 Interval estimation of contrasts in ANOVA (one-way layout) . . 244
6.9 Interval estimation of contrasts in one-way MANOVA . . . 254
Exercises 260

7. RANK PROCEDURES IN FACTORIAL EXPERIMENTS

7.1 Introduction 266
7.2 The method of n rankings 267
7.3 Ranking after alignment 286
7.4 Asymptotically distribution-free procedures based on robust estimators of the parameters 308
7.5 Rank order tests for ordered alternatives in randomized blocks . 323
7.6 Multiple comparison of treatments 328
7.7 Rank order tests for interactions in factorial experiments . . 331
7.8 Analysis of covariance based on general rank scores . . . 338
Exercises 340

8. RANK TESTS FOR INDEPENDENCE

8.1 Introduction 243
8.2 Formulation of a class of association parameters . . . 345
8.3 Permutationally distribution-free tests for $H_0^{(q)}$ 347
8.4 Asymptotic normality of $\mathbf{T}_n$ for arbitrary $F(\mathbf{x})$ 354
8.5 Testing independence of two sets of variates 358
8.6 Pairwise independence 360
8.7 Another model for dependence between two sets of variables . 361
8.8 Parametric theory 362
8.9 Asymptotic relative efficiency 365
Exercises. 371
8.10 Tests for independence in bivariate populations 373

9. RANK TESTS FOR HOMOGENEITY OF DISPERSION MATRICES

9.1 Introduction 377
9.2 Preliminary notions 378
9.3 Permutation tests for $H_0^{(1)}$ 381
9.4 Asymptotic permutation distribution of L_N 385
9.5 Asymptotic normality of $\mathbf{S}_N^{(k)}$, $k = 1, \ldots, c$ for arbitrary $F_{(1)}, \ldots, F_{(c)}$. 387
9.6 Nonparametric tests for $H_0^{(2)}$ 393
9.7 Tests for $H_0^{(3)}$ 398
Exercises. 399

10. APPENDICES

10.1 Introduction 400
10.2 Appendix 1: the two-sample case 400
10.3 Appendix 2: the one-sample case 405
10.4 Appendix 3: aligned rank order statistics 407
10.5 The proof of Theorem 3.6.6 408

REFERENCES 415

INDEX 433

Nonparametic Methods in Multivariate Analysis

CHAPTER 1

A Brief Outline of the Material Covered in the Book

1.1. INTRODUCTION

The primary purpose of this book is to present a systematic and comprehensive account of some aspects of the statistical analysis of multivariate observations without explicitly assuming forms for the underlying probability distributions. We assume that the reader is familiar with parametric multivariate statistical analysis, where the underlying distributions are mostly assumed to be multivariate normal. For fundamental and detailed accounts of the parametric theory, the reader is referred to Anderson (1958), Rao (1965, Chapter 8), and Roy (1957), among others.

In multivariate statistical analysis, we are usually concerned with a set of stochastic vectors. For example, we may be interested in the scores or grades in different subjects of a set of school children; blood pressure, EKG, body temperature, and heart rates of a set of patients; and so on. The random variables constituting the stochastic vectors are supposed to have some (joint) probability distribution on some appropriate-dimensional Euclidean space. In the parametric case, frequently, we assume that these probability distributions are appropriate-dimensional multivariate normal. However, for well-known reasons, this assumption has some limitations. In practice, the forms of the underlying distributions are very seldom known, and also, in many practical situations, there may be distinct indication that the underlying distributions are non-normal. In such cases, the use of the standard parametric procedures is subject to criticism regarding validity and optimality.

In univariate theory, though there are suitable families of non-normal distributions (for which optimum inference procedures exist), during the past three decades a considerable amount of work has been done on the development of statistical inference procedures which remain valid for broad families of underlying distributions. These are conventionally known as nonparametric or distribution-free procedures. In multivariate theory, the need for such procedures is felt even more strongly. First, unlike the univariate case, there are few well-known multivariate non-normal distributions, and for such distributions, suitable inference procedures are not available. Second, the possibility of departure from normality in the multivariate case becomes manifold because of the increase in the number of variates and their interrelationships. The different variates of a stochastic vector may not be marginally normally distributed, and even if they are so, their joint distribution can still be quite different from a multi-normal one.

We attempt to present here the multivariate analogues of the univariate nonparametric procedures for a class of problems. These procedures are valid for certain broad families of underlying distributions and yet are reasonably efficient for normal distributions too. For lack of space and other reasons, certain related problems cannot be dealt with here. However, these are briefly outlined in section 1.3 for convenience of reading and completeness of coverage. For the univariate nonparametric procedures, we suggest our chapter 3 and a recent book by Hájek and Šidák (1967), which together cover a wide spectrum of the theory.

1.2. THE ORGANIZATION OF THIS BOOK

We expect that the reader of this book is familiar with the theory of probability and statistical inference (at the standard graduate level). Thus, it is taken for granted that the reader is also familiar with advanced calculus and real analysis. For the convenience of the reader, some of the basic tools in probability (and measure) theory, needed frequently in chapters 3–10, are briefly (mostly without proof) considered in chapter 2. Since the derivations of these results are given in many standard textbooks on probability theory (referred to in chapter 2), we believe that the interested reader can always look at these books for proofs as and when necessary.

Chapter 3 provides a basic survey of several important aspects of nonparametric inference, mostly related to univariate models. The various nonparametric tests and estimates are either based on Hoeffding's U-statistics (and their extensions) or on suitable families of rank order statistics. The distribution theory of U-statistics has been studied very thoroughly and elegantly by Hoeffding (1948a), and, since then, extended in various directions by many workers. Our section 3.2 provides a detailed account of the distribution theory of U-statistics and of some allied statistics. The distribution theory of various families of rank order statistics is studied in sections 3.6 and 3.7. Section 3.3 is concerned with the study of the theory of permutation tests for the four basic hypotheses of invariance, and the corresponding asymptotic theory is presented in the next section. Section 3.5 deals with the large sample power of nonparametric (and in particular permutation) tests. In the rest of the chapter, the concepts of Pitman-Noether and of Bahadur efficiencies of tests are discussed. In passing, we may remark that while dealing with the theory of rank order statistics (in section 3.6), we have mainly considered the results of Chernoff and Savage (1958) along with the various extensions and modifications of this work by various authors. We take this opportunity to refer to the highly original recent works of Hájek (1968) and Pyke and Shorack (1968a). We believe that both these papers have great potential in the theory of rank order statistics. Since, to date, not much work (particularly in the multivariate area) has been done in this area, we do not stress these approaches in the present work.

In chapter 4, we consider the theory of rank tests for the multivariate single sample location problem. The hypothesis of sign invariance of section 3.3 is extended to the multivariate case, and permutationally distribution-free sign tests and a general class of rank order tests are constructed. These tests may be regarded as the nonparametric competitors of the classical Hotelling T^2-test used for the normal theory model. The asymptotic power and efficiency of these tests are also studied in detail.

Chapter 5 is devoted to the study of multivariate multisample tests for the location and scale problems. In this context, multivariate extensions of the median procedure of Brown and Mood (1951), the rank sum procedures of Wilcoxon (1945), Mann and Whitney (1947) and Kruskal and Wallis (1952), the Fisher-Yates normal scores test, and a variety of scale tests are considered. These tests are shown to be permutationally (conditionally) distribution-free. The asymptotic power properties and relative efficiencies (with respect to the normal theory likelihood ratio tests) are also presented. Finally, the problem of analysis of covariance in one-way layout models is considered, and some rank order tests by Quade (1967) and the present authors are studied.

In chapter 6, we consider the problem of robust (point as well as interval) estimation in some linear models based on suitable rank statistics. Specifically, the following problems are studied: (i) estimation of the location (vector) of a symmetric (multivariate) distribution, (ii) estimation of the difference between the location (vectors) of two multivariate distributions, and (iii) estimation of contrasts in one-way ANOVA and MANOVA. For this purpose, the rank statistics developed in chapters 4 and 5 are used to derive the estimates. The problem of nonparametric simultaneous tests and confidence bands is also studied.

Chapter 7 is completely devoted to the study of nonparametric techniques in two-factor or more than two-factor designs. First, we consider the method of n-rankings which leads to genuinely distribution-free procedures for testing the hypothesis of no treatment effects in two-factor ANOVA and MANOVA models. Our general procedure contains, as special cases, the well-known rank sum procedure of Friedman (1937) and the median procedure of Brown and Mood (1951). The efficiency results are presented in detail. Since the method of n rankings does not utilize the information contained in the inter-block comparisons, it usually entails some loss in efficiency. This loss can be recovered by the second procedure: ranking after alignment. Here, the nuisance parameters are eliminated first by suitable intrablock transformations and then the overall ranking is made on the aligned observations. This procedure is shown to be conditionally distribution-free and usually more efficient than the first one. Finally, the estimation procedure of chapter 6 is extended to the two-way layout problem, and asymptotically distribution-free procedures based on these robust estimates are considered. Several other problems are also studied, namely, (i) tests for interactions in three factor experiments, (ii) analysis of covariance in two-way layouts, (iii) tests for ordered alternatives in two-way layouts, and (iv) robust estimation in incomplete block designs.

Chapter 8 examines the interrelationships of the different variates of a stochastic $p(\geq 2)$-vector. Two basic problems are considered in this context. First, we consider the problem of testing independence of different subsets of a stochastic vector. Under suitable groups of transformations, permutationally invariant tests are studied. Next, we consider some nonparametric analogues of the measures of association based on product moments. The role of such measures in testing theory is stressed. Asymptotic power properties are also studied.

In chapter 9, we develop a class of rank order tests for the homogeneity of dispersion matrices, with or without assuming the identity of the location vectors. In the former case, some permutationally distribution-free tests are developed and their large sample properties are studied, while in the latter case, only asymptotically distribution-free tests are considered. At the end we briefly outline the problem of testing simultaneously the homogeneity of the location parameters and of the dispersion matrices.

Chapter 10 (appendix) deals with the derivations of some of the results used in the previous chapters.

1.3. A BRIEF REVIEW OF SOME ALLIED PROBLEMS IN NONPARAMETRIC MULTIVARIATE ANALYSIS

In chapters 4 to 9, we consider some of the basic problems in nonparametric multivariate analysis. There are certain other problems of interest which we have not been able to consider in this book for lack of space. Without going into the technical details, we give below a very brief outline of these problems and their appropriate references.

(a) Paired Comparison Models. In a sense, these models are the multi-sample analogues of the one-sample models considered in chapter 4. We illustrate them briefly as follows:

Consider $t(\geq 2)$ treatments in an experiment involving paired comparisons. Suppose that for the pair (i, j) of treatments $(1 \leq i < j \leq t)$, the n_{ij} encounters yield n_{ij} responses on each of $p(\geq 1)$ attributes. The parameters of interest are the set of preference parameters for each attribute. Let $\alpha_i^{(1)}, \ldots, \alpha_i^{(p)}$ be the levels of the p attributes for the treatment $i(1 \leq i \leq t)$, and let

$$\pi_{ij}^{(k)} = P\{\alpha_i^{(k)} \prec \alpha_j^{(k)}\}, \qquad 1 \leq i < j \leq t, \quad 1 \leq k \leq p, \tag{1.3.1}$$

where $\prec$ indicates a preference in the direction of the sign. We want to test the null hypothesis that

$$H_0\colon \pi_{ij}^{(k)} = \tfrac{1}{2}, \quad \text{for all} \quad 1 \leq i < j \leq t \quad \text{and} \quad 1 \leq k \leq p; \tag{1.3.2}$$

i.e., the t treatments do not differ among themselves, in respect to each of the p attributes. For $t = 2$, (1.3.2) reduces to the sign-test problem considered in chapter 4. For $t > 2$, the usual chi-square type tests (based on the multinomial distributions relating to the 2^p possible outcomes of the p preferences for each pair) contain spurious degrees of freedom, and hence are usually less efficient than other tests which specifically test (1.3.2) against possible alternatives with various restrictions on the $\pi_{ij}^{(k)}$ (and thereby eliminate the spurious degrees of freedom). The task of formulating these alternative hypotheses and appropriate tests for these hypotheses constitutes the problem of multivariate paired comparisons. Work in this direction has been done by Sen and David (1968) and Davidson and Bradley (1969), among others.

Often the preferences can be expressed quantitatively. Thus, here the n_{ij} encounters yield the observed differences $\mathbf{Z}_{ij,\alpha} = (Z_{ij,\alpha}^{(1)}, \ldots, Z_{ij,\alpha}^{(p)})'$, $\alpha = 1, \ldots, n_{ij}$, $1 \leq i < j \leq t$. In such a case, one can extend the rank order tests of chapter 4 to provide suitable tests for the hypothesis (1.3.2), which in this case reduces to the hypothesis that all the $\mathbf{Z}_{ij,\alpha}$ are distributed (diagonally) symmetrically about $\mathbf{0}$. Tests for this hypothesis are due to Shane and Puri (1969) and, Puri and Shane (1970).

Paired comparison designs are special cases of general incomplete block designs with blocks of size 2. A natural extension of the class of tests considered above to the case of some incomplete block design is due to Sen (1969d).

(b) Robust Estimation and Testing in a Simple Regression Model. Consider a sequence of stochastic $(p \times N)$ matrices

$$\mathbf{E}_N = (\mathbf{X}_1, \ldots, \mathbf{X}_N); \quad \mathbf{X}_\alpha = (\mathbf{X}_\alpha^{(1)}, \ldots, \mathbf{X}_\alpha^{(p)})', \quad \alpha = 1, \ldots, N, \tag{1.3.3}$$

where the $\mathbf{X}_\alpha$ are independently distributed according to continuous distributions $F_\alpha(\mathbf{x})$, $\mathbf{x} \in R^p$, and

$$F_\alpha(\mathbf{x}) = F(\mathbf{x} - \boldsymbol{\beta}_0 - \boldsymbol{\beta}_1 c_\alpha), \qquad 1 \leq \alpha \leq N, \tag{1.3.4}$$

where $\boldsymbol{\beta}_0 = (\beta_0^{(1)}, \ldots, \beta_0^{(p)})'$ and $\boldsymbol{\beta}_1 = (\beta_1^{(1)}, \ldots, \beta_1^{(p)})'$ are unknown parameters, and $\mathbf{c}_N = (c_1, \ldots, c_N)$ is a (known) vector of constants. The two basic problems are (i) the estimation of $\boldsymbol{\beta}_0$ and $\boldsymbol{\beta}_1$ and (ii) tests for $\boldsymbol{\beta}_0$ and $\boldsymbol{\beta}_1$. The estimation problem, based on suitable (multivariate) rank statistics, is considered in detail in Sen and Puri (1969). These extend the results of chapter 6 to a wider class of problems. The testing problem is a special case of the following more general problem.

(c) General Linear Hypotheses. For the stochastic matrix $\mathbf{E}_N$ in (1.3.3), consider the general model

$$F_\alpha(\mathbf{x}) = F(\mathbf{x} - \boldsymbol{\beta}_0 - \boldsymbol{\beta}\mathbf{c}_\alpha), \qquad 1 \leq \alpha \leq N, \tag{1.3.5}$$

where $\boldsymbol{\beta}$ is a $p \times q$ matrix of unknown parameters and $\mathbf{c}_\alpha = (c_\alpha^{(1)}, \ldots, c_\alpha^{(p)})'$, $1 \leq \alpha \leq N$, are known vectors of regression constants. We are interested in testing the hypotheses

$$H_0^{(1)}: \boldsymbol{\beta} = \mathbf{0}, \qquad H_0^{(2)}: \boldsymbol{\beta}_0 = \mathbf{0}, \quad \text{and} \quad H_0^{(3)}: \boldsymbol{\beta}_0 = \mathbf{0}, \quad \boldsymbol{\beta} = \mathbf{0}. \tag{1.3.6}$$

Tests for $H_0^{(1)}$, based on suitable rank statistics, are due to Puri and Sen (1969a), while those for $H_0^{(2)}$ and $H_0^{(3)}$ are due to Hušková (1970). All these procedures are based on the recent fundamental work of Hájek (1968). The corresponding estimation procedures for $\boldsymbol{\beta}_0$ and $\boldsymbol{\beta}$ follow along the lines of the recent work of Jurečková (1967, 1969).

(d) Regression Problems for Grouped Data. Consider the same model as in (1.3.5), but the case where the $\mathbf{X}_\alpha$, $\alpha = 1, \ldots, N$ are not observable. We are given a set of finite or countable number of class intervals (rectangles in p-dimensional Euclidean space), and have information only as to the different class intervals to which the different X_i belong. Precisely, for each i, we know the class interval to which X_i belongs and also the corresponding $\mathbf{c}_i$, $1 \leq i \leq N$. It is indeed possible to derive a class of rank order tests which are natural extensions of those in (c), and which are usually much more efficient than the classical chi-square type tests based on Pearsonian "goodness of fit" criteria. For details, we refer the reader to Sen (1967a) and Ghosh (1969).

(e) Regression Problems with Stochastic Predictors. In (b) and (c), we have considered the case where the regression constants are nonstochastic. It is also possible to extend the theory to the case of stochastic predictors, so that for the conditional distribution of the primary variate given the others, we have models essentially similar to (1.3.4) or (1.3.5). This requires certain modifications and extensions of the theory considered in (*b*) and (*c*). The details are worked out by Ghosh and Sen (1971).

REFERENCES

Anderson (1958), Brown and Mood (1951), Chernoff and Savage (1958), Davidson and Bradley (1969), Friedman (1937), Ghosh (1969), Ghosh and Sen (1971), Hájek (1968), Hoeffding (1948a), Hušková (1970), Jurečková (1967, 1969), Kruskal and Wallis (1952), Mann and Whitney (1947), Puri and Sen (1969a,b,c), Puri and Shane (1970), Pyke and Shorack (1968a), Quade (1967), Rao, C. R. (1965), Roy (1957), Sen (1967a, 1969d), Sen and David (1968), Sen and Puri (1969), Shane and Puri (1969), Wilcoxon (1945).

CHAPTER 2

Preliminaries

2.1. INTRODUCTION

This chapter is intended to outline in a convenient form the basic concepts and the tools of probability theory which are used in the statistical theory developed in the subsequent chapters. A prerequisite is some familiarity with real variable theory, such as ideas of measure and measurable functions. Essential definitions are given and the theorems are stated with no or very few proofs. They should be taken as a rapid review rather than an exposition. The reader interested in the details of proofs and comprehensive study of the theory of probability should refer to books by Kolmogorov (1933), Doob (1967), Gnedenko and Kolmogorov (1954), Loève (1963), Feller (1965), Krickeberg (1965), Moran (1968), Neveu (1965), Lamperti (1966), and Tucker (1967).

2.2. PROBABILITY SPACES

Let Ω be any nonempty set of points ω, and suppose $\mathcal{A}$ is a Borel field of subsets of Ω. This means that $\mathcal{A}$ is a collection of subsets which contains the empty set $\varnothing$, and is closed under the complements, and under unions of at most countably many of its members. Let P be a non-negative function defined on $\mathcal{A}$ such that (i) $P(A) \geq 0$ for every $A \in \mathcal{A}$, (ii) $P(\Omega) = 1$, and (iii) $P(\bigcup_{n=1}^{\infty} A_n) = \sum_{n=1}^{\infty} P(A_n)$ provided $A_n \in \mathcal{A}$ and $A_n \cap A_m = \varnothing$ for all $n \neq m$. P is called a probability measure and $(\Omega, \mathcal{A}, P)$ is called a probability space. It follows that

$$P(\varnothing) = 0, \quad P\left(\bigcup_{n=1}^{\infty} A_n\right) \leq \sum_{n=1}^{\infty} P(A_n), \quad P(A_1) \leq P(A_2) \quad \text{if} \quad A_1 \subset A_2$$

$$P(\liminf A_n) \leq \liminf P(A_n) \leq \limsup P(A_n) \leq P(\limsup A_n),$$

and, if $\lim A_n$ exists, then $P(\lim A_n) = \lim P(A_n)$.

We shall denote by R the real line $(-\infty, \infty)$, and by R^n the n-dimensional real Euclidean space. R^n is the set of all ordered n-tuples $\mathbf{x} = (x_1, \ldots, x_n)$;

equivalently, R^n is the product space $\prod_{i=1}^{n} R_i$ of n real lines $R_i = (-\infty < x_i < +\infty)$. Now, in R^n, consider a class $\mathcal{F}$ of open sets or closed sets or n-dimensional closed or open rectangles or the field generated by the finite unions of the sets of the type $a_i \leq x_i < b_i$, where $-\infty \leq a_i < b_i \leq \infty$. The Borel field (or the minimal σ-field) containing $\mathcal{F}$ is the same in each case; and we shall denote it by $\mathcal{A}_n$.

Random Variable (r.v.)

A real-valued function $X(\omega)$ defined on Ω is called a r.v. if, for every Borel set A in the real line R, the set $\{\omega\colon X(\omega) \in A\}$ is in $\mathcal{A}$. The function X may be regarded as a mapping of Ω into R, and this may be written as $\Omega \xrightarrow{X} R$. The set $\{\omega\colon X(\omega) \in A\}$ is called the inverse image of A, and is denoted by $X^{-1}(A)$. With this notation X is a r.v. if and only if $X^{-1}[(-\infty, x)] \in \mathcal{A}$ for every x.

Distribution Function (d.f.)

If X is a r.v., its d.f. is defined by $F(x) = P[X < x]$ for all $x \in R$. It follows that $F(x)$ is nondecreasing and continuous from the left on R, with $F(-\infty) = 0$, $F(+\infty) = 1$. Conversely, every function F with the above properties is the d.f. of a r.v. on some probability space. Note also that $F(x)$, being a function of bounded variation, has at most a countable number of discontinuities. Furthermore, every d.f. $F(x)$ determines two functions $F_c(x)$ and $F_d(x)$ such that $F_c(x)$ is continuous, $F_d(x)$ is a step function, and $F(x) = F_c(x) + F_d(x)$. This is the well-known decomposition theorem (Loève, 1963). In fact $F_c(x)$ can further be decomposed as $F_c(x) = F_{c_1}(x) + F_{c_2}(x)$ where $F_{c_1}(x)$ is absolutely continuous and $F_{c_2}(x)$ has derivative equal to zero almost everywhere. The d.f. $F(x)$ is said to be absolutely continuous, if there exists a Borel measurable function $f(x)$ defined over R such that $F(x) = \int_{-\infty}^{x} f(t)\, dt$ for all real x. The function f is called the density of F. If $f(x)$ is continuous at x, $F'(x) = f(x)$.

Random Vectors

The collection of n single-valued real functions $X_1(\omega), \ldots, X_n(\omega)$ mapping Ω into R^n is called an n-dimensional random vector or an n-dimensional r.v. if the set $\{\omega\colon X_1(\omega) < x_1, \ldots, X_n(\omega) < x_n\} \in \mathcal{A}$ for all $\mathbf{x} = (x_1, \ldots, x_n)$. As in the case of one-dimensional r.v., we define the distribution function of the random vector $\mathbf{X} = (X_1, \ldots, X_n)$ as $F(x_1, \ldots, x_n) = P\left[\bigcap_{i=1}^{n} [X_i < x_i]\right]$ where $-\infty < x_i < \infty$, $i = 1, \ldots, n$. F is nondecreasing, continuous at least from the left in each of its arguments, $\lim_{x_i \to -\infty} F(x_1, \ldots, x_n) = 0$ for each $i = 1, \ldots, n$, and $\lim (\min x_i \to \infty) F(x_1, \ldots, x_n) = 1$. Conversely, every d.f. $F(\mathbf{x})$ with the foregoing properties is the d.f. of a random vector on some probability space. Note also that $F(\mathbf{x})$ is continuous everywhere except over the union of a countable set of hyperplanes of the form $x_i = c$, $1 \leq i \leq n$. The d.f. of X_i is $F_i(x) = F(\infty, \ldots, \infty, x, \infty, \ldots, \infty)$ where x is in the ith position, and is called the marginal distribution of X_i. As in the one-dimensional case, the joint d.f. F of an n-dimensional r.v. $\mathbf{X} = (X_1, \ldots, X_n)$ is said to be *absolutely continuous* if there exists a Borel measurable function $f(x_1, \ldots, x_n)$ defined over R^n such that

$$F(x_1, \ldots, x_n) = \int_{-\infty}^{x_1} \cdots \int_{-\infty}^{x_n} f(u_1, \ldots, u_n)\, du_1, \ldots, du_n.$$

The function $f(x_1, \ldots, x_n)$ is called a density of $F(x_1, \ldots, x_n)$. If $f(x_1, \ldots, x_n)$ is continuous in all the variables, then $\partial^n F/(\partial x_1, \ldots, \partial x_n) = f(x_1, \ldots, x_n)$. F is said to be discrete if there exists a countable subset $S \subset R^n$ such that $\mu_F(S) = 1$ where μ_F is the Lebesgue-Stieltjes measure over R^n determined by F. Note also that a multivariate d.f. is discrete if and only if every one-dimensional marginal distribution function is discrete.

We shall use the notation $EX = \int x\, dF(x)$ (where $F(x)$ is the d.f. of X and the integral is of the Stieltjes type) for this integral, and it will be called the expected value or the mean value of X. When we speak of EX, it is understood that $\int |x|\, dF(x)$ is finite; otherwise EX does not exist. Let $Y = g(X)$ be a Borel function of X, and denote by $G(y)$ the d.f. of Y. Then $EY = \int y\, dG(y) = \int g(x)\, dF(x)$. If we are dealing with an n-dimensional r.v. $\mathbf{X} = (X_1, \ldots, X_n)$ and $Y = g(X_1, \ldots, X_n)$ is a Borel function of $(X_1, \ldots, X_n)$, then $EY = \int y\, dG(y) = \int g(x_1, \ldots, x_n)\, dF(x_1, \ldots, x_n)$. In particular, if $\mathbf{\Sigma} = (\sigma_{ij})$ is the covariance matrix of $F(x_1, \ldots, x_n)$, and if $\mathbf{a} = (a_1, \ldots, a_n)'$, $\mathbf{b} = (b_1, \ldots, b_n)'$, $\mathbf{X} = (X_1, \ldots, X_n)'$ and $\boldsymbol{\mu} = (\mu_1, \ldots, \mu_n)'$ where $\mathbf{a}$ and $\mathbf{b}$ are the vectors of constants, and $\boldsymbol{\mu} = E\mathbf{X}$, then

$$E(\mathbf{a}'\mathbf{X}) = \mathbf{a}'\boldsymbol{\mu}, \qquad \operatorname{var}(\mathbf{a}'\mathbf{X}) = \mathbf{a}'\mathbf{\Sigma}\mathbf{a},$$

$$\operatorname{var}(\mathbf{b}'\mathbf{X}) = \mathbf{b}'\mathbf{\Sigma}\mathbf{b}, \qquad \operatorname{cov}(\mathbf{a}'\mathbf{X}, \mathbf{b}'\mathbf{X}) = \mathbf{a}'\mathbf{\Sigma}\mathbf{b} = \mathbf{b}'\mathbf{\Sigma}\mathbf{a}.$$

We now give a few results regarding the moments of a distribution function. (Most of these results may also be found in Loève, 1963, and Rao, 1965.)

(*a*) *If X is a non-negative r.v. and if EX exists, then*

$$EX = \int_0^\infty [1 - F(x)]\, dx.$$

(*b*) **Chebychev Theorem.** *Let $g(x)$ be a non-negative function of the random variable X. Then for every $k > 0$,*

$$P[g(X) \geq k] \leq E[g(X)]/k.$$

Special cases of (*b*) are (*c*) and (*d*), where

(*c*) **Chebychev Inequality.** *Let X be a r.v. with mean μ and variance σ^2. Then*

$$P[|x-\mu| \geq k\sigma] \leq 1/k^2, \qquad k > 0.$$

(*d*) *Let X be a non-negative random variable with mean μ. Then for every $\epsilon > 0$, $P[X \geq \epsilon\mu] \leq 1/\epsilon$.*

Multivariate generalizations of the Chebychev inequality have been given by Birnbaum, Raymond, and Zuckerman (1947), Lal (1955), Olkin and Pratt (1958), Marshall and Olkin (1960), and Birnbaum and Marshall (1961), among others.

(*e*) **Cantelli Inequality.**

$$P(X - EX \leq \lambda) \leq \frac{\text{var } X}{\lambda^2 + \text{var } X} \qquad \text{if } \lambda < 0$$

$$\geq 1 - \frac{\text{var } X}{\lambda^2 + \text{var } X} \qquad \text{if } \lambda \geq 0.$$

(*f*) **Berge Inequality.** *Let X and Y be r.v.'s such that $EX = \mu$, $EY = \eta$,* var $X = \sigma^2$, var $Y = \tau^2$, cov $(X, Y) = \rho\sigma\tau$. *Then*

$$P\left[\max\left(\frac{|X-\mu|}{\sigma}, \frac{|Y-\eta|}{\tau}\right) \geq \epsilon\right] \leq \frac{1+\sqrt{1-\rho^2}}{\epsilon^2}.$$

(*g*) *Let X and Y be two r.v.'s with means μ and η, variances σ^2 and τ^2, and d.f.'s F and G respectively. Then*

$$F(x+\mu) - F(-x+\mu) \geq G(x+\eta) - G(-x+\eta)$$

for each x implies that $\sigma^2 \leq \tau^2$.

(*h*) *If $0 < k < k'$ and $E|X|^{k'} < \infty$, then $EX^k < \infty$.*

(*i*) **C_r-Inequality.** *If X and Y are any r.v.'s dependent or not, then*

$$E|X + Y|^r \leq c_r[E|X|^r + E|Y|^r]$$

where

$$c_r = 1 \quad \text{if} \quad 0 < r \leq 1 \quad \text{and} \quad c_r = 2^{r-1} \quad \text{if} \quad r > 1.$$

(*j*) **Holder's Inequality.** *If $k > 1$ and $1/k + 1/k' = 1$, then*

$$E|XY| \leq [E|X|^k]^{1/k} \cdot [E|Y|^{k'}]^{1/k'}.$$

Holder's inequality with $k = k' = 2$ is called

(*k*) **Schwarz Inequality.** $[E|XY|]^2 \leq EX^2 \cdot EY^2$.

(*l*) **Minkowski Inequality.** *If $k \geq 1$, then*

$$[E|X+Y|^k]^{1/k} \leq [E|X|^k]^{1/k} + [E|Y|^k]^{1/k}.$$

(*m*) log $E(|X|^k)$ *is a convex (from below) function of k. Suppose that this* *function exists for some* $k > 0$. *Then it exists for all smaller nonzero values of* k, *and approaches* 0 *as* $k \to 0$. *Hence* $(1/k)E|X|^k$ *is nondecreasing. Therefore*

(*n*) $[E|X|^k]^{1/k}$ *is a nondecreasing function of* k *for* $k > 0$.

(*o*) **Liapounov's Inequality.** *Let* $0 < a < b < c$. *Then*

$$[E|X|^b]^{c-a} \leq [E|X|^c]^{b-a} \cdot [E|X|^a]^{c-b}.$$

(*p*) **Jensen's Inequality.** *If* $g(x)$ *is a convex function and* EX *exists, then* $g[EX] \leq E[g(X)]$.

(*q*) **Ali-Chan Inequality.** *Let* $F(x)$ *be symmetric, continuous, and strictly increasing, and let* $X_{(i)}$ *represent the* ith *order statistic in a sample of size* n *from* $F(x)$. *Then for* $i \geq (n+1)/2$

$$E(X_{(i)}) \geq G(i/(n+1)) \quad \textit{if } F \textit{ is unimodal* and}$$
$$E(X_{(i)}) \leq G(i/(n+1)) \quad \textit{if } F \textit{ is U-shaped†}$$

where $x = G(u)$ *is the inverse function of* $F(x) = u$.

For proof see Ali and Chan (1965).

(*r*) *Let* $\mathbf{X} = (X_1, \ldots, X_p)$ *be a r.v. such that* $EX_i = 0$ *and*

$$EX_iX_j = \sigma_{ij}; \quad i, j = 1, \ldots, p.$$

Let $a_1, \ldots, a_n$ *be any real numbers. Then the necessary and sufficient condition that there exists a distribution of* $\mathbf{X}$ *with second-order moments* σ_{ij} *is that* $Q = \sum_{i,j} a_i a_j \sigma_{ij} = E(\sum a_i X_i)^2 \geq 0$.

(*s*) **Kolmogorov Inequality.** *Let* X_i, $i = 1, \ldots, n$, *be* n *independent r.v.'s such that* $EX_i = 0$ *and* var $X_i = \sigma_i^2 < \infty$. *Then*

$$P\left[\max_{1 \leq i \leq n} \left|\sum_{k=1}^{i} X_k\right| > \epsilon\right] < \frac{1}{\epsilon^2} \sum_{k=1}^{i} \sigma_k^2.$$

Kolmogorov's inequality is a special case of

(*t*) **Hájek-Renyi Inequality.** *Let* X_i, $i = 1, 2, \ldots$, *be independent r.v.'s such that* $EX_i = 0$, *and* var $X_i = \sigma_i^2 < \infty$. *Let* $c_1 \geq c_2 \geq \cdots$ *be a nonincreasing sequence of positive constants. Then, for any* $\epsilon > 0$

$$P\left[\max_{m \leq i \leq n} c_i \left|\sum_{k=1}^{i} X_k\right| \geq \epsilon\right] \leq \frac{1}{\epsilon^2}\left[c_m^2 \sum_{k=1}^{m} \sigma_k^2 + \sum_{k=m+1}^{n} c_k^2 \sigma_k^2\right]$$

and

(*u*) **Birnbaum-Marshall Inequality.** *Let* X_i, $i = 1, \ldots, n$, *be r.v.'s such that*

$$E(|X_k| \mid X_1, \ldots, X_{k-1}) \geq \lambda_k |X_{k-1}|$$

* F is said to be *unimodal* if there exists at least one real number c such that $F(x)$ is convex for $x < c$ and concave for $x > c$.

† F is said to be *U-shaped* if there exists at least one real number c such that $F(x)$ is concave for $x < c$ and convex for $x > c$.

where $\lambda_k \geq 0$, $k = 2, 3, \ldots, n$. *Let* $a_k > 0$, $b_k = \max(a_k, a_{k+1}\lambda_{k+1}, a_{k+2}\lambda_{k+1}\lambda_{k+2}, \ldots, a_n \prod_{i=k+1}^{n} \lambda_i)$, $k = 1, \ldots, n$, $b_{n+1} = 0$, *and let* $X_0 = 0$. *If* $r \geq 1$ *is such that* $E|X_k|^r < \infty$, $k = 1, 2, \ldots, n$, *then*

$$P\left\{\max_{1\leq k\leq n} a_k |X_k| \geq 1\right\} \leq \sum_{k=1}^{n} (b_k^r - \lambda_{k+1}^r b_{k+1}^r) E|X_k|^r$$

$$= \sum_{k=1}^{n} b_k^r (E|X_k|^r - \lambda_k^r E|X_{k-1}|^r).$$

(*v*) **Birnbaum-Marshall Lemma.** *Let* $\{X_t, t \geq 0\}$ *be a separable process.* Suppose that* f *is a positive function on* $[0, \infty)$ *having at most countably many discontinuities, and let* $\tau > 0$. *If* S *is a countable set dense in* $[0, \infty)$ *satisfying the definition of separability and containing the set of discontinuities of* f *as well as* 0 *and* τ, *then* $\{\omega: \sup_{t\in[0,\tau]} [|X_t(\omega)|/f(t)] < 1\}$ *is measurable and*

$$P\left\{\sup_{t\in[0,\tau]} \frac{|X_t|}{f(t)} < 1\right\} = \lim_{k\to\infty} P\left\{|X_t| \leq \frac{(k-1)f(t)}{k} \quad \text{for all} \quad t \in S \cap [0, \tau]\right\}.$$

For the proofs of (*u*) and (*v*) along with some important results on multivariate Chebychev inequalities, see Birnbaum and Marshall (1961).

2.3. CONVERGENCES

Given a sequence of r.v.'s, there are several modes of convergence of X_n to a limiting r.v. X.

(*a*) X_n *converges in probability* to X, and we write $X_n \xrightarrow{p} X$, if, for every $\epsilon > 0$, $P[|X_n - X| \geq \epsilon] \to 0$.

(*b*) X_n *converges almost surely* to X, and we write $X_n \xrightarrow{\text{a.s.}} X$, if $P[\lim_{n\to\infty} X_n = X] = 1$ or, equivalently, for every $\epsilon > 0$,

$$P\left\{\bigcup_{k\geq n} [|X_k - X| \geq \epsilon]\right\} \to 0 \quad \text{as} \quad n \to \infty.$$

(*c*) X_n *converges to* X *in the rth mean*, and we write $X_n \xrightarrow{r} X$, for $r > 0$ if $E|X_n - X|^r \to 0$. Convergence in the rth mean for $r = 2$ is called *convergence in quadratic mean.*

All these convergences are probabilistic. This means, that if X_n, $n = 1, 2, \ldots, X$ has the same distribution as X_n', $n = 1, 2, \ldots, X'$, then $X_n \to X$ in any of the above senses implies that $X_n' \to X'$ in the same sense.

* For definition, see section 2.11.

Also, note that the class of r.v.'s is complete under the operation of convergence in any of the above senses, that is, for any sequence of r.v.'s $\{X_n\}$

$$(d)\ X_n - X_m \xrightarrow{p} 0 \quad \text{as } m, n \to \infty \Leftrightarrow X_n \xrightarrow{p} X \quad \text{for some r.v. } X$$

$$X_n - X_m \xrightarrow{\text{a.s.}} 0 \quad \text{as } m, n \to \infty \Leftrightarrow X_n \xrightarrow{\text{a.s.}} X \quad \text{for some r.v. } X$$

$$X_n - X_m \xrightarrow{r} 0 \quad \text{as } m, n \to \infty \Leftrightarrow X_n \xrightarrow{r} X \quad \text{for some r.v. } X.$$

The following are some well-known facts.

(*e*) If $X_n \xrightarrow{\text{a.s.}} X$, then $X_n \xrightarrow{p} X$. If $X_n \xrightarrow{p} X$, then there is a subsequence $X_{n_k} \xrightarrow{\text{a.s.}} X$ as $k \to \infty$, with

$$\sum_{k=1}^{\infty} P[|X_{n_k} - X| \geq 2^{-k}] < \infty$$

(*f*) If $X_n \xrightarrow{r} X$, then $X_n \xrightarrow{p} X$; if $X_n \xrightarrow{r} X$, then $X_n \xrightarrow{r'} X$ where $0 < r' < r$.

(*g*) $X_n \xrightarrow{r} X \not\Rightarrow X_n \xrightarrow{\text{a.s.}} X \not\Rightarrow X_n \xrightarrow{r} X$ for any r.

An r.v. X is said to be a.s. bounded if $P[|X| > k] = 0$ for some finite k. A sequence of r.v.'s $\{X_n\}$ is a.s. bounded, if each X_n is a.s. bounded with the same k. Thus

(*h*) If a sequence of r.v.'s $\{X_n\}$ is a.s. bounded, and $X_n \xrightarrow{p} X$, then $X_n \xrightarrow{r} X$ for every $r > 0$.

This is a special case of

(*i*) If $|X_n| < Y$ for each n, where Y is another r.v. such that $E\,|Y|^r < \infty$ for $r > 0$, then $X_n \xrightarrow{p} X \Rightarrow X_n \xrightarrow{r} X$, and $E\,|X|^r < \infty$.

The r.v.'s whose rth absolute moments are finite are said to form the *space* $\mathfrak{L}_r$ over some probability space. We say that $X \in \mathfrak{L}_r$ if $E\,|X|^r < \infty$. $\mathfrak{L}_0$ is the space of all r.v.'s and $\mathfrak{L}_\infty$ is the space of all a.s. bounded r.v.'s. As in Loève (1963), we shall denote by L'_∞ the space of r.v.'s a.s. bounded by 1. Thus

$$(j)\quad L_0 \supset L_r \supset L_s \supset L_\infty \supset L'_\infty, \qquad 0 \leq r \leq s \leq \infty.$$

We may summarize relations between the convergences as follows:

$$X_n \xrightarrow{\text{a.s.}} X \Rightarrow X_n \xrightarrow{p} X \Rightarrow X_{n_k} \xrightarrow{\text{a.s.}} X$$

$$\text{with} \quad \sum_{k=1}^{\infty} P[|X_{n_k} - X| \geq 2^{-k}] < \infty$$

$$X_n \xrightarrow{r} X \Rightarrow X_n \xrightarrow{p} X; \quad X_n \xrightarrow{r} X \Rightarrow X_n \xrightarrow{r'} X, \quad 0 < r' < r.$$

Analogously, let $\mathbf{X} = (X_1, \ldots, X_p)$ and $\mathbf{X}_n = (X_{1n}, \ldots, X_{pn}), n = 1, 2, \ldots$ be a sequence of p-dimensional r.v.'s. Then

(*k*) $\mathbf{X}_n \xrightarrow{p} \mathbf{X}$ if $P[|\mathbf{X}_n - \mathbf{X}| \geq \epsilon] \to 0$ as $n \to \infty$ for every $\epsilon > 0$.

$$\mathbf{X}_n \xrightarrow{\text{a.s.}} \mathbf{X} \quad \text{if} \quad P\left[\lim_{n\to\infty} \mathbf{X}_n = \mathbf{X}\right] = 1.$$

(*l*) If $\mathbf{X}_n \xrightarrow{\text{a.s.}} \mathbf{X}$, then $\mathbf{X}_n \xrightarrow{p} \mathbf{X}$. If $\mathbf{X}_n \xrightarrow{p} \mathbf{X}$, then there exists a subsequence $\mathbf{X}_{n_k}$ of $\mathbf{X}_n$ such that $\mathbf{X}_{n_k} \xrightarrow{\text{a.s.}} \mathbf{X}$ as $k \to \infty$.

(*m*) Let $\mathbf{X}, \mathbf{X}_n$, $n = 1, \ldots$ be a sequence of p-dimensional r.v.'s such that $\mathbf{X}_n \xrightarrow{p} \mathbf{X}$. Let $g(\mathbf{X})$ be a continuous function of $\mathbf{X}$ in R^p. Then $g(\mathbf{X}_n) \xrightarrow{p} g(\mathbf{X})$.

2.4. SUMS OF INDEPENDENT RANDOM VARIABLES

This section will contain some of the most well-known and frequently used results on the convergence of the partial sums of sequences of r.v.'s. Let $\{X_n\}$ be a sequence of independent r.v.'s and let $S_n = X_1 + \cdots + X_n$. Then the series $\sum_{n=1}^{\infty} X_n$ converges a.s. if the sequence $\{S_n\}$ of partial sums converges a.s. Also $\sum_{n=1}^{\infty} X_n$ converges in probability (or in the rth mean) if there exists a r.v. X such that $S_n \xrightarrow{p} X$ (or $S_n \xrightarrow{r} X$).

The general problem of convergence of $\sum X_n$ is dealt with easily if we assume that X_n have finite second moments. Applying the Chebychev inequality, we obtain

$$P\left[\left|\sum_{n=k}^{N}\{X_n - EX_n\}\right| \geq \epsilon\right] \leq \epsilon^{-2}\sum_{n=k}^{N} \operatorname{var} X_n \tag{2.4.1}$$

and hence if

$$\sum_{n=1}^{\infty} \operatorname{var} X_n < \infty, \quad \text{then} \quad \sum_{k=1}^{n}(X_k - EX_k) \xrightarrow{p} 0.$$

This result is sufficiently strengthened because of the following theorem due to Kolmogorov (1928, 1930).

Theorem 2.4.1. (Kolmogorov's Inequalities). *Let $\{X_n\}$ be a sequence of independent r.v.'s with finite variances, and let $S_n = \sum_{i=1}^{n} X_i$. Then for every $\epsilon > 0$,*

$$P\left[\max_{1\leq i\leq n} |S_i - ES_i| \geq \epsilon\right] \leq \epsilon^{-2}\sum_{i=1}^{n} \operatorname{var} X_i. \tag{2.4.2}$$

If, in addition, there exists a constant c such that $|X_i| \leq c > 0$ for all i, then

$$P\left[\max_{1\leq i\leq n} |S_i - ES_i| \geq \epsilon\right] \geq 1 - (\epsilon + 2c)^2\left\{\sum_{i=1}^{n} \operatorname{var} X_i\right\}^{-1} \tag{2.4.3}$$

Using Theorem 2.4.1 we obtain (Loève, 1963)

Theorem 2.4.2 *If $\sum_{n=1}^{\infty} \operatorname{var} X_n$ converges, then $\sum_{k=1}^{n} (X_k - EX_k)$ converges a.s.*

Theorem 2.4.1 also yields

Theorem 2.4.3. *If $|X_n| < c$ for all n where c is some positive constant, and if $\sum \operatorname{var} X_n$ diverges, then $\sum (X_n - EX_n)$ diverges a.s.*

Corollary 2.4.3.1. *If* $\sum_{n=1}^{\infty} \operatorname{var} X_n < \infty$, *and if* $\sum_{n=1}^{\infty} EX_n$ *converges, then* $\sum_{n=1}^{\infty} X_n$ *converges a.s.*

Corollary 2.4.3.2. *If* $P[\,|X_n| \le c] = 1$ *for all n and some* $c > 0$ *and if* $\sum X_n$ *converges a.s., then* $\sum \operatorname{var} X_n < \infty$ *and* $\sum_{n=1}^{\infty} EX_n$ *converges.*

The above results depend upon the existence of moments. If no assumption is made regarding moments, the method of truncation offers a very useful tool for establishing a.s. convergence. X_n is said to be truncated at a value $c > 0$, if it is replaced by a r.v. Y_n defined by

$$Y_n = \begin{cases} X_n & \text{if } |X_n| < c \\ 0 & \text{if } |X_n| \ge c. \end{cases}$$

Theorem 2.4.4 (Kolmogorov's Three Series Criterion). *The series* $\sum X_n$ *of independent summands converges a.s. to a r.v. if, and only if, for a fixed* $c > 0$, *the three series*

$$\sum P[|X_n| \ge c], \quad \sum \operatorname{var} Y_n, \quad \text{and} \quad \sum EY_n$$

all converge. (Y_n *is* X_n *truncated at* $c > 0$.)

In discussing the a.s. convergence of infinite series of r.v.'s it is assumed that the reader knows the theory of measure on an infinite-dimensional product space. We now state the result that if $\sum X_n$ does not converge a.s., then it must diverge a.s. This follows from the well-known Borel-Cantelli lemma.

Theorem 2.4.5. Borel-Cantelli Lemma. *Let* $\{A_n\}$ *be a sequence of events. Then*

(*a*) $\sum P(A_n) < \infty \Rightarrow P(\limsup A_n) = 0$;

(*b*) $\sum P(A_n) = \infty$ *and the events* A_n *are independent* $\Rightarrow P(\limsup A_n) = 1$.

Corollary 2.4.5. *If the events* A_n *are independent and* $A_n \to A$ *then* $P(A) = 0$ *or* 1.

An important application of Theorem 2.4.5 is

Theorem 2.4.6. *Let* $\{X_n\}$ *be a sequence of r.v.'s, and* $\{c_n\}$ *a sequence of constants. Then*

(*a*) $\sum P[|X_n| > c_n] < \infty \Rightarrow P[|X_n| > c_n$ *infinitely often*$] = 0$

(*b*) $\sum P[|X_n| > c_n] = \infty$ *and* X_n *are independent* $\Rightarrow P[|X_n| > c_n$ *infinitely often*$] = 1$.

It may be mentioned that if the X_i, $i = 1, 2, \ldots$ are bounded a.s., then much sharper bounds are available (see Hoeffding, 1963, and Stieger, 1969).

2.5. LAWS OF LARGE NUMBERS

Consider a sequence of partial sums $S_n = \sum_{k=1}^n X_k$ of independent r.v.'s $\{X_n\}$. Given two sequences of real numbers a_n and $b_n \to \infty$, we say that the sequence S_n is stable in probability or a.s. if $S_n/b_n - a_n \xrightarrow{p} 0$ or $S_n/b_n - a_n \xrightarrow{\text{a.s.}} 0$. The simplest examples are

(*a*) **Bernoulli Law of Large Numbers.** *Let* $\{X_n\}$ *be independent and identically distributed r.v.'s with* $P[X_i = 1] = p$ *and* $P[X_i = 0] = 1 - p$. *Then*

$$\frac{S_n}{n} - p \xrightarrow{p} 0 \quad \text{as} \quad n \to \infty. \tag{2.5.1}$$

In fact a stronger result is

(*b*) **Borel Strong Law of Large Numbers.** *Let* $\{X_n\}$ *be independent and identically distributed r.v.'s with* $P[X_i = 1] = p$ *and* $P[X_i = 0] = 1 - p$. *Then*

$$\frac{S_n}{n} - p \xrightarrow{\text{a.s.}} 0 \quad \text{as} \quad n \to \infty. \tag{2.5.2}$$

When the convergence is in probability, we say that the weak law of large numbers holds; when it is a.s., the strong law of large numbers holds.

If the independent r.v.'s have finite variances, the weak law of large numbers is easily generalized (by an application of Chebychev's inequality) as follows:

(*c*) *Let the independent r.v.'s* X_n *have finite variances, and let* b_n *be a sequence of positive numbers such that* $b_n[\sum_{i=1}^n \operatorname{var} X_i]^{-1/2} \to \infty$; *then*

$$b_n^{-1} \sum_{i=1}^n [X_i - EX_i] \xrightarrow{p} 0. \tag{2.5.3}$$

A special case of (*c*) is the well-known

(*d*) **Chebychev Theorem.** *Let* X_i, $i = 1, 2 \ldots$ *be independent r.v.'s such that* $EX_i = \mu_i$, $\operatorname{var} X_i = \sigma_i^2$. *Then*

$$\lim_{n\to\infty} \frac{1}{n^2} \sum_{i=1}^n \sigma_i^2 = 0 \Rightarrow \bar{X}_n - \bar{\mu}_n \xrightarrow{p} 0 \tag{2.5.4}$$

where

$$\bar{X}_n = \sum_{i=1}^n X_i/n, \quad \text{and} \quad \bar{\mu}_n = \sum_{i=1}^n \mu_i/n.$$

The results (*c*) and (*d*) hold even if the variables are pairwise uncorrelated.

If $\mu_i = \mu$ and $\sigma_i^2 = \sigma^2$, the assumptions of (*d*) are automatically satisfied, and we have $\bar{X}_n \xrightarrow{p} \mu$. In fact this result holds even without the existence of the second moment.

(*e*) **Khinchine's Law of Large Numbers.** *If X_n are independent and identically distributed r.v.'s, then*

$$EX_i = \mu \text{ exists} \Rightarrow \bar{X}_n \xrightarrow{p} \mu. \tag{2.5.5}$$

A stronger version of this theorem is the well-known

(*f*) **Kolmogorov's Strong Law of Large Numbers.** *Let X_n be independent and identically distributed r.v.'s. Then $\bar{X}_n \xrightarrow{\text{a.s.}}$ a constant μ, if and only if EX_n exists and is equal to μ.*

Note that if $E|X_n| = \infty$, then $\bar{X}_n$ diverges a.s.

(*g*) **Kolmogorov's Sufficient Condition for the Strong Law of Large Numbers.** *Let X_n be independent r.v.'s such that $EX_i = \mu_i$, $\operatorname{var} X_i = \sigma_i^2$. Then*

$$\sum_{i=1}^{\infty} \frac{\sigma_i^2}{i^2} < \infty \Rightarrow \frac{1}{n} \sum_{i=1}^{n} (X_i - \mu_i) \xrightarrow{\text{a.s.}} 0. \tag{2.5.6}$$

In passing, we may mention

(*h*) *Let X_n be independent r.v.'s. Then $X_n \xrightarrow{\text{a.s.}}$ some r.v. $X \Rightarrow P[X = c] = 1$ for some constant c.*

(*i*) *Let X_i be independent r.v.'s such that $EX_i = \mu_i$. Then*

$$n^{-(1+\delta)} \sum_{i=1}^{n} E|X_i|^{1+\delta} \to 0 \quad \text{as} \quad n \to \infty \tag{2.5.7}$$

$$\text{for some} \quad 0 < \delta \leq 1 \Rightarrow \frac{1}{n} \sum_{i=1}^{n} (X_i - \mu_i) \xrightarrow{p} 0.$$

(*j*) *Let X_n be independent r.v.'s all having the same d.f. $F(x)$, and let $\bar{X}_n = \sum_{i=1}^{n} X_i/n$. Then a necessary and sufficient condition for the existence of constant $\mu_1, \mu_2, \ldots$ such that $\bar{X}_n - \mu_n \xrightarrow{p} 0$ is that*

$$\int_{|x|>t} dF(x) = o(t^{-1}) \quad \text{as} \quad t \to \infty. \tag{2.5.8}$$

2.6. CONVERGENCE OF DISTRIBUTION FUNCTIONS

We say that a sequence F_n of d.f.'s converges *weakly* to a d.f. F, and we write $F_n \xrightarrow{W} F$, if $F_n \to F$ on the continuity set $C(F)$ of F.

If the sequence X_n of r.v.'s has the d.f.'s $F_n(x)$, we say that X_n converges in distribution (or in law) to a r.v. X with the d.f. $F(x)$, and we write $\mathfrak{L}(X_n) \to \mathfrak{L}(X)$, if $F_n(x) \xrightarrow{W} F(x)$ as $n \to \infty$.

The d.f. $F(x)$ is called the limiting or the asymptotic d.f. of X_n. Limiting distributions play an important role in statistical applications. We state below some of the important results in deriving limit distributions.

(*a*) $F_n \xrightarrow{W} F$ if, and only if, $F_n \to F$ on a set D dense on the real line $R = (-\infty, \infty)$.

(*b*) **Weak Compactness Theorem.** *Every sequence of d.f.'s is weakly compact, that is, there is a subsequence which tends to a function (not necessarily a d.f.) at all continuity points of the latter.*

(*c*) **Helly-Bray Lemma.** *Let $g(x)$ be a continuous function, and let $F_n(x) \to F(x)$ for all values x in $[a, b]$ for which $F(x)$ is continuous including at $x = a$ and $x = b$. Then*

$$\int_a^b g(x)\, dF_n(x) \to \int_a^b g(x)\, dF(x). \tag{2.6.1}$$

An extension of (c) is

(*d*) **Extended Helly-Bray Lemma.** *Let $g(x)$ be bounded and continuous on $R = (-\infty, \infty)$. Then $F_n \xrightarrow{W} F \Rightarrow$*

$$\int_{-\infty}^{\infty} g(x)\, dF_n(x) \to \int_{-\infty}^{\infty} g(x)\, dF(x). \tag{2.6.2}$$

The following theorem concerns the convergence of d.f.'s to the convergence of the corresponding moments.

(*e*) **Frechet-Shohat Theorem.** *If for each n, the moments $\mu_{r_n} = \int x^r\, dF_n(x)$, $r = 1, 2, \ldots$, of $F_n(x)$ exist, and if $\lim_{n\to\infty} \mu_{r_n} = \mu_r$, $r = 1, 2, \ldots$, then μ_r, $r = 1, 2, \ldots$ are the moments of a d.f. Furthermore, if $F(x)$ is the only d.f. having the moments μ_r, $r = 1, 2, \ldots$, then $F_n(x) \xrightarrow{W} F(x)$ as $n \to \infty$.*

The following theorem provides a sufficient condition for the uniqueness of the determination of a distribution by its moments.

(*f*) **Uniqueness Theorem.** *If the moments μ_r of a d.f. $F(x)$ are finite, and are such that the series $\sum_{r=1}^{\infty} \mu_r(t^r/r!)$ is absolutely convergent for some $t > 0$, then $F(x)$ is the only d.f. having moments μ_r.*

With the help of this theorem, it is easy to establish that the convergence of moments to the moments of a normal distribution implies that the limiting distribution is normal.

The next theorem concerns the convergence of distributions and the convergence of the corresponding characteristic functions. The characteristic function $\varphi_n(t)$ of a d.f. $F_n(x)$ is defined as $\varphi_n(t) = \int e^{itx}\, dF_n(x)$ for real t, and is in general a complex-valued function.

(*g*) **Theorem (Levy).** *If two d.f.'s have the same characteristic function, they are identical.*

(*h*) **Theorem (Levy-Cramér).** *Let $F_n(x)$ be a sequence of d.f.'s with corresponding characteristic functions $\varphi_n(t)$. Then a necessary and sufficient condition that $F_n(x)$ converges weakly to a d.f. $F(x)$ is that, for every real t, $\varphi_n(t)$ converges to a limit $\varphi(t)$ which is continuous at $t = 0$. $\varphi(t)$ is the characteristic function of $F(x)$.*

We shall now give multivariate analogs of the above theorems. First we introduce some notations. Let $F(\mathbf{x})$ be a function defined over R^p. Let $a_i < b_i$, $1 \leq i \leq p$ be p pairs of real numbers. Set $\mathbf{a} = (a_1, \ldots, a_p)$, $\mathbf{b} = (b_1, \ldots, b_p)$ and $[\mathbf{a}, \mathbf{b}) = \{\mathbf{x} = (x_1, \ldots, x_p)\colon\ a_i \leq x_i < b_i;\ 1 \leq i \leq p\}$. $[\mathbf{a}, \mathbf{b})$ is called a cell in R^p. Let $\Delta_{k,p}$ denote the set of $\binom{p}{k}$ p-tuples $(z_1, \ldots, z_p)$ where each z_i is either a_i or b_i and such that exactly k of the z_i are a_i. Then $\Delta = \bigcup_{k=0}^{p} \Delta_{k,p}$ is the set of 2^p vertices of the cell $[\mathbf{a}, \mathbf{b})$ in R^p. We shall denote an arbitrary vertex by δ.

Let F_n be a sequence of d.f.'s over R^p. We say that F_n converges weakly to a d.f. F, and we write $F_n(\mathbf{x}) \xrightarrow{W} F(\mathbf{x})$, if $F_n(\mathbf{x}) \to F(\mathbf{x})$ on the continuity set $C(F)$ of F and such that

(i) F is continuous from the right in each coordinate.

(ii) $\mu_F[\mathbf{a}, \mathbf{b}) = \sum_{k=0}^{p} (-1)^k \sum_{\delta \in \Delta_{k,p}} F(\boldsymbol{\delta}) \geq 0$ for every cell $[\mathbf{a}, \mathbf{b})$ in R^p.

(*i*) *If $F_n(\mathbf{x})$ is a sequence of d.f.'s over R^p, and if $F_n \xrightarrow{W} F$, then F is continuous everywhere over R^p except over a countable union of hyperplanes $x_i = c$.*

(*j*) *If $F_n(\mathbf{x})$ is a sequence of d.f.'s over R^p, and if $F_n(\mathbf{x})$ converges pointwise over a dense subset of R^p, then there exists one and only one d.f. $F(\mathbf{x})$ over R^p such that $F_n(\mathbf{x}) \xrightarrow{W} F(\mathbf{x})$.*

(*k*) **Helly-Bray Theorem.** *If F_n is a sequence of d.f.'s over R^p, and $F_n \xrightarrow{W} F$, and if $I = [\mathbf{a}, \mathbf{b}]$ is a bounded, closed cell in R^p whose boundaries contain no discontinuities of F, and if g is a continuous function over I, then*

$$\int_I g(\mathbf{x})\, dF_n(\mathbf{x}) \to \int_I g(\mathbf{x})\, dF(\mathbf{x}) \quad \text{as} \quad n \to \infty.$$

There are also multivariate analogs of theorems (*e*) and (*f*), but for our purposes, it suffices to note that the moments of a multivariate normal distribution determine the distribution uniquely. Next we state

(*l*) **Continuity Theorem.** *Let $\varphi_n(\mathbf{t})$ be a sequence of characteristic functions whose d.f.'s are $F_n(\mathbf{x})$, $\mathbf{x} \in R^p$. If $\varphi_n(\mathbf{t}) \to \varphi(\mathbf{t})$ which is continuous at $(0, \ldots, 0)$, then $\varphi(\mathbf{t})$ is the characteristic function of a d.f. $F(\mathbf{x})$, and $F_n(\mathbf{x}) \xrightarrow{W} F(\mathbf{x})$.*

This theorem has many applications. Some of the frequently used ones are the following:

(*m*) **Sverdrup's Theorem.** *Let $\mathbf{X}_n = (X_{1n}, \ldots, X_{pn})$, $n = 1, 2, \ldots$ be an infinite sequence of random vectors and let $g(x_1, \ldots, x_p)$ be a real-valued function with the p-dimensional space as its domain. Assume that (i) the limit of the probability distribution of $\mathbf{X}_n$ (as $n \to \infty$) is the probability distribution of a random vector $\mathbf{X} = (X_1, \ldots, X_p)$,* (ii) *there is a closed set S in the p-dimensional space such that (a) $g(x_1, \ldots, x_p)$ is a jointly continuous function of $(x_1, \ldots, x_p)$ for all $(x_1, \ldots, x_p) \in S$, (b) $P[\mathbf{X} \in S] = 1$, (c) for any $\epsilon > 0$,* $\lim_{n\to\infty} P[\mathbf{X}_n \in S_\epsilon] = 1$, *where S_ϵ is the ϵ-neighborhood of S. Then the limit of the probability distribution of $g(\mathbf{X}_n)$ is the probability distribution of $g(\mathbf{X})$.*

Taking S as the whole space we have the following special case of the above theorem.

Corollary. *Let $\mathbf{X}_n = (X_{1n}, \ldots, X_{pn})$, $n = 1, 2, \ldots$ be an infinite sequence of random vectors and let $g(x_1, \ldots, x_p)$ be a real-valued continuous function with the p-dimensional space as its domain. Then under the assumption* (i) *of the above theorem, the limit of the probability distribution of $g(\mathbf{X}_n)$ is the probability distribution of $g(X)$.*

For the proof of the above theorem together with some applications, see Sverdrup (1952).

(*n*) If $\mathbf{X}$, $\mathbf{X}_n$, $n = 1, 2, \ldots$ is a sequence of p-dimensional r.v.'s, and if $\mathbf{X}_n \underset{\text{a.s.}}{\xrightarrow{p}} \mathbf{X}$, then $\mathcal{L}(\mathbf{X}_n) \to \mathcal{L}(\mathbf{X})$.

(*o*) *If F, F_n, $n = 1, 2, \ldots$ is a sequence of d.f.'s and $F_n \xrightarrow{W} F$ where F is continuous, then* $\lim_{n\to\infty} \sup_x |F_n(x) - F(x)| = 0$, *that is, $F_n(x) \to F(x)$ uniformly in x, as $n \to \infty$.*

(*p*) *Let $\{X_n, Y_n\}$, $n = 1, 2, \ldots$ be a sequence of r.v.'s. Then $|X_n - Y_n| \xrightarrow{p}$ 0, $\mathcal{L}(X_n) \to \mathcal{L}(Y) \Rightarrow \mathcal{L}(Y_n) \to \mathcal{L}(Y)$; that is, the limiting distribution of Y_n exists, and is the same as that of Y.*

(*q*) *Let $\{X_n, Y_n\}$, $n = 1, 2, \ldots$ be a sequence of r.v.'s. Then*

(i) $\mathcal{L}(X_n) \to \mathcal{L}(X)$, $\quad Y_n \xrightarrow{p} 0 \Rightarrow X_n Y_n \xrightarrow{p} \mathbf{0}$

(ii) $\mathcal{L}(X_n) \to \mathcal{L}(X)$, $\quad Y_n \xrightarrow{p} a \Rightarrow \mathcal{L}(X_n + Y_n) \to \mathcal{L}(X + a)$

$$\Rightarrow \mathcal{L}(X_n Y_n) \to \mathcal{L}(aX)$$

$$\Rightarrow \mathcal{L}(X_n/Y_n) \to \mathcal{L}(X/a) \quad \text{if} \quad a \neq 0.$$

(iii) $\mathcal{L}(X_n) \to \mathcal{L}(X)$, $\quad Y_n \xrightarrow{p} a \Rightarrow \mathcal{L}(X_n, Y_n) \to \mathcal{L}(X, a)$.

(*r*) *Let* $\mathbf{X} = (X_1, \ldots, X_p)$ *and* $\mathbf{X}_n = (X_{1n}, \ldots, X_{pn})$, $n = 1, 2, \ldots$ *be a sequence of p-dimensional r.v.'s, and let for any real* $\boldsymbol{\lambda} = (\lambda_1, \ldots, \lambda_p)$, $\mathcal{L}(\boldsymbol{\lambda}\mathbf{X}'_n) \to \mathcal{L}(\boldsymbol{\lambda}\mathbf{X}')$. *Suppose* $F(\mathbf{x})$ *is the d.f. of* $\mathbf{X}$. *Then the limiting distribution of* $\mathbf{X}_n$ *exists, and is equal to* $F(\mathbf{x})$.

(*s*) *Let g be a continuous function. Then*

$$X_n - Y_n \xrightarrow{p} 0, \qquad \mathcal{L}(Y_n) \to \mathcal{L}(Y) \Rightarrow g(X_n) - g(Y_n) \xrightarrow{p} 0.$$

(*t*) **Scheffe's Theorem.** *Let* $f_n(x)$ *be the density of the r.v.* X_n *and let* $f_n(x) \to f(x)$, *as* $n \to \infty$. *Then*

(i) $f(x)$ *is a density function* $\Rightarrow \int_R |f_n(x) - f(x)|\, dx \to 0;$

(ii) $|f_n(x)| < g(x)$ *and* $\int g(x)\,dx$ *exists* $\Rightarrow f(x)$ *is a density function, and*

$$\int_R |f_n(x) - f(x)|\, dx \to 0.$$

Let $X_1, \ldots, X_n$ be independent r.v.'s distributed according to the d.f. $F(x)$; and let $F_n(x)$ be the empirical c.d.f. (cumulative distribution function) of $X_1, \ldots, X_n$, i.e., $F_n(x) = n^{-1}$[number of $X_i \le x$, $1 \le i \le n$]. Then, according to Borel's strong law of large numbers, $F_n(x) \xrightarrow{\text{a.s.}} F(x)$ for fixed x. A stronger result is

(*u*) **Glivenko-Cantelli Theorem.** $P[\sup_{-\infty<x<\infty} |F_n(x) - F(x)| \to 0] = 1$, *that is,* $F_n(x) \to F(x)$, *uniformly in* x, *with probability* 1.

2.7. THE CENTRAL LIMIT THEOREM

The expression "the central limit theorem" refers to a body of theorems, each of which deals with the conditions under which sums of independent r.v.'s are asymptotically normally distributed. These theorems have played a very fruitful role in the development of probability theory as well as applications. We state a few of them.

(*a*) **Lindeberg-Levy Theorem.** *Let* X_n, $n = 1, 2, \ldots$ *be a sequence of i.i.d. (independent and identically distributed) r.v.'s, such that* $EX_n = \mu$, *and* var $X_n = \sigma^2$ *exist. Then, as* $n \to \infty$, *the distribution of*

$$Y_n = \sqrt{n}\,(\bar{X}_n - \mu)/\sigma, \quad \text{where} \quad \bar{X}_n = \sum_{i=1}^{n} X_i/n$$

tends to the normal distribution with mean 0 and variance 1.

If we assume that the third moments exist, then we can drop the assumption that the X_i have the same d.f.

(*b*) **Liapounoff Theorem.** *Let $X_n, n = 1, 2, \ldots$ be a sequence of independent r.v.'s. Let* $EX_n = \mu_n$, var $X_n = \sigma_n^2 \neq 0$, *and* $E|X_n - \mu_n|^3 = \beta_n$ *exist for each n. Let*

$$B_n = \left(\sum_{i=1}^{n} \beta_i\right)^{1/3}, \qquad C_n = \left(\sum_{i=1}^{n} \sigma_i^2\right)^{1/2}$$

Then, if $B_n/C_n \to 0$ as $n \to \infty$, the d.f. of

$$Y_n = \left(\sum_{i=1}^{n} X_i - \sum_{i=1}^{n} \mu_i\right)/C_n$$

tends to the normal distribution with mean 0 *and variance* 1.

A very powerful result in this direction is

(*c*) **Lindeberg-Feller Theorem.** *Let X_n, $n = 1, 2, \ldots$ be a sequence of independent r.v.'s and let $G_n(x)$ be the d.f. of X_n.*

Let $EX_n = \mu_n$ and var $X_n = \sigma_n^2 \neq 0$ *exist. Then*

$$\lim_{n\to\infty} \max_{1\le i\le n} \frac{\sigma_i}{C_n} = 0, \tag{2.7.1}$$

$$\lim_{n\to\infty} P[Y_n \le x] = \int_{-\infty}^{x} \frac{1}{\sqrt{2\pi}} e^{-t^2/2}\, dt \tag{2.7.2}$$

if and only if, for every $\epsilon > 0$,

$$\lim_{n\to\infty} \frac{1}{C_n^2} \sum_{i=1}^{n} \int_{|x-\mu_i|>\varepsilon C_n} (x - \mu_i)^2\, dG_i(x) = 0 \tag{2.7.3}$$

where Y_n and C_n are defined in (*b*).

Note: If the r.v.'s X_i are identically distributed, then (2.7.3) reduces to the assertion that

$$\int_{|x-\mu|>\varepsilon\sigma\sqrt{n}} (x - \mu)^2\, dG(x) \to 0$$

which is obviously true since the variance is finite. If the distributions vary but σ_i^2 is bounded away from zero, then the uniform integrability of $(X_i - \mu_i)^2$ is sufficient for (2.7.3).

From the Lindeberg-Feller theorem follows

(*d*) *Let X_n, $n = 1, 2, \ldots$ be a sequence of independent uniformly bounded r.v.'s; that is, there exists a constant $\alpha > 0$ such that for every $n, P[|X_n| \geq \alpha] = 0$, and suppose that* var $X_n \neq 0$ *for every n. Then the necessary and sufficient condition that* (2.7.2) *holds is that* $\lim_{n\to\infty} C_n^2 = \infty$.

In order to make practical applications of limit theorems it is essential to estimate the rates of convergence. The following theorem, due to Berry and Esséen, gives such information.

(*e*) **Berry-Esséen Theorem.** *Let X_n, $n = 1, 2, \ldots$ be independent r.v.'s with a common distribution having mean* 0, *variance* σ^2, *and finite third moment. Then for all* x *and* n,

$$(2.7.4) \qquad |P(\sqrt{n}\,\bar{X}_n \le x) - \Phi_\sigma(x)| \le \frac{33}{4}\frac{E\,|X_1|^3}{\sigma^3\sqrt{n}};$$

$$\Phi_\sigma(x) = \int_{-\infty}^{x} \frac{1}{\sqrt{2\pi}\,\sigma} e^{-t^2/2\sigma^2}\,dt.$$

A generalization of the Berry-Esséen theorem is

(*f*) **Esséen Theorem.** *Let X_n, $n = 1, 2, \ldots$ be independent r.v.'s with a common distribution having mean* 0, *variance* σ^2, *and finite absolute* $2 + \delta'$ *moment* $\beta_{2+\delta'}$, $0 < \delta' \le 1$; *then for all* x *and* n,

$$(2.7.5) \qquad |P(\sqrt{n}\,\bar{X}_n \le x) - \Phi_\sigma(x)| < c(\delta')\left[\frac{\rho_{2+\delta'}}{n^{\delta'/2}} + \frac{\rho_{2+\delta'}^{1/\delta'}}{n^{1/2}}\right]$$

where

$$\rho_{2+\delta'} = \beta_{2+\delta'}/\sigma^{2+\delta'}.$$

Another modification of the central limit theorem is to investigate the convergence of the density of $\sqrt{n}\,\bar{X}_n$, if it exists, to that of $\Phi(x)$. This does not follow from the mere existence of a density of X_i, but the following is true.

(*g*) *If X_n, $n = 1, 2, \ldots$ are independent with a common distribution F having mean* 0 *and variance* σ^2, *and if F has a bounded density, then*

$$(2.7.6) \qquad \lim_{n\to\infty}\ \sup_{-\infty<x<\infty} \left|\frac{d}{dx}P(\sqrt{n}\,\bar{X}_n \le x) - \frac{1}{\sqrt{2\pi}\,\sigma}e^{-x^2/2\sigma^2}\right| = 0$$

(*h*) **Multivariate Central Limit Theorem I (Lindeberg-Levy).** *Let $\mathbf{X}_n = (X_{1n}, \ldots, X_{pn})$, $n = 1, 2, \ldots$ be a sequence of independent p-dimensional r.v.'s, with the same distribution having the second-order moment matrix $M = (m_{jk})$ where the determinant $|M| \ne 0$. Let $F_n(z_1, \ldots, z_p)$ be the d.f. of the r.v. $\mathbf{Z}_n = (Z_{1n}, \ldots, Z_{pn})$ where $Z_{jn} = \sum_{i=1}^{n} (X_{ji} - E(X_{ji}))/\sqrt{nm_{jj}}$, $j = 1, \ldots, p$; then $F_n(z_1, \ldots, z_p)$ converges to the p-variate normal distribution with mean* $\mathbf{0}$ *and the second-order moment matrix $\mathbf{N} = (\eta_{jk})$ where*

$$\eta_{jk} = \begin{cases} 1 & \text{if } j = k \\ \dfrac{m_{jk}}{\sqrt{m_{jj}m_{kk}}} & \text{if } j \ne k. \end{cases}$$

The generalization of Liapounoff's theorem (*b*) to the multidimensional case was given by Bernstein (1926); the generalization of the Lindeberg-Feller theorem can be found in Gnedenko (1944), and for the multivariate extension of Esséen's theorem, see Bhattacharyya (1968, 1970).

(*i*) **Multivariate Central Limit Theorem II (Wald and Wolfowitz, 1944).** *Let F_n denote the joint d.f. of the p-dimensional r.v. $\mathbf{X}_n = (X_{1n}, \ldots, X_{pn})$, $n = 1, 2, \ldots$ and $F_{\lambda n}$, the d.f. of the linear function $\sum_{i=1}^{p} \lambda_i X_{in}$. Then the necessary and sufficient condition that $F_n \to$ a p-variate d.f. F is that $F_{\lambda n}$ converges to a limit d.f. for each vector $\boldsymbol{\lambda}$.*

(*j*) **Multivariate Central Limit Theorem III.** *Let $\mathbf{X}_n$, $n = 1, 2, \ldots$ be a sequence of independent p-dimensional r.v.'s, such that $E(\mathbf{X}_n) = \mathbf{0}$,* cov $(\mathbf{X}_n) =$ *$\boldsymbol{\Sigma}_n$. Suppose that as $n \to \infty$, $1/n \sum_{i=1}^{n} \boldsymbol{\Sigma}_i \to \boldsymbol{\Sigma} \neq \mathbf{0}$, and for every $\epsilon > 0$,*

$$\frac{1}{n} \sum_{n=1}^{n} \int_{\|x\| > \epsilon\sqrt{n}} \|x\|^2 d\, F_i \to 0$$

where F_i is the d.f. of $\mathbf{X}_i$, and $\|x\|$ is the Euclidean norm of the vector $\mathbf{X}$. Then the r.v. $\sum_{i=1}^{n} \mathbf{X}_i/\sqrt{n}$ converges to the p-dimensional normal distribution with mean $\mathbf{0}$ and dispersion matrix $\boldsymbol{\Sigma}$.

2.8. DISTRIBUTION OF QUADRATIC FORMS

Let $X_1, \ldots, X_n$ be n independent r.v.'s each having $N(0, 1)$ distribution. We consider the distribution of the quadratic form $Q = \mathbf{X}'\mathbf{A}\mathbf{X}$ where $\mathbf{X}' = (X_1, \ldots, X_n)$, and $\mathbf{A}$ is a symmetric matrix. (If $\mathbf{A}$ is not symmetric, we could consider the distribution of $\frac{1}{2}\mathbf{X}'(\mathbf{A} + \mathbf{A}')\mathbf{X}$.) Since $\mathbf{A}$ is symmetric there exists an orthogonal transformation $\mathbf{X} = C\mathbf{Y}$ such that $\mathbf{C}'\mathbf{A}\mathbf{C}$ is a diagonal matrix with diagonal elements $\lambda_1, \ldots, \lambda_n$ which are the characteristic roots of the matrix $\mathbf{A}$. Thus

$$\mathbf{X}'\mathbf{A}\mathbf{X} = \mathbf{Y}'C'\mathbf{A}C\mathbf{Y} = \sum_{k=1}^{n} \lambda_k Y_k^2 \tag{2.8.1}$$

when $\mathbf{Y}' = (Y_1, \ldots, Y_n)$, and the Y_k, $k = 1, \ldots, n$ are independent r.v.'s each having normal distribution with zero mean and variance 1. Since the characteristic function of $\lambda_k Y_k^2$ is $(1 - 2i\lambda_k t)^{-1/2}$, it follows that the characteristic function of $\mathbf{X}'\mathbf{A}\mathbf{X}$ is

$$\prod_{k=1}^{n} (1 - 2i\lambda_k t)^{-1/2} = |I - 2it\mathbf{A}|^{-1/2}. \tag{2.8.2}$$

Thus $\mathbf{X}'\mathbf{A}\mathbf{X}$ has a gamma type distribution, if and only if, all the nonzero λ_k are equal.

Now suppose we have two quadratic forms $Q_1 = \mathbf{X}'\mathbf{A}\mathbf{X}$ and $Q_2 = \mathbf{X}'\mathbf{B}\mathbf{X}$ in the same set of normal variates. Then their joint characteristic function is

$$E\{\exp(it_1\mathbf{X}'\mathbf{A}\mathbf{X} + it_2\mathbf{X}'\mathbf{B}\mathbf{X})\}. \tag{2.8.3}$$

Suppose that t_1 and t_2 are real. Then (2.8.3) equals $E\{\exp i\mathbf{X}'(t_1\mathbf{A} + t_2\mathbf{B})\mathbf{X}\}$ where $t_1\mathbf{A} + t_2\mathbf{B}$ is a real symmetric matrix. It follows that for real t_1 and t_2, the joint characteristic function is obtained from (2.8.2) by replacing $t\mathbf{A}$ by $t_1\mathbf{A} + t_2\mathbf{B}$, and is therefore

$$|\mathbf{I} - 2it_1\mathbf{A} - 2it_2\mathbf{B}|^{-1/2}. \tag{2.8.4}$$

From this result, we can deduce some important criteria for the independence of the distributions of Q_1 and Q_2. Clearly, for this to hold, it is necessary and sufficient that

$$|\mathbf{I} - 2it_1\mathbf{A}|\,|\mathbf{I} - 2it_2\mathbf{B}| = |\mathbf{I} - 2it_1\mathbf{A} - 2it_2\mathbf{B}|, \tag{2.8.5}$$

a result first obtained by Cochran (1934). Since $(\mathbf{I} - 2it_1\mathbf{A})(\mathbf{I} - 2it_2\mathbf{B}) = \mathbf{I} - 2it_1\mathbf{A} - 2it_2\mathbf{B} - 4t_1t_2\mathbf{AB}$, it is sufficient that

$$\mathbf{AB} = 0. \tag{2.8.6}$$

The fact that this is also a necessary condition was established by Craig (1943).

We can summarize the above discussion in the form of the following theorem.

Theorem 2.8.1. *Let* $\mathbf{X}'\mathbf{A}\mathbf{X}$ *and* $\mathbf{X}'\mathbf{B}\mathbf{X}$ *have chi-square distributions with degrees of freedom (D.F.)* r_1 *and* r_2 *respectively. The necessary and sufficient condition that they are independently distributed is that* $\mathbf{AB} = \mathbf{0}$.

Cochran (1934) has given a number of theorems which deal with the sum of quadratic forms in normal r.v.'s. (See also James, 1952.)

We consider n independent r.v.'s X_i having $N(\mu_i, 1)$ distributions, $i = 1, \ldots, n$. Let $\mathbf{X}' = (X_1, \ldots, X_n)$, and $\boldsymbol{\mu}' = (\mu_1, \ldots, \mu_n)$.

Theorem 2.8.2. *Let* Q_i, $i = 1, \ldots, k$ *be k quadratic forms with ranks* r_i, $i = 1, \ldots, k$ *such that* $\mathbf{X}'\mathbf{X} = \sum_{i=1}^k Q_i$. *Then the necessary and sufficient condition that* (i) *for each i,* Q_i *has noncentral chi-square distribution with D.F.'s* r_i *and noncentrality parameter* λ_i, *and* (ii) $Q_1, \ldots, Q_k$ *are independent, is that* $n = \sum_{i=1}^k r_i$, *in which case* $\lambda_i = \boldsymbol{\mu}'\mathbf{A}_i\boldsymbol{\mu}$ *if* $Q_i = \mathbf{X}'\mathbf{A}_i\mathbf{X}$ *and* $\sum_{i=1}^n \mu_i^2 = \sum_{i=1}^k \lambda_i$.

In what follows, we shall assume $\mu_i = 0$ for all $i = 1, \ldots, n$ for simplicity of presentation. The results remain valid, however, if instead of the central chi-square distribution, we have noncentral chi-square distribution.

Theorem 2.8.3. *The necessary and sufficient condition that* $\mathbf{X}'\mathbf{A}\mathbf{X}$ *has a chi-square distribution is that* $\mathbf{A}^2 = \mathbf{A}$. *That is, A is idempotent; in which case the D.F. of the distribution is* rank $\mathbf{A}$ = trace $\mathbf{A}$.

Theorem 2.8.4. *Let* $X'X = Q_1 + Q_2$, *and let* Q_1 *have chi-square distribution with* r_1 *D.F. Then* Q_2 *has chi-square distribution with* $(n - r_1)$ *D.F.*

Theorem 2.8.5. *Let $Q = Q_1 + Q_2$, and let Q and Q_1 have chi-square distribution with D.F. r_1 and r_2 respectively. Suppose that Q_2 is non-negative. Then Q_2 has chi-square distribution with $r_1 - r_2$ D.F.*

Theorem 2.8.6. *Let $\mathbf{X}'\mathbf{X} = \sum_{i=1}^k \mathbf{X}'\mathbf{A}_i\mathbf{X}$. Then the necessary and sufficient condition that* (i) *$\mathbf{X}'\mathbf{A}_i\mathbf{X}$ has noncentral chi-square distribution with D.F. $r_i =$* rank *A_i and noncentrality parameter λ_i, $i = 1, \ldots, k$ and* (ii) *$X'\mathbf{A}_iX$, $i = 1, \ldots, k$ are independent is that either*

$$\mathbf{A}_i^2 = \mathbf{A}_i, \qquad i = 1, \ldots, k \tag{2.8.7}$$

or

$$\mathbf{A}_i\mathbf{A}_j = 0 \tag{2.8.8}$$

for all $i \neq j$.

Consider now the case when $X_1, \ldots, X_n$ are correlated. Let $\mathbf{Y}' = (Y_1, \ldots, Y_n)$ be a r.v. with mean $\boldsymbol{\mu}$ and dispersion matrix $\boldsymbol{\Sigma}$. Consider the linear transformation $\mathbf{Y} = \mathbf{C}\mathbf{X}$ so that the components of $\mathbf{X}$ are independent normal. The quadratic forms in $\mathbf{Y}$ transform to the quadratic forms in $\mathbf{X}$ and the theorems 2.8.1 to 2.8.6 apply. For details see Rao (1965).

2.9. SOME RESULTS ON MULTIVARIATE NORMAL DISTRIBUTION

Let $\mathbf{X}' = (X_1, \ldots, X_p)$ and $\mathbf{Y}' = (Y_1, \ldots, Y_q)$ be two r.v.'s and let E, D, C, V denote the expectation, dispersion matrix, covariance, and variance respectively. Then

$$\begin{aligned} E\mathbf{X}' &= [EX_1, \ldots, EX_p], \qquad \mathbf{D}(\mathbf{X}) = \begin{pmatrix} V(X_1), \ldots, C(X_1, X_p) \\ C(X_p, X_1), \ldots, V(X_p) \end{pmatrix}, \\ \mathbf{C}(\mathbf{X}, \mathbf{Y}) &= \begin{pmatrix} C(X_1, Y_1), \ldots, C(X_1, Y_q) \\ \cdots \quad \cdots \quad \cdots \\ C(X_p, Y_1), \ldots, C(X_p, Y_q) \end{pmatrix} = [\mathbf{C}(\mathbf{Y}, \mathbf{X})]' \end{aligned} \tag{2.9.1}$$

and if $\mathbf{a}$ and $\mathbf{b}$ are column vectors, $V(\mathbf{a}'\mathbf{X}) = \mathbf{a}'D(\mathbf{X})\mathbf{a}$, $C(\mathbf{a}'\mathbf{X}, \mathbf{b}'\mathbf{Y}) = \mathbf{a}'C(\mathbf{X}, \mathbf{Y})\mathbf{b}$.

Let $\mathbf{X}_1, \ldots, \mathbf{X}_k$ be k p-dimensional r.v.'s and $a_1, \ldots, a_k$ be k fixed constants. Then

$$\mathbf{D}\left(\sum_{i=1}^k a_i\mathbf{X}_i\right) = \begin{cases} \sum_{i=1}^k a_i^2\mathbf{D}(\mathbf{X}_i) & \text{if the } X_i \text{ are uncorrelated} \\ \left(\sum_{i=1}^k a_i^2\right)\mathbf{D}(\mathbf{X}) & \text{if the } X_i \text{ are uncorrelated and have the same dispersion matrix.} \end{cases} \tag{2.9.2}$$

Next, let $\mathbf{X}$ be a p-dimensional r.v. and B be a $(q \times p)$ matrix; then

$$E(\mathbf{BX}) = \mathbf{B}E\mathbf{X}$$
(2.9.3)
$$\mathbf{D}(\mathbf{BX}) = \mathbf{BD}(\mathbf{X})\mathbf{B}'$$

Let $\mathbf{X} = (X_1, \ldots, X_p)'$ have a p-variate nonsingular normal distribution $N_p(\boldsymbol{\mu}, \Sigma)$; that is, its density is

(2.9.4) $$f(x) = (2\pi)^{-p/2}|\Sigma|^{-\frac{1}{2}}e^{-\frac{1}{2}(\mathbf{X}-\boldsymbol{\mu})'\Sigma^{-1}(\mathbf{X}-\boldsymbol{\mu})}$$

Then the following results hold. For proofs, we refer to Rao (1965).

(i) *Let* $\mathbf{a} = (a_1, \ldots, a_p)'$ *be a vector of constants. Then* $\mathbf{a}'\mathbf{X}$ *has* $N_1(\mathbf{a}'\boldsymbol{\mu}, \mathbf{a}'\boldsymbol{\Sigma}\mathbf{a})$ *distribution.*

(ii) *The characteristic function of* $\mathbf{X}$ *is* $\varphi_{\mathbf{X}}(\mathbf{t}) = \exp(i\mathbf{t}'\boldsymbol{\mu} - \frac{1}{2}\mathbf{t}'\boldsymbol{\Sigma}\mathbf{t})$.

(iii) *If, for every* $\mathbf{a}$, $\mathbf{a}'\mathbf{X}$ *has* $N_1(\mathbf{a}'\boldsymbol{\mu}, \mathbf{a}'\boldsymbol{\Sigma}\mathbf{a})$ *distribution, then* $\mathbf{X}$ *has* $N_p(\boldsymbol{\mu}, \boldsymbol{\Sigma})$ *distribution. If* $\boldsymbol{\Sigma}$ *is a diagonal matrix, then the components of* $\mathbf{X}$ *are independent.*

(iv) *Let* $\mathbf{X} = (\mathbf{X}_1, \mathbf{X}_2)'$ *and let*

$$\boldsymbol{\Sigma} = \begin{pmatrix} \boldsymbol{\Sigma}_{11}\boldsymbol{\Sigma}_{12} \\ \boldsymbol{\Sigma}_{21}\boldsymbol{\Sigma}_{22} \end{pmatrix} = \begin{pmatrix} D(\mathbf{X}_1), C(\mathbf{X}_1, \mathbf{X}_2) \\ C(\mathbf{X}_2, \mathbf{X}_1), D(\mathbf{X}_2) \end{pmatrix}$$

Then $\mathbf{X}_1$ *and* $\mathbf{X}_2$ *are independent if and only if* $\boldsymbol{\Sigma}_{12} = \mathbf{0}$.

(v) *If the subvectors* $\mathbf{X}_1, \ldots, \mathbf{X}_k$ *of* $\mathbf{X}$ *are pairwise independent, they are mutually independent.*

(vi) $\mathbf{X}$ *has* $N_p(\boldsymbol{\mu}, \boldsymbol{\Sigma})$ *distribution, if and only if,* $\mathbf{X} = \boldsymbol{\mu} + \mathbf{BG}$, $\mathbf{BB}' = \boldsymbol{\Sigma}$ *where* $\mathbf{B}$ *is a* $p \times m$ *matrix of* rank m, *and* $\mathbf{G}$ *has* $N_m(\mathbf{0}, \mathbf{I})$ *distribution.*

(vii) *If* $\mathbf{X}$ *has p-variate normal distribution, the marginal distribution of any subset of q components of* $\mathbf{X}$ *is q-variate normal.*

(viii) *Let* $\mathbf{X}$ *have* $N_p(\boldsymbol{\mu}, \boldsymbol{\Sigma})$ *distribution; and let* $\mathbf{C}$ *be* $q \times p$ *matrix. Then* $\mathbf{Y} = \mathbf{CX}$ *has* $N_q(\mathbf{C}\boldsymbol{\mu}, \mathbf{C}\boldsymbol{\Sigma}\mathbf{C}')$ *distribution.*

(ix) *Let* $\mathbf{X}_1' = (X_1, \ldots, X_n)$, $\mathbf{X}_2' = (X_{r+1}, \ldots, X_p)$ *be two subsets of* $\mathbf{X}'$ *and* $\boldsymbol{\Sigma}_{11}, \boldsymbol{\Sigma}_{12}, \boldsymbol{\Sigma}_{22}$ *be the partitions of* $\boldsymbol{\Sigma}$ *as in* (iv). *Then the conditional distribution of* $\mathbf{X}_2$ *given* $\mathbf{X}_1$ *is* $N_{p-r}(\boldsymbol{\mu}_2 + \Sigma_{21}\Sigma_{11}^{-1}(\mathbf{X}_1 - \boldsymbol{\mu}_1), \boldsymbol{\Sigma}_{22} - \boldsymbol{\Sigma}_{21}\boldsymbol{\Sigma}_{11}^{-1}\boldsymbol{\Sigma}_{12})$ *where* $E(\mathbf{X}_i) = \boldsymbol{\mu}_i$, $i = 1, 2$.

(x) *Let* $\mathbf{X}_i$ *have* $N_p(\boldsymbol{\mu}_i, \boldsymbol{\Sigma}_i)$ *distributions for* $i = 1, \ldots, n$ *and let* $\mathbf{X}_i$ *be all independent. Then for fixed constants* a_i, $i = 1, \ldots, n$,

$$\mathbf{Y} = \sum_{i=1}^{n} a_i\mathbf{X}_i \quad \textit{has} \quad N_p(\Sigma a_i\boldsymbol{\mu}_i, \Sigma a_i^2\boldsymbol{\Sigma}_i) \quad \textit{distribution.}$$

(xi) *Let* $\mathbf{X}_i$, $i = 1, \ldots, n$ *be independent r.v.'s each having* $N_p(\boldsymbol{\mu}, \boldsymbol{\Sigma})$ *distribution. Then*

$$\frac{1}{n}\sum_{i=1}^{n} X_i \quad \textit{has} \quad N_p\left(\boldsymbol{\mu}, \frac{1}{n}\boldsymbol{\Sigma}\right) \quad \textit{distribution.}$$

(xii) *Let* $\mathbf{X}$ *have* $N_p(\boldsymbol{\mu}, \boldsymbol{\Sigma})$ *distribution. Then the necessary and sufficient condition that* $(\mathbf{X} - \boldsymbol{\mu})'\mathbf{A}(\mathbf{X} - \boldsymbol{\mu})$ *has chi-square distribution with k degrees of freedom is that* $\boldsymbol{\Sigma}(\mathbf{A}\boldsymbol{\Sigma}\mathbf{A} - \mathbf{A})\boldsymbol{\Sigma} = 0$ *in which case* $k = \text{trace}(\mathbf{A}\boldsymbol{\Sigma})$.

(xiii) *Let* $\mathbf{X}_1$ *and* $\mathbf{X}_2$ *be two independent p-dimensional r.v.'s such that* $\mathbf{X} = \mathbf{X}_1 + \mathbf{X}_2$ *has p-variate normal distribution. Then both* $\mathbf{X}_1$ *and* $\mathbf{X}_2$ *have p-variate normal distributions.*

(xiv) *Let* $\mathbf{X}_1, \ldots, \mathbf{X}_n$ *be* n (≥ 2) *independent p-dimensional r.v.'s. Let*

$$\mathbf{Y}_1 = \sum_{i=1}^{n} a_i \mathbf{X}_i \quad \textit{and} \quad \mathbf{Y}_2 = \sum_{i=1}^{n} b_i \mathbf{X}_i$$

where the a_i *and* b_i *are given constants. Let* $\mathbf{a}' = (a_1, \ldots, a_n)$, $b' = (b_1, \ldots, b_n)$. *Then*

(a) $\mathbf{X}_i$ *are i.i.d. p-variate normal r.v.'s* and $\mathbf{a}'\mathbf{b} = 0 \Rightarrow \mathbf{Y}_1$ *and* $\mathbf{Y}_2$ *are independently distributed.*

(*b*) $\mathbf{Y}_1$ *and* $\mathbf{Y}_2$ *are independently distributed* $\Rightarrow \mathbf{X}_i$ *has p-variate normal distribution for each i such that* $a_i b_i \neq 0$ *and the* $\mathbf{X}_i$ *are not necessarily identically distributed.*

2.10. MARTINGALES

The concept of martingales is due to Paul Levy. Doob realized its great importance, and developed a systematic theory. Hájek (1961) showed that there are cases in rank test theory which can be studied by means of the theory of martingales. This section is devoted to a brief study of martingales and their properties. For details the reader is referred to Doob (1967), Loève (1963), and Feller (1965).

Let $\{X_\tau, \tau \in T\}$ be a family of r.v.'s on a set T ordered by the relation $<$, and let $\mathcal{B}_\tau = \mathcal{B}\{X_{t'}, t' < \tau\}$ be the sub-σ-field of events induced by the subfamily of all the r.v.'s $X_{t'}$ with $t' < \tau$. Suppose that $E|X_\tau| < \infty$ for all $\tau \in T$. Then the family is said to be a

(i) *martingale* if $X_s = E[X_t \mid \mathcal{B}_s]$ a.s. for $s < t$, $s \in T$ and $t \in T$.
(ii) *submartingale* if $X_s \leq E[X_t \mid \mathcal{B}_s]$ a.s. for $s < t$, $s \in T$ and $t \in T$.
(iii) *reverse martingale* if it is a martingale under the reverse ordering of T,
(iv) *reverse submartingale* if it is a submartingale under the reverse ordering of T.

From the definitions, it follows that a sequence of r.v.'s $\{X_n\}$ with finite expectations is a martingale if $E[X_{n+1} \mid X_n, \ldots, X_1] = X_n$ a.s. for $n = 1, 2, \ldots$; submartingale if $E[X_{n+1} \mid X_n, \ldots, X_1] \geq X_n$ for $n = 1, 2, \ldots$; reverse martingale if $E[X_n \mid X_{n+1}, X_{n+2}, \ldots] = X_{n+1}$ a.s. for $n = 1, 2, \ldots$, and reverse submartingale if $E[X_n \mid X_{n+1}, X_{n+2}, \ldots] \geq X_{n+1}$ for $n = 1, 2, \ldots$.

If $X_1, X_2, \ldots, X$ is a martingale, it is said to be closed on the right by X and on the left by X_1. X is called the nearest closing r.v. for $\{X_n\}$, if for any other closing r.v. Y, the sequence $X_1, \ldots, X_n, \ldots, X, Y$ is a martingale.

It follows that (i) if $\{X_n\}$ is a sequence of independent r.v.'s with zero expectations, then the sequence $\{S_n\}$ where $S_n = \sum_{i=1}^n X_i$ is a martingale, (ii) if the joint density of $(X_1, \ldots, X_n)$ is either p_n or q_n, then the sequence of likelihood ratios $\{q_n/p_n\}$ is a martingale, (iii) if $\{X_n\}$ is a submartingale, then $\{X_n^+\}$ where $X_n^+ = X_n I_{[X\geq 0]}$ is a submartingale; if $\{X_n\}$ is a martingale, then for every $r \geq 1$, $\{|X_n|^r\}$ is a submartingale, (iv) if $\mathcal{B}$ is a sub-σ-field of events and g is any function continuous on R and convex from below, then $g(E[X \mid \mathcal{B}]) \leq E[g(X) \mid \mathcal{B}]$; if g is $\uparrow$ and $X' \leq E[X \mid \mathcal{B}]$ a.s., then $g(X') \leq E[g(X) \mid \mathcal{B}]$ a.s.

Theorem 2.10.1. *Let $\{X_n\}$ be a martingale. Then*

(*a*) *Inequalities:*
$EX_1 = EX_2 = \cdots; EX_1^+ \leq EX_2^+ \cdots; E|X_1|^r \leq E|X_2|^r \leq \cdots.$

(*b*) *Convergence: If* $\lim EX_n^+ < \infty$ *or* $\lim EX_n^- < \infty$, *then there exists a r.v. X such that $X_n \xrightarrow{\text{a.s.}} X < \infty$ or $> -\infty$ respectively. If* $\lim E|X_n| < \infty$, *then $X_n \xrightarrow{\text{a.s.}} X$ finite.*

(*c*) *Closure: The martingale is closed on the right by a r.v. $Y \in L_r$, $r \geq 1$ if and only if the $|X_n|^r$ are uniformly integrable; then $X_n \to X$ a.s. or in the* r*th mean and X is the nearest of the closing r.v.'s.*

Theorem 2.10.2. *Let $\{X_n\}$ be a submartingale. Then*

(*a*) *Inequalities:*

$$EX_1 \leq EX_2 \leq \cdots; EX_1^+ \leq EX_2^+ \leq \cdots;$$

$$X_n \geq 0 \text{ a.s.} \Rightarrow EX_1^r \leq EX_2^r \leq \cdots.$$

(*b*) *Convergence: If* $\sup_n E|X_n| < \infty$, *then there exists a r.v. X such that $X_n \to X$ a.s. and* $E|X| \leq \sup_n E|X_n|$.

(*c*) *Closure: If the $|X_n|^r$, $r \geq 1$ are uniformly integrable, then there exists a r.v. X such that $X_n \to X$ a.s. with $X \in L_r$ and X is the nearest of closing r.v.'s. If every $X_n \geq 0$ a.s. then the X_n^r are uniformly integrable, if and only if there is a closing on the right r.v. $Y \in \mathfrak{L}_r$.*

Theorem 2.10.3. Martingale Reversed Sequence. *Let $\{X_n\}$ be a reversed martingale. Then*

(*a*) *Inequalities:* $\cdots = EX_2 = EX_1; \cdots \leq EX_2^+ \leq EX_1^+; \cdots \leq E|X_2|^r \leq E|X_1|^r$.

(*b*) *Convergence: If $EX_1^+ < \infty$ or $EX_1^- < \infty$, then there exists a r.v. X such that $X_n \xrightarrow{\text{a.s.}} X$ with $X < \infty$ or $> -\infty$ respectively.*

(c) Closure: If $E|X_1|^r < \infty$, then $X_n \to X$ a.s. or in the rth *mean* $(r \geq 1)$ *with $X \in \mathfrak{L}_r$ and X is the nearest of the closing r.v.'s.*

Theorem 2.10.4. Submartingale Reversed Sequence. *Let $\{X_n\}$ be a reverse submartingale. Then*

(a) Inequalities: $\cdots \leq EX_2 \leq EX_1$; $\cdots \leq EX_2^+ \leq EX_1^+$. *Every $X_n \geq 0$ a.s.* $\Rightarrow \cdots \leq EX_2^r \leq EX_1^r$, $r \geq 1$.

(b) Convergence: If $EX_1^+ < \infty$, then $X_n \xrightarrow{\text{a.s.}} X < \infty$.

(c) Closure: If $|X_n|^r$ are uniformly integrable, then there exists a r.v. X such that $X_n \to X$ a.s. or in the rth *mean* $(r \geq 1)$ *with $X \in \mathfrak{L}_r$ and X is the nearest of the closing r.v.'s.*

For the proofs of the above theorems, see Loève (1963).

These theorems can be used to prove zero-one laws, the Kolmogorov inequality, the strong laws of large numbers and various interesting results connected with the sums of independent r.v.'s. (Cf. Loève, 1963, Chapter 29; Chow 1960, Steiger 1969; Kounias and Weng, 1969.)

2.11. WEAK CONVERGENCE OF SOME STOCHASTIC PROCESSES

Let $F_n^*(x)$, the empirical c.d.f., be the proportion of the sample observations lying to the left of x, and let the empirical process be defined by $\{F_n^*(x): \infty < x < \infty\}$. This process plays a very fundamental role in the development of the theory in some of the subsequent chapters. The Glivenko-Cantelli theorem [cf. section 6] stresses that if the sample random variables are independent and identically distributed according to a distribution $F(x)$, then $F_n^*(x) \to F(x)$ a.s., as $n \to \infty$ for all x: $-\infty < x < \infty$; the result remains true even when the random variables are not identically distributed, provided we replace $F(x)$ by the average c.d.f. $\bar{F}_{(n)}(x) = n^{-1} \sum_{i=1}^n F_i(x)$, where $F_1, \ldots, F_n$ are the respective underlying distributions. Our main concern is to study the asymptotic behavior of the process

$$\{n^{1/2}[F_n^*(x) - \bar{F}_{(n)}(x)]; -\infty < x < \infty\}.$$

For identically distributed random variables, extensive study of this sample process has been made by *(a)* the Russian school, led by A. N. Kolmogorov and N. V. Smirnov, and *(b)* the American school, led by J. L. Doob and followed by quite a number of others. The theory is vastly unified by the subsequent fundamental works of Prohorov (1956) and Skorohod (1956). The problem can now be characterized as a particular case of probability measures in complete separable metric spaces, their existence, equivalence, and convergence. We are not in a position to present a detailed account of this rich contribution, mainly because the material constitutes a great bulk and has already been treated in depth in Billingsley (1968) and Parthasarathy (1967). We shall briefly sketch only the results relevant to our sample processes.

Let X be *separable* (i.e., every pair of distinct points of X have disjoint neighborhoods), and let $X \times Y$ be the Cartesian product of X and Y. If a function d: $X \times Y \to R$ (the real line) exists, such that (*a*) $d(x, y) \geq 0$, where $d(x, y) = 0$ iff $x = y$; (*b*) $d(x, y) = d(y, x)$; and (*c*) $d(x, z) \leq d(x, y) + d(y, z)$ for all $x, y, z \in X$; then X is a *metric space* with the *metric d*. (X, d) is separable when X is separable under the metric d. Also, X is a complete metric space, when, in addition, $d(x_m, x_n) \to 0$ as $m, n \to \infty$ implies that $d(x_n, x_\infty) \to 0$ as $n \to \infty$ for some $x_\infty = x \in X$.

Consider now the space $\mathbf{Z}$ of all real-valued continuous functions $Z(t)$, $t \in I$, where I is the unit (closed) interval $\{t\colon 0 \leq t \leq 1\}$. If we define the metric

$$d(x, y) = \sup_{t \in I} |x(t) - y(t)| \quad \text{where} \quad x(\cdot), y(\cdot) \in \mathbf{Z}, \tag{2.11.1}$$

then $\mathbf{Z}$ is a complete separable metric space. Consider now a stochastic process $\{X(t)\colon t \in I\}$, defined on some probability space $(\Omega, \mathcal{A}, P)$. Thus, $\{X(t)\colon t \in I\}$ is an indexed (possibly uncountable) family of real-valued random variables on Ω; i.e., given $\omega \in \Omega$, $X(t, \omega)$ is a real-valued function on I and is called a *realization* of the stochastic process; the set of realizations is thus composed of the realizations for various possible $\omega \in \Omega$. We shall consider stochastic processes $\{X(t)\colon t \in I\}$ whose realizations (also called sample paths) are all continuous on I. The probability measure μ corresponding to this stochastic process is defined by $\mu(A) = P[\{\omega\colon X(t, \omega) \in A\}]$, for every $A \in \mathcal{B}$, where $\mathcal{B}$ is the smallest σ-field of subsets of $\mathbf{Z}$ with respect to which the mappings $\omega \to X(t, \omega)$, $t \in I$, are all measurable. Note that, with this definition, the events, such as $\sup_{t \in I} |X(t, \omega)| < c$ (c real), are also measurable. We are concerned with the existence (of one), equivalence (of two or more), and convergence (of a sequence) of such measures on $\mathbf{Z}$.

Existence of a Measure on Z

Denote by $\boldsymbol{\tau}_m = \{t_1, \ldots, t_m\} (\subset [0, 1])$ an arbitrary finite subset of I, and let T_m be the set of all possible $\boldsymbol{\tau}_m \subset I$. Also, let

$$P(\mathbf{x}; \boldsymbol{\tau}_m) = P\{X(t_i) \leq x_i,\ i = 1, \ldots, m\}; \qquad \boldsymbol{\tau}_m \in T_m, \quad \mathbf{x} \in R^m; \tag{2.11.2}$$

be the probability distribution of the m-vector $[X(t_1), \ldots, X(t_m)]$, $m \geq 1$. Let μ be a probability measure defined on $\mathbf{Z}$, and let $\mu(\cdot; \boldsymbol{\tau}_m)$ be the corresponding finite (m-) dimensional probability law induced by μ. If such a measure μ exists, then, of course,

$$P(\mathbf{x}; \boldsymbol{\tau}_m) = \mu(\{\omega: X(t_i, \omega) < x_i, i = 1, \ldots, m\}) \tag{2.11.3}$$

for all $\boldsymbol{\tau}_m \in T_m$ and $\mathbf{x} \in \mathfrak{R}^m$, but the converse is not necessarily true. However, it can be shown that (cf. Hájek and Šidák, 1967, p. 175) that if $P(\mathbf{x}; \boldsymbol{\tau}_m)$ satisfies the condition that for every $\epsilon > 0$,

$$\lim_{\delta \to 0} \inf_{\boldsymbol{\tau}_m \in T_m} P\Big(\max_{\substack{|t-t'| < \delta \\ t, t' \in I}} |X(t') - X(t)| < \epsilon \Big) = 1, \tag{2.11.4}$$

then and only then there exists a measure μ on $\mathbf{Z}$, such that (2.11.3) holds for all $\boldsymbol{\tau}_m \in T_m$ and $\mathbf{x} \in R^m$, $m \geq 1$. A. N. Kolmogorov has considered a very simple and easily verifiable sufficient condition for (2.11.4). Instead of the simple form given in Hájek and Šidák (1967, p. 177), we consider the following more general version, which will be required in the sequel.

Theorem 2.11.1. *If there exist positive constants K, r, and s, such that for all $(t, t') \in I$,*

$$E\,|X(t') - X(t)|^r = \iint_{R^2} |u - v|^r \, dP(u, v; t, t') \leq K\,|t' - t|^{1+s}, \tag{2.11.5}$$

where K does not depend on (t, t'), then (2.11.4) *holds, and hence there exists a measure μ for which* (2.11.3) *holds.*

An elegant proof of this theorem is given in Parthasarathy (1967, p. 216). For illustration, consider the following two stochastic processes.

(i) Let $\{\xi(t): 0 \leq t \leq \infty\}$ be a stochastic process, with

$$E\xi(t) = 0, \qquad E\{\xi(s)\xi(t)\} = s, \qquad 0 \leq s \leq t \leq \infty, \tag{2.11.6}$$

and for every $\boldsymbol{\tau}_m (\in T_m)$, the m-dimensional distribution $P(\mathbf{X}; \boldsymbol{\tau}_m)$ is Gaussian with null mean vector and covariance matrix, specified by (2.11.6). It follows then that

$$E[\xi(t) - \xi(s)]^4 = 3[E\{\xi(t) - \xi(s)\}^2]^2 = 3(t - s)^2, \tag{2.11.7}$$

for all $0 \leq s \leq t \leq \infty$. Hence, (2.11.5) holds with $K = 3$, $r = 4$, $s = 1$. Thus, for this process, there exists a probability measure on the space of all continuous functions. The process is known as the *Brownian movement process* and the probability measure as the *Wiener measure*. Note that, here, instead of $I = [t: 0 \leq t \leq 1]$, we have considered the entire positive part of the real line, i.e., $(0, \infty)$, so that in (2.11.1) through (2.11.5), I may be replaced by $I^* = (0, \infty)$.

(ii) Let $\{Z(t): 0 \leq t \leq 1\}$ be a stochastic process with

$$EZ(t) = 0 \quad \text{and} \quad E[Z(s) \cdot Z(t)] = s(1 - t), \qquad 0 \leq s \leq t \leq 1, \tag{2.11.8}$$

and for every $\boldsymbol{\tau}_m (\in T_m)$, the m-dimensional distribution $P(\mathbf{x}; \boldsymbol{\tau}_m)$ is Gaussian. Then, we have

$$(2.11.9) \quad E[Z(t) - Z(s)]^2 = (t - s)(1 - t + s), \qquad 0 \leq s \leq t \leq 1;$$

$$(2.11.10) \quad E[Z(t) - Z(s)]^4 = 3[E(Z(t) - Z(s))^2]^2 \leq 3(t - s)^2, \qquad 0 \leq s \leq t \leq 1,$$

and hence, (2.11.5) again holds with $K = 3$, $r = 4$, and $s = 1$. This insures the existence of a probability measure on $\mathbf{Z}$. The process $\{Z(t)\colon 0 \leq t \leq 1\}$ is known as the *Brownian bridge*.

It follows from the results of Doob (1949) that the two processes can be linked by

$$(2.11.11) \qquad \xi(t) = (t + 1)Z\left(\frac{t}{t + 1}\right), \qquad 0 \leq t < \infty.$$

We will have the occasion to use both the processes.

Equivalence of Two Measures on Z

Suppose now that μ and ν are two probability measures on $\mathbf{Z}$. As before, let $\boldsymbol{\tau}_m = (t_1, \ldots, t_m)$ be a finite subset of I and let T_m be the class of all possible $\boldsymbol{\tau}_m$, $m \geq 1$. The probability measures on the m-dimensional vector space induced by μ and ν are denoted by $\mu(\cdot; \boldsymbol{\tau}_m)$ and $\nu(\cdot; \boldsymbol{\tau}_m)$, respectively. Then, it can be verified that a necessary and sufficient condition that on $\mathbf{Z}$, $\boldsymbol{\mu} = \boldsymbol{\nu}$ is that

$$(2.11.12) \qquad \mu(\cdot; \boldsymbol{\tau}_m) = \nu(\cdot; \boldsymbol{\tau}_m), \quad \text{for every} \quad \boldsymbol{\tau}_m \in T_m, \quad m \geq 1.$$

This concept is then utilized in the main results of this section.

Convergence of a Sequence of Probability Measures on Z

Corresponding to the sequence $\{n\}$ of positive integers, consider a sequence $\{X_n(t)\colon t \in I\}$ of stochastic processes. Suppose that for each n, $\{X_n(t)\colon t \in I\}$ satisfies the condition of Theorem 2.11.1 (or the weaker condition (2.11.4)), so that the process can realize in the space $\mathbf{Z}$ of all continuous functions on I. Also, let $[\xi(t)\colon t \in I]$ be a stochastic process with a suitable probability measure defined on I. For any $\boldsymbol{\tau}_m \subset I$, we denote the m-dimensional distribution functions of $[X_n(t_1), \ldots, X_n(t_m)]$ and $[\xi(t_1), \ldots, \xi(t_m)]$ by $P_n(\mathbf{x}; \boldsymbol{\tau}_m)$ and $\pi(\mathbf{x}; \boldsymbol{\tau}_m)$ respectively. In order that for $[X_n(t)\colon t \in I]$ a probability measure exists, we require (2.11.4), and in order that this measure converges (as $n \to \infty$) to the measure corresponding to $[\xi(t)\colon t \in I]$, we require that $P_n(\mathbf{x}; \boldsymbol{\tau}_m) \to \pi(\mathbf{x}; \boldsymbol{\tau}_m)$ for every $\boldsymbol{\tau}_m \in T_m$ and all $m \geq 1$. Combining these, we have the following.

Theorem 2.11.2. *The sequence of stochastic processes* $\{X_n(t): t \in I\}$ *weakly converges to* $[\xi(t): t \in I]$ *in the Prohorov sense, that is, the metric* (2.11.1) *defined on* $X_n(t) - \xi(t)$, $t \in I$, *converges in law to zero as* $n \to \infty$, *if and only if*

$$\text{(2.11.13)} \quad \text{(i)} \quad \lim_{\delta \to 0} \liminf_{n} P\left\{ \max_{|t-s|<\delta} |X_n(t) - X_n(s)| < \epsilon \right\} = 1,$$

$$\text{(2.11.14)} \quad \text{(ii)} \quad \lim_{n \to \infty} P_n(\mathbf{x}; \boldsymbol{\tau}_m) = \pi(\mathbf{x}; \boldsymbol{\tau}_m), \quad \text{for every} \quad \boldsymbol{\tau}_m \in T_m, \quad m \geq 1.$$

Note that a sufficient condition for (2.11.13) is specified by (2.11.5) and for (2.11.14) it is sufficient to show that for every real $\boldsymbol{\lambda}_{(m)} = (\lambda_1, \ldots, \lambda_m)$, the distribution of $\sum_{j=1}^m \lambda_j X_n(t_j)$ converges to the distribution of $\sum_{j=1}^m \lambda_j \xi(t_j)$, for all $\boldsymbol{\tau}_m \in T_m$ and $m \geq 1$.

Let us now consider two sequences of stochastic processes $\{X_n(t): t \in I\}$ and $\{Y_n(t): t \in I\}$, both defined on the same probability space. Suppose that for both these processes, (2.11.5) holds, ensuring their realizations on $\mathbf{Z}$. We denote by $P_n(\mathbf{x}; \boldsymbol{\tau}_m)$ and $P_n^*(\mathbf{x}; \boldsymbol{\tau}_m)$ the m-dimensional distributions of the vectors $[X_n(t_1), \ldots, X_n(t_m)]$ and $[Y_n(t_1), \ldots, Y_n(t_m)]$, respectively. Then, from Theorem 2.11.2, we may deduce the following.

Theorem 2.11.3. *The two sequences of stochastic processes* $\{X_n(t): t \in I\}$ *and* $\{Y_n(t): t \in I\}$ *are asymptotically weakly equivalent in the sense that* $d(X_n, Y_n)$ *defined by* (2.11.1) *converges in law to zero as* $n \to \infty$, *if and only if* (2.11.13) *holds for each of these sequences and*

$$\text{(2.11.15)} \quad \lim_{n \to \infty} |P_n(\mathbf{x}; \boldsymbol{\tau}_m) - P_n^*(\mathbf{x}; \boldsymbol{\tau}_m)| = 0, \quad \text{for every} \quad \boldsymbol{\tau}_m \in T_m, \quad m \geq 1.$$

Some Results Connected with Gaussian Processes

We have already noted that for the Brownian movement or the Brownian bridge process, the events, such as $\sup_{t \in I} X(t) < c$, and $\sup_{t \in I} |X(t)| < c$, are measurable. We specify their probability laws for subsequent use.

Theorem 2.11.4 (Doob). *Let* $[\xi(t): 0 \leq t < \infty]$ *be a separable Brownian movement process with* $E\xi(t) = 0$ *and* $E[\xi(s)\xi(t) \mid s \leq t] = s$, $0 \leq s \leq t < \infty$. *Then, for every* α, β, λ *and* T (*all* > 0),

$$\text{(2.11.16)} \quad \begin{aligned} P\left\{ \sup_{0 \leq t \leq T} Z(t) \geq \lambda \right\} &= 2P\{Z(T) \geq \lambda\} \\ &= \sqrt{2/\pi T} \int_\lambda^\infty \left(\exp\left\{ \frac{-x^2}{2T} \right\} \right) dx; \end{aligned}$$

$$\text{(2.11.17)} \quad P\left\{ \sup_{0 \leq t < \infty} [Z(t) - \alpha - \beta t] \geq 0 \right\} = e^{-2\alpha\beta};$$

$$\text{(2.11.18)} \quad P\left\{ \sup_{0 \leq t < \infty} |Z(t)|/(\alpha + \beta t) \geq 1 \right\} = 2 \sum_{m=1}^{\infty} (-1)^{m-1} e^{-2m^2\alpha\beta}.$$

For proof, see Doob (1949, pp. 397–401). Thus, if we use the relation (2.11.11), then from Theorem 2.11.4, we arrive at the following.

Theorem 2.11.5. *Let* $[Z(t)\colon 0 \le t \le 1]$ *be a Brownian bridge on* I*; i.e.,* $EZ(t) = 0$ *and* $E[Z(s)Z(t) \mid s \le t] = s(1-t)\colon 0 \le s \le t \le 1$. *Then, for every* $\lambda > 0$,

$$P\Big\{\sup_{t\in I} Z(t) > \lambda\Big\} = e^{-2\lambda^2}, \tag{2.11.19}$$

$$P\Big\{\sup_{t\in I} |Z(t)| > \lambda\Big\} = 2\sum_{m=1}^{\infty}(-1)^{m-1}e^{-2m^2\lambda^2}. \tag{2.11.20}$$

Limiting Behavior of the Empirical Distribution Function

We are now in a position to study the limiting behavior of the empirical process, considered in the beginning of this section.

Let $X_1, X_2, \ldots, X_n, \ldots$ be independent r.v.'s with continuous d.f.'s $F_1, F_2, \ldots, F_n, \ldots$ respectively. Define $c(u) = 1$ or 0 according as $u \ge$ or < 0, and let

$$F_n^*(x) = n^{-1}\sum_{i=1}^{n} c(x - X_i), \qquad -\infty < x < \infty. \tag{2.11.21}$$

Thus, $F_n^*(n)$ is the empirical c.d.f. and it is an unbiased estimate:

$$\bar{F}_{(n)}(x) = n^{-1}\sum_{i=1}^{n} F_i(x), \quad \text{for every} \quad n \ge 1 \quad \text{and all} \quad -\infty < x < \infty. \tag{2.11.22}$$

Note that $F_n^*(x)$ is a step function with n jumps of $1/n$, at the n points $X_{(1)} < \cdots < X_{(n)}$, the ordered values of $X_1, \ldots, X_n$, whereas $\bar{F}_{(n)}(x)$ is continuous. Our interest is to study the *empirical process*

$$n^{1/2}[F_n^*(x) - \bar{F}_{(n)}(x)]; \qquad -\infty < x < \infty. \tag{2.11.23}$$

(a) Reduction to a Process on $I = [0, 1]$.* We define

$$Y_{ni} = \bar{F}_{(n)}(X_i),\ G_{ni}(t) = F_i[\bar{F}_{(n)}^{-1}(t)] = P\{Y_{ni} \le t\},\ i = 1, \ldots, n; \tag{2.11.24}$$

since $0 \le Y_{ni} \le 1$ for every i, the $G_{ni}(t)$ are all defined on I. Also,

$$\bar{G}_{(n)}(t) = n^{-1}\sum_{i=1}^{n} G_{ni}(t) = t, \qquad 0 \le t \le 1.$$

Define the empirical c.d.f.

$$G_n^*(t) = n^{-1}\sum_{i=1}^{n} c(t - Y_{ni}) = F_n^*[\bar{F}_{(n)}^{-1}(t)], \qquad 0 \le t \le 1. \tag{2.11.25}$$

* For an alternative approach based on $D[0, 1]$ methods, refer to Skorohod (1956), Parthasarathy (1967) or Rao (1969).

Then, we can write (2.11.23) equivalently as

$$V_n(t) = n^{1/2}[G_n^*(t) - t]; \qquad 0 \leq t \leq 1. \tag{2.11.26}$$

Since $V_n(t)$ has n discontinuities (i.e., jumps of $n^{-1/2}$), it is not continuous a.e. (almost everywhere) on I, and hence, does not belong to **Z**. However, the difficulty can be avoided as follows. Define another process $[V_n^*(t), t \in I]$ by

$$V_n^*(t) = V_n\left(\frac{k-1}{n}\right) + [nt - (k-1)]\left[V_n\left(\frac{k}{n}\right) - V_n\left(\frac{k-1}{n}\right)\right], \tag{2.11.27}$$
$$\frac{k-1}{n} \leq t < \frac{k}{n},$$

for $k = 1, \ldots, n$. Note that $[V_n^*(t), t \in I]$ has continuous sample paths a.e. We show that such a process can actually be realized in **Z**, the space of all continuous functions on I. Moreover, we also show that as $n \to \infty$, the two processes $[V_n(t); t \in I]$ and $[V_n^*(t), t \in I]$ converge to each other, in the sense that $\sup_{t \in I} |V_n(t) - V_n^*(t)| \to 0$, in probability, as $n \to \infty$.

From (2.11.25) and (2.11.26), it follows that for all $0 \leq s \leq t \leq 1$,

$$EV_n(t) = 0, \qquad E[V_n(s)V_n(t)] = n^{-1}\sum_{i=1}^{n} G_{ni}(s)[1 - G_{ni}(t)]. \tag{2.11.28}$$

Hence, for all $0 \leq s \leq t \leq 1$,

$$\begin{aligned} E[V_n(t) - V_n(s)]^2 &= n^{-1}\sum_{i=1}^{n} [G_{ni}(t) - G_{ni}(s)][1 - G_{ni}(t) + G_{ni}(s)] \\ &\leq n^{-1}\sum_{i=1}^{n} [G_{ni}(t) - G_{ni}(s)] \\ &= t - s, \end{aligned} \tag{2.11.29}$$

and by some routine computations, we have

(2.11.30)

$$E[V_n(t) - V_n(s)]^4 \leq 3(t-s)^2, \quad \text{for every} \quad \{F_i\} \quad \text{and} \quad 0 \leq s \leq t \leq 1.$$

Hence, from (2.11.27), (2.11.29), and (2.11.30), we obtain

$$E[V_n^*(t) - V_n^*(s)]^4 \leq M(t-s)^2, \quad \text{for every } \{F_i\}, \quad \text{where } M \leq \infty. \tag{2.11.31}$$

Hence, using Theorem 2.11.1, it follows that for every $\epsilon > 0$,

$$\liminf_{\delta \to 0} \inf_{n} P\left\{\max_{|t-s|<\delta} |V_n^*(t) - V_n^*(s)| < \epsilon\right\} = 1. \tag{2.11.32}$$

Again, using (2.11.28) and (2.11.27), we obtain

$$\sup_{0 \leq t \leq 1} |V_n^*(t) - V_n(t)| \leq 2n^{-1/2} + \max_{1 \leq k \leq n}\left|V_n^*\left(\frac{k}{n}\right) - V_n^*\left(\frac{k-1}{n}\right)\right|, \tag{2.11.33}$$

and hence, from (2.11.32) and (2.11.33), it follows that for every $\epsilon > 0$,

$$\liminf_n P\left\{ \sup_{0 \le t \le 1} |V_n(t) - V_n^*(t)| < \epsilon \right\} = 1. \tag{2.11.34}$$

Thus, for large n, the process $[V_n(t), t \in I]$ may be replaced (in the sense of weak convergence) by $[V_n^*(t), t \in I]$. The later process can now be used to derive some useful results, considered below.

(b) Kolmogorov-Smirnov Bound. We shall now consider a celebrated result by Kolmogorov for identically distributed random variables. In this case, $F_i = F$, for every $i \Rightarrow G_{ni}(t) = t$: $0 \le t \le 1$, for every i. Hence, from (2.11.28), we have

$$EV_n(t) = 0, \qquad E[V_n(s)V_n(t)] = s(1-t); \qquad 0 \le s \le t \le 1, \quad \text{for every } i. \tag{2.11.35}$$

Since $G_n^*(t)$ is the average of n independent and identically distributed random variables $c(t - Y_{ni})$, $i = 1, \ldots, n$ (where $Y_{ni} = F(X_i)$, $i = 1, \ldots, n, \ldots$ do not depend on n), by the classical central limit theorem (cf. section 7), it follows that for every $m (\ge 1)$ and $\boldsymbol{\tau}_m = \{t_1, \ldots, t_m\} \subset [0, 1]$, the joint distribution of $[V_n(t_1), \ldots, V_n(t_m)]$ is asymptotically multinormal with null mean vector and dispersion matrix with elements

$$t_i(1 - t_j), \qquad 1 \le i \le j \le m. \tag{2.11.36}$$

Hence, using (2.11.32), (2.11.34), and Theorem 2.11.2, it follows that $[V_n(t)\colon t \in I]$ weakly converges to the Brownian bridge on I. Thus, if we define the Brownian bridge $[Z(t)\colon t \in I]$ as in (2.11.8) to (2.11.10), and let

$$B^+ = \sup_{0 \le t \le 1} Z(t), \qquad B^- = \sup_{0 \le t \le 1} [-Z(t)], \qquad B = \sup_{0 \le t \le 1} |Z(t)|\ ; \tag{2.11.37}$$

$$D_n^+ = \sup_{0 \le t \le 1} V_n(t), \qquad D_n^- = \sup_{0 \le t \le 1} [-V_n(t)], \qquad D_n = \sup_{0 \le t \le 1} |V_n(t)|, \tag{2.11.38}$$

then as $n \to \infty$,

$$D_n^+ \to B^+, \quad D_n^- \to B^-, \quad D_n \to B \quad \text{in law.} \tag{2.11.39}$$

Hence, upon making use of Theorem 2.11.5 and (2.11.39), we arrive at the following:

Theorem 2.11.6. *For every $\lambda > 0$,*

$$\begin{aligned} &\lim_{n\to\infty} P\left\{ \sup_{-\infty < x < \infty} n^{1/2}[F_n^*(x) - F(x)] \ge \lambda \right\} \\ &\qquad = \lim_{n\to\infty} P\left\{ \sup_{-\infty < x < \infty} n^{1/2}[F(x) - F_n^*(x)] \ge \lambda \right\} = e^{-2\lambda^2}, \end{aligned} \tag{2.11.40}$$

$$\lim_{n\to\infty} P\left\{ \sup_{-\infty < x < \infty} n^{1/2} |F_n^*(x) - F(x)| \ge \lambda \right\} = 2\sum_{m=1}^{\infty} (-1)^{m-1} e^{-2m^2\lambda^2}. \tag{2.11.41}$$

Smirnov (1939a) and Birnbaum and Tingey (1951) studied the distribution of D_n^+ and D_n^- for finite n. Wald and Wolfowitz (1939) and subsequently Massey (1950) provided a small sample solution for the distribution of D_n. The proofs of the results due to Birnbaum and Tingey (1951), and Massey (1950) are given in Wilks (1962). The reader is also referred to Kolmogorov (1933) for some earlier results in this area.

Consider now the two-sample case. Let $F_m(x)$ and $G_n(x)$ denote the empirical c.d.f.'s of the first and the second samples of sizes m and n respectively. Let $D_n^+ = \sup_x (F_m(x) - G_n(x))$, $D_n^- = \sup_x (G_n(x) - F_m(x))$, and $D_n = \sup_x |F_m(x) - G_n(x)|$. The reader is referred to Hájek and Šidák (1967) for the asymptotic distribution theory of D_n^+, D_n^-, and D_n. Theorem 2.11.6 remains valid for $n_0^{1/2}D_n^+$, $n_0^{1/2}D_n^-$ and $n_0^{1/2}D_n^-$ (where $n_0 = mn/(m+n)$), when the samples are from the same distribution; the original proof is due to Smirnov (1939). For $m = n$ finite, we refer to Massey (1951) and, Gnedenko and Korolink (1951). See also Wilks (1962, pp. 454–459). For other results in this area, the reader is referred to Anderson and Darling (1952), Birnbaum (1952, 1953a, 1953b), Birnbaum and Pyke (1958), Kiefer (1959), and Birnbaum and Hall (1960).

(c) Conservative Property of the Kolmogorov-Smirnov Bound. We shall now show that if the random variables are not identically distributed, then (2.11.40) and (2.11.41) hold, provided we replace $F(x)$ by $\bar{F}_{(n)}(x)$, and the equality signs by $\leq$ signs. This result is proved in Sen, Bhattacharyya, and Suh (1969).

In this case, the multivariate central limit theorem leads to the asymptotic multinormality of the m-vector $[V_n(t_1), \ldots, V_n(t_m)]$ (for every $\boldsymbol{\tau}_m \subset I$), with the covariance matrix having elements

$$n^{-1}\sum_{\alpha=1}^{n} G_{n\alpha}(t_i)[1 - G_{n\alpha}(t_j)], \qquad 1 \leq i \leq j \leq m. \tag{2.11.42}$$

For simplicity of notations, let $G_{n\alpha}(t_i) = t_{i\alpha}$, $\alpha = 1, \ldots, n$, $i = 1, \ldots, m$. Consider then a sequence of Gaussian processes $[Z_n(t)\colon t \in I]$, where $EZ_n(t) = 0$, $0 \leq t \leq 1$, and $E[Z_n(s)Z_n(t) \mid s \leq t]$ is given by (2.11.42) where we put $t_i = s$ and $t_j = t$. This sequence of Gaussian processes can be conceived of as an average of n independent Gaussian processes. Since, by definition, $E[Z_n(t) - Z_n(s)]^2 = n^{-1}\sum_{\alpha=1}^{n}(t_\alpha - s_\alpha)(1 - t_\alpha + s_\alpha) \leq (t-s)$ and $E[Z_n(t) - Z_n(s)]^4 = 3\{E[Z_n(t) - Z_n(s)]^2\}^2 \leq 3(t-s)^2$, again by Theorem 2.11.1, the existence of such a process follows. It follows that for every $m(\geq 1)$, $\boldsymbol{\tau}_m \subset I$ and real $\lambda_1, \ldots, \lambda_m$, the distributions of $\sum_{j=1}^{m}\lambda_j V_n(t_j)$ and $\sum_{j=1}^{m}\lambda_j Z_n(t_j)$ are asymptotically the same, and hence, it follows that $[V_n(t)\colon t \in I]$ converges in law to $[Z_n(t)\colon t \in I]$ as $n \to \infty$. Hence, it suffices to show that

$$\lim_{n\to\infty} P\left\{\sup_{0\leq t\leq 1} |Z_n(t)| > \lambda\right\} \leq 2\sum_{m=1}^{\infty}(-1)^{m-1}e^{-2\lambda^2 m^2}. \tag{2.11.43}$$

Consider now the Brownian bridge $[Z(t): t \in I]$ with $EZ(t) = 0$,

$$E[Z(s)Z(t) \mid s \le t] = s(1 - t), \qquad 0 \le s \le t \le 1.$$

For finitely many points $0 < t_1 < \cdots < t_m < 1$, let $\boldsymbol{\Sigma}_n^{(m)}$ and $\boldsymbol{\Sigma}^{(m)}$ be the covariance matrices of $[Z_n(t), t \in I]$ and $[Z(t), t \in I]$, respectively. It is then easy to verify that

$$\text{(2.11.44)} \quad \boldsymbol{\Sigma}^{(m)} - \boldsymbol{\Sigma}_n^{(m)} = \left(\left(n^{-1}\sum_{\alpha=1}^{n}(t_{i\alpha} - t_i)(t_{j\alpha} - t_j)\right)\right) = \boldsymbol{\Sigma}_n^{*(m)},$$

where $\boldsymbol{\Sigma}^{(m)}$ is positive definite, while $\boldsymbol{\Sigma}_n^{(m)}$ and $\boldsymbol{\Sigma}_n^{*(m)}$ are at least positive semidefinite. Also, by Theorem 2.11.6,

$$\text{(2.11.45)} \qquad \lim_{m\to\infty} P\left\{\max_{1\le j\le m} |Z(t_j)| \ge \lambda\right\} \le 2\sum_{k=1}^{\infty}(-1)^{k-1}e^{-2k^2\lambda^2}.$$

So the desired result will follow if we can show that for every $m(\ge 1)$ and $0 \le t^{(1)} < \cdots < t^{(m)} \le 1$,

$$\text{(2.11.46)} \quad P\left\{\max_{1\le j\le m} |Z_n(t_j)| > \lambda\right\} \le P\left\{\max_{1\le j\le m} |Z(t_j)| > \lambda\right\}, \qquad \lambda > 0,$$

and this is a direct consequence of the following lemma.

Lemma 2.11.7. *Let C_p be a p-dimensional convex space symmetric about the origin. Consider two p-vectors $\mathbf{X}$ and $\mathbf{Y}$ both distributed normally with null means and dispersion matrices $\mathfrak{B}_1$ and $\mathfrak{B}_2$ respectively, where $\mathfrak{B}_1$ is positive definite, $\mathfrak{B}_2$ and $\mathfrak{B}_3 = \mathfrak{B}_1 - \mathfrak{B}_2$ are at least positive semidefinite. Then*

$$\text{(2.11.47)} \qquad P\{\mathbf{X} \in C_p\} \le P\{\mathbf{Y} \in C_p\},$$

where the equality sign holds only when $\mathfrak{B}_3$ is a null matrix.

Proof: There exists a nonsingular matrix $\mathbf{D}$ such that $\mathbf{D}\mathfrak{B}_1\mathbf{D}' = \mathbf{I}_p$ and $\mathbf{D}\mathfrak{B}_2\mathbf{D}' = \boldsymbol{\nu}_p = \text{diag}(\nu_1, \ldots, \nu_p)$, where $\nu_1, \ldots, \nu_p$ are the characteristic roots of $\mathfrak{B}_1^{-1}\mathfrak{B}_2$. By the hypothesis of the lemma, $0 \le \nu_1, \ldots, \nu_p \le 1$. Under the mapping $\mathbf{X} \to \mathbf{DX} = \mathbf{X}^*$, $\mathbf{Y} \to \mathbf{DY} = \mathbf{Y}^*$, let C_p^* be the image of C_p. Since C_p is convex, so is C_p^*. Let then $\nu_1, \ldots, \nu_q > 0$, while the remaining $p - q$ (≥ 0) of the ν's are set equal to zero. If $q = 0$, the proof is trivial, so it will be assumed that $q \ge 1$. Then, for the q-dimensional distribution of $\mathbf{X}_q^*$ and $\mathbf{Y}_q^*$ (the first q components of $\mathbf{X}^*$ and $\mathbf{Y}^*$), let C_{pq}^* be the intersection of C_p^* with the q-dimensional hyperplane $X_{q+1}^* = \cdots = X_p^* = 0$. Since $\nu_1, \ldots, \nu_q$ are all less than or equal to unity, and $X_1^*, \ldots, X_q^*$ (or $Y_1^*, \ldots, Y_q^*$) are all independent, well-known results on the multivariate normal distribution yield that

$$\text{(2.11.48)} \qquad P\{\mathbf{X}_q^* \in C_{pq}^*\} \le P\{\mathbf{Y}_q^* \in C_{pq}^*\},$$

where the equality sign holds *iff* $\nu_1 = \cdots = \nu_q = 1$. Since $C^*_{pq} \subset C^*_p$, it follows that $P\{\mathbf{X}^*_q \in C^*_{pq}\} \geq P\{\mathbf{X}^* \in C^*_p\} = P\{\mathbf{X} \in C_p\}$, while by virtue of $\nu_{q+1} = \cdots = \nu_p = 0$, $Y^*_{q+1} = \cdots = Y^*_p = 0$, with probability one. Hence, $P\{\mathbf{Y}^*_q \in C^*_{pq}\} = P\{\mathbf{Y}^* \in C^*_p\} = P\{\mathbf{Y} \in C_p\}$. Hence,

$$P\{\mathbf{Y} \in C_p\} \geq P\{\mathbf{X} \in C_p\}, \tag{2.11.49}$$

where the equality sign holds iff $\nu_1 = \cdots = \nu_q = 1$ and $C^*_{pq} \equiv C^*_p$, i.e., when $\mathcal{B}_3$ is null. Q.E.D. ◁

We summarize the above results as follows:

Theorem 2.11.8. *For every $\lambda > 0$,*

$$\liminf_n P\left\{\sup_{-\infty < x < \infty} n^{1/2}[F^*_n(x) - \bar{F}_{(n)}(x)] > \lambda\right\} \leq e^{-2\lambda^2}, \tag{2.11.50}$$

$$\liminf_n P\left\{\sup_{-\infty < x < \infty} n^{1/2}\, |F^*_n(x) - \bar{F}_{(n)}(x)| > \lambda\right\} \leq 2\sum_{k=1}^{\infty}(-1)^{k-1}e^{-2k^2\lambda^2}, \tag{2.11.51}$$

where the equality sign holds when $F_i = F$, for every $i \geq 1$.

Note that the actual magnitude of the left-hand side of (2.11.50) or (2.11.51) may be substantially less than their right-hand sides when the c.d.f.'s are not all identical. For example, let $F_i(x) = [x - i + 1]$, $i - 1 \leq x \leq i$, $F_i(x) = 0$, $x \leq i - 1$ and $F_i(x) = 1$, $x \geq i$, for $i = 1, 2, \ldots$. Then, for any n, $\bar{F}_{(n)}(x) = n^{-1}x$: $0 \leq x \leq n$, while $\bar{F}_{(n)}(x) = 0$, for every $x \leq 0$ and $\bar{F}_{(n)}(x) = 1$ for every $x \geq n$. In this case,

$$\sup_{-\infty < x < \infty} n^{1/2}\, |F^*_n(x) - \bar{F}_{(n)}(x)| < n^{-1/2}, \tag{2.11.52}$$

with probability one, and hence, for every $\lambda > 0$, the left-hand sides of (2.11.50) and (2.11.51) actually converge to zero as $n \to \infty$.

(d) Some Bounds on $[V_n(t)\colon t \in I]$. We consider the following theorem which provides better tail bounds for $[V_n(t)\colon t \in I]$.

Theorem 2.11.9. *Let $q(t)$ be a non-negative function, defined on I, such that (a) $q(t)$ is $\uparrow$ in t for $0 \leq t \leq \frac{1}{2}$ and $\downarrow$ in t: $\frac{1}{2} < t < 1$, and (b) $[q(t)]^{-1}$ is square integrable on I. If $F_1 = \cdots = F_n = F$, then for all $n \geq 1$,*

$$\begin{aligned} &P\{n^{1/2}\, |F^*_n(x) - F(x)| \leq q(F(x)),\ -\infty < x < \infty\} \\ &\qquad \geq 1 - \int_0^1 [q(t)]^{-2}\, dt. \end{aligned} \tag{2.11.53}$$

Proof: Define $G^*_n(t)$ as in (2.11.25) and let $V_n(t) = n^{1/2}[G^*_n(t) - t]$, $t \in I$. Then, we are to show that under the hypothesis of the theorem,

$$P\left\{\sup_{0 \leq t \leq 1} |V_n(t)|/q(t) > 1\right\} < \int_0^1 [q(t)]^{-2}\, dt. \tag{2.11.54}$$

Define (for an arbitrary positive integer N, to be chosen later on)

$$t_j = j/N, \qquad 0 \le j \le N; \tag{2.11.55}$$

$$Z_j = n^{1/2}[G_n^*(t_j) - t_j] = V_n(t_j), \qquad 0 \le j \le N. \tag{2.11.56}$$

By definition, $Z_0 = 0$, with probability one, and

$$\begin{aligned} E(Z_j \mid Z_0, Z_1, \ldots, Z_{j-1}) &= E(Z_j \mid Z_{j-1}) = [(1 - t_j)/(1 - t_{j-1})]Z_{j-1} \\ &= (N - j)/(N - j + 1)Z_{j-1} = \psi_j Z_{j-1}, \qquad \text{(say) } 1 \le j \le N. \end{aligned} \tag{2.11.57}$$

Hence, Jensen's inequality leads to

$$E(|Z_j| \mid Z_0, \ldots, Z_{j-1}) \ge \psi_j \, |Z_{j-1}|, \qquad 1 \le j \le N. \tag{2.11.58}$$

Define

$$1/q(t_j) = a_j, \qquad b_j = \max\left(a_j, a_{j+1}\psi_{j+1}, \ldots, a_{[N/2]} \prod_{k=j+1}^{[N/2]} \psi_k\right), \tag{2.11.59}$$

for $1 \le j \le [N/2]$, where $[s]$ is the integral part of s. Since $q(t)$ is $\uparrow$ in t on $0 \le t \le \frac{1}{2}$, $q^{-1}(t)$ is $\downarrow$ in t on $0 \le t \le \frac{1}{2}$, and hence, a_j is $\downarrow j$: $1 \le j \le [N/2]$. Thus,

$$a_j = b_j, \qquad j = 1, \ldots, N/2, \qquad \text{as} \quad \psi_j < 1 \quad \text{for every} \quad j. \tag{2.11.60}$$

Hence, by the Birnbaum-Marshall lemma (see section 2)

$$\begin{aligned} P\left\{\max_{1 \le j \le [N/2]} |Z_j| \, a_j > 1\right\} &\le \sum_{j=1}^{[N/2]} b_j^2\{E(Z_j^2) - \psi_j^2 E(Z_{j-1}^2)\} \\ &\le \frac{1}{N} \sum_{j=1}^{[N/2]} a_j^2 \le \int_0^{1/2} [q(t)]^{-2} \, dt, \end{aligned} \tag{2.11.61}$$

as

$$E(Z_j^2) = j(N - j)/N^2, \qquad 1 \le j \le N. \tag{2.11.62}$$

Consequently, by using again Birnbaum-Marshall lemma

$$P\left\{\sup_{0 \le t \le 1/2} |V_n(t)|/q(t) > 1\right\} < \int_0^{1/2} [q(t)]^{-2} \, dt. \tag{2.11.63}$$

By symmetry, we have

$$P\left\{\sup_{1/2 \le t \le 1} |V_n(t)|/q(t) > 1\right\} < \int_{1/2}^{1} [q(t)]^{-2} \, dt, \tag{2.11.64}$$

and hence, the theorem follows from (2.11.54), (2.11.63), and (2.11.64). Q.E.D. ◁

For an alternative version of the proof of the theorem, see Pyke and Shorack (1968a).

If $F_1, \ldots, F_n$ are not all identical, we can generalize (2.11.53) at the cost of a mild restriction on $q(t)$, namely, that for every $h > 0$, $0 < t < 1$,

$$|\{t(1-t)\}^{1/2}\{q(t+h)/q(t) - 1\}| < Kh; \qquad K < \infty. \tag{2.11.65}$$

In chapter 4, we shall take $q(t)$ as

$$q(t) = c\{t(1-t)\}^{\frac{1}{2}-\delta}, \qquad 0 < \delta \leq \tfrac{1}{2}, \quad c > 0, \tag{2.11.66}$$

and for this, it is easy to verify that (2.11.65) holds. Proceeding then as in the proof of our Theorem 2.11.8 [along the lines of Theorem 5.1 of Sen (1970a)], we obtain the following theorem.

Theorem 2.11.10 *For the stochastic process* $[V_n(t)\colon t \in I]$, *defined by* (2.11.26), *if* $q(t)$ *satisfies* (2.11.65) *and the conditions of Theorem* 2.11.9, *then*

$$\lim_{n\to\infty} P\left\{\sup_{t\in I} |V_n(t)|/q(t) > 1\right\} \leq \int_0^1 [q(t)]^{-2}\, dt. \tag{2.11.67}$$

In particular, for $q(t)$, defined by (2.11.66), we have

$$\lim_{n\to\infty} P\left\{\sup_{t\in I} |V_n(t)|/\{t(1-t)\}^{1/2-\delta} > c\right\} \leq \frac{1}{c^2}\int_0^1 [t(1-t)]^{-1+2\delta}\, dt, \tag{2.11.68}$$

and hence, by proper choice of c, we have

$$\sup_{t\in I} V_n(t)/\{t(1-t)\}^{1/2-\delta} = O_p(1). \tag{2.11.69}$$

Thus, even if $F_1, \ldots, F_n$ are not identical,

$$\sup_{-\infty<x<\infty} \frac{n^{1/2}[F_n^*(x) - \bar{F}_{(n)}(x)]}{\{\bar{F}_{(n)}(x)[1-\bar{F}_{(n)}(x)]\}^{1/2-\delta}}, \quad \delta \leq \tfrac{1}{2}, \tag{2.11.70}$$

is bounded in probability.

2.12. SOME RESULTS IN MEASURE THEORY

The purpose of this section is to give a brief review of some of the basic results in measure theory. For proofs, we refer to Halmos (1950).

Let Ω be the basic set of points ω (in probability theory, the set of elementary events).

Let τ be any class of sets. The smallest field (σ-field) containing τ is called the field (σ-field) generated by τ. The following result is well known.

Proposition 2.12.1. *For any class* τ *of subsets of* Ω, *there exists a unique field* (σ*-field*) *generated by* τ.

A sequence A_n of sets is said to be monotone if it is either nondecreasing: $A_1 \subset A_2 \subset \cdots$ and we then write $A_n\uparrow$; or if it is nonincreasing: $A_1 \supset A_2 \supset \cdots$ and we then write $A_n\downarrow$.

Proposition 2.12.2. *Every monotone sequence is convergent, and* $\lim A_n = \bigcup_{n=1}^{\infty} A_n$ *or* $\bigcap_{n=1}^{\infty} A_n$ *according* $A_n\uparrow$ *or* $A_n\downarrow$.

We say that a class τ of subsets is *monotone* if it is closed under formation of limits of monotone sequences, that is, if $A_n \in \tau$ and $A_n \uparrow A$, or $A_n \downarrow A$ then $A \in \tau$.

Proposition 2.12.3 (Monotone Class Theorem). *A σ-field is a monotone field and conversely. The smallest monotone class M and the smallest σ-field $\mathcal{A}$ over the same field τ coincide.*

A set function is a function whose domain is a class of sets. A set function μ on a class of sets τ is called *finitely additive* if

$$\mu\left(\bigcup_{i=1}^{n} A_i\right) = \sum_{i=1}^{n} \mu(A_i) \tag{2.12.1}$$

where $A_1, \ldots, A_n \in \tau$ and are pairwise disjoint.

μ is called *σ-additive* if, for any sequence of pairwise disjoint sets $A_i \in K$, $i = 1, 2, \ldots$

$$\mu\left(\bigcup_{i=1}^{\infty} A_i\right) = \sum_{i=1}^{\infty} \mu(A_i). \tag{2.12.2}$$

Definition 2.12.1. *A measure μ is an extended real-valued, non-negative and σ-additive set function defined on σ-field $\mathcal{A}$ of sets, satisfying $\mu(\phi) = 0$. A measure μ is finite if $\mu(\Omega) < \infty$; a measure is a probability measure if $\mu(\Omega) = 1$.*

A measure μ is called σ-finite if there exists a sequence of sets $\{A_n\}$ in $\mathcal{A}$ such that $\Omega = \bigcup A_n$ and $\mu(A_n) < \infty$, $n = 1, 2, \ldots$.

$(\Omega, \mathcal{A}, \mu)$ is called a measure space. If $\mu(\Omega) < \infty$, the measure space is finite; if $\mu(\Omega) = 1$, then $(\Omega, \mathcal{A}, \mu)$ is called a probability space.

We restrict ourselves to σ-finite measures. The extension problem for measures is: Suppose we are given a finitely additive measure μ_0 on a field $\mathcal{F}_0$, when does there exist a measure μ on the σ-field generated by $\mathcal{F}_0$ which agrees with μ_0 on $\mathcal{F}_0$?

A measure has certain continuity properties. μ is said to be continuous *from below* or *above* according as

$$\mu(\lim A_n) = \lim \mu(A_n)$$

for every sequence $A_n\uparrow$, or for every sequence $A_n\downarrow$, such that $\mu(A_n) < \infty$ for some n.

Proposition 2.12.4. *Let μ be a measure on the σ-field $\mathcal{A}$.*

(a) If $A_n \downarrow A$, $A_n \in \mathcal{A}$ and if $\mu(A_n) < \infty$ for some n, then

$$\lim_{n\to\infty} \mu(A_n) = \mu\left(\lim_{n\to\infty} A_n\right).$$

(b) *If* $A_n \uparrow A$, $A_n \in \mathcal{A}$, *then*

$$\lim_{n\to\infty} \mu(A_n) = \mu(A).$$

Note that continuity at φ reduces to continuity from above.

Extension Theorem 2.12.1. *A measure μ on a field τ can be extended to a measure on the smallest σ-field over τ. If, moreover, μ is σ-finite, then the extension is uniquely determined, and is σ-finite.*

Some important measurable spaces are $R^{(1)}$: the real line, $\mathcal{A}_1$: the smallest σ-field containing all intervals; $\mathcal{R}_k$: the k-dimensional Euclidean space, $\mathcal{A}_k$: the smallest σ-field containing all k dimensional rectangles.

Let S be a bounded set on $R^{(1)}$. Take $\Omega = (a, b)$ and let $(a, b) \supset S$. Write $S^c = (a, b) - S$. Let U be a finite or countable union of intervals such that

$$S \subset U \subset (a, b). \tag{2.12.3}$$

Represent U as a union of disjoint intervals, and take as $L(U)$ the sum of their lengths. Then the greatest lower bound of $L(U)$ extended over all U satisfying (2.12.3) is called the *outer measure* of S. We denote it by $\bar{L}(S)$. The expression $\underline{L}(S) = b - a - \bar{L}(S^c)$ is called the *inner measure* of S. The set S is Lebesgue-*measurable* (or L-measurable) if $\underline{L}(S) = \bar{L}(S)$. We denote the common value by $L(S)$. Let now S be an unbounded set on $R^{(1)}$. S is L-measurable if, for every $x > 0$, the intersection $[-x, x] \cap S$ is L-measurable. Every Borel set on $R^{(1)}$ is L-measurable. The converse is not true. There exist L-measurable sets which are not Borel sets. However, if a set S is L-measurable and is not a Borel set, it can be represented as $S = S_1 \cup S_2$ where S_1 is a Borel set and S_2 is L-measurable with $L(S_2) = 0$.

Consider two spaces Ω, R and a mapping $\Omega \xrightarrow{X} R$. The inverse image of a set $B \subset R$ is defined as $X^{-1}(B) = \{\omega \in \Omega.\ X(\omega) \in B\}$. The inverse image preserves all set operations, that is,

$$X^{-1}(\bigcup_i A_i) = \bigcup_i (X^{-1}(A_i))$$

and

$$X^{-1}(A - B) = X^{-1}(A) - X^{-1}(B); \qquad X^{-1}\left(\bigcap_i A_i\right) = \bigcap_i (X^{-1}(A_i)).$$

Definition. *Consider two measurable spaces $(\Omega, \mathcal{A})$ and $(R, \mathcal{B})$. A mapping $\Omega \xrightarrow{X} R$ is called measurable if the inverse image of every set in $\mathcal{B}$ is in $\mathcal{A}$.*

Proposition 2.12.5. *Let $\tau \subset \mathcal{B}$ be such that the minimal σ-field over τ is the Borel field $\mathcal{B}$. Then $\Omega \xrightarrow{X} R$ is measurable if the inverse image of every set in τ is in $\mathcal{A}$.*

Definition. *A function X defined on Ω to $R^{(1)}$ is called a measurable function if it is a measurable map from $(\Omega, \mathcal{A})$ to $(R^{(1)}, \mathcal{B}_1)$.*

The measurable functions on $(\Omega, \mathcal{A})$ are called $\mathcal{A}$-measurable functions.

Proposition 2.12.6. *The class of $\mathcal{A}$-measurable functions is closed under pointwise convergence; that is, if X_n are $\mathcal{A}$-measurable, and if $\lim_{n\to\infty} X_n(\omega) = X(\omega)$ exists, then X is $\mathcal{A}$-measurable.*

The indicator of a set $A \subset \Omega$ is the function

$$I_A(\omega) = \begin{cases} 1 & \text{if} \quad \omega \in A \\ 0 & \text{if} \quad \omega \in A^c. \end{cases}$$

A *simple function* is any finite linear combination of indicator functions: $X = \sum_j x_j I_{A_j}(\omega)$ where $A_j \in \mathcal{A}$.

Let $(\Omega, \mathcal{A}, \mu)$ be a measure space. Let X be a non-negative $\mathcal{A}$-measurable function. Then the integral on Ω of a non-negative simple function $X = \sum_{j=1}^m x_j I_{A_j}$ is defined by $\int_\Omega X\, d\mu = \sum_{j=1}^m x_j \mu(A_j)$. Now let X_n be a non-decreasing sequence of non-negative simple functions which converges to X. Then the integral on Ω of X is defined by $\int_\Omega X\, d\mu = \lim \int_\Omega X_n\, d\mu$.

Note that this limit may be infinite. For any $\mathcal{A}$-measurable function X, suppose that $\int_\Omega |X|\, d\mu < \infty$. In this case, define $X^+ = X \cdot I_{[X \geq 0]}$ and $X^- = -XI_{[X<0]}$. Then the integral on Ω of a $\mathcal{A}$-measurable function X is defined by

$$\int_\Omega X\, d\mu = \int_\Omega X^+\, d\mu - \int_\Omega X^-\, d\mu.$$

(*a*) The elementary properties of the integral are: Let $\int X\, d\mu$, $\int Y\, d\mu$, $\int X\, d\mu + \int Y\, d\mu$ exist. Then

(i) $\int (X + Y)\, d\mu = \int X\, d\mu + \int Y\, d\mu$; $\int_{A+B} X\, d\mu = \int_A X\, d\mu + \int_B X\, d\mu$; $\int cX\, d\mu = c \int X\, d\mu$.

(ii) $X \geq 0 \Rightarrow \int X\, d\mu \geq 0$; $\quad X \geq Y \Rightarrow \int X\, d\mu \geq \int Y\, d\mu$; $X = Y$ a.e. $\Rightarrow \int X\, d\mu = \int Y\, d\mu$

(iii) *X integrable $\Leftrightarrow |X|$ integrable $\Rightarrow X$ a.e. finite; $|X| \leq Y$ integrable $\Rightarrow X$ integrable; X and Y integrable $\Rightarrow X + Y$ integrable.*

(*b*) *Let X_n and Y_n be two nondecreasing sequences of non-negative simple functions having the same limit. Then*

$$\lim \int X_n\, d\mu = \lim \int Y_n\, d\mu.$$

(*c*) *Convergence theorems:*

(i) **Monotone Convergence Theorem 2.12.2.** *If $0 \leq X_n \uparrow X$, then $\int X_n\, d\mu \uparrow \int X\, d\mu$.*

(ii) **Theorem 2.12.3.** *If X is integrable, then $\int_A |X|\, d\mu \to 0$ as $\mu(A) \to 0$.*

(iii) **Fatou-Lebesgue Theorem 2.12.4.** *Let Y and Z be integrable functions. If $Y \leq X_n$ or $X_n \leq Z$, then*

$$\int \liminf X_n\, d\mu \leq \liminf \int X_n\, d\mu, \quad \limsup \int X_n\, d\mu \leq \int \limsup X_n\, d\mu.$$

If $Y \leq X_n \uparrow X$ or $Y \leq X_n \leq Z$ and $X_n \xrightarrow{\text{a.e.}} X$, then

$$\lim_{n\to\infty} \int X_n\, d\mu = \int X\, d\mu.$$

(iv) **Dominated Convergence Theorem 2.12.5.** *If $|X_n| \leq Y$ a.e. with Y integrable and if $X_n \xrightarrow{\text{a.e}} X$ or $X_n \xrightarrow{\mu} X$, then*

$$\int X_n\, d\mu \to \int X\, d\mu.$$

A consequence of this theorem is the following theorem.

(v) *Let $X(\omega, \theta)$ be a measurable function of ω for each θ and let $X(\omega, \theta) \to X(\omega, \theta_0)$ as $\theta \to \theta_0$ for each ω. Furthermore, let $|X(\omega, \theta)| < Y(\omega)$ for all ω and θ in any interval including θ_0, and let $Y(\omega)$ be integrable over a set S. Then*

$$\int_S X(\omega, \theta)\, d\mu \to \int_S X(\omega, \theta_0) \quad \text{as } \theta \to \theta_0$$

(vi) **Fubini Theorem 2.12.6.** *Let $(\Omega, \mathcal{A}_i, \mu_i)$, $i = 1, 2$, be two finite measure spaces. If the function $X(\omega_1, \omega_2)$ defined on $\Omega_1 \times \Omega_2$ is measurable with respect to $\mathcal{A}_1 \times \mathcal{A}_2$ and is either non-negative or integrable on $\Omega_1 \times \Omega_2$ with respect to the measure $\mu_1 \times \mu_2$, then*

$$\int_{\Omega_1 \times \Omega_2} X\, d(\mu_1 \times \mu_2) = \int_{\Omega_1} d\mu_1 \int_{\Omega_2} X\, d\mu_2 = \int_{\Omega_1} d\mu_2 \int_{\Omega_2} X\, d\mu_1.$$

(vii) **Radon-Nikodym Theorem.** *Let μ and γ be two σ-finite measures on $(\Omega, \mathcal{A})$. Then there exists a measurable function X such that*

$$\gamma(A) = \int_A X\, d\mu \quad \textit{for all} \quad A \in \mathcal{A} \tag{2.10.1}$$

if and only if γ is absolutely continuous with respect to μ (i.e., $\mu(A) = 0 \Rightarrow \gamma(A) = 0$ for all $A \in \mathcal{A}$; we write this as $\gamma \ll \mu$). (Another way of denoting (2.10.1) is to say that the Radon-Nikodym derivative of γ with respect to μ exists and equals X; i.e., $\partial\gamma/\partial\mu = X$.)

PRINCIPAL REFERENCES

Ali and Chan (1965), Anderson and Darling (1952), Bernstein (1926), Billingsley (1968), Bhattacharyya (1968a), Birnbaum, Raymond and Zuckerman (1947), Birnbaum and Tingey (1951), Birnbaum (1952, 1953a, 1953b), Birnbaum and Hall (1960), Birnbaum and Pyke (1958), Chow (1960), Cochran (1934), Craig (1943), Doob (1949, 1967), Feller (1965), Gnedenko (1944), Gnedenko and Koroljuk (1951), Gnedenko and Kolmogorov (1954), Hájek and Rènyi (1955), Hájek and Šidàk (1967), Halmos (1950), Hoeffding (1963), James (1952), Kiefer (1959), Kolmogorov (1928, 1930, 1933, 1953, 1954), Kounias and Weng (1969), Krickeberg (1965), Lal (1955), Lamperti (1966), Loève (1963), Massey (1950, 1951), Moran (1968), Neveu (1965), Olkin (1960), Olkin and Pratt (1958), Parthasarathy (1967), Prohorov (1956), Rao, C. R. (1965), Rao, J. S. (1969), Sen (1970a), Sen, Bhattacharyya and Suh (1969), Skorohod (1956), Smirnov (1939), Stieger (1969), Tucker (1967), Wald and Wolfowitz (1939), Wilks (1962).

CHAPTER 3

A Survey of Nonparametric Inference

3.1. INTRODUCTION

The theory of nonparametric methods is essentially concerned with the development of statistical inference procedures without making any explicit assumption regarding the functional form of the probability distribution of the sample observations. For the past thirty years various nonparametric procedures have been considered by various workers. The beauty of these procedures is that they remain valid for a wide class of probability distributions and thus increase the scope of the inferences to be made. In the parametric case, the probability law is usually of specified form and involves only a set of finite (or countable) number of unknown (real-valued) parameters in which we are interested. Thus the parameter space is an appropriate dimensional Euclidean space. On the contrary, in the nonparametric case, the probability law is unspecified, and as a result, the parameter space is the space of all distributions (or some proper subspace of that). Thus, in the nonparametric case, the basic formulation of the problem often requires a different approach.

The developments in nonparametric methods are mostly confined to the fields of theory of estimation and of testing of hypotheses, for small as well as large sample sizes. However, the performance characteristics of these methods are usually studied only for large samples. The study of their small-sample properties (efficiency, power, etc.) often requires quite cumbersome computational schemes, and the prospect of developments along this line depends appreciably on modern computing facilities. Even the development of the asymptotic theory of nonparametric inference is somewhat spotty and piecemeal as compared to its parametric counterpart. We shall find it convenient to classify the general body of the theory of nonparametric inference under the following topics:

(*a*) concept of permutation tests, allied permutation theory, and the asymptotic properties of permutation tests;

(*b*) estimation of regular functionals of distribution functions and its role in nonparametric inference;

(*c*) distribution theory of rank order statistics and its use;

(*d*) certain goodness of fit tests; and

(*e*) nonparametric tolerance regions.

The basic ideas concerning permutation tests originated from R. A. Fisher (1935). For relatively rigorous mathematical treatment of permutation tests the reader is referred to Pitman (1937a,b; 1938), Welch (1937), Scheffé (1943b), and Lehmann and Stein (1949). Study of the asymptotic theory of permutation tests has been done by Wald and Wolfowitz (1944), Noether (1949b), Hoeffding (1951a, 1952), Dwass (1953, 1955b), Motoo (1957), Hájek (1961), and Sen (1965b, 1966c), among others. Estimation of regular functionals of distribution functions has been considered by Halmos (1946), von Mises (1947), and Hoeffding (1948a) among others. The latter paper provides a very systematic account of the theory of the so-called U-*statistics*, on which further works are due to Lehmann (1951a), Sukhatme (1957, 1958a, b), Sukhatme and Sethuraman (1959), Sen (1960, 1963a, 1965b), and Nandi and Sen (1963), among others. Many nonparametric tests and estimates are based on U-statistics, and to mention a few important ones, we may refer to the papers of Kendall (1938), Wilcoxon (1945), Mann and Whitney (1947), Kruskal and Wallis (1952), Sukhatme (1957, 1958a, b), Tamura (1960) and Bhapkar (1961b). However, these results do not apply to another important class of nonparametric statistics, *rank order statistics*, of which van der Waerden's (1952/53, 1953a) test and Fisher and Yates' (1938) test are notable examples. This class also includes the tests of Mood (1950, 1954), the signed rank test of Wilcoxon (1949) and Kruskal and Wallis (1952), and Fraser's (1957a) normal score tests, as particular cases. We may also mention the work of Lehmann (1953), Hoeffding (1953), and Terry (1952) in this connection. A very elegant theory of the asymptotic distribution of rank order statistics has been established by Chernoff and Savage (1958). This has been generalized in different directions by Govindarajulu (1960), Klotz (1962), Hodges and Lehmann (1963), Bhuchongkul (1964), Bhuchongkul and Puri (1965), Lehmann (1963d), Puri (1964, 1965a, 1965b, 1965c, 1967, 1969), Mehra and Puri (1967), Puri and Sen (1967a, 1967b, 1968, 1969b), Sen (1970a,e), and Sen and Puri (1967), among others. We shall deal with the multivariate versions of some of these results in subsequent chapters.

Well-known nonparametric "goodness of fit" tests are due to Karl Pearson (1900), Kolmogorov (1933), Cramér and von Mises (cf. Cramér, 1946, pp. 450–452), and Kuiper (1960), among others. In chapter 2 (section 2.11), we considered briefly the large sample distribution of the one-sample Kolmogorov-Smirnov statistic. We also cited some references on the small-sample distribution of this statistic as well as its two-sample and multisample analogs. A brief outline of the Cramér-von Mises statistic is given in Cramér (1946, pp. 450–452), and an excellent review of both these statistics is given by Darling (1957). A later expository paper by Barton and Mallows (1965) also provides some additional literature on these topics. For the Kuiper test, see Kuiper (1960) and Abrahamson (1967).

Finally, nonparametric tolerance regions have been studied by Wilks (1941), Wald (1943a), Tukey (1948), and Fraser (1951, 1953), among others. Since this branch of nonparametric inference has been more or less exhaustively studied and is contained in many standard textbooks, we shall not discuss it here.

In this chapter, we shall attempt to survey the univariate nonparametric procedures which have some relevance to the multivariate procedures to be considered in the subsequent chapters.

3.2. DISTRIBUTION THEORY OF REGULAR FUNCTIONS OF DISTRIBUTION FUNCTIONS

In the nonparametric case, the probability distribution of the sample observations is of unspecified form, and it belongs to some class of unknown distributions (the class of all nondegenerate distributions, the class of all continuous distributions, etc.). Consequently, in the nonparametric case, the parameters are usually defined as certain functionals of the distribution functions. A formulation of such functionals is considered in the next paragraph. However, we note that prudent choice of such a functional is an important task and plays a fundamental role in many problems of nonparametric inference. In addition, we are faced with the problem of finding suitable estimators of such functionals and of studying their various properties. For this study, we shall mainly consider the results of Halmos (1946), von Mises (1947) and Hoeffding (1948a) along with their generalizations by others. We shall consider in detail the single-sample situation, but as and when necessary, we shall also consider briefly the case of several samples.

Let $X_1, \ldots, X_n$ be n independent and identically distributed (real- or vector-valued) random variables (i.i.d.r.v.) having a cumulative distribution function (c.d.f.) $F(x)$. Let R^k stand for the $k(\geq 1)$-dimensional Euclidean space, and let us assume that $X_i \in R^k$, for some $k \geq 1$. Let then $\mathcal{F}$ be a class of c.d.f.'s in the k-dimensional Euclidean space, and let $\theta(F)$ be a real-valued functional of the c.d.f. F whose domain is $\mathcal{F}$ and whose range is contained in the real line R. We may note that $\theta(F)$ may also be vector-valued, having $s(\geq 1)$ components. However, for simplicity of presentation, we shall assume that $s = 1$. $\theta(F)$ is called a *regular functional* over $\mathcal{F}$, if for all $F \in \mathcal{F}$, $\theta(F)$ admits an unbiased estimator, say $\phi(X_1, \ldots, X_n)$, that is,

$$\int \cdots \int_{R^{kn}} \phi(x_1, \ldots, x_n)\, dF(x_1) \cdots dF(x_n) = \theta(F) \quad \text{for all} \quad F \in \mathcal{F}. \tag{3.2.1}$$

If (3.2.1) holds, we also say that $\theta(F)$ is estimable. If $\theta(F)$ is estimable, the smallest sample size (say m) for which (3.2.1) holds is called the *degree* (over $\mathcal{F}$) of $\theta(F)$, and $\phi(X_1, \ldots, X_m)$ is called the *kernel* of $\theta(F)$. Without any loss of generality we may assume that a kernel is symmetric in its m arguments (as otherwise, we may consider the following kernel which is symmetric in $X_1, \ldots, X_m$:

$$\phi_0(X_1, \ldots, X_m) = \frac{1}{m!} \sum \phi(X_{\alpha_1}, \ldots, X_{\alpha_m}), \tag{3.2.2}$$

where the summation extends over all possible ($m!$) permutations $(\alpha_1, \ldots, \alpha_m)$ of $(1, \ldots, m)$). It readily follows that if $\theta_i(F)$, $i = 1, \ldots, r(\geq 1)$, are all regular functionals, then any polynomial in them is also so. In particular, if m_i is the degree of $\theta_i(F)$, $i = 1, \ldots, r$, then the degrees of $\sum_1^k \theta_i(F)$ and $\prod_1^k \theta_i(F)$ are less than or equal to $\max(m_1, \ldots, m_k)$ and $\sum_i^k m_i$, respectively. Let us consider the following simple examples.

Example 3.2.1. Let $\mu(F) = \int_R x\, dF(x)$ and $\sigma^2(F) = \int_R (x - \mu)^2\, dF(x)$, where x is real-valued. We then note that

$$\sigma^2(F) = \frac{1}{2} \iint_{R^2} (x_1 - x_2)^2\, dF(x_1)\, dF(x_2), \tag{3.2.3}$$

so that $\sigma^2(F)$ is an estimable parameter of degree 2 for the class of c.d.f.'s, for which the second moment exists.

In general, $\mu_k(F) = \int_R (x - \mu)^k\, dF(x)$, $k \geq 1$, cumulants, Gini's mean difference (i.e., $\iint_{R^2} |x_1 - x_2|\, dF(x_1)\, dF(x_2)$), etc., are all estimable parameters over appropriate classes of c.d.f.'s.

Example 3.2.2. Suppose $X = (X^{(1)}, X^{(2)})$ has the bivariate c.d.f. $F(x_1, x_2)$. X_1 and X_2 are said to be *concordant* if $X_1^{(1)} - X_2^{(1)}$ and $X_1^{(2)} - X_2^{(2)}$ have the same sign. Let $\theta(F)$ be the *probability of concordance*. If we consider the kernel

$$\phi(X_1, X_2) = \begin{cases} 1, & \text{if } X_1, X_2 \text{ are concordant,} \\ 0, & \text{otherwise;} \end{cases} \tag{3.2.4}$$

then it is easily seen that $E_F\{\phi(X_1, X_2)\} = \theta(F)$ for all F. Thus $\theta(F)$ is an estimable parameter of degree 2.

Later on, we shall consider some more examples of this type.

Once the kernel $\phi(X_1, \ldots, X_m)$ of $\theta(F)$ is obtained, one may proceed in the following two ways. First, let us define the empirical c.d.f. $F_n(x)$

$$F_n(x) = \frac{1}{n} [\text{number of } X_i \leq x,\ i = 1, \ldots, n] \tag{3.2.5}$$

and with the definition of $\theta(F)$ in (3.2.1), consider the statistic

$$\theta(F_n) = \int_{R^{kn}} \cdots \int \phi(x_1, \ldots, x_m)\, dF_n(x_1) \cdots dF_n(x_m) \tag{3.2.6}$$
$$= n^{-m} \sum_{\alpha_1=1}^{n} \cdots \sum_{\alpha_m=1}^{n} \phi(X_{\alpha_1}, \ldots, X_{\alpha_m}).$$

Such an estimator is quite simple to define, but may not be unbiased (e.g., if in (3.2.3) we use $F_n(x)$ instead of $F(x)$, the resulting statistic will be

$$\frac{1}{n}\sum_1^n X_i^2 - \left(\frac{1}{n}\sum_1^n X_i\right)^2$$

which is a biased estimator of $\sigma^2(F)$). Some studies of the asymptotic distribution theory of such functionals of the empirical c.d.f.'s have been made by von Mises (1947) and Fillipova (1961). We shall not enter into detailed discussion of these functionals, but will refer briefly to their relation to the other class of unbiased estimators which we consider now.

Corresponding to a (symmetric) kernel $\phi(X_1, \ldots, X_m)$ of $\theta(F)$, we define a symmetric unbiased estimator (or *U-statistic*) by

$$U_n = U(\mathbf{X}_n) = \binom{n}{m}^{-1} \sum_S \phi(X_{\alpha_1}, \ldots, X_{\alpha_m}), \tag{3.2.7}$$

where $\mathbf{X}_n = (X_1, \ldots, X_n)$ and the summation S extends over all possible $1 \le \alpha_1 < \cdots < \alpha_m \le n$. Clearly

$$E_F\{U(\mathbf{X}_n)\} = \theta(F) \quad \text{for all} \quad F \in \mathcal{F} \quad \text{and} \quad n \ge m. \tag{3.2.8}$$

Let us also write

$$\phi_c(x_1, \ldots, x_c) = E_F\{\phi(x_1, \ldots, x_c, X_{c+1}, \ldots, X_m)\}, \tag{3.2.9}$$

$$\psi_c(X_1, \ldots, X_c) = \phi_c(X_1, \ldots, X_c) - \theta(F); \tag{3.2.10}$$

$$\zeta_c(F) = E_F\{\psi_c^2(X_1, \ldots, X_c)\}, \quad \text{for all} \quad c = 0, 1, \ldots, m, \tag{3.2.11}$$

where naturally $\phi_0 = \theta(F)$ and $\zeta_0(F) = 0$ for all F. From (3.2.9), (3.2.10), and (3.2.11) it readily follows that $\zeta_c(F)$, $1 \le c \le m$ are also regular functionals and $\zeta_c(F)$ is nondecreasing in c: $1 \le c \le m$; $\zeta_m(F)$ is the variance of the kernel itself, and thus if $\zeta_m(F) < \infty$, $\zeta_c(F) < \infty$ for all $c = 1, \ldots, m$. Further, if for some $d \ge 1$, $\zeta_d(F_0) = 0$ then $\zeta_c(F_0) = 0$ for all $c \le d$, where F_0 is any point of $\mathcal{F}$. Hoeffding (1948a) has obtained the following interesting inequalities

$$0 \le \zeta_c \le \frac{c}{d}\zeta_d \quad \text{for all} \quad 1 \le c < d \le m, \tag{3.2.12}$$

the proof of which is left as an exercise. Now, it follows from (3.2.7) that

$$\sigma^2(U_n) = \binom{n}{m}^{-2} \sum_S \sum_{S^*} \operatorname{cov}\{\phi(X_{\alpha_1}, \ldots, X_{\alpha_m}), \phi(X_{\beta_1}, \ldots, X_{\beta_m})\}, \tag{3.2.13}$$

where the summation S^* extends over all possible $1 \leq \beta_1 < \cdots < \beta_m \leq n$. Now for each $\phi(X_{\alpha_1}, \ldots, X_{\alpha_m})$ there will be $\binom{m}{c}\binom{n-m}{m-c}$ terms of the form $\phi(X_{\beta_1}, \ldots, X_{\beta_m})$ for which exactly c of the β's are the same as c of the α's, where c may range from 0 to m. Thus, using (3.2.9), (3.2.10), and (3.2.11), we get from (3.2.13)

$$\sigma^2(U_n) = \binom{n}{m}^{-1} \sum_{c=1}^{m} \binom{m}{c}\binom{n-m}{m-c} \zeta_c(F). \tag{3.2.14}$$

Again, using (3.2.12), (3.2.14), and a few simple algebraic adjustments, we can show that

$$n\sigma^2(U_n) \text{ is decreasing in } n; \quad \sigma^2(U_m) = \zeta_m(F); \tag{3.2.15}$$

and

$$\lim_{n\to\infty} n\sigma^2(U_n) = m^2\zeta_1(F).$$

If now for some $d \geq 0$ and $F_0 \in \mathcal{F}$,

$$\zeta_0(F_0) = \cdots = \zeta_d(F_0) = 0, \qquad \zeta_{d+1}(F_0) > 0, \tag{3.2.16}$$

$\theta(F)$ will be called *stationary of order d* for $F = F_0 \in \mathcal{F}$. It then follows from (3.2.15) that $n\sigma^2(U_n)$ tends to a positive limit for $F = F_0$, only if $\theta(F)$ is stationary of order zero. In fact, in a majority of the cases, we shall see that $\theta(F)$ is stationary of order zero and our subsequent study will be concerned with such parameters. However, for clarification of ideas, we consider the following counterexample.

Example 3.2.3. Let X_i, $i = 1, \ldots, n$ be i.i.d.r.v. having expectation $\mu(F)$ and let $\theta(F) = [\mu(F)]^k$, where k is a positive integer. Naturally, $\theta(F)$ is estimable, as $\phi(X_1, \ldots, X_k) = \prod_1^k X_i$ is an unbiased estimator of it for all F. Let F_0 be a c.d.f. for which $\mu(F_0) = 0$. It is easy to show that $\zeta_1(F_0) = \cdots = \zeta_{k-1}(F_0) = 0$, $\zeta_k(F_0) > 0$, if $\int x^2\, dF_0(x) > 0$, that is, $\theta(F)$ is stationary of order $k - 1$ for $F = F_0$.

Some similar examples are given as exercises at the end of this chapter.

If $\theta(F)$ is stationary of order zero, $n^{1/2}[U_n - \theta(F)]$ will have a nondegenerate distribution with zero mean and limiting variance $m^2\zeta_1(F)(> 0)$. In general, if $\theta(F)$ is stationary of order d, $n^{(1+d)/2}[U_n - \theta(F)]$ will have a nondegenerate distribution with zero mean and limiting variance

$$\left\{(d+1)!\binom{m}{d+1}^2 \zeta_{d+1}(F)\right\},$$

but it may not be normal when $d > 0$.

In the theory of unbiased estimation, the U-statistics play a very fundamental role. First, if there exists an unbiased estimator $f(X_1, \ldots, X_n)$ of $\theta(F)$, and U_n is the corresponding U-statistic, then

$$\text{var}_F \{U_n\} \leq \text{var}_F \{f(X_1, \ldots, X_n)\}, \tag{3.2.17}$$

where the equality sign holds if and only if $f(x_1, \ldots, x_n) = U_n$ a.e. The same result also holds for the risks using any convex loss function. The proof of this result follows readily from the Rao-Blackwell theorem (cf. Fraser, 1957b, pp. 141–142) and hence is omitted. Second, if X_i, $i = 1, \ldots, n$ are all real-valued, the order statistic

$$t(\mathbf{X}_n) = \{X_{(1)} \leq \cdots \leq X_{(n)}\} \tag{3.2.18}$$

is a symmetric function of $\mathbf{X}_n = (X_1, \ldots, X_n)$ and conversely, any symmetric function of the sample observations can be written as a function of the order statistics. Furthermore, the order statistics $t(\mathbf{X}_n)$ are complete for the class of distributions over R^n corresponding to each coordinate having the same distribution $F(x)$ which is (i) any discrete distribution, or (ii) any absolutely continuous distribution. The proofs of these results are contained in Halmos (1946) and Fraser (1957b, pp. 28–31). Extensions of these results to the multisample case are due to Fraser (1953b). Now, if $t(\mathbf{X}_n)$ is complete for the class of c.d.f.'s $\mathcal{F}$, then U_n is the unique minimum variance unbiased (m.v.u.) estimator as well as the minimum risk (with any convex loss function) estimator of $\theta(F)$ for $F \in \mathcal{F}$; for the proof, see Fraser (1957, p. 142).

The above results give some justification for using U-statistics in the theory of estimation. However, its importance should not be *overemphasized*, since the scope is only confined to unbiased estimation. U-statistics are quite often used in nonparametric methods, especially as suitable estimators or test statistics. For such purposes, we often require the knowledge of the exact sampling distribution of U-statistics. In many nonparametric problems, under suitable hypotheses of invariance, the exact sampling distribution of U-statistics can be obtained by direct enumerations of several equally likely realizations. Examples are the exact sampling distributions (under the null hypotheses) of the Wilcoxon-Mann-Whitney statistic or Kendall's rank correlation coefficient (cf. Owen, 1962; Hájek, 1969). However, unlike the parametric case, in general, the small-sample distribution of the U-statistic may not be very neat or explicit in form. We consider below the following theorems on the large-sample distribution of U-statistics. In this connection, we shall consider specifically regular functionals which are stationary of order zero.

Theorem 3.2.1 (Hoeffding). *If $\theta(F)$ is stationary of order zero and $\zeta_m(F) < \infty$,*

$$\lim_{n\to\infty} P_F\{n^{1/2}[U_n - \theta(F)]/m[\zeta_1(F)]^{1/2} \leq x\} = \int_{\infty}^{x} \frac{1}{\sqrt{2\pi}} e^{-t^2/2}\, dt,$$

for all real x.

Proof: Let us define

$$Y_n = n^{-1/2} \sum_{i=1}^{n} \psi_1(X_i), \tag{3.2.19}$$

where by (3.2.10) $\psi_1(X)$ has mean zero and variance $\zeta_1(F) > 0$. Thus from the classical central limit theorem (under Lindeberg's condition) (cf. chapter 2, p. 22) it readily follows that

$$\mathcal{L}(Y_n/[\zeta_1(F)]^{1/2}) \to N(0, 1). \tag{3.2.20}$$

Essentially by the same arguments as in (3.2.13) and (3.2.14) it is easily shown that

$$E_F(Y_n^2) = \zeta_1(F),\ (n/m^2)E_F[U_n - \theta(F)]^2 = \zeta_1(F) + O(n^{-1}), \tag{3.2.21}$$

and
$$n^{1/2}E_F\{Y_n[U_n - \theta(F)]\} = m\zeta_1(F) + O(n^{-1}).$$

Consequently,

$$E\{n^{1/2}[U_n - \theta(F)] - mY_n\}^2 = O(n^{-1}), \tag{3.2.22}$$

and hence using Chebychev's lemma, we obtain

$$n^{1/2}[U_n - \theta(F)] - mY_n = O_p(n^{-1/2}). \tag{3.2.23}$$

The rest of the proof is straightforward, and is therefore omitted. ◁

Theorem 3.2.2 (Hoeffding). *If $\zeta_m(F) < c_1 < \infty$, $\zeta_1(F) > c_2 > 0$, and $E_F\{|\psi_1(X)|^3\} < c_3 < \infty$, all uniformly in $F \in \mathcal{F}$, then the convergence in Theorem* 3.2.1 *is uniform in x ($-\infty < x < \infty$) and $F(\in \mathcal{F})$.*

Proof: Let us denote by $G_n(x)$ and $H_n(x)$ the c.d.f.'s of $Y_n/[\zeta_1(F)]^{1/2}$ and $n^{1/2}[U_n - \theta(F)]/m[\zeta_1(F)]^{1/2}$ respectively, and let $\Phi(x)$ be the standardized normal c.d.f. Then, it follows from (3.2.21) and the Berry-Esséen theorem (chapter 2, p. 24) that if $\zeta_1(F) > c_2 > 0$ and $E_F\{|\Psi_1(X)|^3\} < c_3 < \infty$, uniformly in $F \in \mathcal{F}$, then

$$|G_n(x) - \Phi(x)| < cn^{-1/2}, \tag{3.2.24}$$

$$c < \infty, \quad \text{uniformly in } x\ (-\infty < x < \infty) \quad \text{and} \quad F \in \mathcal{F}.$$

Also (3.2.22) and (3.2.23) hold uniformly in $F \in \mathcal{F}$ (as $\zeta_m(F) < c_1 < \infty$ uniformly in $F \in \mathcal{F}$). Thus, if we write $n^{-1/2}[U_n - \theta(F)] = mY_n + R_n$, we have $E_F(R_n^2) = O(n^{-1})$, uniformly in $F \in \mathcal{F}$. Hence, using the straightforward inequalities

$$G_n(x - \epsilon) - P\{|R_n| > \epsilon[\zeta_1(F)]^{1/2}\} \leq H_n(x) \leq G_n(x + \epsilon) + P\{|R_n| > \epsilon[\zeta_1(F)^{1/2}]\}, \tag{3.2.25}$$

and

$$P\{|R_n| > \epsilon[\zeta_1(F)]^{1/2}\} \leq P\{|R_n| > \epsilon c_1^{1/2}\} = O\left(\frac{1}{n\epsilon^2 c_1}\right) < \epsilon', \tag{3.2.26}$$

(where ϵ' is also arbitrary for $n \geq n_0(\epsilon)$), we get from (3.2.25) and (3.2.26) that for $n > n_0(\epsilon)$

$$|G_n(x \pm \epsilon) - H_n(x)| \leq \epsilon', \quad \text{uniformly in } x \text{ and } F. \tag{3.2.27}$$

Again, $|\Phi(x \pm \epsilon) - \Phi(x)| \leq \epsilon/\sqrt{2\pi}$, uniformly in x. Consequently, from (3.2.24), (3.2.27), and the last inequality

$$|H_n(x) - \Phi(x)| < \delta, \tag{3.2.28}$$

uniformly in x $(-\infty < x < \infty)$ and $F \in \mathfrak{F}$, where δ is also arbitrarily small. Hence the theorem. ◁

The following theorem states the stochastic relationship between $\theta(F_n)$ and U_n.

Theorem 3.2.3 (Hoeffding). *For any regular functional $\theta(F)$, if the variance of $\theta(F_n)$ exists, then $|\theta(F_n) - U_n| - O_p(n^{-1})$. Hence, if $\theta(F)$ is stationary of order zero, $n^{1/2}[\theta(F_n) - \theta(F)]$ and $n^{1/2}[U_n - \theta(F)]$ have the same limiting distribution.*

Proof: It readily follows from (3.2.6) and (3.2.7) that

$$\theta(F_n) = U_n + R_n, \quad \text{where} \quad E_F(R_n^2) = O(n^{-2}). \tag{3.2.29}$$

Consequently by Chebychev's lemma, $|R_n| = O_p(n^{-1})$. Hence the first part of the theorem. The second part of the theorem follows readily from the fact that

$$n^{1/2}\{|[\theta(F_n) - \theta(F)] - [U_n - \theta(F)]|\} = n^{1/2}|R_n| = O_p(n^{-1/2}), \tag{3.2.30}$$

and that for $\zeta_1(F) > 0$, $n^{1/2}[U_n - \theta(F)]$ has a nondegenerate limiting distribution. Hence the theorem. ◁

Corollary 3.2.3.1. *If $\theta(F)$ is stationary of order zero, the variance of $\theta(F_n)$ exists, and $\zeta_m(F) < \infty$, then $n^{1/2}[\theta(F_n) - \theta(F)]/m[\zeta_1(F)]^{1/2}$ converges in law to a normal distribution with zero mean and unit variance. Furthermore, if the conditions of Theorem 3.2.3 hold, this convergence is uniform in $F \in \mathfrak{F}$.*

The proof is an immediate consequence of the preceding three theorems and some simple algebraic manipulations.

Corollary 3.2.3.2. *If $\theta(F) = (\theta_1(F), \ldots, \theta_s(F))$ is a vector of regular functionals and $U_n = (U_n^{(1)}, \ldots, U_n^{(s)})$ is the corresponding vector of U-statistics, then under the conditions of Theorem 3.2.2,*

$$(n^{1/2})[U_n^{(i)} - \theta_i(F)]/m_i, \qquad i = 1, \ldots, s, \tag{3.2.31}$$

have (jointly) asymptotically a multinormal distribution with null mean vector and dispersion matrix with elements

$$\tau_{ij} = \zeta_1^{(i,j)}(F) = E_F\{\psi_1^{(i)}(x_\alpha)\psi_1^{(j)}(x_\alpha)\}, \qquad i, j = 1, \ldots, s. \tag{3.2.32}$$

The same result also applies to the vector

$$n^{1/2}[\theta_i(F_n) - \theta_i(F)]/m_i, \qquad i = 1, \ldots, s, \tag{3.2.33}$$

provided the variances of all $\theta_i(F_n)$ $(i = 1, \ldots, s)$ exist.

In actual practice we often need to estimate the values of $\zeta_1(F)$, defined by (3.2.11). We have already noted that $\zeta_c(F)$, $1 \leq c \leq m$, are all estimable parameters. An unbiased estimator of $\zeta_c(F) + \theta^2(F)$ is

$$U_{c,n} = \left\{\binom{n}{2m-c}\binom{2m-c}{c}\binom{2m-2c}{m-c}\right\}^{-1} \sum\nolimits_c \phi(X_{\alpha_1}, \ldots, X_{\alpha_m}) \cdot \phi(X_{\beta_1}, \ldots, X_{\beta_m}), \tag{3.2.34}$$

where the summation $\sum_c$ extends over all possible $1 \leq \alpha_1 < \cdots < \alpha_m \leq n$, $1 \leq \beta_1 < \cdots < \beta_m \leq n$ with the restriction that $\alpha_i = \beta_i$, $i = 1, \ldots, c$, and $\alpha_i \neq \beta_j$ for any other (i, j), for $c = 1, \ldots, m$. Since $\theta^2(F)$ is also a regular functional (with kernel $\phi(X_1, \ldots, X_m)\phi(X_{m+1}, \ldots, X_{2m})$), an unbiased symmetric estimator of $\zeta_c(F)$ can readily be obtained from (3.2.34). However, sometimes it may be computationally more convenient to consider the following type of estimator which may not be strictly unbiased. Let us write for this purpose

$$V_n^{(1)}(X_i) = \binom{n-1}{m-1}^{-1} \sum\nolimits_i \phi(X_i, X_{\alpha_2}, \ldots, X_{\alpha_m}), \qquad i = 1, \ldots, n, \tag{3.2.35}$$

where the summation $\sum_i$ extends over all possible $1 \leq \alpha_2 < \cdots < \alpha_m \leq n$, with $\alpha_j \neq i$. It then follows readily that

$$U_n = \frac{1}{n}\sum_{i=1}^{n} V_n^{(1)}(X_i). \tag{3.2.36}$$

Again, essentially by the same arguments as in (3.2.13) and (3.2.14), it can be shown that if $\zeta_m(F) < \infty$

$$E_F[V_n^{(1)}(X_i) - \theta(F)]^2 = \zeta_1(F) + O(n^{-1}), \qquad E_F\{\psi_1^2(X_i)\} = \zeta_1(F),$$

$$E_F\{\psi_1(X_i)[V_n^{(1)}(X_i) - \theta(F)]\} = \zeta_1(F) + O(n^{-1}), \tag{3.2.37}$$

and hence we readily obtain

$$E_F\{[V_n^{(1)}(X_i) - \theta(F)] - \psi_1(X_i)\}^2 = O(n^{-1});$$

that is,

$$\{[V_n^{(1)}(X_i) - \theta(F)] - \psi_1(X_i)\}^2 \xrightarrow{p} 0, \tag{3.2.38}$$

uniformly in $i = 1, \ldots, n$.

Also by Khinchine's law of large numbers (chapter 2, p. 18)

$$\frac{1}{n}\sum_{i=1}^{n} \psi_1^2(X_i) \xrightarrow{p} \zeta_1(F) \quad \text{as } n \to \infty. \tag{3.2.39}$$

Let us consider the following lemma.

Lemma 3.2.4. *Let $\{a_i\}$ and $\{b_i\}$ be two sequences of random variables (not necessarily independent or identically distributed), such that*

$$\text{(i)} \quad \frac{1}{n}\sum_{i=1}^{n} a_i^2 \xrightarrow{p} A > 0, \quad \textit{and} \quad \text{(ii)} \quad \frac{1}{n}\sum_{i=1}^{n} (a_i - b_i)^2 \xrightarrow{p} 0.$$

Then

$$\frac{1}{n}\sum_{i=1}^{n} b_i^2 \xrightarrow{p} A > 0.$$

The proof is simple and is omitted (cf. Sen, 1960, p. 4). Now, from (3.2.38), we readily obtain

$$\frac{1}{n}\sum_{i=1}^{n} \{[V_n(X_i) - \theta(F)] - \psi_1(X_i)\}^2 \xrightarrow{p} 0, \tag{3.2.40}$$

and hence from (3.2.39), (3.2.40), and Lemma 3.2.4, we obtain

$$\frac{1}{n}\sum_{i=1}^{n} [V_n(X_i) - \theta(F)]^2 \xrightarrow{p} \zeta_1(F) > 0. \tag{3.2.41}$$

Let us now define

$$S_1^2 = \frac{1}{n-1}\sum_{i=1}^{n} [V_n(X_i) - U_n]^2. \tag{3.2.42}$$

It then readily follows that

$$S_1^2 = \frac{1}{n-1}\left\{\sum_{i=1}^{n} [V_n(X_i) - \theta(F)]^2\right\} - \frac{n}{n-1}[U_n - \theta(F)]^2. \tag{3.2.43}$$

Since by (3.2.14), $\sigma^2(U_n) = O(n^{-1})$, $U_n - \theta(F) \xrightarrow{p} 0$, and so from (3.2.41), (3.2.42), and (3.2.43) we arrive at the following

Theorem 3.2.5. *If $\zeta_m(F) < \infty$, then S_1^2, defined by* (3.2.42), *converges in probability to $\zeta_1(F)$, defined by* (3.2.11) as $n \to \infty$.

If we define similarly

$$V_n^{(c)}(X_{i_1}, \ldots, X_{i_c})$$

(3.2.44)
$$= \binom{n-c}{m-c}^{-1} \sum_{i_1, \ldots, i_c} \phi(X_{i_1}, \ldots, X_{i_c}, X_{\alpha_{c+1}}, \ldots, X_{\alpha_m}),$$

where the summation $\sum_{i_1, \ldots, i_c}$ extends over all $1 \leq \alpha_{c+1} < \cdots < \alpha_m \leq n$, with $\alpha_j \neq i_k$, $k = 1, \ldots, c$, for $i_1 \neq i_2 \neq i_c = 1, \ldots, n$, and

(3.2.45)
$$S_c^2 = \binom{n}{c}^{-1} \sum_{i \leq i_1 < \cdots < i_c \leq n} [V_n^{(c)}(X_{i_1}, \ldots, X_{i_c}) - U_n]^2$$

then proceeding precisely as above we arrive at the following:

Corollary 3.2.5.1. *If* $\zeta_m(F) < \infty$, *then* S_c^2, *defined by* (3.2.45), *stochastically converges to* $\zeta_c(F)$, *for all* $c = 1, \ldots, m$.

From (3.2.14) we may readily conclude that U_n is a consistent estimator of $\theta(F)$. However, this restricts the proof to the class of distributions for which $\zeta_m(F)$ is finite. Further relaxations of this regularity condition are due to Sen (1960), Hoeffding (1961), and Berk (1966). We state the following theorem on the convergence of U-statistics.

Theorem 3.2.6. (Hoeffding-Berk). *If* $\theta(F)$ *is estimable,* U_n *converges to* $\theta(F)$ *almost surely as* $n \to \infty$.

Proof: Let $\mathbf{Y}_n = (X_{(1)}, \ldots, X_{(n)})$ be the order statistics associated with $X_1, \ldots, X_n$, and let $\mathcal{F}_n$ be the σ-field generated by $(\mathbf{Y}_n, X_{n+1}, \ldots, \ldots)$. Then, for any $1 \leq i_1 < \cdots < i_m \leq n$, $E\{\phi(X_{i_1}, \ldots, X_{i_m}) \mid \mathcal{F}_n\}$ is the same, and hence, equal to $E[U_n \mid \mathcal{F}_n] = U_n$ almost surely (a.s.). Hence, $E[U_{m+j} \mid \mathcal{F}_n] = U_n$ a.s. for all $j = 0, 1, \ldots, n - m$. Thus, $\{U_n, \mathcal{F}_n\}$ forms a reverse martingale sequence. The proof of the theorem now follows by using Theorem 2.10.3. ◁

Sproule (1969) has proved that S_1^2, defined by (3.2.42), converges almost surely to $\zeta_1(F)$. For related results on $\theta(F_n)$, see Ghosh and Sen (1970).

We have so far considered the case where $X_1, \ldots, X_n$ have the common c.d.f. $F(x)$. We shall now consider certain extensions of the results derived so far. These will be in the following directions:

(i) $\{X_1, \ldots, X_n\}$ have distributions which are not all identical,

(ii) $\{X_1, \ldots, X_n\}$ is a realization from an m-dependent stationary stochastic process, and

(iii) $\{X_1, \ldots, X_n\}$ are drawn without replacement from a finite universe (of size $N > n$).

Consider first the case where the distributions of $X_1, \ldots, X_n$ are not all identical. Let $F_\alpha(x)$ be the c.d.f. of X_α, $\alpha = 1, \ldots, n$. Following Hoeffding

(1948a) let

$$\theta_{\alpha_1,\ldots,\alpha_m} = E\{\phi(X_{\alpha_1},\ldots,X_{\alpha_m})\}; \tag{3.2.46}$$

$$\begin{aligned}\psi_{c(\alpha_1,\ldots,\alpha_c)\beta_1,\ldots,\beta_{m-c}}&(x_{\alpha_1},\ldots,x_{\alpha_c}) \\ &= E\{\phi(x_{\alpha_1},\ldots,x_{\alpha_c},X_{\beta_1},\ldots,X_{\beta_{m-c}})\} - \theta_{\alpha_1,\ldots,\alpha_c,\beta_1,\ldots,\beta_{m-c}};\end{aligned} \tag{3.2.47}$$

$$\begin{aligned}\zeta_{c(\alpha_1,\ldots,\alpha_c)\beta_1,\ldots,\beta_{m-c};\gamma_1,\ldots,\gamma_{m-c}} &= E\{\psi_{c(\alpha_1,\ldots,\alpha_c)\beta_1,\ldots,\beta_{m-c}}(X_{\alpha_1},\ldots,X_{\alpha_c}) \\ &\quad \times \psi_{c(\alpha_1,\ldots,\alpha_c)\gamma_1,\ldots,\gamma_{m-c}}(X_{\alpha_1},\ldots,X_{\alpha_c})\},\end{aligned} \tag{3.2.48}$$

for $c = 1,\ldots,m$, and all possible α's, β's, and γ's.

$$\begin{aligned}\zeta_{c,n} &= \left\{\binom{n}{2m-c}\binom{2m-c}{c}\binom{2m-2c}{m-c}\right\}^{-1} \\ &\quad \times \sum \zeta_{c(\alpha_1,\ldots,\alpha_c)\beta_1,\ldots,\beta_{m-c};\gamma_1,\ldots,\gamma_{m-c}},\end{aligned} \tag{3.2.49}$$

where the summation extends over all possible $1 \le \alpha_1 < \cdots < \alpha_c \le n$, $1 \le \beta_1 < \cdots < \beta_{m-c} \le n$, $1 \le \gamma_1 < \cdots < \gamma_{m-c} \le n$ with $\alpha_i \ne \beta_i$, $\alpha_i \ne \gamma_j$ and $\beta_i \ne \gamma_j$; for $c = 1,\ldots,m$.

Then the variance of U_n, defined by (3.2.7), will be equal to

$$\sigma^2(U_n) = \binom{n}{m}^{-1}\sum_{c=1}^{m}\binom{m}{c}\binom{n-m}{m-c}\zeta_{c,n}. \tag{3.2.50}$$

Further, let

$$\bar{\psi}_{1(i)}(X_i) = \binom{n-1}{m-1}^{-1}\sum{}' \psi_{1(i)\alpha_2,\ldots,\alpha_m}(X_i), \qquad i = 1,\ldots,n, \tag{3.2.51}$$

where the summation $\sum'$ extends over all $1 \le \alpha_2 < \cdots < \alpha_m \le n$, with $\alpha_j \ne i$. Then we have the following theorem.

Theorem 3.2.6. (Hoeffding). *Suppose there is a number A such that for every $n = 1, 2, \ldots,$*

(i) $$\int\cdots\int \phi^2(x_1,\ldots,x_m)\,dF_{\alpha_1}(x_1)\cdots dF_{\alpha_m}(x_m) < A$$

(for all $1 \le \alpha_1 < \cdots < \alpha_m \le n$*);*

(ii) $E\,|\bar{\psi}^3_{1(i)}(X_i)| < \infty$, *for all* $i = 1,\ldots,n$,

and

(iii) $$\lim_{n\to\infty}\sum_{i=1}^{n} E\,|\bar{\psi}^3_{1(i)}(X_i)| \Big/ \left\{\sum_{i=1}^{n} E\{\bar{\psi}^2_{1(i)}(X_i)\}\right\}^{3/2} = 0.$$

Then, as $n \to \infty$, the limiting distribution of $[U_n - E(U_n)]/\sigma(U_n)$ is normal with mean zero and variance one.

The proof is similar to that of Theorem 3.2.2, and hence is left as an exercise to the reader. The same result also holds for $\theta(F_n)$ defined by (3.2.6). We shall now attach some simple form to the mean and variance of U_n, when

F_α's are not all identical. For this, let us write as in Sen (1969b)

(3.2.52) $$\bar{F}_{(n)}(x) = \frac{1}{n}\sum_{i=1}^{n} F_i(x), \quad \text{for} \quad n = 1, 2, \ldots,$$

and we denote the quantities $\theta(F)$ and $\zeta_c(F)$ for $F = \bar{F}_{(n)}$ by $\theta(\bar{F}_{(n)})$ and $\zeta_c(\bar{F}_{(n)})$ respectively. Also let

$$\theta(F_i, \bar{F}_{(n)}) = \int\left[\iint \phi(x_1, x_2, \ldots, x_m)\, d\bar{F}_{(n)}(x_2) \cdots d\bar{F}_{(n)}(x_m)\right] dF_i(x_1),$$
$$i = 1, \ldots, n;$$

$$\Delta_n^2 = \frac{1}{n}\sum_{i=1}^{n} [\theta(F_i, \bar{F}_{(n)}) - \theta(\bar{F}_{(n)})]^2 \geq 0.$$

Theorem 3.2.7. *Under the conditions of Theorem* 3.2.6, $n^{1/2}[U_n - \theta(\bar{F}_{(n)})]/m[\zeta_1(\bar{F}_{(n)}) - \Delta_n^2]^{1/2}$ *has asymptotically a normal distribution with zero mean and unit variance.*

Proof: It is sufficient to show that

(i) $n^{1/2}\,|E(U_n) - \theta(\bar{F}_{(n)})| = o(1)$, and (ii) $|\zeta_{1,n} - \{\zeta_1(\bar{F}_n) - \Delta_n^2\}| \to 0$

as $n \to \infty$. Now,

$$E(U_n) \doteq \binom{n}{m}^{-1} \sum_{1\leq\alpha_1<\cdots<\alpha_m\leq n} \int \cdots \int \phi(x_1, \ldots, x_m) \times dF_{\alpha_1}(x_1) \cdots dF_{\alpha_m}(x_m)$$

(3.2.53) $$= \frac{1}{n(n-1)\cdots(n-m+1)} \sum_{\alpha_1\neq\cdots\neq\alpha_m=1}^{n} \int \cdots \int \phi(x_1, \ldots, x_m) \times dF_{\alpha_1}(x_1) \cdots dF_{\alpha_m}(x_m).$$

Using (3.2.1) and (3.2.52), we note that after some simplifications, $\theta(\bar{F}_n)$ can be written as

(3.2.54) $$\theta(\bar{F}_{(n)}) = n^{-m} \sum_{\alpha_1=1}^{n} \cdots \sum_{\alpha_m=1}^{n} \int \cdots \int \phi(x_1, \ldots, x_m)\, dF_{\alpha_1}(x_1) \cdots dF_{\alpha_m}(x_m).$$

Comparing (3.2.53) and (3.2.54), it is easily seen that

(3.2.55) $$|\theta(\bar{F}_{(n)}) - E(U_n)| = O(n^{-1}).$$

Now, using (3.2.46) through (3.2.49), we may write $\zeta_{1,n}$ as

(3.2.56) $$n^{-[2m-1]} \sum\nolimits_1^{**}\Big\{\int \cdots \int \phi(x_1, \ldots, x_m)\phi(x_m, \ldots, x_{2m-1}) \times dF_{i_1}(x_m) \prod_{1}^{m-1} dF_{j_l}(x_l)\, dF_{k_l}(x_{m+l}) - \theta_{i_1, j_1, \ldots, j_{m-1}} \cdot \theta_{i_1, k_1, \ldots, k_{m-1}}\Big\},$$

where $p^{[q]} = p \cdots (p - q + 1)$ and the summation $\sum_1^{**}$ extends over all possible $1 \leq i_1 \neq j_1 \neq \cdots \neq j_{m-1} \neq k_1 \neq \cdots \neq k_{m-1} \leq n$. Now, as in (3.2.53) and (3.2.54) it can be shown that

$$\left| n^{-[2m-1]} \sum_1^{**} \theta_{i_1, j_1, \ldots, j_{m-1}} \theta_{i_1, k_1, \ldots, k_{m-1}} - \frac{1}{n} \sum_{i=1}^{n} \theta^2(F_i; \bar{F}_{(n)}) \right| = O(n^{-1}), \tag{3.2.57}$$

where $\theta(F_i; \bar{F}_{(n)})$ is defined just after (3.2.52). Also, by definition

$$(1/n) \sum_{i=1}^{n} \theta(F_i; \bar{F}_{(n)}) = \theta(\bar{F}_{(n)}; \bar{F}_{(n)}) = \theta(\bar{F}_{(n)}). \tag{3.2.58}$$

Finally, using (3.2.10), (3.2.11), and (3.2.12), we obtain

$$\begin{aligned} \zeta_1(\bar{F}_{(n)}) + \theta^2(\bar{F}_{(n)}) = n^{-(2m-1)} \sum_1^0 \int \cdots \int \phi(x_1, \ldots, x_m) \\ \times \phi(x_m, \ldots, x_{2m-1}) \, dF_{i_1}(x_m) \prod_1^{m-1} dF_{j_l}(x_l) \, dF_{k_l}(x_{m+l}), \end{aligned} \tag{3.2.59}$$

where the summation $\sum_1^0$ extends over all possible $1 \leq i_1 \leq n$, $1 \leq j_l \leq n$, $1 \leq k_l \leq n$ for $l = 1, \ldots, m - 1$. Hence, we get from (3.2.56), (3.2.57), and (3.2.58) and some simple algebraic manipulations that

$$|\zeta_{1,n} - \{\zeta_1(\bar{F}_n) - \Delta_n^2\}| = O(n^{-1}). \tag{3.2.60}$$

This completes the proof of the theorem. ◁

Let us now consider the case where the sample observations are not all independent. Two important dependent processes are the so-called m-dependent stationary stochastic process and the symmetric dependent process, the latter being particularly simplified in the sampling scheme of drawing (without replacement) a random sample from a finite universe. All the results concerning the unbiasedness, consistency, and asymptotic normality of the U-statistics, obtained earlier in this section, can be easily extended to the case of m-dependent processes. We shall not enter into the detailed discussion of these results. The interested reader is referred to the papers by Sen (1963c, 1965c) which give a somewhat detailed account of such processes. We have, however, included a few exercises to familiarize the reader with this topic. The sampling distribution theory of U-statistics in samples (drawn without replacement) from a finite universe has been studied in detail by Nandi and Sen (1963), and this has an important role in the permutational distribution theory of U-statistics. We shall take up this subject in a subsequent section.

The concept of regular functions and the use of U-statistics have been further extended to the case of more than one sample by Lehmann (1951a) among others. For simplicity of presentation, we shall only consider the two-sample case and leave the general case of $c(\geq 2)$ samples as an exercise.

Let $X_1, \ldots, X_{n_1}$, and $Y_1, \ldots, Y_{n_2}$ be two independent samples of sizes n_1 and n_2, drawn from two populations having c.d.f.'s $F(x)$ and $G(x)$ respectively. Then $\theta = \theta(F, G)$, regarded as a functional of both the c.d.f.'s F and G, is said to be regular (over the product set $\{(\mathcal{F} \times \mathcal{G}): F \in \mathcal{F}, G \in \mathcal{G}\}$) if there exists a statistic $\phi(X_1, \ldots, X_{n_1}, Y_1, \ldots, Y_{n_2})$ such that

$$\int \cdots \int \phi(x_1, \ldots, x_{n_1}, y_1, \ldots, y_{n_2})\, dF(x_1) \cdots dF(x_{n_1})\, dG(y_1) \cdots dG(y_{n_2}) = \theta(F, G), \tag{3.2.61}$$

for all $(F, G) \in (\mathcal{F} \times \mathcal{G})$. The smallest sample sizes (say, m_1 and m_2) for which (3.2.61) holds may be regarded as the *degree vector* of $\theta(F, G)$, and the corresponding statistic as *kernel*. Here also, we may without any loss of generality assume that $\phi(X_1, \ldots, X_{m_1}, Y_1, \ldots, Y_{m_2})$ is symmetric in its first m_1 arguments as well as in its last m_2 arguments, though the roles of the two sets need not be symmetric. Let us first consider a few simple examples.

Example 3.2.4. Let $\theta(F, G) = P\{X < Y\}$ when X and Y have the c.d.f.'s F and G respectively. Obviously,

$$\theta(F, G) = \int_{-\infty}^{\infty} \int_{-\infty}^{\infty} \phi(x, y)\, dF(x)\, dG(y), \tag{3.2.62}$$

where

$$\phi(x, y) = \begin{cases} 1, & \text{if } x < y, \\ 0, & \text{otherwise.} \end{cases} \tag{3.2.63}$$

Thus, $\theta(F, G)$ is an estimable parameter for all $(F, G) \in (\mathcal{F}, \mathcal{G})$, where $\mathcal{F}(\mathcal{G})$ is the class of all distributions on the real line.

Example 3.2.5. Let $\theta(F, G) = \int_{-\infty}^{\infty} [F(x) - G(x)]^2\, d[\{F(x) + G(x)\}/2]$. Thus $\theta(F, G)$ may be regarded as a *distance function* for the two c.d.f.'s $F(x)$ and $G(x)$, and $\theta(F, G) > 0$ if $F(x)$ and $G(x)$ do not agree at least on a set of points of measure nonzero (cf. Lehmann, 1951a). Let us consider the statistic $\phi(X_1, X_2, Y_1, Y_2)$ defined below:

$$\phi(X_1, X_2, Y_1, Y_2) = \begin{cases} 1, & \text{if } \max(X_1, X_2) < \min(Y_1, Y_2), \\ & \text{or } \min(X_1, X_2) > \max(Y_1, Y_2); \\ 0, & \text{otherwise.} \end{cases} \tag{3.2.64}$$

If now the X's and Y's have respectively the c.d.f.'s F and G, then

$$\begin{aligned} E\phi(X_1, X_2, Y_1, Y_2) &= \int_{-\infty}^{\infty} [1 - G(x)]^2 2F(x)\, dF(x) \\ &\quad + \int_{-\infty}^{\infty} [1 - F(x)]^2 2G(x)\, dG(x) \end{aligned} \tag{3.2.65}$$

and after some essential simplification (3.2.65) reduces to

$$(3.2.66)\quad \frac{1}{3} + 2\int_{-\infty}^{\infty} [F(x) - G(x)]^2 \, d\,\frac{[F(x) + G(x)]}{2} \quad \text{for all} \quad (F, G) \in (\mathcal{F}, \mathcal{G}).$$

Thus, the *kernel*

$$\psi(X_1, X_2, Y_1, Y_2) = \begin{cases} \frac{1}{3}, & \text{if } \min(X_1, X_2) > \max(Y_1, Y_2) \\ & \text{or } \max(X_1, X_2) < \min(Y_1, Y_2) \\ -\frac{1}{6}, & \text{otherwise,} \end{cases}$$

is an unbiased and symmetric (in X_1, X_2 and Y_1, Y_2) estimator of $\theta(F, G)$ for all pairs of continuous c.d.f.'s (on the real lines).

Now corresponding to the kernel $\phi(X_1, \ldots, X_{m_1}, Y_1, \ldots, Y_{m_2})$ of $\theta(F, G)$, we define the generalized U-statistic as

$$(3.2.67)\quad U_{n_1,n_2} = \binom{n_1}{m_1}^{-1}\binom{n_2}{m_2}^{-1} \sum \phi(X_{\alpha_1}, \ldots, X_{\alpha_{m_1}}, Y_{\beta_1}, \ldots, Y_{\beta_{m_2}})$$

where the summation extends over all possible $1 \leq \alpha_1 < \cdots < \alpha_{m_1} \leq n_1$, $1 \leq \beta_1 < \cdots < \beta_{m_2} \leq n_2$. By way of summary, let us briefly recapitulate the basic properties of U_{n_1,n_2}. First let

$$(3.2.68)\quad \begin{aligned} \theta(F_{n_1}, G_{n_2}) &= n_1^{-m_1} n_2^{-m_2} \\ &\times \sum_{\alpha_1=1}^{n_1} \cdots \sum_{\alpha_{m_1}=1}^{n_1} \sum_{\beta_1=1}^{n_2} \cdots \sum_{\beta_{m_2}=1}^{n_2} \phi(X_{\alpha_1}, \ldots, X_{\alpha_{m_1}}, Y_{\beta_1}, \ldots, Y_{\beta_{m_2}}). \end{aligned}$$

Then

$$(3.2.69)\quad |U_{n_1,n_2} - \theta(F_{n_1}, G_{n_2})| = O_p(n^{-1}); \qquad n = \min(n_1, n_2)$$

provided the variance of $\theta(F_{n_1}, G_{n_2})$ exists.

The U-statistic U_{n_1,n_2} is the m.v.u. estimator of $\theta(F, G)$ and is unique if the order statistic (corresponding to the two samples) is complete in the extended sense of Fraser (1953b). Also, let

$$(3.2.70)\quad \begin{aligned} &\psi_{cd}(x_1, \ldots, x_c, y_1, \ldots, y_d) \\ &= E\{\phi(x_1, \ldots, x_c, X_{c+1}, \ldots, X_{m_1}, y_1, \ldots, y_d, Y_{d+1}, \ldots, Y_{m_2}) \\ &\qquad - \theta(F, G)\}; \end{aligned}$$

$$(3.2.71)\quad \zeta_{cd} = E\{\psi_{cd}^2(X_1, \ldots, X_c, Y_1, \ldots, Y_d)\},$$

for $c = 0, \ldots, m_1$, $d = 0, \ldots, m_2$. (By definition $\zeta_{00} = 0$.) Then by the same technique as in Theorem 3.2.2, it can be shown that

$$(3.2.72)\quad \text{(i)} \quad \left| U_{n_1,n_2} - \frac{m_1}{n_1}\sum_{\alpha=1}^{n_1} \psi_{10}(X_\alpha) - \frac{m_2}{n_2}\sum_{\beta=1}^{n_2} \psi_{01}(Y_\beta) - \theta(F, G) \right| = O_p(n^{-1}),$$

$$(3.2.73)\quad \text{(ii)} \quad \mathcal{L}\left(\left[\left(\frac{m_1}{n_1}\right) \sum_{\alpha=1}^{n_1} \psi_{10}(X_\alpha) + \left(\frac{m_2}{n_2}\right) \sum_{\beta=1}^{n_2} \psi_{01}(Y_\beta) \right] \Big/ \gamma_{n_1,n_2} \right) \to N(0, 1),$$

where

$$\gamma^2_{n_1,n_2} = \left(\frac{m_1^2}{n_1}\right)\zeta_{10} + \left(\frac{m_2^2}{n_2}\right)\zeta_{01}. \tag{3.2.74}$$

Thus from (3.2.72), (3.2.73), and (3.2.74),

$$\mathcal{L}([U_{n_1,n_2} - \theta(F, G)]/\gamma_{n_1,n_2}) \to N(0, 1). \tag{3.2.75}$$

The uniformity of the convergence in (3.2.75) can be established under conditions similar to that of Theorem 3.2.2, while the consistency of U_{n_1,n_2} as an estimate of $\theta(F, G)$ will follow as in Theorem 3.2.6. Let us also define

$$V_n^{(1,0)}(X_i) = \binom{n_1 - 1}{m_1 - 1}^{-1}\binom{n_2}{m_2}^{-1} \sum\nolimits_i \phi(X_i, X_{\alpha_2}, \ldots, X_{\alpha m_1}, Y_{\beta_1}, \ldots, Y_{\beta m_2}), \tag{3.2.76}$$

where the summation $\sum_i$ extends over all possible $1 \le \alpha_2 < \cdots < \alpha_{m_1} \le n_1$, $1 \le \beta_1 < \cdots < \beta_{m_2} \le n_2$ with $\alpha_j \ne i$, $i = 1, \ldots, n_1$. In a similar manner, we may define

$$V_n^{(0,1)}(Y_j) = \binom{n_1}{m_1}^{-1}\binom{n_2 - 1}{m_2 - 1}^{-1} \sum\nolimits_j \phi(X_{\alpha_1}, \ldots, X_{\alpha_{m_1}}, Y_j, Y_{\beta_2}, \ldots, Y_{\beta_{m_2}}), \tag{3.2.77}$$

where the summation $\sum_j$ extends over all possible $1 \le \alpha_1 < \cdots < \alpha_{m_1} \le n_1$, $1 \le \beta_2 < \cdots < \beta_{m_2} \le n_2$, with $\beta_i \ne j$, for $i = 1, \ldots, n_2$. Also let

$$S_{10}^2 = \frac{1}{n_1 - 1}\sum_{i=1}^{n_1} [V_n^{(1,0)}(X_i) - U_{n_1,n_2}]^2, \tag{3.2.78}$$

$$S_{01}^2 = \frac{1}{n_2 - 1}\sum_{j=1}^{n_2} [V_n^{(0,1)}(Y_j) - U_{n_1,n_2}]^2. \tag{3.2.79}$$

Then, by the same method as in (3.2.35) through (3.2.43), we conclude

$$S_{10}^2 \xrightarrow{p} \zeta_{10} \quad \text{and} \quad S_{01}^2 \xrightarrow{p} \zeta_{01}. \tag{3.2.80}$$

We consider some further results on U-statistics in section 3.3.

3.3. THEORY OF PERMUTATION TESTS

For a wide class of nonparametric problems, the underlying null hypothesis relates to the invariance of the joint distribution of the sample observations under certain finite groups of transformations which map the sample space onto itself. Let us consider the following examples, to which we will refer throughout the section.

(I) Let $X_1, \ldots, X_n$ be n independent real-valued random variables, distributed according to the cumulative distribution functions (c.d.f.'s) $F_1(x), \ldots, F_n(x)$, respectively. These c.d.f.'s are not necessarily all identical or continuous. We consider the null hypothesis

$$H_{0,1}: F_i(-x) = 1 - F_i(x - o) \quad \text{for all} \quad i = 1, \ldots, n, \qquad x \in R^1, \tag{3.3.1}$$

where R^k stands for the k-dimensional Euclidean space ($k \geq 1$). $H_{0,1}$ implies that the joint distribution of $\mathbf{E}_n = (X_1, \ldots, X_n)$ remains invariant under the $M_1 = 2^n$ transformations $G_n^{(1)} = \{g_{n,\alpha}, \alpha = 1, \ldots, n\}$ where

$$g_n \mathbf{E}_n = [(-1)^{j_1} X_1, \ldots, (-1)^{j_n} X_n], \qquad j_i = 0, 1; \quad i = 1, \ldots, n. \tag{3.3.2}$$

This may be termed the hypothesis of *sign invariance.*

(II) Let $X_i = (X_{1i}, X_{2i})$, $i = 1, \ldots, n$, be n independent and identically distributed random variables (i.i.d.r.v.), distributed according to a bivariate c.d.f. $F(x_1, x_2)$, which need not be continuous. The null hypothesis $H_{0,2}$ states that

$$H_{0,2}: F(x_1, x_2) = F(x_1, \infty) F(\infty, x_2) \quad \text{for all} \quad (x_1, x_2) \in R^2. \tag{3.3.3}$$

This implies that the joint distribution of $\mathbf{E}_n = (X_1, \ldots, X_n)$ remains invariant under the $M_2 = n!$ permutations of the type

$$\mathbf{E}_n(\mathbf{R}_n) = \{(X_{1i}, X_{2R_i}), i = 1, \ldots, n\}, \tag{3.3.4}$$

where $\mathbf{R}_n = (R_1, \ldots, R_n)$ is any permutation of $(1, \ldots, n)$. This may be termed the hypothesis of *matching invariance.*

(III) Let $X_{ij}, j = 1, \ldots, n_i$ be n_i i.i.d.r.v. distributed according to a c.d.f. $F_i(x)$, for $i = 1, \ldots, c(\geq 2)$. The different sets of random variables are assumed to be mutually stochastically independent and the c.d.f.'s $F_1, \ldots, F_c$ may or may not be continuous. The null hypothesis of interest is that

$$H_{0,3}: F_1(x) \equiv \cdots \equiv F_c(x) \quad \text{for all} \quad x \in R^1. \tag{3.3.5}$$

This implies that the joint distribution of $\mathbf{E}_N = (X_{11}, \ldots, X_{cn_c})$ (where $\sum_{i=1}^{c} n_i = N$) remains invariant under the $M_3 = N!$ permutations of the coordinates of $\mathbf{E}_N$ among themselves. Equivalently, this implies that the joint distribution of $\mathbf{E}_N$ remains invariant under all possible partitionings of $\mathbf{E}_N$ into c subsets of sizes $n_1, \ldots, n_c$ respectively. $H_{0,3}$ may be termed the hypothesis of *invariance under partitioning into subsets.*

(IV) Let $\mathbf{X}_i = (X_{1i}, \ldots, X_{pi})$, $i = 1, \ldots, n$ be n independent random variables distributed according to the p variate c.d.f.'s $F_i(x_1, \ldots, x_p)$, $i = 1, \ldots, n$, respectively, where $p(\geq 2)$ is a positive integer. The null hypothesis of interest is

$$H_{0,4}: F_i(x_1, \ldots, x_p) = F_i(x_{j_1}, \ldots, x_{j_p}), \quad \text{for all} \quad i = 1, \ldots, n, \qquad (x_1, \ldots, x_p) \in R^p, \tag{3.3.6}$$

where $(j_1, \ldots, j_p)$ is any permutation of $(1, \ldots, p)$. This implies that the joint distribution of $E_n = (\mathbf{X}_1, \ldots, \mathbf{X}_n)$ remains invariant under the $M_4 = (p!)^n$ permutations on the p intravector coordinates within each of the n vectors $\mathbf{X}_i$, $i = 1, \ldots, n$. This may be termed the hypothesis of *interchangeability*.

The multivariate generalizations of these problems are easy to conceive. For example, (I) and (III) may be readily generalized to vector-valued random variables, (II) and (IV) to the situation where each element X_{ij} has k components. Further, in some problems of design of experiments, we are faced with a problem as in (IV), where p is not necessarily a constant, and may vary over $i = 1, \ldots, n$. Finally, we also have a situation as in (II), where one of the variates (say X_1) is a nonstochastic ordered variable. In this case, the problem is often termed that of *randomness in a series of observations*.

In what follows we shall assume that N is a positive integer, and let Z_N be a random variable which may assume values z_N, which are points in a space $\mathfrak{Z}_N$. The probability distribution of Z_N, defined on an additive class $\mathcal{A}_N$ of subsets A_N of $\mathfrak{Z}_N$, is denoted by $P(A_N) = P\{Z_N \in A_N\}$. We define a partition Π_N of $\mathfrak{Z}_N$, which is a class of mutually exclusive subsets S of $\mathfrak{Z}_N$, such that every point z_N of $\mathfrak{Z}_N$ belongs to one of the subsets S. The set of all points which lie in the same subset S, containing any typical point z_N, is denoted by $T(z_N)$, and the number of points of $T(z_N)$ by $M(z_N)$. The partition Π_N is assumed to be such that

(i) $M(z_N)$ is finite for all finite N, and

(ii) if S_M stands for the set theoretic union of all those sets $S \in \Pi_N$ which contains exactly M elements then there exist mutually exclusive subsets $S_M^{(i)}$, $i = 1, \ldots, M$, which are measurable and are such that every element S of Π_N has one and only one element in common with each $S_M^{(i)}$, $i = 1, \ldots, M$. For such a partition Π_N of $\mathfrak{Z}_N$, we define the *null hypothesis* (H_0) *of invariance as that of the invariance of the distribution of* Z_N *under* Π_N.

Now let us consider a *test function* $\phi_N(z_N)$ $(0 \leq \phi_N \leq 1)$, which is defined in such a manner that

$$\sum_{z' \in T(z_N)} \phi_N(z') = M(z_N) \cdot \alpha \qquad (0 < \alpha < 1), \quad \text{for all} \quad z_N \in \mathfrak{Z}_N, \tag{3.3.7}$$

where α is the size of the test. It is then easy to check that

$$E_{H_0}\{\phi_N(Z_N)\} = E_{H_0}[E\{\phi_N(Z_N) \mid T(Z_N)\}] = \alpha. \tag{3.3.8}$$

Thus, $\phi_N(Z_N)$ will be a strictly distribution-free similar size α test for H_0. $\phi_N(Z_N)$ will be termed, henceforth, a *permutation test* as it is really based on the consideration of all possible permutations of Z_N over $T(Z_N)$. Condition (3.3.7) is known as the $S(\alpha)$ *structure of tests* (cf. Scheffé, 1943b). Thus, by constitution *permutation tests are similar size* α *tests having the* $S(\alpha)$ *structure*. It is easy to verify that all the four hypotheses of invariance, considered in

(I)–(IV), satisfy the regularity conditions for Π_N, stated above, and hence, admit of the existence of permutation (similar) tests.

The idea of permutation tests is due to R. A. Fisher (1935) and a rigid formulation of the basic concepts is due to Scheffé (1943b). The theory is also developed by Lehmann and Stein (1949) and Hoeffding (1952). These papers also contain a useful account of the various small- and large-sample properties of permutation tests, which we shall consider only briefly. For detailed study, the reader is referred to the papers mentioned above.

Theorem 3.3.1. **(Scheffé, Lehmann and Stein).** *For testing a hypothesis of invariance, any test of structure $S(\alpha)$ is similar and of size α* $(0 < \alpha < 1)$.

The proof follows readily from (3.3.7) and (3.3.8).

Theorem 3.3.2. **(Lehmann and Stein).** *Let H be the hypothesis of invariance under the partition Π_N and let g be a probability density function not in H. For any Z in S_M denote by $Z^{(1)}, \ldots, Z^{(M)}$ the M points of $T(Z)$ arranged so that $g(Z^{(1)}) \geq g(Z^{(2)}) \geq \cdots \geq g(Z^{(M)})$. For testing H against g a most powerful test of size α is given by*

$$\phi(Z) = \begin{cases} 1 & \text{if} \quad g(Z) > g(Z^{(1+[M\alpha])}), \\ a & \text{if} \quad g(Z) = g(Z^{(1+[M\alpha])}), \\ 0 & \text{if} \quad g(Z) < g(Z^{(1+[M\alpha])}), \end{cases} \tag{3.3.9}$$

where $\sum_{i=1}^{M} \phi(Z^{(i)}) = M\alpha$, $0 \leq \alpha \leq 1$, and a may depend on Z through $T(Z)$. (Here $[M\alpha]$ is the largest integer less than or equal to $M\alpha$.)

The proof follows from the fundamental Neyman-Pearson lemma (cf. Lehmann, 1959, p. 65).

The above results give some justifications of permutation tests. Among the class of nonparametric tests too, permutation tests have some merits. First, these tests are valid under regularity conditions which are usually less stringent than those required with other nonparametric tests for the same hypothesis. For example, permutation tests are even applicable for discrete or categorical sample spaces where other rank tests may not be strictly distribution-free. Secondly, in multivariate problems, even under the hypothesis of invariance and for continuous parent distributions, the usual nonparametric tests available for univariate problems, when extended, are found to depend on the unknown distributions. On the other hand, in most of the cases, it is not difficult to formulate suitable permutation tests for these problems, which are genuinely distribution-free.

In spite of all these advantages, the permutation tests have two drawbacks. First, the construction of the exact permutation test, satisfying (3.3.7), deems the knowledge of all the $M(z_N)$ points for any $z_N \in \mathfrak{Z}_N$. In most of the cases, $M(z_N)$ increases at a very fast rate with the increase in N (e.g., in problems (I)–(IV), $M(z_N)$ increases either exponentially or even at a faster rate with N). This makes the scheme of evaluation of the exact test $\phi(z)$ in (3.3.7) prohibitively laborious for large N. To simplify the situation various attempts have been made by various workers, and mostly these relate to some simple limiting distributions of suitable test statistics under various types of permutation models. These limiting distributions are then proposed to construct large sample permutation tests. These will be discussed in detail in section 3.4. The second drawback is caused by the fact that permutation tests are, by construction, conditional tests, and in the majority of the cases, the evaluation of the exact power function of the tests seems to be considerably difficult. This calls for the study of large sample power properties of permutation tests and their relations with those of some unconditional tests. This will be the subject matter of section 3.5.

3.4. PERMUTATIONAL LIMIT THEOREMS

In most of the cases, the test function $\phi(Z_N)$ in (3.3.7) is based on the values of some suitable test statistic $t_N = t(Z_N)$. Thus, in such a problem, we require to know the permutation distribution of t_N in order that $\phi(Z_N)$ may be constructed. The problem of evaluating the exact permutation distribution of t_N becomes increasingly laborious as N increases. The earlier attempts by Pitman (1937a, 1937b, 1938), Welch (1937), and others, consist in evaluating the first four moments of the exact permutation distribution of t_N and showing that these converge to the corresponding moments of some well-known distribution (normal, chi-square, etc.) as N increases. The conclusion of the convergence of the true permutation distribution of t_n to the aimed limiting distribution as claimed by the convergence of only the first four moments, though found to be true in most cases, cannot be statistically justified for obvious reasons.

The first statistically sound attempt on this line was due to Wald and Wolfowitz (1944) who established an elegant permutational central limit theorem on a class of statistics, which we shall call *linear permutation statistics*. Further extensions of this theorem are due to Madow (1948), Noether (1949b), Hoeffding (1951a), Dwass (1953, 1955b), Fraser (1956) and Hájek (1961) among others. Hoeffding (1951a) has established another central limit theorem on a relatively more general type of statistics. This statistic will be termed the *bilinear permutation statistic*. Motoo (1957) has relaxed the conditions of the above theorem to some extent. An important subclass of linear and bilinear permutation statistics is the class of *rank order statistics*, considered by Dwass (1956), Chernoff and Savage (1958), Bhuchongkul (1964), Puri (1964), Klotz (1962), Puri and Sen (1969b) and many others. Sen (1965b, 1966c, 1967e) has considered the permutation theory of a class of *generalized U-statistics* with their various applications. We shall review the theory in this order and also consider various illustrations pertaining to the problems (I)–(IV) considered in section 3.3.

Before the main theorems are presented, let us consider the different sets of regularity conditions offered by different workers and also their mutual implications. Let us consider a nondecreasing sequence of positive integers $\{N_\nu\}$, defined for each positive integer ν, such that

$$\lim_{\nu\to\infty} N_\nu = \infty. \tag{3.4.1}$$

For each ν, we also define a sequence $H_\nu = (h_{\nu 1}, \ldots, h_{\nu N_\nu})$ of N_ν real numbers, and let

$$\bar{h}_\nu = N_\nu^{-1} \sum_{i=1}^{N_\nu} h_{\nu i}; \tag{3.4.2}$$

$$\mu_{r,\nu}(H) = N_\nu^{-1} \sum_{i=1}^{N_\nu} (h_{\nu i} - \bar{h}_\nu)^r, \qquad r = 2, 3, \ldots; \tag{3.4.3}$$

$$\nu_{r,\nu}(H) = N_\nu^{-1} \sum_{i=1}^{N_\nu} |h_{\nu i} - \bar{h}_\nu|^r, \qquad r \geq 0; \tag{3.4.4}$$

$$R_\nu(H) = \operatorname*{range}_{i=1,\ldots,N_\nu} [h_{\nu i}] = \max_i h_{\nu i} - \min_j h_{\nu j}. \tag{3.4.5}$$

Then, we consider the following conditions:

Condition I (Wald-Wolfowitz):

$$\mu_{r,\nu}(H)/\{\mu_{2,\nu}(H)\}^{r/2} = O(1) \quad \text{for all} \quad r = 3, 4, \ldots. \tag{3.4.6}$$

Condition II (Noether):

$$\mu_{r,\nu}(H)/\{\mu_{2,\nu}(H)\}^{r/2} = o(N_\nu^{r/2-1}) \quad \text{for all} \quad r = 3, 4, \ldots. \tag{3.4.7}$$

Condition II is shown by Hoeffding (1951a) to be equivalent to either of the following two: For some $r > 2$

$$\nu_{r,\nu}(H)/\{\mu_{2,\nu}(H)\}^{r/2} = o(N_\nu^{r/2-1}), \tag{3.4.8}$$

or

$$R_\nu(H)/\{\mu_{2,\nu}(H)\}^{1/2} = o(N_\nu^{1/2}). \tag{3.4.9}$$

Condition III (Hájek): If $\{n_\nu\}$ is any nondecreasing sequence of positive integers such that

$$\lim_{\nu\to\infty} n_\nu = \infty \quad \text{but} \quad \lim_{\nu\to\infty} n_\nu/N_\nu = 0, \tag{3.4.10}$$

then

$$\left[\max_{1\le i_1<\cdots<i_{n_\nu}\le N_\nu}\left\{\sum_{j=1}^{n_\nu}[h_{\nu i_j}-\bar{h}_\nu]^2\right\}\Big/\mu_{2,\nu}(H)\right]=o(N_\nu). \tag{3.4.11}$$

This condition may be put in the form

$$\left[\lim_{\nu\to\infty} n_\nu=\infty\right] \text{ implies } \frac{1}{\mu_{2,\nu}(H)}\int_{|x-\bar{h}_\nu|>\{n_\nu\mu_{2,\nu}(H)\}^{\frac{1}{2}}}(x-\bar{h})^2\,dG_\nu(x)=o(1), \tag{3.4.12}$$

which may also be expressed in terms of two more or less similar conditions given by Hájek (1961, p. 517).

Let us now consider another double sequence $D_\nu=\{d_{\nu ij},\ 1\le i,j\le N_\nu\}$, defined for each ν, and we state the following conditions on D_ν.

Condition IV (Hoeffding):

$$\frac{1}{N_\nu^2}\sum_{i=1}^{N_\nu}\sum_{j=1}^{N_\nu}|d_{\nu ij}|^r\Big/\left\{\frac{1}{N_\nu^2}\sum_{i=1}^{N_\nu}\sum_{j=1}^{N_\nu}d_{\nu ij}^2\right\}^{r/2}=o(N_\nu^{r/2-1}), \tag{3.4.13}$$

for $r=3,4,\ldots$; (3.4.13) will be satisfied if

$$\left\{\max_{1\le i,j\le N_\nu}d_{\nu ij}^2\right\}\Big/\left\{\frac{1}{N_\nu^2}\sum_{i=1}^{N_\nu}\sum_{j=1}^{N_\nu}d_{\nu ij}^2\right\}=o(N_\nu). \tag{3.4.14}$$

Condition V (Motoo): Let

$$d_\nu^2=N_\nu^{-1}\sum_{i=1}^{N_\nu}\sum_{j=1}^{N_\nu}d_{\nu ij}^2; \tag{3.4.15}$$

then for any positive $\epsilon>0$

$$N_\nu^{-1}\sum_{|d_{\nu ij}|>\epsilon d_\nu}d_{\nu ij}^2/d_\nu^2\to 0 \quad\text{as}\quad \nu\to\infty. \tag{3.4.16}$$

Now, using the results of Hoeffding (1951a), Noether (1949b) and Motoo (1957), it is easy to verify that

$$(\mathrm{I})\Rightarrow(\mathrm{II}),\qquad (\mathrm{IV})\Rightarrow(\mathrm{V}). \tag{3.4.17}$$

Again, if A_ν and B_ν are two sequences of real numbers, defined as in H_ν, and if D_ν has elements $d_{\nu ij}=a_{\nu i}b_{\nu j}$ for all (i,j), then on denoting by $[X, Y]$ the event that A_ν satisfies the condition X and B_ν the condition Y, it can be shown (cf. Hájek, 1961) that

$$[\mathrm{I},\mathrm{I}]\Rightarrow[\mathrm{I},\mathrm{II}]\Rightarrow[\mathrm{IV}]\Rightarrow[\mathrm{V}], \tag{3.4.18}$$

$$[\mathrm{II},\mathrm{III}]\Rightarrow[\mathrm{V}]\Rightarrow[\mathrm{II},\mathrm{II}]. \tag{3.4.19}$$

Now, corresponding to a sequence $A_\nu = (a_{\nu 1}, \ldots, a_{\nu N_\nu})$ of N_ν real numbers, we define a random variable $\mathbf{X}_\nu = (X_{\nu 1}, \ldots, X_{\nu N_\nu})$ which takes on each permutation of $(a_{\nu 1}, \ldots, a_{\nu N_\nu})$ with the same probability $1/N_\nu!$. Also let $\mathbf{B}_\nu$ be a second sequence of N_ν real numbers $(b_{\nu 1}, \ldots, b_{\nu N_\nu})$. Then, we define a *linear permutation statistic* L_ν as

$$L_\nu = \sum_{i=1}^{N\nu} b_{\nu i} X_{\nu i}. \tag{3.4.20}$$

We denote by $\mathcal{T}_\nu$ the permutational probability distribution of L_ν, generated by the $N_\nu!$ equally likely permutations of X_ν on A_ν. Then, it is easy to verify that

$$\begin{aligned} E\{L_\nu \mid \mathcal{T}_\nu\} &= N_\nu \bar{a}_\nu \bar{b} = \lambda_\nu \quad \text{(say)}, \\ V\{L_\nu \mid \mathcal{T}_\nu\} &= [N_\nu^2/(N_\nu - 1)]\mu_{2,\nu}(A)\mu_{2,\nu}(B) = \sigma_\nu^2 \quad \text{(say)}, \end{aligned} \tag{3.4.21}$$

where $\bar{a}_\nu$, $\bar{b}_\nu$, $\mu_{2,\nu}(A)$ and $\mu_{2,\nu}(B)$ are defined as in (3.4.2) and (3.4.3). Let us also denote by $\mathcal{L}_{\mathcal{T}}(Z_N) \to N(0, 1)$ the statement that under the permutation model, the random variable Z_N converges in law to a normal distribution with zero mean and unit variance.

Now, let us consider the following theorem which was established first by Wald and Wolfowitz (1944) under [I, I], and was subsequently relaxed to [II, I] by Noether (1949b) and Hoeffding (1951a).

Theorem 3.4.1 (Wald-Wolfowitz-Noether-Hoeffding). *If B_ν satisfies condition* I *in* (3.4.6) *and A_ν satisfies condition* II *in* (3.4.7), *then*

$$\mathcal{L}_{\mathcal{T}}([L_\nu - \lambda_\nu]/\sigma_\nu) \to N(0, 1). \tag{3.4.22}$$

The proof of this theorem is based on the evaluation of the rth moment of L_ν under the permutations model for all $r = 1, 2, 3, \ldots$. Without any loss of generality, we may standardize A_ν and B_ν in such a manner that $\lambda_\nu = 0$ and $\sigma_\nu^2 = N_\nu^2/(N_\nu - 1) \simeq N_\nu$.

We introduce the following notations: Let us define for H_ν the following symmetric functions

$$H(\gamma_1, \ldots, \gamma_m) = \sum_m h_{\nu i_1}^{\gamma_1} \cdots h_{\nu i_m}^{\gamma_m}, \tag{3.4.23}$$

where the summation $\sum_m$ extends over all permutations of $(i_1, \ldots, i_m)$ of m integers chosen from $(1, \ldots, N_\nu)$. Then from (3.4.20), we have

$$\begin{aligned} E\{([L_\nu - \lambda_\nu]/\sigma_\nu)^r \mid \mathcal{T}_\nu\} &= \left(\frac{N_\nu^2}{N_\nu - 1}\right)^{-r/2} E\left\{\left(\sum_{\alpha=1}^{N_\nu} b_{\nu i} X_{\nu i}\right)^r \mid \mathcal{T}_\nu\right\} \\ &= N_\nu^{-r/2} \sum_{i_1=1}^{N_\nu} \cdots \sum_{i_r=1}^{N_\nu} b_{\nu i_1} \cdots b_{\nu i_r} \\ &\quad \times E(X_{\nu i_1} \cdots X_{\nu i_r} \mid \mathcal{T}_\nu). \end{aligned} \tag{3.4.24}$$

Let us consider the term on the right-hand side of (3.4.24). Suppose $i_1, \ldots, i$ are not all distinct and $i_1, \ldots, i_m$ $(m \leq r)$ are the distinct integers which are repeated $\gamma_1, \ldots, \gamma_m$ times respectively. Thus $\gamma_1 + \cdots + \gamma_m = r$. Then

$$(3.4.25) \qquad E\{X_{\nu i_1}^{\gamma_1} \cdots X_{\nu i_m}^{\gamma_m} \mid \mathfrak{F}_\nu\} = (N_\nu^{[m]})^{-1} A_\nu(\gamma_1, \ldots, \gamma_m),$$

where $N_\nu^{[m]} = N_\nu(N_\nu - 1) \cdots (N_\nu - m + 1)$. Also, the coefficient of (3.4.25) in the expansion (3.4.24) is nothing but

$$(3.4.26) \qquad B_\nu(\gamma_1, \ldots, \gamma_m) \cdot W(r; \gamma_1, \ldots, \gamma_m),$$

where $W(r; \gamma_1, \ldots, \gamma_m)$ is the number of ways in which the r suffixes $i_1, \ldots, i_r$ can be tied up into m distinct groups of numbers $\gamma_1, \ldots, \gamma_m$, respectively. Now, under the condition I on B_ν and condition II on A_ν, it can be verified (cf. Wald and Wolfowitz, 1944; Noether, 1949b) that

$$(3.4.27) \qquad N_\nu^{-m} A_\nu(\gamma_1, \ldots, \gamma_m) B_\nu(\gamma_1, \ldots, \gamma_m) = o(N_\nu^{r/2}),$$

unless $m = r/2$ and $\gamma_1 = \cdots = \gamma_m = 2$, and in this case

$$(3.4.28) \qquad N_\nu^{-r} A_\nu(2, \ldots, 2) B_\nu(2, \ldots, 2) = 1 + o(1).$$

Further $W(r, 2, \ldots, 2) = (2k - 1)(2k - 3) \cdots 3 \cdot 1$ if $r = 2k$, and otherwise $W(r, \gamma_1, \ldots, \gamma_m)$ is finite. Consequently, from (3.4.24), (3.4.25), (3.4.26), (3.4.27), and (3.4.28), we get

$$(3.4.29) \qquad \begin{aligned} E\{([L_\nu - \lambda_\nu]/\sigma_\nu)^r \mid \mathfrak{F}_\nu\} &= \frac{r!}{2^r \left(\frac{r}{2}\right)!}, \quad && \text{if } r \text{ is even,} \\ &= o(1), && \text{if } r \text{ is odd, for } r = 1, 2, 3, \ldots. \end{aligned}$$

Hence the theorem. ◁

The above theorem deals with sufficient conditions for (3.4.21) to hold. Hájek (1961) has obtained a necessary and sufficient condition for (3.4.21) to hold. For this, let us consider a sequence $U_1, \ldots, U_{N_\nu}$ of independent random variables each having uniform distribution over the interval [0, 1], and let $Z_1 < \cdots < Z_{N_\nu}$ be the order statistics associated with U_i's. We take a nondecreasing sequence $a_{\nu 1} \leq \cdots \leq a_{\nu N_\nu}$ of real numbers and put

$$(3.4.30) \quad a_\nu(\lambda) = a_{\nu i} \quad \text{for} \quad (i - 1)/N_\nu < \lambda \leq i/N_\nu, \qquad i = 1, \ldots, N_\nu.$$

The function $a_\nu(\lambda)$ will be called a quantile function of $a_{\nu 1} \leq \cdots \leq a_{\nu N_\nu}$. As $(i - 1)/N_\nu < i/(N_\nu + 1) < i/N_\nu$, we have

$$(3.4.31) \qquad a_{\nu i} = a_\nu(i/N_\nu) = a_\nu(i/(N_\nu + 1)), \qquad 1 \leq i \leq N_\nu.$$

Furthermore,

(3.4.32)
$$\bar{a}_\nu = \frac{1}{N_\nu}\sum_{i=1}^{N_\nu} a_{\nu i} = \int_0^1 a_\nu(\lambda)\,d\lambda,$$
$$\sigma^2 = \frac{1}{N}\sum_{i=1}^{N_\nu}(a_{\nu i} - \bar{a}_\nu)^2 = \int_0^1 [a_\nu(\lambda) - \bar{a}_\nu]^2\,d\lambda.$$

Then, in a very interesting way, it has been shown by Hájek (1961) that L_ν, defined by (3.4.20), is (asymptotically) equivalent in the quadratic mean to the statistic

$$T_\nu = \sum_{i=1}^{N_\nu}(b_{\nu i} - \bar{b}_\nu)a_\nu(U_i) + \bar{b}_\nu \sum_{i=1}^{N_\nu} a_{\nu i}, \tag{3.4.33}$$

where $a_\nu(\lambda)$ denotes the quantile functions of $a_{\nu 1} \le \cdots \le a_{\nu N_\nu}$. Since $a_\nu(U_i)$, $i = 1, \ldots, N_\nu$ are i.i.d.r.v., the classical central limit theorem under Lindeberg's condition may be applied to T_ν to prove its asymptotic normality. Hence, after some simplifications, we arrive at the following.*

Theorem 3.4.2 (Hájek). *If both A_ν and B_ν satisfy condition* II, *then $\mathcal{L}_\mathcal{J}([L_\nu - \lambda_\nu]/\sigma_\nu) \to N(0, 1)$, if and only if condition* V *is satisfied by D_ν, where $d_{\nu ij} = a_{\nu i}b_{\nu j}$ for $i, j = 1, \ldots, N_\nu$.*

Hájek has also considered another useful theorem when the roles of A_ν and B_ν (in L_ν) may not be symmetric.

Theorem 3.4.3 (Hájek). *If A_ν and B_ν both satisfy condition* II, *then A_ν will satisfy condition* V *in conjunction with B_ν, if and only if, A_ν satisfies condition* III.

Let us now consider the case of *bilinear permutation statistics*. Following Hoeffding (1951a), we define a sequence of N_ν^2 real numbers by

$$B_\nu^* = \{b_\nu(i, j), 1 \le i, j \le N_\nu\}, \tag{3.4.34}$$

where N_ν satisfies (3.4.1). Let then $(R_{\nu 1}, \ldots, R_{\nu N_\nu})$ be any permutation of $(1, \ldots, N_\nu)$. We define a random variable $R_\nu = (R_{\nu 1}, \ldots, R_{\nu N_\nu})$ which takes on each permutation of $(1, \ldots, N_\nu)$ with the same probability $1/N_\nu!$, and let

$$\mathcal{C}_\nu = \sum_{i=1}^{N_\nu} b_\nu(i, R_{\nu i}) \tag{3.4.35}$$

* An extension of this theorem to the problem of sign invariance is due to Ghosh (1969).

be our desired statistic. It may be noted that L_ν defined in (3.4.20) is a particular case of $\mathcal{T}_\nu$ where $b_\nu(i,j) = a_{\nu i} b_{\nu j}$ for all (i, j). Let us then define

$$\xi_\nu = N_\nu^{-2} \sum_{i=1}^{N_\nu} \sum_{j=1}^{N_\nu} b_\nu(i,j), \tag{3.4.36}$$

$$d_{\nu ij} = b_\nu(i,j) - N_\nu^{-1}\left\{\sum_{g=1}^{N_\nu} b_\nu(g,j) + \sum_{h=1}^{N_\nu} b_\nu(i,h)\right\} + \xi_\nu \quad \text{for} \quad i, j = 1, \ldots, N_\nu, \tag{3.4.37}$$

and let

$$D_\nu = \{d_{\nu ij};\ 1 \le i, j \le N_\nu\}. \tag{3.4.38}$$

It is then easy to show that

$$E\{\mathcal{T}_\nu \mid \mathcal{S}_\nu\} = N_\nu \xi_\nu \tag{3.4.39}$$

$$\text{var}\{\mathcal{T}_\nu \mid \mathcal{S}_\nu\} = \frac{1}{N_\nu - 1} \sum_{i=1}^{N_\nu} \sum_{j=1}^{N_\nu} d_\nu^2(i,j). \tag{3.4.40}$$

The following theorem is an extension of Hoeffding's (1951a) original theorem by Motoo (1957), the proof of which will not be considered here.

Theorem 3.4.4 (Hoeffding-Motoo). *If* D_ν *in* (3.4.38) *satisfies condition* V($\Leftarrow$IV), *then*

$$\mathcal{L}_{\mathcal{S}}([\mathcal{T}_\nu - N_\nu \xi_\nu]/[\text{var}(\mathcal{T}_\nu \mid \mathcal{S}_\nu)]^{1/2}) \to N(0, 1). \tag{3.4.41}$$

It may be noted that Theorems 3.4.1 and 3.4.2 can be readily generalized to the case of more than one L_ν and Theorem 3.4.4 to the case of more than one $\mathcal{T}_\nu$ (cf. Hájek, 1961; Hoeffding, 1952). In either case, we require a nonsingularity condition on the asymptotic permutation covariance matrix of the set of permutation statistics for which the multinormality is desired. Some other permutational central limit theorems are due to Dwass (1953, 1955b) and Hoeffding (1952).

Before we proceed to consider the application of these theorems in some specific problems, we would like to emphasize a distinction between two types of permutation tests: *component randomization tests* and *rank permutation tests* (cf. Wilks, 1962, p. 462). For the first type of tests, the test function $\phi(Z_N)$ in (3.3.8) is based on the permutation distribution of some statistic $t_N = t(Z_N)$ on the $M(Z_N)$ equally likely points in the set $T(Z_N)$, where $M(Z_N)$ and $T(Z_N)$ are defined in section 1. In such a case, usually the statistic t_N is chosen to be some parametrically optimum test statistic. Though these permutation tests are quite simple to define, in actual practice they impose some problems, particularly when the sample size is not very small. In order that we may use Theorems 3.4.1–3.4.4 for the asymptotic permutation distribution of such a t_n, we require t_n to be either some linear or bilinear statistic. Moreover, t_n depends explicitly on the values of the sample observations. Hence, even if t_n is linear or bilinear, the sequences A_ν, B_ν, or B_ν^* may contain random variables, and hence any condition on them, such as conditions I to V, will not, in general, hold with certainty, even if the sample size is very large. However, if A_ν or B_ν is composed of i.i.d.r.v., the conditions I–V will hold, in probability, under certain conditions on the parent distributions. For this we have the following simple theorem.

Theorem 3.4.5 (Hoeffding). *If H_ν is composed of i.i.d.r.v. with a nonzero variance and a finite absolute moment of order $2 + \delta$ for some $\delta > 0$, then condition II holds, with probability one, as $\nu \to \infty$.*

In many problems, H_ν is not composed of i.i.d.r.v. but is composed of $c(\geq 2)$ independent sets of i.i.d.r.v., where the distribution may vary from one set to another in any arbitrary manner. Let $n_\nu^{(1)}, \ldots, n_\nu^{(c)}$ be c positive integers defined for each positive integer ν, and let us assume that

$$\sum_{k=1}^{c} n_\nu^{(k)} = N_\nu \to \infty \quad \text{as} \quad \nu \to \infty; \tag{3.4.42}$$

$$\lim_{\nu \to \infty} n_\nu^{(k)} = \infty \quad \text{for all} \quad k = 1, \ldots, c(\geq 2), \tag{3.4.43}$$

though $n_\nu^{(k)}/N_\nu$ may or may not tend to any limit λ_k $(0 < \lambda_k < 1)$ as $\nu \to \infty$. Let us then in the kth set have $n_\nu^{(k)}$ i.i.d.r.v. distributed according to a c.d.f. $F_k(x)$, $k = 1, \ldots, c$. Then the following theorem is a straightforward generalization of Theorem 3.4.5.

Theorem 3.4.6. *If H_ν is composed of c independent sets of $n_\nu^{(1)}, \ldots, n_\nu^{(c)}$ i.i.d.r.v. distributed according to c.d.f.'s $F_1, \ldots, F_c$, respectively, and if these c.d.f.'s have all nonzero variances and finite moments up to the order $2 + \delta$, for some $\delta > 0$, then under (3.4.42) and (3.4.43), H_ν satisfies condition II, with probability one, as $\nu \to \infty$.*

Again in many cases both A_ν and B_ν are stochastic vectors, and it can be shown that both satisfy condition II, with probability one. In such a case, in order that we may use Theorem 3.4.2 we require to show that condition V holds, in probability, for $d_{\nu ij} = a_{\nu i} \cdot b_{\nu j}$; $i, j = 1, \ldots, N_\nu$. If A_ν is composed of i.i.d.r.v. distributed according to a c.d.f. $F(x)$ having a finite and nonzero variance σ_1^2, and similarly B_ν is composed of i.i.d.r.v. distributed according to a c.d.f. $G(y)$ having a finite and nonzero variance σ_2^2, then it is easy to verify that $N_\nu^{-1} d_\nu^2$ (defined in (3.4.15)) converges, with probability one, to $\sigma_1^2\sigma_2^2 > 0$. Hence, it can be shown that (3.4.16) holds, with probability one. This leads to the following.

Theorem 3.4.7. *If A_ν and B_ν are independent and contain i.i.d.r.v. distributed according to c.d.f.'s F and G, respectively, both having (nonzero) finite second-order moments, then condition* V *holds, with probability one, for the sequence* $D_\nu = (a_{\nu i} b_{\nu j}, 1 \leq i, j \leq N_\nu)$. *Similarly, if both F and G have finite moments up to the order* $2 + \delta$ *for some* $\delta > 0$, *then* D_ν *satisfies condition* IV, *with probability one, as* $\nu \to \infty$.

It may be noted that for this theorem we need not assume that A_ν and B_ν are stochastically independent. In fact, if $C_\nu = (A_\nu, B_\nu) = \{(a_{\nu i}, b_{\nu i}), i = 1, \ldots, N_\nu\}$ is a sequence of N_ν (bivariate) i.i.d.r.v. distributed according to a c.d.f. $F(x, y)$, then we have the following.

Theorem 3.4.8. *The conclusions of Theorem* 3.4.7 *also hold if the c.d.f.* $F(x, y)$ *satisfies the moment conditions imposed on* $F(x)$ *and* $G(y)$ *in Theorem* 3.4.7.

As the proofs of these theorems are not difficult but only lengthy, they are omitted. Also, both Theorems 3.4.7 and 3.4.8 may be further generalized as in Theorem 3.4.6. Because of their similarity of formulation, these are not considered.

It is thus seen that for component randomization tests based on certain types of permutation statistics, the theorems considered earlier can be used to prove the asymptotic normality of the permutation distribution of such statistics. Naturally, here the class of statistics is somewhat narrow in the sense that many other statistics do not belong to this class. Later, we shall see how a much wider class of statistics can be used in permutation tests and how their asymptotic normality can be proved by some further theorems. Before that, we will consider briefly the *rank permutation tests.* These tests possess two distinctive advantages: (i) the test procedure remains invariant under any monotonic transformation of the variates and (ii) if the parent c.d.f.'s are continuous, so that the possibility of ties may be ignored, in probability, often the permutation distribution of such a statistic becomes identical with the unconditional null distribution of the same. (An exception of this feature is the multivariate case, where even under the null hypothesis of invariance, the permutation distribution and the unconditional null distribution will be usually different from each other.) An important class of rank permutation tests is the class of *rank order tests* considered by Hoeffding (1951b), Terry (1952), Dwass (1956), Chernoff and Savage (1958), Govindarajulu (1960), Klotz (1962), Sen (1970a), Puri (1964, 1965a), Bhuchongkul (1964), Sen and Govindarajulu (1966), Govindarajulu, Le Cam, and Raghavachari (1967), and Puri and Sen (1969b) among many others. These tests are essentially based on statistics of the type (3.4.20), where both the sequences A_ν and B_ν are composed of elements which are explicit functions of ranks $1, \ldots, N_\nu$. For these statistics, all the Theorems 3.4.1–3.4.4 considered earlier hold under very mild restrictions on the parent c.d.f.'s and the rank functions. However, in many multivariate problems or in problems where the parent c.d.f.'s are not continuous, these rank order statistics will have null distribution depending on the unknown parent distribution, and the simplest way of rendering these strictly distribution-free is to resort to the permutation principle. However, all rank tests are not based on statistics belonging to the class of linear or bilinear rank order statistics. For these tests the theorems considered earlier may not be applicable. For this reason, we shall consider the following development which covers a wide class of nonparametric test statistics.

The U-statistics, as explained in section 3.2, play an important role in the theory of nonparametric inference. These may be used for both component randomization and rank permutation tests, and moreover they are applicable to distributions which are not necessarily continuous. In particular, if U-statistics are also rank statistics and the parent distributions are continuous, then usually under a suitable hypothesis of invariance, the distributions of U-statistics do not depend on the parent distribution. However, this statement again does not apply to multivariate situations. Also, when the parent distributions are not necessarily continuous or when the U-statistics are not explicit functions of ranks, their distributions will depend on the unknown parent distribution, and a convenient way of overcoming this difficulty is to adopt the permutation principle. Sen (1965b, 1966c, 1967c) has developed the permutation distribution theory of U-statistics under conditions which are quite similar to the ones required for the unconditional distribution theory, as has been imposed by Hoeffding (1948a).

To start with, let us assume that there is a finite universe, having N_ν units whose values are $a_{\nu 1}, \ldots, a_{\nu N_\nu}$, respectively, and we denote by $A_\nu = (a_{\nu 1}, \ldots, a_{\nu N_\nu})$. In a random sample of size n_ν drawn (without replacement) from this finite universe, let $X_{\nu 1}, \ldots, X_{\nu n_\nu}$ be the n_ν observations. We define a kernel as in (3.2.1) and note that the regular functional θ will be a function of A_ν, and hence, for simplicity, it is denoted by θ_ν. We define a U-statistic $U_\nu (= U_{n_\nu})$ as in (3.2.7). It can be shown (cf. Nandi and Sen, 1963) that U_ν will be the m.v.u. estimator of θ_ν. Further let $n_\nu \to \infty$ as $\nu \to \infty$,

$$(3.4.44) \quad n_\nu / N_\nu = \lambda_\nu \to \lambda\colon\ 0 \le \lambda < 1 \text{ as } \nu \to \infty, \quad (N_\nu \to \infty \text{ as } \nu \to \infty)$$

and define

$$(3.4.45) \qquad \phi_1(X_{\nu 1}) = \binom{N_\nu - 1}{m - 1}^{-1} \sum{}^1 \phi(X_{\nu 1}, X_{\nu 2}, \ldots, X_{\nu m}),$$

where the summation $\sum^1$ extends over all possible choices of $X_{\nu 2}, \ldots, X_{\nu m}$

from the remaining $(N_\nu - 1)$ values in A_ν, and let

(3.4.46a) $$\zeta_{1,N} = \frac{1}{N_\nu} \sum_{i=1}^{N_\nu} [\phi_1(a_{\nu i}) - \theta_\nu]^2;$$

then essentially by the same technique as in (3.2.23) and (3.2.24) it can be shown that if (3.4.44) holds and if

(3.4.46b) $$E_\nu\{\phi^2(X_{\nu 1}, \ldots, X_{\nu m})\} < \infty, \qquad \text{uniformly in } \nu,$$

then

$$n_\nu^{1/2}\left\{[U_\nu - \theta_\nu] - \frac{m}{n_\nu}\sum_{i=1}^{n_\nu}[\phi_1(X_{\nu i}) - \theta_\nu]\right\} \xrightarrow{p} 0.$$

(For details, see Theorems 4.1, 4.2 and 4.3 of Nandi and Sen, 1963.) Further, if $E_\nu\{|\phi|^{2+\delta}\} < \infty$ for some $\delta > 0$, it readily follows from Theorem 3.4.1 (as in this case, all possible $\binom{N_\nu}{n_\nu}$ choices of n_ν units from N_ν units are equally likely; or in other words, we can write $\sum_{i=1}^{n_\nu} [\phi_1(X_{\nu i}) - \theta_\nu]$ in the same form as in (3.4.25) with $b_{\nu i}$ as $+1$ or 0) that the standardized form of $\sum_{i=1}^{n_\nu} [\phi_1(X_{\nu i}) - \theta_\nu]$ has asymptotically (i.e., as $\nu \to \infty$) a normal distribution. Hence, from the asymptotic equivalence of U_ν and $1/n_\nu \sum_{i=1}^{n_\nu} \phi_1(X_{\nu i})$, we readily arrive at the following.

Theorem 3.4.9 (Nandi-Sen). *If* (3.4.43) *and* (3.4.46) *hold, and for some* $\delta > 0$, $E_\nu\{|\phi_1(X_{\nu i})|^{2+\delta}\} < \infty$, *uniformly in* ν, *then*

(3.4.47) $$\mathcal{L}\left(\frac{n_\nu^{1/2}[U_\nu - \theta_\nu]}{m\{(1-\lambda_\nu)\zeta_{1,\nu}\}^{1/2}}\right) \to N(0, 1).$$

Let us consider next the situation which seems to be appropriate to problems (I) and (III) in section 3.3. Let $X_{i1}, \ldots, X_{in_i}$ be n_i i.i.d.r.v. distributed according to a c.d.f. $F_i(x)$, for $i = 1, \ldots, c(\geq 2)$, and let $Z_N = (X_{11}, \ldots, X_{cn_c})$. We have in mind the null hypothesis (3.3.5) and the finite group of transformations in (III) of section 3.3, which leaves the joint distribution of Z_N invariant. Let then

(3.4.48) $$\phi(X_{1\alpha_{11}}, \ldots, X_{1\alpha_{1m_1}}, \ldots, X_{c\alpha_{c1}}, \ldots, X_{c\alpha_{cm_c}})$$

be a *kernel* of some *estimable parameter* $\theta = \theta(F_1, \ldots, F_c)$, where $(m_1, \ldots, m_c)$ are all non-negative integers, at least one of them being strictly positive. ϕ is assumed to be symmetric in the m_i arguments of the ith set for $i = 1, \ldots, c$, though the roles of the c different sets may not be symmetric. The U-statistic corresponding to ϕ in (3.4.48) is given by

(3.4.49) $$U_N = \prod_{i=1}^{c} \binom{n_i}{m_i}^{-1} \sum_S \phi(X_{1\alpha_{11}}, \ldots, X_{1\alpha_{1m_1}}, \ldots, X_{c\alpha_{c1}}, \ldots, X_{c\alpha_{cm_c}}),$$

where the summation S extends over all possible $1 \le \alpha_{i1} < \cdots < \alpha_{im_i} \le n_i$, for $i = 1, \ldots, c$. U_N is an unbiased estimator of $\theta(F_1, \ldots, F_c)$. (Usually, we have more than one estimable parameter in the formulation of the problem; however, for simplicity, we consider specifically the case of a single U-statistic.) In order to induce the nonparametric structure of the hypothesis, we shall assume that

$$\theta(F_1, \ldots, F_c) = \theta^0 \text{ (known)} \quad \text{for all} \quad F_1 \equiv \cdots \equiv F_c \equiv F \text{ (unknown)}. \tag{3.4.50}$$

We shall see later on that in most of the problems, we can select θ in such a manner that (3.4.50) holds. Also, let ϕ^* be the completely symmetric form of ϕ in (3.4.48), i.e., ϕ^* is obtained by taking the average overall possible permutations of $m_1 + \cdots + m_c (= m$, say) coordinates of ϕ in (3.4.48). Further, let γ_N^2 be the (unconditional) variance of U_N in (3.4.49) when $H_{0,3}$ in (3.3.5) holds. We also assume that $n_i = n_\nu^{(i)}$, $i = 1, \ldots, c$ are non-decreasing in ν and $N = N_\nu$ satisfies (3.4.1) along with

$$\lim_{\nu \to \infty} n_\nu^{(i)}/N_\nu = \lambda_i: \ 0 < \lambda_i < 1 \quad \text{for} \quad i = 1, \ldots, c. \tag{3.4.51}$$

Then, the following two conditions will be imposed on θ and ϕ.

Condition VI:

$$\phi^*(z_1, \ldots, z_m) = \theta^0 \quad \text{for all} \quad (z_1, \ldots, z_m) \in R^m. \tag{3.4.52}$$

Condition VII:

$$\lim_{N \to \infty} N\gamma_N^2 = \gamma^2(F, \lambda_1, \ldots, \lambda_c) > 0. \tag{3.4.53}$$

It may be noted that (3.4.53) is also implicit in the asymptotic normality of the unconditional distribution of U-statistics (cf. Hoeffding, 1948a), while (3.4.52) will usually follow from (3.4.50), and will not be a serious restriction in a majority of the problems.

Let us denote the permutational variance of U_N in (3.4.49) (conditioned on a given Z_N) by $\sigma_U^2(Z_N)$, and note that $\sigma_U^2(Z_N)$ is also a linear function of several U-statistics in the N observations in Z_N (cf. Sen, 1965a, b; 1966c). If N is large, then $N\sigma_U^2(Z_N)$ converges stochastically to a single U-statistic, i.e., the contribution of other terms in $\sigma_U^2(Z_N)$ can be made arbitrarily small.

The sampling distribution of U-statistics in random samples drawn without replacement from a finite universe, which has been studied in the preceding theorem, has been extended further by Sen (1965b, 1966c) to consider the permutation distribution of U-statistics which we present below (without proof).

Theorem 3.4.10. *Under $H_{0,3}$ defined in* (3.3.5), *if the kernel ϕ in* (3.4.48) *has a finite second-order moment and if* (3.4.52) *holds then $\sigma_U^2(Z_N)$ is an unbiased estimator of γ_N^2. If in addition to* (3.4.52), *ϕ has a finite fourth-order moment and the order statistic associated with Z_N is complete then $\sigma_U^2(Z_N)$ is the minimum variance unbiased (m.v.u.) estimator of γ_N^2.*

If the c.d.f.'s $F_1(x), \ldots, F_c(x)$ are not all identical, we define

$$\bar{F}_{\{N\}}(x) = \sum_{i=1}^{c} (n_i/N)F_i(x), \tag{3.4.54}$$

and note that (3.4.54) converges to $\bar{F}(x) = \sum_{i=1}^{c} \lambda_i F_i(x)$, as $N \to \infty$.

Theorem 3.4.11. *If ϕ in* (3.4.48) *possesses a finite fourth-order moment for all $F_1, \ldots, F_c$ then $\sigma_U^2(Z_N)$ is a consistent estimator of $\gamma^2(\bar{F}, \lambda_1, \ldots, \lambda_c)$. Further, if $\gamma^2(\bar{F}, \lambda_1, \ldots, \lambda_c) > 0$, then it is also asymptotically the m.v.u. estimator of the same.*

Theorem 3.4.12. *The permutation distribution of $t_N = \{U_N - \theta^0\}/\sigma_U(Z_N)$ (under the $N!$ equally likely permutations of the coordinates of Z_N) asymptotically, in probability, reduces to a normal distribution with zero mean and unit variance, provided conditions* VI *and* VII *hold and either* (i) $H_{0,3}$ *in* (3.3.5) *holds and ϕ in* (3.4.48) *has a finite absolute moment of order $2 + \delta$, $\delta > 0$, or* (ii) *ϕ has a finite fourth-order moment.*

We shall see later on how these theorems can also be used in the one-sample situation, as considered in problem (I) of section 3.3. We would like to present here some more theorems applicable to the problem of matching invariance defined in section 3.3. In this problem, we have the null hypothesis (3.3.3) and consequent on this, we have the $n!$ equally likely permutations of the type (3.3.4). Thus, if we have any estimable parameter $\theta = \theta(F)$, where F is the bivariate c.d.f. in (3.3.3), then conditioned on the observed sample point $\mathbf{E}_n$ (defined just after (3.3.3)), we have $n!$ equally likely sample points, from each of which we can calculate a U-statistic for θ. Thus, we will have a permutation distribution of this U-statistic, generated by the $n!$ equally likely permutations in (3.3.4). Here also the permutation variance of the U-statistic can be calculated from the sample (cf. Sen, 1967e, (3.11) and (3.14)) and is denoted by $\sigma_U^2(E_N)$, while the variance of the U-statistic under the assumption that (3.3.3) holds and $F(x, \infty) = G(x)$, $F(\infty, y) = H(y)$, is denoted by $\gamma_n^2(G, H)$. Then by analogy with conditions VI and VII, we impose the following.

Condition VIII: If ϕ^* is the completely symmetric form of $\phi = \phi([X_{11}, X_{22}], \ldots, [X_{1m}, X_{2m}])$, obtained by taking average over the $m!$ permutations of $X_{21}, \ldots, X_{2m}$ among the m positions, then

$$\phi^* = \theta^0 \text{ (known)} \quad \text{for all} \quad X_1, \ldots, X_m \in R^m. \tag{3.4.55}$$

Condition IX:

$$\lim_{n\to\infty} n\gamma_n^2(G, H) = \gamma(G, H) > 0. \tag{3.4.56}$$

Theorem 3.4.13. *Under $H_{0,2}$ defined in* (3.3.3), *$\sigma_U^2(E_n)$ is an unbiased estimator of $\gamma_n^2(G, H)$ provided the kernel ϕ has a finite second-order moment. If further, ϕ has a finite fourth-order moment, no matter whether $H_{0,2}$ in* (3.3.3) *holds or not, $n\,|\sigma_U^2(E_n) - \gamma_n^2(G, H)| \xrightarrow{p} 0$.*

Theorem 3.4.14. *Under the conditions* (3.4.55), (3.4.56) *and the existence of the fourth-order moment of the kernel ϕ the permutation distribution (i.e., over the $n!$ possible permutations in* (3.3.4)) *of $(U_n - \theta^0)/\sigma_U(E_n)$, asymptotically, in probability, reduces to a normal distribution with zero mean and unit variance.*

The proofs of these two theorems are given in Sen (1967e). It may be noted that the permutational distribution theory of U-statistics, as has been sketched above, remains valid even when the parent distribution is not continuous or the U-statistic is not a rank statistic. On the other hand, the unconditional distribution of U-statistics is found to be strictly distribution-free only for continuous parent distributions and for kernels which are invariant under monotonic transformations of the observations.

Let us now consider certain examples illustrating the uses of Theorems 3.4.1–3.4.14 in actual practice. We first consider problem III of section 3.3, and for simplicity, we take $c = 2$. We consider the two-sample test by Pitman (1937a), which is based on the statistic

$$t = \frac{\bar{X}_1 - \bar{X}_2}{\{s^2(1/n_1 + 1/n_2)\}^{1/2}}, \tag{3.4.57}$$

where $\bar{X}_1 = n_1^{-1}\sum_1^{n_1} X_{1i}$, $\bar{X}_2 = n_2^{-1}\sum_1^{n_2} X_{2i}$, and

$$s^2 = (n_1 + n_2 - 2)^{-1}\left\{\sum_1^{n_1}(X_{1i} - \bar{X}_1)^2 + \sum_1^{n_2}(X_{2i} - \bar{X}_2)^2\right\}.$$

It is easy to check that t is a strictly monotonically increasing function of $t^* = (\bar{X}_1 - \bar{X}_2)/\{s_0^2(1/n_1 + 1/n_2)\}^{1/2}$, where

$$(n_1 + n_2)s_0^2 = \sum_1^{n_1} X_{1i}^2 + \sum_1^{n_2} X_{2i}^2 - (n_1\bar{X}_1 + n_2\bar{X}_2)^2/(n_1 + n_2). \tag{3.4.58}$$

Also, by definition, s_0^2 remains invariant under the $(n_1 + n_2)!$ permutations of the coordinates of $(X_{11}, \ldots, X_{1n_1}, X_{21}, \ldots, X_{2n_2})$ among themselves. Thus, in order to study the permutation distribution theory of t^* (or equivalently of t), it is sufficient to study the same for $\bar{X}_1 - \bar{X}_2$. Now, under the permutational model, we can write

$$\bar{X}_1 - \bar{X}_2 = \sum_{\alpha=1}^{N} C_{N\alpha} Z_\alpha, \qquad N = n_1 + n_2, \tag{3.4.59}$$

where $(Z_1, \ldots, Z_N)$ takes on each permutation of $(X_{11}, \ldots, X_{2n_2})$ with equal probability $1/N!$, and

$$C_{N\alpha} = \begin{cases} 1/n_1, & \text{if } Z_\alpha \text{ is a first sample observation,} \\ -1/n_2, & \text{otherwise, for } \alpha = 1, \ldots, N. \end{cases} \tag{3.4.60}$$

Then, it is easy to verify that $(C_{N1}, \ldots, C_{NN})$ satisfies condition I, and $(Z_1, \ldots, Z_N)$ satisfies condition II, in probability, if $E(|Z|^{2+\delta}) < \infty$, for some $\delta > 0$. Thus, under the existence of the moments of $X_{1\alpha}$ and $X_{2\alpha}$ up to the order $2 + \delta$, for some $\delta > 0$, we have no difficulty in applying Theorem 3.4.1 to use the asymptotic normality of t^* under the permutation model. Let us next consider the well-known rank sum test by Wilcoxon (1945) and Mann and Whitney (1947). In this case, the test is based on the statistic U which may be expressed as a monotonic function of the difference of the average ranks of the two-sample observations. In this setup, if all the $X_{1\alpha}$ and $X_{2\alpha}$'s are distinct, then the ranks are the distinct integers $1, \ldots, N$, and we may write U as in (3.4.59), with the only change that $(Z_1, \ldots, Z_N)$ takes on each permutation of $(1, \ldots, N)$ with equal probability $1/N!$. It is then easy to verify that $(1, \ldots, N)$ satisfies condition II and hence, we may again use Theorem 3.4.1. If however, some of the observations are tied, we may again work with the kernel

$$\phi(X_{1\alpha}, X_{2\beta}) = \begin{cases} 1 & \text{if } X_{1\alpha} < X_{2\beta} \\ \frac{1}{2} & \text{if } X_{1\alpha} = X_{2\beta} \\ 0 & \text{if } X_{1\alpha} > X_{2\beta} \end{cases} \tag{3.4.61}$$

and show that conditions VI and VII hold for this kernel, and hence, we may use Theorem 3.4.12 for the asymptotic normality of U under the permutation model. This problem has been considered by Putter (1955) and Sen (1965a). The problem of scales in the two-sample case may be tackled in a similar manner.

Let us then consider problem (I) of section 3.3. We may consider either of the following two approaches. The first one is based on the basic permutational argument as considered in (3.3.2). Thus, if we consider the t-test considered by Fisher (1935), under the sign-invariant transformation (3.3.2), we may write our statistic as

$$t = \sum_{i=1}^{n} C_i X_i \bigg/ \left\{ \sum_{i=1}^{n} (C_i X_i)^2 \right\}^{1/2}, \tag{3.4.62}$$

where C_i is a random variable which can take only two values ± 1, with probability $\frac{1}{2}$ for each, $i = 1, \ldots, n$. Here for simplicity, let us assume that $X_1, \ldots, X_n$ all have a common distribution. Thus, the denominator of t is sign-invariant, while the numerator (under the permutation model) is the sum of n independent and identically distributed random variables, and has mean zero, variance $\sum_1^n X_i^2$ and absolute $(2 + \delta)$th-order $(\delta > 0)$ moment equal to $\sum_1^n |X_i|^{2+\delta}$. Thus, if $E\,|X_i|^{2+\delta} < \infty$, it is easily seen by an application of the Barry-Esséen theorem (chapter 2, p. 24) that in probability, t has asymptotically a normal (permutation) distribution. In the case of Wilcoxon's signed-rank test, we use a statistic similar to (3.4.62) with X_i replaced by the rank R_i of $|X_i|$ (among $|X_1|, \ldots, |X_n|$), $i = 1, \ldots, n$, so that $(R_1, \ldots, R_n)$ is a permutation of $(1, \ldots, n)$. Consequently, again the Barry-Esséen theorem may be used to prove the desired asymptotic normality. The second approach to this one-sample problem is based upon the following reduction to the two-sample problem. Under the null hypothesis that X has a distribution symmetric about $x = 0$, the two conditional distributions

$$(3.4.63) \qquad G(x) = P\{X \le x \mid X > 0\}, \qquad H(x) = P\{X \ge -x \mid X < 0\}$$

are identical. Thus, if we form two samples, one with the positive values among $(X_1, \ldots, X_n)$ and the other with the negative values among $(X_1, \ldots, X_n)$, then conditioned on a given m (the number of positive observations), we are faced essentially with a two-sample problem, with the sample sizes m and $n - m$ respectively. Hence, we may use the procedures discussed earlier.

Let us then consider problem (II) of section 3.3. In this case, if we rank $(X_{i1}, \ldots, X_{in})$ among themselves and assign the ranks $R_{i1}, \ldots, R_{in}$ respectively (for $i = 1, 2$), then a very simple test (due to Hotelling and Pabst, 1936) may be based on

$$(3.4.64) \qquad Z_n = \sum_{j=1}^{n} R_{1j}R_{2j} - n(n+1)^2/4,$$

and the asymptotic normality would then readily follow from Theorem 3.4.1. However, we are often interested in using some other suitable measure of association as for example, the difference-sign covariance, and moreover, the distribution of (X_{1i}, X_{2i}) may not be continuous, so that the possibility of ties may not be ignored, in probability. In such a situation, we may work with a class of U-statistics which satisfies conditions X and XI and then use Theorem 3.4.14 to derive the desired asymptotic normality of the permutation distribution. The details of these are omitted. Incidentally, Pitman's (1937b) test for independence based on the product moment correlation coefficient also belongs to this latter class, and hence may be tackled by Theorem 3.4.14. It may also be tackled by Theorem 3.4.5 and Theorem 3.4.6.

Finally, problem (IV) of section 3.3 can be more appropriately tackled by some multivariate procedures, and will be discussed in Chapter 7.

3.5. ASYMPTOTIC POWER OF NONPARAMETRIC TESTS

The study of the power properties of nonparametric tests poses certain problems. First, we have already noted in section 3.4 that under suitable hypotheses of invariance, the distribution of the nonparametric test statistic does not depend on the parent population. But when the null hypothesis is not true, the sampling distribution of nonparametric statistics depends on the parent distribution in some way or other. Second, the permutation tests considered in the preceding two sections are essentially conditional tests, and the study of their power properties demands the knowledge of the unconditional distribution of the allied test statistics. The evaluation of the exact unconditional non-null distribution often becomes quite a laborious task, if not impracticable. For this reason, the growth of the literature on the small-sample power properties of nonparametric tests is not at all satisfactory. However, in large samples, under certain conditions, such a study can be made, and we shall discuss these. We shall mainly consider the permutation tests considered in sections 3.3 and 3.4 and shall indicate how the results can be generalized to a wide class of nonparametric tests. In this context, we classify permutation tests into three types. First, some permutation tests are really based on the permutation distribution of some well-known statistics, which are also used in the parametric case when the parent distributions are specified. As examples of these, we may refer to the two-sample test by Pitman (1937a) considered in (3.4.57), the one sample test by Fisher (1935) considered in (3.4.62), or the test for independence in the bivariate case, considered by Pitman (1937b). All these statistics are also used for the corresponding optimum parametric tests based upon the assumption of normality of the parent distributions. These we will call *Type I permutation tests*. The second type of permutation tests is the class of all rank permutation and rank order tests, considered by Hotelling and Pabst (1936), Kendall (1938), Lehmann (1951a), Wilcoxon (1945), Mann and Whitney (1947), Hoeffding (1948a, 1948b, 1952), Terry (1952), Dwass (1953, 1955b, 1956), Chernoff and Savage (1958), Puri (1964, 1965a, 1965b) and Puri and Sen (1969a) among others. These will be referred to as *Type II permutation tests*. Finally we shall also consider permutation tests which are based on a class of U-statistics, whether rank tests or value permutation tests, though they may not correspond to well-known test statistics used in the parametric case. These will be called *Type III permutation tests*. Each type presents some difficulties in the study of the exact power properties of the tests.

Let us first consider the Type I permutation tests. In this case, the test statistics have usually some well-known distribution even when the null hypothesis is not true. For example, the statistic t in (3.4.57) has the noncentral Student's t distribution with $n_1 + n_2 - 2$ d.f. when the hypothesis $H_{0,3}$ does not hold, but the two c.d.f.'s F_1 and F_2 are normal and differ only by a translation parameter. On the other hand, the critical value of such a permutation test statistic is itself a random variable in the sense that it usually depends on the observation vector, held fixed for the permutation model. This makes the study of the exact power properties impracticable in most cases. However, it has been shown by Hoeffding (1952) that under certain conditions, the large-sample power properties of Type I permutation tests can be studied.

As in section 3.1, let $P(A_N) = P\{Z_N \in A_N\}$ be defined on an additive class $\mathcal{A}_N$ of subsets A_N of $\mathfrak{Z}_N$, and lst $\mathcal{G}_N$ be a finite group of transformations g_N of $\mathfrak{Z}_N$ onto itself which also maps P onto itself. Let H be the hypothesis of invariance of $P(A_N)$ under the group of transformations in $\mathcal{G}_N$, there being M_N such elements in $\mathcal{G}_N$. Let $t_N(Z_N)$ be a real-valued function of Z_N on $\mathfrak{Z}_N$. For any $z_N \in \mathfrak{Z}_N$, let

$$t_N^{(1)}(z_N) \leq \cdots \leq t_N^{(M_N)}(z_N) \tag{3.5.1}$$

be the ordered values of $t_N(g_N z_N)$ for $g \in_N \mathcal{G}_N$. Then, given the level of significance α $(0 < \alpha < 1)$, one can find a critical value $t_N^{(k_N)}(z_N)$ and the test function $\phi_N(z_N)$ follows:

$$\phi_N(Z_N) = \begin{cases} 1, & \text{if} \quad t_N(Z_N) > t_N^{(k_N)}(z_N), \\ a_N(z_N), & \text{if} \quad t_N(Z_N) = t_N^{(k_N)}(z_N), \\ 0, & \text{if} \quad t_N(Z_N) < t_N^{(k_N)}(z_N), \end{cases} \tag{3.5.2}$$

where k_N and $a_N(z_N)$ are so chosen that $E_H\{\phi_N(Z_N)\} = \alpha$. (3.5.2) relates to one-sided alternatives, but the modifications for the two-sided test are evident. In the asymptotic case, we will take N to be indefinitely large and assume that

$$\lim_{N\to\infty} M_N = \infty, \qquad \lim_{N\to\infty} k_N = \infty, \quad \text{and} \quad \lim_{N\to\infty} (k_N/M_N) = \alpha: \quad 0 < \alpha < 1. \tag{3.5.3}$$

Let now

$$F_N(t; z_N) = P\{t_N(Z_N) \leq t \mid H, Z_N = z_N\} \tag{3.5.4}$$

be the conditional c.d.f. of $t_N(Z_N)$ (under H), given $Z_N = z_N$. Also let

$$F_N(t) = P\{t_N(Z_N) \leq t \mid H\} = E_H\{(F(t; Z_N)\}, \tag{3.5.5}$$

be the unconditional c.d.f. of $t_N(Z_N)$. If $F_N(t)$ is specified (as will be the case for Type I permutation tests), the unconditional test based on $F_N(t)$ may be characterized by a test function $\phi_N^*(Z_N)$, which is usually parametric in nature, and is given by

$$\phi_N^*(Z_N) = \begin{cases} 1, & \text{if} \quad t_N(Z_N) > t_{N,\alpha}, \\ a_N, & \text{if} \quad t_N(Z_N) = t_{N,\alpha}, \\ 0, & \text{if} \quad t_N(Z_N) < t_{N,\alpha}, \end{cases} \tag{3.5.6}$$

where $t_{N,\alpha}$ and a_N are two constants (independent of Z_N), chosen in such a manner that $E_H\{\phi_N^*(Z_N)\} = \alpha$. In a majority of the cases dealing with Type I permutation tests, ϕ_N^* is usually some optimal (one-sided) parametric test based on the specified (or assumed) nature of the c.d.f. $F_N(t)$ in (3.5.5). Hoeffding (1952) has proved the asymptotic power equivalence of the two sequences of tests $\{\phi_N\}$ and $\{\phi_N^*\}$ defined in (3.5.2) and (3.5.6), respectively, for some class of alternative distributions, which we denote by $\{P_N(A)\}$.

Theorem 3.5.1 (Hoeffding). *Suppose that for a sequence of distributions* $\{P_N\}$, $F_N(t; Z_N) \xrightarrow{p} F(t)$, *at every point of continuity* t *of* $F(t)$, *where* $F(t)$ *is a c.d.f., and that* $F(t) = 1 - \alpha$ *has a unique solution* $t = t_\alpha$, *which is also a point of continuity of* $F(t)$, *then* $t_N^{(k_N)}(Z_N) \xrightarrow{p} t_\alpha$. *Also, if there exists a function* $H(t)$ *also continuous at* $t = t_\alpha$, *such that*

$$P\{t_N(Z_N) \le t \mid P_N\} \to H(t) \quad \text{as} \quad N \to \infty \tag{3.5.7}$$

at every point of continuity of $H(t)$, *then*

$$E\{\phi_N(Z_N) \mid P_N\} \to 1 - H(t_\alpha) \quad \text{as} \quad N \to \infty. \tag{3.5.8}$$

Again, if for the same sequence $\{P_N\}$ *of distributions,* $F_N(t) \to F(t)$, *at every point of continuity of* $F(t)$, *so that* $t_{N\alpha} \to t_\alpha$, *then*

$$E\{\phi_N^*(Z_N) \mid P_N\} \to 1 - H(t_\alpha) \quad \text{as} \quad N \to \infty. \tag{3.5.9}$$

Proof: If $F_N(t; Z_N) \xrightarrow{p} F(t)$ at every point of continuity of $F(t)$ and if $F(t) = 1 - \alpha$ has a unique solution $t = t_\alpha$, which is also a point of continuity of $F(t)$, then it is easy to verify using (3.5.2) and (3.5.6) that $t^{(k_N)}(Z_N) \xrightarrow{p} t_\alpha$, while (3.5.8) really follows from (3.5.7) and the stochastic convergence of $t^{(k_N)}(Z_N)$ to t_α. The last part of the theorem is simple and hence, its proof is omitted. ◁

Thus, it follows from the above theorem that for the sequence of alternatives $\{P_N\}$, the two sequences of tests $\{\phi_N\}$ and $\{\phi_N^*\}$ are asymptotically power equivalent. This result is particularly important when $\{\phi_N^*\}$ is a sequence of some optimum tests, so that the use of the same test statistic under the permutation setup results also in an asymptotically optimum test. Now, regarding the conditions under which $F_N(t; Z_N) \xrightarrow{p} F(t)$, Hoeffding (1952) has presented another interesting theorem.

Theorem 3.5.2. (Hoeffding). *A necessary and sufficient condition that* $F_N(t; Z_N) \xrightarrow{p} F(t)$ *at every point of continuity of* $F(t)$ *is that*

$$\begin{aligned} &P\{t_N(G_N Z_N) \le t\} \to F(t) \\ &P\{t_N(G_N Z_N) \le t, t_N(G_N' Z_N) \le t\} \to F^2(t), \end{aligned} \tag{3.5.10}$$

where G_N *and* G_N' *are random variables whose values are the* M_N *elements*

g_N of $\mathcal{G}_N$ (each element having the same probability $1/M_N$), and where G_N, G'_N, and Z_N are independent.

Proof: By the definitions of $t_N^{(k_N)}(Z_N)$ and $F_N(t; Z_N)$,

$$P\{t_N^{(k_N)}(Z_N) \leq y\} = P\{F_N(y, Z_N) \geq k_N/M_N\} \tag{3.5.11}$$

for every real y. Let y be a point of continuity of $F(y)$. Since, by assumption $k_N/M_N \to 1 - \alpha = F(\lambda)$, and $y < \lambda$ implies $F(y) < F(\lambda)$, the right-hand side of (3.5.11) tends to 0 if $y > \lambda$. Similarly it tends to 1 if $y < \lambda$. Hence $t_N^{(k_N)}(Z_N) \to \lambda$ in probability. ◁

A sufficient condition for a sequence of random variables to converge in probability to a constant c is that their means and variances converge, respectively, to c and 0. If the random variables are uniformly bounded, the condition is also necessary. Hence $F_n(y, X) \to F(y)$ in probability if and only if

$$EF_N(y, Z_N) \to F(y), \qquad EF_N(y, Z_N)^2 \to F(y)^2. \tag{3.5.12}$$

We can write

$$F_N(y, x) = M_N^{-1} \sum_g C(gx),$$

where $C(x) = 1$ or 0 according as $t_N(x) \leq y$ or $> y$. Hence

$$EF_N(y, Z_N) = M^{-1} \sum_g P\{t_N(gZ_N) \leq y\} \tag{3.5.13}$$

$$EF_N(y, Z_N)^2 = M^{-2} \sum_g \sum_{g'} P\{t_N(gZ_N) \leq y, t_N(g'Z_N) \leq y\}. \tag{3.5.14}$$

Let G be the random transformation defined earlier, let G' have the same distribution as G, and let G, G', and Z_N be mutually independent. Then equations (3.5.13) and (3.5.14) can be written as

$$EF_N(y, Z_N) = P\{t_N(GZ_N) \leq y\}, \tag{3.5.15}$$

and

$$EF_N(y, Z_N)^2 = P\{t_N(GZ_N) \leq y, t_N(G'Z_N) \leq y\}. \tag{3.5.16}$$

Note that $t_N(GZ_N)$ and $t_N(G'Z_N)$ are identically distributed, but not independent (except in the trivial case when the random variable $F_N(y, Z_N)$ has variance 0). Equations (3.5.15) and (3.5.16) imply that (3.5.12) is satisfied if $t_N(GZ_N)$ has the limiting distribution function $F(y)$, and ($t_N(GZ_N)$ and $t_N(G'Z_N)$) are independent in the limit.

It may be noted that the function $t_N(Z_N)$ in $\phi_N(Z_N)$ may be replaced by another function $S_N(Z_N)$, such that for any $z_N \in \mathfrak{Z}_N$ and for any two elements

g_N and g'_N of $\mathcal{G}_N$

$$(3.5.17) \qquad S_N(g_N z_N) - S_N(g'_N z_N) \quad \text{and} \quad t_N(g_N z_N) - t_N(g'_N z_N)$$

have the same sign. Thus, if the conditions of the above two theorems are not satisfied with $t_N(z_N)$, they may possibly be satisfied with some $S_N(z_N)$. In particular we have the following equivalence theorem.

Theorem 3.5.3. (Hoeffding). *Let $F_N^*(t; z_N)$ be the conditional c.d.f. (given z_N) of $S_N(z_N)$ (under H) which is defined as*

$$(3.5.18) \qquad S_N(z_N) = C_N(z_N)t_N(z_N) + d_N(z_N),$$

where $C_N(Z_N)$ and $d_N(Z_N)$ are invariant under $\mathcal{G}_N$; and $C_N(Z_N)$ and $d_N(Z_N)$ converge in probability to 1 *and* 0 *respectively; then*

$$F_N^*(t; Z_N) \xrightarrow{p} F(t) \quad \text{if and only if} \quad F_N(t; Z_N) \xrightarrow{p} F(t).$$

The proof of this theorem is simple and is left as Exercise 3.5.1.

Using the above theorems Hoeffding (1952) has shown that the permutation test (3.4.62) is asymptotically power equivalent to the one-sided Student's t-test against shift in location. He has also considered the permutation analog of the classical analysis of variance tests (considered by Pitman, 1938, and Welch, 1937) and has shown that the permutation tests and the classical tests are asymptotically power equivalent, for Pitman's type of translation alternatives. He has further considered the two-sample location test (3.4.57) by Pitman (1937b) and has proved a general theorem which has been extended by Dwass (1955b) to a general class of two-sample problems. For this, let us define $\mathbf{B}_N = (b_{N1}, \ldots, b_{NN})$ to be a sequence of real numbers, and let $\mathbf{Z}_i = (Z_{i1}, \ldots, Z_{in_i})$ be n_i i.i.d.r.v. distributed according to a c.d.f. $G_i(x)$, for $i = 1, \ldots, c\ (\geq 2)$, all these c stochastic vectors are assumed to be independent. Let $N = \sum_{i=1}^{c} n_i$, and $\mathbf{Z}_N = (\mathbf{Z}_1, \ldots, \mathbf{Z}_c)$. Let us then define

$$(3.5.19) \qquad \int x\, dG_i(x) = \mu_i, \qquad \int (x - \mu_i)^2\, dG_i(x) = \sigma_i^2: \qquad 0 < \sigma_i^2 < \infty, \quad i = 1, \ldots, c;$$

$$(3.5.20) \qquad G(x) = \sum_{i=1}^{c} \lambda_i G_i(x),$$

where we assume that

$$(3.5.21) \qquad \lim_{N\to\infty} n_i/N = \lambda_i\colon\ 0 < \lambda_i < 1, \quad \sum_{i}^{c} \lambda_i = 1.$$

Also, let $G^*(x)$ be the c.d.f. obtained from $G(x)$ by a linear transformation of the independent variable, so that the first two moments of $G^*(x)$ are 0 and 1, respectively, and let $G_N(x; Z_N)$ be the empirical c.d.f. of $\mathbf{Z}_N$, i.e.,

$$(3.5.22) \qquad G_N(x; \mathbf{Z}_N) = \frac{1}{N}(\text{number of } Z_{ij} \leq x).$$

Theorem 3.5.4. (Hoeffding-Dwass). *If* $G_N(x; Z_N) \xrightarrow{p} G(x)$ *at every point of continuity or* $G(x)$, *and either* B_N *satisfies condition* II *of section* 3.4 *or* $G^*(x)$ *is normal, then* $L_N = \sum_{i=1}^{N} b_{Ni}Z_{Ni}$ *has asymptotically a normal distribution when standardized.*

The proof is simple and is sketched in a paper by Dwass (1955b).

The above theorem may be used to show that Pitman's (1937a) two-sample location test (3.4.57) is asymptotically power equivalent to the classical Student's t-test. Hoeffding (1952) has also considered Pitman's (1937b) test for independence in bivariate samples (based on the correlation coefficient), and has obtained its asymptotic power equivalence with the classical test for no correlation. Thus for Type I permutation tests the above theorems may be used to study the asymptotic power properties.

Let us now consider the Type II permutation tests. Mostly, these are based on certain rank statistics, and if the parent distribution is continuous (so that the possibility of ties can be ignored, in probability), the permutation distribution of such a statistic agrees with its unconditional null distribution. (However, in many multivariate problems, this is not true.) Hence, in such a case, the critical values of the statistics (in some suitable test functions) do not depend on the conditioning variable held fixed. On the other hand, as the statistics are usually functions of ranks of the observations, the exact distribution when the null hypothesis (of invariance) is not true, is, in most of the cases, quite involved, and the process of evaluating the exact power of the tests is usually prohibitively tedious. However, for certain types of rank tests, the power function of the tests can be elegantly studied for large samples. The two most notable classes of such rank statistics are (i) rank statistics which are functions of U-statistics and (ii) the rank order statistics. For U-statistics, Hoeffding (1948a) has proved a very elegant theorem (cf. Theorem 3.2.1) on the asymptotic normality of the standardized form, and this may be readily used to study the asymptotic power properties of rank tests based on U-statistics. Further developments on this line are due to Lehmann (1951a), Sukhatme (1958), among many others. For rank order statistics, Chernoff and Savage (1958) have established a very useful limit theorem for a class of rank order statistics which has the same structure as our linear permutation statistics in section 3.4, and is somewhat more specialized in the sense that the sequence A_ν (cf. section 3.4) consists of only elements 0 and 1. The Chernoff-Savage theorem has been used to study the asymptotic power and power efficiency of a class of two-sample tests. The extensions of this celebrated theorem are due to Puri [(1964), (1965a), (1965b)], Puri and Sen [(1966), (1967b), (1969b)], Govindarajulu (1960), Govindarajulu, Le Cam, and Raghavachari (1967), Bhuchongkul (1964), and Sen (1970a), among others. In the next section, we shall consider in detail these rank order statistics, their asymptotic distribution theory and the role of such statistics in nonparametric inference. In fact, throughout this book, special emphasis will be laid on the use of such rank order statistics in multivariate analysis. However, the scope of these rank tests is somewhat limited only to the class of continuous distributions. In actual practice, often we have discrete sample spaces, and in section 3.4, we have considered four theorems (Theorems 3.4.11 to 3.4.14) which related to the permutation distribution theory of a certain class of U-statistics under various types of invariance models. In such a case, again the critical values of the test statistics are random variables and we are faced with the same situation as with Type I permutation tests, discussed earlier. We may, however, note certain points here. First, as we are dealing with U-statistics, under the conditions VI–IX, the asymptotic normality of the unconditional distributions of such statistics will follow from the general theorems by Hoeffding (1948a) and others discussed in section 3.2. Hence, as in Theorem 3.5.1, we require only to show that the critical values of the permutation distributions of such U-statistics (which are random variables) converge in probability to some constants, which are the critical values of the limiting unconditional null distribution of the same U-statistics. Now, using Theorems 3.4.12 and 3.2.2, and (3.2.74), it can easily be shown (for details cf. Sen, 1965a, b) that Theorem 3.5.1 is also applicable to the permutation tests based on U-statistics, provided the conditions VI–IX of section 3.4 hold. Thus, asymptotically, the power of the permutation tests based on U-statistics can be traced under certain conditions from the (unconditional) asymptotic normality of U-statistics. Exercise 3.5.2 clarifies this situation.

3.6. DISTRIBUTION THEORY OF RANK ORDER STATISTICS

In sections 3.3 and 3.4 we considered the distribution theory of U-statistics and some other nonparametric statistics. In section 3.5, we noted that an important class of nonparametric tests is based on the so-called rank order statistics. The permutation distribution theory of such rank order statistics has been studied in section 3.4, while in section 3.5, we noted that the study of the power properties of such tests demands the knowledge of the distribution theory of rank order statistics in the non-null case. The study of the distribution theory of rank order statistics in the non-null case was initiated by Dwass (1955b, 1956). However, one of the most significant theorems on this topic is due to Chernoff and Savage (1958). These authors developed a new theorem for the asymptotic normality of a class of rank order statistics for the two-sample problem under very mild conditions on the distributions of the samples. Their conditions on the score generating functions are rather stringent, though they are satisfied by most of the well-known rank order statistics. However, in this chapter we present their theorem under conditions which are less stringent than theirs.* In 1968, Pyke and Shorack provided a new approach to the Chernoff-Savage theorem. Their methods† are based on the concept of weak convergence in metric spaces. They are of great potential for the further development of the rank order theory, but are not discussed here since so far they have not been applied to the problems of multivariate analysis considered in this book. Neither the Chernoff-Savage (1958) nor the Pyke-Shorack (1968a) methods have been applied to the problems concerning regression analysis. Hájek (1962) considered a class of rank order statistics for the linear regression model. He provided a limit theorem under very mild conditions on the score-generating functions, but imposed the restriction of contiguity on the distributions of the samples. This approach has already been given an excellent coverage in Hájek and Šidák (1967) and is not discussed here. Hájek (1968) has also proved a very general limit theorem for the asymptotic normality of a class of rank order statistics when each observation may have a different distribution. The results he obtained have been extended in various directions in the univariate and multivariate problems by Hoeffding (1968), Jurečková (1967, 1969), Hušková (1970), Puri and Sen (1969a), and Sen and Puri (1969). These results are in the direction of regression analysis, and will be considered in a separate book.

For simplicity of presentation we shall first consider the two-sample problem (see problem (III) of section 3.3) and subsequently extend the results to the case of the one-sample problem (see problem (I) of section 3.3) as well as the c (≥ 2)-sample problem.

3.6.1. The Two-sample Case

Let $X_1, \ldots, X_m$ and $Y_1, \ldots, Y_n$ be two independent samples from absolutely continuous c.d.f.'s $F(x)$ and $G(x)$ respectively. Let $N = m + n$, $\lambda_N = m/N$, and we assume that for all N, the inequalities $0 < \lambda_0 \leq \lambda_N \leq 1 - \lambda_0 < 1$ hold for some fixed $\lambda_0 \leq \frac{1}{2}$.

Let $F_m(x)$ and $G_n(x)$ be the sample cumulative distribution functions of $X_1, \ldots, X_m$ and $Y_1, \ldots, Y_n$, respectively. Then $H_N(x) = \lambda_N F_m(x) + (1 - \lambda_N) G_n(x)$ is the combined sample cumulative distribution function. The combined population cumulative distribution function is $H(x) = \lambda_N F(x) + (1 - \lambda_N) G(x)$.

Define $Z_{N,i} = 1$, if the ith smallest observation in the combined sample is an X observation, and otherwise define $Z_{N,i} = 0$. Now consider the statistics

* For another proof of the Chernoff-Savage (1958) theorem under relaxed conditions on scores, see Govindarajulu, Le Cam, and Raghavachari (1967).

† The reader is also referred to Pyke (1970), and Hájek (1970), "Discussion on Pyke's paper" in Nonparametric Techniques in Statistical Inference, Cambridge University Press, England. Editor: M. L. Puri.

of the form

$$mT_N = \sum_{i=1}^{N} E_{N,i} Z_{N,i} \tag{3.6.1}$$

where the $E_{N,i}$ are given numbers. An equivalent representation of T_N is

$$T_N = \int_{-\infty}^{\infty} J_N\left[\frac{N}{N+1} H_N(x)\right] dF_m(x) \tag{3.6.2}$$

where $E_{N,i} = J_N(i/(N+1))$. While J_N need be defined only at $1/(N+1), \ldots, N/(N+1)$, we shall extend its domain of definition to $(0, 1]$ by letting J_N be constant on $(i/(N+1), (i+1)/(N+1)]$. Throughout this section, K will be used as a generic constant which may depend on J_N but will not depend on $F(x)$, $G(x)$, m, n, or N. Statements involving o_p, O_p will always be uniform in $F(x)$, $G(x)$, $H(x)$, and λ_N in the interval $0 < \lambda_0 \leq \lambda_N \leq 1 - \lambda_0 < 1$.

The statistics of the form (3.6.1) are usually considered for the problem of testing that F and G are identical. Examples of the statistics of the form (3.6.1) are (i) Wilcoxon's two-sample test with $E_{N,i} = i$, (ii) the Fisher-Yates-Hoeffding-Terry normal scores two-sample test with $E_{N,i}$ as the expected value of the ith order statistic of a sample of size N from the standard normal distribution, (iii) Van der Waerden's two-sample test with $E_{N,i} = \Phi^{-1}(i/(N+1)]$ where

$$\Phi(x) = \int_{-\infty}^{x} \frac{1}{\sqrt{2\pi}} e^{-t^2/2}\, dt,$$

(iv) the Ansari-Bradley two-sample test with

$$E_{N,i} = \left|\frac{i}{N+1} - \frac{1}{2}\right|,$$

(v) Mood's two-sample test with

$$E_{N,i} = \left(\frac{i}{N+1} - \frac{1}{2}\right)^2,$$

(vi) a test with

$$E_{N,i} = \left\{\Phi^{-1}\left(\frac{i}{N+1}\right)\right\}^2,$$

(vii) a test with $E_{N,i}$ as the expected value of the square of the ith order statistic of a sample of size N from the standard normal distribution.

Examples (i) to (iii) are the tests for location; (iv) to (vii) are (if it is assumed that X and Y have the same median) tests for scale.

We make the following assumptions.

(*a*) $\lim_{N\to\infty} J_N(u) = J(u)$ exists for $0 < u < 1$ and is not constant,

(*b*) $\int_{-\infty}^{\infty}\left[J_N\left[\frac{N}{N+1}H_N(x)\right] - J\left[\frac{N}{N+1}H_N(x)\right]\right] dF_m(x) = o_p(N^{-1/2})$,

(*c*) $|J^{(i)}(u)| = |d^iJ(u)/du^i| \leq K[u(1-u)]^{-i-1/2+\delta}$, $i = 0, 1$; for some $\delta > 0$.

Then we have the following theorem.

Theorem 3.6.1.* **(Chernoff-Savage).** *Under the assumptions* (*a*), (*b*), *and* (*c*), *and for fixed* $F(x)$, $G(x)$, *and* λ_N,

$$\lim_{N\to\infty} P\left(\frac{T_N - \mu_N}{\sigma_N} \leq t\right) = \int_{-\infty}^{t} \frac{1}{\sqrt{2\pi}} e^{-x^2/2}\, dx, \tag{3.6.3}$$

where

$$\mu_N = \int_{-\infty}^{\infty} J[H(x)]\, dF(x) \tag{3.6.4}$$

and

$$\begin{aligned} N\sigma_N^2 = 2(1-\lambda_N)\Bigg\{ &\iint_{-\infty<x<y<\infty} G(x)[1-G(y)]J'[H(x)]J'[H(y)] \\ &\times dF(x)\, dF(y) + \frac{1-\lambda_N}{\lambda_N} \iint_{-\infty<x<y<\infty} F(x)[1-F(y)] \\ &\times J'[H(x)]J'[H(y)]\, dG(x)\, dG(y)\Bigg\} \end{aligned} \tag{3.6.5}$$

provided $\sigma_N \neq 0$.

Remarks: In equations (3.6.4) and (3.6.5) we put subscripts N on μ and σ^2 to emphasize that these depend on F, G, and λ_N and are meaningful in the more general case where F, G, and λ_N are not fixed. Corollary 3.3.1 will extend this theorem to obtain convergence to normality uniformly with respect to F, G, and λ_N for a broad range of F, G, and λ_N.

To facilitate the proof of Corollary 3.3.1, we will regard F, G, and λ_N as variable throughout the proof of Theorem 3.6.1 except where it is specified otherwise.

Assumption (*a*) is likely to be filled whenever one speaks of a sequence of tests. In the special case when $E_{N,i} = E(i/(N+1))$, then $J_N = E = J$ and assumption (*b*) will automatically be satisfied. Section 10.5 deals with sufficient conditions under which assumptions (*a*) and (*b*) hold. We may

* The assumptions of this theorem are milder than those of Chernoff and Savage (1958).

mention that assumption (*c*) is the basic assumption. It has two functions: it limits the growth of the coefficients $E_{N,i}$ and it supplies certain smoothness properties.

Proof of Theorem 3.6.1: We rewrite T_N as

$$(3.6.6)\qquad T_N = \int_{-\infty}^{\infty} J_N\left[\frac{N}{N+1} H_N\right] dF_m(x)$$
$$= \int_{-\infty}^{\infty}\left\{J_N\left[\frac{N}{N+1} H_N\right] - J\left[\frac{N}{N+1} H_N\right]\right\} dF_m(x)$$
$$+ \int_{-\infty}^{\infty} J\left[\frac{N}{N+1} H_N\right] dF_m(x).$$

In the second integral, we write $dF_m = d(F_m - F + F)$,

$$(3.6.7)\quad J\left[\frac{N}{N+1} H_N\right] = J[H] + (H_N - H)J'[H] - \frac{H_N}{N+1} J'[H]$$
$$+ \left\{J\left[\frac{N}{N+1} H_N\right] - J[H]\right.$$
$$\left. - \left(\frac{N}{N+1} H_N - H\right) J'[H]\right\}.$$

Using (3.6.7), the expression for T_N becomes

$$(3.6.8)\qquad T_N = \mu_N + B_{1N} + B_{2N} + \sum_{i=1}^{4} C_{iN}$$

where μ_N is defined in (3.6.4) and

$$(3.6.9)\quad B_{1N} = \int_{-\infty}^{\infty} J[H]\, d[F_m(x) - F(x)]$$

$$(3.6.10)\quad B_{2N} = \int_{-\infty}^{\infty} (H_N - H)J'[H]\, dF(x)$$

$$(3.6.11)\quad C_{1N} = -\frac{1}{N+1}\int_{-\infty}^{\infty} H_N J'[H]\, dF_m(x)$$

$$(3.6.12)\quad C_{2N} = \int_{-\infty}^{\infty} (H_N - H)J'[H]\, d[F_m(x) - F(x)]$$

$$(3.6.13)\quad C_{3N} = \int_{-\infty}^{\infty}\left\{J\left[\frac{N}{N+1} H_N\right] - J[H]\right.$$
$$\left. - \left(\frac{N}{N+1} H_N - H\right) J'[H]\right\} dF_m(x)$$

$$(3.6.14)\quad C_{4N} = \int_{-\infty}^{\infty}\left\{J_N\left[\frac{N}{N+1} H_N\right] - J\left[\frac{N}{N+1} H_N\right]\right\} dF_m(x).$$

The μ_N, B, and C terms represent the "constant," "first-order random," and "higher-order random" portions respectively of T_N. The proof of this theorem is accomplished by showing that (i) the μ_N term is nonrandom and finite, (ii) $B_{1N} + B_{2N}$ has normal distribution in the limit, and (iii) the C terms are all $o_p(N^{-1/2})$. We shall establish these facts in the following lemmas.

Lemma 3.6.1. *$\int_{-\infty}^{\infty} J[H(x)]\, dF(x)$ is finite.*

Proof: Using assumption (c) and the fact that $dH \geq \lambda_0\, dF$, we get

$$\left| \int_{-\infty}^{\infty} J[H(x)]\, dF(x) \right| \leq K \int_0^1 [H(1-H)]^{\delta - 1/2} dH$$

$$\leq K \int_0^1 \frac{dH}{[H(1-H)]^{1/2}} < \infty. \qquad \triangleleft$$

Lemma 3.6.2. *$B_{1N} + B_{2N}$ has normal distribution in the limit.*

Proof: Integrating B_{2N} by parts, and using the fact that

$$\int_{-\infty}^{\infty} d[F_m(x) - F(x)] = 0,$$

we obtain

$$B_{1N} + B_{2N} = (1 - \lambda_N)\left\{ \int_{-\infty}^{\infty} B(x)\, d[F_m(x) - F(x)] - \int_{-\infty}^{\infty} B^*(x)\, d[G_n(x) - G(x)] \right\}, \tag{3.6.15}$$

where

$$B(x) = \int_{x_0}^{x} J'[H(y)]\, dG(y) \tag{3.6.16}$$

and

$$B^*(x) = \int_{x_0}^{x} J'[H(y)]\, dF(y) \tag{3.6.17}$$

and

$$\lambda_N B^*(x) + (1 - \lambda_N) B(x) = J[H(x)] - J[H(x_0)]$$

with x_0 determined somewhat arbitrarily, say by $H(x_0) = \frac{1}{2}$.

Thus

$$B_{1N} + B_{2N} = (1 - \lambda_N)\left\{ \frac{1}{m} \sum_{i=1}^{m} [B(X_i) - EB(X)] - \frac{1}{n} \sum_{i=1}^{n} [B^*(Y_i) - EB^*(Y)] \right\}, \tag{3.6.18}$$

where E represents expectation and X and Y have the F and G distributions respectively. $\triangleleft$

The two summations given by (3.6.18) involve independent samples of identically distributed random variables and we may apply the central limit theorem to show that $B_{1N} + B_{2N}$ when properly normalized has a normal distribution in the limit. The central limit theorem applies if the variances of $B(X)$ and $B^*(Y)$ are finite and at least one is positive.

First, we shall find a bound on the $(2 + \delta')$th moments of $B(X)$ and $B^*(Y)$ where $0 < \delta' \leq 1$.

Using assumption (c),

$$|B(X)| = \left|\int_{x_0}^{x} J'[H(y)]\, dG(y)\right| \leq K[H(x)[1 - H(x)]]^{\delta - 1/2}.$$

Next, choose $\delta' > 0$ such that $(2 + \delta')(\delta - \frac{1}{2}) > -1$; then

$$E\{|B(X)|\}^{2+\delta'} \leq K \int_{-\infty}^{\infty} [H(x)[1 - H(x)]]^{(2+\delta')(\delta - 1/2)}\, dF(x)$$

$$\leq K \int_{0}^{1} [H(1 - H)]^{(2+\delta')(\delta - 1/2)}\, dH < \infty.$$

(Note that in the second inequality, we have made use of the fact that $dG \leq (1/\lambda_0)\, dH$.)

Similarly, we can establish that $E\{|B^*(Y)|\}^{2+\delta'} < \infty$. The proof follows provided $B(X)$ and $B^*(Y)$ do not both have zero variances.

We shall now find the variance of $B_{1N} + B_{2N}$. From (3.6.18),

$$\text{(3.6.19)} \quad \operatorname{var}(B_{1N} + B_{2N}) = (1 - \lambda_N)^2\left\{\frac{1}{m} \operatorname{var} B(X) + \frac{1}{n} \operatorname{var} B^*(Y)\right\}.$$

Now since

$$B(X) - EB(X) = \int_{-\infty}^{\infty} B(x)\, d[F_1(x) - F(x)]$$

$$= -\int_{-\infty}^{\infty} [F_1(x) - F(x)]J'[H(x)]\, dG(x).$$

Therefore

$$\text{(3.6.20)} \quad \operatorname{var} B(X) = E\left\{\int_{-\infty}^{\infty}\int_{-\infty}^{\infty} [F_1(x) - F(x)][F_1(y) - F(y)] \cdot J'[H(x)]J'[H(y)]\, dG(x)\, dG(y)\right\}$$

$$= 2 \iint_{-\infty < x < y < \infty} F(x)[1 - F(y)]J'[H(x)]J'[H(y)] \times dG(x)\, dG(y).$$

(Note that the application of Fubini's theorem (chapter 2, p. 47) permits the interchange of expectation and integral.)

Similarly, the variance of $B^*(Y)$ is given by

$$(3.6.21) \qquad \operatorname{var} B^*(Y) = 2 \iint_{-\infty<x<y<\infty} G(x)[1-G(y)]J'[H(x)] \times J'[H(y)]\, dF(x)\, dF(y).$$

Hence from (3.6.19), (3.6.20), and (3.6.21)

$$(3.6.22) \qquad \operatorname{var}(B_{1N}+B_{2N}) = \sigma_N^2, \qquad \text{defined by (3.6.5).}$$

Lemma 3.6.3. $\sqrt{N}\, C_{iN} \to 0$ *in probability uniformly in* F, G, *and* λ_N, *and for all* $i = 1, 2, 3, 4$.

The proof of this lemma is given in the Appendix (cf. section 10.2).

We have thus proved that the C terms are of higher order uniformly in F, G, and λ_N, the A term is finite and nonrandom, and $B_{1N} + B_{2N}$ is the sum of two independent terms each of which is the average of random variables with mean zero and finite second moments. Theorem 3.6.1 follows.

We now extend the proof of Theorem 3.6.1 to the case where F, G, and λ_N are not fixed, so that the asymptotic normality holds uniformly with respect to F, G, and λ_N.

Theorem 3.6.3. *If the assumptions of Theorem* 3.6.1 *are satisfied, and* F, G, *and* λ_N $(0 < \lambda_0 \le \lambda_N \le 1 - \lambda_0 < 1)$ *are chosen so that* $B(X)$ *and* $B^*(Y)$ *have variances bounded away from zero, then the asymptotic normality of* T_N *is uniform with respect to* F, G, *and* λ_N.

Proof: Since the A term is finite (Lemma 3.6.1) and the C terms are all $o_p(N^{-1/2})$ uniformly in F, G, and λ_N (Lemma 3.6.3), it suffices to establish the uniform convergence for $B_{1N} + B_{2N}$. For this it suffices to bound $\rho_{2+\delta'} = \beta_{2+\delta'}/\sigma^{2+\delta'}$ for $B(X)$ and $B^*(Y)$ (cf. Esséen's theorem, chapter 2, p. 24). Since we have shown in Theorem 3.6.1 that the $2 + \delta'$ moments of $B(X)$ and $B^*(Y)$ are bounded, all that is required is to bound the variances of $B(X)$ and $B^*(Y)$ away from zero. The theorem follows. ◁

Problem 3.6.3 gives sufficient conditions under which $B(X)$ and $B^*(Y)$ have variances bounded away from zero.

Corollary 3.6.3. *If the assumptions of Theorem* 3.6.1 *are satisfied,* $0 < \lambda_0 \le \lambda_N \le 1 - \lambda_0 < 1$, *and* $F(x) = \Psi(x - \theta_N)$, $G(x) = \Psi(x - \phi_N)$ *where* Ψ *has a density* ψ, *then (a) the asymptotic normality of* T_N *is uniform with* *respect to* θ_N *and* ϕ_N *for* $\phi_N - \theta_N$ *in some neighborhood of zero; (b) if* $\phi_N - \theta_N \to 0$,

$$\lim_{N\to\infty} \frac{\lambda_N N \sigma_N^2}{1-\lambda_N} = 2 \iint_{0<x<y<1} x(1-y)J'(x)J'(y)\,dx\,dy \tag{3.6.23}$$

$$= \int_0^1 J^2(x)\,dx - \left(\int_0^1 J(x)\,dx\right)^2.$$

The proof of the corollary is given as an exercise (Exercise 3.6.4).

3.6.2. One-Sample Case

We have in mind problem (I) of section 3.4 and for this we consider the following simplified situation.

Let $X_1, \ldots, X_n$ be n independent and identically distributed random variables, each having an absolutely continuous cumulative distribution function (c.d.f.) $F(x)$, defined on the real line $\{-\infty < x < \infty\}$. Let $Z_{n,\alpha}$ be equal to 1 if the αth smallest observation among the values $|X_i|$, $i = 1, \ldots, n$, is from a positive X and let $Z_{n,\alpha}$ be equal to zero otherwise (for $\alpha = 1, \ldots, n$). Then we consider one-sample rank order statistics of the form

$$T_n = \frac{1}{n}\sum_{\alpha=1}^{n} Z_{n,\alpha}E_{n,\alpha}, \tag{3.6.24}$$

where $E_{n,\alpha} = J_n(\alpha/(n+1))$, $\alpha = 1, \ldots, n$ are functions of the ranks α $(= 1, \ldots, n)$ and are explicitly known. The function $J_n(u)$ is defined only at $u = \alpha/(n+1)$, $\alpha = 1, \ldots, n$, but we may extend its domain of definition to (0, 1) by letting it have constant values over $(\alpha/(n+1), (\alpha+1)/(n+1))$, $\alpha = 1, \ldots, n$. We define by $F_n(x)$ the sample c.d.f. of $X_1, \ldots, X_n$. Thus

$$F_n(x) = \frac{1}{n}(\text{number of } X_\alpha \le x); \tag{3.6.25}$$

also, let

$$H_n(x) = \frac{1}{n}(\text{number of } |X_\alpha| \le x, \alpha = 1, \ldots, n) \tag{3.6.26}$$

$$= \begin{cases} F_n(x) - F_n(-x-) & \text{if } x \ge 0 \\ 0 & \text{otherwise,} \end{cases}$$

$$H(x) = \begin{cases} F(x) - F(-x) & \text{if } x \ge 0 \\ 0 & \text{otherwise.} \end{cases} \tag{3.6.27}$$

Further we assume that (i) the conditions (a) and (c) of Theorem 3.6.1 are satisfied and (ii)

$$\int_0^\infty \left\{J_n\left[\frac{n}{n+1}H_n(x)\right] - J\left[\frac{n}{n+1}H_n(x)\right]\right\} dF_n(x) = o_p(n^{-1/2}).$$

We rewrite T_n as

$$T_n = \int_0^\infty J_n\left[\frac{n}{n+1} H_n(x)\right] dF_n(x), \tag{3.6.28}$$

and define

$$\mu_n = \int_0^\infty J[H(x)]\, dF(x); \tag{3.6.29}$$

$$\begin{aligned} \sigma_n^2 = \int_0^\infty J^2[H(x)]\, dF(x) - \left[\int_0^\infty J[H(x)]\, dF(x)\right]^2 \\ + 2\Bigg[\iint\limits_{0<x<y<\infty} H(x)[1 - H(y)]J'[H(x)]J'[H(y)]\, dF(x)\, dF(y) \\ - \iint\limits_{0<x<y<\infty} H(x)J'[H(x)]J[H(y)]\, dF(x)\, dF(y) \\ + \iint\limits_{0<x<y<\infty} J[H(x)][1 - H(y)]J'[H(y)]\, dF(x)\, dF(y)\Bigg]. \end{aligned} \tag{3.6.30}$$

Theorem 3.6.4. *Under the conditions* (i) *and* (ii)

$$\lim_{n\to\infty} P\{n^{1/2}[T_n - \mu_n]/\sigma_n \le x\} = (2\pi)^{-1/2}\int_{-\infty}^{x} e^{-t^2/2}\, dt$$

provided $\sigma_n \neq 0$.

Proof: Writing $F_n(x) = F(x) + [F_n(x) - F(x)]$, making use of condition (ii) and finally by expansion of $J[nH_n(x)/(n+1)]$ around $J[H(x)]$, we get after a little simplification that

$$T_n = B_{1n} + B_{2n} + \sum_{i=1}^{4} C_{in}, \tag{3.6.31}$$

where

$$B_{1n} = \int_0^\infty J[H(x)]\, dF_n(x), \tag{3.6.32}$$

$$B_{2n} = \int_0^\infty [H_n(x) - H(x)]J'[H(x)]\, dF(x), \tag{3.6.33}$$

$$C_{1n} = \frac{-1}{n+1}\int_0^\infty H_n(x)J'[H(x)]\, dF_n(x), \tag{3.6.34}$$

$$C_{2n} = \int_0^\infty [H_n(x) - H(x)]J'[H(x)]\, d[F_n(x) - F(x)], \tag{3.6.35}$$

$$\begin{aligned} C_{3n} = \int_0^\infty \Bigg\{J\left[\frac{n}{n+1} H_n(x)\right] - J[H(x)] \\ - \left[\frac{n}{n+1} H_n(x) - H(x)\right] J'[H(x)]\Bigg\}\, dF_n(x), \end{aligned} \tag{3.6.36}$$

$$C_{4n} = \int_0^\infty \left\{J_n\left[\frac{n}{n+1} H_n(x)\right] - J\left[\frac{n}{n+1} H_n(x)\right]\right\} dF_n(x). \tag{3.6.37}$$

◁

It is proved in the Appendix 2 of chapter 10 that the terms C_{in} ($i = 1, 2, 3, 4$) are all $o_p(n^{-1/2})$. Thus, we require only to show that $n^{1/2}(B_{1n} + B_{2n} - \mu_n)$ has asymptotically a normal distribution with zero mean and variance σ_n^2, where μ_n and σ_n^2 are defined by (3.6.29) and (3.6.30) respectively. Let $c(u)$ be equal to 1 or 0 according as u is positive or not. Then, from (3.6.32) and (3.6.33) we obtain

$$B_{1n} + B_{2n} = \frac{1}{n}\sum_{\alpha=1}^{n} B(X_\alpha), \tag{3.6.38}$$

where

$$B(X_\alpha) = J[H(|X_\alpha|)]c(X_\alpha) + \int_0^\infty [c(x - |X_\alpha|) - H(x)]J'[H(x)]\,dF(x). \tag{3.6.39}$$

By an application of Fubini's theorem, it readily follows from (3.6.39) that

$$E_F(B(X)) = \mu_n \quad \text{and} \quad \operatorname{var}_F(B(X)) = \sigma_n^2, \tag{3.6.40}$$

defined by (3.6.29) and (3.6.30), respectively.

Also, by condition (c), $E_F|B(X)|^{2+\delta} < \infty$, for some $\delta > 0$, and hence $\sigma_n^2 < \infty$. Since σ_n^2 is assumed to be positive, and by (3.6.38), $B_{1n} + B_{2n} - \mu_n$ is the average of n i.i.d.r.v. having zero mean and variance $\sigma_n^2 > 0$, the proof follows. (Note that if $F(x)$ does not depend on n, the condition $E_F|B(x)|^2 < \infty$ is sufficient for the asymptotic normality of T_n.) ◁

Corollary 3.6.4.1. *If* (i) *the assumptions of Theorem* 3.6.4 *hold,* (ii) $F(x)$ *is replaced by a sequence of c.d.f.'s* $\{F_{(n)}(x)\}$, *where* $F_{(n)}(x)$ *is symmetric about* $\theta/n^{1/2}$, θ *being real and finite,* (iii) *the function* $J_n(\alpha/(n+1))$ *is the expected value of the* αth *order statistic in a sample of size n from a distribution* $\psi(x)$, *where*

$$\psi(x) = \psi^*(x) - \psi^*(-x),\ x \geq 0, \text{ and } = 0 \text{ for } x < 0, \tag{3.6.41}$$

$\psi^*(x)$ *being symmetric about* $x = 0$, *then*

$$\lim_{n\to\infty} \sigma_n^2 = \frac{1}{4}\int_0^1 J^2(u)\,du. \tag{3.6.42}$$

The proof is an immediate consequence of Theorem 3.6.4, and is therefore omitted.

Remark: The statistics of the form (3.6.24) are usually considered for the problem of testing that the distribution function $F(x)$ is symmetric about zero. Examples of the test statistics of the form (3.6.24) are (i) the sign test with $E_{n,\alpha} = 1$, $\alpha = 1, \ldots, n$;† (ii) the Wilcoxon signed rank test with $E_{n,\alpha} = \alpha$ for $\alpha = 1, \ldots, n$; and (iii) the test with $E_{n,\alpha}$ as the expected value of the αth order statistic of a sample of size n from a chi distribution with one degree of freedom. This test is called the one-sample Fisher-Yates or normal scores test. For some related results, we refer to Pyke and Shorack (1968b), Hušková (1970) and Sen (1970a,e), and Sen and Ghosh (1971).

3.6.3. Multisample Case

We shall now generalize the results of section 3.6.1 to the case of c (≥ 2) independent samples.

Let $X_{i1}, \ldots, X_{in_i}$ be n_i i.i.d.r.v. having a continuous c.d.f. $F_i(x)$, for $i = 1, \ldots, c$. Let $N = \sum_{i=1}^{c} n_i$ and $\lambda_{i,N} = n_i/N$, $i = 1, \ldots, c$, and assume that for all N, the inequalities $0 < \lambda_0 \leq \lambda_{1,N}, \ldots, \lambda_{c,N} \leq 1 - \lambda_0 < 1$ hold for some $\lambda_0 \leq 1/c$. Let

$$F_{n_i}^{(i)}(x) = \frac{1}{n_i}\,[\text{number of } X_{i\alpha} \leq x;\ \alpha = 1, \ldots, n_i], \qquad i = 1, \ldots, c \tag{3.6.43}$$

be the sample cumulative distribution functions of the c samples. Define

$$H_N(x) = \sum_{i=1}^{c} \lambda_{i,N} F_{n_i}^{(i)}(x) \tag{3.6.44}$$

as the combined sample c.d.f. The combined population c.d.f. is

$$H(x) = \sum_{i=1}^{c} \lambda_{i,N} F_i(x). \tag{3.6.45}$$

Let $Z_{N,i}^{(j)} = 1$, if the ith smallest of $N = \sum_1^c n_i$ observations is from the jth sample, and otherwise let $Z_{N,i}^{(j)} = 0$, $i = 1, \ldots, N$; $j = 1, \ldots, c$. Then, consider the statistics

$$\begin{aligned} T_{N,j} &= \frac{1}{n_j} \sum_{i=1}^{N} E_{N,i} Z_{N,i}^{(j)} \\ &= \int J_N\left[\frac{N}{N+1} H_N(x)\right] dF_{n_j}^{(j)}(x); \qquad j = 1, \ldots, c, \end{aligned} \tag{3.6.46}$$

where $E_{N,i} = J_N(i/(N+1))$. As in section 3.6.1, we extend the domain of definition of J_N to $(0, 1)$ by letting it have constant value on $(i/(N+1), (i+1)/(N+1)]$.

† In this case assumption (i) does not hold and the supplied proof does not apply. However, using the well-known binomial distribution, the desired result can be readily obtained.

Before proving the asymptotic normality of the statistics $T_{N,j}, j = 1, \ldots, c$, we state a few elementary results (see Exercise 3.6.11):

(3.6.47) $$H \geq \lambda_{i,N} F_i \geq \lambda_0 F_i; \qquad i = 1, \ldots, c$$

(3.6.48) $$(1 - F_i) \leq (1 - H)/\lambda_{i,N} \leq (1 - H)/\lambda_0; \qquad i = 1, \ldots, c$$

(3.6.49) $$F_i(1 - F_i) \leq H(1 - H)/\lambda_{i,N}^2 \leq H(1 - H)/\lambda_0^2; \qquad i = 1, \ldots, c$$

(3.6.50) $$dH \geq \lambda_{i,N}\, dF_i \geq \lambda_0\, dF_i$$

(3.6.51) Let (a_N, b_N) be the interval $S_{N\epsilon}$ where

$$S_{N\epsilon} = \{x\colon\ H(x)[1 - H(x)] > \eta_\epsilon \lambda_0 / N\};$$

then, η_ϵ can be chosen independent of F_i and $\lambda_{i,N}$; $i = 1, \ldots, c$, such that $P\{X_{ij} \in S_{N\epsilon};\ i = 1, \ldots, c;\ j = 1, \ldots, n_i\} \geq 1 - \epsilon$.

Theorem 3.6.5. *If* (i) *the assumptions* (*a*) *and* (*c*) *of Theorem* 3.6.1 *are satisfied, and*

(ii) $$\int_{-\infty}^{\infty} \left[J_N\left[\frac{N}{N+1} H_N(x)\right] - J\left[\frac{N}{N+1} H_N(x)\right]\right] dF_{n_j}^{(j)}(x) = o_p(N^{-1/2})$$

for all $j = 1, \ldots, c$,

then the random vector $N^{1/2}[T_{N,1} - \mu_{N,1}, \ldots, T_{N,c} - \mu_{N,c}]$ *where*

(3.6.52) $$\mu_{N,j} = \int J[H_j(x)]\, dF_j(x)$$

has a limiting normal distribution with mean vector zero and variance covariance matrix $\boldsymbol{\Sigma} = ((\sigma_{N,jk}))$ *where*

(3.6.53) $$\sigma_{N,jj} = N\sigma_{N,j}^2 = 2\sum_{\substack{i=1\\ i\neq j}}^{c} \iint_{-\infty<x<y<\infty} A_i(x, y)\, dF_j(x)\, dF_j(y)$$

$$+ \frac{2}{\lambda_j} \sum_{\substack{i=1\\ i\neq j}}^{c} \lambda_i^2 \iint_{-\infty<x<y<\infty} A_j(x, y)\, dF_i(x)\, dF_i(y)$$

$$+ \frac{1}{\lambda_j} \sum_{\substack{i,k=1\\ i\neq k, i\neq j\\ k\neq j}}^{c} \left[\iint_{-\infty<x<y<\infty} A_j(x, y)\, dF_i(x)\, dF_i(y) \right.$$

$$\left. + \iint_{-\infty<y<x<\infty} A_j(y, x)\, dF_i(x)\, dF_k(y) \right]$$

and

$$\sigma_{N;jj'} = -\sum_{i=1}^{c} \lambda_i \Bigg[\iint_{-\infty < x < y < \infty} A_j(x, y)\, dF_i(x)\, dF_{j'}(y) + \iint_{-\infty < y < x < \infty} A_j(y, x)\, dF_i(x)\, dF_{j'}(y) \Bigg]$$
$$- \sum_{i=1}^{c} \lambda_i \Bigg[\iint_{-\infty < x < y < \infty} A_{j'}(x, y)\, dF_i(x)\, dF_j(y) + \iint_{-\infty < y < x < \infty} A_{j'}(y, x)\, dF_i(x)\, dF_j(y) \Bigg]$$
$$- \sum_{i=1}^{c} \lambda_i \Bigg[\iint_{-\infty < x < y < \infty} A_i(x, y)\, dF_j(x)\, dF_{j'}(y) + \iint_{-\infty < y < x < \infty} A_i(y, x)\, dF_j(x)\, dF_{j'}(y) \Bigg] \tag{3.6.54}$$

where

$$A_j(u, v) = F_j(u)[1 - F_j(v)]J'[H(u)]J'[H(v)]. \tag{3.6.55}$$

Proof: Proceeding as in Theorem 3.6.1, we can write

$$T_{N,j} = \mu_{N,j} + B_{1N,j} + B_{2N,j} + \sum_{i=1}^{4} C_{iN,j} \tag{3.6.56}$$

where

$$\mu_{N,j} = \int J[H(x)]\, dF_j(x) \tag{3.6.57}$$

$$B_{1N,j} = \int_{-\infty}^{\infty} J[H(x)]\, d[F_{n_j}^{(j)}(x) - F_j(x)] \tag{3.6.58}$$

$$B_{2N,j} = \int_{-\infty}^{\infty} [H_N(x) - H(x)]J'[H(x)]\, dF_j(x) \tag{3.6.59}$$

$$C_{1,N}^{(j)} = -\frac{1}{N+1} \int_{-\infty}^{\infty} H_N(x)J'[H(x)]\, dF_{n_j}^{(j)}(x) \tag{3.6.60}$$

$$C_{2,N}^{(j)} = \int_{-\infty}^{\infty} (H_N(x) - H(x))J'[H(x)]\, d(F_{n_j}^{(j)}(x) - F_j(x)) \tag{3.6.61}$$

$$C_{3,N}^{(j)} = \int_{-\infty}^{\infty} \left\{ J\left[\frac{N}{N+1} H_N(x)\right] - J[H(x)] - \left(\frac{N}{N+1} H_N(x) - H(x)\right) J'[H(x)] \right\} dF_{n_j}^{(j)}(x) \tag{3.6.62}$$

$$C_{4,N}^{(j)} = \int_{-\infty}^{\infty} \left[J_N\left[\frac{N}{N+1} H_N(x)\right] - J\left[\frac{N}{N+1} H_N(x)\right]\right] dF_{n_j}^{(j)}(x). \tag{3.6.63}$$

Proceeding as in Lemma 3.3.3 we can show (see Exercise 3.6.12) that the C terms are all $o_p(N^{-1/2})$. The difference $N^{1/2}[T_{N,1} - \mu_{N,1}, \ldots, T_{N,c} - \mu_{N,c}] - N^{1/2}[B_{1N,1} + B_{2N,1}, \ldots, B_{1N,c} + B_{2N,c}]$ tends to zero in probability and so the vectors $N^{1/2}[T_{N,1} - \mu_{N,1}, \ldots, T_{N,c} - \mu_{N,c}]$ and $N^{1/2}[B_{1N,1} + B_{2N,1}, \ldots, B_{1N,c} + B_{2N,c}]$ possess the same limiting distribution, if they have one at all. Now proceeding as in Lemma 3.6.2, we can write $B_{1N,j} + B_{2N,j}$ as

$$(3.6.64)\quad B_{1N,j} + B_{2N,j} = -\sum_{\substack{i=1\\ j\neq i}}^{c}\left[\lambda_i \frac{1}{n_i}\sum_{k=1}^{n_i}\{B_j(X_{ik}) - EB_j(X_i)\}\right]$$
$$+ \frac{1}{n_j}\sum_{k=1}^{n_j}\{J[H(X_{jk})] - \lambda_j B_j(X_{jk}) - E[J[H(X_j)]$$
$$- \lambda_j B_j(X_j)]\}$$

where

$$(3.6.65)\qquad B_j(x) = \int_{x_0}^{x} J'[H(y)]\,dF_j(y)$$

with x_0 determined somewhat arbitrarily, say by $H(x_0) = \frac{1}{2}$. E represents the expectation and $X_1, \ldots, X_c$ have the $F_1, \ldots, F_c$ distributions respectively. Now, let us denote

$$(3.6.66)\qquad B_{ij}^*(X_{ik}) = -\lambda_i\{B_j(X_{ik}) - EB_j(X_{ik})\}\quad \text{for}\quad i \neq j$$

and

$$(3.6.67)\quad B_{jj}^*(X_{jk}) = [\{J[H(X_{jk}) - \lambda_j B_j(X_{jk})]\} - E[J[H(X_j)] - \lambda_j B_j(X_j)]].$$

Then we can rewrite (3.6.64) as

$$(3.6.68)\qquad B_{1N,j} + B_{2N,j} = \sum_{i=1}^{c}\left\{\frac{1}{n_i}\sum_{k=1}^{n_i} B_{ij}^*(X_{ik})\right\}.$$

The proof of the theorem now follows by applying the central limit theorem to each of the c independent vectors

$$(3.6.69)\quad \frac{1}{n_i}\sum_{k=1}^{n_i}[B_{i1}^*(X_{ik}), B_{i2}^*(X_{ik}), \ldots, B_{ic}^*(X_{ik})];\qquad i = 1, \ldots, c.$$

To compute the variance of $B_{1N,j,} + B_{2Nj}$ (cf. (3.6.60)), note that

$$-\lambda_i \frac{1}{n_i}\sum_{k=1}^{n_i}\{B_j(X_{ik}) - EB_j(X_i)\} = -\lambda_i \int B_j(x)\,d[F_{n_i}^{(i)}(x) - F_i(x)]$$
$$= \lambda_i \int_{-\infty}^{\infty}[F_{n_i}^{(i)}(x) - F_i(x)]J'[H(x)]\,dF_j(x),$$
$$i = 1, \ldots, c;\quad i \neq j,$$

has mean zero, and variance

$$(3.6.70)\qquad E\left\{\lambda_i\int_{-\infty}^{\infty}[F_{n_i}^{(i)}(x)-F_i(x)]J'[H(x)]\,dF_j(x)\right\}^2$$

$$= E\left\{\lambda_i^2\int_{-\infty}^{\infty}\int_{-\infty}^{\infty}[F_{n_i}^{(i)}(x)-F_i(x)][F_{n_i}^{(i)}(y)-F_i(y)]\right.$$

$$\left.\times\, J'[H(x)]J'[H(y)]\,dF_j(x)\,dF_j(y)\right\}$$

$$= \frac{2\lambda_i}{N}\iint_{-\infty<x<y<\infty} A_i(x,y)\,dF_j(x)\,dF_j(y)$$

$$\text{for}\quad i=1,\ldots,c;\quad i\neq j,$$

where $A_j(u,v)$ is defined by (3.6.55). By a similar argument, the variance of

$$\frac{1}{n_j}\sum_{k=1}^{n_j}\{J[H(X_{jk})]-\lambda_jB_j(X_{jk})-E[J[H(X_j)]-\lambda_jB_j(X_j)]\}$$

that is, $\int_{-\infty}^{\infty}\{J[H(x)]-\lambda_jB_j(x)\}\,d[F_{m_j}^{(j)}(x)-F^{(j)}(x)]$ is given by

$$(3.6.71)\qquad \frac{2}{N\lambda_j}\sum_{\substack{i=1\\ i\neq j}}^{c}\iint_{-\infty<x<y<\infty} A_j(x,y)\,dF_i(x)\,dF_i(y)$$

$$+\frac{1}{N\lambda_j}\sum_{\substack{i=1\\ i\neq k, i\neq j}}^{c}\sum_{\substack{k=1\\ k\neq j}}^{c}\left[\iint_{-\infty<x<y<\infty} A_j(x,y)\,dF_i(x)\,dF_k(y)\right.$$

$$\left.+\iint_{-\infty<y<x<\infty} A_j(y,x)\,dF_i(x)\,dF_k(y)\right],$$

where $A_j(x,y)$ and $A_j(y,x)$ are given by (3.6.55).

Adding the c terms given by (3.6.70) and (3.6.72), we obtain $\sigma_{N,j}^2$ as defined in (3.6.53). The covariance results stated in (3.6.54) are obtained analogously. The theorem follows. ◁

We conclude this Section with the following theorem whose proof is given in the Appendix (Section 10.5).

Theorem 3.6.6. *If* $J_N(i/(N+1))$ *is either* $J(i/(N+1))$ *or is the expected value of the* ith *order statistic of a sample of size* N *from a population with c.d.f.* $F=J^{-1}$, *for* $i=1,\ldots,N$, *then assumptions* (*a*) *and* (*b*) *holds whenever* (*c*) *holds.*

3.7. LOCALLY MOST POWERFUL RANK TESTS

In this section we consider a general type of the two-sample problem for which rank tests can be applied and show how we can obtain the rank tests most powerful for a specific simple alternative.

Let $X_1, \ldots, X_m$ and $Y_1, \ldots, Y_n$ $(N = m + n)$ be independent random samples from populations with unknown cumulative distribution functions F_X and F_Y respectively. We wish to test

$$H_0\colon F_X = F_Y \tag{3.7.1}$$

against

$$H_1\colon F_X = G_\theta, \qquad F_Y = G_\phi, \qquad \theta, \phi \in R, \tag{3.7.2}$$

where G_θ is a specified family of cumulative distribution functions (one for each θ); R is an interval on the real line; θ and ϕ are specified, and are very close to some specified value ϕ_0; and $\theta \neq \phi$.

Definition 3.7.1. *If $\beta_N(\theta, \phi)$ denotes the power of a size α rank test of H_0 against H_1, then the rank test with power $\beta_N^*(\theta, \phi)$ is said to be the locally most powerful rank test (l.m.p.r.t.) for $\theta > \phi$, if $\beta_N^*(\theta, \phi) \geq \beta_N(\theta, \phi)$ uniformly in N, and for all θ and ϕ in some sufficiently small neighborhood of ϕ_0; that is, $\theta, \phi \in [\phi_0 - \epsilon_N, \phi_0 + \epsilon_N]$, $\epsilon_N > 0$.*

The l.m.p.r.t. for $\theta < \phi$ is defined in a similar manner. We note that the neighborhood of ϕ_0 is allowed to vary with N.

Suppose that the following regularity conditions are satisfied.

(*A*) $G_\theta(x)$ has a density function $g_\theta(x)$, which, along with $\partial g_\theta(x)/\partial\theta$, is continuous with respect to θ for $\phi_0 - a \leq \theta \leq \phi_0 + a$, $a > 0$, for almost all x. There exist functions $M_0(x)$ and $M_1(x)$, integrable over $(-\infty, +\infty)$, such that

$$g_\theta(x) \leq M_0(x), \qquad |\partial g_\theta(x)/\partial\theta| \leq M_1(x), \qquad \phi_0 - a \leq \theta \leq \phi_0 + a;$$

(*B*) $g_\theta(x) > 0$ if and only if $g_\phi(x) > 0$;

(*C*) $|J^{(i)}(H)| = |d^iJ/dH^i| \leq K[H(1 - H)]^{-i-1/2+\delta}$, for $i = 0, 1$, and for some $\delta > 0$, where K is a constant, and where

$$J(G_{\phi_0}(x)) = \frac{\partial}{\partial\theta} \log g_\theta(x)\bigg|_{\theta=\phi_0};$$

$$(D) \qquad 0 < \lim_{N\to\infty} m/n = r < \infty.$$

Theorem 3.7.1 (Capon). *The l.m.p.r.t. of H_0 against H_1 is based on the statistic*

$$\tau_N = \frac{1}{m}\sum_{i=1}^{N} a_{Ni}Z_{Ni} \tag{3.7.3}$$

where $Z_{N,i} = 1$, if the ith *smallest observation in the combined sample is an*

X observation and $Z_{N,i} = 0$ *otherwise; and where*

$$a_{Ni} = E_{\phi_0\phi_0}\left[\frac{\partial}{\partial\theta}\log g_\theta(Z_i)\bigg|_{\theta=\phi_0}\right], \tag{3.7.4}$$

that is, a_{Ni} *is the expected value of a certain function* $(\partial/\partial\theta)\log g_\theta(x)\,|_{\theta=\phi_0}$ *of the* ith *smallest observation* Z_i *of a sample of size* N *from the cumulative distribution function* G_{ϕ_0}.

To prove this theorem it is necessary to know the distribution of $Z_{N1}, \ldots, Z_{NN}$ under H_1. If the condition (B) is satisfied, then the joint distribution, under H_1, of Z_{Ni}, $i = 1, \ldots, N$, is given by

$$\begin{aligned} P_{\theta\phi}(Z_{N1} = z_{N1}, \ldots, Z_{NN} = z_{NN}) &= \binom{N}{n}^{-1} E_{\phi\phi}\left\{\prod_{i=1}^{N}\left[\frac{g_\theta(Z_i)}{g_\phi(Z_i)}\right]^{z_{Ni}}\right\} \\ &= \binom{N}{n}^{-1} E_{\theta\theta}\left\{\prod_{i=1}^{N}\left[\frac{g_\phi(Z_i)}{g_\theta(Z_i)}\right]^{y_{Ni}}\right\} \end{aligned} \tag{3.7.5}$$

where Z_i is the ith smallest of the N observations, $y_{Ni} = 1 - z_{Ni}$, and $E_{\phi\phi}$ indicates that the expectation is taken under the assumption that $F_X = F_Y = G_\phi$. Under H_0,

$$P_{\phi_0\phi_0}(Z_{N1} = z_{N1}, \ldots, Z_{NN} = z_{NN}) = \binom{N}{n}^{-1}. \tag{3.7.6}$$

We now expand $P_{\theta\phi}$ as

$$\begin{aligned} P_{\theta\phi} = P_{\phi_0\phi_0} &+ (\theta - \phi_0)\frac{\partial P_{\theta\phi}}{\partial\theta}\bigg|_{\theta=\phi=\phi_0} + (\phi - \phi_0)\frac{\partial P_{\theta\phi}}{\partial\phi}\bigg|_{\theta=\phi=\phi_0} \\ &+ o(|\theta - \phi_0| + |\phi - \phi_0|). \end{aligned} \tag{3.7.7}$$

This extension is valid provided $P_{\theta\phi}$ has continuous partial derivatives in a neighborhood of $\theta = \phi = \phi_0$ and this follows from condition (A).

As a consequence of condition (A), we may interchange differentiation and expectation to obtain from equation (3.7.5)

$$\begin{aligned} \frac{\partial P_{\theta\phi}}{\partial\theta}\bigg|_{\theta=\phi=\phi_0} &= \binom{N}{n}^{-1} E_{\theta_0\phi_0}\left\{\sum_{i=1}^{N}\left(\frac{\partial}{\partial\theta}g_\theta(Z_i)\bigg|_{\theta=\phi_0}\right)(g_{\phi_0}(Z_i))^{-1}Z_{Ni}\right\} \\ &= \binom{N}{n}^{-1}\sum_{i=1}^{N} E_{\theta_0\phi_0}\left(\frac{\partial}{\partial\theta}\log g_\theta(Z_i)\bigg|_{\theta=\phi_0}\right)Z_{Ni} \\ &= \binom{N}{n}^{-1}\sum_{i=1}^{N} a_{Ni}Z_{Ni}. \end{aligned} \tag{3.7.8}$$

Similarly

$$\frac{\partial P_{\theta\phi}}{\partial\theta}\bigg|_{\theta=\phi=\phi_0} = \binom{N}{n}^{-1}\sum_{i=1}^{N} a_{Ni}Y_{Ni} \tag{3.7.9}$$

where a_{Ni} is defined in (3.7.4).

Substituting equations (3.7.8) and (3.7.9) in (3.7.7), we obtain

$$(3.7.10)\quad P_{\theta\phi} = \binom{N}{n}^{-1}\Big[1 + (\theta - \phi)\sum_{i=1}^{N} a_{Ni}Z_{Ni} + (\phi - \phi_0)\sum_{i=1}^{N} a_{Ni} + o(|\theta - \phi_0| + |\phi - \phi_0|)\Big].$$

Now using the Neyman-Pearson fundamental lemma, we find that the most powerful rank test rejects H_0 when

$$1 + (\theta - \phi)\sum_{i=1}^{N} a_{Ni}Z_{Ni} + (\phi - \phi_0)\sum_{i=1}^{N} a_{Ni} + o(|\theta - \phi| + |\phi - \phi_0|) > C.$$

Thus, if $\theta > \phi$, the l.m.p.r.t. rejects H_0, when

$$(3.7.11)\qquad \tau_N = \frac{1}{m}\sum_{i=1}^{N} a_{Ni}Z_{Ni} > C$$

and, if $\theta < \phi$, the l.m.p.r.t. rejects H_0, when

$$(3.7.12)\qquad \tau_N = \frac{1}{m}\sum_{i=1}^{N} a_{Ni}Z_{Ni} < C.$$

This proves the theorem.

Theorem 3.7.1. generalizes the results obtained by Hoeffding (1951b), Terry (1952), Lehmann (1953), and Savage (1956) for different functional alternatives.

We now give applications of the above theorem to some specific cases. In each example, it is easy to verify that the conditions (A), (B), and (C) are satisfied.

Example 3.7.1. Normal Case (Translation Alternatives). Let

$$g_\theta(x) = (2\pi\sigma^2)^{-1/2} e^{-(x-\theta)^2/2\sigma^2}, \qquad \theta, \phi, \phi_0 \in (-\infty, +\infty).$$

Then

$$\frac{\partial}{\partial\theta}\log g_\theta(x)\Big|_{\theta=\phi_0} = \frac{x - \phi_0}{\sigma} \quad \text{and} \quad J(v) = \Phi^{-1}(v), \qquad 0 \le v \le 1,$$

where

$$\Phi(x) = (2\pi)^{-1/2}\int_{-\infty}^{x} e^{-t^2/2}\, dt,$$

and $\Phi^{-1}(v)$ is the inverse function of $\Phi(x)$. Thus the l.m.p.r.t. is based on the statistic

$$(3.7.13)\qquad C_N = \frac{1}{M}\sum_{i=1}^{N} E(\xi_i)Z_{Ni}$$

where ξ_i is the ith smallest observation of a sample of size N from the standard normal distribution. The test based on C_N was proposed originally by Fisher and Yates (1938), and shown to be locally most powerful by Hoeffding (1951b) and Terry (1952). The asymptotic normality of C_N was established in Theorem 3.6.1.

Example 3.7.2. Normal Case (Scalar Alternatives). Let

$$g_\theta(x) = (2\pi\theta)^{-\frac{1}{2}} e^{-\frac{1}{2}(x-\xi)^2/\theta}, \qquad \theta, \phi, \phi_0 \in (0, \infty).$$

Then

$$\frac{\partial}{\partial\theta} \log g_\theta(x) \bigg|_{\theta=\phi_0} = -\frac{1}{2}\phi_0^{-1} + \frac{1}{2}\left(\frac{x-\xi}{\phi_0}\right)^2$$

and

$$J(v) = \tfrac{1}{2}\phi_0^{-1}[\{\Phi^{-1}(u)\}^2 - 1], \qquad 0 \le u \le 1.$$

Thus the l.m.p.r.t. is based on the statistic

$$K_N = \frac{1}{m}\sum_{i=1}^{N} E(\xi_i^2) Z_{Ni} \tag{3.7.14}$$

where ξ_i is the ith smallest observation of a sample of size N from the standard normal distribution. The asymptotic normality of K_N is established in Theorem 3.6.1. See also Raghavachari (1965a) and Puri (1968) for the case when the location parameters are different.

Example 3.7.3. Exponential Case (Scalar Alternatives). Let

$$G_\theta(x) = 1 - e^{-\theta x}, \qquad x \ge 0,$$

and

$$G_\theta(x) = 0 \qquad \text{otherwise}, \qquad \theta, \phi, \phi_0 \in (0, \infty).$$

In this case the l.m.p.r.t. is based on the statistic

$$S_N = \frac{1}{m}\sum_{i=1}^{N} E_{Ni} Z_{Ni} \tag{3.7.15}$$

where

$$E_{Ni} = \sum_{u=N-i+1}^{N} u^{-1} \tag{3.7.16}$$

is the expected value of the ith smallest observation of a sample of size N from the exponential distribution $G_1(x)$.

The test based on the statistic S_N was proposed originally by Savage (1956). The asymptotic normality of S_N is established in Theorem 3.6.1.

Example 3.7.4. The Wilcoxon-Mann-Whitney Rank Sum Test. Let

$$h_{\theta\phi}(u) = (1 - \theta + \phi)u + (\theta - \phi)u^2;$$
$$\theta, \phi, \phi_0 \in (-\infty, \infty), \quad 0 < \theta - \phi < 1.$$

Then

$$J(u) = 2u - 1, \qquad 0 \leq u \leq 1,$$

and, the l.m.p.r.t. is based on the statistic

$$V_N = \frac{1}{m} \sum_{i=1}^{N} E(U_i) Z_{Ni} \tag{3.7.17}$$

where U_i is the ith smallest observation of a sample of size N from the uniform distribution on (0, 1). Since $E(U_i) = i/(N + 1)$,

$$V_N = \frac{1}{m(N + 1)} \sum_{i=1}^{N} iZ_{Ni}.$$

The test based on V_N is called the Wilcoxon-Mann-Whitney or rank sum test. The asymptotic normality of V_N is established in Theorem 3.6.1.

3.8. THE ASYMPTOTIC EFFICIENCY OF TESTS

Suppose for the problem of testing a simple hypothesis that the value of a parameter θ is θ_0 against the alternatives that $\theta > \theta_0$ or $\theta < \theta_0$ or $\theta \neq \theta_0$; we have two or more tests available. Which test is it preferable to use? Obviously, the decision has to be made on grounds involving whatever characteristics the situation seems to demand the test should have. The simplest way to compare two tests is by direct examination of their power functions. Suppose that a test $A^{(1)}$ of size α must be based on N_1 observations to attain a certain power β against the alternative θ, and a test $A^{(2)}$, also of level α, requires N_2 observations to produce the same power at the same alternative. We define the relative efficiency of $A^{(1)}$ with respect to $A^{(2)}$ as the ratio N_2/N_1. The complete comparison of $A^{(1)}$ with $A^{(2)}$ would therefore require evaluation of this "power efficiency function" for all values of its three arguments. In practice the computation of this power function is extremely difficult, and so we have to depend on some criterion, preferably a single numerical measure, which is easily computable for making a choice between tests. Such an approach may be unsatisfactory for tests based on finite samples. We shall therefore confine ourselves to the asymptotic results.

Now suppose we define our measure of test efficiency as the limit of the relative efficiency as the sample sizes tend to infinity. Then we encounter another difficulty. For, in most cases, the power of the tests under consideration tends to one against any fixed alternative θ ($\neq \theta_0$). Such a test is called a consistent test. (A test which is not consistent has a poor performance in the sense that it does not enable us to detect the alternative with certainty if it is true, as the sample size tends to infinity.) Now if the two tests under consideration are both size α consistent tests and we compare them against some fixed alternative value of θ, then their powers will always tend to one as the sample sizes tend to infinity. Thus, unless we know the rates of convergence to the limit distribution, the limiting value of the relative efficiency may not serve as a criterion for distinguishing between tests. Thus, in order to obtain an appropriate asymptotic measure of test efficiency we shall consider a sequence of alternative hypotheses in which θ approaches the test value θ_0 as the sample size tends to infinity, and the power remains bounded away from one. This concept of asymptotic relative efficiency was first developed by Pitman (1948). An exposition of his work together with some generalizations was given by Noether (1955) and Hoeffding and Rosenblatt (1955).

We shall first discuss the Pitman-Noether measure of asymptotic relative efficiency. Later on we shall consider briefly some alternative measures which have been proposed recently.

3.8.1. The Pitman-Noether Efficiency

Let $\beta_N^{(1)}(\theta)$ and $\beta_N^{(2)}(\theta)$ denote the power functions of two tests $A^{(1)}$ and $A^{(2)}$, based on the same set of N observations, against a parametric family of alternatives labeled by θ, and let θ_0 be the value of θ specified by the hypothesis. We assume that both the tests are at the same level of significance α. Consider a sequence of alternatives θ_N and a sequence $N^* = h(N)$ such that

$$\lim_{N\to\infty} \beta_N^{(1)}(\theta_N) = \lim_{N\to\infty} \beta_{N^*}^{(2)}(\theta_N) \tag{3.8.1}$$

where the two limits existing are neither zero nor one. The then asymptotic relative efficiency (A.R.E.) of the test $A^{(2)}$ with respect to the test $A^{(1)}$ is defined as

$$e_{A^{(2)},A^{(1)}} = \lim_{N\to\infty} N/N^* \tag{3.8.2}$$

provided this limit exists and is independent of the particular sequences $\{\theta_N\}$ and $\{h(N)\}$ satisfying (3.8.1).

Usually the N observations constitute a sample, or are divided into $c \geq 2$ samples of sizes n_i; $i = 1, \ldots, c$, with $\sum_1^c n_i = N$. In the latter case, we assume that n_i/N tends to some limit λ_i, $0 < \lambda_i < 1$, as $N \to \infty$. In most of the problems $e_{A^{(2)},A^{(1)}}$ is independent of λ_i, $i = 1, \ldots, k$.

Let $A_N(x)$ be a test for the hypothesis H_0: $\theta = \theta_0$ against the alternative H_1: $\theta > \theta_0$ based on the first N observations, and let the critical region be $T_N \geq \lambda_{N,\alpha}$. Let further θ_N be a sequence of alternatives such that

$$\theta_N = \theta_0 + \frac{k}{N^\delta} \tag{3.8.3}$$

where k and δ are some finite positive constants independent of N.

Then the following theorem gives the limiting power of the test $A_N(x)$.

Theorem 3.8.1. *Suppose*

(*a*) $\lim_{N\to\infty} P_{\theta_0}(T_N \geq \lambda_{N,\alpha}) = \alpha$, *where* α $(0 < \alpha < 1)$ *is a fixed value;*
(*b*) *there exist functions* $\mu_N(\theta)$ *and* $\sigma_N(\theta)$ *such that*

$$\lim_{N\to\infty} P\left\{\frac{T_N - \mu_N(\theta)}{\sigma_N(\theta)} \leq x\right\} = \Phi(x) \tag{3.8.4}$$

uniformly in θ, $\theta_0 \leq \theta \leq \theta_0 + \epsilon$ *where* ϵ *is any positive number and* $\Phi(x)$ *is the standardized normal distribution function;*

(*c*) *for some positive integer* m,

$$\mu_N'(\theta_0) = \cdots = \mu_N^{(m-1)}(\theta_0) = 0, \qquad \mu_N^{(m)}(\theta_0) > 0;$$

$$(d) \quad \lim_{N\to\infty} \frac{\mu^{(m)}(\theta_N)}{\mu_N^{(m)}(\theta_0)} = 1, \qquad \lim_{N\to\infty} \frac{\sigma_N(\theta_N)}{\sigma_N(\theta_0)} = 1;$$

$$(e) \quad \lim_{N\to\infty} N^{-m\delta} \frac{\mu_N^{(m)}(\theta_N)}{\sigma_N(\theta_0)} = c > 0^*.$$

Then the limiting power of the test $A_N(x)$ *is*

$$1 - \Phi\left(\lambda_\alpha - \frac{k^m c}{m\,!}\right) \tag{3.8.5}$$

where $1 - \Phi(\lambda_\alpha) = \alpha$.

Proof: By definition, the limiting power of the test is

$$\lim_{N\to\infty} P_{\theta_N}(T_N \geq \lambda_{N,\alpha}) = \lim_{N\to\infty} \Phi\left(-\frac{\lambda_{N,\alpha} - \mu_N(\theta_N)}{\sigma_N(\theta_N)}\right) = \Phi(-\xi), \tag{3.8.6}$$

where

$$\xi = \lim_{N\to\infty} \frac{\lambda_{N,\alpha} - \mu_N(\theta_N)}{\sigma_N(\theta_N)}. \tag{3.8.7}$$

Now, expanding $\mu_N(\theta_N)$ about θ_0 in a Taylor series, we obtain

$$\mu_N(\theta_N) = \mu_N(\theta_0) + \frac{k^m}{N^{\delta m} m\,!}\mu_N^{(m)}(\theta), \qquad \theta_0 < \hat{\theta} < \theta_N \tag{3.8.8}$$

$$\xi = \lim_{N\to\infty} \frac{\lambda_{N,\alpha} - \mu_N(\theta_0) - \dfrac{k^m \mu_N^{(m)}(\hat{\theta})}{N^{\delta m} m\,!}}{\sigma_N(\theta_N)} \tag{3.8.9}$$

$$= \lim_{N\to\infty}\left\{\frac{\lambda_{N,\alpha} - \mu_N(\theta_0)}{\sigma_N(\theta_0)}\frac{\sigma_N(\theta_0)}{\sigma_N(\theta_N)}\right\} - \lim_{N\to\infty}\left\{\frac{k^m \mu_N^{(m)}(\hat{\theta})}{m\,!\,N^{\delta m}\sigma_N(\theta_0)}\frac{\sigma_N(\theta_0)}{\sigma_N(\theta_N)}\right\}.$$

* The quantity c defined by (e) is called the efficacy of the test $A_N(x)$.

Using the assumptions (*d*) and (*e*), and the fact that $\lim_{N\to\infty} (\lambda_{N\alpha} - \mu_N(\theta_0)/\sigma_N(\theta_0) = \lambda_\alpha$, we obtain

$$\xi = \lambda_\alpha - \frac{k^m c}{m\,!} \tag{3.8.10}$$

which proves the required result.

As a special case of this theorem (that is, for $m = 1$), we have the following theorem.

Theorem 3.8.2. *Suppose, in addition to the assumptions* (*a*) *and* (*b*) *of Theorem* 3.8.1, *the following assumptions are satisfied.*

(*c*) $\mu'_N(\theta_0) > 0$;

(*d*) $\lim_{N\to\infty} \dfrac{\mu'_N(\theta_N)}{\mu'_N(\theta_0)} = 1, \qquad \lim_{N\to\infty} \dfrac{\sigma_N(\theta_N)}{\sigma_N(\theta_0)} = 1;$

(*e*) $\lim_{N\to\infty} N^{-1/2} \dfrac{\mu'_N(\theta_0)}{\sigma_N(\theta_0)} = c > 0.$

Then, the limiting power of the test $A_N(x)$ *is*

$$1 - \Phi(\lambda_\alpha - kc). \tag{3.8.11}$$

From the proofs of the above two theorems, it follows that the assumption (*b*) can be replaced by the somewhat weaker assumption

$$\lim_{N\to\infty} P\left\{\frac{T_N - \mu_N(\theta_N)}{\sigma_N(\theta_N)} \le x\right\} = \Phi(x) \tag{b'}$$

both under the alternative hypothesis H_1 and under the null hypothesis $\theta_N = \theta_0$.

Assume now that we have two level α tests $A^{(1)}$ and $A^{(2)}$ based on the statistics $T^{(1)}$ and $T^{(2)}$, and let N_i, k_i, δ_i, m_i, and c_i be the constants prescribed by the assumptions of Theorem 3.8.1 for the sequence i ($i = 1, 2$). Then the two tests will have the same limiting power against the same sequences of alternatives if

$$\frac{k_1^{m_1} c_1}{m_1\,!} = \frac{k_2^{m_2} c_2}{m_2\,!} \tag{3.8.12}$$

and

$$\frac{k_1}{N_1^{\delta_1}} = \frac{k_2}{N_2^{\delta_2}}\,. \tag{3.8.13}$$

Combining (3.8.12) and (3.8.13), we obtain

$$\frac{N_1^{\delta_1}}{N_2^{\delta_2}} = \frac{k_1}{k_2} = \left(\frac{c_2 m_1\,!}{c_1 m_2\,!}\right)^{1/m_1} k_2^{m_2/m_1 - 1} = \lambda, \qquad \text{say.} \tag{3.8.14}$$

Since λ is a positive constant, N_1/N_2 will tend to a constant as N_1, N_2 tend to infinity if and only if $\delta_1 = \delta_2 = \delta$. Thus by (3.8.2),

$$e_{A^{(2)},A^{(1)}} = \lim_{N_1\to\infty} \frac{N_1}{N_2} = \left(\frac{c_2 m_1!\, k_2^{m_2-m_1}}{c_1 m_2\,!}\right)^{1/m_1\cdot\delta}. \tag{3.8.15}$$

In most practical cases, $m_1 = m_2 = m$, say, then

$$e_{A^{(2)},A^{(1)}} = \left(\frac{c_2}{c_1}\right)^{1/m\delta} = \lim_{N\to\infty}\left(\frac{\mu_{2N}^{(m)}(\theta_0)}{\mu_{1N}^{(m)}(\theta_0)}\,\frac{\sigma_{1N}(\theta_0)}{\sigma_{2N}(\theta_0)}\right)^{1/m\delta}. \tag{3.8.16}$$

Thus, we have proved the following theorem.

Theorem 3.8.3. *If* $\{T^{(1)}\}$ *and* $\{T^{(2)}\}$ *satisfy the assumptions of Theorem* 3.8.1, *and if* $\delta_1 = \delta_2$, $m_1 = m_2$, *then the asymptotic relative efficiency of* $\{A^{(2)}\}$ *with respect to* $\{A^{(1)}\}$ *is given by* (3.8.16).

In many practical applications $m = 1$ and $\delta = \frac{1}{2}$; in which case

$$e_{A^{(2)},A^{(1)}} = \lim_{N\to\infty}\left(\frac{\mu_{2N}'(\theta_0)}{\mu_{1N}'(\theta_0)}\,\frac{\sigma_{1N}(\theta_0)}{\sigma_{2N}(\theta_0)}\right)^2. \tag{3.8.17}$$

The case where the alternative is of the type $\theta_n < \theta_0$ or $|\theta_N| \neq \theta_0$, or the case where $\mu_N^{(m)}(\theta_0) < 0$, can be handled correspondingly.

Example 3.8.1. ARE of Two-Sample rank-order Tests of Location. Let $X_1, \ldots, X_m$ and $Y_1, \ldots, Y_n$ be independent samples from the absolutely continuous cumulative distribution functions $F(x)$ and $F(y - \theta)$ respectively. We are interested in finding the A.R.E. of the tests T_N defined by (3.6.1) with respect to the Student's t-test when F is arbitrary and has finite variance σ^2.

It was shown in Theorem 3.6.1 that $[T_N - \mu_N(\theta)]/\sigma_N(\theta)$ tends to standard normal distribution, where

$$\mu_N(\theta) = \int_{-\infty}^{\infty} J\left[\frac{m}{N}F(x) + \frac{n}{N}F(x-\theta)\right] dF(x) \tag{3.8.18}$$

and, if $m/N \to \lambda$; $0 < \lambda < 1$ and $\theta \to 0$,

$$\sigma_N^2(0) = \frac{n}{mN}A^2 \sim [(1-\lambda)/\lambda N]A^2 \tag{3.8.19}$$

where

$$A^2 = \int_0^1 J^2(x)\,dx - \left(\int_0^1 J(x)\,dx\right)^2. \tag{3.8.20}$$

Now Student's test is based on the statistic

(3.8.21) $$t = \sqrt{\frac{mn}{N}}(\bar{Y} - \bar{X}) \Big/ \sqrt{\frac{\sum (X_i - \bar{X})^2 + \sum (Y_j - \bar{Y})^2}{m + n - 2}}$$

where $\bar{X} = \sum X_i/m$ and $\bar{Y} = \sum Y_i/n$.

The denominator of t converges to σ in probability, and hence t is asymptotically equivalent, in probability, to

(3.8.22) $$t^* = \sqrt{\frac{mn}{N}}(\bar{Y} - \bar{X})/\sigma$$

which has a limiting normal distribution with mean

(3.8.23) $$v_N(\theta) = \sqrt{\frac{mn}{N}} \Big/ \sigma,$$

and variance

(3.8.24) $$\sigma_N^2(\theta) = 1.$$

Thus, assuming that we can differentiate $\mu_N(\theta)$ under the integral sign (see Exercise 3.8.5), and using (3.8.17), we obtain

(3.8.25) $$e_{T_N,t} = \sigma^2\left(\int_{-\infty}^{\infty} \frac{d}{dx} J[F(x)]\, dF(x)\right)^2 \Big/ A^2.$$

Special Cases. (i) Let $J(u) = u$; then the T_N reduces to Wilcoxon's W test. In this case

(3.8.26) $$e_{W,t} = 12\sigma^2\left(\int_{-\infty}^{\infty} f^2(x)\, dx\right)^2$$

Hodges and Lehmann (1956) have shown that $e_{W,t} \geq 0.864$ for all F; $e_{W,t} = 3/\pi \sim 0.955$ when F is normal (0, 1); and $e_{W,t} > 1$ for many non-normal F.

(ii) Let $J(u) = \Phi^{-1}(u)$, then T_N reduces to the normal scores C_N test. In this case

(3.8.27) $$e_{C_N,t} - \sigma^2\left(\int_{-\infty}^{\infty} \frac{f^2(x)\, dx}{\phi\{\Phi^{-1}[F(x)]\}}\right)^2$$

Chernoff and Savage (1958) have shown that $e_{C_N,t} \geq 1$ for all F. Mikulski (1963) has shown that $e_{C_N,t} = 1$ only if F is normal. Thus from the A.R.E. point of view both the Wilcoxon and normal scores procedures appear to be superior compared with the normal theory procedure. No information is yet available on how large sample sizes should be in general for these tests to assert their asymptotic superiority.

Theorem 3.8.2. *Let F be a c.d.f. with density f and variance $\sigma^2 < \infty$. Further assume that $(d/dx)\Phi^{-1}\{F(x)\}$ is bounded as $x \to \pm\infty$. Then $e_{C_N,t} \geq 1$ and $e_{C_N,t} = 1$ if and only if F is normal.*

Proof: (Gastwirth and Wolff, 1968). Without loss of generality we may assume that $\int_{-\infty}^{\infty} xf(x)\,dx = 0$. Applying Jensen's inequality (cf. section 2 of chapter 2), to the function

$$g(X) = \frac{1}{\varphi\{\Phi^{-1}[F(X)]\}/f(X)},$$

we obtain

$$\int_{-\infty}^{\infty} \frac{f^2(x)\,dx}{\varphi\{\Phi^{-1}[F(x)]\}} \geq \left[\int_{-\infty}^{\infty} \varphi\{\Phi^{-1}[F(x)]\}\,dx\right]^{-1}.$$

Integrating the right-hand side by parts, we obtain

$$\int_{-\infty}^{\infty} \varphi\{\Phi^{-1}[F(x)]\}\,dx = x\varphi\{\Phi^{-1}[F(x)]\}\Big|_{-\infty}^{\infty} + \int_{-\infty}^{\infty} x\Phi^{-1}[F(x)]f(x)\,dx$$

Now using the well-known result that $(1 - \Phi(t)/\varphi(t) \to 1$ as $t \to \infty$ and the assumptions of the theorem, it is easy to check that $x\varphi\{\Phi^{-1}[F(x)]\}|_{-\infty}^{\infty} = 0$. The result now follows by applying the Cauchy-Schwartz inequality to $\int_{-\infty}^{\infty} x\Phi^{-1}[F(x)]f(x)\,dx$. ◁

3.8.2. The Asymptotic Relative Efficiency and the Correlation Coefficient

We shall now prove a theorem which shows that the asymptotic relative efficiency of two tests $A^{(1)}$ and $A^{(2)}$ based on the statistics $T^{(1)}$ and $T^{(2)}$ is equal to the square of the limiting correlation coefficient $\rho(T^{(1)}, T^{(2)})$ between $T^{(1)}$ and $T^{(2)}$.

Theorem 3.8.3.* *Let ζ be the class of all tests of H_0 satisfying the assumptions of Theorem* 3.8.2 *and suppose that ζ contains $A^{(1)}$ and $A^{(2)}$. Further suppose that*

(I) *for every given α and k, $A^{(1)}$ is at least as powerful as any other test in* ζ, i.e., $c_1 \geq c_2 > 0$ *for all $A^{(2)} \in \zeta$*

(II) *the joint distribution of*

$$\frac{T_N^{(1)} - \mu_N^{(1)}(\theta)}{\sigma_N^{(1)}(\theta)} \quad \textit{and} \quad \frac{T_N^{(2)} - \mu_N^{(2)}(\theta)}{\sigma_N^{(2)}(\theta)}$$

* The proof of this theorem is due to van Eeden (1963). For the special case dealing with rank order statistics, see Hájek (1962).

tends to a bivariate normal distribution uniform in θ, *or* $\theta_0 \leq \theta \leq \theta_0 + \epsilon$.

(III) $\lim_{N\to\infty} \rho(T_N^{(1)}, T_N^{(2)} \mid \theta_0) = \lim_{N\to\infty} \rho(T_N^{(1)}, T_N^{(2)} \mid \theta_N) = \rho$ say,

then

$$e_{A^{(2)},A^{(1)}} = \rho^2. \tag{3.8.28}$$

Proof: Consider the tests $A(\lambda)$ based on the test statistic

$$T_N^{(\lambda)} = \lambda[T_N^{(1)}/\sigma_N^{(1)}(\theta_0)] + (1-\lambda)[T_N^{(2)}/\sigma_N^{(2)}(\theta_0)] \tag{3.8.29}$$

where λ is a constant independent of N. It is clear that $T_N^{(\lambda)} \in \zeta$ for all λ. Furthermore,

$$\begin{aligned} c_\lambda &= \lim_{N\to\infty} \frac{\mu'_{N,\lambda}(\theta_0)}{N^{1/2}\sigma_{N,\lambda}(\theta_0)} \\ &= [\lambda c_1 + (1-\lambda)c_2]/[\lambda^2 + (1-\lambda)^2 + 2\lambda(1-\lambda)\rho]^{1/2}. \end{aligned} \tag{3.8.30}$$

Since $T_N(\lambda) \in \zeta$ and $c_1 \geq c_2$, for all $A_N^{(2)} \in \zeta$, it follows that

$$c_\lambda \leq c_1 \quad \text{for every} \quad \lambda, \tag{3.8.31}$$

that is,

$$\begin{aligned} &\lambda^2[c_2^2 - c_1^2 - 2c_1(c_2 - \rho c_1)] \\ &\quad - 2\lambda[c_2^2 - c_1^2 - c_1(c_2 - \rho c_1)] + c_2^2 - c_1^2 \leq 0, \quad \text{for every} \quad \lambda. \end{aligned} \tag{3.8.32}$$

Thus $c_1^2(c_2 - \rho c_1)^2 \leq 0$. Hence

$$\rho = c_2/c_1 = (e_{A^{(2)},A^{(1)}})^{1/2}. \tag{3.8.33}$$

This proves the theorem. ◁

3.8.3. Non-normal Cases

So far we have confined ourselves to the case where the tests are based on asymptotically normally distributed statistics. However, the careful examination of the theory will show that in defining the asymptotic relative efficiency we did not make any specific use of the normality assumption. The theory is applicable to the cases where the limiting form of the two distributions is analytically the same, though not normal, provided the two power functions can be made asymptotically the same function by an appropriate choice of the two sample sizes.

In the subsequent chapters we will have occasion to compare tests based on statistics having asymptotically chi-square distributions.

Let T_{Nj}, $j = 1, \ldots, p$, be p statistics with means $\mu_{Nj}(\theta)$ and covariances

$\sigma_{Ni}(\theta)\sigma_{Nj}(\theta)\lambda_{Nij}(\theta)$. It is assumed that for $\theta_N = \theta_0 + k/N^\delta$,

(a) $\mu'_{Nj}(\theta_0) = \cdots = \mu_{Nj}^{(m-1)}(\theta_0) = 0, \qquad \mu_{Nj}^{(m)}(\theta_0) \neq 0; \quad j = 1, \ldots, p;$

(b) $\lim_{N\to\infty} \dfrac{\mu_{Nj}^{(m)}(\theta_N)}{\mu_{Nj}^{(m)}(\theta_0)} = \lim_{N\to\infty} \dfrac{\sigma_{Nj}(\theta_N)}{\sigma_{Nj}(\theta_0)} = \lim_{N\to\infty} \dfrac{\lambda_{Nij}(\theta_N)}{\lambda_{Nij}(\theta_0)} = 1; \quad i, j = 1, \ldots, p;$

(c) $\lim_{N\to\infty} \dfrac{\mu_{Nj}^{(m)}(\theta_0)}{N^{m\delta}(\sigma_{Nj}(\theta_0))} = c_j \neq 0;$

(d) the limiting distributions of

$$\frac{T_{N,j} - \mu_{Nj}(\theta_N)}{\sigma_{Nj}(\theta_N)}, \qquad j = 1, \ldots, p;$$

is a non-degenerate multivariate normal distribution as N tends to infinity. This means that the distribution of

$$(3.8.34) \quad \mathfrak{L}_N = \left\{\frac{T_{N,1} - \mu_{N,1}(\theta_0)}{\sigma_{N1}(\theta_0)}, \ldots, \frac{T_{N,p} - \mu_{N,p}(\theta_0)}{\sigma_{N,p}(\theta_0)}\right\}' \cdot \mathbf{\Lambda}_N^{-1}(\theta_0)\left\{\frac{T_{N,1} - \mu_{N,1}(\theta_0)}{\sigma_{N1}(\theta_0)}, \ldots, \frac{T_{N,p} - \mu_{N,p}(\theta_0)}{\sigma_{Np}(\theta_0)}\right\}$$

is asymptotically a chi square with p degrees of freedom when $\theta = \theta_0$. Here $\mathbf{\Lambda}_N(\theta) = (\lambda_{Nij}(\theta))$. $\mathfrak{L}_N$ may also be written as

$$(3.8.35) \quad \mathfrak{L}_N = \left\{\frac{T_{N,j} - \mu_{N,j}(\theta_N)}{\sigma_{Nj}(\theta_0)} - \frac{\mu_{N,j}(\theta_0) - \mu_{Nj}(\theta_N)}{\sigma_{N,j}(\theta_0)}, j = 1, \ldots, p\right\}' \cdot \mathbf{\Lambda}_N^{-1}(\theta_0)\left\{\frac{T_{N,j} - \mu_{N,j}(\theta_N)}{\sigma_{Nj}(\theta_0)} - \frac{\mu_{N,j}(\theta_0) - \mu_{Nj}(\theta_N)}{\sigma_{N,j}(\theta_0)}, \; j = 1, \ldots, p\right\}$$

Since

$$(3.8.36) \quad \lim_{N\to\infty} \frac{\mu_{Nj}(\theta_N) - \mu_{Nj}(\theta_0)}{\sigma_{Nj}(\theta_0)} = \lim_{N\to\infty} \frac{k^m}{m!\, n^m} \frac{\mu_j^{(m)}(\theta)}{\sigma_{Nj}(\theta_0)}, \qquad \theta_0 < \theta < \theta_N, \\ = \frac{k^m c_j}{m!}$$

using (b), (c), and (d), we find that $\mathfrak{L}_N$ is asymptotically equivalent to

$$(3.8.37) \quad \mathfrak{L}_N^* = \left\{\frac{T_{N,j} - \mu_{N,j}(\theta_N)}{\sigma_{Nj}(\theta_N)} + \frac{k^m c_j}{m!}, \; j = 1, \ldots, p\right\}' \cdot \mathbf{\Lambda}_N^{-1}(\theta_N)\left\{\frac{T_{N,j} - \mu_{N,j}(\theta_N)}{\sigma_{Nj}(\theta_N)} + \frac{k^m c_j}{m!}, \; j = 1, \ldots, p\right\}.$$

From assumption (*d*), it follows that when $\theta = \theta_N$, $\mathfrak{L}_N^*$ has asymptotically a noncentral chi-square distribution with p degrees of freedom and noncentrality parameter

$$\Delta_N = \frac{k^{2m}}{(m!)^2}\, \mathbf{c}'\mathbf{\Lambda}^{-1}(\theta_0)\mathbf{c} \tag{3.8.38}$$

where

$$\mathbf{c} = (c_1, \ldots, c_p)' \quad \text{and} \quad \mathbf{\Lambda}(\theta_0) = \lim_{N\to\infty} \mathbf{\Lambda}_N(\theta_N) = \lim_{N\to\infty} \mathbf{\Lambda}_N(\theta_0). \tag{3.8.39}$$

Let us now consider two tests $A^{(1)}$ and $A^{(2)}$ based on sets of p statistics $T_{Nj}^{(1)}$, $j = 1, \ldots, p$, and $T_{Nj}^{(2)}$, $j = 1, \ldots, p$, satisfying assumptions (*a*) to (*d*) above, with constants c_{1j} and c_{2j} given by (*c*) and correlation matrices $\mathbf{\Lambda}_{1N}(\theta)$ and $\mathbf{\Lambda}_{2N}(\theta)$. Now if $\delta_1 \neq \delta_2$ then one test is unquestionably superior to the other. We therefore consider the case where $\delta_1 = \delta_2 = \delta$. The two tests will then have the same limiting power if

$$\frac{k_1^{2m_1}}{(m_1!)^2}\, \mathbf{c}_1'\mathbf{\Lambda}_1^{-1}(\theta_0)\mathbf{c}_1 = \frac{k_2^{2m_2}}{(m_2!)^2}\, \mathbf{c}_2'\mathbf{\Lambda}_2^{-1}(\theta_0)\mathbf{c}_2 \tag{3.8.40}$$

The two sequences of alternatives will be the same if

$$\frac{k_1}{N_1^\delta} = \frac{k_2}{N_2^\delta}. \tag{3.8.41}$$

It follows then that

$$\frac{N_1}{N_2} \sim \left(\frac{k_1}{k_2}\right)^{1/\delta} = \left(\frac{\mathbf{c}_2'\mathbf{\Lambda}_2^{-1}(\theta_0)\mathbf{c}_2}{\mathbf{c}_1'\mathbf{\Lambda}_1^{-1}(\theta_0)\mathbf{c}_1}\right)^{(2m\delta)^{-1}} \tag{3.8.42}$$

The most common case is when $m = 1$, $\delta = \frac{1}{2}$; in which case

$$e_{A^{(2)},A^{(1)}} = \frac{\mathbf{c}_2'\mathbf{\Lambda}_2^{-1}(\theta_0)\mathbf{c}_2}{\mathbf{c}_1'\mathbf{\Lambda}_1^{-1}(\theta_0)\mathbf{c}_1}. \tag{3.8.43}$$

When $p = 1$, it is easily seen that (3.8.43) reduces to (3.8.17) so that the present theorem is a direct generalization of the Pitman-Noether theorem 3.8.1 to more than one dimension.

The case where μ_{Nj}, σ_{Nj}, and λ_{Nj} depend on more than one parameter, say $\theta_1, \ldots, \theta_c$, may be reduced to the one considered above by putting $\theta_i = \theta_{i0} + k_i/n^{1/2}$. In this case (3.8.43) will involve all the k_i and the unique answer regarding the asymptotic efficiency of $A^{(2)}$ with respect to $A^{(1)}$ may not be possible. However, one can sometimes get useful information, (see e.g., chapters 4 and 5) on the performance of tests by considering the bounds on the ratio of two quadratic forms. In this respect, the following theorem due to Courant is often found useful.

Theorem (Courant). *If* $\mathbf{XAX'}$ *and* $\mathbf{XBX'}$ *are two positive definite quadratic forms in p variables, then the supremum and the infinum of the ratio* $\mathbf{XAX'}/\mathbf{XBX'}$ *with respect to* $\mathbf{X} \neq \mathbf{0}$ *are given by the largest and smallest characteristic roots of* AB^{-1}, *and*

$$\inf_{\mathbf{X}} \frac{\mathbf{XAX'}}{\mathbf{XBX'}} \leq |\mathbf{AB}^{-1}|^{1/p} \leq \sup_{\mathbf{X}} \frac{\mathbf{XAX'}}{\mathbf{XBX'}}.$$

3.8.4. Bahadur Efficiency

In this section, we indicate briefly a different approach to the problems of measuring asymptotic efficiency, due to Bahadur (1960a, b, c).

Suppose we are given a set of probability measures $\{P_\theta, \theta \in \Omega\}$ defined over an arbitrary space S of points s. Let Ω_0 be a subset of Ω, and let H_0 denote the hypothesis that $\theta \in \Omega_0$. To test H_0, suppose we have two sequences of tests $A^{(i)}$, $i = 1, 2$; $N = 1, 2, \ldots$; based on the test statistics $T_N^{(i)}$; $i = 1, 2$; $N = 1, 2, \ldots$, which satisfy the following assumptions.

(I) There exist continuous cumulative distribution functions $F_i(x)$; $i = 1, 2$, such that for each $\theta \in \Omega_0$

$$\lim_{N\to\infty} P_\theta\{T_N^{(i)} \leq x\} = F_i(x), \quad \text{for every} \quad x; \tag{3.8.44}$$

(II) there exist constants a_i, $0 < a_i < \infty$, such that

$$-\log\,[1 - F_i(x)] = \frac{a_i x^2}{2}\,[1 + o(1)], \quad \text{as} \quad x \to \infty, \quad i = 1, 2; \tag{3.8.45}$$

(III) there exist functions $b_i(\theta)$, $0 < b_i(\theta) < \infty$ defined on $\Omega - \Omega_0$, such that for every $\epsilon > 0$,

$$\lim_{N\to\infty} P_\theta\left(\left|\frac{T_N^{(i)}}{N^{1/2}} - b_i(\theta)\right| > \epsilon\right) = 0, \qquad \text{for each} \quad \theta \in \Omega - \Omega_0, \quad \text{and} \quad i = 1, 2. \tag{3.8.46}$$

Then the asymptotic efficiency of $\{A^{(1)}\}$ relative to $\{A^{(2)}\}$ is defined as

$$e^*_{A^{(1)},A^{(2)}} = c_1^*(\theta)/c_2^*(\theta) \quad \text{at} \quad \theta \in \Omega - \Omega_0, \tag{3.8.47}$$

where

$$c_i^*(\theta) = a_i b_i^2(\theta); \qquad i = 1, 2. \tag{3.8.48}$$

A sequence $\{T_N^{(i)}\}$ satisfying the conditions I to III is called a standard sequence for testing H_0 and $c_i(\theta)$ is called the asymptotic slope of the tests based on $\{T_N^{(i)}\}$.

Now under H_0, the limiting distribution of $T_N^{(i)}$ is F_i, and otherwise $T_N^{(i)} \to \infty$ in probability. It follows that for large N, large values of $T_N^{(i)}$ are significant for rejecting H_0. Accordingly, for any given s in S, $1 - F_i(T_N^{(i)}(s))$ may be regarded as an approximate level attained by $T_N^{(i)}$.

To justify the above definition of efficiency, we consider the comparison of levels attained for a given sample size N. In the given situation the test based on $T_N^{(i)}$ is inferior to the test based on $T_N^{(j)}$ if the level attained by $T_N^{(i)}$ exceeds the level attained by $T_N^{(j)}$. Now let us regard the level attained by $T_N^{(i)}$ in a given case as a random variable defined on S. Denote

$$K_N^{(i)} = -2 \log [1 - F_i(T_N^{(i)})]. \tag{3.8.49}$$

Since $K_N^{(i)}$ is a monotone transformation of $T_N^{(i)}$, then for any test of H_0 based on $T_N^{(i)}$, there exists an equivalent test of H_0 based on $K_N^{(i)}$, $i = 1, 2$, all N. Thus comparison of $\{T_N^{(i)}\}$, $i = 1, 2$ is the same as the comparison of $\{K_N^{(i)}\}$, $i = 1, 2$.

Now by assumption I, for each $\theta \in \Omega_0$,

$$\lim_{N\to\infty} P_\theta\{K_N^{(i)} > x\} = e^{-x/2} \quad \text{for every} \quad x > 0; \quad i = 1, 2. \tag{3.8.50}$$

On the other hand, a direct consequence of assumption III is that

$$\lim_{N\to\infty} K_N^{(i)}/N = a_i b_i^2(\theta) \quad \text{for} \quad \theta \in \Omega - \Omega_0; \quad i = 1, 2. \tag{3.8.51}$$

Hence the optimal rejection regions for H_0 are of the form

$$K_N^{(i)} \geq c a_i b_i^2(\theta) N, \qquad 0 < c < 1, \tag{3.8.52}$$

where N is large. Let us now suppose that for given N, there exists a number $n(N)$ such that

$$e^{-\frac{1}{2}\{c a_i b_i{}^2(\theta)\cdot N\}} = e^{-\frac{1}{2}\{c a_j b_j{}^2(\theta)\cdot n(N)\}}, \tag{3.8.53}$$

that is, the type-one probabilities of error (in the asymptotic sense) are equal and $n(N) \to \infty$ as $N \to \infty$. Then

$$e^*_{A_N{}^{(1)}, A_N{}^{(2)}} = \frac{n(N)}{N} = \frac{a_1 b_1^2(\theta)}{a_2 b_2^2(\theta)}, \qquad \theta \in \Omega - \Omega_0, \tag{3.8.54}$$

that is, in large samples, the Bahadur efficiency is equal to the limit of the inverse ratio of the sample sizes needed for the two tests to have the same level of significance.

Under certain regularity conditions, Bahadur (1960) has shown that the limit of (3.8.54) as $\theta \to \theta_0$ is the Pitman efficiency evaluated at $\theta = \theta_0$.

An exposition of Bahadur's work together with some extensions is given by Gleser (1964) and Abrahamson (1967).

There are a few other measures of interest of asymptotic efficiency of tests due to Chernoff (1952), Hodges and Lehmann (1956), Rao (1965) and Walsh

(1946) among others. They are relatively more specialized than those considered in this section, and we do not intend to pursue their study. In passing we might mention that the study of the Bahadur efficiency necessitates the study of large deviations. This field has been pursued only very recently, and in this connection we refer to the works of Bahadur and Ranga Rao (1960), Bahadur (1967), Sanov (1957), Sethuraman (1964), Hoeffding (1965, 1966), and Hoadley (1967), among others.

EXERCISES

Section 3.2

3.2.1. Show that $\mu_k = E(X - \mu)^k$, where $\mu = E(X)$, $k = 1, 2, \ldots$, are all estimable parameters. Hence, or otherwise, obtain the U-statistic corresponding to μ, μ_2, and μ_3. (Hoeffding, 1948a.)

3.2.2. Let $\zeta_c(F)$, $c = 1, \ldots, m$ be defined as in (3.2.11). Prove that for all $1 \le c < d \le m$, $0 \le \zeta_c \le (c/d)\zeta_d$. (Hoeffding, 1948a.)

3.2.3. Let $\sigma^2(U_n)$ be defined as in (3.2.14). Show that (i) $n\sigma^2(U_n)$ is decreasing in n and (ii) $\lim_{n\to\infty} n\sigma^2(U_n) = m^2\zeta_1(F)$. (Hoeffding, 1948a.)

3.2.4. Let F be the class of all Cauchy distributions, viz.,

$$dF(x) = (\lambda/\pi)\, dx/\{\lambda^2 + (x - \theta)^2\}, \qquad -\infty < x < \infty$$

$[\lambda > 0, -\infty < \theta < \infty]$. Show that θ is estimable and has degree 3, though the kernel has no finite variance. (Sen, 1960.)

3.2.5. Let $D(x, y) = F(x, y) - F(x, \infty)F(\infty, y)$ where (X, Y) has the bivariate c.d.f. $F(x, y)$, and let

$$\Delta = \Delta(F) = \iint_{R^2} D^2(x, y)\, dF(x, y) \ge 0.$$

Show that $0 \le \Delta \le \frac{1}{30}$. (Hoeffding, 1948b.)

3.2.6. (Continued.) Let $c(u) = 1$ if $u \ge 0$ and $c(u) = 0$, otherwise. Also let $\psi(x_1, x_2, x_3) = c(x_1 - x_2) - c(x_1 - x_3)$, $\phi(x_1, y_1, \ldots, x_5, y_5) = \frac{1}{4}\psi(x_1, x_2, x_3)\psi(x_1, x_4, x_5)\psi(y_1, y_2, y_3)\psi(y_1, y_4, y_5)$ and let

$$u_n = \frac{1}{n \cdots (n-4)} \sum \psi(X_{\alpha_1}, Y_{\alpha_1}, \ldots, X_{\alpha_5}, Y_{\alpha_5}),$$

where the summation $\sum$ extends over all possible $\alpha_1 \ne \cdots \ne \alpha_5 = 1, \ldots, n$. Show that (i) $E(U_n) = \Delta(F)$, (ii) U_n is stationary of order 1 if $\Delta(F) = 0$. (Hoeffding, 1948b.)

3.2.7. Let Ω and Ω^* be the class of continuous and absolutely continuous bivariate c.d.f.'s. Show that there do not exist rank tests of independence which are unbiased on any significance level with respect to the classes Ω or Ω^*. (Hoeffding, 1948b.)

3.2.8. Let $\mathbf{X} = (X^{(1)}, X^{(2)})$ have the bivariate c.d.f. $F(x_1, x_2)$. Two observations $\mathbf{X}_\alpha$ and $\mathbf{X}_\beta$ are said to be *concordant* (or *discordant*) if $X_\alpha^{(1)} - X_\beta^{(1)}$ and $X_\alpha^{(2)} - X_\beta^{(2)}$ have the same (or opposite) signs. Let P_c and P_d be respectively the probability of concordance and discordance, and let $\tau = P_c - P_d = 2P_c - 1$. Justify the use of τ as a measure of association of $\mathbf{X}^{(1)}$ and $\mathbf{X}^{(2)}$. Show that τ is an estimable parameter of degree 2 and the corresponding U-statistic is

$$t_n = 4\binom{n}{2}^{-1} \sum_{1 \le \alpha < \beta \le n} c(X_\alpha^{(1)} - X_\beta^{(1)})c(X_\alpha^{(2)} - X_\beta^{(2)}) - 1.$$

Obtain the variance of t_n and show that if $(X^{(1)}, X^{(2)})$ are independent $\sigma^2(t_n) = 2(2n + 5)/9n(n - 1)$. (Hoeffding, 1948a.)

3.2.9. If in the sample $\{(X_\alpha^{(1)}, X_\alpha^{(2)}),\ \alpha = 1, \ldots, n\}$ all $X_\alpha^{(1)}$'s (and $X_\alpha^{(2)}$'s) are different, the Spearman's rank correlation coefficient is defined by

$$K_n' = \frac{12}{n(n^2 - 1)} \sum_{\alpha=1}^{n} \left(R_\alpha^{(1)} - \frac{n+1}{2}\right)\left(R_\alpha^{(2)} - \frac{n+1}{2}\right),$$

where $R_\alpha^{(k)}$ is the rank of $X_\alpha^{(k)}$ among all $X_1^{(k)}, \ldots, X_n^{(k)}$ for $\alpha = 1, \ldots, n$, $k = 1, 2$. Also, the grade correlation coefficient is defined by

$$\rho_g = 3 \int_{-\infty}^{\infty} \int_{-\infty}^{\infty} [2F(x, \infty) - 1][2F(\infty, y) - 1]\, dF(x, y).$$

Show that K_n' stochastically converges to ρ_g. Obtain the exact variance of K_n' when $X^{(1)}, X^{(2)}$ are independent. Obtain also the unbiased estimator of ρ_g. (Hoeffding, 1948a.)

3.2.10. Suppose we have $c\ (\geq 2)$ independent samples of sizes $n_1, \ldots, n_c$ drawn from distributions $F_1(x), \ldots, F_c(x)$, respectively. Let $\theta(F)$ be a regular functional, and $U_1, \ldots, U_c$ be the U-statistics from the 1st, $\ldots$, cth sample, respectively. Also, let U_0 be the U-statistic corresponding to the pooled sample size $N (= \sum_1^c n_i)$.

(i) Show that U_0 is the minimum variance unbiased estimator of $\theta(F)$ if $F_1 = \cdots = F_c = F$. (ii) If $\theta(F)$ is stationary of order zero, then $\bar{U} = \sum_{i=1}^c n_i U_i/N$ is asymptotically equivalent to U_0. (iii) If $\theta(F)$ is stationary of order higher than zero, then the efficiency of $\bar{U}$ (with respect to U_0) cannot exceed $1/c$. (Sen, 1967b.)

3.2.11. (Continued.) If $F_1, \ldots, F_c$ are not all identical, let

$$\bar{F}_N(x) = \sum_1^c n_i F_i(x)/N$$

and let $n_i/N \to \lambda_i$: $0 < \lambda_i < 1$ for all $i = 1, \ldots, c$;

$$\bar{\theta}(F) = \sum_{i=1}^{c} \lambda_i \theta(F_i) \quad \text{and} \quad \theta(\bar{F}) = \lim_{N\to\infty} \theta(\bar{F}_N).$$

(i) Show that $\bar{U}$ is the m.v.u. estimator of $\bar{\theta}(F)$.

(ii) U_0 is asymptotically the m.v.u. estimator of $\theta(\bar{F})$.

(iii) The bias of U_0 is at most of the order N^{-1}.

(iv) Give an example where $\bar{\theta}(F) = \theta(\bar{F})$ and another example where $\bar{\theta}(F) \neq \theta(\bar{F})$. (Sen, 1967b.)

3.2.12. Suppose $\{X_i\}$ is an m-dependent stationary stochastic process, with marginal c.d.f. $F(x)$. A regular functional $\theta(F)$ is said to be a *nonserial parameter* if it depends only on $F(x)$ (but not on the joint c.d.f.'s of two or more dependent random variables). A kernel $f(X_{\alpha_1}, \ldots, X_{\alpha_k})$ will be called a nonserial statistic if $|\alpha_i - \alpha_j| > m$ for all $i \neq j = 1, \ldots, k$.

The corresponding symmetric (unbiased) estimator is defined by

$$U_n^0 = \binom{n - km + m}{k}^{-1} \sum_{S_0} f(X_{\alpha_1}, \ldots, X_{\alpha_k})$$

where the summation S_0 extends over all possible $\alpha_i \neq \cdots \neq \alpha_k = 1, \ldots, n$, such that $|\alpha_i - \alpha_j| > m$ for all $i \neq j = 1, \ldots, k$. If we define U_n as in (3.2.7) show that

(i) $E(U_n - U_n^0) = O(n^{-1})$,

(ii) $n^{1/2}[U_n - U_n^0] \xrightarrow{p} 0$. (Sen, 1963c.)

3.2.13. (Continued.) If we define $\psi_1(X_\alpha)$ as in (3.2.10) (where the kernel is a nonserial statistic), and let

$$\zeta_{1\cdot h} = E\{\psi_1(X_\alpha)\psi_1(X_{\alpha+h})\}, \qquad h = 0, \ldots, m,$$

$$\zeta_1^0 = \zeta_{1\cdot 0} + 2\sum_{h=1}^{m} \zeta_{1\cdot h},$$

then show that if the kernel has a finite second-order moment, while $E|\psi_1(X_\alpha)|^3 < \infty$, $n^{1/2}[U_n - E(U_n^0)]$ has asymptotically a normal distribution with zero mean and variance $k^2\zeta_1$. (Sen, 1963c.)

3.2.14. Generalize the results of Exercise 3.2.10 in the light of Exercise 3.2.13. (Sen, 1965c.)

3.2.15. Define a kernel as in (3.4.47) and the corresponding U-statistic as in (3.4.48), and obtain (i) the variance of the U-statistic, and (ii) deduce its asymptotic normality.

Section 3.3

3.3.1. For problem (I) of section 3.3 show that the sign test is a similar size α test for all continuous $F_1, \ldots, F_n$.

3.3.2. Show that the sign test for $p = \frac{1}{2}$ is unbiased for two-sided alternatives if the tails are symmetrically truncated.

3.3.3. Obtain a confidence interval for the population quantile using the sign test for the same quantile.

3.3.4. If the c.d.f.'s $F_1, \ldots, F_n$ are not necessarily continuous, obtain the permutation test based on the sign test statistic. (Putter, 1955.)

3.3.5. Obtain the tests based on t_n and K_n' (in Exercises 3.2.9 and 3.2.10) for testing matching invariance in problem (II) of section 3.3 against appropriate alternatives.

3.3.6. *Median test.* In problem (III) of section 3.3, suppose we have two samples of sizes n_1 and n_2 with observations $X_1, \ldots, X_{n_1}$ and $Y_1, \ldots, Y_{n_2}$, respectively. Let Z_N be the ath order statistic of the combined sample of size $N (= m + n)$, and let m_1 and m_2 be the number of observations of the first and the second samples having values not exceeding that of Z_N. Show that if the two samples are from the same distribution,

$$P(m_1, m_2) = \binom{n_1}{m_1}\binom{n_2}{m_2}\bigg/\binom{N}{a}, \qquad 0 \leq m_1 \leq a.$$

Hence, obtain a suitable test for the equality of the two distributions. Obtain the large-sample test procedure. (Mood, 1950.)

3.3.7. (Continued.) Generalize the test in Exercise 3.3.6 for a set of more than one quantile of the pooled sample. (Massey, 1951c.)

3.3.8. (Continued.) Generalize the tests in Exercises 3.3.6 and 3.3.7 to the case of c (≥ 2) samples. (Massey, 1951c; Mood, 1950.)

3.3.9. For the problem of Exercise 3.3.6, let $X_{(r_1)} < \cdots < X_{(r_k)}$ be any k specified quanlties of the first sample, and let $n_{20}, n_{21}, \ldots, n_{2k}$ be the number of observations of the second sample belonging to the $k + 1$ cells formed by these k quantiles. Obtain the distribution of $n_{20}, \ldots, n_{2k}$ for arbitrary distributions of the two-sample observations. Hence, or otherwise, obtain the null distribution of $n_{20}, \ldots, n_{2k}$. Develop an exact, as well as large-sample test, for the equality of the two distributions based on $n_{20}, \ldots, n_{2k}$, and study the consistency of the test. (Sen, 1962a.)

3.3.10. (i) Generalize the test in Exercise 3.3.9 to the case of c (>2) independent samples and obtain the large sample distribution test criterion. (Sen, 1962a.)

3.3.11. Let $X_{(1)} = -\infty < X_{(1)} < X_{(2)} < \cdots < X_{(n_1)} < X_{(n_1+1)} = \infty$ be the order statistic of the first sample. Thus, these ordered variables form $n_1 + 1$ nonoverlapping and contiguous cells (if the possibility of ties is neglected, which is granted by the assumed continuity of the parent distribution). The number of observations of the second sample belonging to these cells are denoted by $r_0, r_1, \ldots, r_{n_1}$ respectively. Show that if the two distributions are identical, the joint distribution of r_j's is

$$\frac{n_1!\, n_2!}{(n_1 + n_2)!} \quad \text{where} \quad r_j \geq 0, \quad \sum_{j=0}^{n2} r_j = n_2.$$

Let now, S_i of the r_j's be equal to i for $i = 0, \ldots, n_2$, so that $S_0 + \cdots + S_{n_2} = n_1 + 1$, $S_1 + 2S_2 + \cdots + n_2 S_{n_2} = n_2$. Show that under the hypothesis of equality of the two distributions of the two-sample observations, the probability function of $S_0, \ldots, S_{n_2}$ is given by

$$\frac{n_1!\, n_2!}{(n_1 + n_2)!} \cdot \frac{(n_1 + 1)!}{S_0!\, S_1! \cdots S_{n_2}!}.$$

Hence, or otherwise, obtain the distribution of S_0. Show that for large

n_1, n_2 for which $n_2 = \rho n_1 + o(1)$, $0 < \rho < \infty$,

$$E(S_0) = n_1/(1 + \rho) + o(n_1),$$
$$V(S_0) = n_1\rho^2/(1 + \rho)^3 + o(n_1),$$

and $[S_0 - E(S_0)]/[V(S_0)]^{1/2}$ has asymptotically the normal distribution with zero mean and unit variance. (Wilks, 1961.)

3.3.12. (Continued.) Show that under the hypothesis of equality of distributions, and for any fixed k,

$$\sum_{i=0}^{k} (S_i - n_1 a_i)^2/n_1 a_i + \frac{u^2 + v^2}{n_1\rho^2(1 + \rho)a_k}$$

has asymptotically a chi-square distribution with $k + 1$ D.F., where $a_i = \rho^i/(1 + \rho)^{i+1}$ and

$$u = \sum_{t=0}^{k} (S_i - n_1 a_i)(i - \rho - k - 1)$$
$$u = \sqrt{\rho}\,(1 + \rho)\sum_{i=0}^{k} (S_i - n_1 a_i).$$

Study the consistency of the test. (Wilks, 1961.)

3.3.13. *Run test*. Let us arrange the two-sample observations in order of magnitude, and let u be the total number of runs. (A run is defined by a sequence of observations from the same sample, viz., $XXYXYYX$ will have 5 runs.) Obtain the exact distribution of V_N, the number of runs in samples of sizes (n_1, n_2) when the parent distributions are identical. Show that asymptotically $[V_N - 2n_1\rho/(1 + \rho)](1 + \rho)^{3/2}/2\rho n_1^{1/2}$ has a normal distribution with zero mean and unit variance, where $\rho = n_1/n_2 : 0 < \rho < 1$.
(Wald and Wolfowitz, 1941.)

3.3.14. *Wilcoxon-Mann-Whitney test*. In the combined sample of size $N\,(= n_1 + n_2)$, let the ranks of the first sample observation be denoted by $R_1, \ldots, R_{n_1}$, respectively. Let then

$$V_n = \frac{1}{n_1}\sum_{i=1}^{n_1} R_i - \frac{N+1}{2}.$$

Show that V_n is linearly related to that of a generalized U-statistic. Hence, or otherwise, obtain the asymptotic (null as well as non-null) distribution of V_n. (Lehmann, 1951a.)

3.3.15. Consider Exercise 3.3.14 when the parent distributions are not necessarily continuous, and suggest some permutation tests.
(Putter, 1955; Chanda, 1963; Sen, 1965b.)

3.3.16. For the two-sample scale problem, consider the parameter

$$\theta(F, G) = P\{|X_1 - X_2| > |Y_1 - Y_2|\},$$

where the X's and Y's have the c.d.f.'s F and G, respectively. Obtain an unbiased test for H_0: $F = G$ against $\theta(F, G) > \frac{1}{2}$ (or $<\frac{1}{2}$). Show that the test based on the corresponding generalized U-statistic is not distribution-free, even asymptotically. (Sukhatme, 1957.)

3.3.17. Obtain the permutation variance of the generalized U-statistic corresponding to $\theta(F, G)$ in Exercise 3.3.16, and hence, or otherwise, obtain a permutationally distribution free test for the hypothesis in Exercise 3.3.16.

(Sen, 1965b.)

3.3.18. Consider the kernel

$$\phi(X_1, X_2, Y_1, Y_2) = \begin{cases} 1, & \text{if} \quad X_1 \le (Y_1, Y_2) \le X_2 \quad \text{or} \quad X_2 \le (Y_1, Y_2) \le X_1, \\ & \text{or} \quad Y_1 \le (X_1, X_2) \le Y_2 \quad \text{or} \quad Y_2 \le (X_1, X_2) \le Y_1, \\ 0, & \text{otherwise.} \end{cases}$$

Show that the corresponding U-statistic provide a suitable test for the two-sample scale problem. (Tamura, 1960.)

3.3.19. Consider the kernel

$$\phi(X_i, X_j) = \begin{cases} 1 & \text{if} \quad X_i + X_j > 0, \\ 0 & \text{if} \quad X_i + X_j \le 0, \end{cases}$$

where $X_1, \ldots, X_n$ are i.i.d.r.v. having a continuous c.d.f. $F(x)$. Obtain the corresponding U-statistic and its exact variance when $F(x)$ is symmetric about zero. (Tukey, 1949; Walsh, 1949.)

3.3.20. Define by R_i the rank of $|X_i|$ among all $(X_1|, \ldots, |X_n)$, $i = 1, \ldots, n$, and let S_i be the sign of X_i, $i = 1, \ldots, n$. Then show the relationship between $W_n = \sum_i^n S_i R_i$ and the U-statistic in Exercise 3.3.19.

(Wilcoxon, 1949; Walsh, 1957.)

3.3.21. Use Exercises 3.3.19 and 3.3.20 for the one-sample location test.

3.3.22. *Sukhatme's scale test.* Let $X_1, \ldots, X_m$ and $Y_1, \ldots, Y_n$ represent two independent samples of sizes m and n of independent observations from two populations with continuous c.d.f.'s $F(x)$ and $F(\theta x)$ respectively. Then for testing the hypothesis $\theta = 1$, Sukhatme's test is based on the statistic

$$S = \sum_{i=1}^{n} \sum_{\substack{j,k=1 \\ j \ne k}}^{m} Q(X_j, X_k, Y_i) + 2 \sum_{\substack{i,j=1 \\ i \ne j}}^{n} \sum_{k=1}^{m} Q(X_k, Y_i, Y_j)$$

$$+ \frac{m + n - 2}{2} \sum_{i=1}^{n} \sum_{j=1}^{m} k(X_j, Y_i),$$

where

$$Q(u, v, w) = \begin{cases} 1 & \text{if} \quad 0 < u < w, \quad 0 < v < w, \\ & \text{or} \quad w < u < 0, \quad w < v < 0, \\ 0 & \text{otherwise,} \end{cases}$$

and

$$k(u, u) = \begin{cases} 1 & \text{if} \quad 0 < u < v \quad \text{or} \quad v < u < 0, \\ 0 & \text{otherwise.} \end{cases}$$

Prove that S has asymptotically a normal distribution. (Sukhatme, 1958b.)

Section 3.4

3.4.1. Prove the results (3.4.21). (Wald and Wolfowitz, 1944.)

3.4.2. Prove the results (3.4.24) and (3.4.25). (Wald and Wolfowitz, 1944.)

3.4.3. Show that L_ν, defined by (3.4.20), is stochastically equivalent to T_ν, defined by (3.4.33). (Hájek, 1961.)

3.4.4. Obtain the proofs of Theorems 3.4.3, 3.4.4, and 3.4.5. (Hájek, 1961; Motoo, 1957.)

3.4.5. Show that in the univariate case if the parent c.d.f. is continuous and a generalized U-statistic is a sole function of the ranks of the observations, then the permutation distribution of the U-statistic (under the hypothesis of invariance under partitioning into subsets) coincides with its exact null distribution. Verify this with the aid of Exercises 3.3.14 and 3.3.18.

3.4.6. By means of some counterexamples show that for the same hypothesis of invariance under partitioning into subsets, the conclusions of Exercise 3.4.5 may not hold for multivariate problems. (Chatterjee and Sen, 1964.)

3.4.7. Show that in the bivariate case (with continuous c.d.f.) the conclusion of Exercise 3.4.5 also holds for the hypothesis of matching invariance. (Scheffé, 1943b.)

3.4.8. A generalized U-statistic U_N defined in (3.2.67) is said to be of type A if the corresponding kernel $\phi(X_{\alpha_1}, \ldots, X_{\alpha_{m_1}}, Y_{\beta_1}, \ldots, Y_{\alpha_{m_2}})$ satisfies the following:

$$\frac{1}{(m_1 + m_2)!} \sum_{S^*} \phi(X_{\alpha_1}, \ldots, X_{\alpha_{m_1}}, Y_{\beta_1}, \ldots, Y_{\beta_{m_2}}) = \text{constant}$$

for all $(X_{\alpha_1}, \ldots, Y_{\beta_{m_2}})$, where the summation extends over all possible $(m_1 + m_2)!$ permutations of the $m_1 + m_2$ arguments in the $(m_1 + m_2)$ ordered positions of ϕ. Show that the covariance of $\phi(X_{\alpha_1}, \ldots, X_{\alpha_{m_1}}, Y_{\beta_1}, \ldots, Y_{\beta_{m_2}})$ and $\phi(X_{\alpha_{m_1+1}}, \ldots, X_{\alpha_{2m_1}}, Y_{\beta_{m_2+1}}, \ldots, Y_{\beta_{2m_2}})$, under the permutation model of problem (III) of section 3.3 is exactly equal to zero. (Sen, 1965b.)

Section 3.5

3.5.1. Obtain the proofs of Theorems 3.5.3 and 3.5.4. (Hoeffding, 1952; Dwass, 1953, 1955b.)

3.5.2. Let a U-statistic U_N be defined as in (3.2.67) and suppose we want to test the hypothesis of the equality of the two c.d.f.'s F and G. Show that if $F(x)$ and $G(x)$ are replaced by two sequences of c.d.f.'s $\{F_N(x)\}$ and $\{G_N(x)\}$, such that both $F_N(x)$ and $G_N(x)$ converge to a common c.d.f. $H(x)$ at all points of continuity of the latter, then for this sequence of alternatives, the permutation test based on U_N is asymptotically power-equivalent to the unconditional test based on the same U_N (though the latter may not be exactly distribution-free). (Sen, 1965b.)

3.5.3. Use the results of Exercise 3.5.2 to study the asymptotic power equivalence of the tests in Exercises 3.3.14 and 3.3.15 and of Exercises 3.3.16 and 3.3.17.

3.5.4. For Exercise 3.5.2 show that if for some permutation test based on a statistic T_N, the conditions for the applicability of Pitman efficiency hold, then the permutation test based on the generalized U-statistic corresponding to T_N has a Pitman efficiency which cannot be smaller than that of T_N.

(Sen, 1965b.)

Section 3.6

3.6.1. Verify that the assumptions (*a*), (*b*), and (*c*) of section 3.6.1 hold whenever $F\,(=J^{-1})$ is a normal, logistic, uniform, or exponential c.d.f. and $J_N(i/(N+1))$ is the expected value of the ith order statistic of a sample of size N from the distribution F. (Chernoff and Savage, 1958.)

3.6.2. If $F_1(x)$ is the sample c.d.f. based on a sample of size 1 and $F(x)$ is the corresponding population c.d.f., and if $x < y$, then

(i) $$E\{[F_1(x) - F(x)][F_1(y) - F(y)]\} = F(x)[1 - F(y)],$$

(ii) $$E\{d[F_1(x) - F(x)]\, d[F_1(y) - F(y)]\} = -dF(x)\, dF(y),$$

and

(iii) $$E\{[F_1(x) - F(x)]\, d[F_1(x) - F(x)]\} = [1 - F(x)]\, dF(x).$$

(Chernoff and Savage, 1958.)

3.6.3. If $J(u)$ is monotonic and $F(x)$, $G(x)$, and λ_N are restricted to a set for which $|J'(H(x))| > 0$ has a positive measure with respect to $F(x)$ and $G(x)$, then $B(X)$ and $B^*(Y)$, defined by (3.6.16) and (3.6.17), have variances bounded away from zero.

3.6.4. Prove Corollary 3.6.3. (Chernoff and Savage, 1958.)

3.6.5. *Two-sample location test.* Let $X_1, \ldots, X_m$ be i.i.d.r.v. having the c.d.f. $F(x)$ and $Y_1, \ldots, Y_n$ be i.i.d.r.v. having the c.d.f. $G(x)$. The null hypothesis states that $F = G$, while the set of translation type of alternatives state that $G(x) = F(x + \delta)$, $\delta \neq 0$.

(*a*) *Wilcoxon-Mann-Whitney test.* (see also Exercise 3.3.14.) This test is based on the statistic $W = (\sum_{i=1}^{N} iZ_{n,i})/mN$ where $N = m + n$ and $Z_{N,i}$ is defined in section 3.6.1. Show that if F and G are absolutely continuous

$$\mathcal{L}([W - \mu_N(W)]/\sigma_N(W)) \to N(0, 1)$$

where $\mu_N(W) = \int_{-\infty}^{\infty} H(x)\, dF(x)$ and

$$\sigma_N^2(W) = \frac{2(1-\lambda_N)}{\lambda_N}\Bigg[\iint\limits_{-\infty<x<y<\infty} G(x)[1 - G(y)]\, dF(x)\, dF(y) + \frac{1-\lambda_N}{\lambda_N}\iint\limits_{-\infty<x<y<\infty} F(x)[1 - F(y)]\, dG(x)\, dG(y)\Bigg]$$

Under the null hypothesis, $\mu_N(W)$ and $\sigma_N^2(W)$ reduce to $\frac{1}{2}$ and $(1 - \lambda_N)/12N\lambda_N$, respectively (as compared to the exact expressions $\frac{1}{2}$ and $n(N + 1)/12mN^2$).*

* For exact tables of small-sample significant probabilities and some allied recursive relations, see Mann and Whitney (1947) and Owen (1962).

(*b*) *Median and quantile tests.* (Mathiesen, 1943; Mood, 1950.) See Exercises 3.3.6 through 3.3.10.

(*c*) *The normal score test.* This test is based on the statistic

$$V = \frac{1}{m} \sum_{i=1}^{N} E_{N,i} Z_{N,i}$$

where $E_{N,i}$ is the expected value of the ith order statistic of a sample of size N drawn from a standardized normal distribution $\Phi(x)$. Show that V satisfies all the conditions needed for the asymptotic normality of Theorem 3.6.1 and the asymptotic mean and variance of V are

$$\mu_N = \int_{-\infty}^{\infty} \Phi^{-1}(H(x))\, dF(x)$$

and

$$\sigma_N^2(V) = \frac{2(1 - \lambda_N)}{\lambda_N} \left[\iint_{-\infty < x < y < \infty} [A(x, y)\, dF(x)\, dF(y) + B(x, y)\, dG(x)\, dG(y) \right]$$

where

$$A(x, y) = G(x)[1 - G(y)]/\phi[\Phi^{-1}(H(x))]\phi[\Phi^{-1}(H(y))],$$

$$B(x, y) = F(x)[1 - F(y)]/\phi[\Phi^{-1}(H(x))]\phi[\Phi^{-1}(H(y))].$$

Under the hypothesis $F = G$, the above expressions reduce to 0 and $(1 - \lambda_N)/N\lambda_N$, respectively. (Chernoff and Savage, 1958.)

(*d*) *Van der Waerden's test.* This is based on

$$V^* = \frac{1}{m} \sum_{i=1}^{V} \Phi^{-1}[i/(N + 1)] Z_{N,i}$$

(van der Waerden, 1953a.)

Show that the statistics V^* and V (in (*c*)) are asymptotically equivalent.

3.6.6. *One-sample location tests.* Let $X_1, \ldots, X_n$ be i.i.d.r.v. having the continuous c.d.f. $F(x)$.

(*a*) *Sign test.* H_0: $F(x)$ has the median $x = 0$. See Exercises 3.3.1 and 3.3.2.

If further $F(x)$ can be written as $F_0(x - \theta)$ where θ is the median of $F(x)$ and $F_0(u)$ is symmetric about $u = 0$, then for testing H_0: $\theta = 0$, consider the following tests.

(*b*) *Signed rank test.* This test is based on the statistic $W^* = n^{-1} \sum_{i=1}^{n} iZ_{N,i}$ where $Z_{N,i}$ is defined in section 3.6.2. With the help of Theorem 3.6.2 obtain the asymptotic distribution of W^* and the expressions for its asymptotic mean and variance. Show that the exact mean and variance of W^* (under H_0: $\theta = 0$) are $n + 1/4$ and $(n + 1)(2n + 1)/24n$, respectively.

(Wilcoxon, 1949; Tukey, 1949.)

(c) *Normal score test* (Fraser, 1957a). This test is based on the statistic

$$V^* = \frac{1}{n} \sum_{i=1}^{n} E_{n,i} Z_{n,i}$$

where $E_{n,i}$ is the expected value of the ith order statistic of a sample of size n drawn from the chi distribution with 1 d.f. Obtain the asymptotic distribution of V^* and the expressions for its mean and variance with the aid of Theorem 3.6.4.

3.6.7. *Two-sample scale tests (locations known).* Let $X_1, \ldots, X_m$ be i.i.d.r.v. having the continuous c.d.f. $F(x)$ and $Y_1, \ldots, Y_n$ be i.i.d.r.v. having the continuous c.d.f. $G(x) = F(\theta x)$. The null hypothesis is H_0: $\theta = 1$.

(a) *Mood's test.* This test is based on

$$M^* = \frac{1}{mN^2} \sum_{i=1}^{n} \left(r_i - \frac{N+1}{2}\right)^2$$

where r_i is the rank of the ith X observation in the pooled sample. Show that the statistic M^* belongs to the family of the Chernoff-Savage type of rank order statistics (cf. section 3.6.1), and hence or otherwise, obtain the asymptotic mean, variance, and normality of M^* for arbitrary F and G. Show that under H_0, the mean and variance of the asymptotic distribution reduce to $\frac{1}{12}$ and $(1 - \lambda_N)/180\lambda_N$, respectively.

(Mood, 1954; Chernoff and Savage, 1958.)

(b) *The simplest rank order test** is based on

$$B = \frac{1}{m} \sum_{i=1}^{N} E_{Ni} Z_{Ni} \text{ where } E_{Ni} = \frac{1}{2} + \frac{1}{2N} - \left|\frac{1}{2} + \frac{1}{2N} - \frac{i}{N}\right|$$

Show that E_{Ni} satisfies the assumptions of Theorem 3.6.1. Hence B has asymptotically a normal distribution with mean $\mu_N(B)$ and variance $\sigma_N^2(B)$ given by (3.6.4) and (3.6.5) respectively, where

$$J(u) = \tfrac{1}{2} - |\tfrac{1}{2} - u|.$$

Under the hypothesis H_0: $\theta = 1$,

$$\mu_N(B) = \tfrac{1}{4}; \qquad \sigma_N^2(B) = \frac{(1 - \lambda_N)}{48N\lambda_N}.$$

(Ansari and Bradley, 1960.)

(d) *The normal score test* is based on

$$D = \frac{1}{m} \sum_{i=1}^{N} E_{N,i} Z_{N,i}$$

where $E_{N,i} = [\Phi^{-1}(i/(N+1))]^2$. Obtain its asymptotic mean, variance and normality with the aid of Theorem 3.6.1. (Capon, 1961; Klotz, 1962.)

* Tables for the significant values of B are due to Ansari and Bradley (1960), Barton and David (1958) and Siegel and Tukey (1960).

For other scale tests by Sukhatme (1957), Lehmann (1951a), and Tamura (1960), see Exercises 3.3.16–3.3.18.

3.6.8. *A lemma on the modified U-statistic.* Let $X_1, \ldots, X_n$ be n i.i.d.r.v. having the c.d.f. $F(x - v)$. Let $\phi(u_1, \ldots, u_s)$ with $s \leq n$ be a real-valued symmetric function of its arguments such that if $W(x_1, x_2, \ldots, x_s, t) = \phi(x_1 - t, x_2 - t, \ldots, x_s - t)$ where $A(t - v) = E\phi(X_1 - t, \ldots, X_s - t)$, the following conditions are satisfied.

$$(a) \quad |W(x_1, x_2, x_s, r)| \leq M_1, \quad \text{and} \quad E|W(X_1, X_2, \ldots, X_s; r + h) - W(X_1, X_2, \ldots, X_s; r)| \leq M_2 \cdot h$$

where M_1 and M_2 are fixed constants. There exists a sequence $\{r_j\}$ such that for each set of x's

$$(b) \quad \sup_{0 \leq r_j \leq k} |W(x_1, \ldots, x_s, r_j) - W(x_1, \ldots, x_s, 0)| = \sup_{0 \leq r \leq k} |W(x_1, \ldots, x_s, r) - W(x_1, \ldots, x_s, 0)|.$$

Further, let $\hat{v}(x_1, \ldots, x_n)$ be an estimate of ξ such that given $\epsilon > 0$, there exists a number b such that for n sufficiently large, $P\{|\hat{v} - v| \geq b/\sqrt{n}\} \leq \epsilon$.

Define

$$U_n = \binom{n}{s}^{-1} \sum \phi(X_{\alpha_1} - v, \ldots, X_{\alpha_s} - v),$$

the summation being taken over all subscripts α such that $1 \leq \alpha_1 < \alpha_2 < \cdots < \alpha_s \leq n$ and

$$L_n = \binom{n}{s}^{-1} \sum [\phi(X_{\alpha_1} - \hat{v}, \ldots, X_{\alpha_s} - \hat{v}) - A(\hat{v} - v)].$$

Then $\sqrt{n}\, L_n$ and $\sqrt{n}\,(U_n - E_n)$ have the same limiting distributions as $n \to \infty$. (Sukhatme, 1958a.)

3.6.9. *A lemma for modified rank order statistics.* Let $X_1, \ldots, X_n$ be i.i.d.r.v. having the c.d.f. $F(x - v)$ where v is the location parameter. Let $\hat{v} = \hat{v}(X_1, \ldots, X_n)$ be a consistent estimator of v such that $n^{1/2}(\hat{v} - v)$ is bounded in probability as $n \to \infty$. Let $M(x)$ be a real-valued function defined over $(-\infty, +\infty)$ such that the derivative $M'(x)$ exists, $E(M'(X - v)) = 0$ where X has the c.d.f. $F(x - v)$, and $|M'(x - v)| \leq kT(x)$ uniformly in t for $|t| \leq c$ where c and k are constants. Also assume that $E[T(X)]^2 < \infty$. Then $L_n = n^{-1/2} \sum_{i=1}^n \{M(X_i - v) - M(X_i - \hat{v})\} \to 0$ in probability as $n \to \infty$. (Raghavachari, 1965a.)

3.6.10. *Two-sample scale tests when the locations are unknown.* Let $X_1, \ldots, X_m$ be i.i.d.r.v. having the c.d.f. $F(x - \xi)$ and $Y_1, \ldots, Y_n$ be i.i.d.r.v. having the c.d.f. $G(x - \eta) = F[\theta(x - \xi)]$. Let $\hat{\xi} = \xi(X_1, \ldots, X_m)$ and $\hat{\eta} = \hat{\eta}(Y_1, \ldots, Y_n)$ be consistent estimators of ξ and η, such that $N^{1/2}(\hat{\xi} - \xi)$ and $N^{1/2}(\hat{\eta} - \eta)$ are bounded in probability (where m/N is bounded away from 0 and 1). Consider the centered observations $X_i^* = X_i - \hat{\xi}$, $i = 1, \ldots, m$ and $Y_i^* = Y_i - \hat{\eta}$, $i = 1, \ldots, n$.

(*a*) Show that Mood's test (cf. Exercise 3.6.6(*a*)) based on these centered observations is asymptotically distribution-free under H_0: $\theta = 1$.

(*b*) Show that Lehmann's test (cf. Exercise 3.3.16) is not distribution-free even asymptotically (under H_0: $\theta = 1$).

(*c*) Show that if (i) $f(x) = F'(x)$ and $g(x) = G'(x)$ are symmetric about the origin and (ii) $f(x)/\phi[\Phi^{-1}(F(x))]$ and $g(x)/\phi[\Phi^{-1}(G(x))]$ are bounded, the asymptotic distribution of the normal score statistic (cf. Exercise 3.6.6(*c*)) based on the centered observations and the same statistic based on the observations $X_i - \xi$, $Y_j - \eta$ are the same. Also show that if (ii) holds then $h(x)/\phi[\Phi^{-1}(H(x))]$ is bounded, where $H(x) = (m/N)F(x) + (n/N)G(x)$.

(Raghavachari, 1965a.)

3.6.11. Prove Corollary 3.6.4.1. (Puri and Sen, 1969b.)

3.6.12. Prove the assertions (6) to (10) of section 3.6.3. (Puri, 1964.)

3.6.13. Prove that the *C*-terms in Theorems 3.6.5 are all $o_p(N^{-1/2})$. (Puri, 1964.)

3.6.14. If for all i, $\lim_{N\to\infty} n_i/N = \lambda_i$ exists ($i = 1, \ldots, c$) and there exist c sequences of c.d.f.'s $\{F_i(x) = F_{N_i}(x)\}$ $i = 1, \ldots, c$ such that $F_{N_i}(x)$ converges to an absolutely continuous c.d.f. $F(x)$ at all points of continuity of the latter, then under the assumptions (and notations) of Theorem 3.6.5, the random vector $[n_1^{1/2}(T_{N,1} - \mu_{N,1}), \ldots, n_c^{1/2}(T_{N,c} - \mu_{N,c})]$ has a limiting normal distribution with a null mean vector and a covariance matrix with elements $A^2[\delta_{ij} - (\lambda_i\lambda_j)^{1/2}]$, where A^2 is defined in (3.8.20) and δ_{ij} is the Kronecker delta. (Puri, 1964.)

3.6.15. *Multisample location tests.* Let $X_{i1}, \ldots, X_{in_i}$ be i.i.d.r.v. having the c.d.f. $F_i(x)$ for $i = 1, \ldots, c$ and the null hypothesis to be tested relates to $F_1 = \cdots = F_c$ against the set of alternatives that they differ only by locations.

(*a*) *The Kruskal-Wallis test* is based on the statistic

$$H = \frac{12}{N(N+1)} \sum_{i=1}^{c} n_i\left(\bar{R}_i - \frac{N+1}{2}\right)^2, \qquad N = \sum_{1}^{c} n_i,$$

where $\bar{R}_i$ is the average rank of the n_i observations of the ith sample among the N pooled sample observations. Prove that if $F_i(x)$, $i = 1, \ldots, c$, are all absolutely continuous and (i) $F_i(x) = F(x + N^{-1/2}\theta_i)$, $i = 1, \ldots, c$, (ii) $\lim_{N\to\infty} n_i/N = \lambda_i$ $(0 < \lambda_i < 1)$ exists and (iii)

$$\lim_{N\to\infty} \int_{-\infty}^{\infty} N^{-1/2}[F(x + N^{-1/2}t) - F(x)]\, dF(x)$$

exists and is finite, then the limiting distribution of H is noncentral chi square with $c - 1$ degrees of freedom and the noncentrality parameter

$$\Delta_H = 12\left[\int_{-\infty}^{\infty} f^2(x)\, dx\right]^2 \sum_{k=1}^{c} \lambda_k(\theta_k - \bar{\theta})^2 \quad \text{where} \quad \bar{\theta} = \sum_{1}^{c} \lambda_k\theta_k.$$

(Kruskal, 1952; Andrews, 1954; Puri, 1964.)

(*b*) *Brown and Mood's median test.* (See Exercise 3.3.8.) Prove that under the assumptions (i) and (ii) of (*a*) and the existence of the density function at the population median (of F), the median test statistic has asymptotically a noncentrality parameter

$$\Delta_M = 4[F'(a)]^2 \sum_{i=1}^{c} \lambda_k(\theta_k - \bar{\theta})^2 \quad \text{where} \quad F(a) = \tfrac{1}{2}.$$

(Mood, 1954; Andrews, 1954.)

(*c*) *The normal score test* is based on the statistic

$$V = \sum_{i=1}^{c} n_i T_{N,i}^2 / A^2$$

where $T_{N,i}$ is given by (3.6.46).

Prove that if (i) and (ii) of (*a*) hold and if

$$\lim_{N\to\infty} N^{1/2} \int_{-\infty}^{\infty} \left[J\left\{\sum_1^c \lambda_k F(x + N^{-1/2}\{\theta_\alpha - \theta_i\})\right\} - J\{F(x)\}\right] dF(x)$$

exists and is finite, then $V^{(c)}$ has asymptotically a noncentral chi-square distribution with $c - 1$ D.F. and the noncentrality parameter

$$\Delta_V = \left[\int_{-\infty}^{\infty} \frac{d}{dx} J\{F(x)\}\, dF(x)\right]^2 \sum_1^c \lambda_k(\theta_k - \bar{\theta})^2,$$

where $J\{F(x)\} = \Phi^{-1}\{F(x)\}$, $\Phi(x)$ being the standardized normal c.d.f.

(Puri, 1964.)

3.6.16. *Multisample scale problem.* In the setup of Exercise 3.6.15, we want to test H_0: $F_1 = \cdots = F_c$ against scale alternatives. Define $T_{N,i}$, $i = 1, \ldots, c$ as in (3.6.46) and let $\mathcal{L}_N = \sum_{k=1}^c n_k(T_{N,k} - \bar{E}_N)^2/A^2$, where A^2 is given by (3.8.20), and $\bar{E}_N = 1/N \sum_{\alpha=1}^N E_{N,\alpha}$.

(*a*) *Mood's test.* Consider the scores $E_{N,\alpha} = (\alpha/(N+1) - \frac{1}{2})^2$, $\alpha = 1, \ldots, N$ and denote the corresponding $\mathcal{L}_N$ by $\mathcal{L}_N^{(1)}$. Show that if (i) $F_i(x) = F(x[1 + N^{-1/2}\nu_i])$, $i = 1, \ldots, c$, and (ii)

$$x \frac{d}{dx} J[F(x)] \quad \text{is bounded}$$

then $\mathcal{L}_N^{(1)}$ has asymptotically the noncentral chi-square distribution with $c - 1$ D.F. and the noncentrality parameter

$$\Delta_M = 180 \sum_{k=1}^{c} \lambda_k(\nu_k - \bar{\nu})^2 \left[\int_{-\infty}^{\infty} x f(x)[2F(x) - 1]\, dF(x)\right]^2,$$

where $\bar{\nu} = \sum_1^c \lambda_k \nu_k$, the λ_k's being the same as in Exercise 3.6.14.

(*b*) The normal score scale test is based on the statistic $\mathcal{L}_N$ with scores $E_{N,\alpha} = [\Phi^{-1}(\alpha/N + 1)]^2$, $\alpha = 1, \ldots, N$. Show that the normal score statistic has asymptotically the noncentral chi-square distribution with $c - 1$ D.F. and the noncentrality parameter

$$\Delta_v = 2 \sum_{k=1}^{c} \lambda_k(\nu_k - \bar{\nu})^2 \left[\int_{-\infty}^{\infty} \frac{x\Phi^{-1}(F(x)) f(x)\, dF(x)}{\phi[\Phi^{-1}(F(x))]}\right]^2.$$

(*c*) Also consider the extension of the test (*b*) in Exercise 3.6.7 to the c-sample case.

(*d*) Finally, as in the two-sample case consider the scale problem when the locations are unknown. Here also consider consistent estimates of locations and work with the observations centered at the respective estimated locations. Show that under assumptions similar to the ones in Exercise 3.6.10 the modified rank order test in (*a*) and (*b*) will also be asymptotically distribution-free. (Puri, 1965a, 1968.)

3.6.17. *An alternative class of multisample location and scale tests.* Let the first sample order statistic be denoted by $X_{1(0)} = -\infty, X_{1(1)}, \ldots, X_{1(n_1)}, X_{1(n_1+1)} = \infty$ and let us define the nonoverlapping and contiguous cells $\{I_j: j = 0, \ldots, n_1\}$ by $I_j: X_{1(j)} < x \leq X_{1(j+1)}, j = 0, \ldots, n_1$. Also let $r_{k,j}$ denote the number of observations of the kth sample belonging to the jth cell I_j, for $k = 1, \ldots, c, j = 0, \ldots, n_1$; so that

$$\sum_{j=0}^{n_1} r_{k,j} = n_k \quad \text{for} \quad k = 1, \ldots, c; \qquad r_{1,j} = \begin{cases} 1, & \text{for } j = 0, \ldots, n_1 - 1 \\ 0, & \text{for } j = n_1. \end{cases}$$

Finally let us define a sequence of real numbers $\{a(j, n), j = 0, \ldots, n\}$ for each positive integer n, and assume that this sequence satisfies the following three conditions.

(C.1) For each n and all $j = 0, \ldots, n$,

$$a(j, n) \leq K\left\{\frac{(j+1)(n+1-j)}{(n+1)^2}\right\}^{\delta-\frac{1}{2}}, \quad \text{for some} \quad \delta > 0, \quad K > 0.$$

Then, let

$$\bar{a}_n = \sum_{j=0}^{n} a(j, n)/(n+1), \qquad \sigma_{a_n}^2 = \sum_{j=0}^{n} \{a^2(j, n)/(n+1)\} - \bar{a}_n^2.$$

It can be shown that $\bar{a}_n$ and $\sigma_{a_n}^2$ are both bounded, even when $n \to \infty$.

(C.2) $\bar{a}_n$ as well as $\sigma_{a_n}^2$ converges to some finite limit as $n \to \infty$, and we denote by

$$\bar{a} = \lim_{n=\infty} \bar{a}_n, \qquad \sigma_a^2 = \lim_{n\to\infty} \sigma_{a_n}^2.$$

(C.3) For each finite n as well as when $n \to \infty$, $\sigma_{a_n}^2$ is positive, that is,

$$\sigma_{a_n}^2 > 0 \quad \text{for all} \quad n, \quad \text{and} \quad \sigma_a^2 > 0.$$

Condition (C.3) implies that $a(j, n)$ is not a constant. Let us now define

$$S_{N,k} = \sum_{j=0}^{n_1} a(j, n_1) r_{k,j}/n_k, \qquad k = 1, \ldots, c, \quad N = \sum_{k=1}^{c} n_k,$$

and

$$\bar{S}_N = \sum_{J=0}^{n_1} a(j, n_1)\left(\sum_{k=1}^{c} r_{k,j}\right) \Big/ N = \sum_{k=1}^{c} n_k S_{N,k}/N.$$

Then our proposed test is based on $T_N = \sum_{k=1}^{c} n_k [S_{N,k} - \bar{S}_N]^2/\sigma_{a_{n_1}}^2$.

(*a*) Show that under H_0: $F_1 = \cdots = F_c$, T_N has asymptotically the chi-square distribution with $c - 1$ d.f.

(*b*) For the sequence of alternatives $\{H_N\}$; H_N: $F_k(x) = F_{k,N}(x) = F(x + N^{-1/2}[\alpha_k + \beta_k x])$, $k = 1, \ldots, c$, show that T_N has asymptotically the noncentral chi-square distribution with $c - 1$ D.F., and the noncentrality parameter $\Delta = \sum_1^c \lambda_k(\theta_k - \bar{\theta})^2/\sigma_a^2$, where

$$\theta_k = \int_{-\infty}^{\infty} (\alpha_k + \beta_k x) b(F(x)) f^2(x)\, dx, \qquad k = 1, \ldots, c,$$

$\bar{\theta} = \sum_1^c \lambda_k \theta_k$ and $b(u) = \lim_{n_1 \to \infty} n_1 b(j, n_1)$ where $j/n_1 \to u: 0 < u < 1$ and $b(j, n_1) = a(j, n_1) - a(j + 1, n_1)$. Hence, or otherwise, show that the tests in Exercises 3.6.15 and 3.6.16 have the same noncentral chi-square distribution if $a(j, n_1)$ is made to correspond with $J_N(j/(n + 1))$ (i.e., the same weight function (but for different sample sizes) is used).

(Sen, 1963a; Sen and Govindarajulu, 1966.)

Section 3.7

3.7.1. Prove the assertion (3.7.5). (Hoeffding, 1951b.)

3.7.2. (*a*) Prove that the statistic τ_N given by (3.7.11) has asymptotically the normal distribution with mean

$$\mu_{\theta\phi}(\tau_N) = \int_{-\infty}^{\infty} J(H_{\theta\phi}(x))\, dG_\theta(x)$$

and variance

$$\sigma_{\theta\phi}^2(\tau_N) = \frac{2n}{N^2}\left\{ \iint\limits_{-\infty<x<y<\infty} G_\phi(x)[1 - G_\phi(y)] J'[H_{\theta\phi}(x)] J'[H_{\theta\phi}(y)]\, dG_\theta(x)\, dG_\theta(y) \right.$$

$$\left. + \frac{n}{m} \iint\limits_{-\infty<x<y<\infty} G_\theta(x)[1 - G_\theta(y)] J'[H_{\theta\phi}(x)] J'[H_{\theta\phi}(y)]\, dG_\phi(x)\, dG_\phi(y) \right\}$$

provided $\sigma_{\theta\phi}^2(\tau_N) \neq 0$ where

$$H_{\theta\phi}(x) = \frac{n}{N} G_\phi(x) + \frac{m}{n} G_\theta(x)$$

$$J(G_{\phi_0}(x)) = \left.\frac{\partial \log g_\theta(x)}{\partial \theta}\right|_{\theta=\phi_0}.$$

Under the hypothesis H_0: $\theta = \phi = \phi_0$

$$\sigma_{\phi_0}^2(\tau_N) = \frac{n}{mN} A^2$$

where A^2 is given by (3.8.20),

(*b*) Prove that $\sigma_{\theta\phi}^2(\tau_N)$ is continuous at the point (ϕ_0, ϕ_0) uniformly in N. (Capon, 1961.)

3.7.3. (*a*) Prove that the Wilcoxon two-sample test is (i) l.m.p.r.t. when $G_\theta(x) = 1/(1 + e^{-(x-\theta)})$, (ii) unbiased against one-sided alternatives.

(*b*) Prove that the similar properties hold for the one-sample Wilcoxon test. (Lehmann, 1953; Capon, 1961.)

3.7.4. (*a*) Prove that the two-sample normal scores test is (i) l.m.p.r.t. against normal alternatives, (ii) unbiased against one-sided alternatives.

(*b*) Prove that similar properties hold for the one-sample normal scores test. (Capon, 1961.)

3.7.5. Let $X_1, \ldots, X_N$ be a sample from the double exponential distribution with density $\frac{1}{2}e^{-|x-\theta|}$. Prove that for testing $\theta \leq 0$ against $\theta > 0$, the sign test is (i) locally most powerful, and (ii) unbiased. (Capon, 1961.)

Section 3.8

3.8.1. Show that if the number of quantiles is increased with the sample sizes, the quantile test in Exercise 3.3.10 is consistent against any divergence of the *c* distribution. (Sen, 1962a.)

3.8.2. Study the consistency of the run test in Exercise 3.3.13. (Wolfowitz, 1946.)

3.8.3. Show that the Wilcoxon test (in Exercise 3.3.14) is consistent against the set of alternatives that $P\{X < Y\} \neq \frac{1}{2}$ (or $>$ or $<\frac{1}{2}$). (Dantzig, 1951.)

3.8.4. Show that the rank order tests (considered in Exercises 3.6.5 through 3.6.14) are all consistent.

3.8.5. Let (i) F be a continuous c.d.f., differentiable in each of the open intervals $(-\infty, a_1), (a_1, a_2), \ldots, (a_{s-1}, a_s), (a_s, \infty)$, with the derivative of F bounded in each of these intervals, and (ii) the function $(d/dx)\{J[F(x)]\}$ be bounded as $x \to \pm\infty$; then

$$\lim_{N\to\infty} N^{1/2}[\mu_N(\theta) - \mu_N(0)] = \int_{-\infty}^{\infty} \theta \frac{d}{dx}\{J[F(x)]\}\, dF(x)$$

where $\mu_N(\theta)$ is the same as μ_N in (3.6.4) with $G(x) = F(x - \theta N^{-1/2})$. (Hodges and Lehmann, 1961; Puri, 1964.)

3.8.6. (*a*) Show that for any absolutely continuous distribution function $F(x)$ with variance σ^2, the A.R.E. of the sign test with respect to the t test is $4\sigma^2\{f(0)\}^2$ where f is the density of $F(x)$. When F is a $N(0, 1)$ distribution, this reduces to $2/\pi$.

(*b*) Show that if F possesses a unimodal density in the sense that $0 \leq |x| \leq |x'| \Rightarrow f(x') < f(x)$, then $4\sigma^2\{f(0)\}^2 \geq \frac{1}{3}$. (Hodges and Lehmann, 1956.)

3.8.7. (*a*) Show that for the two-sample location problem (Exercise 3.6.5), if the c.d.f. $F(x)$ has a continuous density function $f(x)$ and a finite variance σ^2, then the A.R.E. of the Wilcoxon's test with respect to Student's t-test is

$$e_{W,t} = 12\sigma^2\left\{\int_{-\infty}^{\infty} f^2(x)\, dx\right\}^2.$$

(*b*) Prove that $e_{W,t}$ is also the A.R.E. of the Wilcoxon's signed-rank test (cf. Exercise 3.6.6(*b*)) with respect to Student's t-test.

(*c*) Show that $e_{W,t}$ is also the A.R.E. of the Kruskal-Wallis H test (cf. Exercise 3.6.15(*a*)) with respect to the one-way analysis of variance test.

(*d*) Prove that $e_{W,t} \geq 0.864$, for all F, and the lower bound is attained for the density $f(x) = b(a^2 - x^2)$ for $-a \leq x \leq a$ and $f(x) = 0$ otherwise. (Andrews, 1954; Hodges and Lehmann, 1956.)

3.8.8. Show that the quantile tests in Exercises 3.3.8 and 3.3.10 are asymptotically power equivalent for the same local alternatives provided the same set of quantiles is used in both cases. (Sen, 1962a.)

3.8.9. (*a*) For the two-sample location problem, if the parent c.d.f. $F(x)$ possesses a continuous density function $f(x)$ and a finite variance σ^2, show that the A.R.E. of the normal score test with respect to the t-test is

$$e_{V,t} = \sigma^2\left[\int_{-\infty}^{\infty} \frac{d}{dx}\,\Phi^{-1}\{F(x)\}\,dF(x)\right]^2.$$

(*b*) Show that $e_{V,t}$ is also the A.R.E. of the one-sample normal score test with respect to the t-test.

(*c*) Also show that $e_{V,t}$ is the A.R.E. of the c sample normal score test with respect to the one-way analysis of variance test.

(Chernoff and Savage, 1958; Puri, 1964; Puri and Sen, 1969b.)

3.8.10. (*a*) With the aid of Exercises 3.8.7 and 3.8.9 show that the A.R.E. of the Wilcoxon's test with respect to the normal score test is

$$e_{W,V} = 12\left[\int_{-\infty}^{\infty} f^2(x)\,dx\right]^2 \Big/ \left[\int_{-\infty}^{\infty} \frac{f^2(x)}{\phi[\Phi^{-1}(F(x))]}\,dx\right]^2.$$

(*b*) Show that $e_{W,V} \leq 6/\pi = 1.91$ for all F.

(*c*) Compute $e_{W,V}$ for the case when F is (i) rectangular, (ii) exponential, (iii) normal, (iv) logistic, (v) double exponential, and (vi) Cauchy.

(Hodges and Lehmann, 1961.)

3.8.11. (*a*) Show that for the two-sample scale problem (Exercises 3.6.7(*a*) and 3.6.10(*a*)), if the c.d.f. $F(x)$ has a continuous density function $f(x)$ and a finite variance σ^2, then the A.R.E. of the Mood's test with respect to the F test is

$$e_{M^*,F} = 45(\nu_2 + 2)\left(\int_{-\infty}^{\infty} xf(x)[2F(x) - 1]\,df(x)\right)^2$$

where

$$\nu_2 = [E[x - Ex]^4/\{E(x - Ex)^2\}^2] - 3$$

(*b*) Show that $e_{M^*,F}$ is also the A.R.E. of the c-sample Mood's test with respect to the Bartlett likelihood ratio test for the homogeneity of variances.

(Puri, 1965a, 1968.)

3.8.12. (*a*) Show that for the two-sample scale problem (Exercise 3.6.7(*b*)), the A.R.E. of the simplest rank order test with respect to the F test is

$$e_{B,F} = 12(\nu_2 + 2)\left(\int_{-\infty}^{0} xf^2(x)\,dx - \int_{0}^{\infty} xf^2(x)\,dx\right)^2.$$

(*b*) Show that $e_{B,F}$ is also the A.R.E. of the c-sample simplest rank test with respect to the Bartlett likelihood ratio test for the homogeneity of variances. (Ansari and Bradley, 1960; Puri, 1965a.)

3.8.13. Show that (cf. Exercise 3.3.22), the A.R.E. of the two-sample Sukhatme test with respect to the F test is

$$e_{S,F} = \tfrac{720}{61}(\nu_2 + 2)\left(2\int_{-\infty}^{\infty} xF(x)f^2(x)\,dx - \int_{-\infty}^{\infty} xf^2(x)\,dx\right)^2.$$

In particular, if $f(x) = e^{-x^2/2}/\sqrt{2\pi}$, $e_{S,F} = 0.69$; if

$$f(x) = 1(-\tfrac{1}{2} \le x \le \tfrac{1}{2}), \qquad e_{S,F} = 0.80;$$

and if

$$f(x) = \tfrac{1}{2}e^{-|x|}, \qquad e_{S,F} = 1.03.$$

(Sukhatme, 1958.)

3.8.14. (*a*) Show that for the two-sample scale problems (Exercises 3.6.7(*d*) and 3.6.10(*c*)), the A.R.E. of the normal scores test with respect to the F-test is

$$e_{D,F} = \left(\int \frac{x\Phi^{-1}[F(x)]f(x)\,dF(x)}{\phi[\Phi^{-1}[F(x)]]}\right)^2.$$

(*b*) Show that $e_{D,F}$ is also the A.R.E. of the c-sample normal scores test with respect to Bartlett's likelihood ratio test for the homogeneity of variances. (Puri, 1965a.)

(*c*) Show that $e_{D,F}$, which takes the value 1 if F is the standard cumulative normal distribution function, can otherwise take any value between 0 and ∞. (Klotz, 1962; Raghavachari, 1965b.)

3.8.15. *Efficiency comparisons for different densities of c-sample scale tests.* Prove the results given in the accompanying table.

Density	$e_{B,D}$	$e_{M^*,D}$	e_{B,M^*}
Exponential	0	0	0.600
Rectangular	0	0	0.806
Normal	0.608	0.760	0.800
Logistic	0.750	0.896	0.837
Double exponential	0.744	0.900	0.860
Cauchy	1.783	1.670	1.068

3.8.16. Let X be a random variable with a continuous symmetric distribution $F(x; \theta)$ with mean θ, and let $X_1, \ldots, X_n$ be a sample from this distribution. Let $U_1, \ldots, U_n$ be the ordered absolute values of the observations. Let, for for $i = 1, \ldots, n$; let $Z_{n,i}$ be defined by

$$Z_{n,i} = \begin{cases} 1, & \text{if } U_i \text{ corresponds to a positive observation,} \\ -1, & \text{if } U_i \text{ corresponds to a negative observation,} \end{cases}$$

and consider tests for H_0: $\theta = 0$ based on test statistics of the form

$$t_n = \sum_{i=1}^{n} a_{n,i} Z_{n,i}, \qquad t'_n = \sum_{i=1}^{n} a'_{n,i} Z_{n,i}$$

(where $a_{n,i}$ and $a'_{n,i}$ are given functions of i and n) and

$$\bar{X}_n = n^{-1} \sum_{i=1}^{n} X_i = n^{-1} \sum a_{n,i} U_i.$$

Prove that the correlation coefficient under H_0 of t_n and t'_n is

$$\rho(t_n, t'_n; H_0) = \left(\sum_{i=1}^{n} a_{n,i} a'_{n,i} \Big/ \left[\sum_{i=1}^{n} a^2_{n,i} \sum_{i=1}^{n} a'^2_{n,i} \right]^{1/2} \right)$$

and the correlation coefficient under H_0 of t_n and $\bar{X}_n$ is

$$\rho(t_n, \bar{X}_n; H_0) = \left(\sum_{i=1}^{n} a_{n,i} E U_i / \sigma \left[n \sum_{i=1}^{n} a^2_{n,i} \right]^{1/2} \right)$$

where E denotes the expectation and σ^2 is the variance of X.

(van Eeden, 1963.)

3.8.17. (Continued.) (*a*) Let X have a normal distribution with mean θ and variance 1. The best tests in this case are the tests based on the mean and the normal scores test. Using Theorem 3.8.3, show that the asymptotic relative efficiency of the sign test with respect to the normal scores test or the t-test is $2/\pi$.

(*b*) Let X have a double exponential distribution with density $\frac{1}{2}e^{-|x-\theta|}$. The best test in this case is the sign test. Show that the A.R.E. of the best test with respect to the normal scores test is $\pi/2$. (van Eeden, 1963.)

3.8.18. Consider the two-sample location problem as stated in Exercise 3.6.5. Show that no invariant test of $\delta = 0$ vs $\delta = \delta_N$ can have greater efficacy than the likelihood ratio test for testing $\delta = 0$ vs $\delta = \delta_N$ (for the case where the likelihood ratio test is informative). (Chernoff and Savage, 1958.)

3.8.19. (Continued.) Show that the efficacy of the likelihood ratio test L of $\delta = 0$ vs $\delta = \delta_N$ (cf. Theorem 3.2.1) is given by

$$E = E_{\phi_0} \left(\frac{\partial \log g_\theta(x)}{\partial \theta} \bigg|_{\theta = \phi_0} \right)^2.$$

(Capon, 1961.)

3.8.20. (Continued.) *Asymptotically efficient test.* We shall say that a test is asymptotically efficient for the sequence of alternatives $\{\delta_N\}$, if its efficacy achieves the upper bound (E). Using this definition show that the locally most powerful rank order tests considered in Theorem 3.7.1 are asymptotically efficient. (Capon, 1961.)

3.8.21. For the one-sample location problem, obtain the exact Bahadur efficiency of the sign test with respect to the t-test. (Bahadur, 1960b.)

PRINCIPAL REFERENCES

Abrahamson (1967), Andrews (1954), Ansari and Bradley (1960), Bahadur (1960a, b, c, 1967), Bahadur and Ranga Rao (1960), Barton and Mallows (1965), Berk (1966), Bhapkar (1961b), Bhuchongkul (1964), Bhuchongkul and Puri (1965), Capon (1961), Chanda (1963), Chatterjee (1966a), Chatterjee and Sen (1964), Chernoff (1952), Chernoff and Savage (1958), Cramér (1946), Dantzig (1951), Darling (1957), David and Barton (1958), Dwass (1953, 1955b, 1956), Fillipova (1961), Fisher (1935), Fisher and Yates (1938), Fraser (1951, 1953, 1954, 1956, 1957a, 1957b), Gastwirth and Wolff (1968), Ghosh and Sen (1970), Gleser (1964), Govindarajulu (1960), Govindarajulu, Le Cam, and Raghavachari (1967), Hájek (1961, 1962, 1968a, b, 1969), Hájek and Šidák (1967), Halmos (1946), Hoadley (1967), Hodges and Lehmann (1956, 1961, 1963), Hoeffding (1948a, 1948b, 1951a, b, 1952, 1953, 1965, 1966, 1968), Hoeffding and Rosenblatt (1955), Hotelling and Pabst (1936), Hušková (1970), Jurečková (1967, 1969), Kendall (1938), Klotz (1962), Kolmogorov (1933), Kruskal and Wallis (1952), Kuiper (1960), Lehmann (1951a, 1953, 1963), Lehmann and Stein (1949), Madow (1948), Mann and Whitney (1947), Massey (1951a, b, c), Mathisen (1943), Mehra and Puri (1967), Mikulski (1963), Mood (1949, 1950, 1954), Motoo (1957), Nandi and Sen (1963), Noether (1949a, b, 1955), Owen (1962), Pearson (1900), Pitman (1937a, 1937b, 1938, 1948), Puri (1964, 1965a, 1965b, 1965c, 1967, 1968), Puri and Sen (1966, 1967a, 1967b, 1968a, 1969a, b) Putter (1955), Pyke and Shorack (1968a, b), Raghavachari (1965a, b) Rao (1965), Sanov (1957), Savage (1956), Scheffé (1943b), Sen (1960, 1962a, 1963a, 1965a, 1965b, 1966c, 1967e, 1970a, e), Sen and Ghosh (1971), Sen and Govindarajulu (1966), Sen and Puri (1967), Sethuraman (1964), Siegel and Tukey (1960), Sproule (1969), Sukhatme (1956, 1957, 1958), Sukhatme and Sethuraman (1959), Tamura (1960), Terry (1952), Tukey (1948, 1949), van der Waerdon (1952/53, 1953a), van Eeden (1963), von Mises (1947), Wald (1943), Wald and Wolfowitz (1941, 1944), Walsh (1946, 1949, 1957), Welch (1937, 1938), Wilcoxon (1945, 1949), Wilks (1941, 1961), Wolfowitz (1946).

CHAPTER 4

Rank Tests For The Multivariate Single-Sample Location Problems

4.1. INTRODUCTION

In this chapter, we consider the following two hypotheses testing problems. First, we may desire to test whether several (vector-valued) random variables have specified median vectors. Such tests should be valid for distributions not necessarily all identical nor symmetric about the respective median vectors. Second, we may desire to test whether the sample observations are drawn from (unknown) distributions with a specified median vector, assuming of course that all the distributions are diagonally symmetric about their respective median vectors. For the first problem, multivariate generalizations of the univariate sign test are considered. For the second problem, a general class of rank order tests is considered and it includes the multivariate generalizations of the Wilcoxon signed rank test and the normal scores test (cf. Exercise 3.6.5) as special cases. The large-sample power and power efficiency of this class of rank order tests are also studied. The development of all these multivariate tests depends on the formulation of suitable hypotheses of sign invariance and these characterize the existence of (conditionally) distribution-free tests for both the problems. The sign-invariant permutation distribution theory of U-statistics (cf. section 3.2) is also developed and incorporated in the construction of suitable permutation tests based on U-statistics.

For convenience of presentation, we shall start with a simple bivariate sign test of location by Chatterjee (1966b). More complicated sign tests are also referred to subsequently, and these are followed by a general class of rank order tests.

4.2. A BIVARIATE SIGN TEST FOR LOCATION

Let $\mathbf{X}_\alpha = (X_{1\alpha}, X_{2\alpha})$, $\alpha = 1, \ldots, n$, be n independent stochastic vectors having continuous cumulative distribution functions (c.d.f.'s) $F_1(\mathbf{x}), \ldots, F_n(\mathbf{x})$, $\mathbf{x} \in R^2$, respectively. The problem is to test whether $F_1, \ldots, F_n$ have

n specified pairs of (marginal) medians (assumed to be uniquely defined). By suitably choosing the origins, we may assume that the pair of hypothetical medians for each $\mathbf{X}_\alpha$ is $\mathbf{0} = (0, 0)$, $\alpha = 1, \ldots, n$. Then the null hypothesis is

$$H_0\colon F_\alpha(0, \infty) = F_\alpha(\infty, 0) = \tfrac{1}{2}, \quad \text{for all} \quad \alpha = 1, \ldots, n, \tag{4.2.1}$$

where $F_1, \ldots, F_n$ are otherwise arbitrary. For each $\mathbf{X}_\alpha$, the events (i) $X_{1\alpha} \leq 0$, $X_{2\alpha} \leq 0$, and (ii) $X_{1\alpha} > 0$, $X_{2\alpha} > 0$ are called *concordance* of the first and second kind, and the events (iii) $X_{1\alpha} \leq 0$, $X_{2\alpha} > 0$ and (iv) $X_{1\alpha} > 0$, $X_{2\alpha} \leq 0$ as *discordance* of the first and second kinds respectively, $\alpha = 1, \ldots, n$. Also, let γ_α be the probability of concordance of $(X_{1\alpha}, X_{2\alpha})$, and assume that

$$0 < \gamma_\alpha < 1 \quad \text{for all} \quad \alpha = 1, \ldots, n. \tag{4.2.2}$$

Finally, let us denote for X_α, the conditional probability of a concordance (discordance) of the first kind given concordance (discordance) by $\theta_\alpha(\tau_\alpha)$, for $\alpha = 1, \ldots, n$. Then, (4.2.1) may equivalently be written as

$$H_0\colon \theta_\alpha = \tau_\alpha = \tfrac{1}{2} \quad \text{for all} \quad \alpha = 1, \ldots, n. \tag{4.2.3}$$

The sign test to be considered below is based on the following principle. Among the n observations $\mathbf{X}_\alpha$, $\alpha = 1, \ldots, n$, let $C_i(D_i)$ be the number of concordances (discordances) of the ith kind, $i = 1, 2$. Also, let $C = C_1 + C_2$ and $D = D_1 + D_2$. Then $C + D = n$. Under (4.2.3), C_1 and C_2 (as well as D_1 and D_2) should be stochastically equal whereas if (4.2.1) does not hold, (4.2.3) cannot hold, and hence, $C_1 - C_2$ or $D_1 - D_2$ will be stochastically different from zero. So, it is suggested that the test may be based on $C_1 - C_2$ and $D_1 - D_2$. However, even when (4.2.3) holds, the joint distribution of (C_1, C_2, D_1, D_2) depends on F_α, $\alpha = 1, \ldots, n$ through the unknown values of $F_\alpha(0, 0)$, $\alpha = 1, \ldots, n$ (cf. Exercise 4.2.1). Therefore the following conditional probability law is used to construct a test which is conditionally distribution-free. We consider the conditional distribution of C_1, C_2, D_1, and D_2, given which pairs among $(X_{1\alpha}, X_{2\alpha})$, $\alpha = 1, \ldots, n$ are concordant and which discordant. Let $c(0 \leq c \leq n)$ be any integer. We consider any partition

$$(i_1, \ldots, i_c), \quad (i_{c+1}, \ldots, i_n) \quad \text{where} \quad i_1 < \cdots < i_c, \quad i_{c+1} < \cdots < i_n, \tag{4.2.4}$$

of the set of numbers $1, \ldots, n$ into two disjoint subsets containing c and $n - c$ numbers respectively. (If $c = 0$ or n, the first or second subset is of course empty.) Let $\mathcal{E}_{i_1 \ldots i_c}$ be the event that among $\mathbf{X}_1, \ldots, \mathbf{X}_n$, the i_1th, $\ldots$, i_cth pairs are concordant and the rest are discordant. Then

$$P\{\mathcal{E}_{i_1 \ldots i_c}\} = \gamma_{i1} \cdots \gamma_{ic}(1 - \gamma_{i_{c+1}}) \cdots (1 - \gamma_{i_n}). \tag{4.2.5}$$

The probability that $\mathcal{E}_{i_1\ldots i_c}$ will occur and there will be just c_1 concordances of the first kind and just d_1 discordances of the first kind $(0 \le c_1 \le c, 0 \le d_1 \le n - c)$ is

$$(4.2.6)\quad \gamma_{i_1}\cdots\gamma_{i_c}(1-\gamma_{i_{c+1}})\cdots(1-\gamma_{i_n})\cdot\{(1-\theta_{i_1})\cdots(1-\theta_{i_c}) \Sigma_1 \Pi_1 [\theta_{i_j}/(1-\theta_{i_j})]\}\cdot\{(1-\tau_{i_{c+1}})\cdots(1-\tau_{i_n}) \Sigma_2 \Pi_2 [\tau_{i_j}/(1-\tau_{i_j})]\}$$

where Π_1 denotes a product over a subset of c_1 of the values $1, 2, \ldots, c$ of j, and Σ_1 denotes the sum over all the $\binom{c}{c_1}$ such subsets; similarly, Π_2 denotes a product over a subset of d_1 of the values $c + 1, \ldots, n$ of j, and Σ_2 denotes the sum over all the $\binom{n-c}{d_1}$ such subsets. From (4.2.5) and (4.2.6) the required conditional distribution of C_1 and D_1 is given by

$$(4.2.7)\quad P\{C_1 = c_1, D_1 = d_1 \mid \mathcal{E}_{i_1\ldots i_c}\} = (1-\theta_{i_1})\cdots 1-\theta_{i_c}) \Sigma_1 \Pi_1 [\theta_{i_j}/(1-\theta_{i_j})]\}\cdot\{(1-\tau_{i_{c+1}})\cdots(1-\tau_{i_c}) \Sigma_2 \Pi_2 [\tau_{i_j}/(1-\tau_{i_j})]\},$$
$$0 \le c_1 \le c, \qquad 0 \le d_1 \le n - c.$$

Under H_0 given by (4.2.3), whatever be $(F_1, \ldots, F_n)$, (4.2.7) gives

$$(4.4.8)\quad P\{C_1 = c_1, D_1 = d_1 \mid \mathcal{E}_{i_1\ldots i_c}, H_0\} = P\{C_1 = c_1, D_1 = d_1 \mid C = c, H_0\} = \binom{c}{c_1}\binom{n-c}{d_1}2^{-n},$$
$$0 \le c_1 \le c, \qquad 0 \le d_1 \le n - c.$$

Thus, under H_0, given $C = c$, C_1 and D_1 are independently distributed as binomial random variables with parameters $(c, \frac{1}{2})$ and $(n - c, \frac{1}{2})$ respectively. Hence, for testing H_0, it seems reasonable to use the statistic

$$(4.2.9)\quad T = (4/C)(C_1 - C/2)^2 + [4/(n - C)][D_1 - (n - C)/2]^2.$$

(For $C = 0$ or n, one of the terms in T is absent.) Given $C = c$, the conditional distribution of T under H_0 would be clearly distribution-free. Let $F_n(t \mid c)$ denote the c.d.f. of this distribution as obtained by summing (4.2.8) over those combinations (c_1, d_1) for which the value of T does not exceed t. For any $\epsilon(0 < \epsilon < 1)$ let $t_{n,\epsilon}(c)$ be the value of t for which $1 - F_n(t \mid c) \le \epsilon < 1 - F_n(t - 0 \mid c)$. We then define a critical function $\phi(t, c)$ which assumes the value 0 for $t < t_{n,\epsilon}(c)$ and 1 for $t > t_{n,\epsilon}(c)$. For $t = t_{n,\epsilon}(c)$ we take $\phi(t,c) = a_{n,\epsilon}(c)$ where $0 \le a_{n,\epsilon}(c) < 1$ is chosen such that

$$(4.2.10)\quad E\{\phi(T, C) \mid C = c, H_0\} = \epsilon.$$

It follows then from the results of chapter 3 (cf. section 3.3] that $\phi(T, c)$ is a strictly size ϵ test for H_0 in (4.2.3). This is a conditional test and is randomized in nature. A nonrandomized test may be as follows:

$$\text{(4.2.11)} \qquad \text{reject } H_0 \text{ if } T > t_{n,\epsilon}(c), \quad \text{otherwise} \quad \text{accept } H_0;$$

whatever be $F_1, \ldots, F_N$, the size of this test $\leq \epsilon$.

Although the above development of the test is based on heuristic reasoning, it is of interest to note that, from considerations of invariance and sufficiency, we are led to a class of similar tests. The problem of testing the hypothesis (4.2.3) remains invariant under all transformations $X'_{1\alpha} = g_\alpha^{(1)}(X_{1\alpha})$, $X'_{2\alpha} = g_\alpha^{(2)}(X_{2\alpha})$, $\alpha = 1, \ldots, n$, with strictly increasing continuous functions $g_\alpha^{(i)}(x)$ for which $g_\alpha^{(i)}(0) = 0$, $i = 1, 2$, and all permutations of the n pairs $(X_{1\alpha}, X_{2\alpha})$, $\alpha = 1, \ldots, n$. Therefore, if we restrict ourselves to invariant tests, we need consider only tests based on the set of maximal invariants C_1, D_1, C. The joint distribution of these is obtained by summing (4.2.6) over all the $\binom{n}{c}$ partitions (4.2.4). From this, the power of any critical function $\psi(c_1, d_1, c)$ is seen to be continuous in the parameters, so that a test $(c_1, d_1 x)$ for (4.2.3) is unbiased only if it is similar. Now it may be noted that C is a sufficient statistic for the family of distributions of C_1, D_1, and C under H_0. Further, as the induced family of distributions of C includes all binomial distributions with parameter n, the statistic C may be shown to be complete (cf. Lehmann, 1959; p. 131). Thus $\psi(c_1, d_1, c)$ would give a similar level ϵ test for H_0, if and only if

$$\text{(4.2.12)} \quad E\{\psi(C_1, D_1, C) \mid C = c, H_0\} = \epsilon \quad \text{for all} \quad c, \qquad 0 \leq c \leq n.$$

Thus we are led to consider the class of critical functions which satisfy (4.2.12). By (4.2.10), the critical function $\phi(t, c)$ belongs to this class. The choice of the particular critical function $\phi(t, c)$ is made from considerations of practical convenience and is to some extent justified by its properties of unbiasedness and consistency.

To apply these tests in practice, we require knowledge of $t_{n,\epsilon}(c)$ and $a_{n,\epsilon}(c)$. For small n, these may be found easily with the help of binomial tables. We now show that, for large n, we may approximate $F_n(t \mid c)$ by the c.d.f. $F_{\chi_2^2}(t)$ of the χ_2^2 distribution and use the upper ϵ point $\chi^2_{2,\epsilon}$ of the latter for $t_{n,\epsilon}(c)$.

From (4.2.8) and (4.2.9), by the DeMoivre-Laplace limit theorem, we get that $F_n(t \mid c) \to F_{\chi_2^2}(t)$ if c and $n - c$ both tend to infinity. Now, writing $C = \sum_{\alpha=1}^n Z_\alpha$, $n - C = \sum_{\alpha=1}^n (1 - Z_\alpha)$ where Z_α assumes the value 1 or 0 according as $(X_{1\alpha}, X_{2\alpha})$ is concordant or discordant, by Kolmogorov's three-series criterion (cf. Theorem 2.4.4) we see that, as $n \to \infty$, if

$$\sum_{\alpha=1}^{\infty} \gamma_\alpha = \infty, \qquad \sum_{\alpha=1}^{\infty} (1 - \gamma_\alpha) = \infty, \tag{4.2.13}$$

both C and $N - C$ tend to ∞ almost surely. Also, (4.2.2) implies (4.2.13). Hence, assuming (4.2.13), we have

$$t_{n,\epsilon}(C) \xrightarrow{\text{a.s.}} \chi^2_{2,\epsilon}, \qquad a_{n,\epsilon}(C) \xrightarrow{\text{a.s.}} 0. \tag{4.2.14}$$

Therefore for large n, the randomized as well as the nonrandomized test reduces to the test:

$$\text{reject } H_0 \text{ if } T > \chi^2_{2,\epsilon}; \quad \text{otherwise} \quad \text{accept } H_0. \tag{4.2.15}$$

The test is shown by Chatterjee (1966b) to be consistent and unbiased against certain alternatives. We leave these as Exercises 4.2.2 and 4.2.3.

In the particular case when $\mathbf{X}_\alpha$, $\alpha = 1, \ldots, n$ are all identically distributed, the statistic T defined in (4.2.9) is also proposed by Bennett (1964) for testing the hypothesis that the common median vector is a null vector. His test is, however, an asymptotically distribution-free test based on the rejection rule (4.2.15). Chatterjee's test not only generalizes Bennett's test for distributions not necessarily all identical but also exhibits its unbiasedness and distribution freeness under a simple conditional approach. The asymptotic power properties of Chatterjee's sign test will be considered in a later section. In passing, we may remark that Hodges (1955) has considered an association invariant sign test on which some further work has been done by Klotz (1964), and Joffe and Klotz (1962). The test is strictly distribution-free but neither the null nor the non-null distribution (for any suitable sequence of alternative hypotheses) of the test statistic appears to be simple. The null distribution is of course tabulated by Joffe and Klotz (1962) for sample sizes up to 30. But there is little scope for studying the Pitman efficiency of such a test. The only study of the efficiency of this test was made by Klotz (1962), and he was concerned with the Bahadur efficiency. Further, since it is a sign test, it is anticipated that as in the univariate case, for nearly normal c.d.f.'s, its efficiency may be appreciably low. Blumen (1958) has also considered an association invariant sign test which is subject to the same criticisms as the Hodges sign test. These sign tests will not be considered in this book. The multivariate generalization of Chatterjee's sign test, actually based on certain U-statistics, is considered in section 4.6.

4.3. THE HYPOTHESIS OF SIGN INVARIANCE AND THE BASIC PERMUTATION PRINCIPLE

Consider now the second problem. Let $\mathbf{X}_\alpha = (X_{1\alpha}, \ldots, X_{p\alpha})'$, $\alpha = 1, \ldots, n$ $(p \geq 1)$, be n independent random variables having continuous c.d.f.'s $F_\alpha(\mathbf{x})$, $\mathbf{x} \in R^p$, $\alpha = 1, \ldots, n$, respectively. We shall write

$$(4.3.1) \quad F_\alpha(\mathbf{x}) = F_\alpha(\mathbf{x}; \boldsymbol{\theta}), \qquad \boldsymbol{\theta} = (\theta_1, \ldots, \theta_p)', \qquad \alpha = 1, \ldots, n,$$

where $\boldsymbol{\theta}$ stands for a location (vector) parameter. In this section, we assume that $F_\alpha(\mathbf{x}; \boldsymbol{\theta})$ is diagonally symmetric about $\boldsymbol{\theta}$, for all $\alpha = 1, \ldots, n$; the diagonal symmetry is defined as follows:

If for a random variable $\mathbf{Z}$ ($\in R^p$), the vectors $\mathbf{Z}$ and $(-1)\ \mathbf{Z}$ both have the common c.d.f. $F(\mathbf{x})$, $x \in R^p$, then $F(\mathbf{x})$ is said to be diagonally symmetric about $\mathbf{0}$. If $F(\mathbf{x})$ is absolutely continuous having a continuous density function $f(\mathbf{x})$, then diagonal symmetry of $F(\mathbf{x})$ around $\mathbf{0}$ is equivalent to

$$(4.3.2) \qquad f((-1)\mathbf{x}) = f(\mathbf{x}) \quad \text{for all} \quad \mathbf{x} \in R^p.$$

Under this setup, we frame the null hypothesis as

$$(4.3.3) \quad H_0\colon \boldsymbol{\theta} = \mathbf{0}, \quad \text{where} \quad F_1, \ldots, F_n \text{ are all diagonally symmetric about } \boldsymbol{\theta}.$$

We are interested in the alternative hypothesis

$$(4.3.4) \quad K\colon \boldsymbol{\theta} \neq \mathbf{0}, \quad F_\alpha \text{ is diagonally symmetric about } \boldsymbol{\theta}, \quad \alpha = 1, \ldots, n.$$

In the particular case of $p = 1$, and $F_1 = \cdots = F_N = F$, we have considered some rank order tests for H_0 in (4.3.3) in section 3.6.2. Here, we considered the tests for the general case $p \geq 1$ and where F_α's are not necessarily identical. In the univariate case, under the null hypothesis (4.3.3), the distributions of suitable rank tests (cf. section 3.6.2) do not depend on the unknown (common) c.d.f. F. But this does not hold in the multivariate case, unless $F_\alpha(\mathbf{x})$ is equal to the product of the marginal c.d.f.'s, for all $\alpha = 1, \ldots, n$ (cf. Exercise 4.3.1). To overcome this difficulty, we consider the basic sign invariance principle which leads to conditionally distribution-free tests.

Let us denote the sample point by

$$(4.3.5) \qquad \mathbf{Z}_n = (\mathbf{X}_1, \ldots, \mathbf{X}_n),$$

and the sample space of $\mathbf{Z}_n$ by $\mathcal{Z}_n$. Consider the following finite group $\mathcal{G}_n$ of transformations $\{g_n\}$ given by

$$(4.3.6) \quad g_n\mathbf{Z}_n = ((-1)^{j_1}\mathbf{X}_1, \ldots, (-1)^{j_n}\mathbf{X}_n); \qquad j_\alpha = 0, 1; \qquad \alpha = 1, \ldots, n,$$

where $(-1)\mathbf{X} = (-X_1, \ldots, -X_n)$. Thus $\mathcal{G}_n$ has 2^n distinct elements $\{g_n\}$. Now, under H_0 in (4.3.3), the joint distribution of $\mathbf{Z}_n$ remains invariant under the group of transformations $\mathcal{G}_n$, which maps the sample space onto itself. Hence, for any point $\mathbf{z}_n \in Z_n$, there exists a set $S(\mathbf{z}_n)$ of 2^n points obtained by applying the group of transformations $\mathcal{G}_n$ in (4.3.6), such that when (4.3.3) holds, the conditional distribution of $\mathbf{Z}_n$, given $\mathbf{Z}_n \in S(\mathbf{z}_n)$ is uniform over the 2^n (conditionally equally likely) realizations. This leads to a permutational (conditional) probability measure $\mathcal{T}_n$ which attaches equal probability mass (i.e., 2^{-n}) to all the 2^n elements of $S(\mathbf{z}_n)$.

Let us now consider a test function $\phi(\mathbf{Z}_n)$, chosen in such a manner that

$$(4.3.7) \quad E\{\phi(\mathbf{Z}_n) \mid \mathcal{T}_n\} = \epsilon \quad (0 < \epsilon < 1), \quad \text{the desired level of significance.}$$

The possibility of (4.3.7) is ensured by the completely known measure $\mathcal{T}_n$. Now, (4.3.7) implies that

$$(4.3.8) \qquad E\{\phi(\mathbf{Z}_n) \mid H_0 \text{ in } (4.3.3)\} = E\{E[\phi(\mathbf{Z}_n) \mid \mathcal{T}_n] \mid H_0\} = \epsilon,$$

and hence, $\phi(\mathbf{Z}_n)$ is a similar test of size ϵ. Since (4.3.8) holds for all diagonally symmetric $F_1, \ldots, F_n$, the existence of the distribution-free tests for H_0 in (4.3.3) is thus proved. Now, in actual practice, we select $\phi(\mathbf{Z}_n)$ with special attention to the class of alternative hypotheses. Moreover, for convenience of carrying out the test actually we prefer to base $\phi(\mathbf{Z}_n)$ on a single-valued statistic $T(\mathbf{Z}_n)$. With this end in view, in section 4.4, we consider a class of rank order tests for H_0 in (4.3.3), while in section 4.6, we consider a class of tests based on appropriate U-statistics.

4.4. A CLASS OF (LINEAR) RANK ORDER STATISTICS AND THE ALLIED TESTS

We rank the n elements in each row of $\mathbf{Z}_n$ in the increasing order of their absolute values; we get a $p \times n$ rank matrix

$$(4.4.1) \qquad \mathbf{R}_n = \begin{pmatrix} R_{11} & \cdots & R_{1n} \\ \cdot & \cdots & \cdot \\ R_{p1} & \cdots & R_{pn} \end{pmatrix}$$

where $R_{j\alpha}$ is the rank of $|X_{j\alpha}|$ among all $|X_{j1}|, \ldots, |X_{jn}|$, $\alpha = 1, \ldots, n$; by virtue of the assumed continuity of F_α's, the possibility of ties is neglected, in probability. For every $i = 1, \ldots, p$, we replace the ranks $1, \ldots, n$ in the ith row of $\mathcal{R}_n$ by a set of general scores $\{E_{n,\alpha}^{(i)}, \alpha = 1, 2, \ldots, n\}$ which is a set of n real constants, we get a $p \times n$ matrix of general scores $\mathbf{E}_n$ corresponding to $\mathcal{R}_n$:

$$(4.4.2) \qquad \mathbf{E}_n = \begin{pmatrix} E_{n,R_{11}}^{(1)} & \cdots & E_{n,R_{1n}}^{(1)} \\ \cdot & \cdots & \cdot \\ E_{n,R_{p1}}^{(p)} & \cdots & E_{n,R_{pn}}^{(p)} \end{pmatrix}.$$

Later on, we shall specify certain conditions of $\mathbf{E}_n$. We consider now the univariate rank order statistics coordinatewise:

$$(4.4.3) \qquad T_n^{(j)} = \sum_{\alpha=1}^{n} E_{n,R_{j\alpha}}^{(j)} C_{j\alpha}, \qquad j = 1, 2, \ldots, p,$$

and let $\mathbf{T}_n = (T_n^{(1)}, \ldots, T_n^{(p)})'$, where $C_{j\alpha} = +1$ or -1 according as $X_{j\alpha} > 0$ or <0 respectively. $T_n^{(j)}$ is thus the difference of the sum of the scores $E_{n,\alpha}^{(j)}$ for which $X_{j\alpha} > 0$ and the sum of those for which $X_{j\alpha} < 0$.

We first consider the mean and dispersion matrix of $\mathbf{T}_n$ under $\mathcal{T}_n$. We denote by $\mathbf{C}_\alpha = (C_{1\alpha}, \ldots, C_{p\alpha})'$, $\alpha = 1, \ldots, n$. Now, the group of transformation $\mathcal{G}_n$ leaves $\mathbf{E}_n$ invariant, while it attaches equal probability (under $\mathcal{T}_n$) to the two likely realizations $(-1)^j\mathbf{C}_\alpha$, $j = 1, 2$, for all $\alpha = 1, \ldots, n$. Hence, $E(\mathbf{C}_\alpha \mid \mathcal{T}_n) = \mathbf{0}$, $E(\mathbf{C}_\alpha\mathbf{C}_\alpha' \mid \mathcal{T}_n) = \mathbf{C}_\alpha\mathbf{C}_\alpha'$, and $\mathbf{C}_\alpha$, $\alpha = 1, \ldots, n$ are all stochastically independent under $\mathcal{T}_n$. Hence, we obtain that

$$E\{\mathbf{T}_n \mid \mathcal{T}_n\} = \mathbf{0}, \tag{4.4.4}$$

$$E\{\mathbf{T}_n\mathbf{T}_n' \mid \mathcal{T}_n\} = n\mathbf{V}_n; \qquad \mathbf{V}_n = ((v_{n,jk}))_{j,k=1,\ldots,p}, \tag{4.4.5}$$

where

$$v_{n,jk} = (1/n)\sum_{\alpha=1}^{n} E_{n,R_{j\alpha}}^{(j)} E_{n,R_{k\alpha}}^{(k)} C_{j\alpha}C_{k\alpha}; \qquad j, k = 1, \ldots, p. \tag{4.4.6}$$

Note that $v_{n,jj} = (1/n)\sum_{\alpha=1}^{n}\{E_{n,\alpha}^{(j)}\}^2$, $j = 1, \ldots, p$ are all nonstochastic, but $v_{n,jk}, j \neq k = 1, \ldots, p$ are all stochastic and depend on $\mathbf{Z}_n$. We assume that $\mathbf{V}_n$ is positive definite. ($\mathbf{V}_n$, being a covariance matrix, will be positive semidefinite at least. If $\mathbf{V}_n$ is singular, then we can work with the highest-order nonsingular minor of $\mathbf{V}_n$ and work with the corresponding variates.) Then, we consider the following positive definite quadratic form

$$S_n = \frac{1}{n}\{\mathbf{T}_n\mathbf{V}_n^{-1}\mathbf{T}_n'\}, \quad \text{where} \quad \mathbf{V}_n^{-1}\mathbf{V}_n = \mathbf{I}_p. \tag{4.4.7}$$

If $\mathbf{T}_n$ is stochastically different from $\mathbf{0}$, S_n will be large. Hence it seems natural to use the following test function:

$$\phi(\mathbf{z}_n) = \begin{cases} 1 & \text{if } S_n > S_{n,\epsilon}(\mathbf{z}_n) \\ a_{n,\epsilon} & \text{if } S_n = S_{n,\epsilon}(\mathbf{z}_n) \\ 0 & \text{if } S_n < S_{n,\epsilon}(\mathbf{z}_n) \end{cases} \tag{4.4.8}$$

where $S_{n,\epsilon}(\mathbf{z}_n)$ and $a_{n,\epsilon}$ are so chosen that $E\{\phi(\mathbf{Z}_n) \mid \mathcal{T}_n\} = \epsilon$, $0 < \epsilon < 1$. Thus, $\phi(\mathbf{Z}_n)$ will be a similar size ϵ test for H_0 in (4.3.3).

For small samples, $S_{n,\epsilon}(\mathbf{z}_n)$ and $a_{n,\epsilon}$ are to be determined from the actual permutational c.d.f. of S_n, while for large samples, we shall simplify (4.4.8) considerably. It is worth considering here several specific forms of S_n for specific $\mathbf{E}_n$. First, if $E_{n,\alpha}^{(j)} = 1$ for all $\alpha = 1, \ldots, n$, $j = 1, \ldots, p$, S_n in (4.4.7) leads to the multivariate sign tests and it includes T in (4.2.9) as a particular case when $p = 2$. This score being a constant for all α, does not satisfy the conditions of section 4.4.2, and will be studied in section 4.4.4. Second, if $E_{n,\alpha}^{(j)} = \alpha/(n+1)$, $1 \leq \alpha \leq n$, $j = 1, \ldots, p$, then S_n reduces to the multivariate generalization of the one-sample Wilcoxon signed rank statistic. Third, if $E_{n,\alpha}^{(j)}$ is the expected value of the αth smallest observation of a sample of size n from a chi-distribution with 1 d.f., $\alpha = 1, \ldots, n$; $j = 1, \ldots, p$, then S_n is the multivariate one-sample normal scores statistics. The relative performances of these statistics will be considered in section 4.6.

4.4.1. Large-Sample Permutation Distribution of S_n

Let us denote the marginal c.d.f. of $X_{j\alpha}$ by $F_{j,\alpha}(x)$ and that of $(X_{j\alpha}, X_{k\alpha})$ by $F_{jk,\alpha}(x, y)$ for $j \neq k = 1, \ldots, p$. Define

$$(4.4.9) \quad F_{j(n)}^{*} = (1/n)\sum_{\alpha=1}^{n} F_{j,\alpha}; \qquad F_{jk(n)}^{*} = (1/n)\sum_{\alpha=1}^{n} F_{jk,\alpha}, \quad j \neq k = 1, \ldots, p,$$

$$(4.4.10) \quad H_{j,\alpha}(x) = F_{j,\alpha}(x) - F_{j,\alpha}(-x) \quad \text{if} \quad x \geq 0 \quad \text{and is 0, otherwise,}$$

$$(4.4.11) \qquad H_{j(n)}^{*} = (1/n)\sum_{\alpha=1}^{n} H_{j,\alpha}, \qquad j = 1, \ldots, p;$$

$$(4.4.12) \quad H_{jk,\alpha}(x, y) = F_{jk,\alpha}(x, y) + F_{jk,\alpha}(-x, y) + F_{jk,\alpha}(x, -y) + F_{jk,\alpha}(-x, -y), \qquad x \geq 0, \quad y \geq 0$$

$$(4.4.13) \qquad H_{jk(n)}^{*} = (1/n)\sum_{\alpha=1}^{n} H_{jk,\alpha}.$$

Also, define the corresponding empirical c.d.f.'s as

$$(4.4.14) \qquad F_{nj}(x) = (1/n)\sum_{\alpha=1}^{n} c(x - X_{j\alpha}), \qquad j = 1, \ldots, p;$$

$$(4.4.15) \quad F_{njk}(x, y) = (1/n)\sum_{\alpha=1}^{n} c(x - X_{j\alpha})c(y - X_{k\alpha}), \qquad j \neq k = 1, \ldots, p,$$

where $c(u)$ is 1 or 0 according as $u \geq 0$ or not. Let then

$$(4.4.16) \qquad H_{nj}(x) = F_{nj}(x) - F_{nj}(-x-) \quad \text{for} \quad x \geq 0, \qquad = 0, \quad x < 0;$$

$$(4.4.17) \quad H_{njk}(x, y) = F_{njk}(x, y) + F_{njk}(-x-, y) + F_{njk}(x, -y-) + F_{njk}(-x-, -y-),$$

for $j \neq k = 1, \ldots, p$. Finally, as in section 3.6.2, we let

$$(4.4.18) \qquad E_{n,\alpha}^{(j)} = J_{n,j}\left(\frac{\alpha}{n+1}\right), \qquad 1 \leq \alpha \leq n, \quad i \leq j \leq p,$$

where $J_{n,j}$ satisfies the assumptions I, II, and III of section 3.6.2, for each $j = 1, \ldots, p$. In addition, we make the following assumptions.

(IV) For each j (= 1, . . . , p)

$$\lim_{n\to\infty}\int_0^1 [J_{n,j}(u) - J_j(u)]^2\, du = 0, \tag{4.4.19}$$

where $J_j(u) = \lim_{n\to\infty} J_{n,j}(u)$: $0 < u < 1$ is defined by assumption I and is assumed to be not a constant. Let us then define

$$\boldsymbol{\nu}_n = ((\nu_{njk}))_{j,k=1,\dots,p}, \tag{4.4.20}$$

where

$$\nu_{njk} = \int_0^\infty\int_0^\infty J_j(H^*_{j(n)}(x))J_k(H^*_{k(n)}(y))\, dH^*_{jk(n)}(x, y). \tag{4.4.21}$$

(V) $\text{Inf}_n\, |\boldsymbol{\nu}_n| > 0$, i.e., $\boldsymbol{\nu}_n$ is positive definite for all n.

We shall first consider the following.

Theorem 4.4.1. *Under the assumptions* I–IV, $\mathbf{V}_n \sim_p \boldsymbol{\nu}_n$, *where* $\mathbf{V}_n$ *and* $\boldsymbol{\nu}_n$ *are defined by* (4.4.6) *and* (4.4.21), *respectively.*

Proof: Using (4.4.6) and (4.4.18), we write (for $j \neq k$)

$$v_{n,jk} = \int_0^\infty\int_0^\infty J_{n,j}\left(\frac{n}{n+1} H_{nj}(x)\right) J_{n,k}\left(\frac{n}{n+1} H_{nk}(y)\right) dH_{njk}(x, y),$$

which after using (4.4.19) reduces to

$$\int_0^\infty\int_0^\infty J_j\left(\frac{n}{n+1} H_{nj}(x)\right) J_k\left(\frac{n}{n+1} H_{nk}(y)\right) dH_{njk}(x, y) + o(1). \tag{4.4.22}$$

The first term of (4.4.22) can be written as

$$\begin{aligned}
&\int_0^\infty\int_0^\infty J_j(H^*_{j(n)}(x))J_k(H^*_{k(n)}(y))\, dH_{njk}(x, y) \\
&\quad+ \int_0^\infty\int_0^\infty \left[J_j\left(\frac{n}{n+1} H_{nj}(x)\right) - J_j(H^*_{j(n)}(x))\right] \\
&\quad\times J_k\left(\frac{n}{n+1} H_{nk}(y)\right) dH_{njk}(x, y) + \int_0^\infty\int_0^\infty J_j(H^*_{j(n)}(x)) \\
&\quad\times \left[J_k\left(\frac{n}{n+1} H_{nk}(y)\right) - J_k(H^*_{k(n)}(y))\right] dH_{njk}(x, y).
\end{aligned} \tag{4.4.23}$$

Now, the first term of (4.4.23) involves an average over n independent random variables, viz., $[J_j(H^*_{j(n)}(X_{j\alpha}))J_k(H^*_{k(n)}(X_{k\alpha}))]$, $\alpha = 1, \dots, n$. Hence, using a law of large numbers (see (2.5.7)) in conjunction with assumption III of section 3.6.2, it readily follows that it is asymptotically equivalent, in probability, to ν_{njk} in (4.4.21). Thus, we only require to show that the two other terms of (4.4.23) both converge in probability to zero, as $n \to \infty$. Now, applying the Cauchy-Schwarz inequality, we obtain that the absolute value of the second integral of (4.4.23) is bounded above by

$$(4.4.24)\quad \left\{\left[\int_0^\infty \left[J_j\left(\frac{n}{n+1} H_{nj}(x)\right) - J_j(H^*_{j(n)}(x))\right]^2 dH_{nj}(x)\right] \times \left[\int_0^\infty J_k^2\left(\frac{n}{n+1} H_{nk}(y)\right) dH_{nk}(y)\right]\right\}^{1/2}$$

where the second term converges to a finite quantity by assumption III. Now, let $S_{n,\epsilon} = \{x : \epsilon < H^*_{j(n)}\, x < 1 - \epsilon\}$, $\epsilon > 0$, and $S^*_{n,\epsilon}$ be the complementary set. Then, by assumption III, $J \in L_r$, $r > 2$, and hence, by the weak law of large numbers, for every $\epsilon > 0$, there exists an $\eta(> 0)$, such that

$$(4.4.25)\quad \int_{S^*_{n,\epsilon}} [J(nH_{nj}/(n+1)) - J(H^*_{j(n)})]^2\, dH_{nj} \le 2\left[\int_{S^*_{n,\epsilon}} J^2(nH_{nj}/(n+1))\, dH_{nj} + \int_{S^*_{n,\epsilon}} J^2(H^*_{j(n)})\, dH_{nj}\right] < \eta,$$

in probability, as $n \to \infty$. Again, using (2.11.70) (with $\bar{F}_n$ and F_n^* being replaced by $H^*_{j(n)}$ and H_{nj} respectively,) and the first derivative condition in assumption III, we obtain that

$$(4.4.26)\quad \int_{S_{n,\epsilon}} [J(nH_{nj}/(n+1)) - J(H^*_{j(n)})]^2\, dH_{nj} < \eta,$$

in probability, as $n \to \infty$. Hence, (4.4.24) is $o_p(1)$ as $n \to \infty$. In a similar manner, it follows that the third integral of (4.4.23) also converges to zero, in probability, as $n \to \infty$. Consequently, $v_{n,jk} \sim_p \nu_{njk}$ for all $j \neq k = 1, \ldots, p$. Also, $v_{n,jj} = \int_0^1 J^2_{n,j}(u)\, du \to \int_0^1 J_j^2(u)\, du = \nu_{n,jj}$ for all $j = 1, \ldots, p$, as $n \to \infty$. Hence the theorem. ◁

It follows from (4.4.9), (4.4.12), (4.4.13), and (4.4.21) that

$$(4.4.27)\quad \nu_{njk} = \int_0^\infty \int_0^\infty J_j(H^*_{j(n)}(x)) J_k(H^*_{k(n)}(y))\, d[F^*_{jk(n)}(x, y) + F^*_{jk(n)}(-x, -y) + F^*_{jk(n)}(-x, y) + F^*_{jk(n)}(x, -y)]$$

Consider now the following: (i) there exists a $J^*(u)$ $(0 < u < 1)$, such that

$$(4.4.28)\quad J_j(u) = J_j^*\left(\frac{1+u}{2}\right) \quad \text{and} \quad J_j^*(u) + J_j^*(1-u) = \text{constant}, \quad 0 < u < 1,\ \ j = 1, \ldots, p,$$

and (ii) for the sequence $\{F_\alpha(\mathbf{x}), \alpha = 1, \ldots, n\}$, $F^*_{j(n)}$ converges to a limiting c.d.f. $F^*_j(x)$ which is symmetric about 0, and $F^*_{jk(n)}$ converges to a limiting (bivariate) c.d.f. F^*_{jk} for all $j \neq k = 1, \ldots, p$. Then

$$\text{(4.4.29)} \quad \nu_{n,jk} \to \nu^*_{jk} = \int_{-\infty}^{\infty}\int_{-\infty}^{\infty} J^*_j(F^*_j(x))J^*_k(F^*_k(y))\, dF^*_{jk}(x, y), \qquad j, k = 1, \ldots, p.$$

Also, as a special case, it follows that under H_0 in (4.3.3) and for (4.4.28)

$$\text{(4.4.30)} \quad \nu_{n,jk} = \nu^*_{njk} = \int_{-\infty}^{\infty}\int_{-\infty}^{\infty} J^*_j(F^*_{j(n)}(x))J^*_k(F^*_{k(n)}(y))\, dF^*_{jk(n)}(x, y),$$

for all $j, k = 1, \ldots, p$. (4.4.29) will hold particularly when $F_\alpha(\mathbf{x})$ is diagonally symmetric about $n^{-1/2}c_\alpha\boldsymbol{\theta}$, $\alpha = 1, \ldots, n$, where $c_1, \ldots, c_n$ all lie in a finite interval; $\boldsymbol{\theta} = (\theta_1, \ldots, \theta_p)'$ has real elements, and further the average c.d.f. $F^*_n = (1/n)\sum_{\alpha=1}^n F_\alpha$ converges in the limit to a c.d.f. F^*. It also follows from Theorem 4.4.1 that under the assumptions I–V, $\mathbf{V}_n$ is positive definite, in probability.

Theorem 4.4.2. *Under the assumptions* I–V, *and* $\mathcal{T}_n$, *as* $n \to \infty$

$$\text{(4.4.31)} \qquad \mathcal{L}_{\mathcal{T}_n}(n^{-1/2}\mathbf{T}_n) \xrightarrow{p} N(\mathbf{0}, \boldsymbol{\nu}_n),^*$$

and also, the permutation distribution of S_n *converges asymptotically, in probability, to a chi-square distribution with* p *degrees of freedom (D.F).*

Proof: (4.4.31) will follow, if it can be shown that for any non-null $\boldsymbol{\delta}$,

$$\text{(4.4.32)} \qquad \mathcal{L}_{\mathcal{T}_n}(n^{-1/2}\boldsymbol{\delta}'\mathbf{T}_n/[\boldsymbol{\delta}'\boldsymbol{\nu}_n\boldsymbol{\delta}]^{1/2}) \xrightarrow{p} N(0, 1).$$

Now, from (4.4.3) and assumption II of section 3.6.2, we have

$$\text{(4.4.33)} \qquad n^{-1/2}\boldsymbol{\delta}'\mathbf{T}_n = n^{-1/2}\sum_{\alpha=1}^{n}\left[\sum_{j=1}^{p}\delta_j c_\alpha^{(j)} J_j\left(\frac{R_{j\alpha}}{n+1}\right)\right] + o_p(1).$$

Let us set $Y_\alpha = \sum_{j=1}^p \delta_j C_\alpha^{(j)} J_j(R_{j\alpha}/(n+1))$, $\alpha = 1, \ldots, n$. Under $\mathcal{T}_n$, $\boldsymbol{\delta}$ and $J_j(R_\alpha^{(j)}/(n+1))$, $\alpha = 1, \ldots, n$, $j = 1, \ldots, p$ are all invariant, while the vectors $\mathbf{C}_\alpha$, $\alpha = 1, \ldots, n$ are all stochastically independent, each having only two equally likely realizations, namely $(-1)^\beta\mathbf{C}_\alpha$, $\beta = 0, 1$; $\alpha = 1, \ldots, n$. Hence, under $\mathcal{T}_n$, $Y_1, \ldots, Y_n$ are all independent and each Y_i can have only two equally likely realizations. Hence, (4.4.32) will follow, if we can show that

* $\mathcal{L}(\mathbf{X}_n) \to N(\mathbf{a}_n, \mathbf{B}_n)$ means that for any non-null $\boldsymbol{\delta}$, as $n \to \infty$, the distribution of $\boldsymbol{\delta}'(\mathbf{X}_n - \mathbf{a}_n)/[\boldsymbol{\delta}'\mathbf{B}_n\boldsymbol{\delta}]^{1/2}$ converges to a standard normal distribution.

for $Y_1, \ldots, Y_n$, the following holds: for some $r > 2$

$$(4.4.34)\quad \lim_{n\to\infty} n^{-(r-2)/2}\left\{\frac{\frac{1}{n}\sum_{\alpha=1}^{n} E\{|Y_\alpha|^r \mid \mathfrak{S}_n\}}{\left[\frac{1}{n}\sum_{\alpha=1}^{n} E\{Y_\alpha^2 \mid \mathfrak{S}_n\}\right]^{r/2}}\right\} = 0, \quad \text{in probability.}$$

Now, by definition of Y_α's and by Theorem 4.4.1 and assumption V, $(1/n)\sum_{\alpha=1}^{n} E(Y_\alpha^2 \mid \mathfrak{S}_n) \sim_p \boldsymbol{\delta}'\boldsymbol{\nu}_n\boldsymbol{\delta}$ is positive. So, (4.4.34) will follow, if we can show that for some $r > 2$, $(1/n)\sum_{\alpha=1}^{n} E\{\,|Y_\alpha|^r \mid \mathfrak{S}_n\}$ is finite. To show this we note that for $r > 2$

$$(4.4.35)\quad \frac{1}{n}\sum_{\alpha=1}^{n} E\{|Y_\alpha|^r \mid \mathfrak{S}_n\} = \frac{1}{n}\sum_{\alpha=1}^{n} E\left\{\left|\sum_{j=1}^{p}\delta_j C_\alpha^{(j)} J_j\left(\frac{R_{j\alpha}}{n+1}\right)\right|^r \mid \mathfrak{S}_n\right\}$$

$$\leq p^{r-1}\sum_{j=1}^{p}|\delta_j|^r\left\{\frac{1}{n}\sum_{\alpha=1}^{n} E\left\{\left|J_j\left(\frac{R_{j\alpha}}{n+1}\right)\right|^r \mid \mathfrak{S}_n\right\}\right.$$

$$= p^{r-1}\sum_{j=1}^{p}|\delta_j|^r\left\{\frac{1}{n}\sum_{\alpha=1}^{n}\left|J_j\left(\frac{\alpha}{n+1}\right)\right|^r < \infty,\right.$$

for all $r \leq 2 + \delta$,

where $2 + \delta$ is defined by assumption III of section 3.6.2, and by the same assumption $(1/n)\sum_{\alpha=1}^{n} |J_j(\alpha/(n+1))|^{2+\delta}$ is finite for all n. Hence (4.4.34) holds, and this proves (4.4.32) and (4.4.31). The second part of the theorem is a direct consequence of (4.4.31) and (xii) of section 2.9. Q.E.D.

4.4.2. Large-Sample (Unconditional) Distribution of S_n

In this section, we shall extend Theorem 4.4.2 to the unconditional model. For the sake of convenience, we shall work with the statistic $\mathbf{T}_n^*$ where

$$(4.4.36)\quad \mathbf{T}_n^* = (T_{n,1}^*, \ldots, T_{n,p}^*)' = \frac{1}{2n}[\mathbf{T}_n + \bar{\mathbf{E}}_n],$$

where

$$(4.4.37)\quad \bar{\mathbf{E}} = (\bar{E}_{n,1}, \ldots, \bar{E}_{n,p}); \qquad \bar{E}_{n,j} = (1/n)\sum_{\alpha=1}^{n} E_{n,\alpha}^{(j)}, \qquad j = 1, \ldots, p.$$

As in section 3.6.2, we can write

$$(4.4.38)\quad T_{n,j}^* = (1/n)\sum_{\alpha=1}^{n} E_{n,\alpha}^{(j)} Z_{n\alpha}^{(j)}, \qquad j = 1, \ldots, p,$$

where $Z_{n\alpha}^{(j)}$ is 1 or 0 according as the αth smallest observation among

$|X_{j1}|, \ldots, |X_{jn}|$ corresponds to a positive X_{ji} or not, $\alpha = 1, \ldots, n$; $j = 1, \ldots, p$. We introduce the following notations:

$$\text{(4.4.39)} \quad \mu_{nj,\alpha} = \int_0^\infty J_j[H^*_{j(n)}(x)]\, dF_{j,\alpha}(x), \quad j = 1, \ldots, p; \qquad \alpha = 1, \ldots, n,$$

$$\text{(4.4.40)} \quad \mu^*_{nj} = (1/n)\sum_{\alpha=1}^{n} \mu_{nj,\alpha}, \quad j = 1, \ldots, p; \qquad \boldsymbol{\mu}^*_n = (\mu^*_{n1}, \ldots, \mu^*_{np})'.$$

It may be noted that

$$\text{(4.4.41)} \qquad \mu^*_{nj} = \int_0^\infty J_j[H^*_{j(n)}(x)]\, dF^*_{j(n)}(x), \qquad j = 1, \ldots, p.$$

In the sequel, we denote by $F_{jk,\alpha}(x, y)$ a bivariate c.d.f., which for $j = k$ means $F_{j,\alpha}([\min\{x, y\}])$. A similar notation is used for $H_{jk,\alpha}$ and H^*_{jk}.

Theorem 4.4.3. *Under the conditions* I–III *of section* 3.6.2, $n^{1/2}(\mathbf{T}^*_n - \boldsymbol{\mu}^*_n)$ *has asymptotically a multinormal distribution with a null mean vector and dispersion matrix* $\boldsymbol{\Gamma}_n = ((\gamma_{n,jk}))_{j,k=1,\ldots,p}$, *defined by* (4.4.55), *provided* $\boldsymbol{\Gamma}_n$ *is positive definite for N.*

Proof: We write $T^*_{n,j}$ equivalently as

$$\text{(4.4.42)} \quad T^*_{n,j} = \int_0^\infty J_{n,j}\left(\frac{n}{n+1} H_{nj}(x)\right) dF_{nj}(x), \qquad j = 1, \ldots, p.$$

Also, we write

$$\text{(4.4.43)} \quad J_{n,j}\left(\frac{n}{n+1} H_{nj}\right) = J_j\left(\frac{n}{n+1} H_{nj}\right) + \left[J_{n,j}\left(\frac{n}{n+1} H_{nj}\right) - J_j\left(\frac{n}{n+1} H_{nj}\right)\right],$$

$$\text{(4.4.44)} \quad \begin{aligned} J_j\left(\frac{n}{n+1} H_{nj}\right) &= J_j(H^*_{j(n)}) + [H_{nj} - H^*_{j(n)}]J'_j[H^*_{j(n)}] \\ &\quad - \frac{1}{n+1} H_{nj}J'_j(H^*_{j(n)}) + \left\{J_j\left(\frac{n}{n+1} H_{nj}\right)\right. \\ &\quad \left. - J_j(H^*_{j(n)}) - \left[\frac{n}{n+1} H_{nj} - H^*_{j(n)}\right] J'_j(H^*_{j(n)})\right\} \end{aligned}$$

and finally, $dF_{nj} = dF^*_{j(n)} + d[F_{nj} - F^*_{jn}]$, $j = 1, \ldots, p$. Then, from the preceding three equations we have

$$T^*_{n,j} = (1/n)\sum_{\alpha=1}^{n} B_{nj}(X_{j\alpha}) + \sum_{r=1}^{4} C_{r,nj}, \tag{4.4.45}$$

where $B_{nj}(X_{j\alpha}) = B^{(1)}_{nj}(X_{j\alpha}) + B^{(2)}_{nj}(X_{j\alpha})$, and

$$B^{(1)}_{nj}(X_{j\alpha}) = J_j[H^*_{j(n)}(X_{j\alpha})]c(X_{j\alpha}); \tag{4.4.46}$$

$$B^{(2)}_{nj}(X_{j\alpha}) = \int_0^\infty [c(x - |X_{j\alpha}|) - H_{j,\alpha}(x)]J'_j[H^*_{j(n)}(x)]\, dF^*_{j(n)}(x), \tag{4.4.47}$$

($c(u)$ is 1 or 0 according as u is ≥ 0 or not), and

$$\begin{aligned} C_{1,nj} &= \int_0^1 \left[J_{n,j}\left(\frac{n}{n+1} H_{nj}\right) - J_j\left(\frac{n}{n+1} H_{nj}\right)\right] dF_{nj} \\ &= o_p(n^{-\frac{1}{2}}), \qquad \text{by assumption II;} \end{aligned} \tag{4.4.48}$$

$$C_{2,nj} = [1/(n+1)]\int_0^\infty H_{nj}J'_j(H^*_{j(n)})\, dF_{nj}, \tag{4.4.49}$$

$$C_{3,nj} = \int_0^\infty [H_{nj} - H^*_{j(n)}]J'_j(H^*_{j(n)})\, d[F_{nj} - F^*_{j(n)}], \tag{4.4.50}$$

$$\begin{aligned} C_{4,nj} = \int_0^\infty \bigg\{ & J_j\left(\frac{n}{n+1} H_{nj}\right) - J_j(H^*_{j(n)}) \\ & - \left[\frac{n}{n+1} H_{nj} - H^*_{j(n)}\right] J'_j(H^*_{j(n)})\bigg\}\, dF_{nj}, \end{aligned} \tag{4.4.51}$$

for $j = 1, \ldots, p$. As in section 3.6.2, we shall show that $C_{r,nj}$, $r = 1, 2, 3, 4$ constitute the higher-order terms, and $(1/n)\sum_{\alpha=1}^{n} B_{nj}(X_{j\alpha})$ constitutes the principal term. In fact, it will be shown in section 10.2 that $C_{r,nj} = o_p(n^{-\frac{1}{2}})$, for all $r = 1, 2, 3, 4$ and $j = 1, \ldots, p$. Hence, $n^{\frac{1}{2}}(\mathbf{T}^*_n - \mathbf{u}^*_n)$ and $[n^{-\frac{1}{2}}\{\sum_{\alpha=1}^{n} B_{nj}(X_{j\alpha}) - n\mu^*_{nj}\}, j = 1, \ldots, p]'$ have the same limiting distribution, if they have any at all. Thus, to prove the theorem, it suffices to show that the n stochastically independent vectors $\mathbf{B}_n(\mathbf{X}_\alpha) = [B_{n1}(X_{1\alpha}), \ldots, B_{np}(X_{p\alpha})]'$, $\alpha = 1, \ldots, n$ satisfy the conditions of the central limit theorem in the multivariate case. With this end in view, we consider first the moments of these variables. By definition in (4.4.46) and (4.4.47)

$$E\{B^{(1)}_{nj}(X_{j\alpha})\} = \int_0^\infty J_j[H^*_{nj}(x)]\, dF_{j,\alpha}(x) = \mu_{nj,\alpha}, \tag{4.4.52}$$

$$E\{B^{(2)}_{nj}(X_{j\alpha})\} = 0, \quad \text{for all } j = 1, \ldots, p; \qquad \alpha = 1, \ldots, n. \tag{4.4.53}$$

Also, upon writing $H^0_{jk,\alpha}(x, y) = P\{|X_{j\alpha}| \le x, |X_{k\alpha}| \le y\} = F_{jk,\alpha}(x, y) - F_{jk,\alpha}(-x, y) - F_{jk,\alpha}(x, -y) + F_{jk,\alpha}(-x, -y)$, $x \ge 0$, $y \ge 0$, it follows that

$$
\begin{aligned}
\gamma_{njk,\alpha} &= \operatorname{cov}[B_{nj}(X_{j\alpha}), B_{nk}(X_{k\alpha})] \\
&= \sum_{r=1}^{2}\sum_{s=1}^{2} \operatorname{cov}[B^{(r)}_{nj}(X_{j\alpha}), B^{(s)}_{nk}(X_{k\alpha})] \\
&= \int_0^\infty \int_0^\infty J_j[H^*_{j(n)}(x)]J_k[H^*_{k(n)}(y)]\, dF_{jk,\alpha}(x, y) - \mu_{nj,\alpha}\mu_{nk,\alpha} \\
&\quad + \int_{x=0}^\infty \int_{y=0}^\infty \int_{z=-\infty}^\infty J_j[H^*_{j(n)}(y)][c(x - |z|) - H_{k,\alpha}(x)] \\
&\quad \times J'_k[H^*_{k(n)}(x)]\, dF^*_{k(n)}(x)\, dF_{jk,\alpha}(y, z) \qquad (4.4.54) \\
&\quad + \int_{x=0}^\infty \int_{y=0}^\infty \int_{z=-\infty}^\infty J_k[H^*_{k(n)}(y)][c(x - |z|) - H_{j,\alpha}(x)] \\
&\quad \times J'_j[H^*_{j(n)}(x)]\, dF^*_{j(n)}(x)\, dF_{kj,\alpha}(y, z) \\
&\quad + \int_0^\infty \int_0^\infty [H^0_{jk,\alpha}(x, y) - H_{j,\alpha}(x)H_{k,\alpha}(y)]J'_j[H^*_{j(n)}(x)] \\
&\quad \times J'_k[H_{k(n)}(y)]\, dF^*_{j(n)}(x)\, dF^*_{k(n)}(y)
\end{aligned}
$$

for $j, k = 1, \ldots, p$, $\alpha = 1, \ldots, n$. Let then

$$
(4.4.55) \quad \mathbf{\Gamma}_n = ((\gamma_{njk})); \qquad \gamma_{njk} = (1/n)\sum_{\alpha=1}^{n} \gamma_{njk,\alpha}, \qquad j, k = 1, \ldots, p.
$$

Thus, from the preceding four equations, we obtain that

$$
(4.4.56) \qquad E\left\{(1/n)\sum_{\alpha=1}^{n} B_{nj}(X_{j\alpha})\right\} = \mu^*_{nj}, \qquad j = 1, \ldots, p;
$$

$$
(4.4.57) \qquad n \operatorname{cov}\left\{(1/n)\sum_{\alpha=1}^{n} B_{nj}(X_{j\alpha}), \quad (1/n)\sum_{\alpha=1}^{n} B_{nk}(X_{k\alpha})\right\} = \gamma_{njk},
$$

for $j, k = 1, \ldots, p$. Now, by (4.4.41), μ^*_{nj} depends on $(F_1, \ldots, F_n)$ only through the c.d.f. $F^*_{j(n)}$, $j = 1, \ldots, p$. But, $\mathbf{\Gamma}_n$ depends on $F_1, \ldots, F_n$ in a rather involved way. We shall show next that $\mathbf{\Gamma}_n$ satisfies certain interesting matrix inequalities. For this, let

$$
\begin{aligned}
\gamma_{jk}(F^*_{(n)}) = & \int_0^\infty \int_0^\infty J_j(H^*_{j(n)}(x)) J_k(H^*_{k(n)}(y))\, dF^*_{jk(n)}(x, y) - \mu^*_{nj}\mu^*_{nk} \\
& + \int_{x=0}^\infty \int_{y=0}^\infty \int_{z=-\infty}^\infty J_j[H^*_{j(n)}(y)][c(x - |z|) - H^*_{k(n)}(x)] \\
& \times J'_k[H^*_{k(n)}(x)]\, dF^*_{k(n)}(x)\, dF^*_{jk(n)}(y, z) \\
& + \int_{x=0}^\infty \int_{y=0}^\infty \int_{z=-\infty}^\infty J_k[H^*_{k(n)}(y)][c(x - |z|) - H^*_{j(n)}(x)] \\
& \times J'_j[H^*_{j(n)}(x)]\, dF^*_{j(n)}(x)\, dF^*_{kj(n)}(y, z) \\
& + \int_0^\infty \int_0^\infty [H^{0*}_{jk(n)}(x, y) - H^*_{j(n)}(x) H^*_{k(n)}(y)] J'_j[H^*_{j(n)}(x)] \\
& \times J'_j[H^*_{k(n)}(y)]\, dF^*_{j(n)}(x)\, dF^*_{k(n)}(y),
\end{aligned} \tag{4.4.58}
$$

where $H^{0*}_{jk(n)} = (1/n) \sum_{\alpha=1}^n H^0_{jk,\alpha}$; $j, k = 1, \ldots, p$. Let then

$$\mathbf{\Gamma}(F^*_{(n)}) = ((\gamma_{jk}(F^*_{(n)}))), \tag{4.4.59}$$

so that $\mathbf{\Gamma}(F^*_{(n)})$ is the value of $\mathbf{\Gamma}_n$ when $F_1 \equiv \cdots \equiv F_n \equiv F^*_{(n)}$. Also, let

$$\beta^{(1)}_{nj,\alpha} = \mu_{nj,\alpha} - \mu^*_{nj}, \quad \beta^{(2)}_{nj,\alpha} = \int_0^\infty [H_{j\alpha}(x) - H^*_{j(n)}(x)] J'_j[H^*_{j(n)}(x)]\, dF^*_{j(n)}(x), \tag{4.4.60}$$

for $j = 1, \ldots, p$, $\alpha = 1, \ldots, n$. Thus, $(1/n) \sum_{\alpha=1}^n \beta^{(i)}_{nj,\alpha} = 0$ for $i = 1, 2$; $j = 1, \ldots, p$. Finally, let

$$\eta_{jk,n} = (1/n) \sum_{\alpha=1}^n [\beta^{(1)}_{nj,\alpha} + \beta^{(2)}_{nj,\alpha}][\beta^{(1)}_{nk,\alpha} + \beta^{(2)}_{nk,\alpha}], \qquad j, k = 1, \ldots, p; \tag{4.4.61}$$

$$\mathbf{H}_n = ((\eta_{jk,n})). \tag{4.4.62}$$

Note that by definition $\mathbf{H}_n$ is a positive semidefinite matrix; it ceases to be positive definite only when $\beta^{(1)}_{nj,\alpha} + \beta^{(2)}_{nj,\alpha} = 0$ for all $\alpha = 1, \ldots, n$ and for at least one j $(= 1, \ldots, p)$.

Lemma 4.4.4. $\mathbf{\Gamma}_n = \mathbf{\Gamma}(F^*_{(n)}) - \mathbf{H}_n$; $0 \leq |\mathbf{\Gamma}(F^*_{(n)})| < \infty$, *for all* $F_1, \ldots, F_n$.

To prove the lemma, we consider first the following identities whose proofs are evident and therefore omitted.

$$
\begin{aligned}
& (1/n) \sum_{\alpha=1}^n [c(x - |z|) - H_{k,\alpha}(x)]\, dF_{jk,\alpha}(y, z) \\
& \qquad = [c(x - |z|) - H^*_{k(n)}(x)]\, dF^*_{jk(n)}(y, z) - \\
& \qquad\qquad (1/n) \sum_{\alpha=1}^n [H_{k,\alpha}(x) - H^*_{k(n)}(x)]\, d[F_{jk,\alpha}(x) - F^*_{jk(n)}(x)];
\end{aligned} \tag{4.4.63}
$$

$$(4.4.64)\quad \begin{aligned}&(1/n)\sum_{\alpha=1}^{n}[H^{0}_{jk,\alpha}(x,y) - H_{j,\alpha}(x)H_{k,\alpha}(y)]\\ &\quad = [H^{0*}_{jk(n)}(x,y) - H^{*}_{j(n)}(x)H^{*}_{k(n)}(y)]\\ &\quad\quad - (1/n)\sum_{\alpha=1}^{n}[H_{j,\alpha}(x) - H^{*}_{j(n)}(x)][H_{k,\alpha}(y) - H^{*}_{k(n)}(y)];\end{aligned}$$

$$(4.4.65)\quad (1/n)\sum_{\alpha=1}^{n}\mu_{nj,\alpha}\mu_{nk,\alpha} - \mu^{*}_{nj}\mu^{*}_{nk} = (1/n)\sum_{\alpha=1}^{n}(\mu_{nj,\alpha} - \mu^{*}_{nj})(\mu_{nk,\alpha} - \mu^{*}_{nk});$$

$$(4.4.66)\quad \int_{y=0}^{\infty}\int_{z=-\infty}^{\infty} J_j[H^{*}_{n(j)}(y)]\, d[F_{jk,\alpha}(y,z) - F^{*}_{jk(n)}(y,z)] = \beta^{(1)}_{nj,\alpha},$$

for all $j, k = 1, \ldots, p$. The first part of the lemma then follows from (4.4.54) and (4.4.58) after using (4.4.9) and (4.4.59) through (4.4.66). Since $\mathbf{\Gamma}(F^{*}_{(n)})$ is a covariance matrix, the proof of the second part follows directly by showing that $\gamma_{jj}(F^{*}_{(n)}) < \infty$ for all $j = 1, \ldots, p$. This directly follows from (4.4.58) by noting that for $j = k$, $y = z$ and

$$(4.4.67)\quad dF^{*}_{j(n)}(x) \le dH^{*}_{j(n)}(x) \quad \text{for all} \quad x \ge 0;$$

$$(4.4.68)\quad \iint\limits_{0<u<v<1} u^{r}(1-v)^{s}\,|J_j^{(r)}(u)|\,|J_j^{(s)}(v)|\,du\,dv < \infty \quad \text{for all} \quad r, s = 0, 1,$$

where (4.4.68) follows from assumption III of section 3.6.2. We shall next prove the following lemma. ◁

Lemma 4.4.5. *Under assumption* III *of section* 3.6.2,

$$(1/n)\sum_{\alpha=1}^{n}E\{|B_{nj}(X_{j\alpha})|^{2+\delta}\} < \infty,$$

for all $j = 1, \ldots, p$ *and uniformly in* $\{\mathcal{F}_n\} = \{(F_1, \ldots, F_n): F_i \in \mathcal{F},\ i = 1, \ldots, n\}$, $\mathcal{F}$ *being the class of all continuous p-variate c.d.f.'s.*

Proof: By virtue of (4.4.46) and (4.4.47), we require to show that

$$(4.4.69)\quad (1/n)\sum_{\alpha=1}^{n} E\{|B^{(i)}_{nj}(X_{j\alpha})|^{2+\delta}\} < \infty \quad \text{for} \quad i = 1, 2; \qquad j = 1, \ldots, p.$$

Now, by definition in (4.4.46),

$$(4.4.70)\quad \begin{aligned}(1/n)\sum_{\alpha=1}^{n}E\{|B^{(1)}_{nj}(X_{j\alpha})|^{2+\delta}\} &\le \int_0^1 |J_j(H^{*}_{j(n)}(x)|^{2+\delta}\, dF^{*}_{j(n)}(x)\\ &\le \int_0^1 |J_j(u)|^{2+\delta}\, du < \infty,\end{aligned}$$

as $dF^{*}_{j(n)} \le dH^{*}_{j(n)}$ and assumption III holds. Let now Y_{nj} be a random variable (independent of $X_{j1}, \ldots, X_{jn}$) following the c.d.f. $F^{*}_{j(n)}(x)$. Define

$$(4.4.71)\quad d_n(X_{j\alpha}, Y_{nj}) = \begin{cases}[c(Y_{nj} - |X_{j\alpha}|) - H_{j\alpha}(Y_{nj})]J'_j[H^{*}_{j(n)}(Y_{nj})], & Y_{nj} > 0\\ 0, & Y_{nj} \le 0.\end{cases}$$

Then, it is easy to verify that

$$B_{nj}^{(2)}(X_{j\alpha}) = E\{d_n(X_{j\alpha}, Y_{nj}) \mid X_{j\alpha}\}, \quad \alpha = 1, \ldots, n. \tag{4.4.72}$$

Consequently, by straightforward computations we obtain that

$$\begin{aligned} E\{|B_{nj}^{(2)}(X_{j\alpha})|^{2+\delta}\} &\le E\{[E(|d_n(X_{j\alpha}, Y_{nj})|^{1+\delta/2} \mid X_{j\alpha})]^2\} \\ &= 2E\Bigg\{ \iint_{0<x<y<\infty} |\{c(x - |X_{j\alpha}|) - H_{j\alpha}(x)\} \\ &\quad \times \{c(y - |X_{j\alpha}|) - H_{j\alpha}(y)\} J_j'[H_{j(n)}^*(x)] \\ &\quad \times J_j'[H_{j(n)}^*(y)]\,|^{1+\delta/2}\, dF_{j(n)}^*(x)\, dF_{j(n)}^*(y) \Bigg\} \\ &\le 6 \iint_{0<x<y<\infty} H_{j\alpha}(x)[1 - H_{j\alpha}(y)]\,|\, J_j'[H_{j(n)}^*(x)] \\ &\quad \times J_j'[H_{j(n)}^*(y)]\,|^{1+\delta/2}\, dF_{j(n)}^*(x)\, dF_{j(n)}^*(y), \end{aligned} \tag{4.4.73}$$

as

$$\begin{aligned} E[|\{c(x - |X_{j\alpha}|) - H_{j\alpha}(x)\}\{c(y - |X_{j\alpha}|) - H_{j\alpha}(y)\}\,|^{1+\delta/2} \mid x < y] \\ = H_{j\alpha}(x)[\{1 - H_{j\alpha}(x)\}\{1 - H_{j\alpha}(y)\}]^{1+\delta/2} \\ + [H_{j\alpha}(y) - H_{j\alpha}(x)][H_{j\alpha}(x)\{1 - H_{j\alpha}(y)\}]^{1+\delta/2} \\ + [1 - H_{j\alpha}(y)][H_{j\alpha}(x)H_{j\alpha}(y)]^{1+\delta/2} \\ \le 3H_{j\alpha}(x)[1 - H_{j\alpha}(y)], \end{aligned} \tag{4.4.74}$$

since $\delta > 0$ and $y > x$. Hence, using (4.4.64) (with $j = k$), we obtain from (4.4.73) that

$$\begin{aligned} (1/n)\sum_{\alpha=1}^{n} E\{|B_{nj}^{(2)}(X_{j\alpha})|^{2+\delta}\} \\ = 6\Bigg[\iint_{0<x<y<\infty} H_{j(n)}^*(x)[1 - H_{j(n)}^*(y)] \\ \times |J_j'[H_{j(n)}^*(x)]J_j'[H_{j(n)}^*(y)]|^{1+\delta/2}\, dF_{j(n)}^*(x) F_{j(n)}^*(y) \\ - (1/2n)\sum_{\alpha=1}^{n}\Bigg\{\int_0^1 [H_{j\alpha}(x) - H_{j(n)}^*(x)]\,|J_j'[H_{j(n)}^*(x)]|^{1+\delta/2} \\ \times dF_{j(n)}^*(x)\Bigg\}^2\Bigg] \\ \le 6 \iint_{0<u<v<1} u(1 - v)\,|J_j'(u)J_j'(v)|\, du\, dv^{\,1+\delta/2} < \infty, \end{aligned} \tag{4.4.75}$$

by (4.4.67) and assumption III.

Hence the lemma. ◁

We return now to the proof of Theorem 4.4.3. It remains to show that for any non-null $\mathbf{d}$, $\mathbf{d}'\mathbf{B}_n(\mathbf{X}_\alpha)$, $\alpha = 1, \ldots, n$ satisfy the Lindeberg condition of the classical central limit theorem. Now, by assumption $\mathbf{\Gamma}_n$ is positive definite, hence

$$n^{-1}\sum_{\alpha=1}^{n} \text{var}\,(\mathbf{d}'\mathbf{B}_n(\mathbf{X}_\alpha)) = \mathbf{d}'\mathbf{\Gamma}_n\,\mathbf{d} > 0. \tag{4.4.76}$$

Also, by Lemma 4.4.5, we obtain that

$$\begin{aligned} n^{-1}\sum_{\alpha=1}^{n} E\,\{|\mathbf{d}'(\mathbf{B}_n(\mathbf{X}_\alpha) - E\mathbf{B}_n(\mathbf{X}_\alpha))|^{2+\delta}\} &\le 2^{1+\delta}n^{-1}\sum_{\alpha=1}^{n} E\{\,|\mathbf{d}'\mathbf{B}_n(\mathbf{X}_\alpha)|^{2+\delta}\} \\ &\le (2p)^{1+\delta}\sum_{j=1}^{p} |d_j|^{2+\delta}\left\{(1/n)\sum_{\alpha=1}^{n} E\,|B_{nj}(X_{j\alpha})|^{2+\delta}\right\} < \infty. \end{aligned} \tag{4.4.77}$$

Consequently, for any non-null $\mathbf{d}$,

$$\lim_{n\to\infty} n^{-\delta/2}\left\{\frac{[n^{-1}\sum_{\alpha=1}^{n} E\,|\mathbf{d}'(\mathbf{B}_n(\mathbf{X}_\alpha) - E\mathbf{B}_n(\mathbf{X}_\alpha))|^{2+\delta}]}{[n^{-1}\sum_{\alpha=1}^{n} \text{var}\,(\mathbf{d}'\mathbf{B}_n(\mathbf{X}_\alpha))]^{1+\delta/2}}\right\} = 0, \tag{4.4.78}$$

where by assumption III, $\delta > 0$. Hence the theorem. ◁

Theorem 4.4.3 generalizes the results of Sen (1970a) to the multivariate case and of Sen and Puri (1967) to the case when the c.d.f.'s are not all identical. For some related results, the reader is also referred to Hušková (1970). When $F_1(\mathbf{x}) \equiv \cdots \equiv F_n(\mathbf{x}) \equiv F(\mathbf{x})$, it follows from (4.4.60) that $\beta^{(i)}_{nj,\alpha} = 0$ for $i = 1, 2$ and $j = 1, \ldots, p$, and hence, from Lemma 4.4.4, we obtain the following:

Lemma 4.4.6. *If $F_1 \equiv \cdots \equiv F_n \equiv F$, then $\mathbf{\Gamma}_n = \mathbf{\Gamma}(F)$, where $\mathbf{\Gamma}(F)$ is defined by* (4.4.59) *with $F^*_{(n)} \equiv F$.*

In fact, we have a stronger result, as follows:

Lemma 4.4.7. *If F_α have univariate marginals all symmetric about* 0, *for $\alpha = 1, \ldots, n$, then $\mathbf{H}_n = \mathbf{0}^{p\times p}$ for all $F_1, \ldots, F_n$.*

Proof: If $F_{j\alpha}$ is symmetric about 0, then $H_{j\alpha}(x) = 2F_{j\alpha}(x) - 1$, $x \ge 0$, for all $\alpha = 1, \ldots, n$. Hence $H^*_{j(n)}(x) = 2F^*_{j(n)}(x) - 1$ for all $x \ge 0$, $j = 1, \ldots, p$. Consequently, writing $\beta^{(1)}_{nj,\alpha}$ as $\int_0^\infty J_j[H^*_{j(n)}(x)]\,d[F_{j,\alpha}(x) - F^*_{j(n)}(x)] = \frac{1}{2}\int_0^\infty J_j[H^*_{j(n)}(x)]\cdot d[H_{j,\alpha}(x) - H^*_{j(n)}(x)]$, then integrating it by parts and finally adding to $\beta^{(2)}_{nj,\alpha}$, we obtain that $\beta^{(1)}_{nj,\alpha} + \beta^{(2)}_{nj,\alpha} = 0$ for all $j = 1, \ldots, p$, $\alpha = 1, \ldots, n$. Hence, by (4.4.61), $\mathbf{H}_n = \mathbf{0}$. Q.E.D.

It also follows that when $F_{j\alpha}$'s are all symmetric about 0,

$$\mu^*_{nj} = \int_0^\infty J_j[H^*_{j(n)}(x)]\,dF^*_{j(n)}(x) = \frac{1}{2}\int_0^1 J_j(u)\,du, \qquad j = 1, \ldots, p, \tag{4.4.79}$$

$$\gamma_{jj}(F^*_{(n)}) = \frac{1}{4}\int_0^1 J_j^2(u)\,du, \qquad j = 1, \ldots, p \tag{4.4.80}$$

are all independent of $F^*_{(n)}$. However, $\gamma_{jk}(F^*_{(n)})$ depends on the unknown c.d.f. $F^*_{(n)}$ for all $j \neq k = 1, \ldots, p$. We consider next the following lemma, whose proof we leave as Exercise 4.4.3.

Lemma 4.4.8. *If F_α, $\alpha = 1, \ldots, n$ are all diagonally symmetric about* $\mathbf{0}$, *then* $\mathbf{\Gamma}(F^*_{(n)}) \equiv \frac{1}{4}\mathbf{v}_n$, *where $\mathbf{v}_n$ is defined by* (4.4.20) *and* (4.4.21).

We are now in a position to consider also a class of asymptotically distribution-free tests for H_0 in (4.3.3) based on $\mathbf{T}^*_n$ in (4.4.36). Let $\hat{\mathbf{\Gamma}}(F^*_{(n)})$ be any consistent estimate of $\mathbf{\Gamma}(F^*_{(n)})$ [in the sense that $\hat{\mathbf{\Gamma}}(F^*_{(n)}) \sim_p \mathbf{\Gamma}(F^*_{(n)})$ as $n \to \infty$]. Then, using Theorem 4.4.3, Lemma 4.4.6, and (4.4.79), we propose the test statistic

$$S^*_n = n(\mathbf{T}^*_n - \boldsymbol{\mu}^*_0)'[\hat{\mathbf{\Gamma}}(F^*_{(n)})]^{-1}(\mathbf{T}^*_n - \boldsymbol{\mu}^*_0), \tag{4.4.81}$$

where

$$\boldsymbol{\mu}^*_0 = (\mu^*_{10}, \ldots, \mu^*_{p0}); \qquad \mu^*_{j0} = \frac{1}{2}\int_0^1 J_j(u)\, du, \qquad j = 1, \ldots, p. \tag{4.4.82}$$

It follows from Theorem 4.4.3 that under H_0 in (4.3.3), S^*_n has asymptotically a χ^2 distribution with p d.f., where of course, by virtue of Lemma 4.4.8 and Theorem 4.4.1, $\mathbf{\Gamma}(F^*_{(n)})$ is assumed to be positive definite. We shall now study the large sample properties of the tests based on S_n in (4.4.7) and S^*_n in (4.4.81), and show that they are asymptotically power-equivalent.

For this purpose, we consider the following sequence of admissible alternative hypotheses $\{H_n\}$ which specifies that

$$H_n\colon\ F_\alpha(\mathbf{x}) = F_{\alpha,n}(x) = F_{\alpha,0}(\mathbf{x} - n^{-1/2}\mathbf{d}_\alpha), \qquad \mathbf{d}_\alpha = (d_{1\alpha}, \ldots, d_{p\alpha})', \quad \alpha = 1, \ldots, n, \tag{4.4.83}$$

where the $d_{\alpha j}$'s are all real and finite and $F_{\alpha,0}$, $\alpha = 1, \ldots, n$ are all diagonally symmetric about $\mathbf{0}$. Furthermore, it is assumed that the score $E^{(j)}_{n,\alpha}$ is the expected value of the αth smallest observation of a sample of size n drawn from a distribution $\Psi_j(x)$, where

$$\Psi_j(x) = \begin{cases} 2\Psi^*_j(x) - 1, & x \geq 0, \\ 0, & x < 0, \end{cases} \tag{4.4.84}$$

and $\Psi^*_j(x) + \Psi^*_j(-x) = 1$ for all x. (4.4.84) implies that $J_j(0) = 0$ and

$$J_j(u) = \Psi_j^{-1}(u) = \Psi_j^{*-1}\left(\frac{1+u}{2}\right) = J^*_j\left(\frac{1+u}{2}\right), \qquad 0 < u < 1, \quad j = 1, \ldots, p. \tag{4.4.85}$$

Thus, in this case (4.4.28) holds. We also denote $F^*_{j(n)}$ as in (4.4.9) and let

$$f^*_{j(n)} = \left(\frac{d}{dx}\right)F^*_{j(n)} = \left(\frac{1}{n}\right)\sum_{\alpha=1}^{n}\left(\frac{d}{dx}\right)F_{j,\alpha}(x), \qquad j = 1, \ldots, p, \tag{4.4.86}$$

and we assume that

$$f^*_{j(n)}(x)J^{*\prime}[F^*_{j(n)}(x)] \text{ is bounded as } x \to \pm\infty, \quad j = 1, \ldots, p. \tag{4.4.87}$$

Finally, we let

$$\begin{aligned} \mathbf{c}^*_{(n)} &= (c^*_{1(n)}, \ldots, c^*_{p(n)})'; \\ c^*_{j(n)} &= \left(\frac{1}{2n}\right)\sum_{\alpha=1}^{n} d_{j\alpha}\int_{-\infty}^{\infty} J^{*\prime}_j[F^*_{j(n)}(x)][f^*_{j(n)}(x)]^2\,dx, \end{aligned} \tag{4.4.88}$$

$j = 1, \ldots, p$; it is of course assumed that the F_α's are all absolutely continuous with continuous density functions. Then the following corollary is an immediate consequence of Theorem 4.4.3, Lemma 4.4.8, (4.4.85), and (4.4.30).

Corollary 4.4.3.1. *Under the assumptions of Theorem* 4.4.3 *and the conditions* (4.4.83), (4.4.85), *and* (4.4.87), $n^{1/2}(\mathbf{T}_n - \boldsymbol{\mu}^*_n)$ *(where* μ^*_j *is defined by* (4.4.82)) *has asymptotically a p-variate normal distribution with mean vector* $\frac{1}{2}\mathbf{c}^*_{(n)}$ *and dispersion matrix* $\frac{1}{4}\boldsymbol{\nu}^*_n$, *where* $\boldsymbol{\nu}^*_n$ *has the elements* ν^*_{njk}, *defined by* (4.4.30). *In particular, when* $F_1 \equiv \cdots \equiv F_n = F$, $\mathbf{c}^*_{(n)} = (\bar{d}_{1(n)}B_1, \ldots, \bar{d}_{p(n)}B_p)'$ *and* ν^*_{njk} *is given by* (4.4.29), *where*

$$\bar{d}_{j(n)} = \frac{1}{n}\sum_{\alpha=1}^{n} d_{j\alpha} \quad \text{and} \quad B_j = \int_{-\infty}^{\infty} (d/dx)J^*_j[F_j(x)]\,dF_j(x), \tag{4.4.89}$$

$$j = 1, \ldots, p.$$

Corollary 4.4.3.2. *The limiting distribution of* $n^{1/2}(\mathbf{T}_n - \boldsymbol{\mu}^*_n)$ *is singular iff one or more* $J^*_j[F^*_{j(n)}(X_{j\alpha})]$ *can be expressed as a linear function of* $J^*_k[F^*_{k(n)}(X_{k\alpha})]$, $k \neq j = 1, \ldots, p$.

The proof is left as Exercise 4.4.4.

We shall now consider the large-sample distribution of S_n in (4.4.7), when the null hypothesis (4.3.3) may not hold. To justify the limiting distribution theory, it will be assumed in the sequel that as $n \to \infty$ $F^*_{(n)}(\mathbf{x}) = (1/n)\sum_{\alpha=1}^{n} F_\alpha(\mathbf{x})$ converges to a limiting c.d.f. $F^*(\mathbf{x})$. Under fairly general conditions, such an assumption may be made, and the following models justify the same.

(I) Homogeneous c.d.f.'s. This is the conventional and usually adopted model, where we assume that $F_1 \equiv \cdots \equiv F_n \equiv F$, so that $F^*_{(n)} \equiv F^* \equiv F$.

(II) Outliers model. Suppose that out of the n observations $\mathbf{X}_\alpha$, $\alpha = 1, \ldots, n$, n_1 are from the c.d.f. $F(\mathbf{x})$, while the remaining $n - n_1 = n_2$ are from different c.d.f.'s (i.e., outliers), where $n_1/n \to 1$ as $n \to \infty$. In this case also $F^*_{(n)}$ converges to F as $n \to \infty$.

(III) Finite mixture model. Suppose that n_j of the observations are drawn from the c.d.f. G_j, $j = 1, \ldots, t\ (\geq 1)$, where $\sum_{j=1}^t n_j = n$ and further $n_j/n \to \lambda_j$: $0 \leq \lambda_j \leq 1$, $j = 1, \ldots, t$. Then, $F^*_{(n)}(\mathbf{x}) \to \sum_{j=1}^t \lambda_j G_j(\mathbf{x}) = \bar{G}(\mathbf{x})$ (say). This model may arise in two different ways (i) when we are combining $t\ (\geq 1)$ independent samples into a combined sample and (ii) when in model (II), we assume that the proportion of the outliers is not negligible, but the outliers come from several homogenous c.d.f.'s.

(IV) Convergence model. Suppose $\{F_\alpha\}$ stands for a sequence of c.d.f.'s such that $\lim_{\alpha\to\infty} F_\alpha = F$ exists. Then obviously, $F^*_{(n)}(\mathbf{x})$ also converges to $F(\mathbf{x})$ as $n \to \infty$.

There may be other models too. Also, when we consider the sequence of alternative hypotheses $\{H_n\}$ in (4.4.83), we make the following assumption for simplification of the results:

(4.4.90) either $\mathbf{d}_\alpha = \mathbf{d}$ for all α and F_α's possibly differ,

or $F_\alpha = F$ for all α and $\bar{d}_{j(n)}$, defined by (4.4.89), converges to $\bar{d}_j$,

as $n \to \infty$, $j = 1, \ldots, p$.

Further, we define B_j, $j = 1, \ldots, p$, as in (4.4.87) and let

(4.4.91) $\mathbf{c}^* = (c_1^*, \ldots, c_p^*)'$; $\quad c_j^* = d_j B(F_j^*)$ or $\bar{d}_j B_j$, $j = 1, \ldots, p$,

where $B(F_j^*)$ is defined as in (4.4.89) with F replaced by F^*. Finally, let $\boldsymbol{\nu}^* = ((\nu_{jk}^*))$, where ν_{jk}^* is defined by (4.4.29). Then we have the following

Theorem 4.4.9. *If* (i) $\lim_{n\to\infty} F^*_{(n)} \equiv F^*$ *exists and* (ii) *the conditions of corollary* 4.4.3.1 *are satisfied, then the limiting distribution of* S_n *in* (4.4.7) *is noncentral chi square with p D.F. and the noncentrality parameter*

$$\Delta_S = \mathbf{c}^{*\prime} T^{-1} \mathbf{c}^*. \tag{4.4.92}$$

The proof of the theorem follows from (4.4.29), Corollary 4.4.3.1 and some straightforward computations. The details are left as Exercise 4.4.5.

Now, in (4.4.80), we have considered $\hat{\boldsymbol{\Gamma}}$ to be any consistent estimate of $\boldsymbol{\Gamma}(F^*_{(n)})$. Since, under (4.4.83), $\boldsymbol{\Gamma}_n$ in (4.4.55) is asymptotically equivalent to $\boldsymbol{\Gamma}(F^*_{(n)})$, of which $\hat{\boldsymbol{\Gamma}}$ is a consistent estimator, it follows from Theorem 4.4.9 that S_n^* has asymptotically the same distribution as of S_n, and $S_n \sim_p S_n^*$. Hence, the permutation test based on S_n and the unconditional test based on S_n^* are asymptotically power equivalent (for $\{H_n\}$). Further, the choice of $\hat{\boldsymbol{\Gamma}}$ is of no importance in the limit.

4.5. TESTS BASED ON U-STATISTICS

In this section, we develop the permutation distribution theory of U-statistics under the sign-invariant transformations in section 4.3. This leads to suitable permutation tests when the parent distributions are not necessarily identical nor continuous.

Consider the group $\mathcal{G}_n$ of transformations in (4.3.6), and the permutational probability measure $\mathcal{T}_n$ induced by the conditional law of $\mathbf{E}_n$ over the 2^n equally likely realizations. We shall consider the distribution theory of U-statistics under $\mathcal{T}_n$. For an (extended) *regular functional* $\theta(F_1, \ldots, F_m)$ of *degree* m (≥ 1), denote the *symmetric kernel* by $\phi(\mathbf{X}_1, \ldots, \mathbf{X}_m)$, so that the corresponding U-statistic (cf. section 3.2)

$$U_n = \binom{n}{m}^{-1} \sum_C \phi(\mathbf{X}_{i_1}, \ldots, \mathbf{X}_{i_m}); \tag{4.5.1}$$

$$C = \{(i_1, \ldots, i_n)\colon 1 \leq i_1 < \cdots < i_n \leq n\}$$

is an unbiased estimate of $\binom{n}{m}^{-1} \sum_C \theta(F_{i_1}, \ldots, F_{i_m})$. We shall specifically consider the class of functionals for which $\theta(F_{i_1}, \ldots, F_{i_m}) = \theta_0$, uniformly in $F_i \in \mathcal{F}$, for all $i = 1, \ldots, n$ where $\mathcal{F}$ is the class of all sign-invariant c.d.f.'s. It will be seen later on that the above condition holds for a broad class of functionals. Thus U_n will be an unbiased estimator of θ_0 if $F_i \in \mathcal{F}_0$, for all i. It is easy to show that the above condition also implies that

$$E\{\phi(\mathbf{X}_{i_1}, \ldots, \mathbf{X}_{i_m}) \mid \mathcal{T}_n\} = \theta_0 \quad \text{for all} \quad 1 \leq i_1 < \cdots < i_m \leq n, \tag{4.5.2}$$

and hence, $E\{U_n \mid \mathcal{T}_n\} = \theta_0$. Let us denote by

$$\begin{aligned} \binom{n}{2m-c} \frac{(2m-c)!}{c!\,(m-c)!\,(m-c)!}\, 2^{(2m-c)}\{\zeta_c(\mathcal{T}_n) + \theta_0^2\} \\ = \textstyle\sum_c^* \phi((-1)^{i_1}\mathbf{X}_{j_1}, \ldots, (-1)^{i_m}\mathbf{X}_{j_m}) \\ \times\, \phi((-1)^{i_{m-c+1}}\mathbf{X}_{j_{m-c+1}}, \ldots, (-1)^{i_{2m-c}}\mathbf{X}_{j_{2m-c}}), \end{aligned} \tag{4.5.3}$$

where the summation $\sum_c^*$ extends over all possible choices of distinct $j_1, \ldots, j_m$ and $i_k = 0, 1$ for $k = 1, \ldots, 2m - c$. Obviously, by (4.5.2) $\zeta_0(\mathcal{T}_n) = 0$. Then, by routine computations along the lines of section 3.2, it follows that

$$\sigma^2_{\mathcal{T}_n} = V\{U_n \mid \mathcal{T}_n\} = \binom{n}{m}^{-1} \sum_{c=1}^{m} \binom{m}{c}\binom{n-m}{m-c} \zeta_c(\mathcal{T}_n). \tag{4.5.4}$$

Now, defining $\zeta_{c,n}$ as in (3.2.49), it follows that the unconditional variance of U_n has the same expression as in (4.5.4) where $\zeta_{c,n}$ replaces $\zeta_c(\mathcal{T}_n)$ for all

$c = 1, \ldots, m$; this is denoted by σ_n^2. Let us denote by

$$F^*_{(n)}(\mathbf{x}) = (1/n)\sum_{\alpha=1}^{n} F_\alpha(\mathbf{x}), \quad \text{so that} \quad F^*_{(n)} \in \mathcal{F} \quad \text{when} \quad F_\alpha \in \mathcal{F}, \forall\alpha. \tag{4.5.5}$$

When $F_1 \equiv \cdots \equiv F_n \equiv F$, σ_n^2 is denoted by $\sigma_n^2(F)$, in order to indicate its possible dependence on F. If $F_i \in \mathcal{F}$, we denote by G_i the c.d.f. of $(-1)\mathbf{X}_\alpha$, and $\bar{F}_\alpha = \frac{1}{2}[F_\alpha + G_\alpha]$, $\alpha = 1, \ldots, n$. Also, let

$$\bar{F}^*_{(n)}(\mathbf{x}) = (1/n)\sum_{\alpha=1}^{n} \bar{F}_\alpha(\mathbf{x}), \quad \text{so that} \quad \bar{F}^*_{(n)} \in \mathcal{F}. \tag{4.5.6}$$

Then, using the results of section 3.2, it can be shown that

$$\text{(i)} \quad n[\sigma^2_{\mathcal{T}_n} - \sigma_n^2(\bar{F}^*_{(n)})] \xrightarrow{p} 0, \quad \text{for all} \quad F_1, \ldots, F_n; \tag{4.5.7}$$

$$\text{(ii)} \quad n[\sigma_n^2 - \sigma_n^2(F^*_{(n)})] \xrightarrow{p} 0, \quad \text{for all} \quad F_1, \ldots, F_n; \tag{4.5.8}$$

and hence,

$$\text{(iii)} \quad n[\sigma^2_{\mathcal{T}_n} - \sigma_n^2] \xrightarrow{p} 0 \quad \text{as } n \to \infty, \quad \text{for all} \quad F_1, \ldots, F_n. \tag{4.5.9}$$

In particular, when $F_1 \equiv \cdots \equiv F_n \equiv F \in \mathcal{F}$ and the kernel ϕ has finite fourth-order moment, $\sigma_{\mathcal{T}_n}$ is the minimum variance unbiased estimator of $\sigma_n^2(F)$. This gives some justification of using the permutation variance of U_n as an estimate of the true variance (when the latter is unknown), even if the permutation principle is not used for the construction of the test.

Now, the 2^n conditionally equally likely realizations of $\mathbf{E}_N$ under $\mathcal{T}_n$ generate the permutation distribution of U_n. For small values of n, one may venture to construct exact permutation tests based on U_n with the aid of this known conditional distribution. However, the labor involved becomes prohibitive as n increases. This leads to the problem of finding the large-sample permutation distribution of U_n, for which we have the following theorem.

Theorem 4.5.1. *If* (i) $E\{|\phi((-1)^{i_1}\mathbf{X}_{\alpha_1}, \ldots, (-1)^{i_m}\mathbf{X}_{\alpha_1})|^r\} < \infty$, *for some* $r < 2$, *uniformly in* $i_k = 0, 1$, $k = 1, \ldots, m$. $1 \le \alpha_1 < \cdots < \alpha_m \le n$, *and* (ii) $n\sigma_n^2(F^*_{(n)})$ *is strictly positive, then the permutation distribution of* $n^{1/2}(U_n - \theta_0)/[m\{\zeta_1(\mathcal{T}_n)\}^{1/2}]$ *is, asymptotically, in probability, a standard normal distribution.*

Proof: Under $\mathcal{T}_n$, $\mathbf{X}_1, \ldots, \mathbf{X}_n$ are all stochastically independent variables and each can have only two (conditionally) equally likely values with probability $\frac{1}{2}$. Let then

$$\begin{aligned} V(\mathbf{x}_i) &= E\left\{\binom{n-1}{m-1}^{-1} \textstyle\sum_i \phi(\mathbf{x}_i, \mathbf{X}_{j_2}, \ldots, \mathbf{X}_{j_m}) \mid \mathcal{T}_n\right\} \\ &= 2^{-(m-1)}\binom{n-1}{m-1}^{-1} \textstyle\sum_i^* \phi(\mathbf{x}_i, (-1)^{i_2}\mathbf{X}_{j_2}, \ldots, (-1)^{i_m}\mathbf{X}_{j_m}), \end{aligned} \tag{4.5.10}$$

where the summation $\sum_i$ extends over all $1 \leq j_2 < \cdots < j_m \leq n$ (but $j_k \neq i$), and $\sum_i^*$ in addition over all $i_k = 0, 1, k = 2, \ldots, m$. Then, essentially by the same projection technique as in section 3.2 (but considering the conditional law $\mathcal{T}_n$), it can be shown that under $\mathcal{T}_n$, $n^{1/2}(U_n - \theta_0)$ and $mn^{-1/2}\sum_{\alpha=1}^n [V(X_\alpha) - \theta_0]$ both have the same limiting law, if they have any at all. So, we require to show that the random variables $V(X_\alpha)$, $\alpha = 1, \ldots, n$, satisfy the Liapounoff condition of the central limit theorem (cf. section 2.7). Now, by virtue of (4.5.2), $V(\mathbf{X}_\alpha) + V((-1)\mathbf{X}_\alpha) = 2\theta_0$ for all $\alpha = 1, \ldots, n$. Thus,

$$(4.5.11)\quad E\left\{(1/n)\sum_{\alpha=1}^n |V(\mathbf{X}_\alpha) - \theta_0|^k \mid \mathcal{T}_n\right\} = (1/n)\sum_{\alpha=1}^n |V(\mathbf{X}_\alpha) - \theta_0|^k.$$

Using then the structural convergence properties of U-statistics, studied in detail in section 3.2, and generalizing them, of course, to the case where $F_1, \ldots, F_n$ are not necessarily identical, it can be shown that

$$(4.5.12)\qquad (1/n)\sum_{\alpha=1}^n [V(\mathbf{X}_\alpha) - \theta_0]^2 \sim_p \zeta_1(F^*_{(n)}); \qquad \zeta_1(F^*_{(n)}) > 0.$$

Also, by (4.5.9), $\zeta_1(\mathcal{T}_n) \sim_p \zeta_1(F^*_{(n)})$. Thus, both the left-hand side of (4.5.12) and $\zeta_1(\mathcal{T}_n)$ are positive, in probability. Again, by condition (i) of the theorem, the expected value of $(1/n)\sum_{\alpha=1}^n |V(\mathbf{X}_\alpha) - \theta_0|^r$ exists for some $r > 2$, and hence, by Chebychev's inequality, $(1/n)\sum_{\alpha=1}^n |V(\mathbf{X}_\alpha) - \theta_0|^r$ is bounded, in probability. Thus,

$$(4.5.13)\quad \sum_{\alpha=1}^n E\{|V(\mathbf{X}_\alpha) - \theta_0|^r \mid \mathcal{T}_n\} \Big/ \left[\sum_{\alpha=1}^n E\{|V(\mathbf{X}_\alpha) - \theta_0|^2 \mid \mathcal{T}_n\}\right]^{r/2} = o_p(1).$$

Hence the theorem. ◁

In the multivariate case, we usually require more than one U-statistic. Hence, we shall recapitulate briefly (without proofs) the vector generalization of the results derived above. For some positive integer $t\ (\geq 1)$ and $\mathbf{F}_n = (F_1, \ldots, F_n)$, let $\boldsymbol{\theta}(\mathbf{F}_n) = (\theta_1(\mathbf{F}_n), \ldots, \theta_t(\mathbf{F}_n))$ be an estimable parameter vector whose kernel is $\boldsymbol{\phi} = (\phi_1, \ldots, \phi)$ where ϕ_k is of degree $m_k\ (\geq 1)$, $k = 1, \ldots, t$. The corresponding U-statistic vector is denoted by $\mathbf{U}_n = (U_{n,1}, \ldots, U_{n,t})'$, and is defined in the same manner as in (4.5.1), and (4.5.2) is assumed to be true coordinatewise; the corresponding vector is denoted by $\boldsymbol{\theta}_0 = (\theta_{0,1}, \ldots, \theta_{0,t})'$. Precisely as in (4.5.3), we define the permutational covariances $\zeta_{c(k,q)}(\mathcal{T}_n)$, $0 \leq c \leq \min(m_k, m_q)$, $k, q = 1, \ldots, t$. Then, the permutational covariance of $U_{n,k}$ and $U_{n,q}$ can be expressed as

$$(4.5.14)\quad \sigma_{n(k,q)}(\mathcal{T}_n) = \binom{n}{m_k}^{-1}\sum_{c=1}^{m_k}\binom{m_q}{c}\binom{n-m_q}{m_k-c}\zeta_{c(k,q)}(\mathcal{T}_n), \quad k, q = 1, \ldots, t.$$

We denote by $n^{-1}\boldsymbol{\Sigma}(\mathcal{T}_n)$ the covariance matrix with elements in (4.5.14), and the corresponding unconditional covariance matrix is denoted by $n^{-1}\boldsymbol{\Sigma}_n$.

The true covariance matrix of $n^{1/2}(\mathbf{U}_n - \boldsymbol{\theta}_0)$ when $F_1 \equiv \cdots \equiv F_n \equiv F$, is denoted by $\boldsymbol{\Sigma}_n(F)$ and $F^*_{(n)}$ is defined as in (4.5.5). Then, it can be shown that (4.5.7), (4.5.8), and (4.5.9) hold when the scalar elements therein are replaced by the corresponding matrices. Also, Theorem 5.4.1 can be generalized to obtain the asymptotic multinormality of $n^{1/2}(\mathbf{U}_n - \boldsymbol{\theta}_0)$ (under $\mathcal{T}_n$). Under the assumption that $\boldsymbol{\Sigma}_n(F^*_{(n)})$ is positive definite, $\boldsymbol{\Sigma}(\mathcal{T}_n)$ will also be so, in probability, and, we denote by $\boldsymbol{\Sigma}^{-1}(\mathcal{T}_n)$ the reciprocal matrix of $\boldsymbol{\Sigma}(\mathcal{T}_n)$. We define now

$$S_n^0 = n(\mathbf{U}_n - \boldsymbol{\theta}_0)'\boldsymbol{\Sigma}^{-1}(\mathcal{T}_n)(\mathbf{U}_n - \boldsymbol{\theta}_0) \tag{4.5.15}$$

Our proposed tests are based on suitable S_n. It follows from the above discussion that

$$\mathcal{L}_{\mathcal{T}_n}(S_n^0) \xrightarrow{p} \chi_t^2, \tag{4.5.16}$$

where χ_t^2 has the chi-square distribution with t d.f. Here also, for small samples, the permutation distribution of $\mathbf{U}_n$ can be used to determine the critical value of S_n, when n is small.

We note that $\boldsymbol{\Sigma}(\mathcal{T}_n)$ stochastically converges to $\boldsymbol{\Sigma}_n(F^*_{(n)})$, and hence, using the results of section 3.2, it follows that under H_0 in (4.3.3) (i.e., $F_\alpha \in \mathcal{F}$, $\forall\alpha$), S_n has asymptotically (unconditionally) a chi-square distribution with t D.F.

Now, as in section 4.4, for the study of the asymptotic power properties of the permutation or the large-sample (unconditional) tests (which are asymptotically equivalent), we consider a sequence of n-tuplets of c.d.f.'s, say, $\{F_{n1}(\mathbf{x}), \ldots, F_{nn}(\mathbf{x})\}$, where $F_{n\alpha}(\mathbf{x})$ converges to some fixed c.d.f. $F_\alpha(\mathbf{x}) \in \mathcal{F}$ as $n \to \infty$ (at all points of continuity of the latter), for all α, where $F^*_{(n)}(\mathbf{x}) = (1/n)\sum_{\alpha=1}^n F_\alpha(\mathbf{x})$ also converges to some nondegenerate c.d.f. $F^*(\mathbf{x})$ as $n \to \infty$. In this connection, we again refer to the four models considered in section 4.4, where this convergence is justified. Then, it follows that

$$[\boldsymbol{\Sigma}(\mathcal{T}_n) - \boldsymbol{\Sigma}_n(F^*_{(n)})] \xrightarrow{p} 0, \tag{4.5.17}$$

and that $\lim_{n\to\infty} \boldsymbol{\Sigma}_n(F^*_{(n)}) = \boldsymbol{\Sigma}(F^*)$ exists and is positive definite. Let us finally assume that $\mathbf{F}_n = (F_{n1}, \ldots, F_{nn})$ is such that

$$\lim_{n\to\infty} n^{\frac{1}{2}}[\boldsymbol{\theta}(\mathbf{F}_n) - \boldsymbol{\theta}_0] = \boldsymbol{\delta} = (\delta_1, \ldots, \delta_t)' \quad \text{exists}, \tag{4.5.18}$$

where δ_j's are all real and finite. Then, it follows from (4.5.15) and the results of section 3.2 that under the above sequence of alternative hypotheses, S_n has asymptotically a noncentral chi-square distribution with t D.F. and the noncentrality parameter

$$\Delta_{S^0} = \boldsymbol{\delta}'\boldsymbol{\Sigma}^{-1}(F^*)\boldsymbol{\delta}. \tag{4.5.19}$$

It is easy to show that the sequence of alternatives in (4.4.82) is contained in (4.5.18) under very mild restrictions on the F_α's.

We shall now consider certain specific tests.

(I) The Multivariate Sign Test. Let us consider the regular functionals

$$\theta_{i,\alpha}^{(1)} = P\{X_{i\alpha} < 0\} + \tfrac{1}{2}P\{X_{i\alpha} = 0\}, \quad i = 1, \ldots, p; \quad \alpha = 1, \ldots, n. \tag{4.5.20}$$

Under the null hypothesis that $F_\alpha \in \mathcal{F}\, \forall \alpha$, $\theta_{i\alpha,}^{(1)} = \frac{1}{2}$ for all $i = 1, \ldots, p$, $\alpha = 1, \ldots, n$. We define the usual sign function $c(u)$ to be 1, $\frac{1}{2}$, or 0 according as u is $>$, $=$, or < 0. Then the corresponding U-statistics are

$$U_{n,i}^{(1)} = (1/n)\sum_{\alpha=1}^{n} c(X_{i\alpha}), \quad i = 1, \ldots, p; \qquad \mathbf{U}_n^{(1)} = (U_{n,1}^{(1)}, \ldots, U_{n,p}^{(1)})'. \tag{4.5.21}$$

According to (4.5.4), the permutational covariance matrix of $\mathbf{U}_n^{(1)}$ has the elements

$$v_{jk}(\mathcal{S}_n) = (1/n^2)\left[\sum_{\alpha=1}^{n} c(X_{j\alpha})c(X_{k\alpha}) - n/4\right], \qquad j, k = 1, \ldots, p. \tag{4.5.22}$$

(For $j = k$, $\sum_{\alpha=1}^{n} [c(X_{j\alpha})]^2 - n/4 = n_j/4$, where n_j is the number of $X_{j\alpha}$'s which are different from zero, $j = 1, \ldots, p$.) The construction of the test statistic follows as in (4.5.15) and is left as Exercise 4.5.1. Let us denote by F_j^* and F_{jk}^* the univariate and bivariate c.d.f.'s of the jth and (j, k)th variates corresponding to the limiting c.d.f. $F^* = \lim_{n\to\infty} F_{(n)}^*$. Let then

$$\nu_{jk}^* = \begin{cases} \frac{1}{4}, & j = k = 1, \ldots, p, \\ F_{jk}^*(0, 0) - \frac{1}{4}, & j \neq k = 1, \ldots, p, \end{cases} \tag{4.5.23}$$

and let

$$\boldsymbol{\nu}^{*-1} = ((\nu^{*jk})) = (\nu_{jk}^*)^{-1}. \tag{4.5.24}$$

Finally, we consider the sequence of alternative hypothesis in (4.4.82) with the further simplifications in (4.4.88), and define

$$\mathbf{c}^{**} = (c_1^{**}, \ldots, c_p^{**})'; \qquad c_j^{**} = \bar{d}_j f_j(0) \quad \text{or} \quad d_j f_j^*(0), \qquad j = 1, \ldots, p, \tag{4.5.25}$$

where $\bar{d}_j$ and d_j are defined by (4.4.88) and $f_j(0)$ is the (marginal) density function of $F_j(x)$ at $x = 0$, $f_j^*(0)$ is the same function for $F_j^*(x)$, $j = 1, \ldots, p$. Then, from the discussion preceding (4.5.19), it follows that under the above sequence of alternative hypotheses, the multivariate sign statistic, based on $\mathbf{U}_n^{(1)}$ in (4.5.21) and having the structure in (4.5.15), has asymptotically a noncentral chi-square distribution with p D.F. and the noncentrality parameter

$$\Delta_M = \mathbf{c}^{**\prime}\boldsymbol{\nu}^{*-1}\mathbf{c}^{**}, \tag{4.5.26}$$

where $\boldsymbol{\nu}^*$ and $\mathbf{c}^{**}$ are defined by (4.5.23) and (4.5.25), respectively. For the particular case $p = 2$, the result gives the asymptotic power of Chatterjee's sign test considered in section 4.2. Some other problems are left as Exercises 4.5.2, 4.5.3, and 4.5.4.

(II) The Extended Signed Rank Test. Let us consider the parameters

$$\theta^{(2)}_{i,\alpha\beta} = P\{X_{i\alpha} + X_{i\alpha} > 0\} + \tfrac{1}{2}P\{X_{i\alpha} + X_{i\beta} = 0\}, \quad i = 1, \ldots, p, \quad \alpha \neq \beta = 1, \ldots, n. \tag{4.5.27}$$

If $F_\alpha \in \mathcal{F}$, $\forall\alpha$, then $\theta^{(2)}_{i,\alpha\beta} = \frac{1}{2}$ for all $\alpha \neq \beta = 1, \ldots, n$, $i = 1, \ldots, p$, while $\theta^{(2)}_{i,\alpha\alpha} = \theta^{(1)}_{i,\alpha}$, defined by (4.5.20). On the other hand, if F_α is symmetric about $u \neq 0$, then (4.5.27) will not be equal to $\frac{1}{2}$. The U-statistic corresponding to (4.5.27) is $\mathbf{U}^{(2)}_n = (U^{(2)}_{n,1}, \ldots, U^{(2)}_{n,p})'$, where

$$U^{(2)}_{n,j} = \binom{n}{2}^{-1} \sum_{\alpha<\beta} c(X_{j\alpha} + X_{j\beta}), \quad j = 1, \ldots, p. \tag{4.5.28}$$

Also, we define the Wilcoxon signed rank statistics by

$$\mathbf{W}_n = (W_{n,1}, \ldots, W_{n,p})'; \quad W_{n,j} = [1/n(n+1)]\sum_{\alpha=1}^{n} S(X_{j\alpha})R_{j\alpha}, \quad j = 1, \ldots, p, \tag{4.5.29}$$

where

$$R_{j\alpha} = \frac{1}{2} + \sum_{\beta=1}^{n} c(|X_{j\alpha}| - |X_{j\beta}|), \quad S(X_{j\alpha}) = c(X_{j\alpha})^{-\frac{1}{2}}. \tag{4.5.30}$$

It is then easy to show that

$$\mathbf{W}_n = [\tfrac{1}{2}(n+1)][(n-1)\{\mathbf{U}^{(2)}_n - \tfrac{1}{2}\mathbf{1}_p\} + 2\{\mathbf{U}^{(1)}_n - \tfrac{1}{2}\mathbf{1}_p\}], \tag{4.5.31}$$

where $\mathbf{U}^{(1)}_n$ is defined by (4.5.21) and $\mathbf{1}_p = (1, \ldots, 1)'$. Thus, we may either work with $\mathbf{W}_n$ or with $\mathbf{U}^{(2)}_n$. From the computational standpoint, $\mathbf{W}_n$ possesses some minor advantages. The permutational covariance matrix of $n^{1/2}\mathbf{W}_n$ has the elements $v_{jk}(\mathcal{T}_n)$ specified by

$$v_{jk}(\mathcal{T}_n) = (1/n)\sum_{\alpha=1}^{n} R_{j\alpha}R_{k\alpha}S(X_{j\alpha})S(X_{k\alpha}), \quad j, k = 1, \ldots, p. \tag{4.5.32}$$

The actual construction of the test follows along (4.5.15) and is left as Exercise 4.5.5. This test statistic generalizes the multivariate version of the Wilcoxon signed rank statistic of section 4.4 to the case where $F_1, \ldots, F_n$ are not necessarily continuous, i.e., where ties among $|X_{j1}|, \ldots, |X_{jn}|$, $j = 1, \ldots, p$ may not be ignored. Some other tests are considered as exercises.

4.6. EFFICIENCY CONSIDERATIONS

We shall now compare the performances of the different tests studied in the preceding sections. If $F_1 \equiv \cdots \equiv F_n \equiv F$ is a multinormal c.d.f., the most powerful invariant (under nonsingular linear transformations) test for H_0 in (4.3.3) is based on the Hotelling's T_n^2 statistic defined as

$$T_n^2 = n\bar{\mathbf{X}}_n'\mathbf{S}_n^{-1}\bar{\mathbf{X}}_n;$$

(4.6.1)
$$\bar{\mathbf{X}}_n = \left(\frac{1}{n}\right)\sum_{\alpha=1}^{n}\mathbf{X}_\alpha,$$

$$\mathbf{S}_n = \frac{1}{(n-1)}\sum_{\alpha=1}^{n}(\mathbf{X}_\alpha - \bar{\mathbf{X}})(\mathbf{X}_\alpha - \bar{\mathbf{X}})'.$$

Suppose now, that the null hypothesis H_0 in (4.3.3) holds but F_α's are not all identical or normal, but have finite moments up to the order $2 + \delta$, $\delta < 0$. Define by $\mathbf{\Sigma}_\alpha$ the covariance matrix of $\mathbf{X}_\alpha$, $\alpha = 1, \ldots, n$, and let $\mathbf{\Sigma}_{(n)}^* = (1/n)\sum_{\alpha=1}^{n}\mathbf{\Sigma}_\alpha$. Using then the multivariate central limit theorem, it follows that under H_0 in (4.3.3),

(4.6.2)
$$\mathcal{L}(n^{1/2}\bar{\mathbf{X}}_n) \rightarrow N(\mathbf{0}, \mathbf{\Sigma}_{(n)}^*).$$

Also writing

(4.6.3)
$$\mathbf{S}_n = \frac{n}{n-1}\left\{\frac{1}{n}\sum_{\alpha=1}^{n}\mathbf{X}_\alpha\mathbf{X}_\alpha' - \bar{\mathbf{X}}_n\bar{\mathbf{X}}_n'\right\},$$

using (4.6.2) and the law of large numbers in (2.5.7), it follows that $S_n \sim_p \mathbf{\Sigma}_{(n)}^*$. Consequently, some routine computations yield that when $F_1, \ldots, F_n$ have all finite moments up to the order $2 + \delta$, $\delta > 0$, then

(4.6.4)
$$\mathcal{L}(T_n^2 \mid H_0) \rightarrow \chi_p^2.$$

In this sense, Hotelling's T_n^2 test has also the same rejection rule as the tests in the preceding two sections. We shall now consider the sequence of alternative hypotheses in (4.4.82) (under the further simplifications in (4.4.88)) and study the asymptotic distribution of T_n^2. Since, for such a sequence of alternative hypotheses,

(4.6.5)
$$\mathcal{L}(n^{1/2}\bar{\mathbf{X}}_n \mid H_n) \rightarrow N(\bar{\mathbf{d}}, \mathbf{\Sigma}^*),$$

(where $\bar{\mathbf{d}} = (\bar{d}_1, \ldots, \bar{d}_p)'$ has the elements defined by (4.4.88), and $\mathbf{\Sigma}^*$ is the covariance matrix of $F^* = \lim_{n\to\infty} F_{(n)}^*$), and even for such alternatives $\mathbf{S}_n \sim_p \mathbf{\Sigma}_{(n)}^* \rightarrow \mathbf{\Sigma}^*$, it follows that T_n^2 has asymptotically a noncentral chi-square distribution with p d.f. and a noncentrality parameter

(4.6.6)
$$\Delta_{T^2} = \bar{\mathbf{d}}'(\mathbf{\Sigma}^*)^{-1}\bar{\mathbf{d}}.$$

Thus, T_n^2 and all the other test statistics in sections 4.4 and 4.5 have asymptotically the same distribution (namely, noncentral χ^2 with p D.F.) differing only in their noncentrality parameters, and a comparison of the noncentrality parameters reveals their asymptotic efficiencies (cf. section 3.8).

Hence, denoting by e_{T_n,T_n^*} the asymptotic relative efficiency (A.R.E.) of a test T_n with respect to a second test T_n^*, we have from Theorem 4.4.1 and (4.6.6) that

$$e_{S_n,T_n^2} = \left[\frac{(\mathbf{c}^{*\prime}\boldsymbol{\nu}^{*-1}\mathbf{c}^*)}{(\bar{\mathbf{d}}'(\boldsymbol{\Sigma}^*)^{-1}\bar{\mathbf{d}})}\right]. \tag{4.6.7}$$

It is quite clear that the A.R.E. in (4.6.7), in general, depends not only on $\boldsymbol{\nu}^*$ and $\boldsymbol{\Sigma}$ but also on $\mathbf{c}^*$ and $\bar{\mathbf{d}}$. Consider the two situations in (4.4.90). In the first situation, where $d_\alpha = d$ for all α's but the F_α's possibly differ, (4.6.7) reduces to

$$e^{(1)}_{S_n,T_n^2} = \left[\frac{(\mathbf{d}'(\mathbf{T}^*)^{-1}\,\mathbf{d})}{(\mathbf{d}'\,\boldsymbol{\Sigma}^{*-1}\,\mathbf{d})}\right], \tag{4.6.8}$$

where

$$\mathbf{T}^* = ((\tau^*_{jk})); \qquad \tau^*_{jk} = \nu^*_{jk}/[B(F_j^*)B(F_k^*)], \qquad j, k = 1, \ldots, p, \tag{4.6.9}$$

and $B(F_j^*)$'s are defined by (4.4.89). Thus, in this case, the A.R.E. depends on $\mathbf{d}$, $\mathbf{T}^*$, and $\boldsymbol{\Sigma}^*$. In the second case, when $F_\alpha \equiv F$ but $\mathbf{d}_\alpha$'s are not all equal, (4.6.7) reduces to

$$e^{(2)}_{S_n,T_n^2} = \left[\frac{(\bar{\mathbf{d}}' \quad ^{-1}}{(\bar{\mathbf{d}}'\,\boldsymbol{\Sigma}^{-1}\,\bar{\mathbf{d}})}\right], \tag{4.6.10}$$

where $\boldsymbol{\Sigma}$ is the common covariance matrix and $\mathbf{T} = \mathbf{T}^* \,|_{F^* \equiv F}$. Thus, in this case, the A.R.E. depends on $\bar{\mathbf{d}}$, $\mathbf{T}$, and $\boldsymbol{\Sigma}$. In any case, unlike the univariate situation (cf. section 3.8), the A.R.E. is not independent of $\mathbf{d}$ or $\bar{\mathbf{d}}$, in general. However, using the Courant theorem on the extremum of the ratio of two positive definite quadratic forms, we obtain the following.

Theorem 4.6.1. *The maximum and minimum values of* $e^{(1)}_{S_n,T_n^2}$ *are given by the maximum and minimum eigenvalues of* $\boldsymbol{\Sigma}^*(\mathbf{T}^*)^{-1}$. *Similarly, the maximum and minimum values of* $e^{(2)}_{S_n,T_n^2}$ *are given by the maximum and minimum eigenvalues of* $\boldsymbol{\Sigma}\,\mathbf{T}^{-1}$.

The proof is simple and left as Exercise 4.6.1.

(The asymptotic relative efficiency of the S_n^0-test in section 4.5 (when $t = p$) with respect to T_n^2 has essentially a similar expression, as can be obtained from (4.5.19) and (4.6.6).)

Thus, it may be of some interest to study the bounds for $e^{(j)}_{S_n,T_n^2}$, $j = 1, 2$, suggested in Theorem 4.6.1 for various special cases of S_n and particular F^* or F. The following interesting cases are considered for illustrations.

(I) The Multivariate Sign Test (M_n Test) (cf. section 4.5). We define ν^*_{jk} as in (4.5.23), so that

$$\tau^*_{jk} = \nu^*_{jk}/[f_j^*(0)f_k^*(0)], \qquad j, k = 1, \ldots, p. \tag{4.6.11}$$

Thus, τ^*_{jk} depends on $F^*_{jk}(0, 0)$, $f^*_j(0)$ and $f^*_k(0)$, $j, k = 1, \ldots, p$; and in general, the relation between $\mathbf{T}^*$ and $\mathbf{\Sigma}^*$ may be quite involved, even when $F_1 \equiv \cdots \equiv F_n \equiv F$. When F (or F^*) is a multivariate normal c.d.f. with a null mean vector and dispersion matrix $\mathbf{\Sigma}$ (or $\mathbf{\Sigma}^*$), then

$$f^*_j(0) = (2\pi)^{-1/2}\sigma_{jj}^{-1/2}, \qquad j = 1, \ldots, p, \tag{4.6.12}$$

$$4F^*_{jk}(0, 0) - 1 = \rho^{(M)}_{jk} = (2/\pi)\sin^{-1}(\sigma_{jk}/[\sigma_{jj}\sigma_{kk}]^{1/2}), \qquad j \neq k = 1, \ldots, p, \tag{4.6.13}$$

and hence, writing $\rho_{jk} = \sigma_{jk}/[\sigma_{jj}\sigma_{kk}]^{1/2}$, $j, k = 1, \ldots, p$; $\mathbf{R} = ((\rho_{jk}))$ and $\mathbf{R}^{(M)} = ((\rho^{(M)}_{jk}))$, it follows that

$$\mathbf{\Sigma}\, T^{-1} = (2/\pi)(([\sigma_{jj}\sigma_{kk}]^{1/2}\rho_{jk}))(([\sigma_{jj}\sigma_{kk}]\rho^{1/2(M)}_{jk}))^{-1}, \tag{4.6.14}$$

and hence, the maximum and the minimum eigenvalues of (4.6.14) are the maximum and the minimum eigenvalues of $(2/\pi)\mathbf{R}[R^{(M)}]^{-1}$, respectively. These depend only on $\mathbf{R}$ (by virtue of (4.6.13)), and hence, allowing $\mathbf{R}$ to vary over its domain (such that $|R| > 0$), the bounds can be deduced. When $p = 2$, the maximum eigenvalues of $(2/\pi)\mathbf{R}(\mathbf{R}^{(M)})^{-1}$ lie between $2/\pi$ and 0.72 (cf. Exercise 4.6.3), where the upper limit 0.72 is attained for $\rho_{12} \simeq \pm 0.70$, while the lower limit $(2/\pi)$ is attained when $\rho_{12} = 0$ or ± 1. For $p > 2$, this becomes a function of more than one ρ_{jk}, and except for certain special cases (cf. Exercises 4.6.4, 4.6.5), no general bounds are available. For non-normal F (or F^*), (4.6.13) does not hold, and as a result, it becomes quite difficult to attach bounds to the A.R.E.

(II) The Multivariate Signed Rank Test (cf. section 4.4). This is a special case of S_n when $E^{(j)}_{n,\alpha} = \alpha/(n + 1)$, $1 \leq \alpha \leq n$, $j = 1, \ldots, p$. In this case, ν^*_{jk}, defined by (4.4.29), reduces to

$$\nu^*_{jj} = \tfrac{1}{3}, \qquad j = 1, \ldots, p, \tag{4.6.15}$$

$$\nu^*_{jk} = 4\int_{-\infty}^{\infty}\int_{-\infty}^{\infty} [F^*_j(x) - \tfrac{1}{2}][F^*_k(y) - \tfrac{1}{2}]\, dF^*_{jk}(x, y), \qquad j \neq k = 1, \ldots, p. \tag{4.6.16}$$

Also

$$B(F^*_j) = \int_{-\infty}^{\infty} (d/dx)(2F^*_j(x) - 1)\, dF^*_j(x) = 2\int_{-\infty}^{\infty} f^*_j(x)\, dF^*_j(x). \tag{4.6.17}$$

Consequently, $\mathbf{T}^*$, defined by (4.6.9), has the elements

$$\tau^*_{jk} = \begin{cases} \frac{1}{12}\left[\int_{-\infty}^{\infty} f^*_j(x)\, dF^*_j(x)\right]^2, & j = k = 1, \ldots, p, \\ \rho^g_{jk}\Big/\left\{12\left[\int_{-\infty}^{\infty} f^*_j(x)\, dF^*_j(x)\right]\left[\int_{-\infty}^{\infty} f^*_k(x)\, dF^*_k(x)\right]\right\}, & j \neq k = 1, \ldots, p, \end{cases} \tag{4.6.18}$$

where ρ^g_{jk} is the grade correlation defined by

$$\rho^g_{jk} = 12\int_{-\infty}^{\infty}\int_{-\infty}^{\infty} [F^*_j(x) - \tfrac{1}{2}][F^*_k(y) - \tfrac{1}{2}]\, dF^*_{jk}(x, y), \tag{4.6.19}$$
$$j, k = 1, \ldots, p; \qquad R^g = ((\rho^g_{jk})).$$

Thus, in this case, the maximum and minimum A.R.E. of the $S_n(R)$ test (where $S_n(R)$ stands for the signed rank test) relative to the T^2_n test are specified by the corresponding eigenvalues of $\mathbf{\Sigma}^*\mathbf{T}^{*-1}$. In particular when F^* (or F) is multinormal,

$$\int_{-\infty}^{\infty} f^*_j(x)\, dF^*_j(x) = [2\sqrt{\pi}\, \sigma_{jj}^{1/2}]^{-1}, \qquad j = 1, \ldots, p, \tag{4.6.20}$$

$$\rho^g_{jk} = (6/\pi)\sin^{-1}[\sigma_{jk}/2\{\sigma_{jj}\sigma_{kk}\}^{1/2}] = (6/\pi)\sin^{-1}(\tfrac{1}{2}\rho_{jk}), \tag{4.6.21}$$

$j, k = 1, \ldots, p$, so that the maximum and minimum characteristic roots of $\mathbf{\Sigma}^*\mathbf{T}^{*-1}$ reduces to that of $\mathbf{R}(\mathbf{R}^g)^{-1}$. For $p = 2$, it is easy to show that the two characteristic roots of $\mathbf{R}(\mathbf{R}^g)^{-1}$ are

$$(3/\pi)(1 + \rho_{12})/(1 + \rho^g_{12}) \quad \text{and} \quad (3/\pi)(1 - \rho_{12})/(1 - \rho^g_{12}). \tag{4.6.22}$$

As such, maximizing and minimizing over the range of variation of ρ_{12} $(-1 < \rho < 1)$ along with the use of (4.6.21) leads to the following:

The maximum A.R.E. is bounded below by $3/\pi$ and above by 0.965, the lower bound is attained when $\rho_{12} = 0$ or $\rho_{12} \to \pm 1$, while the upper bound is attained when $\rho_{12} \to \pm 0.78$. Similarly, the minimum A.R.E. is bound below by 0.87 and above by $3/\pi$; the lower bound is attained when $\rho_{12} \to \pm 1$, while the upper bound is attained when $\rho_{12} \to 0$. Both the A.R.E. functions are symmetric about $\rho_{12} = 0$.

For $p > 2$, the bounds for the A.R.E. depend on more than one ρ_{jk} and, in general, are quite complicated. However, the following results can be proved and these are left as Exercises 4.6.5 and 4.6.7. (I) the minimum A.R.E. can be arbitrarily close to zero, of course in the limiting degenerate case where $\mathbf{\Sigma}^*$ is nonsingular while $\mathbf{T}^*$ converges to a singular matrix. (II) The maximum A.R.E. is always bounded below by one.

(III) The Multivariate Normal Scores Test (cf. section 4.4). In this special case, $J^*_j(u) = \Phi^{-1}(u)$ $0 < u < 1, j = 1, \ldots, p$, where $\Phi(x)$ is the standard normal c.d.f. The corresponding S_n statistic is denoted by $S_n(\Phi)$. Hence, ν^*_{jk}, defined by (4.4.29), reduces to

$$\nu^*_{jk} = \begin{cases} 1, & j = k = 1, \ldots, p, \\ \displaystyle\int_{-\infty}^{\infty}\int_{-\infty}^{\infty} \Phi^{-1}(F^*_j(x))\Phi^{-1}(F^*_k(y))\, dF^*_{jk}(x, y), & j \neq k = 1, \ldots, p \end{cases} \tag{4.6.23}$$

We write $\nu^*_{jk} = \rho^{(\Phi)}_{jk}$ for all $j, k = 1, \ldots, p$, and $\mathbf{R}^{(\Phi)} = ((\rho^{(\Phi)}_{jk}))$. Also, we note that

$$(4.6.24) \quad B(F^*_j) = \int_{-\infty}^{\infty} (d/dx)\Phi^{-1}[F^*_j(x)]\, dF^*_j(x), \qquad j = 1, \ldots, p.$$

It readily follows that if F^* is a multinormal c.d.f., $\rho^{(\Phi)}_{jk} = \rho_{jk}$ for all $j, k = 1, \ldots, p$ and $B(F^*_j) = (\sigma_{jj})^{-1/2}$ for all $j = 1, \ldots, p$. Hence, $\mathbf{\Sigma}^* = \mathbf{T}^*$. Consequently, $\mathbf{\Sigma}^*\mathbf{T}^{*-1} = \mathbf{I}_p$, the identity matrix of order p. Consequently, the A.R.E. of the normal scores test relative to the T^2_n test is equal to 1 for all $\mathbf{\Sigma}^*$. In the univariate case, in chapter 3, we noted that this A.R.E. is bounded below by 1, where the lower bound is attained only for normal c.d.f.'s. But, in the multivariate case, for non-normal c.d.f.'s such a lower bound is not known to be available. However, since for normal F^*, $S_n(\phi)$ and T^2_n tests are asymptotically power equivalent, it follows from the A.R.E. of the $S_n(R)$ test that

$$(4.6.25) \qquad \inf_{F^* \in \mathbf{\Phi}} \inf_{\mathbf{d}^*} e_{S_n(\Phi), S_n(R)} \geq 1,$$

$$(4.6.26) \qquad \sup_{F^* \in \mathbf{\Phi}} \sup_{\mathbf{d}^*} e_{S_n(\Phi), S_n(R)} = \begin{cases} 1/0.87 = 1.15, & p = 2 \\ \infty, & p > 2, \end{cases}$$

where $\mathbf{\Phi}$ is the class of all p-variate nonsingular normal c.d.f.'s.

Some other problems are considered as Exercises 4.6.5 and 4.6.7.

EXERCISES

Section 4.2

4.2.1. Define C_i, D_i $(i = 1, 2)$ as in section 2 and obtain their exact distribution. Show that even when F_α, $\alpha = 1, \ldots, N$, have all null median vector, this distribution depends on F_α, $\alpha = 1, \ldots, N$.

4.2.2. Define γ_α, θ_α, and τ_α as in section 4.2, and consider an infinite sequence $\{F_\alpha\}$ of bivariate c.d.f.'s. Let Ω^* be the subset of the set of all $\{F_\alpha\}$, such that at least one of the following two conditions holds (for Ω^*) as $n \to \infty$:

$$\text{(I)} \qquad \liminf |n^{-1} \sum_1^n \gamma_\alpha(\theta_\alpha - \tfrac{1}{2})| > 0,$$

$$\liminf |n^{-1} \sum_1^n (1 - \gamma_\alpha)(\tau_\alpha - \tfrac{1}{2})| > 0.$$

Then (i) the randomized as well as nonrandomized conditional test based on T in (4.2.9) and the large-sample unconditional test in (4.2.15) are all consistent for $\{F_\alpha\} \in \Omega^*$; (ii) if $F_\alpha \equiv F$ for all α, show that the condition (I) is satisfied by every alternative to H_0 defined in (4.2.3). (Chatterjee, 1966b.)

4.2.3. Consider the set $\Omega^n_{(1)}$ of c.d.f.'s (F_α, $\alpha = 1, \ldots, n$) for which the following two conditions hold:

$$\text{(II)} \qquad \text{either } \theta_\alpha \geq \tfrac{1}{2} \text{ or } \theta_\alpha \leq \tfrac{1}{2} \text{ for each } \alpha = 1, \ldots, n,$$

$$\text{either } \tau_\alpha \geq \tfrac{1}{2} \text{ or } \tau_\alpha \leq \tfrac{1}{2} \text{ for each } \alpha = 1, \ldots, n.$$

Then (i) the randomized conditional test $\phi(T, C)$ is unbiased for all $\{F_\alpha\} \in \Omega^{(1)}_n$; (ii) if $F_\alpha \equiv F$ for each $\alpha = 1, \ldots, n$, (II) holds for all alternatives to H_0 defined in (4.2.3). (Chatterjee, 1966b.)

4.2.4. Show that the nonrandomized conditional test based on T defined in (4.2.9) and the large-sample test based on (4.2.15) can not be unbiased against the alternatives $\Omega^{(1)}_n$. (Chatterjee, 1966.)

Section 4.3

4.3.1. Show that the joint distribution of the ranks of $[|X_{j\alpha}|, \alpha = 1, \ldots, n]$, $j = 1, \ldots, p$, depends on the parent c.d.f.'s $F_1, \ldots, F_n$, even when $F_1 \equiv \cdots \equiv F_n \equiv F$ is diagonally symmetric about $\mathbf{0}$, unless the variables $X_{1\alpha}, \ldots, X_{p\alpha}$ are mutually independent.

4.3.2. Define Hotelling's T_n^2 as in (4.6.1) and show that under $\mathcal{T}_n$ in section 4.3, the distribution of T_n^2 asymptotically, in probability, converges to a chi-square distribution with p D.F., even when $F_1, \ldots, F_n$ are not all identical.

Section 4.4

4.4.1. Prove that $dF^*_{j(n)}(x) \leq dH^*_{j(n)}(x)$ for all $x \geq 0$.

4.4.2. Prove the inequality in (4.4.68).

4.4.3. Prove the Lemma 4.4.8.

4.4.4. Supply the proof of the Corollary 4.4.3.2.

4.4.5. Derive the results in Theorem 4.4.9.

4.4.6. Obtain the first two moments of S_n, defined by (4.4.7), under the permutational model $\mathcal{T}_n$, and show how they differ from the corresponding moments of the limiting distribution of S_n.

Section 4.5

4.5.1. From (4.5.21) and (4.5.22), obtain the multivariate sign test statistic.

4.5.2. Let $F_\alpha(x, y) = F(x/\sigma^{(1)}_\alpha, y/\sigma^{(2)}_\alpha)$, $\alpha = 1, \ldots, n$, where $(1/n)\sum_{\alpha=1}^n \sigma^{(j)}_\alpha = \bar{\sigma}^{(j)}_n \to \bar{\sigma}^{(j)}$ as $n \to \infty$, $j = 1, 2$. Show that the A.R.E. of Chatterjee's sign test with respect to the T_n^2 test is minimum when the $\sigma^{(j)}_\alpha$, $\alpha = 1, \ldots, n$, are all identical, $j = 1, 2$, provided F satisfies some mild restrictions (to be stated).

4.5.3. Generalize the result of 4.5.2 to the p (≥ 2)-variate case.

4.5.4. Show that in Exercises 4.5.2 and 4.5.3, the limit of the covariance matrix $\mathbf{\Gamma}(F^*_{(n)})$ exists for all $\sigma^{(j)}_\alpha$, when H_0 in (4.3.3) holds, and further this limit is independent of all $\sigma^{(j)}_\alpha$'s.

4.5.5. Obtain the test statistic corresponding to the U-statistics in (4.5.28), and also corresponding to the statistics in (4.5.30). Show that for $p = 1$, they reduce (respectively) to the Tukey's signed rank and the Wilcoxon signed rank statistics, when there are no ties.

Section 4.6

4.6.1. Derive the results of Theorem 4.6.1.

4.6.2. Prove (4.6.13).

4.6.3. For the multivariate sign test and multinormal F^*, prove the following: for $p = 2$, the maximum A.R.E. (relative to T_n^2) lies between $2/\pi$ and 0.72 where the minimum value is attained when $\rho_{12} = 0$ or $\rho_{12} \to \pm 1$, while the maximum value is attained when $\rho_{12} = 0.7$. Similarly, the minimum A.R.E. is bounded above by $2/\pi$ (at $\rho_{12} = 0$) and converges to zero as $\rho_{12} \to \pm 1$.

4.6.4. For Exercise 4.6.3, and any $p \geq 2$, the A.R.E. of the sign test relative to the T_n^2 test, $S_n(R)$ test, or $S_n(\Phi)$ test cannot be greater than one. (*Hint:* The proof rests on the result that $((\sin^{-1} \rho_{jk} - \rho_{jk}))$ is positive semidefinite for all $p \geq 2$.)

4.6.5. $(X_{1\alpha}, X_{2\alpha})$ is said to have a totally symmetric (bivariate) c.d.f. if $((-1)^{i_1}X_{1\alpha}, (-1)^{i_2}X_{2\alpha})$ has the same c.d.f. for all $i_j = 0, 1, j = 1, 2$. Show that a sufficient condition for the asymptotic independence of the components of $\mathbf{T}_n^*$ in (4.4.36) is the total symmetry of $(X_{\alpha j}, X_{\alpha k})$, for all $j \equiv k = 1, \ldots, p$, $\alpha = 1, \ldots, n$ (though $F_1, \ldots, F_n$ may not be all identical). For totally symmetric c.d.f.'s show that

$$e_{M_n,T^2} = 4\sum_{j=1}^{p} f_j^{*2}(0)\theta_j^2 \Big/ \left(\sum_{j=1}^{p} \theta_j^2/\sigma_{jj}\right),$$

$$e_{S_n(R),T_n^2} = 12\sum_{j=1}^{p}\left(\int_{-\infty}^{\infty} f_j^{*2}(x)\,dx\right)^2\theta_j^2 \Big/ \left(\sum_{j=1}^{p} \theta_j^2/\sigma_{jj}\right),$$

$$e_{S_n(\Phi),T_n^2} = \sum_{j=1}^{p}\theta_j^2\left(\int_{-\infty}^{\infty} \frac{f_j^{*2}(x)}{\Phi(\Phi^{-1}(F_{[j]}(x)))}\right)^2 \Big/ \left(\sum_{j=1}^{p} \theta_j^2/\sigma_{jj}\right).$$

Also, show that for the entire family $\mathcal{F}$ of totally symmetric absolutely continuous c.d.f's,

$$\inf_{F^*\in\mathcal{F}}\inf_{\mathbf{d}^*} e_{M_n,T_n^2} = 0.33 \qquad \inf_{F^*\in\mathcal{F}}\inf_{\mathbf{d}^*} e_{S_n(R),T_n^2} = 0.86$$

$$\inf_{F^*\in\mathcal{F}}\inf_{\mathbf{d}^*} e_{S_n(\Phi),T_n^2} = 1, \qquad \inf_{F^*\in\mathcal{F}}\inf_{\mathbf{d}^*} e_{S_n^*(\Phi),S_n^*(R)} = \pi/6.$$

(Bickel, 1965; Bhattacharyya, 1967.)

4.6.6. Prove (4.6.21).

4.6.7. Let $\boldsymbol{\Phi}$ be the class of p-variate nonsingular normal c.d.f.'s. Then prove the following results:

(i) $$\inf_{F^*\in\boldsymbol{\Phi}}\inf_{\mathbf{d}^*} e_{S_n(R),T_n^2} = 0 \quad \text{for} \quad p \geq 3.$$

(ii) $$\frac{3}{\pi} \leq e_{S_n(R),T_n^2} \leq 0.965 \quad \text{for} \quad p = 2.$$

(iii) $$\sup_{F^* \in \Phi} \sup_{d^*} e_{S_n(\Phi), S_n(R)}, = \infty \quad \text{for} \quad p \geq 3.$$

(iv) $$\inf_{F^* \in \Phi} \inf_{d^*} e_{S_n(\Phi), S_n(R)} \geq 1 \quad \text{for} \quad p \geq 2.$$

(v) $$\sup_{F^* \in \Phi} \sup_{d^*} e_{S_n(\Phi), M_n} = \infty \quad \text{for} \quad p \geq 3.$$

PRINCIPAL REFERENCES

Anderson (1966). Bennett (1962, 1964), Bhattacharyya (1967), Bickel (1965), Blumen (1958), Chatterjee (1966b), Hodges (1955), Hušková (1970), Joffe and Klotz (1962), Klotz (1962, 1964), Sen (1970a) and Sen and Puri (1967, 1970b).

CHAPTER 5

Multivariate Multisample Rank Tests for Location and Scale

5.1. INTRODUCTION

Consider the problem of testing the identity of c (≥ 2) multivariate distributions $F_1, \ldots, F_c$ based on independent samples from them. When $F_1, \ldots, F_c$ are all multivariate normal distributions, they can differ only in their location (mean) vectors and dispersion matrices. But for non-normal $F_1, \ldots, F_c$, the differences may be due to a variety of reasons, and the identity of location vectors and of dispersion (or association) matrices does not necessarily imply the identity of $F_1, \ldots, F_c$. Consequently, we shall assume that $F_1, \ldots, F_c$ have a common unspecified form but they may differ in their location (or scale) vectors or in their association patterns. In this chapter, suitable rank tests will be studied for the problem of testing the identity of location or of scale vectors of $F_1, \ldots, F_c$ (in the above setup), while in chapter 9 the problem of testing the identity of dispersion matrices will be considered.

It may be noted that in the several-sample multivariate case, the (unconditional) distribution of rank statistics (such as the ones in section 3.6) depends, in general, on the underlying distribution even when the null hypothesis of identity of the distributions holds. To overcome this drawback, the permutational-invariance structure of the distribution of the combined sample observations, studied in section 3.3, is extended here to the multivariate case. This enables us to construct conditional (permutational) tests based on suitable rank statistics, which are the multivariate generalizations of the univariate cases considered in section 3.6. The relationship of such conditional tests and some (analogous) asymptotically distribution-free unconditional tests are also studied here, and in this context, the distribution theory studied in section 3.6 is generalized to the multivariate case. We also discuss the asymptotic relative efficiencies of the proposed rank tests relative to their parametric competitors based on the likelihood-ratio principle.

5.2. FORMULATION OF THE MULTIVARIATE MULTISAMPLE LOCATION AND SCALE PROBLEMS

Let

$$\{\mathbf{X}_\alpha^{(k)} = (X_{1\alpha}^{(k)}, \ldots, X_{p\alpha}^{(k)})', \quad \alpha = 1, \ldots, n_k, k = 1, \ldots, c\}$$

be a set of independent vector-valued random values. The cumulative distribution function (c.d.f.) of $\mathbf{X}_\alpha^{(k)}$ is denoted by $F_k(\mathbf{x})$. The set of admissible hypotheses designates that each $F_k(\mathbf{x})$ belongs to some class of distribution functions Ω. The hypothesis to be tested, say H_0, specifies that

$$F_1(\mathbf{x}) = \cdots = F_c(\mathbf{x}) = F(\mathbf{x}) \quad \text{for all} \quad \mathbf{x}, \quad \text{where} \quad F \in \Omega. \tag{5.2.1}$$

The alternative to H_0 is the hypothesis that each $F_k(\mathbf{x})$ belongs to Ω but that (5.2.1) does not hold. To avoid the problem of ties, it is assumed throughout that the class Ω is the class of all continuous distribution functions.

We shall in this chapter pay particular attention to translation and scale-type alternatives. For translation-type alternatives, we let

$$F_k(\mathbf{x}) = F(\mathbf{x} + \boldsymbol{\delta}_k) \quad \text{for all} \quad k = 1, \ldots, c; \qquad F \in \Omega, \tag{5.2.2}$$

and we are interested in testing

$$H_0^{(1)}: \boldsymbol{\delta}_1 = \cdots = \boldsymbol{\delta}_c = \mathbf{0} \tag{5.2.3}$$

against the alternative that $\boldsymbol{\delta}_1, \ldots, \boldsymbol{\delta}_c$ are not all equal.

Again, let

$$F_k(\mathbf{x}) = F(\mathbf{X}_k^*), \qquad \mathbf{X}_k^* = \left(\frac{x_1 - \mu_1}{\sigma_1^{(k)}}, \ldots, \frac{x_p - \mu_p}{\sigma_p^{(k)}}\right), \qquad F \in \Omega; \tag{5.2.4}$$

and let $\boldsymbol{\sigma}^{(k)} = (\sigma_1^{(k)}, \ldots, \sigma_p^{(k)})$ for all $k = 1, \ldots, c$. Then, under the scale-type alternatives, we are interested in testing

$$H_0^{(2)}: \boldsymbol{\sigma}^{(1)} = \cdots = \boldsymbol{\sigma}^{(c)} = \mathbf{l}, \tag{5.2.5}$$

(when $\mathbf{l}$ is the unit vector) against the alternatives that $\boldsymbol{\sigma}_1, \ldots, \boldsymbol{\sigma}_c$ are not all equal. It may be noted that in (5.2.4), the homogeneity of location (vectors) of $F_1, \ldots, F_c$ is assumed. A more general case, when this is not assumed to be true, will be considered in chapter 9.

5.3. BASIC RANK PERMUTATION PRINCIPLE

Let us rank the N i-variate observations $X_{i\alpha}^{(k)}$, $\alpha = 1, \ldots, n_k$, $k = 1, \ldots, c$ in ascending order of magnitude, and let $R_{i\alpha}^{(k)}$ denote the rank of $X_{i\alpha}^{(k)}$ in this set. The observation vector $\mathbf{X}_\alpha^{(k)} = (X_{1\alpha}^{(k)}, \ldots, X_{p\alpha}^{(k)})'$ then gives rise to the rank vector $\mathbf{R}_\alpha^{(k)} = (R_{1\alpha}^{(k)}, \ldots, R_{p\alpha}^{(k)})'$, $\alpha = 1, \ldots, n_k$, $k = 1, \ldots, c$. The N rank vectors corresponding to the N observation vectors can be represented by the rank matrix

$$\mathbf{R}_N^{p\times N} = \begin{pmatrix} R_{11}^{(1)} & \cdots & R_{1n_1}^{(1)} & \cdots & R_{1n_c}^{(c)} \\ \cdot & & \cdot & & \cdot \\ \cdot & & \cdot & & \cdot \\ \cdot & & \cdot & & \cdot \\ R_{p1}^{(1)} & \cdots & R_{pn_1}^{(1)} & \cdots & R_{pn_c}^{(c)} \end{pmatrix} \tag{5.3.1}$$

Each row of this matrix is a random permutation of the numbers $1, 2, \ldots, N$. Thus, $\mathbf{R}_N^{p\times N}$ is a random matrix which can have $(N!)^p$ possible realizations. We shall say that two rank matrices of the form (5.3.1) are permutationally equivalent if one can be obtained from the other by a rearrangement of its columns. Thus, the matrix $\mathbf{R}_N$ is permutationally equivalent to another matrix $\mathbf{R}_N^*$ which has the same column vectors as in $\mathbf{R}_N$, but they are so arranged that the first row of $\mathbf{R}_N^*$ consists of the numbers $1, 2, \ldots, N$ in the natural order, i.e.

$$\mathbf{R}_N^* = \begin{pmatrix} 1 & 2 & \cdots & N \\ R_{21}^* & R_{22}^* & \cdots & R_{2N}^* \\ \cdot & \cdot & \cdots & \cdot \\ R_{p1}^* & R_{p2}^* & \cdots & R_{pN}^* \end{pmatrix} \tag{5.3.2}$$

Since the p variates $X_{i\alpha}^{(k)}$, $i = 1, \ldots, p$, are, in general, stochastically dependent, the joint distribution of the elements of $\mathbf{R}_N$ (or $\mathbf{R}_N^*$) will depend on the unknown distribution $F\,(\in \Omega)$, even when (5.2.1) holds. Now, $\mathbf{R}_N^*$ can have $(N!)^{p-1}$ possible realizations, as each row of it (other than the first) is a permutation of the numbers $1, \ldots, N$. Let the set of all these realizations be denoted by $\mathcal{R}_N^*$. Thus, the unconditional distribution of $\mathbf{R}_N^*$ over $\mathcal{R}_N^*$ depends, in general, on $F_1, \ldots, F_c$, even when they are identical. However, when $F_1 \equiv \cdots \equiv F_c$, $\mathbf{X}_\alpha^{(k)}$, $\alpha = 1, \ldots, n_k$, $k = 1, \ldots, c$ are all independent and identically distributed random vectors (i.i.d.r.v.), and hence, their joint distribution remains invariant under any permutation of the vectors among themselves. This implies that the conditional distribution of $\mathbf{R}_N$ over the set of $N!$ possible realizations on $S(\mathbf{R}_N^*)$ (which can be obtained by all possible permutations of the columns of $\mathbf{R}_N^*$) will be uniform under H_0: $F_1 \equiv \cdots \equiv F_c \equiv F$. That is,

$$P\{\mathbf{R}_N = \mathbf{r}_N \mid S(\mathbf{R}_N^*), H_0\} = 1/N! \quad \text{for all} \quad \mathbf{r}_N \in S(\mathbf{R}_N^*), \tag{5.3.3}$$

whatever be the parent c.d.f. $F\,(\in \Omega)$. Let us denote by $\mathcal{T}_N$, the conditional (permutational) probability measure generated by the $N!$ (equally likely) possible permutations of the columns of $\mathbf{R}_N^*$. Then, it follows from (5.3.3) that if we consider any statistic which depends explicitly on $\mathbf{R}_N$ then under $\mathcal{T}_N$, it has a completely specified (conditional) distribution. Consequently, we can always construct a test function $\phi(0 \leq \phi \leq 1)$, depending explicitly on $\mathbf{R}_N$, such that $E\{\phi \mid \mathcal{T}_N\} = \epsilon$: $0 < \epsilon < 1$, the level of significance. This, in turn implies that $E\{\phi \mid H_0\} = E[E\{\phi \mid \mathcal{T}_N\}] = \epsilon$. Thus, ϕ will be a similar test of size ϵ.

With this end in view, we shall now consider a class of tests which depends explicitly on $\mathbf{R}_N$ and rests on the probability law $\mathcal{T}_N$. We shall term such tests *permutation tests based on linear rank order statistics.*

5.4. SOME PERMUTATION RANK ORDER TESTS

As in section 3.6, we start with a general class of rank scores defined by explicitly known functions of the ranks $1, \ldots, N$, viz.,

$$E_{N,\alpha}^{(i)} = J_{N(i)}\left(\frac{\alpha}{N+1}\right), \qquad 1 \leq \alpha \leq N, \quad i = 1, \ldots, p. \tag{5.4.1}$$

Now, replacing the ranks $R_{i\alpha}^{(k)}$ in $\mathbf{R}_N$ by $E_{N,R_{i\alpha}^{(k)}}^{(i)}$, for all $i = 1, \ldots, p$, $\alpha = 1, \ldots, n_k$, $k = 1, \ldots, c$, we get a corresponding $p \times N$ matrix of general scores, which we denote by $\mathbf{E}_N$. Thus,

$$\mathbf{E}_N = \begin{pmatrix} E_{N,R_{11}^{(1)}}^{(1)} & \cdots & E_{N,R_{1n_1}^{(1)}}^{(1)} & \cdots & E_{N,R_{1n_c}^{(c)}}^{(1)} \\ \cdots & \cdots & \cdots & \cdots & \cdots \\ \cdots & \cdots & \cdots & \cdots & \cdots \\ E_{N,R_{p1}^{(1)}}^{(p)} & \cdots & E_{N,R_{pn_1}^{(1)}}^{(p)} & \cdots & E_{N,R_{pn_c}^{(c)}}^{(p)} \end{pmatrix} \tag{5.4.2}$$

We then consider the average rank scores for each $i\ (= 1, \ldots, p)$ of the c samples, defined by

$$T_{Ni}^{(k)} = (1/n_k)\sum_{\alpha=1}^{n} E_{N,R_{\alpha}^{(k)}}^{(i)}, \quad k = 1, \ldots, c; \qquad i = 1, \ldots, p. \tag{5.4.3}$$

Various common forms of (5.4.3) will be identified shortly afterwards. For convenience of presentation, we shall write

$$T_{Ni}^{(k)} = \sum_{\alpha=1}^{N} E_{N,\alpha}^{(i)} Z_{Ni,\alpha}^{(k)} / n_k \tag{5.4.4}$$

where the $E_{N,\alpha}^{(i)}$ are given numbers (scores) and $Z_{Ni,\alpha}^{(k)} = 1$ if the αth smallest observation among $X_{i\alpha}^{(k)}$, $\alpha = 1, \ldots, n_k$, $k = 1, \ldots, c$ is from the kth sample, and otherwise $Z_{Ni,\alpha}^{(k)} = 0$. Then, by straightforward computations, we obtain

$$E(T_{Ni}^{(k)} - \bar{E}_N^{(i)} \mid \mathcal{T}_N) = 0 \tag{5.4.5}$$

$$\operatorname{cov}(T_{Ni}^{(k)}, T_{Nj}^{(q)} \mid \mathcal{T}_N) = (N\,\delta_{kq} - n_k)v_{ij}(\mathbf{R}_N^*)/n_k(N-1), \quad i,j = 1,\ldots,p, \quad k = 1,\ldots,c \tag{5.4.6}$$

where δ_{kq} is the Kronecker delta, and

$$v_{ij}(\mathbf{R}_N^*) = \frac{1}{N}\sum_{k=1}^{c}\sum_{\alpha=1}^{n_k} E_{N\alpha,i}^{(k)} E_{N,\alpha j}^{(q)} - \bar{E}_N^{(i)}\bar{E}_N^{(j)} \tag{5.4.7}$$

where $E_{N\alpha,i}^{(k)}$ is the value of $E_{N,S}^{(i)}$ associated with the rank $S = R_{i\alpha}^{(k)}$, and

$$\bar{E}_N^{(i)} = \sum_{\alpha=1}^{N} E_{N,\alpha}^{(i)}/N, \qquad i = 1,\ldots,p. \tag{5.4.8}$$

Under H_0, all the observations $\mathbf{X}_\alpha^{(k)}$, $\alpha = 1,\ldots,n_k$, $k = 1,\ldots,c$ have the same distribution. Hence, for each i, the mean scores $T_{(k)}^{Ni}$, $k = 1,\ldots,c$ should be close to the weighted total mean score $\sum_{k=1}^{c} n_k T_{N,i}^{(k)}/N = \bar{E}_N^{(i)}$. Hence, it seems reasonable to base a test for H_0 on the contrasts between the mean scores $T_{N,i}^{(k)}$. In particular, we may take a set of $p(c-1)$ contrasts $T_{N,i}^{(k)} - \bar{E}_N^{(i)}$, $k = 1,\ldots,c-1$, $i = 1,\ldots,p$ and base our test on these. Under H_0, all these contrasts would stochastically be numerically small. However, for practical convenience, it would be desirable to formulate the test on the basis of a single function of these contrasts. For this reason we choose a function that would reject the numerical largeness of any of the contrasts. A positive definite quadratic form in these contrasts seem to be a natural answer. Now denoting

$$\mathbf{V}(\mathbf{R}_N^*) = ((v_{ij}(\mathbf{R}_N^*)))_{i,j=1,\ldots,p}, \tag{5.4.9}$$

the (conditional) permutational covariance matrix (under H_0) of the above $p(c-1)$ contrasts taken in that order, is

$$\frac{1}{(N-1)}\left(\left(\frac{N}{n_k}\delta_{kq} - 1\right)\right)_{k,q=1,\ldots,c} \otimes \mathbf{V}(\mathbf{R}_N^*) \tag{5.4.10}$$

where $\otimes$ is the Kronecker product of the two matrices. Since a Kronecker product is positive definite when the factor matrices are so, (5.4.10) would be positive definite if $\mathbf{V}(\mathbf{R}_N^*)$ is positive definite. Hence if $\mathbf{V}(R_N^*)$ is positive definite, the inverse of the matrix (5.4.10) is

$$(N-1)\left(\frac{n_k}{N}\delta_{kq} + \frac{n_k n_q}{n_c}\right)_{k,q=1,\ldots,c} \otimes \mathbf{V}^{-1}(\mathbf{R}_N^*) \tag{5.4.11}$$

where

$$\mathbf{V}^{-1}(R_N^*) = ((v_{ij}(\mathbf{R}_N^*)))^{-1}. \tag{5.4.12}$$

Hence, following the structure of the test asymptotically equivalent to the likelihood ratio test based on Lawley-Hotelling's generalized T^2 statistic, we take as our test statistic $\mathfrak{L}_N$,

$$\mathfrak{L}_N = \sum_{k=1}^{c} n_k[(\mathbf{T}_N^{(k)} - \bar{\mathbf{E}}_N)\mathbf{V}^{-1}(\mathbf{R}_N^*)(\mathbf{T}_N^{(k)} - \bar{\mathbf{E}}_N)'] \tag{5.4.13}$$

which is the weighted sum of c quadratic forms in $(\mathbf{T}_N^{(k)} - \bar{\mathbf{E}}_N)$, $k = 1, \ldots, c$ with the same discriminant $V_N^{-1}(\mathbf{R}_N^*)$, where $\mathbf{T}_N^{(k)} = (T_{N1}^{(k)}, \ldots, T_{Np}^{(k)})$ and $\bar{\mathbf{E}}_N = (\bar{E}_N^{(1)}, \ldots, \bar{E}_N^{(p)})$. In deriving the test statistic $\mathfrak{L}_N$, we have assumed $\mathbf{V}(\mathbf{R}_N^*)$ to be positive definite. However, if $\mathbf{V}(\mathbf{R}_N^*)$ is not positive definite, we can work with the highest-order nonsingular minor of $\mathbf{V}(\mathbf{R}_N^*)$, and obtain the statistic similar to $\mathfrak{L}_N$.

From the remarks made earlier, it follows that the conditional distribution of $\mathfrak{L}_N$ given a $\mathbf{R}_N^*$ is the same under H_0 whatever be $F(\mathbf{x}) \in \Omega$. Hence the following randomization test procedure provides an exact level ϵ test of H_0

$$\phi(\mathbf{R}_N) = \begin{cases} 1 & \text{if } \mathfrak{L}_N > \mathfrak{L}_N(\mathbf{R}_N^*), \\ A_{N,\epsilon}(\mathbf{R}_N^*) & \text{if } \mathfrak{L}_N = \mathfrak{L}_N(\mathbf{R}_N^*), \\ 0 & \text{if } \mathfrak{L}_N < \mathfrak{L}_N(\mathbf{R}_N^*), \end{cases} \tag{5.4.14}$$

where the random cutoff point is determined by

$$E\{\phi(\mathbf{R}_N) \mid \mathfrak{T}_N\} = \epsilon. \tag{5.4.15}$$

(5.4.15) implies that the unconditional size of the test is also ϵ. Thus, the permutation test based on $\mathfrak{L}_N$ is strictly distribution-free.

To carry out the test, one would need to study all the $N!$ possible permuted values of $\mathfrak{L}_N$ under $\mathfrak{T}_N$, there being at the most $N!/\prod_{k=1}^{c} n_k!$ distinct values of $\mathfrak{L}_N$ for any given $\mathbf{R}_N^*$.

Except for small values of N and p, an exact application of the permutation test (5.4.14) is difficult since the amount of computation becomes prohibitive. This leads to the study of the large-sample distribution of $\mathfrak{L}_N$, to be considered now. Before that we identify various common tests based on $\mathfrak{L}_N$ in (5.4.13).

(I) *The Multivariate Multisample Median Test.* Here, we put $E_{N,\alpha}^{(i)}$ to be equal to 1 or 0 according as $\alpha \leq [N/2]$ ($= a$, say) or not, for $\alpha = 1, \ldots, N$; $i = 1, \ldots, p$. Thus, $T_{N,i}^{(k)}$ is the sample proportion of the kth-sample ith variate values which are less than or equal to the ath smallest of the combined-sample ith variate values, $i = 1, \ldots, p$; $k = 1, \ldots, c$. It follows from (5.4.7) that in this case

$$v_{ii}(\mathbf{R}_N^*) = \frac{a(N-a)}{N^2} \quad \text{for all} \quad i = 1, \ldots, p \quad \text{and} \quad \mathbf{R}_N^* \in \mathfrak{R}_N^*, \tag{5.4.16}$$

while for $i \neq j$

$$v_{ij}(\mathbf{R}_N^*) = a_{N,ij} - (a/N)^2, \tag{5.4.17}$$

where $a_{N,ij}$ is the proportion of the combined sample observations for which the ith and jth variate values are simultaneously less than or equal to the corresponding ath smallest values, $i \neq j = 1, \ldots, p$. Finally, looking at (5.4.13), (5.4.16), and Exercise 3.3.8, we observe that for $p = 1$, the statistic $\mathfrak{L}_N$ reduces to the univariate several-sample median test statistic by Brown and Mood (1951).

(II) *Rank Sum Test*. Here, we put $E_{N,\alpha}^{(i)} = \alpha/(N+1)$, $1 \leq \alpha \leq N$, $i = 1, \ldots, p$, so that $T_{N,i}^{(k)}$ reduces to $(N+1)^{-1}$ times the average rank of the kth-sample ith variate observations among the combined-sample ith variate observations, for $k = 1, \ldots, c$ and $i = 1, \ldots, p$. The expressions for $v_{ij}(\mathbf{R}_N^*)$ are quite easy to obtain from (5.4.7) by substituting $E_{N,\alpha}^{(i)} = \alpha/(N+1)$, $1 \leq \alpha \leq N$, $i = 1, \ldots, p$. For $p = 1$, $\mathfrak{L}_N$ again reduces to the well-known Kruskal-Wallis test considered in Exercise 3.6.15.

(III) *The Normal Scores Test*. Here we define $E_{N,\alpha}^{(i)}$ to be the expected value of the αth smallest observation of a sample of size N from a standard normal distribution, for $\alpha = 1, \ldots, N$; $i = 1, \ldots, p$. (For the univariate case, see Exercise 3.6.15.)

Consider now the scale problem. Here we may work with (i) $E_{N,\alpha}^{(i)}$ equal to 1 or 0 according as $|\alpha - (N+1)/2| \geq b$ or not, (ii) $E_{N,\alpha}^{(i)}$ to be equal to $|\alpha/(N+1) - \frac{1}{2}|$, (iii) $E_{N,\alpha}^{(i)}$ to be equal to $(\alpha/(N+1) - \frac{1}{2})^2$, or (iv) $E_{N,\alpha}^{(i)}$ as the squares of the normal scores in III (location problem), for all $\alpha = 1, \ldots, N$; $i = 1, \ldots, p$. For the univariate case, see Exercises 3.6.7 and 3.6.16.

Asymptotic Permutation Distribution of $\mathfrak{L}_N$. We introduce first the following notations and assumptions.

Let the marginal c.d.f.'s of $X_{i\alpha}^{(k)}$ and $(X_{i\alpha}^{(k)}, X_{j\alpha}^{(k)})'$ be denoted by $F_{[i]}^{(k)}(x)$ and $F_{[i,j]}^{(k)}(x, y)$ respectively; $i, j = 1, \ldots, p$, $k = 1, \ldots, c$. Let $N = \sum_{k=1}^{c} n_k$ and $\lambda_N^{(k)} = n_k/N$, and assume that for all N, the inequalities

$$0 < \lambda_0 \leq \lambda_N^{(k)} \leq 1 - \lambda_0 < 1, \qquad k = 1, \ldots, c, \tag{5.4.18}$$

hold for some fixed $\lambda_0 \leq 1/c$. Let

$$F_{N[i]}^{(k)}(x) = n_k^{-1}(\text{number of } X_{i\alpha}^{(k)} \leq x, \alpha = 1, \ldots, n_k). \tag{5.4.19}$$

Then $F_{N[i]}^{(k)}(x)$ is the sample c.d.f. of the observations in the ith coordinate of the k sample. Define

$$H_{N[i]}^{(k)}(x) = \sum_{k=1}^{c} \lambda_N^{(k)} F_{N[i]}^{(k)}(x), \qquad i = 1, \ldots, p. \tag{5.4.20}$$

Thus $H_{N[i]}^{(k)}(x)$ is the sample c.d.f. of the observations in the ith coordinate of the combined sample. Let

$$F^{(k)}_{N[i,j]}(x, y) = n_k^{-1}(\text{number of } (X^{(k)}_{i\alpha}, X^{(k)}_{j\alpha}) \leq (x, y)) \tag{5.4.21}$$

$$H_{N[i,j]}(x, y) = \sum_{k=1}^{c} \lambda_N^{(k)} F^{(k)}_{N[i,j]}(x, y) \tag{5.4.22}$$

$$H_{[i,j]}(x, y) = \sum_{k=1}^{c} \lambda_N^{(k)} F^{(k)}_{[i,j]}(x, y). \tag{5.4.23}$$

Even though $H_{[i]}(x)$ and $H_{[i,j]}(x, y)$ depend on N through the λ's, our notation suppresses this fact for convenience.

Now consider the random variables $T^{(k)}_{Ni}$ defined in (5.4.4) where the scores $E^{(i)}_{N,\alpha} = J_{N(i)}(\alpha/N + 1)$ satisfy the assumptions (*a*) and (*b*) of section 3.6.1 for each $i = 1, \ldots, p$, and (*d*)

$$\int_{-\infty}^{\infty}\int_{-\infty}^{\infty}\left[J_{N(i)}\left(\frac{N}{N+1} H_{N[i]}(x)\right)J_{N(j)}\left(\frac{N}{N+1} H_{N[j]}(y)\right) - J_{(i)}\left(\frac{N}{N+1} H_{N[i]}(x)\right)J_{(j)}\left(\frac{N}{N+1} H_{N[j]}(y)\right)\right] dF^{(k)}_{N[i,j]}(x, y) = o_p(1), \tag{5.4.24}$$

$$i \neq j = 1, \ldots, p.$$

It may be noted that (5.4.24) is satisfied if, for all $i = 1, \ldots, p$

$$\int_{-\infty}^{\infty}\left[J_{N(i)}\left(\frac{N}{N+1} H_{N(i)}(x)\right) - J_{(i)}\left(\frac{N}{N+1} H_{N(i)}(x)\right)\right]^2 dF^{(k)}_{N(i)}(x) = o_p(1). \tag{5.4.25}$$

Finally, we shall assume that either assumption (*c*) of section 3.6.1 holds, or the following holds; there exists a finite set of half-open intervals $0 = \alpha_0 \leq u < \alpha_1^{(i)}$, $\alpha_1^{(i)} \leq u < \alpha_2^{(i)}, \ldots, \alpha_{r_i-1}^{(i)} \leq u < 1 = \alpha_{r_i}^{(i)}$, such that $J_{N(i)}(u)$ assumes constant values on these intervals and

$$J_{(i)}(u) = J_{(i),j} \quad \text{for all} \quad u\colon \alpha_{j-1}^{(i)} \leq u < \alpha_j^{(i)}, \qquad j = 1, \ldots, r_i, \tag{5.4.26}$$

where for some $\delta > 0$

$$\sum_{j=1}^{r_i} |J_{(i),j}|^{2+\delta}(\alpha_j^{(i)} - \alpha_{j-1}^{(i)}) < \infty, \quad \text{for all} \quad i = 1, \ldots, p. \tag{5.4.27}$$

With reference to the examples considered above, it is seen that the assumption (*c*) holds for rank sum or normal scores test, while it does not hold for the median test. However, (5.4.26) and (5.4.27) hold for a wider class of tests including the median test or the scale test based on (i). These assumptions all together are satisfied by the majority of the tests used in actual practice.

Now let $F_{[i]}(x)$ and $F_{[i,j]}(x, y)$ be the marginal c.d.f.'s of $X^{(k)}_{i\alpha}$ and $(X^{(k)}_{i\alpha}, X^{(k)}_{j\alpha})$ respectively when each $\mathbf{X}^{(k)}_{\alpha}$, $\alpha = 1, \ldots, n_k$, $k = 1, \ldots, c$ have the same distribution function $F(\mathbf{x})$. Denote

$$(5.4.28)\quad \nu_{ij}(F) = \begin{cases} \int_0^1 J_{(i)}^2(u)\,du - \mu_i^2; \quad \mu_i = \int_0^1 J_{(i)}(u)\,du, \\ \qquad\qquad\qquad\qquad i = j = 1, \ldots, p, \\ \int_{-\infty}^{\infty}\int_{-\infty}^{\infty} J_{(i)}(F_{[i]}(x))J_{(j)}(F_{[j]}(y))\,dF_{[i,j]}(x, y) - \mu_i\mu_j, \\ \qquad\qquad\qquad\qquad i \neq j = 1, \ldots, p \end{cases}$$

and

$$(5.4.29)\qquad \boldsymbol{\nu}(F) = ((\nu_{ij}(F))), \qquad i, j = 1, \ldots, p.$$

Let Ω_0 denote the class of all continuous p-variate c.d.f.'s for which $\boldsymbol{\nu}(F)$ is positive definite, and assume that

$$(5.4.30)\qquad F_k(\mathbf{x}) \in \Omega_0, \qquad k = 1, \ldots, c.$$

It may be noted that (5.4.30) implies that

$$(5.4.31)\qquad H(\mathbf{x}) = \sum_{k=1}^{c} \lambda_N^{(k)} F_k(\mathbf{x}) \in \Omega_0,$$

that is, $\boldsymbol{\nu}(H)$ is positive definite uniformly in $\lambda_N^{(k)}$, $k = 1, \ldots, c$ satisfying (5.4.18). Furthermore, it may be noted that the conditions a, b, c (or its alternate in (5.4.26) and (5.4.27)) on $J_{N(i)}$ are sufficient for the asymptotic multinormality of the joint permutation distribution of $T_{N,i}^{(k)}$'s. (5.4.30) or (5.4.31) is required to ensure the nonsingularity of the above asymptotic distribution, as the same is required, in turn, for the asymptotic distribution of $\mathcal{L}_N$. Finally, (5.4.24) or (5.4.25) is required to establish the convergence (in probability) of the (random) permutation covariance matrix $\mathbf{V}(\mathbf{R}_N^*)$ to $\boldsymbol{\nu}(H)$, which will be required in the sequel.

Theorem 5.4.1. *If $V(\mathbf{R}_N^*)$ is asymptotically nonsingular, then the joint permutation distribution of the $p(c-1)$ random variables $\{T_{Ni}^{(k)} - \bar{E}_N^{(i)};$ $i = 1, \ldots, p;\ k = 1, \ldots, c-1\}$ defined by* (5.4.4) *and* (5.4.8) *is asymptotically a $p(c-1)$-variate normal distribution.*

Proof: As in (3.4.30), we define for each $i\,(= 1, \ldots, p)$, the quantile functions $a_i(\lambda)$: $0 < \lambda < 1$, and let

$$(5.4.32)\qquad T_{N,i}^{*(k)} = n_k^{-1}\sum_{\alpha=1}^{n} a_i(U_{(R_{i\alpha}^{(k)})}^{(i)}), \qquad 1 \le i \le p, \quad 1 \le k \le c,$$

where $U_{(1)}^{(i)} \le \cdots \le U_{(N)}^{(i)}$ are the ordered random variables of a sample of size N from a rectangular c.d.f. on (0, 1), and the vector $\mathbf{U} = (U^{(1)}, \ldots, U^{(p)})'$ of the unordered random variables (corresponding to the ordered $U_{(\alpha)}^{(i)}$, $1 \le i \le p$, $1 \le \alpha \le N$) has the same distribution as of $(F_{[i]}(X_{i\alpha}^{(k)}),$

$i = 1, \ldots, p)'$; $R_{i\alpha}^{(k)}$ are defined by (5.3.1). Then, by Theorem 3.4.2, and the Bonferroni inequality,

$$n_k^{1/2}[T_{N,i}^{(k)} - T_{N,i}^{*(k)}] \xrightarrow{p} 0, \quad \text{for all} \quad i = 1, \ldots, p, \quad k = 1, \ldots, c. \tag{5.4.33}$$

Now, considering any linear compound, viz.,

$$\sum_{i=1}^{p} \sum_{k=1}^{c} d_{ik}\sqrt{n_k}\,(T_{N,i}^{*(k)}), \quad \text{where} \quad \mathbf{d} \neq \mathbf{0}, \tag{5.4.34}$$

we can readily apply the central limit theorems of section 2.7, and conclude that $[n_k^{1/2}(T_{N,i}^{*(k)} - \bar{E}_N^{(i)}),\ i = 1, \ldots, p,\ k = 1, \ldots, c]$ has asymptotically a multinormal distribution. The proof of the theorem is then completed by using (5.4.33). Q.E.D.

Theorem 5.4.2. *If the conditions a, b, c (or its alternate in* (5.4.26) *and* (5.4.27)) *and d hold, then, for all* $F_k(\mathbf{x}) \in \Omega$, $k = 1, \ldots, c$ *and uniformly in* $\lambda_N^{(1)}, \ldots, \lambda_N^{(c)}$ *satisfying* (5.4.18),

$$\mathbf{V}(\mathbf{R}_N^*) \underset{p}{\sim} \boldsymbol{\nu}(H) \quad as \quad N \to \infty. \tag{5.4.35}$$

Proof: From (5.4.8) it follows that

$$\bar{E}_N^{(i)} = \int_{-\infty}^{\infty} J_{N(i)}\left(\frac{N}{N+1} H_{N(i)}\right) dH_{N(i)} \to \int_0^1 J_{(i)}(u)\, du = \mu_i,$$

for all $i = 1, \ldots, p$. Hence, from (5.4.7) and (5.4.24), we obtain

$$\begin{aligned} &v_{ij}(\mathbf{R}_N^*) + \mu_i\mu_j \\ &\quad = \iint_{R^2} J_{(i)}\left(\frac{N}{N+1} H_{N[i]}\right) J_{(j)}\left(\frac{N}{N+1} H_{N[j]}\right) dH_{N[i,j]}(x, y) + o_p(1) \end{aligned} \tag{5.4.36}$$

where R^2 denote the two-dimensional real space. So our problem reduces to that of showing that the integral on the right-hand side of (5.4.36) is stochastically equivalent to

$$A = \iint_{R^2} J_{(i)}(H_{[i]}) J_{(j)}(H_{[j]})\, dH_{[i,j]}(x, y). \tag{5.4.37}$$

Let us first prove the theorem under condition (c). Define

$$I_{N(i)} = \{x\colon N^{-1/3} \leq H_{[i]}(x) \leq 1 - N^{-1/3}\}; \qquad i = 1, \ldots, p$$

$$I_{N(i,j)} = I_{N(i)} \cap I_{N(j)}, \qquad j = 1, \ldots, p.$$

Then, using the Cauchy-Schwarz inequality, for the integral over $R^2 - I_{N(i,j)}$, we can write the right-hand side of (5.4.36) as

$$A_N + B_{1N} + B_{2N} + o_p(1) \tag{5.4.38}$$

where

$$A_N = \iint_{I_{N(i,j)}} J_{(i)}(H_{[i]})J_{(j)}(H_{[j]})\, dH_{N[i,j]}(x, y) \tag{5.4.39}$$

$$B_{1N} = \iint_{I_{N(i,j)}} J_{(i)}\left(\frac{N}{N+1} H_{N[i]}\right) \times \left[J_{(j)}\left(\frac{N}{N+1} H_{N[j]}\right) - J_{(j)}(H_{[j]})\right] dH_{N[i,j]}(x, y) \tag{5.4.40}$$

$$B_{2N} = \iint_{I_{N(i,j)}} J_{(j)}(H_{[j]}) \times \left[J_{(i)}\left(\frac{N}{N+1} H_{N[i]}\right) - J_{(i)}(H_{[i]})\right] dH_{N[i,j]}(x, y). \tag{5.4.41}$$

Now, using condition (*c*), one easily sees that for all points in $I_{N(i,j)}$

$$\left|J_{(i)}\left(\frac{N}{N+1} H_{N[i]}\right) - J_{(i)}(H_{[i]})\right| = \left\{N^{1/2}\left|\frac{N}{N+1} H_{N[i]} - H_{[i]}\right|\right\} \times \left\{N^{-1/2}\left|J'_{(i)}\left(\phi\frac{N}{N+1} H_{N[i]} + (1-\phi)H_{[i]}\right)\right|\right\} = o_p(1), \quad 0 < \phi < 1 \tag{5.4.42}$$

as

$$\sup_x N^{1/2}\,|H_{N[i]}(x) - H_{[i]}(x)| \le \sum_{k=1}^{c}(1/\lambda_N^{(k)})^{1/2} \sup_x \sqrt{n_k}\,|F_{N[i]}^{(k)}(x) - F_{[i]}^{(k)}(x)| = O_p(1).$$

Also, noting that

$$\int_{I_{N(i)}} \left|J_{(i)}\left(\frac{N}{N+1} H_{N[i]}\right)\right| dH_{N[i]}(x) < \infty \tag{5.4.43}$$

for all $i = 1, \ldots, p$ (by condition (*c*)), we obtain from the preceding three equations that

$$|B_{1N}| = o_p(1). \tag{5.4.44}$$

Similarly, it is easily seen that under the stated regularity conditions

$$\int_{I_{N(i)}} |J_{(i)}(H_{[i]}(x))|\, dH_{N[i]}(x) \tag{5.4.45}$$

is bounded in probability for all $i = 1, \ldots, p$. Hence, from (5.4.41), (5.4.42), and (5.4.45) it follows that

$$|B_{2N}| = o_p(1). \tag{5.4.46}$$

Finally, A_N in (5.4.39) can be written as

$$\sum_{k=1}^{c} \lambda_N^{(k)} \iint_{I_{N(i,j)}} J_{(i)}(H_{[i]}(x))J_{(j)}(H_{[j]}(y))\, dF_{N[i,j]}^{(k)}(x, y) = \sum_{k=1}^{c} \lambda_N^{(k)} \left\{ \frac{1}{n_k} \sum_{\alpha=1}^{n_k} g_N(X_{i\alpha}^{(k)}, X_{j\alpha}^{(k)}) \right\} \tag{5.4.47}$$

where

$$g_N(x, y) = \begin{cases} J_{(i)}(H_{[i]}(x))J_{(j)}(H_{[j]}(y)) & \text{if} \quad (x, y) \in I_{N(i,j)} \\ 0, & \text{otherwise} \end{cases} \tag{5.4.48}$$

and, where $\{(X_{i\alpha}^{(k)}, X_{j\alpha}^{(k)});\ \alpha = 1, \ldots, n_k\}$ are independent and identically distributed random variables (i.i.d.i.v.) having the bivariate c.d.f. $F_{[i,j]}^{(k)}(x, y)$. Now by virtue of condition (c), it is easily seen (by using elementary inequalities (cf. chapter 3)) that

$$\sup_N E\{|g_N(X_{i\alpha}^{(k)}, X_{j\alpha}^{(k)})|^{1+\delta}\, |F_{[i,j]}^{(k)}\} < \infty \quad \text{for} \quad k = 1, \ldots, c, \tag{5.4.49}$$

where δ is a positive quantity defined in condition (c). Also, $\{g_N(X_{i\alpha}^{(k)}, X_{j\alpha}^{(k)}),\ \alpha = 1, \ldots, n_k\}$ are n_k independent and identically distributed random variables. Hence, using (5.4.49) we obtain

$$\frac{1}{n_k} \sum_{\alpha=1}^{n_k} g_N(X_{i\alpha}^{(k)}, X_{j\alpha}^{(k)}) \underset{p}{\sim} \iint_{R^2} J_{(i)}(H_{[i]}(x))J_{(j)}(H_{[j]}(y))\, dF_{[i,j]}^{k)}(x, y) \tag{5.4.50}$$

for $k = 1, \ldots, c$, which along with (5.4.47) implies that $A_N \underset{p}{\sim} A$, where A is defined by (5.4.37).

Let us now consider the proof when condition (c) is replaced by (5.4.26) and (5.4.27). We can write the first term on the right-hand side of (5.4.36) as

$$\sum_{r=1}^{r_i} \sum_{s=1}^{r_j} J_{(i),r} J_{(j),s} A_{N,rs}^{(i,j)}, \tag{5.4.51}$$

where $NA_{N,rs}^{(i,j)} =$ {Number of $\mathbf{X}_\alpha^{(k)}$ such that $H_{N[i]}(X_{i\alpha}^{(k)}) \in (\alpha_{r-1}^{(i)}, \alpha_r^{(i)})$, $H_{N[j]}(X_{j\alpha}^{(k)}) \in (\alpha_{s-1}^{(j)}, \alpha_s^{(j)})$ for $\alpha = 1, \ldots, n_k$, $k = 1, \ldots, c$}. Also, let $B_{N,rs}^{(i,j)} =$ $(1/N)$ {number of $\mathbf{X}_\alpha^{(k)}$ such that $H_{[i]}(X_{i\alpha}^{(k)}) \in (\alpha_{r-1}^{(i)}, \alpha_r^{(i)})$, $H_{[j]}(X_{j\alpha}^{(k)}) \in (\alpha_{s-1}^{(j)},$

$\alpha_s^{(j)}$), $\alpha = 1, \ldots, n_k$, $k = 1, \ldots, c\}$. Then, using the inequality following (5.4.42), it readily follows that

$$\sup_{\substack{r=1,\ldots,r_i \\ s=1,\ldots,r_j}} N^{1/2} |A_{N,rs}^{(i,j)} - B_{N,rs}^{(i,j)}| = O_p(1) \quad \text{for all} \quad i \neq j = 1, \ldots, p. \tag{5.4.52}$$

Thus, it follows from (5.4.51) and (5.4.52) that we are only to show that

$$\sum_{r=1}^{r_i} \sum_{s=1}^{r_j} J_{(i),r} J_{(j),s} B_{N,rs}^{(i,j)} \sim_p A, \qquad \text{defined by (5.5.47).} \tag{5.4.53}$$

Now, by definition, $B_{N,rs}^{(i,j)}$ is an average over N independent indicators (i.e., zero-one valued) random variables, and hence, by the law of large numbers

$$|B_{N,rs}^{i,j)} - E(B_{N,rs}^{(i,j)})| \xrightarrow{p} 0 \quad \text{for all} \quad r = 1, \ldots, r_i, \quad s = 1, \ldots, r_j. \tag{5.4.54}$$

Now,

$$E(B_{N,rs}^{(i,j)}) = \iint_{E_{rs}} dH_{[i,j]}(x, y), \tag{5.4.55}$$

where $E_{rs} = \{(x, y)\colon H_{[i]}(x) \in (\alpha_{r-1}^{(i)}, \alpha_r^{(i)}),\ H_{[j]}(y) \in (\alpha_{s-1}^{(j)}, \alpha_s^{(j)})\}$. Consequently, (5.4.53) follows from (5.4.54), (5.4.55), and (5.4.37). Thus

$$v_{ij}(\mathbf{R}_N^*) \sim_p \nu_{ij}(H) \quad \text{for all} \quad i \neq j = 1, \ldots, p. \tag{5.4.56}$$

Also, we note that $v_{ii}(\mathbf{R}_{N_1}^*)$ is independent of $\mathbf{R}_N^*$ and converges as $N \to \infty$ to $\nu_{ii} = \int_0^1 J_{(i)}^2(u)\, du - (\int_0^1 J_{(i)}(u)\, du)^2$, which is also the value of $\nu_{ii}(H)$, as the latter is independent of H. Hence,

$$v_{ii}(\mathbf{R}_N^*) \to \nu_{ii} \quad \text{for all} \quad i = 1, \ldots, p. \tag{5.4.57}$$

Hence the theorem.

The preceding two theorems lead to the following.

Theorem 5.4.3. *Under the conditions of Theorem* 5.4.2, *the permutation distribution of* $\mathfrak{L}_N$ *asymptotically, in probability, reduces to the chi-square distribution with* $p(c - 1)$ *degrees of freedom.*

In section 3.6, we have considered some sufficient conditions for the assumptions (a), (b), and (c) (of section 3.6.1) to hold. We shall show that essentially the same conditions validate (5.4.25) (which implies (5.4.24)). We shall consider first the case where $E_{N,\alpha}^{(i)}$ is the expected value of the αth smallest observation of a sample of size N from a continuous distribution $\psi_i(x)$, and let $\xi_{N,\alpha}^{(i)}$ be defined by $\psi_i(\xi_{N,\alpha}^{(i)}) = \alpha/(N + 1)$ for $\alpha = 1, \ldots, N$. Then, (5.4.25) may be written as

$$(1/n_k) \sum_{j=1}^{n_k} [E_{N,\alpha_j}^{(i)} - \xi_{N,\alpha_j}^{(i)}]^2, \tag{5.4.58}$$

where $1 \leq \alpha_1 < \cdots < \alpha_{n_k} \leq N$ are n_k distinct integers. (5.4.58) is less than

$$(N/n_k)\{(1/N)\sum_{\alpha=1}^{N}[E_{N,\alpha}^{(i)} - \xi_{N,\alpha}^{(i)}]^2\}. \tag{5.4.59}$$

Hence, it suffices to show that (5.4.59) converges to 0 as $N \to \infty$. To do this we consider the following lemma.

Lemma 5.4.4. *Let $\Psi(x)$ be any continuous c.d.f., satisfying*

(i) $\int x\, d\Psi(x) = 0, \qquad \int x^2\, d\Psi(x) < \infty,$

(ii) *$\Psi(x)$ is symmetric about $x = 0$; $\Psi(x)$ is convex for all $x \leq 0$ and concave for all $x > 0$ (or vice versa). Then*

$$\lim_{N\to\infty}(1/N)\sum_{i=1}^{N}[E_{N,i} - \xi_{N,i}]^2 = 0. \tag{5.4.60}$$

Proof: The left-hand side of (5.4.60) can be written as

$$\frac{1}{N}\sum_{i=1}^{N}E_{N,i}^2 + \frac{1}{N}\sum_{i=1}^{N}\xi_{N,i}^2 - \frac{2}{N}\sum_{i=1}^{N}E_{N,i}\xi_{N,i}. \tag{5.4.61}$$

It follows from the Ali and Chan inequality (cf. section 2.2) that if $\Psi(x)$ is concave (convex) for $x \geq 0$, then for $i \geq [(N+1)/2]$, $E_{N,i} \geq (\leq)\, \xi_{N,i}$ and opposite inequalities hold for $i < [(N+1)/2]$. Thus, if $\Psi(x)$ is convex for $x \leq 0$ and concave for $x > 0$, we readily get from the above result that

$$\sum_{i=1}^{N}\xi_{N,i}^2 \leq \sum_{i=1}^{N}\xi_{N,i}E_{N,i} \leq \sum_{i=1}^{N}E_{N,i}^2 \tag{5.4.62}$$

and the opposite inequalities hold if $\Psi(x)$ is concave for $x \leq 0$ and convex for $x \geq 0$. Thus, from (5.4.62), we get

$$\frac{1}{N}\sum_{i=1}^{N}(E_{N,i} - \xi_{N,i})^2 \leq \left|\frac{2}{N}\sum_{i=1}^{N}E_{N,i}^2 - \frac{2}{N}\sum_{i=1}^{N}\xi_{N,i}^2\right| \tag{5.4.63}$$

Again it follows from Hoeffding's results (cf. Exercise 5.4.4) that

$$\frac{1}{N}\sum_{i=1}^{N}E_{N,i}^2 \to \int x^2\, d\Psi(x) < \infty, \quad \text{as} \quad N \to \infty, \tag{5.4.64}$$

and by the fundamental theorem of integral calculus as $N \to \infty$,

$$\frac{1}{N}\sum_{i=1}^{N}\xi_{N,i}^2 \to \int x^2\, d\Psi(x) < \infty. \tag{5.4.65}$$

From (5.4.63), (5.4.64), and (5.4.65) we readily conclude that (5.4.62) converges to zero as $N \to \infty$. Q.E.D.

In particular, if $\Psi(x)$ is a standardized normal c.d.f., it is convex for $x \leq 0$ and concave for $x > 0$, and hence the result of Lemma 5.7.1 holds. The same lemma holds for all unimodal symmetric distribution with finite variance, such as logistic, double exponential, rectangular, and many other c.d.f.'s.

Remark: If condition (*c*) of section 3.6.1 is replaced by (5.4.26) and (5.4.27), the above proof does not hold, but (5.4.25) follows by more elementary treatment (cf. Exercise 5.4.3). Also, for scale problems, usually we take $E_{N,\alpha}^{(i)}$ as the square of the expected value of the αth smallest observation of a sample of size N from the c.d.f. $\Psi_i(x)$ and $J_{(i)}(\alpha/(N+1))$ as $[\xi_{N,\alpha}^{(i)}]^2$. In this case, it is easy to extend the result of Lemma 5.4.4, provided $\int x^4 \, d\Psi(x) < \infty$. (cf. Exercise 5.4.5).

Further, we require to show that $\boldsymbol{\nu}(F)$ in (5.4.29) is positive definite under certain conditions. $\boldsymbol{\nu}(F)$ will cease to be positive definite when there exists one or more linear relations among the variables $J_{(i)}(F_{[i]}(x))$, $i = 1, \ldots, p$. This in turn requires that the scatter of the points of the distribution $F(x_1, \ldots, x_p)$ is confined to a $(p-1)$ or lower-dimensional hypercurve whose equation is

$$\sum_{i=1}^{p} c_i J_{(i)}(F_{[i]}(X_i)) = \text{constant} \tag{5.4.66}$$

(or more than one such equation). Thus, if the c.d.f. $F(x_1, \ldots, x_p)$ is nonsingular in the sense that there is no $(p-1)$ or lower-dimensional subspace of the p-dimensional Euclidean space E_p whose probability mass is exactly equal to unity, then (5.4.66) cannot hold, in probability, and hence $\boldsymbol{\nu}(F)$ would be positive definite. In actual practice this is practically no serious restriction on F.

Finally, in the case of two populations, $\mathfrak{L}_N$ in (5.4.13) can be expressed directly in terms of $\mathbf{T}_N^{(1)}$ or $\mathbf{T}_N^{(2)}$, viz.,

$$\mathfrak{L}_N = (n_1 N/n_2)(\mathbf{T}_N^{(1)} - \bar{E}_N)V^{-1}(\mathbf{R}_N^*)(\mathbf{T}_N^{(1)} - \bar{E}_N)', \tag{5.4.67}$$

where $N = n_1 + n_2$. This follows from the fact that $n_1\mathbf{T}_N^{(1)} + n_2\mathbf{T}_N^{(2)} = N\bar{E}_N$.

By virtue of Theorem 5.4.3, the permutation test procedure based on $\mathfrak{L}_N$ simplifies in large samples to the following rule:

$$\begin{aligned} \text{if } \mathfrak{L}_N &\geq \chi^2_{\epsilon, p(c-1)}, && \text{reject } H_0, \\ &< \chi^2_{\epsilon, p(c-1)}, && \text{accept } H_0, \end{aligned} \tag{5.4.68}$$

where $\chi_{\epsilon,r}$ is the $100(1-\epsilon)\%$ point of a chi-square distribution with r d.f. (degrees of freedom).

In order to study the power properties of the test considered above, we require to study the unconditional distribution of $\mathfrak{L}_N$, under appropriate classes of alternative hypotheses. This, in turn, requires the study of the joint distribution of the rank order statistics defined by (5.4.4) and is considered in the next section.

5.5. ASYMPTOTIC MULTINORMALITY OF T_N, $k = 1, \ldots, c$, $i = 1, \ldots, p$ FOR ARBITRARY $F_1(\mathbf{x}), \ldots, F_c(\mathbf{x})$

We shall consider first the case when assumption (c) holds, and subsequently, we shall prove a second theorem when (c) is replaced by (5.4.26) and (5.4.27).

Theorem 5.5.1. *If $F_k \in \mathscr{F}$ for all $k = 1, \ldots, c$, and if $J_{N(i)}$ satisfies the conditions (a), (b), and (c) of section (3.6.1) for all $i = 1, \ldots, p$, then the stochastic vector with elements $[N^{1/2}(T_{Ni}^{(k)} - \mu_{Ni}^{(k)}), i = 1, \ldots, p; k = 1, \ldots, c]$ has a limiting normal distribution with zero mean vector and covariance matrix $\mathbf{\Sigma} = ((\sigma_{N[i,j]}^{(k,q)}))$, where*

$$\mu_{Ni}^{(k)} = \int_{-\infty}^{+\infty} J_{(i)}(H_{[i]}(x))\, dF_{[i]}^{(k)}(x); \qquad i = 1, \ldots, p; \quad k = 1, \ldots, c. \tag{5.5.1}$$

$$\begin{aligned}
\sigma_{N[i,j]}^{(k,q)} = {} & 2 \sum_{\substack{r=1 \\ r \neq k}}^{c} \lambda_N^{(r)} \iint\limits_{-\infty < x < y < \infty} A_i^{(r)}(x, y)\, dF_{[i]}^{(r)}(x)\, dF_{[i]}^{(r)}(y) \\
& + \frac{2}{\lambda_N^{(k)}} \sum_{\substack{r=1 \\ r \neq k}}^{c} [\lambda_N^{(r)}]^2 \iint\limits_{-\infty < x < y < \infty} A_i^{(k)}(x, y)\, dF_{[i]}^{(r)}(x)\, dF_{[i]}^{(r)}(y) \\
& + \frac{1}{\lambda_N^{(k)}} \sum_{\substack{r=1 \\ s \neq r, r \neq k \\ s \neq k}}^{c} \sum_{s=1}^{c} \lambda_N^{(r)} \lambda_N^{(s)} \Bigg[\iint\limits_{-\infty < x < y < \infty} A_i^{(k)}(x, y)\, dF_{[i]}^{(r)}(x)\, dF_{[i]}^{(s)}(y) \\
& + \iint\limits_{-\infty < y < x < \infty} A_i^{(k)}(y, x)\, dF_{[i]}^{(r)}(x)\, dF_{[i]}^{(s)}(y) \Bigg] \\
& \qquad \text{(if } k = q,\ i = j\text{);}
\end{aligned} \tag{5.5.2}$$

$$\begin{aligned}
& = \sum_{r=1}^{c} \sum_{s=1}^{c} \frac{\lambda_N^{(r)} \lambda_N^{(s)}}{\lambda_N^{(k)}} \int_{-\infty}^{+\infty} \int_{-\infty}^{+\infty} B_{i,j}^{(k)}(x, y)\, dF_{[i]}^{(r)}(x)\, dF_{[j]}^{(s)}(y) \\
& \quad - \sum_{r=1}^{c} \lambda_N^{(r)} \int_{-\infty}^{+\infty} \int_{-\infty}^{+\infty} B_{i,j}^{(k)}(x, y) [dF_{[i]}^{(r)}(x)\, dF_{[j]}^{(k)}(y) \\
& \quad + dF_{[i]}^{(k)}(x)\, dF_{[j]}^{(r)}(y)] \\
& \quad + \sum_{r=1}^{c} \lambda_N^{(r)} \int_{-\infty}^{+\infty} \int_{-\infty}^{+\infty} B_{i,j}^{(r)}(x, y)\, dF_{[i]}^{(k)}(x)\, dF_{[j]}^{(k)}(y) \\
& \qquad \text{(if } k = q,\ i \neq j\text{);}
\end{aligned} \tag{5.5.3}$$

$$
\begin{aligned}
&= -\sum_{r=1}^{c} \lambda_N^{(r)} \Big[\iint_{-\infty<x<y<\infty} A_i^{(k)}(x, y)\, dF_{[i]}^{(r)}(x)\, dF_{[i]}^{(q)}(y) \\
&\quad + \iint_{-\infty<y<x<\infty} A_i^{(k)}(y, x)\, dF_{[i]}^{(r)}(x)\, dF_{[i]}^{(q)}(y) \Big] \\
&\quad - \sum_{r=1}^{c} \lambda_N^{(r)} \Big[\iint_{-\infty<x<y<\infty} A_i^{(q)}(x, y)\, dF_{[i]}^{(r)}(x)\, dF_{[i]}^{(k)}(y) \\
&\quad + \iint_{-\infty<y<x<\infty} A_i^{(q)}(y, x)\, dF_{[i]}^{(r)}(x)\, dF_{[i]}^{(k)}(y) \Big] \\
&\quad + \sum_{r=1}^{c} \lambda_N^{(r)} \Big[\iint_{-\infty<x<y<\infty} A_i^{(r)}(x, y)\, dF_{[i]}^{(k)}(x)\, dF_{[i]}^{(q)}(y) \\
&\quad + \iint_{-\infty<y<x<\infty} A_i^{(r)}(y, x)\, dF_{[i]}^{(k)}(x)\, dF_{[i]}^{(q)}(y) \Big] \\
&\qquad (\text{if } k \neq q,\ i = j);
\end{aligned}
\tag{5.5.4}
$$

$$
\begin{aligned}
&= -\sum_{r=1}^{c} \lambda_N^{(r)} \int_{-\infty}^{+\infty} \int_{-\infty}^{+\infty} B_{i,j}^{(k)}(x, y)\, dF_{[i]}^{(r)}(x)\, dF_{[j]}^{(q)}(y) \\
&\quad - \sum_{r=1}^{c} \lambda_N^{(r)} \int_{-\infty}^{+\infty} \int_{-\infty}^{+\infty} B_{i,j}^{(q)}(x, y)\, dF_{[i]}^{(k)}(x)\, dF_{[j]}^{(r)}(y) \\
&\quad + \sum_{r=1}^{c} \lambda_N^{(r)} \int_{-\infty}^{+\infty} \int_{-\infty}^{+\infty} B_{i,j}^{(r)}(x, y)\, dF_{[i]}^{(k)}(x)\, dF_{[j]}^{(q)}(y) \\
&\qquad (\text{if } k \neq q,\ i \neq j);
\end{aligned}
\tag{5.5.5}
$$

where

$$A_i^{(\alpha)}(t, u) = F_{[i]}^{(\alpha)}(t)[1 - F_{[i]}^{(\alpha)}(u)]J'_{(i)}[H_{[i]}(t)]J'_{(i)}[H_{[i]}(u)] \tag{5.5.6}$$

$$B_{i,j}^{(\alpha)}(t, u) = [F_{[i,j]}^{(\alpha)}(t, u) - F_{[i]}^{(\alpha)}(t)F_{[j]}^{(\alpha)}(u)]J'_{(i)}[H_{[i]}(t)]J'_j[H_{[j]}(u)]. \tag{5.5.7}$$

Proof: As in Theorem 3.6.5, we rewrite $T_{Ni}^{(k)}$ as

$$T_{Ni}^{(k)} = \mu_{Ni}^{(k)} + B_{1N(i)}^{(k)} + B_{2N(i)}^{(k)} + \sum_{r=1}^{4} C_{rN(i)}^{(k)} \tag{5.5.8}$$

where $\mu_{Ni}^{(k)}$ is given by (5.5.1),

$$B_{1N(i)}^{(k)} = \int J_{(i)}(H_{[i]}(x))\, d[F_{N[i]}^{(k)}(x) - F_{[i]}^{(k)}(x)] \tag{5.5.9}$$

$$B_{2N(i)}^{(k)} = \int [H_{N[i]}(x) - H_{[i]}(x)]J'_{(i)}[H_{(i)}(x)]\, dF_{[i]}^{(k)}(x) \tag{5.5.10}$$

and the $C_{rNi}^{(k)}$, $r = 1, 2, 3, 4$ are all $o_p(N^{-1/2})$; for $i = 1, \ldots, p$ and $k = 1, \ldots, c$. The difference $N^{1/2}(T_{Ni}^{(k)} - \mu_{Ni}^{(k)}) - N^{1/2}(B_{1N(i)}^{(k)} + B_{2N(i)}^{(k)})$ tends to zero in probability for all $i = 1, \ldots, p$ and $k = 1, \ldots, c$ and so the vectors

$$[N^{1/2}(T_{Ni}^{(k)} - \mu_{Ni}^{(k)}), \quad i = 1, \ldots, p; \quad k = 1, \ldots, c]$$

and

$$[N^{1/2}(B_{1N(i)}^{(k)} + B_{2N(i)}^{(k)}), \quad i = 1, \ldots, p; \quad k = 1, \ldots, c]$$

have the same limiting distribution if they have one at all. Thus, to prove this theorem, it suffices to show that for any real non-null $\boldsymbol{\delta} = (\boldsymbol{\delta}_1', \ldots, \boldsymbol{\delta}_c')$, the random variable

$$N^{1/2} \sum_{k=1}^{c} \boldsymbol{\delta}_k' [\mathbf{B}_{1N}^{(k)} + \mathbf{B}_{2N}^{(k)}] \tag{5.5.11}$$

(where $\mathbf{B}_{rN}^{(k)} = (B_{rN(1)}^{(k)}, \ldots, B_{rN(p)}^{(k)})'$, $r = 1, 2$), has asymptotically a normal distribution.

Now, as in Theorem 3.6.5, we can rewrite $B_{1N(i)}^{(k)} + B_{2N(i)}^{(k)}$ as

$$\sum_{\substack{q=1\\ \neq k}}^{c} \lambda_N^{(q)} \left\{ \frac{1}{n_q} \sum_{\alpha=1}^{n_q} B_{kq(1)}^{(i)}(X_{i\alpha}^{(q)}) - \frac{1}{n_k} \sum_{\alpha=1}^{n_k} B_{kq(2)}^{(i)}(X_{i\alpha}^{(k)}) \right\} \tag{5.5.12}$$

where

$$B_{kq(1)}^{(i)}(X_{i\alpha}^{(q)}) = \int_{-\infty}^{\infty} [F_{1[i]}^{(q)} - F_{(i)}^{(q)}] J_{(i)}'(H_{[i]}) \, dF_{[i]}^{(k)}(x) \tag{5.5.13}$$

$$B_{kq(2)}^{(i)}(X_{i\alpha}^{(k)}) = \int_{-\infty}^{\infty} [F_{1[i]}^{(k)} - F_{[i]}^{(k)}] J_{(i)}'(H_{[i]}) \, dF_{[i]}^{(q)}(x) \tag{5.5.14}$$

$F_{1[i]}^{(k)}$ being the empirical c.d.f. of $X_{1\alpha}^{(k)}$, whose true c.d.f. is $F_{[i]}^{(k)}(x)$, for $q, k = 1, \ldots, c$; $i = 1, \ldots, p$. If we now define

$$B_{0k}^{(i)}(X_{i\alpha}^{(k)}) = \sum_{\substack{q=1\\ \neq k}}^{c} \delta_q^{(i)} B_{qk(1)}^{(k)}(X_{i\alpha}^{(k)}), \qquad B_{k0}^{(i)}(X_{i\alpha}^{(k)}) = \sum_{\substack{q=1\\ \neq k}}^{c} \lambda_N^{(q)} B_{kq(2)}^{(i)}(X_{i\alpha}^{(k)}), \tag{5.5.15}$$

$$B_k^{(i)}(X_{i\alpha}^{(k)}) = B_{0k}^{(i)}(X_{i\alpha}^{(k)}) - \frac{\delta_k^{(i)}}{\lambda_N^{(k)}} B_{k0}^{(i)}(X_{i\alpha}^{(k)}) \tag{5.5.16}$$

and

$$B_N^{(i)} = \frac{1}{N} \sum_{k=1}^{c} \sum_{\alpha=1}^{n_k} B_k^{(i)}(X_{i\alpha}^{(k)}), \quad i = 1, \ldots, p, \qquad \mathbf{B}_N = (B_N^{(1)}, \ldots, B_N^{(p)})' \tag{5.5.17}$$

then (5.5.11) can be written as $N^{1/2}\mathbf{B}_N$, which is a stochastic p vector with elements $B_N^{(i)}$ $(i = 1, \ldots, p)$ which involve c independent sums of independent and identically distributed random variables $B_k^{(i)}(X_{i\alpha}^{(k)})$, $\alpha = 1, \ldots, n_k$; $k = 1, \ldots, c$. Further, as in Theorem 3.6.5 it can be easily shown that

$$E\{|B_k^{(i)}(X_i^{(k)})|^{2+\delta}\} < \infty \tag{5.5.18}$$

for all $i = 1, \ldots, p$; $k = 1, \ldots, c$, where δ is defined under condition (c).

Hence, the desired asymptotic multinormality follows from (5.5.17) and (5.5.18) through an application of the (vector case) central limit theorem.

Further, it can be shown by routine algebraic manipulations that

$$\begin{aligned}
&\operatorname{cov}\{B^{(i)}_{kq(1)}(X^{(q)}_{i\alpha}),\, B^{(j)}_{k'q'(1)}(X^{(q')}_{j\alpha})\} = 0 \quad \text{if} \quad q \neq q', \\
&= \int_{-\infty}^{+\infty}\int_{-\infty}^{+\infty} B^{(q)}_{i,j}(x, y)\, dF^{(k)}_{[i]}(x)\, dF^{(k')}_{[j]}(y) \quad \text{if} \quad q = q'; \quad i \neq j; \\
(5.5.19) \quad &= \iint_{-\infty<x<y<\infty} A^{(q)}_{i}(x, y)\, dF^{(k)}_{[i]}(x)\, dF^{(k')}_{[i]}(y) \\
&\quad + \iint_{-\infty<y<x<\infty} A^{(q)}_{i}(y, x)\, dF^{(k)}_{[i]}(x)\, dF^{(k')}_{[i]}(y) \quad \text{if} \quad q = q', \quad i = j,
\end{aligned}$$

where $A^{(q)}_{i}(x, y)(A^{(q)}_{i}(y, x))$ and $B^{(q)}_{i,j}(x, y)$ are given by (5.5.6) and (5.5.7) respectively.

$$\begin{aligned}
&\operatorname{cov}\{B^{(i)}_{kq(2)}(X^{(k)}_{i\alpha}),\, B^{(j)}_{k'q'(2)}(X^{(k')}_{j\alpha})\} = 0 \quad \text{if} \quad k \neq k' \\
&= \int_{-\infty}^{+\infty}\int_{-\infty}^{+\infty} B^{(k)}_{i,j}(x, y)\, dF^{(q)}_{[i]}(x)\, dF^{(q')}_{[j]}(y) \quad \text{if} \quad k = k'; \quad i \neq j; \\
(5.5.20) \quad &= \iint_{-\infty<x<y<\infty} A^{(k)}_{i}(x, y)\, dF^{(q)}_{[i]}(x)\, dF^{(q')}_{[i]}(y) \\
&\quad + \iint_{-\infty<y<x<\infty} A^{(k)}_{i}(y, x)\, dF^{(q)}_{[i]}(x)\, dF^{(q')}_{[i]}(y) \quad \text{if} \quad k = k'; \quad i = j.
\end{aligned}$$

Finally

$$\begin{aligned}
&\operatorname{cov}\{B^{(i)}_{kq(1)}(X^{(k)}_{i\alpha}),\, B^{(j)}_{k'q'(2)}(X^{(k')}_{j\alpha})\} = 0, \quad \text{if} \quad q \neq k'; \\
&= \int_{-\infty}^{+\infty}\int_{-\infty}^{+\infty} B^{(q)}_{i,j}(x, y)\, dF^{(k)}_{[i]}(x)\, dF^{(q')}_{[j]}(y) \quad \text{if} \quad q = k'; \quad i \neq j; \\
(5.5.21) \quad &= \iint_{-\infty<x<y<\infty} A^{(q)}_{i}(x, y)\, dF^{(k)}_{[i]}(x)\, dF^{(q')}_{[i]}(y) \\
&\quad + \iint_{-\infty<y<x<\infty} A^{(q)}_{i}(y, x)\, dF^{(k)}_{[i]}(x)\, dF^{(q')}_{[i]}(y) \quad \text{if} \quad q = k'; \quad i = j.
\end{aligned}$$

The rest of the proof follows from (5.5.12)—(5.5.14), (5.5.19)–(5.5.21), and some routine computation. Q.E.D. ◁

We shall now consider the case when condition (c) is replaced by (5.4.26) and (5.4.27). For this, define

$$H_{[i]}(\zeta_s^{(i)}) = \alpha_s^{(i)}, \qquad s = 1, \ldots, r_i - 1, \quad i = 1, \ldots, p, \tag{5.5.22}$$

and assume that at $\zeta_s^{(i)}$, $F_{[i]}^{(k)}(x)$, $k = 1, \ldots, c$ are all absolutely continuous with

$$h_{[i]}(\zeta_s^{(i)}) = (\partial/\partial x)H_{[i]}(x) \mid_{\zeta_s^{(i)}} > 0 \quad \text{for all} \quad s = 1, \ldots, r_i - 1, \quad i = 1, \ldots, p. \tag{5.5.23}$$

Note that for notational simplicity the dependence of $\zeta_s^{(i)}$ on N is understood. Also, define

$$a_{k,s}^{(i)} = \mathrm{f}_{[i]}^{(k)}(\zeta_s^{(i)})/h_{[i]}(\zeta_s^{(i)}), \qquad k = 1, \ldots, c; \quad s = 1, \ldots, r_i - 1; \quad i = 1, \ldots, p. \tag{5.5.24}$$

Thus, by definition

$$0 \le a_{k,s}^{(i)} \le 1/\lambda_0 \quad \text{for all} \quad k, s, \text{and } i. \tag{5.5.25}$$

We define $\mu_{Ni}^{(k)}$ as in (5.5.1), but instead of (5.5.2), we let here

$$\begin{aligned}
\sigma_{N[i,j]}^{(k,q)} = \sum_{s=1}^{r_i-1}\sum_{s'=1}^{r_j-1} &[J_{(i),s} - J_{(i),s+1}][J_{(j),s'} - J_{(j),s'+1}] \\
&\Bigg\{\left(\frac{\delta_{kq}}{\lambda_N^{(k)}} - a_{k,s}^{(i)} - a_{q,s'}^{(j)}\right)[F_{[i,j]}^{(k)}(\zeta_s^{(i)}, \zeta_{s'}^{(j)}) - F_{[i]}^{(k)}(\zeta_s^{(i)})F_{[j]}^{(k)}(\zeta_{s'}^{(j)})] \\
&\quad + a_{k,s}^{(i)}a_{q,s'}^{(j)}[H_{[i,j]}(\zeta_s^{(i)}, \zeta_{s'}^{(j)}) - H_{[i]}(\zeta_s^{(i)})H_{[j]}(\zeta_{s'}^{(j)})] \\
&\quad - a_{k,s}^{(i)}a_{q,s'}^{(j)}\Bigg[\sum_{k'=1}^{c}\lambda_N^{(k')}\{F_{[i]}^{(k')}(\zeta_s^{(i)}) - H_{[i]}(\zeta_s^{(i)})\} \\
&\quad \times \{F_{[j]}^{(k)}(\zeta_{s'}^{(j)}) - H_{[j]}(\zeta_{s'}^{(j)})\}\Bigg]\Bigg\},
\end{aligned} \tag{5.5.26}$$

for $k, q = 1, \ldots, c$, $i, j = 1, \ldots, p$, where δ_{kq} is the Kronecker delta, and where for $i = j$, the bivariate c.d.f. $F_{[i,i]}^{(k)}(x, y)$ (or $H_{[i,i]}(x, y)$) is to be read as $F_{[i]}^{(k)}(x)$ (or $H_{[i]}(x)$), if $x \le y$.

Theorem 5.5.2. *If $J_{N(i)}$ satisfies the conditions* (a) *and* (b) *of section* 3.6.1, *and if* (5.4.26), (5.4.27), *and* (5.5.23) *hold, then the stochastic vector* $[N^{1/2}(T_{Ni}^{(k)} - \mu_{Ni}^{(k)}),\ k = 1, \ldots, c,\ i = 1, \ldots, p]$ *has asymptotically a multi-normal distribution with null mean vector and covariance matrix with elements defined by* (5.5.26).

Outline of the Proof: Define the combined sample quantiles by

$$H_{N[i]}(Z_{N,s}^{(i)}) = \alpha_s^{(i)}, \qquad s = 1, \ldots, r_i - 1, \quad i = 1, \ldots, p. \tag{5.5.27}$$

Then, under (5.5.23), we will show that

$$N^{1/2}\,|Z_{N,s}^{(i)} - \zeta_s^{(i)}| = O_p(1) \quad \text{for all} \quad s = 1, \ldots, r_i - 1, \quad i = 1, \ldots, p \tag{5.5.28}$$

Using (5.5.22) and (5.5.27), it follows that for any $a > 0$,

$$\begin{aligned} &P\{N^{1/2}(Z_{N,s}^{(i)} - \zeta_s^{(i)}) > a\} \\ &= P\left\{\frac{1}{N}\sum_{k=1}^{c}\sum_{\alpha=1}^{n_k} c(\zeta_s^{(i)} + N^{-1/2}a - X_{i\alpha}^{(k)}) \leq \alpha_s^{(i)}\right\} \\ &= P\left\{N^{-1/2}\left[\sum_{k=1}^{c}\sum_{\alpha=1}^{n_k} c(\zeta_s^{(i)} + N^{-1/2}a - X_{i\alpha}^{(k)}) - NH_{[i]}(\zeta_s^{(i)} + N^{-1/2}a)\right]\right. \\ &\qquad \left.\leq N^{1/2}[\alpha_s^{(i)} - H_{[i]}(\zeta_s^{(i)} + N^{-1/2}a)]\right\}, \end{aligned} \tag{5.5.29}$$

where $c(u)$ is 1 or 0 according as u is positive or not. Now, the second term within the parenthesis on the extreme right-hand side of (5.5.29) converges to $-ah_{[i]}(\zeta_s^{(i)})$, while, by the classical central limit theorem, the first term has asymptotically a normal distribution with zero mean and limiting variance

$$\sum_{k=1}^{c} \lambda_N^{(k)} F_{[i]}^{(k)}(\zeta_s^{(i)})[1 - F_{[i]}^{(k)}(\zeta_s^{(i)})] < \tfrac{1}{4}. \tag{5.5.30}$$

Consequently, for any arbitrarily small $\epsilon(>0)$, we can always select a value of a, say a_ϵ and $N \geq N_0(\epsilon)$, such that

$$P\{N^{1/2}(Z_{N,s}^{(i)} - \zeta_s^{(i)}) > a_\epsilon\} < \epsilon/2R, \tag{5.5.31}$$

where $R = \sum_{i=1}^{p}(r_i - 1)$. Similarly, for $N \geq N_0(\epsilon)$,

$$P\{N^{1/2}(Z_{N,s}^{(i)} - \zeta_s^{(i)}) < -a_\epsilon\} < \epsilon/2R. \tag{5.5.32}$$

Then, (5.5.28) follows from (5.5.31), (5.5.32), and the Bonferroni inequality. Next, we shall show that *for any real and finite a*

$$\begin{aligned} n_k^{1/2}\,|[&F_{N[i]}^{(k)}(\zeta_s^{(i)} + N^{-1/2}a) - F_{N[i]}^{(k)}(\zeta_s^{(i)})] \\ &- [F_{[i]}^{(k)}(\zeta_s^{(i)} + N^{-1/2}a) - F_{[i]}^{(k)}(\zeta_s^{(i)})]| = o_p(1), \end{aligned} \tag{5.5.33}$$

for all $k = 1, \ldots, c$; $s = 1, \ldots, r_i - 1$ and $i = 1, \ldots, p$. The proof of (5.5.33) is a direct consequence of (2.11.26), (2.11.27), (2.11.32), (2.3.34), and the Bonferroni inequality.

Now, using the definitions of $H_{N[i]}$ and $H_{[i]}$, (5.5.23) and (5.5.33), it follows by some routine analysis [cf. Sen (1970c)] that for any real and finite a

$$\begin{aligned} N^{1/2}\,|[&F_{N[i]}^{(k)}(\zeta_s^{(i)} + N^{-1/2}a) - F_{N[i]}^{(k)}(\zeta_s^{(i)})] \\ &- a_{k,s}^{(i)}[H_{N[i]}(\zeta_s^{(i)} + N^{-1/2}a) - H_{N[i]}(\zeta_s^{(i)})]| = o_p(1) \end{aligned} \tag{5.5.34}$$

for all $k = 1, \ldots, c$; $s = 1, \ldots, r_i - 1$ and $i = 1, \ldots, p$. Hence, using (5.5.22), (5.5.23), (5.5.27), (5.5.28), and (5.5.34), it follows that

$$N^{1/2} \,|[F_{N[i]}^{(k)}(Z_{N,s}^{(i)}) - F_{N[i]}^{(k)}(\zeta_s^{(i)})] + a_{k,s}^{(i)}[H_{N[i]}(\zeta_s^{(i)}) - H_{[i]}(\zeta_s^{(i)})]| = o_p(1) \tag{5.5.35}$$

for all $k = 1, \ldots, c$; $s = 1, \ldots, r_i - 1$; $i = 1, \ldots, p$. Now, using conditions (*a*) and (*b*) of section 3.6.1, (5.5.1), (5.4.26), and (5.4.27), we obtain that

$$\begin{aligned}
N^{1/2}(T_{Ni}^{(k)} - \mu_{Ni}^{(k)}) &= \sum_{s=1}^{r_i-1}[J_{(i),s} - J_{(i),s+1}]N^{1/2} \\
&\quad \times [F_{N[i]}^{(k)}(Z_{N,s}^{(i)}) - F_{[i]}^{(k)}(\zeta_s^{(i)})] + o_p(1) \\
&= \sum_{s=1}^{r_i-1} [J_{(i),s} - J_{(i),s+1}] \\
&\quad \times \{N^{1/2}[F_{[i]}^{(k)}(Z_{N,s}^{(i)}) - F_{N[i]}^{(k)}(\zeta_s^{(i)})] \\
&\quad + N^{1/2}[F_{N[i]}^{(k)}(\zeta_s^{(i)}) - F_{[i]}^{(k)}(\zeta_s^{(i)})]\} + o_p(1) \\
&= \sum_{s=1}^{r_i-1} [J_{(i),s} - J_{(i),s+1}] \\
&\quad \times \{N^{1/2}[F_{N[i]}^{(k)}(\zeta_s^{(i)}) - F_{[i]}^{(k)}(\zeta_s^{(i)})] \\
&\quad - a_{k,s}^{(i)} \sum_{q=1}^{c} \lambda_N^{(q)} \cdot N^{1/2}[F_{N[i]}^{(q)}(\zeta_s^{(i)}) - F_{[i]}^{(q)}(\zeta_s^{(i)})]\} + o_p(1),
\end{aligned} \tag{5.5.36}$$

where the last term follows from (5.5.35). Now, recalling that $F_{N[i]}^{(k)}(\zeta_s^{(i)}) = (1/n_k) \sum_{\alpha=1}^{n_k} c(\zeta_s^{(i)} - X_{i\alpha}^{(k)})$, $k = 1, \ldots, c$, $s = 1, \ldots, r_i - 1$, $i = 1, \ldots, p$, we observe that (5.5.36) involves a linear function of averages over c independent sets of independent stochastic vectors. Hence, the rest of the proof of the asymptotic normality follows from the classical central limit theorem (for bounded variables) in the vector case. Finally, the computation of the terms in (5.5.26) follows from (5.5.36) on noting that

$$\operatorname{cov}\{c(\zeta_s^{(i)} - X_{i\alpha}^{(k)}), c(\zeta_{s'}^{(j)} - X_{j\beta}^{(q)})\} = \delta_{\alpha\beta} \cdot \delta_{kq} \cdot \{F_{[i,j]}^{(k)}(\zeta_s^{(i)}, \zeta_{s'}^{(j)}) - F_{[i]}^{(k)}(\zeta_s^{(i)})F_{[j]}^{(k)}(\zeta_{s'}^{(j)})\}, \tag{5.5.37}$$

for all $k, q = 1, \ldots, c$; $s = 1, \ldots, r_i - 1$, $s' = 1, \ldots, r_j - 1$; $i, j = 1, \ldots, p$. Q.E.D.

It may be noted that the asymptotic normal distribution derived in Theorems 5.5.1 and 5.5.2 is singular and the rank of this distribution can at most be equal to $p(c - 1)$. If the null hypothesis is true, then it is readily seen that

$$\sigma_{N[i,j]}^{(k,q)} = (\delta_{kq}/\lambda_N^{(k)} - 1)\nu_{ij}(F), \tag{5.5.38}$$

where δ_{kq} is the Kronecker delta and $\nu_{ij}(F)$ is given by (5.4.28). Thus, if $\boldsymbol{\nu}(F)$ is positive definite, the asymptotic distribution of $[N^{1/2}(T_{Ni}^{(k)} - \mu_i^{(k)}),\ i = 1, \ldots, p,\ k = 1, \ldots, c]$ is multivariate normal of rank $p(c - 1)$, and hence, it follows that

$$(5.5.39) \qquad \mathfrak{L}_N^* = \sum_{k=1}^{c} n_k[\mathbf{T}_N^{(k)} - \boldsymbol{\mu}]\boldsymbol{\nu}^{-1}(F)[\mathbf{T}_N^{(k)} - \boldsymbol{\mu}]'$$

has under H_0 asymptotically a chi-square distribution with $p(c - 1)$ degrees of freedom.

If now $\hat{\boldsymbol{\nu}}^{-1}(F)$ is a consistent estimator of $\boldsymbol{\nu}^{-1}(F)$, then

$$(5.5.40) \qquad \hat{\mathfrak{L}}_N = \sum_{k=1}^{c} n_k(T_N^{(k)} - \boldsymbol{\mu})\hat{\nu}^{-1}(F)(T_N^{(k)} - \boldsymbol{\mu})'$$

has also, under H_0, asymptotically a chi-square distribution with $p(c - 1)$ degrees of freedom. Hence an asymptotic test of H_0 may be based on the statistic $\hat{\mathfrak{L}}_N$. We shall call the tests based on $\hat{\mathfrak{L}}_N$ asymptotically distribution-free rank order tests.

5.6. THE LIMITING DISTRIBUTIONS OF RANK ORDER TESTS FOR SEQUENCES OF TRANSLATION AND SCALE ALTERNATIVES

In this section, we shall concern ourselves with a sequence of admissible alternative hypotheses $\{H_N\}$ which specifies that for each $k = 1, \ldots, c$,

$$(5.6.1) \qquad F_k(\mathbf{x}) = F(x_{1N}^{(k)}, \ldots, x_{pN}^{(k)}) \quad \text{with} \quad F \in \Omega_0$$

where

$$(5.6.2) \qquad x_{iN}^{(k)} = (x_i + \theta_i^{(k)}N^{-1/2})(1 + \delta_i^{(k)}N^{-1/2}), \qquad i = 1, \ldots, p.$$

We shall use the notation

$$(5.6.3) \qquad \boldsymbol{\theta}^{(k)} = (\theta_1^{(k)}, \ldots, \theta_p^{(k)}) \quad \text{and} \quad \delta^{(k)} = (\delta_1^{(k)}, \ldots, \delta_p^{(k)})$$

and assume that at least one of the following two equalities

$$\boldsymbol{\theta}^{(1)} = \cdots = \boldsymbol{\theta}^{(c)}, \qquad \boldsymbol{\delta}^{(1)} = \cdots = \boldsymbol{\delta}^{(c)}$$

is not true. It may be noted that if $\boldsymbol{\delta}^{(1)} = \cdots = \boldsymbol{\delta}^{(c)}$, we have the usual location problem, and if $\boldsymbol{\theta}^{(1)} = \cdots = \boldsymbol{\theta}^{(c)}$, we have the scale problem; while if both the equalities hold, then $F_1 \equiv \cdots \equiv F_c$. Furthermore we shall denote as before the marginal c.d.f.'s of the ith variate, and that of the ith and jth variates, corresponding to the distribution function $F(x_1, \ldots, x_p)$ by $F_{[i]}$ and $F_{[i,j]}$ respectively.

Next, let us write

$$(5.6.4)\quad B_{(i)}(F) = \int_{-\infty}^{+\infty} \frac{d}{dx} J_{(i)}(F_{[i]}(x))\, dF_{[i]}(x), \qquad i = 1, \ldots, p,$$

$$(5.6.5)\quad C_{(i)}(F) = \int_{-\infty}^{+\infty} x \frac{d}{dx} J_{(i)}(F_{[i]}(x))\, dF_{[i]}(x), \qquad i = 1, \ldots, p,$$

$$(5.6.6)\quad \eta_i^{(k)} = B_{(i)}(F)(\theta_i^{(k)} - \bar{\theta}_i) + C_i(F)(\delta_i^{(k)} - \bar{\delta}_i), \qquad k = 1, \ldots, c,$$

$$(5.6.7)\quad \boldsymbol{\eta}^{(k)} = (\eta_1^{(k)}, \ldots, \eta_p^{(k)}), \qquad k = 1, \ldots, c;$$

where

$$\bar{\theta}_i = \sum_{k=1}^{c} \lambda^{(k)} \theta_i^{(k)} \quad \text{and} \quad \bar{\delta}_i = \sum_{k=1}^{c} \lambda^{(k)} \delta_i^{(k)} \quad \text{for} \quad i = 1, \ldots, p.$$

It is to be noted that when the condition (c) of section 3.6.1 is replaced by (5.4.26), (5.4.27), and (5.5.23), $\boldsymbol{\eta}^{(k)}$ has to be defined in a slightly different manner. We define $\zeta_s^{(i)}$ as in (5.5.22) with $H_{[i]}$ replaced by $F_{[i]}$, and $h_{[i]}(\zeta_s^{(i)})$ is also replaced by $f_{[i]}(\zeta_s^{(i)})$. Then, $\eta_i^{(k)}$ is to be defined as

$$(5.6.8)\quad \begin{aligned} \eta_i^{(k)} &= \theta_i^{(k)} \sum_{s=1}^{r_i-1} \{J_{(i),s} - J_{(i),s+1}\} f_{[i]}(\zeta_s^{(i)}) \\ &\quad + \delta_i^{(k)} \sum_{s=1}^{r_i-1} \{J_{(i),s} - J_{(i),s+1}\} \zeta_s^{(i)} f_{[i]}(\zeta_s^{(i)}) \end{aligned}$$

for $k = 1, \ldots, c$ $i = 1, \ldots, p$. Then, we have the following.

Theorem 5.6.1. *If* (i) *for all* $k = 1, \ldots, c$, $\lim_{N\to\infty} \lambda_N^{(k)} = \lambda^{(k)}$ *exists and is a positive fraction less than unity, and* (ii) *the conditions of Theorem* 5.5.1 *or Theorem* 5.5.2 *are satisfied then under the hypothesis* $\{H_N\}$ *defined by* (5.6.1) *and* (5.6.2), *the random vector* $[N^{1/2}(T_{Ni}^{(k)} - \mu_{Ni}^{(k)}), i = 1, \ldots, p; k = 1, \ldots, c]$ *has a limiting normal distribution with mean vector zero and covariance matrix* $\Sigma^* = ((\sigma_{[i,j]}^{(k,q)}))$ *where*

$$(5.6.9)\quad \sigma_{[i,j]}^{(k,q)} = \left(\frac{\delta_{kq}}{\lambda^{(k)}} - 1\right) \nu_{ij}(F), \quad i, j = 1, \ldots, p \cdot \quad k, q = 1, \ldots, c,$$

where δ_{kq} *is the usual Kronecker delta and* $\nu_{ij}(F)$ *is given by* (5.4.28).

The proof of this theorem is an immediate consequence of Theorem 5.5.1, Theorem 5.5.2, and some routine algebra, and is therefore omitted.

The following theorem gives the asymptotic distribution of $\hat{\mathfrak{L}}_N$ under the hypothesis $\{H_N\}$.

Theorem 5.6.2. *If* (i) *conditions* (i) *and* (ii) *of Theorem* 5.6.1 *are satisfied.* (ii) $F_{[i]}(x)$ *has a density* $f_{[i]}(x)$, *and for all* q (= 1, . . . , c), $i = 1, \ldots, p$,

$$\text{(iii)}\quad \alpha_i^{(q)} = \lim_{N\to\infty} \sqrt{N}\int_{-\infty}^{\infty}\left[J_{(i)}\left\{\sum_{k=1}^{c}\lambda_N^{(k)}F_{[i]}(x+\theta_i^{(k)}N^{-\frac{1}{2}})(1+\delta_i^{(k)}\cdot N^{-\frac{1}{2}})\right\} - J_{(i)}\{F_{[i]}^{(q)}(x)\}\right] dF_{[i]}^{(q)}(x)$$

exists and is finite, then $\hat{\mathfrak{L}}_N$ *is asymptotically distributed as noncentral chi square with* $p(c-1)$ *degrees of freedom and noncentrality parameter*

$$\Delta = \sum_{k=1}^{c}\lambda^{(k)}[\boldsymbol{\alpha}^{(k)}\boldsymbol{\nu}^{-1}(F)\boldsymbol{\alpha}'^{(k)}] \tag{5.6.10}$$

where

$$\boldsymbol{\alpha}^{(k)} = (\alpha_1^{(k)}, \ldots, \alpha_p^{(k)}), \qquad k = 1, \ldots, c.$$

In many situations, the noncentrality parameter can be computed easily with the aid of the following lemma.

Lemma 5.6.3. *If* $F(\mathbf{x})$ *and the function* $J_{(i)}$ *are such that for every* $i = 1, \ldots, p$ (i) $F_{[i]}(x)$ *is continuous and differentiable in each of the open intervals* $(-\infty, a_{i1}), (a_{i1}, a_{i2}), \ldots, (a_{i2}, \infty)$, (ii) $F_{[i]}(x)$ *has a density* $f_{[i]}(x)$ *which is bounded in each of these intervals, and* (iii) *the function* $(d/dx)\{J_{(i)}[F_{[i]}(x)]\}$ *is bounded as* $x \to \pm\infty$ *then, the limits* $\boldsymbol{\alpha}^{(k)}$ $k = 1, \ldots, c$ *exist and are given by* $\boldsymbol{\alpha}^{(k)} = -\boldsymbol{\eta}^{(k)}$ *where* $\boldsymbol{\eta}^{(k)}$ *is given by* (5.6.7) *and* (5.6.8).

The proof of this lemma is similar to problem 3.6.15, and is therefore omitted. In case the conditions of Lemma 5.6.3 are satisfied, then

$$\Delta_{\mathfrak{L}} = \sum_{k=1}^{c}\lambda^{(k)}[\boldsymbol{\eta}^{(k)}(\boldsymbol{\nu}^{-1}(F))\boldsymbol{\eta}'^{(k)}] \tag{5.6.11}$$

where $\boldsymbol{\eta}^{(k)}$ is given by (5.6.7) and (5.6.8).

Theorem 5.6.4. *If the conditions of Theorem* 5.6.1 *and Lemma* 5.6.3 *are satisfied then under the hypothesis* $\{H_N\}$ *defined by* (5.6.1) *and* (5.6.2) *the statistic* $\mathfrak{L}_N$, *defined by* (5.4.13), *has, asymptotically, a noncentral chi-square distribution with the noncentrally parameter* $\Delta_{\mathfrak{L}}$ *defined in* (5.6.11).

Proof: It follows from Theorem 5.6.1 that under $\{H_N\}$ the rank of the asymptotic normal distribution of $(T_N^{(k)} - \mu_N^{(k)})$, $k = 1, \ldots, c$, is $p(c-1)$. Further, it follows from Theorem 5.4.2 and Theorem 5.6.1 that under the stated conditions

$$\boldsymbol{\nu}(H) \to \boldsymbol{\nu}(F) \text{ as } N \to \infty, \tag{5.6.12}$$

$$\boldsymbol{\nu}(\mathbf{R}_N^*) \underset{p}{\sim} \boldsymbol{\nu}(H) \sim \boldsymbol{\nu}(F). \tag{5.6.13}$$

Finally under $\{H_N\}$ it is easily seen (by using (5.4.1) and (5.5.1)) that

$$(5.6.14) \quad [\mu_{Ni}^{(k)} - \bar{E}_N^{(i)}] \to \eta_i^{(k)} \quad \text{for all} \quad i = 1, \ldots, p; \quad k = 1, \ldots, c.$$

The rest of the proof of the theorem follows from Theorem 5.6.1 and the results of section 2.9 [particularly (xii)]. ◁

It now follows from Theorems 5.4.3, 5.6.1, 5.6.2, and 5 6.34 that under the sequence $\{H_N\}$ of alternative hypotheses

$$(5.6.15) \qquad \mathfrak{L}_N \underset{p}{\sim} \hat{\mathfrak{L}}_N$$

and hence, using the results of section 3.5, it follows that for testing H_0 against the set of alternatives $\{H_N\}$ the permutation test and the asymptotically distribution-free rank test are asymptotically power equivalent. ◁

5.7. MULTIVARIATE TESTS BASED ON THE *U*-STATISTICS

In this section, we shall consider another class of multivariate multisample tests. These tests are all permutation tests and are based on an appropriate generalized *U*-statistic, the theory of which is discussed in chapter 3.

As before, $\mathbf{X}_\alpha^{(k)} = (X_{1\alpha}^{(k)}, \ldots, X_{p\alpha}^{(k)})$, $\alpha = 1, \ldots, n_k$ are independent observations from the c.d.f. $F_k(\mathbf{x})$, $k = 1, \ldots, c$, which need not be continuous. Let us denote

$$(5.7.1) \qquad \mathbf{F} = (F_1, \ldots, F_c), \qquad \mathbf{F} \in \Omega.$$

Let

$$(5.7.2) \qquad \boldsymbol{\theta}(F) = (\theta_1(\mathbf{F}), \ldots, \theta_t(\mathbf{F}))', \qquad t \geq 1$$

be a vector-valued regular functional of $\mathbf{F} \in \Omega$, and assume that

$$(5.7.3) \qquad \boldsymbol{\theta}(\mathbf{F}) = \boldsymbol{\theta}^0 = (\theta_1^0, \ldots, \theta_t^0)' \quad \text{under} \quad H_0 \quad \text{in (5.2.1)}$$

where $\boldsymbol{\theta}^0$ is known and is independent of $F(\mathbf{x})$. Since $\boldsymbol{\theta}(\mathbf{F})$ is estimable for all $\mathbf{F} \in \Omega$, there exists a vector-valued *kernel* $\boldsymbol{\phi} = (\phi_1, \ldots, \phi_t)'$ of $\boldsymbol{\theta}(\mathbf{F})$, where

$$(5.7.4) \quad \phi_i = \phi_i(X_{\alpha_{11}}^{(1)}, \ldots, X_{\alpha_{1m_{i1}}}^{(1)}, \ldots, X_{\alpha_{c1}}^{(c)}, \ldots, X_{\alpha_{cm_{ic}}}^{(c)}), \qquad i = 1, \ldots, t$$

and the *degree* matrix

$$(5.7.5) \qquad \mathbf{m} = \begin{bmatrix} m_{11} & \cdots & m_{1c} \\ \cdot & \cdot & \cdot \\ \cdot & \cdot & \cdot \\ \cdot & \cdot & \cdot \\ m_{t1} & \cdots & m_{tc} \end{bmatrix}$$

have elements m_{ij}, which are all non-negative integers, and where ϕ_i is symmetric in $(X_{\alpha_{k1}}^{(k)}, \ldots, X_{\alpha_{km_{ik}}}^{(k)})$ for each $k = 1, \ldots, c$; though the roles of the c different sets may not be symmetric. Then the generalized U-statistic $U_N = (U_{N1}, \ldots, U_{Nt})$ corresponding to $\boldsymbol{\phi}$ is given by

$$(5.7.6)\quad U_{Ni} = \left[\prod_{i=1}^{c} \binom{n_k}{m_{ik}}^{-1}\right] \cdot \sum_{S_i} \phi_i(X_{\alpha_{11}}^{(1)}, \ldots, X_{\alpha_{cm_{ic}}}^{(c)}), \qquad i = 1, \ldots, t,$$

where the summation S_i extends over all possible $1 \leq \alpha_{k1} < \cdots < \alpha_{km_{ik}} \leq n_k$, $k = 1, \ldots, c$; $i = 1, \ldots, t$, and the subscript N stands for $(\sum_{k=1}^{c} n_k)$, the combined sample size. We shall consider the tests for H_0 based on the statistic $\mathbf{U}_N$.

We combine the c samples together into a combined sample of size $N\,(= \sum_{k=1}^{c} n_k)$, and denote the N observations by a vector with c-tuple elements

$$(5.7.7)\quad \mathbf{Z}_N = (Z_1, \ldots, Z_N); \quad Z_\alpha = (Z_{1\alpha}, \ldots, Z_{p\alpha}), \qquad \alpha = 1, \ldots, N$$

where conventionally, we let

$$(5.7.8)\quad \begin{aligned} Z_\alpha &= X_\alpha^{(1)}, \quad \text{for} \quad \alpha = 1, \ldots, n_1 \\ &= X_{\alpha - n_1 - \cdots - n_{k-1}}^{(k)}, \quad \text{for} \quad \alpha = n_1 + \cdots + n_{k-1} + 1, \ldots, \\ &\qquad\qquad n_1 + \cdots + n_k\,; \qquad k = 2, \ldots, c. \end{aligned}$$

$\mathbf{Z}_N$ will be termed the *collection vector*. Now, under (H_0), $\mathbf{Z}_N$ is composed of N i.i.d. random vectors, and as a result, under (H_0) and given $\mathbf{Z}_N$, all possible permutations of the coordinates of $\mathbf{Z}_N$ are equally likely (conditionally), each one having the conditional probability $1/N!$. Hence, all possible partitioning of the N observations into c subsets of sizes $n_1, \ldots, n_c$ respectively, are conditionally (under H_0 and given $\mathbf{Z}_N$) equally likely, each having the permutational probability

$$(5.7.9)\qquad \binom{N}{n_1, \ldots, n_c}^{-1} = \frac{n_1! \cdots n_c!}{N!}$$

As (5.7.9) is independent of $\mathbf{Z}_N$ as well as of $F(\mathbf{x})$, we use the known permutation distribution of $\mathbf{U}_N$ (which is induced by the completely known probability measure (5.7.9)), and the associated test based on it will be conditionally distribution-free.

Now the formulation of the test procedure depends on the properties of the permutation distribution of $\mathbf{U}_N$. Hence, we will consider first some results on the permutation distribution of $\mathbf{U}_N$; in later sections we will consider the proposed class of tests and their properties.

Permutation Properties of U_N. Let us define $m_i = m_{i1} + \cdots + m_{ic}$, for $i = 1, \ldots, t$, and let

$$(5.7.10)\qquad \phi_i^*(Z_{\alpha_1}, \ldots, Z_{\alpha_{m_i}}) = (m_i!)^{-1} \textstyle\sum_{S_i^*} \phi_i(Z_{\alpha_1}, \ldots, Z_{\alpha_{m_i}}),$$

where the summation S_i^* extends over all the $m_i!$ possible permutations of the coordinates of $\phi_i(\ldots)$, so that ϕ_i^* is the completely symmetric form of ϕ_i, $i = 1, \ldots, t$. Also let

$$\boldsymbol{\phi}^* = (\phi_1^* = \cdots, \phi_t^*)'. \tag{5.7.11}$$

The class of $\boldsymbol{\theta}(\mathbf{F})$ considered here is assumed to satisfy the following two conditions on its estimators $\mathbf{U}_N$:

(5.7.12) (i) $\boldsymbol{\phi}^* = \boldsymbol{\theta}^0$, for all $\mathbf{Z}_N$;

(5.7.13) (ii) the covariance matrix of $N^{1/2}\{\mathbf{U}_N - \boldsymbol{\theta}^0\}$

has asymptotically (as $N \to \infty$ subject to

$$n_k/N \to \lambda_k\colon\ 0 < \lambda_k < 1; \qquad \sum_{k=1}^{c} \lambda_k = 1) \tag{5.7.14}$$

a positive definite limit $\boldsymbol{\Gamma}(F, \boldsymbol{\lambda})$, under H_0; where F is a common (unknown) c.d.f. and $\boldsymbol{\lambda} = (\lambda_1, \ldots, \lambda_c)$.

Let now $\mathcal{T}(\mathbf{Z}_N)$ stand for the permutation probability measure generated by (5.7.15). It is then easily seen that

$$E\{\mathbf{U}_N \mid \mathcal{T}(\mathbf{Z}_N)\} = \boldsymbol{\theta}^0 \quad \text{for all} \quad \mathbf{Z}_N. \tag{5.7.15}$$

Let us also assume that

$$E\{\boldsymbol{\phi}\boldsymbol{\phi}' \mid F \in \omega_0\} < \infty. \tag{5.7.16}$$

We now consider under H_0 the covariance of ϕ_i and ϕ_j (defined in (5.7.4)), when d_k of $\{X_\alpha^{(k)}\}$'s are common between the two sets of $\{X_\alpha^{(k)}\}$'s (one in each of ϕ_i and ϕ_j), for $k = 1, \ldots, c$. We denote the above covariance as

$$\zeta_{d_1 \ldots d_c}^{(i,j)}(F), \quad 0 \le d_k \le \min(m_{ik}, m_{jk}), \quad k = 1, \ldots, c', \quad i, j = 1, \ldots, t. \tag{5.7.17}$$

On the other hand, if we consider the probability measure $\mathcal{T}(\mathbf{Z}_N)$, we denote the same covariance as

$$\zeta_{d_1 \ldots d_c}^{(i,j)}(\mathbf{Z}_N), \quad 0 \le d_k \le \min(m_{ij}, m_{jk}), \quad k = 1, \ldots, d; \quad i, j = 1, \ldots, t. \tag{5.7.18}$$

It may be noted here that the expressions in (5.7.17) are all estimable parameters, while those in (5.7.18) are all random variables, as they depend explicitly on $\mathbf{Z}_N$. Further, the permutation covariances in (5.7.18) can be evaluated very simply as the average of the product of deviations of ϕ_i and ϕ_j from θ_i° and θ_j° respectively, where the average extends over all possible terms for which d_k of the $\{Z_\alpha\}$'s are common in the kth set of $\{Z_\alpha\}$'s of ϕ_i and ϕ_j, for $k = 1, \ldots, c$.

It then follows that

$$\zeta^{(i,j)}_{0\ldots0}(\mathbf{Z}_N) = 0 \quad \text{for all} \quad i, j = 1, \ldots, t. \tag{5.7.19}$$

Also, it follows by simple arguments that

$$\sigma_{ij}(\mathbf{Z}_N) = \operatorname{cov}\{U_{Ni}, U_{Nj} \mid \mathcal{T}(\mathbf{Z}_N)\}$$

$$= \left\{\prod_{k=1}^{c}\binom{n_k}{m_{jk}}^{-1}\right\}\sum_{d_1=0}^{m_{i1}}\cdots\sum_{d_c=0}^{m_{ic}}\prod_{k=1}^{c}\left\{\binom{m_{ik}}{d_k}\binom{n_k - m_{ik}}{m_{jk} - d_k}\right\}\zeta^{(i,j)}_{d_1\ldots d_c}(\mathbf{Z}_N); \tag{5.7.20}$$

$$\sigma_{ij}(F) = \operatorname{cov}\{U_{Ni}, U_{Nj} \mid F_1 = \cdots = F_c = F\}$$

$$= \left\{\prod_{k=1}^{c}\binom{n_k}{m_{jk}}^{-1}\right\}\sum_{d_1=1}^{m_{i1}}\cdots\sum_{d_c=0}^{m_{ic}}\prod_{k=1}^{c}\left\{\binom{m_{ik}}{d_k}\binom{n_k - m_{ik}}{m_{jk} - d_k}\right\}\zeta^{(i,j)}_{d_1\ldots d_c}(F), \tag{5.7.21}$$

for all $i, j = 1, \ldots, t$. Let then

$$\boldsymbol{\Sigma}(\mathbf{Z}_N) = ((\sigma_{ij}(\mathbf{Z}_N)))_{i,j=1,\ldots,t} \quad \text{and} \quad \boldsymbol{\Sigma}(F) = ((\sigma_{ij}(F)))_{i,j=1,\ldots,t}. \tag{5.7.22}$$

With these notations, we will state the following theorems, the proofs of which are straightforward extensions of similar theorems in section 3.5 and are left as exercises.

Theorem 5.7.1. *If* (5.7.16) *holds and* $\mathbf{F} \in \omega_0$, $N|\boldsymbol{\Sigma}(\mathbf{Z}_N) - \boldsymbol{\Sigma}(F)| \xrightarrow[\text{a.s.}]{} 0$, *and further if* (5.7.13) *holds then* $N\boldsymbol{\Sigma}(\mathbf{Z}_N) \xrightarrow[a.s.]{} \boldsymbol{\Gamma}(F, \boldsymbol{\lambda})$. *Moreover, if* $\boldsymbol{\phi}$ *has finite fourth-order moments for all* $\mathbf{F} \in \omega_0$, *and for the c.d.f.* F, *the associated collection vector is complete, and then* $N\boldsymbol{\Sigma}(\mathbf{Z}_N)$ *has uniformly* (*in* $\mathbf{F}$) *the minimum concentration ellipsoid* (*as well as minimum risk with any convex loss function*) *among all unbiased estimators of* $N\boldsymbol{\Sigma}(F)$.

Theorem 5.7.2. *If* $\boldsymbol{\Phi}$ *has finite fourth-order moments for all* $\mathbf{F} \in \Omega$, *then* $N\,|\boldsymbol{\Sigma}(\mathbf{Z}_N) - \boldsymbol{\Sigma}(\bar{F})| \xrightarrow{\mathcal{P}} 0$, *where* $\bar{F} = \sum_{k=1}^{c} \lambda_k F_k$.

Theorem 5.7.3. *If, under* H_0, $\boldsymbol{\phi}$ *has finite moments up to the order* $2 + \delta$, *for some* $\delta > 0$, *or if under the alternative* $\boldsymbol{\phi}$ *has finite fourth-order moments, then subject to* (5.7.13) *and* (5.7.14), *the permutation distribution of* $N^{1/2}\{\mathbf{U}_N - \boldsymbol{\theta}^0\}$ *asymptotically, in probability, reduces to a multinormal distribution with a null mean vector and a dispersion matrix* $\boldsymbol{\Gamma}(\bar{F}, \boldsymbol{\lambda})$. *Further, the permutation distribution of the statistic*

$$T_N = [\mathbf{U}_N - \boldsymbol{\theta}^0]'(\boldsymbol{\Sigma}(\mathbf{Z}_N))^{-1}[\mathbf{U}_N - \boldsymbol{\theta}^0] \tag{5.7.23}$$

asymptotically, in probability, converges to a chi-square distribution with t *degrees of freedom* (*D.F.*).

With the aid of these results we shall now construct a suitable test based on $\mathbf{U}_N$. For small samples one can use the permutation distribution of $\mathbf{U}_N$ (under the model (5.7.9)) to formulate a class of strictly distribution-free tests. Though this problem appears to be deterministic, the labor involved in this evaluation of the exact permutation distribution of $\mathbf{U}_N$ increases almost prohibitively with the increase in the sample sizes. Hence, it seems advisable to use a suitable test statistic and to approximate the permutation distribution of this statistic by some simple law, which simplifies the test for large samples at least. Theorem 5.7.3 suggests that T_N in (5.7.23) may be a suitable test statistic and the chi-square distribution with t d.f. may be used to determine the critical value of T_N for large N.

A second type of asymptotically distribution-free tests may be considered at this stage. Let us define $N\mathbf{\Sigma}(F)$ as in (5.7.22) and let $N\hat{\mathbf{\Sigma}}(F)$ be any consistent estimator of it. We then consider a statistic T_N^* defined as

$$T_N^* = N[\mathbf{U}_N - \boldsymbol{\theta}^0]'(N\,\hat{\mathbf{\Sigma}}(F))^{-1}[\mathbf{U}_N - \boldsymbol{\theta}^0]. \tag{5.7.24}$$

It can then be shown, that under H_0, T_N^* has asymptotically a χ^2 distribution with t D.F., and hence, an asymptotic test procedure may be suggested which is based on the T_N^* rejecting the null hypothesis for large T_N^*.

We shall now establish certain asymptotic power equivalence relations of the two tests based on T_N and T_N^* respectively. This is done for a sequence of alternative hypotheses (including the null hypothesis), for which the power of the tests lie in the open interval (0, 1). For this purpose, let us assume that there is a sequence of values of the pooled sample size N (subject to (5.7.14)), and for each N, we replace $\mathbf{F} \in \Omega$ by $\mathbf{F}_N \in \Omega$, where $\{\mathbf{F}_N\}$ is a sequence of c-tuplets of c.d.f.'s, defined for each N, such that each F_{kN} converges (as $N \to \infty$) to a common $F (k = 1, \ldots, c)$, at all points of continuity of the latter, in such a manner that

$$H_N\colon\ \boldsymbol{\theta}(\mathbf{F}_N) = \boldsymbol{\theta}^0 + N^{-1/2}\boldsymbol{\eta}_N; \qquad \boldsymbol{\eta}_N \to \boldsymbol{\eta} \text{ as } N \to \infty, \tag{5.7.25}$$

where $\boldsymbol{\eta} = (\eta_1, \ldots, \eta_t)$ is a real and finite vector. It may be noted that the formulation of the class of alternative hypotheses $\{H_N\}$ in (5.7.25) is somewhat more general than the usual type of location or scale alternatives in the Pitman sense, where the continuity of $\mathbf{F}$ appears to be more or less essential.

Theorem 5.7.4. *If* $\boldsymbol{\phi}$ *has finite fourth moment for all* $\mathbf{F} \in \Omega$ *and if* (5.7.12) *and* (5.7.13) *hold, then under* $\{H_N\}$ *in* (5.7.25), $T_N \underset{p}{\sim} T_N^*$; *moreover, either of the test statistics has asymptotically (under* $\{H_N\}$*) a noncentral chi-square distribution with t D.F. and the noncentrality parameter*

$$\Delta = \boldsymbol{\eta}'(\mathbf{\Gamma}(F, \boldsymbol{\lambda}))^{-1}\boldsymbol{\eta}. \tag{5.7.26}$$

Proof: It follows that under $\{H_N\}$, $\bar{F}(x) = \sum_{k=1}^{c} \lambda_k F_k(x)$ converges to F as $N \to \infty$, and hence by Theorem 5.7.2,

$$N\mathbf{\Sigma}(\mathbf{Z}_N) \xrightarrow{p} \mathbf{\Gamma}(F, \boldsymbol{\lambda}). \tag{5.7.27}$$

Also $N\hat{\mathbf{\Sigma}}(F)$ estimates consistently $N\mathbf{\Sigma}(F)$ and hence by (5.7.14), it converges in probability to $\mathbf{\Gamma}(F, \boldsymbol{\lambda})$. Hence, we get that under the given conditions

$$|N\mathbf{\Sigma}(\mathbf{Z}_N) - N\hat{\mathbf{\Sigma}}(F)| \xrightarrow{p} \mathbf{0}, \tag{5.7.28}$$

where $\mathbf{0}$ is a $t \times t$ null matrix. Further from (5.7.25), it follows that $N^{1/2}\{\boldsymbol{\theta}(\mathbf{F}_N^*) - \boldsymbol{\theta}^0\}$ is a real vector with elements which are all asymptotically finite. Also, from the well-known property of asymptotic multinormality of U-statistics, it follows that $N^{1/2}\{\mathbf{U}_N - \boldsymbol{\theta}(\mathbf{F}_N)\}$ is a vector which is bounded in probability, and hence,

$$S_N = N[\mathbf{U}_N - \boldsymbol{\theta}^0]'(\mathbf{\Gamma}(F, \boldsymbol{\lambda}))^{-1}[\mathbf{U}_N - \boldsymbol{\theta}^0] \tag{5.7.29}$$

is also bounded in probability. Consequently, from (5.7.27), (5.7.28), (5.7.30), and the definitions of T_N and T_N^*, we get that under $\{H_N\}$, $T_N \underset{p}{\sim} T_N^*$.

Further, it is easily shown that under $\{H_N\}$, the statistic S_N in (5.7.29) has asymptotically a noncentral chi-square distribution with t D.F. and the noncentrality parameter Δ, defined in (5.7.26), and from (5.7.27)–(5.7.29) we get $\{H_N\}$

$$|T_N - S_N| \underset{p}{\sim} |T_N^* - S_N| \xrightarrow{p} 0. \tag{5.7.30}$$

Hence the last part of the theorem. ◁

Hence, using the results of section 3.5, we can conclude that the permutation test based on T_N and the asymptotically distribution-free test based on T_N^* are asymptotically power-equivalent for the sequence of alternatives $\{H_N\}$ in (5.7.25).

5.8. ASYMPTOTIC RELATIVE EFFICIENCY OF RANK ORDER TESTS

Consider first the multivariate multisample location problem. In the parametric theory, it is usually assumed that $F_1, \ldots, F_c$ are all multinormal c.d.f.'s with a common (nonsingular) dispersion matrix $\mathbf{\Sigma}$ and mean vectors possibly different from each other. The commonly used test based on the likelihood ratio criterion has the test statistic $U_N = |N\hat{\mathbf{\Sigma}}_\Omega|/|N\hat{\mathbf{\Sigma}}_\omega|$, where

$$N\hat{\mathbf{\Sigma}}_\Omega = \sum_{k=1}^{c}\sum_{\alpha=1}^{n_k}(\mathbf{X}_\alpha^{(k)} - \bar{\mathbf{X}}^{(k)})(\mathbf{X}_\alpha^{(k)} - \bar{\mathbf{X}}^{(k)})'; \tag{5.8.1}$$

$$N\hat{\mathbf{\Sigma}}_\omega = \sum_{k=1}^{c} n_k(\bar{\mathbf{X}}^{(k)} - \bar{\mathbf{X}})(\bar{\mathbf{X}}^{(k)} - \bar{\mathbf{X}})' + N\hat{\mathbf{\Sigma}}_\Omega; \tag{5.8.2}$$

and

$$\bar{\mathbf{X}}^{(k)} = (1/n_k)\sum_{\alpha=1}^{n_k}\mathbf{X}_\alpha^{(k)}, \quad k = 1, \ldots, c; \qquad \bar{\mathbf{X}} = \sum_{k=1}^{c}(n_k/N)\bar{\mathbf{X}}^{(k)}. \tag{5.8.3}$$

It can be shown (cf. Exercise 5.8.1) that $-N \log U_N$ is asymptotically, in probability, equivalent to

$$(5.8.4) \qquad W_N = \sum_{k=1}^{c} n_k(\bar{\mathbf{X}}^{(k)} - \bar{\mathbf{X}})\mathbf{\Sigma}^{-1}(\bar{\mathbf{X}}^{(k)} - \bar{X})'.$$

Since by the central limit theorem, $n_k^{1/2}(\bar{\mathbf{X}}^{(k)} - E\bar{\mathbf{X}}^{(k)})$, $k = 1, \ldots, c$ are independent random vectors following asymptotically a common multinormal c.d.f. with null mean vector and a dispersion matrix $\mathbf{\Sigma}$ (for all $F_1, \ldots, F_c$ differing only in locations and having finite fourth-order moments), it follows that W_N has asymptotically a noncentral chi-square distribution with $p(c - 1)$ degrees of freedom and the noncentrality parameter $\sum_{k=1}^{c} \lambda^{(k)}(E\bar{\mathbf{X}}^{(k)} - E\bar{\mathbf{X}})\mathbf{\Sigma}^{-1}(E\bar{\mathbf{X}}^{(k)} - E\bar{\mathbf{X}})'$. Thus, under (5.6.2) (with $\boldsymbol{\delta}^{(k)} = \mathbf{0}$, $k = 1, \ldots, c$), W_N has asymptotically a noncentral chi-square distribution with $p(c - 1)$ D.F. and noncentrality parameter

$$(5.8.5) \qquad \Delta_W = \sum_{k=1}^{c} \lambda^{(k)}(\boldsymbol{\theta}^{(k)} - \bar{\boldsymbol{\theta}})' \mathbf{\Sigma}^{-1} (\boldsymbol{\theta}^{(k)} - \bar{\boldsymbol{\theta}}),$$

where $\bar{\boldsymbol{\theta}} = \sum_{k=1}^{c} \lambda^{(k)}\boldsymbol{\theta}^{(k)}$. Without any loss of generality, we may put $\bar{\boldsymbol{\theta}} = \mathbf{0}$. This simplifies the notations to some extent.

Hence, from (5.8.5), Theorem 5.6.2, and (5.6.11), it follows that the asymptotic relative efficiency (A.R.E.) of the class of location tests $\mathfrak{L}_N$ relative to W_N is

$$(5.8.6) \qquad e_{\mathfrak{L},W} = \left[\sum_{k=1}^{c} \lambda^{(k)}\{\boldsymbol{\eta}^{(k)\prime}\boldsymbol{\nu}^{-1}(F)\boldsymbol{\eta}^{(k)}\}\right] \Big/ \left[\sum_{k=1}^{c} \lambda^{(k)}\{\boldsymbol{\theta}^{(k)\prime}\mathbf{\Sigma}^{-1}\boldsymbol{\theta}^{(k)}\}\right],$$

where

$$(5.8.7) \qquad \boldsymbol{\eta}^{(k)} = (\eta_1^{(k)}, \ldots, \eta_p^{(k)})';$$

$\eta_i^{(k)}$ given by (5.6.6) and (5.6.8) with $\delta_i^{(k)} = 0$, for $i = 1, \ldots, p$, $k = 1, \ldots, c$.

Special cases. (I) The rank-sum test $\mathfrak{L}_N(R)$, for which $J_{(i)}(u) = u$: $0 < u < 1$, $i = 1, \ldots, p$. It follows from (5.8.6), (5.8.7), and (5.6.6) that

$$(5.8.8) \qquad e_{\mathfrak{L}_N(R), W_N} = \sum_{k=1}^{c} \lambda^{(k)}(\boldsymbol{\xi}^{*(k)\prime}\boldsymbol{\nu}^{*-1}(F)\boldsymbol{\xi}^{*(k)}) \Big/ \sum_{k=1}^{c} \lambda^{(k)}(\boldsymbol{\theta}^{(k)\prime}\mathbf{\Sigma}^{-1}\boldsymbol{\theta}^{(k)})$$

where

$$(5.8.9) \qquad \boldsymbol{\xi}^{*(k)} = (\xi_1^{*(k)}, \ldots, \xi_p^{*(k)})'; \quad \xi_i^{*(k)} = \theta_i^{(k)} \int_{-\infty}^{\infty} f_i^2(x)\, dx,$$
$$i = 1, \ldots, p, \quad k = 1, \ldots, c;$$

$$(5.8.10) \qquad \nu_{ij}^*(F) = \begin{cases} \frac{1}{12} & \text{if } i = j = 1, \ldots, p, \\ \displaystyle\int_{-\infty}^{\infty}\int_{-\infty}^{\infty} F_{[i]}(x)F_{[j]}(y)\, dF_{[i,j]}(x, y) - \tfrac{1}{4}, & i \neq j = 1, \ldots, p \end{cases}$$

and $\boldsymbol{\nu}^{*-1}(F) = ((\nu_{ij}^{*}(F)))^{-1}$. Since this efficiency is similar to that of the $S_n(R)$ test in the multivariate one-sample problem, for various allied results, the reader is referred to section 4.6.

(II) The normal scores test $\mathfrak{L}_N(\phi)$, for which $J_{(i)}(u) = \Phi^{-1}(u)$, $0 < u < 1$, $i = 1, \ldots, p$, where $\Phi(x)$ is the standard (univariate) normal c.d.f. Here

$$e_{\mathfrak{L}(\phi),W} = \sum_{k=1}^{c} \lambda^{(k)}(\boldsymbol{\delta}^{(k)\prime}\boldsymbol{\tau}^{-1}(F)\,\boldsymbol{\delta}^{(k)}) \Big/ \sum_{k=1}^{c} \lambda^{(k)}(\boldsymbol{\theta}^{(k)\prime}\,\boldsymbol{\Sigma}^{-1}\,\boldsymbol{\theta}^{(k)}). \tag{5.8.11}$$

where

$$\boldsymbol{\delta}^{(k)} = (\delta_1^{(k)}, \ldots, \delta_p^{(k)}), \qquad \delta_i^{(k)} = \theta_i^{(k)} \int_{-\infty}^{\infty} \frac{d}{dx}\Phi^{-1}(F_{[i]}(x))\, dF_{[i]}(x); \tag{5.8.12}$$

$$J_{(i)} = \phi^{-1};$$

$$\tau_{ij}(F) = \begin{cases} 1 & \text{if } \; i = j = 1, \ldots, p, \\ \displaystyle\int_{-\infty}^{\infty}\int_{-\infty}^{\infty} \Phi^{-1}[F_{[i]}(x)]\Phi^{-1}[F_{[j]}(y)]\, dF_{[i,j]}(x, y), & i \neq j = 1, \ldots, p, \end{cases} \tag{5.8.13}$$

and $\boldsymbol{\tau}^{-1}(F) = ((\tau_{ij}(F)))^{-1}$. We note that if $F(\mathbf{x})$ is $N(\mathbf{0}, \boldsymbol{\Sigma})$, the c.d.f. of p-variate normal distribution with mean vector $\mathbf{0}$ and nonsingular covariance matrix $\boldsymbol{\Sigma} = (\rho_{ij}\sigma_i\sigma_j)$, then the asymptotic efficiency of the multivariate normal scores test relative to the likelihood ratio W_N test is 1. Thus, in the family of normal distributions, the efficiency property of the univariate normal scores test with respect to the t-test is preserved in its multisample multivariate extension. The same is not true for the extension of Kruskal-Wallis-Wilcoxon test as is evident from the results of chapter 4. In the univariate case, we have shown in chapter 3 that the asymptotically normal scores test is always at least as good as the Student's t-test or the classical F test for all continuous distribution functions F. The generalization of this result to the multivariate case is quite difficult, since the efficiency not only depends on $B_i(F)$, $i = 1, \ldots, p$ but also on the matrix $\boldsymbol{\tau}(F)$, the sequence $\boldsymbol{\theta}^{(k)}$, $k = 1, \ldots, c$ and $\lambda^{(1)}, \ldots, \lambda^{(c)}$. However, it is easy to check that if the p variates in $F(\mathbf{x})$ are all uncorrelated, then (5.8.11) ≥ 1 for all $F \in \Omega_0$ independent of the direction $\boldsymbol{\theta}^{(k)}$ through which H_0 is approached. On the other hand, if the p variates are correlated, we may find the upper and lower bounds to (5.8.16) by allowing variations over $\boldsymbol{\theta}^{(1)}, \ldots, \boldsymbol{\theta}^{(c)}$ and keeping $\lambda^{(1)}, \ldots, \lambda^{(c)}$ fixed. These bounds evidently depend on the characteristic roots of $\boldsymbol{\tau}^{-1}\boldsymbol{\Sigma}$. For some allied results, see chapter 4 and papers by Chatterjee and Sen (1964) and Bickel (1965).

(III) Median test for which $J_{(i)}(u)$ is 1 or 0 according as u is $\leq \frac{1}{2}$ or not, $i = 1, \ldots, p$. Here

$$e_{\mathfrak{L}(M),W} = \left[\sum_{k=1}^{c} \lambda^{(k)}\boldsymbol{\theta}^{(k)\prime}\mathbf{T}^{*-1}(F)\boldsymbol{\theta}^{(k)}\right] \Big/ \left[\sum_{k=1}^{c} \lambda^{(k)}\boldsymbol{\theta}^{(k)\prime}\,\boldsymbol{\Sigma}^{-1}(F)\boldsymbol{\theta}^{(k)}\right], \tag{5.8.14}$$

where $\mathbf{T}^{*-1}(F) = ((\tau_{ij}^*(F)))^{-1}$ and

$$(5.8.15)\quad \tau_{ij}^*(F) = \begin{cases} 1/4f_{[i]}^2(0), & i = j = 1, \ldots, p, \\ [F_{[i,j]}(0, 0) - F_{[i]}(0)F_{[j]}(0)]/[f_{[i]}(0)f_{[j]}(0)], & \\ & i \neq j = 1, \ldots, p. \end{cases}$$

The A.R.E. is similar to that of the multivariate one-sample sign test relative to the Hotelling's T^2 test, and hence, for details, the reader is referred to section 4.6.

Let us now consider the multivariate multisample scale problem. In the parametric case, under the assumption of nonsingular normal distribution of each of $F_1, \ldots, F_c$, some optimum likelihood ratio test is available for testing the identity of the dispersion matrices of $F_1, \ldots, F_c$. If, however, we are interested only in testing the identity of the variances for each of the p variates, we can have a similar test whose test statistic (say W_N) has asymptotically under H_N in (5.6.1) and (5.6.2) (where $\boldsymbol{\theta}^{(k)} = \mathbf{0}$ for all $k = 1, \ldots, c$), a noncentral chi-square distribution with $p(c - 1)$ D.F. and the noncentrality parameter

$$(5.8.16)\qquad \Delta_W = \sum_{k=1}^{c} \lambda^{(k)}[(\boldsymbol{\delta}^{(k)'})\boldsymbol{\Gamma}^{-1}(\boldsymbol{\delta}^{(k)})]$$

where $\boldsymbol{\Gamma} = ((\gamma_{ij}))_{i,j=1,\ldots,p}$ with

$$(5.8.17)\qquad \gamma_{ij} = 2\sigma_{ij}^2, \qquad i, j = 1, \ldots, p,$$

$\boldsymbol{\Sigma} = ((\sigma_{ij}))$ being the covariance matrix of F.

Hence the asymptotic efficiency of the $\mathfrak{L}_N$ test with respect to the W_N test is

$$(5.8.18)\qquad e_{\mathfrak{L},W} = \Delta_{\mathfrak{L}}/\Delta_W.$$

In particular, when the underlying c.d.f. F is nonsingular and multinormal, the multivariate normal scores test for scale (cf. section 5.4) can be easily shown to be asymptotically as efficient as the W test. However, the A.R.E. of the three other scale tests referred to in section 5.4 depends on $\boldsymbol{\Sigma}$ as well as on $\boldsymbol{\delta}^{(k)}$, $k = 1, \ldots, c$, even if F is multinormal. As in location tests, the bounds for the A.R.E. of these tests are to be obtained by evaluating the maximum and minimum characteristic roots of $\boldsymbol{\Gamma}^{-1}\boldsymbol{\Sigma}$. The same procedure is also needed for the normal scores test when F is nonnormal.

In all the examples (for location and scale problems), we have considered the case where $J_{N(i)}$ does not depend on i $(= 1, \ldots, p)$. In situations where we have strong evidence that the c.d.f.'s $F_{(i)}$, $i = 1, \ldots, p$ have some definite pattern, we may use different scores for different i. But, usually, the cases considered above are used in practice.

5.9. RANK ORDER TESTS FOR THE ANALYSIS OF COVARIANCE PROBLEM

Consider a stochastic vector $\mathbf{Z} = (X_1, X_2, \ldots, X_p)$, where X_1 is the primary variate and $\mathbf{X} = (X_2, \ldots, X_p)$ is a concomitant stochastic vector. Let then $\mathbf{Z}_\alpha^{(k)}$, $\alpha = 1, \ldots, n_k$ be n_k independent and identically distributed random vectors having a continuous c.d.f. $G_k(\mathbf{z})$, $\mathbf{z} \in R^p$, for $k = 1, \ldots, c\ (\geq 2)$. The marginal c.d.f. of $\mathbf{X}_\alpha^{(k)}$ is denoted by $F_k^{(1)}(\mathbf{x})$, $k = 1, \ldots, c$, and it is assumed that

$$F_1^{(1)}(\mathbf{x}) \equiv \cdots \equiv F_c^{(1)}(\mathbf{x}), \qquad \mathbf{x} \in R^{p-1}. \tag{5.9.1}$$

Such an assumption is often found justified in practice (cf. Scheffé, 1959, where the parametric case is thoroughly studied). It is desired to test the null hypothesis that $G_1, \ldots, G_c$ are all identical (assuming of course, the identity of $F_1^{(1)}, \ldots, F_c^{(1)}$), against shift alternatives.

We reduce this problem to the multivariate multisample location problem by specifying further that in this case, we are specifically interested only in difference in locations of the first variate (i.e., $X_{1\alpha}^{(k)}$, $\alpha = 1, \ldots, n_k$, $k = 1, \ldots, c$), while utilizing the information contained in the concomitant variates under the assumption (5.9.1).

We adopt the same notations as in sections 5.3 and 5.4. Since the null hypothesis of identity of $G_1, \ldots, G_c$ retains the permutation-invariance structure in (5.3.3), we have no problem in constructing suitable rank order tests. We define $T_{N,i}^{(k)}$, $i = 1, \ldots, p$, $k = 1, \ldots, c$ as in (5.4.4) and $v_{N,ij}(\mathbf{R}_N^*)$ as in (5.4.9). Now, to utilize the information contained in the concomitant variates, we eliminate from $T_{N,1}^{(k)}$, the variation accounted to regression of $T_{N,1}^{(k)}$ on $T_{N,2}^{(k)}, \ldots, T_{N,p}^{(k)}$. Since the moments of $\{T_{N,1}^{(k)}\}$ are specified by (5.4.5) and (5.4.6) and their asymptotic multinormality by Theorem 5.4.1, we may fit a linear regression. The fitted values are

$$\bar{E}_N^{(1)} - \sum_{i=2}^{p} (V_{N,i1}/V_{N,11})(T_{N,i}^{(k)} - \bar{E}_N^{(i)}), \qquad k = 1, \ldots, c, \tag{5.9.2}$$

where $V_{N,ij}$ is the cofactor of $v_{ij}(\mathbf{R}_N^*)$ in $\mathbf{V}(\mathbf{R}_N^*)$. Thus, the residuals are

$$T_{N,1}^{(k)} - \left\{\bar{E}_N^{(1)} - \sum_{i=2}^{p} (V_{N,i1}/V_{N,11})(T_{N,i}^{(k)} - \bar{E}_N^{(i)})\right\}$$

$$= \sum_{i=1}^{p} (V_{N,i1}/V_{N,11})(T_{N,i}^{(k)} - \bar{E}_N^{(i)}) = T_{N,k}^* \text{ say}, \qquad k = 1, \ldots, c. \tag{5.9.3}$$

From (5.4.6) and (5.9.3), we obtain after some simplifications that

$$\operatorname{cov}(T_{N,k}^*, T_{N,q}^* \mid \mathcal{T}_N) = [(N\delta_{kq} - n_k)/n_k(N-1)]\, |V_N|/V_{N,11}, \tag{5.9.4}$$

for $k, q = 1, \ldots, c$, where $|V_N|$ is the determinant of $\mathbf{V}_N(\mathbf{R}_N^*)$. Thus, following the same arguments as in section 5.4, we may consider the test statistic

$$\mathfrak{L}_N^* = \{V_{N,11}/|V_N|\}\sum_{k=1}^{c} n_k(T_{N,k}^*)^2. \tag{5.9.5}$$

As in (5.4.14) and (5.4.15), the permutation (conditional) distribution of $\mathfrak{L}_N^*$ (under (5.3.3)) can be used to provide a distribution-free test, but the labor involved increases prohibitively as N increases. However, from the results of Theorems 5.4.1 and 5.4.2, it follows that the permutation distribution of $\mathfrak{L}_N^*$ reduces asymptotically in probability to a chi-square distribution with $c - 1$ D.F. Also, it follows from Theorem 5.5.1, Theorem 5.5.2, and (5.5.39) that the unconditional null distribution of $\mathfrak{L}_N^*$ asymptotically reduces to a chi-square distribution with $c - 1$ D.F. Thus, a large-sample test procedure may be constructed which rejects the null hypothesis when $\mathfrak{L}_N^* \geq \chi^2_{\alpha,c-1}$.

For the study of the asymptotic non-null distribution of $\mathfrak{L}_N^*$, we consider here the following sequence $\{H_N^*\}$ of alternative hypotheses, specified by

$$H_N^*\colon\ G_k(\mathbf{z}) = G_{k,N}(\mathbf{z}) = G(z_1 - N^{-1/2}\theta_k, z_2, \ldots, z_p), \quad k = 1, \ldots, c, \tag{5.9.6}$$

where $\theta_1, \ldots, \theta_c$ are all real and finite. Then, we arrive at the following theorem from Theorem 5.6.1, Theorem 5.6.2, and Lemma 5.6.3.

Theorem 5.9.1. *Under* (5.9.6) *and the conditions of Theorem* 5.6.1 *as well as Lemma* 5.6.3, $\mathfrak{L}_N^*$ *has asymptotically a noncentral chi-square distribution with* $c - 1$ *D.F. and the noncentrality parameter*

$$\Delta_{\mathfrak{L}^*} = \nu^{11}(G)\sum_{k=1}^{c}\lambda^{(k)}(\eta_1^{(k)} - \bar{\eta}_1)^2, \tag{5.9.7}$$

where $\nu^{11}(G)$ *is the element in the first row and first column of* $\nu^{-1}(G)$ ($\boldsymbol{\nu}(F)$ *is defined by* (5.4.29) *and* $\boldsymbol{\nu}^{-1}(G) = (\nu(G))^{-1}$, *for* $F \equiv G$), $\eta_1^{(k)}$, $k = 1, \ldots, c$ *are defined by* (5.6.6) *and* (5.6.8), *and* $\bar{\eta}_1 = \sum_{k=1}^{c}\lambda^{(k)}\eta_1^{(k)}$.

Let us now denote the (common) covariance matrix of $Z_\alpha^{(k)}$ by $\mathbf{\Sigma}(G)$, and under the assumption that it is nonsingular, we denote its reciprocal by $\mathbf{\Sigma}^{-1}(G) = ((\sigma^{ij}(G)))$. The classical parametric analysis of covariance test under the assumed multinormality of G is based on the variance-ratio criterion (after the variation due to concomitant variates is eliminated from the between- and within-group mean squares). For details, the reader is referred to Scheffé (1959, chapter 6). It can be shown that if G be any multivariate c.d.f. with finite moments up to the order $2 + \delta$, for some $\delta > 0$, then under (5.9.6), $c - 1$ times this variance ratio criterion has asymptotically a noncentral chi-square distribution with $c - 1$ D.F. and the noncentrality parameter

$$\Delta_{\mathcal{F}} = \sigma^{11}(G)\sum_{k=1}^{c}\lambda^{(k)}(\bar{\theta}_k - \bar{\theta})^2; \qquad \bar{\theta} = \sum_{k=1}^{c}\lambda^{(k)}\theta_k. \tag{5.9.8}$$

(The proof is left as Exercise 5.9.5.)

Thus, the asymptotic relative efficiency (A.R.E.) of the $\mathfrak{L}_N^*$ test with respect to the variance-ratio test is

$$e_{\mathfrak{L}^*,\mathcal{F}} = \Delta_{\mathfrak{L}^*}/\Delta_{\mathcal{F}} = [\nu^{11}(G)/\sigma^{11}(G)]C^2, \tag{5.9.9}$$

where $C = B_{(1)}(G)$, defined by (5.6.4), if condition (c) of section 3.6.1 holds, and it is equal to $\sum_{s=1}^{r_j-1}[J_{(1),x} - J_{(1),s+1}]g_{(1)}(\zeta_s^{(1)})$, if the condition ($c$) is replaced by (5.4.26) and (5.4.27), while F is replaced by G. Now, going back to chapter 3, section 3.8, we observe that the efficiency of the rank order test based on $T_{N,1}^{(k)}$, $k = 1, \ldots, c$ for the ANOVA problem with respect to the classical parametric ANOVA test is equal to

$$e^0 = C^2\sigma_{11}(G)/\nu_{11}(G). \tag{5.9.10}$$

Hence, from (5.9.9) and (5.9.10), we obtain that

$$e_{\mathfrak{L}^*,\mathcal{F}} = e^0[\nu^{11}(G)\cdot\nu_{11}(G)]/[\sigma^{11}(G)\sigma_{11}(G)]. \tag{5.9.11}$$

Since $\nu^{11}(G)\nu_{11}(G) \geq 1$ and $\sigma^{11}(G)\cdot\sigma_{11}(G) \geq 1$, (5.9.11) leads to

$$e^0/[\sigma^{11}(G)\sigma_{11}(G)] \leq e_{\mathfrak{L}^*,\mathcal{F}} \leq e^0[\nu^{11}(G)\nu_{11}(G)]. \tag{5.9.12}$$

Since values as well as bounds for e^0 have already been studied in chapter 3, (5.9.9), (5.9.11), and (5.9.12) may be used to extend these for the analysis of covariance problem.

Special Cases **(i)** ***Rank Sum Procedure,*** *i.e.*, $E_{N,\alpha}^{(i)} = \alpha/(N+1)$, $1 \leq \alpha \leq N$, $i = 1, \ldots, p$. Here, $e^\circ = 12\sigma_{11}(G)(\int_{-\infty}^{\infty} g_{[1]}^2(x)\,dx)^2$. It is bounded below by 0.864 for all continuous G, is equal to $3/\pi$ for normal c.d.f.'s and is greater than 1 for many non-normal c.d.f.'s. However, upper bounds to $\sigma^{11}(G)\sigma_{11}(G)$ (over all possible G) are not easy to obtain, and as a result, no lower bound to (5.9.9) may be obtained. The following cases are of some interest:

(a) *G is a p-variate normal c.d.f.* Let $\mathbf{R} = ((\rho_{ij}))$ stand for the product-moment correlation matrix, and let $P_g = ((\rho_{g.ij}))$ be the matrix of the grade correlation, i.e. $\rho_{g.ij} = (6/\pi)\sin^{-1}(\frac{1}{2}\rho_{ij})$, $i, j = 1, \ldots, p$. Denote by $\mathbf{R}^{-1} = ((\rho^{ij}))$ and $\mathbf{R}_g^{-1} = ((\rho_g^{ij}))$ the corresponding reciprocal matrices. Then, it follows from (5.9.9) and some simplifications that

$$e_{\mathfrak{L}^*,\mathcal{F}} = (3/\pi)\rho_g^{11}/\rho^{11}. \tag{5.9.13}$$

For $p = 2$, (5.9.13) simplifies to $(3/\pi)(1 - \rho_{12}^2)/(1 - \rho_{g12}^2)$. It is equal to $3/\pi$ when $\rho_{12} = 0$ and it tends monotonically (in each direction) to 0.866 as $|\rho| \to \pm 1$. Thus, for one concomitant variate, (5.9.13) is bounded below by

0.866 for normal c.d.f.'s. For $p \geq 3$, the expression in (5.9.13) can be arbitrarily close to zero, of course only in some limiting degenerate cases (cf. Exercise 5.9.6). On the other hand, it cannot be greater than $3/\pi$ for any $p \geq 1$. The proof will follow if we can show that $\rho_g^{11} \leq \rho^{11}$ for all $p \geq 1$, in normal c.d.f.'s. For this, it is sufficient to show that $((2 \sin^{-1} \frac{1}{2}\rho_{ij} - \rho_{ij}))$ is positive semidefinite, and the proof is left as an exercise (cf. Exercise 5.9.7).

(*b*) The primary variate is uncorrelated with (but not necessarily independent of) the concomitant variates. Then $\sigma_{1i} = 0$, $i = 2, \ldots, p$, so that $\sigma^{11}(G)\sigma_{11}(G) = 1$. Since $\nu^{11}(G)\nu_{11}(G) \geq 1$, we therefore obtain that

$$e_{\mathfrak{L}^*,\mathfrak{F}} \geq e^0. \tag{5.9.14}$$

Here, all the bounds applicable to e^0 are also applicable to $e_{\mathfrak{L}^*,\mathfrak{F}}$.

(ii) Normal Scores Procedure. Here, we take $E_{N,\alpha}^{(i)}$ as the expected value of the αth smallest observation of a sample of size N from a standard normal distribution $\Phi(x)$. From the results of chapter 3, it follows that

$$e^0 = \sigma_{11}(G)\left[\int_{-\infty}^{\infty} \{g_{[1]}^2(x)/\phi(\Phi^{-1}(G_{[1]}(x)))\}\, dx\right]^2 \geq 1, \tag{5.9.15}$$

where the equality sign holds only when $G_{[1]}$ is a normal c.d.f. Thus, the A.R.E. is bounded below by $\nu^{11}(G)/\sigma^{11}(G)$. In the particular case (*a*) when G is normal, $\nu^{11}(G)/\sigma^{11}(G) = 1$ (as $\boldsymbol{\nu} = \boldsymbol{\Sigma}$), and as a result, the A.R.E. equals 1, independently of $p \geq 1$ and $\boldsymbol{\Sigma}$. This results establishes the supremacy of the normal scores procedure over the rank sum procedure as the latter has been shown to have an A.R.E. bounded above by $3/\pi$ and below by 0 (for normal c.d.f.'s). Also, in case (*b*), when $\sigma_{12} = 0$, $i = 2, \ldots, p$, the A.R.E. is bounded below by 1.

(iii) Median Procedure. Here $E_{N,\alpha}^{(i)} = 1$ if $\alpha \leq [N/2]$ and is 0, otherwise, and e^0 equals to $4g_{[1]}^2(0)$ (cf. Exercise 3.8.6). As such, the A.R.E. is equal to $4g_{[1]}^2(0)\nu^{11}(G)/\sigma^{11}(G)$. If G is a normal c.d.f., this simplifies to

$$(2/\pi)(\rho_M^{11}/\rho^{11}), \tag{5.9.16}$$

where $\mathbf{R}_M = ((\rho_{Mij}))$, $\mathbf{R}_M^{-1} = ((\rho_M^{ij}))$ and $\rho_{M,ij} = (2/\pi)\sin^{-1}\rho_{ij}$. Here also for $p \geq 2$, (5.9.16) may be arbitrarily close to zero and bounded above by $2/\pi$ (cf. Exercise 5.9.8). Also, in case (*b*) when $\sigma_{1i} = 0$, $i = 2, \ldots, p$, the A.R.E. is equal to $e^0 = 4g_{[1]}^2(0)$.

EXERCISES

Section 5.3

5.3.1. Obtain the unconditional distribution of $\mathbf{R}_N$ in the special case $p = 1$. Show that under H_0: $F_1 \equiv \cdots \equiv F_c$ this agrees with (5.3.3) when F is continuous. Show also that this agreement does not hold when $p \geq 2$.

Section 5.4

5.4.1. Prove (5.4.6) and (5.4.10).

5.4.2. Prove that (5.4.25) implies (5.4.24).

5.4.3. Show that (5.4.25) holds if (5.4.26) and (5.4.27) hold along with conditions (*a*) and (*b*) of section 3.6.1.

5.4.4. Prove (5.4.64) and, in general, the following: $N^{-1}\sum_{i=1}^{N} g(E_{N,i}) \to \int_{-\infty}^{\infty} g(x)\, d\Psi(x)$ whenever the latter exists.

5.4.5. Show that under the conditions of Lemma 5.4.4

$$\lim_{N\to\infty} \left(\frac{1}{N}\right) \sum_{i=1}^{N} [E_{N,i}^2 - \xi_{N,i}^2]^2 = 0,$$

if $\int x^4\, d\Psi(x) < \infty$.

Section 5.5

5.5.1. Show that $B_{1N(i)}^{(k)} + B_{2N(i)}^{(k)}$ is equal to the expression (5.5.12).

5.5.2. Establish (5.5.19), (5.5.20), and (5.5.21).

5.5.3. Prove that under the null hypothesis

$$\sigma_{[Ni,j]}^{(k,q)} = \left(\frac{\delta_{kq}}{\lambda_N^{(k)}} - 1\right) \nu_{ij}(F)$$

where $\nu_{ij}(F)$ is given by (5.4.29).

5.5.4. Prove that under the null hypothesis, the limiting distribution of $\mathcal{L}_N^*$ as well as that of $\mathcal{L}_N$ is chi-square with $p(c - 1)$ D.F.

Section 5.6

5.6.1. Complete the proof of Theorem 5.6.1.

5.6.2. Complete the proof of Theorem 5.6.2.

5.6.3. Complete the proof of Lemma 5.6.3.

5.6.4. Prove the assertion (5.6.12).

Section 5.7

5.7.1. Obtain the test statistic for the multivariate location problem, when $F_1, \ldots, F_c$ are not necessarily continuous, and we desire to use the rank sum procedure. (*Hint:* Express the coordinatewise individual sample rank sums as a function of several *U*-statistics and then use the results of section 5.7.)

5.7.2. For the same problem, study the conditions under which the applied permutation covariance matrix is positive definite.

Section 5.8

5.8.1. Show that for U_N, defined just before (5.8.1), $-N \log U_N$ is asymptotically equivalent to W_N, defined by (5.8.4).

5.8.2. Obtain the proof for the asymptotic distribution of W_N for the scale problem in (5.8.15) and (5.8.16).

5.8.3. Obtain the expression for (5.8.17) when we use the following rank scores: (a) $E_{N,\alpha}^{(i)} = |\alpha/(N+1) - \frac{1}{2}|$, (b) $E_{N,\alpha}^{(i)} = [\alpha/(N+1) - \frac{1}{2}]^2$ and (c) $E_{N,\alpha}^{(i)}$ is 1 or 0 according as $|\alpha - (N+1)/2|$ is $\geq (N+1)/4$ or not.

Section 5.9

5.9.1. Prove that under the permutation model of section 5.3, the fitted values of $T_{N,1}^{(k)}$ from the regressors $T_{N,i}^{(k)}$, $i = 2, \ldots, p$, are given by (5.9.2).

5.9.2. Prove (5.9.4).

5.9.4. Obtain the exact expression for the variance-ratio criterion for the analysis of covariance problem when G is assumed to be a multinormal c.d.f. (Scheffé, 1959, Ch. 6.)

5.9.5. Show that under (5.9.6), $c - 1$ times the variance-ratio criterion has asymptotically a noncentral chi-square distribution with $c - 1$ d.f. and the noncentrality parameter defined by (5.9.8), provided G has finite moments up to the order $2 + \delta$, $\delta > 0$.

5.9.6. Show that for $p = 3$, we can have a sequence of product moment correlation matrix $\mathbf{R}_\nu \to \mathbf{R}$ (nonsingular) as $\nu \to \infty$, such that (for normal G) the corresponding $\mathbf{R}_{\nu g}$ has $\rho_\nu^{11} \to 0$ as $\nu \to \infty$.

5.9.7. Show that $(2 \sin^{-1} \frac{1}{2}\rho_{ij} - \rho_{ij}))$ is positive semidefinite. (*Hint:* Express $\sin^{-1} a$ in a power series of a, and utilize the result that if $((\rho_{ij}))$ is positive definite then $((\rho_{ij}^k))$ is also so for any integer $k \geq 1$.)

5.9.8. Show that (5.9.16) is bounded above by $(2/\pi)$. [*Hint:* Apply the same technique as in Exercise 5.9.7.]

5.9.9. Consider now the case of k primary variates and $p - k$ $(p > k \geqslant 1)$ concomitant variates, for which (5.9.1) holds. (i) Obtain the standard MANOCA (multivariate analysis of covariance) tests for the equality of treatment vectors of the c (k-variate) treatment responses, when the parent c.d.f. is multinormal. (ii) Show that even if the parent c.d.f. is not normal but possesses finite moments up to the second order, a monotonic function of the normal-theory likelihood ratio test statistic has asymptotically chi-square distribution. (iii) Extend the statistic (5.9.5) to this situation, and (iv) study the asymptotic relative efficiency results. (Sen and Puri, 1970a.)

PRINCIPAL REFERENCES

Bhattacharyya (1967), Chatterjee and Sen (1964, 1965, 1966), Puri and Sen (1966, 1969c), Quade (1966, 1967), Sen (1966c, 1970c), Sen and Puri (1970a) and Tamura (1966).

CHAPTER 6

Estimators In Linear Models (One-Way Layouts) Based On Rank Tests

6.1. INTRODUCTION

In linear models, the method of least squares provides minimum variance unbiased (m.v.u.) estimators. But this requires the existence of the moments up to the second order, and besides, the estimates are vulnerable to gross errors. In addition, the associated confidence interval being based on the assumed normality of the parent distribution is sensitive to any departure from normality. Alternative robust estimators are proposed by various workers. The Winsorized means or trimmed means, initially proposed by Tukey (1960) (see also Huber, 1964) are quite commonly in use. Linear functions of order statistics are also proposed by many workers. A nice account of this is available in Sarhan and Greenberg (1962) (see also Gastwirth (1966)). A third approach is based on robust rank tests for linear models (such as the ones considered in the previous three chapters). In this chapter, we shall develop the theory of robust estimation of parameters in linear models, based on suitable rank tests. Specifically, the following problems will be considered:

(*a*) estimation (point as well as interval) of the location vector of a symmetric distribution,

(*b*) estimation of the difference in location vectors of two multivariate distributions, and

(*c*) estimation of contrasts in ANOVA and MANOVA (one-way layout.)

Certain generalizations of these problems will be considered in chapter 7.

6.2. POINT ESTIMATION OF LOCATION OF A SYMMETRIC DISTRIBUTION

For convenience we first consider briefly the point estimation of a location parameter of a symmetric univariate distribution. Suppose that $X_1, \ldots, X_n$ are independently distributed with common c.d.f. $F(x - \theta)$, where F is absolutely continuous and symmetric about zero. The least squares estimator of θ is given by

$$\bar{X}_n = \frac{1}{n}\sum_{i=1}^{n} X_i. \tag{6.2.1}$$

Furthermore, if F is normal, it is well known that $\bar{X}_n$ is the m.v.u. estimator of θ provided that the variance of F is finite.

In this chapter we shall consider the estimators of θ based on the following alignment consideration. Consider a statistic $h(\mathbf{X}_n) = h(X_1, \ldots, X_n)$, nonparametric or otherwise for testing the hypothesis $\theta = 0$ against the alternatives $\theta > 0$. Since, by definition, $X_1 - \theta, \ldots, X_n - \theta$ are independently and identically distributed according to the c.d.f. $F(x)$, the alignment procedure consists in selecting an estimator $\hat{\theta}_n$ of θ in such a manner that the statistic $h(\mathbf{X}_n - \hat{\theta}_n\mathbf{1}_n)$, where $\mathbf{1}_n$ is the n vector $(1, \ldots, 1)$, based on the observations $X_1 - \hat{\theta}_n, \ldots, X_n - \hat{\theta}_n$ is close to $E(h(\mathbf{X}_n) \mid \theta = 0)$. There is either a unique such value of θ which would then serve as estimate, or an interval of such values, in which case the midpoint of the interval provides a natural estimate. To formalize this, suppose that

(6.2.2) $h(\mathbf{x}_n + a\mathbf{1}_n)$ is a nondecreasing function of a for each $\mathbf{x}_n$.

(6.2.3) the distribution of $h(\mathbf{X}_n)$ is symmetric about a fixed point $\mu_n (= E_0(h(\mathbf{X}_n)))$ when $\theta = 0$.

Let

$$\begin{aligned} \theta_n^* &= \sup\{a\colon\ h(x_n - a\mathbf{1}_n) > \mu_n\}, \\ \theta_n^{**} &= \inf\{a\colon\ h(x_n - a\mathbf{1}_n) < \mu_n\}, \end{aligned} \tag{6.2.4}$$

and

$$\hat{\theta}_n = (\theta_n^* + \theta_n^{**})/2. \tag{6.2.5}$$

Then, for suitable functions h, we consider $\hat{\theta}_n$ as estimator of θ.

The estimator $\hat{\theta}_n$ defined in (6.2.5) forms a general class of estimates of θ. A few important cases are as follows:

(i) Let $h(\mathbf{X}_n)$ be the Wilcoxon signed rank statistic. Then the corresponding estimator (to be called the Wilcoxon estimator $\hat{\theta}_{n(R)}$) is the median of the $n(n+1)/2$ values of $(X_\alpha + X_\beta)/2$ for $1 \le \alpha \le \beta \le n$ (cf. Hodges and Lehmann, 1963; and Sen, 1963b);

(ii) Let $h(\mathbf{X}_n)$ be the sign statistic. Then the corresponding estimator (to be called the median estimator $\hat{\theta}_{n(M)}$) is the median of the observations $X_1, \ldots, X_n$.

(iii) Let $h(\mathbf{X}_n)$ be the normal scores statistic. Then the corresponding estimator (to be called the normal scores estimator $\hat{\theta}_{N(\Phi)}$) cannot be expressed as a simple function of $X_1, \ldots, X_n$. An iterative procedure is employed to compute $\hat{\theta}_{n(\Phi)}$ (see Puri and Sen, 1967b, 1970).

We now develop the above ideas to the case where the independent observations are vector-valued, not necessarily identically distributed, but each having the same (unknown) point of symmetry which we wish to estimate. (For the special case of estimation when the underlying distributions are all identical, see Puri and Sen, 1970).

Let $\mathbf{X}_\alpha = (X_{1\alpha}, \ldots, X_{p\alpha})'$, $\alpha = 1, \ldots, n$ be independently distributed according to a p-variate c.d.f. $F_\alpha(\mathbf{x} - \boldsymbol{\theta})$, $\mathbf{x} = (x_1, \ldots, x_p)$, $\boldsymbol{\theta} = (\theta_1, \ldots, \theta_p)$, $\alpha = 1, \ldots, n$, where F_α is absolutely continuous and diagonally symmetric about $\mathbf{0}$. By definition, $\mathbf{X}_\alpha - \boldsymbol{\theta}$, $\alpha = 1, \ldots, n$ are distributed independently according to $F_\alpha(\mathbf{x})$, $\alpha = 1, \ldots, n$ where each $F_\alpha(\mathbf{x})$, $\alpha = 1, \ldots, n$ is diagonally symmetric about $\mathbf{0}$. In chapter 4 (cf. (4.4.80)) we considered a class of rank order tests based on the statistic S_n^* for testing the hypothesis $\boldsymbol{\theta} = \mathbf{0}$ against the alternatives $\boldsymbol{\theta} \neq \mathbf{0}$. It was shown there that S_n^* is stochastically small (precisely bounded in probability) when $\boldsymbol{\theta} = \mathbf{0}$. On the other hand, S_n^* is stochastically large if $\boldsymbol{\theta} \neq \mathbf{0}$. Thus one criterion of selecting an estimator $\hat{\boldsymbol{\theta}}_n = (\hat{\theta}_{n1}, \ldots, \hat{\theta}_{np})$ of $\boldsymbol{\theta}$ may be to select it in such a way that S_n^* based on $\mathbf{X}_\alpha - \hat{\boldsymbol{\theta}}_n$, $\alpha = 1, \ldots, n$ is smallest. Since $S_n^* \geq 0$, its smallest possible value is obtained when $T_{n,j}^*$ defined in (4.4.38) is closest to $\frac{1}{2}\bar{E}_{n,j}$ $(= \mu_{j0}^*)$ for each $j = 1, \ldots, p$; where $\bar{E}_{n,j}$ and μ_{j0}^* are defined in (4.4.37) and (4.4.81), respectively. Consequently, the alignment procedure reduces to that of selecting $\hat{\theta}_{nj}$ in such a manner that the statistic T_{nj}^* based on the observations $X_{j\alpha} - \hat{\theta}_{nj}$, $\alpha = 1, \ldots, n$ is closed to μ_{j0}^* for all $j = 1, \ldots, p$. To formalize the above ideas, let us write

(6.2.6) $$\mathbf{X}_n^{(j)} = (X_{j1}, \ldots, X_{jn}), \quad T_{nj}^* = h_j(\mathbf{X}_n^{(j)}), \qquad j = 1, \ldots, p$$

and denote the statistic T_{nj}^* based on $X_{j\alpha} - a_j$, $\alpha = 1, \ldots, n$ by $h_j(\mathbf{X}_n^{(j)} - a_j\mathbf{1}_n)$. Recalling the definition of T_{nj}^* (see 4.4.38), we define the rank scores $\{E_{n,\alpha}^{(j)}\}$ as follows: Let $E_{n,\alpha}^{(j)}$, $\alpha = 1, \ldots, n$ be the expected value of the αth-order statistic of a sample of size n from a distribution

(6.2.7) $$\psi_j(x) = \psi_j^*(x) - \psi_j^*(-x) \quad \text{if} \quad x \geq 0, \quad \text{and} \quad \psi_j(x) = 0 \quad \text{otherwise}$$

where $\psi_j^*(x)$ is symmetric about zero, for each $j = 1, \ldots, p$.

Then, it is easy to check that

(6.2.8) $$h_j(\mathbf{X}_n^{(j)} - a_j\mathbf{1}_n) \quad \text{is nondecreasing in } a_j \text{ for each } \mathbf{X}_n^{(j)}, \quad j = 1, \ldots, p$$

Let

$$\theta^*_{nj} = \sup\{a_j\colon h_j(\mathbf{X}_n^{(j)} - a_j\mathbf{1}_n) > \mu^*_{0j}\},$$
(6.2.9)
$$\theta^{**}_{nj} = \inf\{a_j\colon h_j(\mathbf{X}_n^{(j)} - a_j\mathbf{1}_n) < \mu^*_{0j}\},$$

and

(6.2.10) $$\hat{\boldsymbol{\theta}}_n = (\hat{\theta}_{n1}, \ldots, \hat{\theta}_{np}), \quad \hat{\theta}_{nj} = \tfrac{1}{2}(\theta^*_{nj} + \theta^{**}_{nj}), \qquad j = 1, \ldots, p.$$

Then we propose $\hat{\boldsymbol{\theta}}_n$ as estimator of the location parameter $\boldsymbol{\theta}$.

The estimator $\hat{\theta}_n$ forms a general class of estimators of $\boldsymbol{\theta}$. A few important cases are as follows:

(i) The Wilcoxon estimator $\hat{\boldsymbol{\theta}}_{n(R)}$ resulting from the Wilcoxon signed rank statistic by taking for $\psi^*_j(x)$ in (6.2.7) the rectangular distribution over $(-1, 1)$. In this case, the estimator $\hat{\theta}_{nj(R)}$ of θ_j can be expressed as the median of the $n(n+1)/2$ values of $(X_{j\alpha} + X_{j\beta})/2$ for $1 \le \alpha \le \beta \le n, j = 1, \ldots, p$.

(ii) The normal scores estimator $\hat{\boldsymbol{\theta}}_{n(\Phi)}$ resulting from the normal scores statistic by taking for $\psi^*_j(x)$ in (6.2.7) the standard normal distribution. The $\hat{\boldsymbol{\theta}}_{n(\Phi)}$ cannot, however, be expressed as a simple function of the X's. For the actual computation of $\hat{\boldsymbol{\theta}}_{n(\Phi)}$, an iterative procedure is employed (cf. Puri and Sen, 1967a, 1970).

To be explicit, we write $\hat{\boldsymbol{\theta}}_n$ as $\hat{\boldsymbol{\theta}}_n(\mathbf{X}_1, \ldots, \mathbf{X}_n)$. Then, upon noting that $h_j(\mathbf{X}_n^{(j)} + a_j\mathbf{1}_n - a_j\mathbf{1}_n) = h_j(\mathbf{X}_n^{(j)})$ we obtain from (6.2.9) and (6.2.10) that

(6.2.11) $$\hat{\boldsymbol{\theta}}_n(\mathbf{X}_1 + \mathbf{a}, \ldots, \mathbf{X}_n + \mathbf{a}) = \hat{\boldsymbol{\theta}}_n(\mathbf{X}_1, \ldots, \mathbf{X}_n) + \mathbf{a} \qquad \text{for all} \quad \mathbf{a} = (a_1, \ldots, a_p).$$

Thus, like the least squares estimator

$$\bar{\mathbf{X}}_n = (\bar{X}_{n1}, \ldots, \bar{X}_{np}), \quad \bar{X}_{nj} = \frac{1}{n}\sum_{\alpha=1}^{n} X_{j\alpha}, \qquad j = 1, \ldots, p,$$

the estimator $\hat{\boldsymbol{\theta}}_n$ is translation invariant.

Theorem 6.2.1. *The distribution of $\hat{\boldsymbol{\theta}}_n$ is diagonally symmetric about $\boldsymbol{\theta}$ if $F_\alpha(\mathbf{x})$ for $\alpha = 1, \ldots, n$ are diagonally symmetric.*

Proof: By definition

(6.2.12) $$h_j(\mathbf{X}_n^{(j)}) + h_j(-\mathbf{X}_n^{(j)}) = 2\mu^*_{0j} \quad \text{for all} \quad j = 1, \ldots, p.$$

Since each $F_\alpha(\mathbf{x})$ is diagonally symmetric, $\mathbf{X}_\alpha - \boldsymbol{\theta}$ and $\boldsymbol{\theta} - \mathbf{X}_\alpha$ have the same c.d.f. $F_\alpha(\mathbf{x})$ for $\alpha = 1, \ldots, n$. Moreover, by virtue of (6.2.11), we may assume without loss of generality that $\boldsymbol{\theta} = \mathbf{0}$. Then $\hat{\boldsymbol{\theta}}_n(\mathbf{X}_1, \ldots, \mathbf{X}_n)$ and $\hat{\theta}(-\mathbf{X}_1, \ldots, -\mathbf{X}_n)$ have the same distribution (as $\boldsymbol{\theta} = \mathbf{0}$). Hence, to prove that $\hat{\boldsymbol{\theta}}_n(\mathbf{X}_1, \ldots, \mathbf{X}_n)$ and $-\hat{\boldsymbol{\theta}}_n(\mathbf{X}_1, \ldots, \mathbf{X}_n)$ have the same distribution, we have only to show that

(6.2.13) $$\hat{\boldsymbol{\theta}}_n(\mathbf{X}_1, \ldots, \mathbf{X}_n) = -\hat{\boldsymbol{\theta}}_n(-\mathbf{X}_1, \ldots, -\mathbf{X}_n).$$

Using (6.2.12), we obtain

$$\begin{aligned}(6.2.14)\quad &\sup\{a_j: h_j(\mathbf{X}_n^{(j)} - a_j\mathbf{1}_n) > \mu_{0j}^*\}\\ &\quad = -\inf\{-a_j: h_j(-\mathbf{X}_n^{(j)} + a_j\mathbf{1}_n) < \mu_{0j}^*,\ j = 1, \ldots, p\}\end{aligned}$$

$$\begin{aligned}(6.2.15)\quad &\inf\{a_j: h_j(\mathbf{X}_n^{(j)} - a_j\mathbf{1}_n) < \mu_{0j}^*\}\\ &\quad = -\sup\{-a_j: h_j(-\mathbf{X}_n^{(j)} + a_j\mathbf{1}_n) > \mu_{0j}^*,\ j = 1, \ldots, p\}.\end{aligned}$$

Consequently, from (6.2.9), (6.2.11), (6.2.14), and (6.2.15) we obtain

$$(6.2.16)\qquad \hat{\theta}_{nj}(\mathbf{X}_n^{(j)}) = -\hat{\theta}_{nj}(-\mathbf{X}_n^{(j)}) \quad \text{for all} \quad j = 1, \ldots, p.$$

The proof follows. ◁

Theorem 6.2.2. *The condition (6.2.8) and the (absolute) continuity of $F_\alpha(\mathbf{x})$ for each $\alpha = 1, \ldots, n$ imply the (absolute) continuity of the joint distribution of $\hat{\boldsymbol{\theta}}_n$.*

Proof: By virtue of (6.2.8), for any fixed numbers $t_2, \ldots, t_n$, $h_j(x_1, x_1 + t_2, \ldots, x_1 + t_n)$ is nondecreasing in x_1. Let $v_j(t_2, \ldots, t_n)$ be such that $h_j(x_1, x_1 + t_2, \ldots, x_1 + t_n)$ is $<$ or $\geq \mu_{0j}^*$ according as x_1 is $<$ or $\geq$ $v_j(t_2, \ldots, t_n)$. Then by (6.2.9),

$$(6.2.17)\qquad \theta_{nj}^{**}(y_1, y_1 + t_2, \ldots, y_1 + t_n) = y_1 - v_j(t_2, \ldots, t_n).$$

If $F_\alpha(\mathbf{x})$ is continuous, the distribution of $\mathbf{X}_n^{(j)}$ is also so. Furthermore,

$$(6.2.18)\quad \theta_{nj}^{**}(\mathbf{X}_n^{(j)}) = c \Leftrightarrow X_{j1} = v_j(X_{j2} - X_{j1}, \ldots, X_{jn} - X_{j1}) + c$$

and hence, the continuity of the joint distribution of $\mathbf{X}_n^{(j)}$ implies that the set of points (s) for which (6.2.18) holds has probability zero. Hence, $P\{\theta_{nj}^{**}(\mathbf{X}_n^{(j)}) = c\} = 0$ for all $j = 1, \ldots, p$ and real c. Similarly $P\{\theta_{nj}^{*}(\mathbf{X}_n^{(j)}) = c\} = 0$ for all $j = 1, \ldots, p$ and real c. These two facts along with (6.2.9) and (6.2.10) imply that $P\{\hat{\theta}_{nj} = c\} = 0$ for all $j = 1, \ldots, p$ and real c. Since the continuity of all marginals implies the continuity of a joint distribution, the proof follows. Suppose now that $F_\alpha(\mathbf{x})$ is absolutely continuous for each $\alpha = 1, \ldots, n$. Then, it is seen from (6.2.18) that for any set A on the real line with Lebesgue measure zero, the set $S_j^* = \{\mathbf{X}_n^{(j)}: \theta_{nj}^{**}(\mathbf{X}_n^{(j)}) \in A\}$ has also Lebesgue measure zero. Hence $P\{S_j^*\} = 0, j = 1, \ldots, p$. The rest of the proof follows as in the case of continuous $F_\alpha(\mathbf{x})$, $\alpha = 1, \ldots, n$. ◁

So far, we have studied the small-sample properties of $\hat{\boldsymbol{\theta}}_n$. We shall now study the asymptotic normality of $\hat{\boldsymbol{\theta}}_n$.

As in section 4.4, we shall make the assumption that

$$F_{(n)}^*(\mathbf{x}) = \frac{1}{n}\sum_{\alpha=1}^{n} F_\alpha(\mathbf{x})$$

converges to a limiting c.d.f. $F^*(\mathbf{x})$ as $n \to \infty$. (The conditions under which this assumption is justified have already been discussed in section 4.4.)

Theorem 6.2.3. *If $\psi_j(x)$, $j = 1, \ldots, p$ satisfy the assumptions of Theorem 4.4.3, then $n^{1/2}(\hat{\boldsymbol{\theta}}_n - \boldsymbol{\theta})$ has asymptotically a p-variate normal distribution with null mean vector and covariance matrix $\mathbf{T}^* = ((\tau_{jk}^*))$ where $\tau_{jk}^*, j, k = 1, \ldots, p$ are defined by* (4.6.9).

Proof: For two p-vectors $\mathbf{x}$ and $\mathbf{y}$, let $\mathbf{x} \leq \mathbf{y}$ denote the coordinate wise inequalities $x_i \leq y_i$ for all $i = 1, \ldots, p$. Then from (6.2.8), we obtain for any $\mathbf{a} = (a_1, \ldots, a_p)$

$$\text{(6.2.19)} \quad \begin{cases} \boldsymbol{\theta}_n^{**} < \mathbf{a} \Rightarrow \mathbf{h}(\mathbf{X}_1 - \mathbf{a}, \ldots, \mathbf{X}_n - \mathbf{a}) < \boldsymbol{\mu}_0^* \Rightarrow \boldsymbol{\theta}_n^{**} \leq \mathbf{a}, \\ \boldsymbol{\theta}_n^* > \mathbf{a} \Rightarrow \mathbf{h}(\mathbf{X}_1 - \mathbf{a}, \ldots, \mathbf{X}_n - \mathbf{a}) > \boldsymbol{\mu}_0^* \Rightarrow \boldsymbol{\theta}_n^* \geq \mathbf{a}, \end{cases}$$

where $\mathbf{h} = (h_1, \ldots, h_p)$ and $\boldsymbol{\mu}_0^* = (\mu_{01}^*, \ldots, \mu_{0p}^*)$. Since $F_\alpha(\mathbf{x})$, $\alpha = 1, \ldots, n$ are absolutely continuous, the distributions of $\boldsymbol{\theta}_n^*$ and $\boldsymbol{\theta}_n^{**}$ are also absolutely continuous (Theorem 6.2.2). Hence, from (6.2.19),

$$\text{(6.2.20)} \qquad P\{\boldsymbol{\theta}_n^{**} < \mathbf{a}\} = P\{\mathbf{h}(\mathbf{X}_1 - \mathbf{a}, \ldots, \mathbf{X}_n - \mathbf{a}) < \boldsymbol{\mu}_0^*\}$$

$$\text{(6.2.21)} \qquad P\{\boldsymbol{\theta}_n^* < \mathbf{a}\} = P\{\mathbf{h}(\mathbf{X}_1 - \mathbf{a}, \ldots, \mathbf{X}_n - \mathbf{a}) \leq \boldsymbol{\mu}_0^*\}.$$

(6.2.20), (6.2.21), and (6.2.9) imply that

$$\text{(6.2.22)} \quad \begin{aligned} P\{\mathbf{h}(\mathbf{X}_1 - \mathbf{a}, \ldots, \mathbf{X}_n - \mathbf{a}) < \boldsymbol{\mu}_0^*\} &\leq P\{\hat{\boldsymbol{\theta}}_n < \mathbf{a}\} \\ &\leq P\{\mathbf{h}(\mathbf{X}_1 - \mathbf{a}, \ldots, \mathbf{X}_n - \mathbf{a}) \leq \boldsymbol{\mu}_0^*\} \end{aligned}$$

From (6.2.17), we obtain

$$\text{(6.2.23)} \quad \begin{aligned} &\lim_{n \to \infty} P_{\boldsymbol{\theta}}\{n^{1/2}(\hat{\boldsymbol{\theta}}_n - \boldsymbol{\theta}) \leq \mathbf{a}\} \\ &\qquad = \lim_{n \to \infty} P_0\{\mathbf{h}(\mathbf{X}_1 - n^{-1/2}\mathbf{a}, \ldots, \mathbf{X}_n - n^{-1/2}\mathbf{a}) \leq \mu_0^*\} \end{aligned}$$

where P_0 indicates that the probability is computed when $\boldsymbol{\theta} = \mathbf{0}$.

Let $\Phi(x_1, \ldots, x_p; \mathbf{m}, \boldsymbol{\Sigma})$ be the p-variate multinormal c.d.f. with mean vector $\mathbf{m}$ and dispersion matrix $\boldsymbol{\Sigma}$. Then from the Corollary 4.4.3.1, it follows that as $n \to \infty$ $\mathcal{L}(n^{1/2}\{\mathbf{h}(\mathbf{X}_1 - n^{-1/2}\mathbf{a}, \ldots, \mathbf{X}_n - n^{-1/2}\mathbf{a}) - \boldsymbol{\mu}_0^*\})$ converges to $\Phi(x_1, \ldots, x_p; (a_1c_1, \ldots, a_pc_p), \frac{1}{4}\boldsymbol{\nu}^*)$ where $c_j = \frac{1}{2}B(F_j^*)$, $B(F_j^*)$ is the same as B_j in (4.4.89) with F replaced by F^*, and $\boldsymbol{\nu}^* = ((\boldsymbol{\nu}_{jk}^*))$ is defined in (4.4.30). Hence the right-hand side of (6.2.23) equals

$$\text{(6.2.24)} \qquad \Phi(a_1c_1, \ldots, a_pc_p; \mathbf{0}, \tfrac{1}{4}\boldsymbol{\nu}^*), \quad \text{for all} \quad \mathbf{a}.$$

The proof now follows from (6.2.23), (6.2.24), and the fact (cf. (4.6.9)) that $\tau_{jk}^* = \nu_{jk}^*/4c_jc_k$ for $j, k = 1, \ldots, p$. Q.E.D. ◁

For the special case of the Wilcoxon and the normal scores estimators, the dispersion matrix T^* reduces to (4.6.18) and (4.6.22), respectively.

In section 6.4 we shall study the asymptotic performance of the estimators $\hat{\boldsymbol{\theta}}_{n(R)}$ and $\hat{\boldsymbol{\theta}}_{n(\Phi)}$ vis-à-vis the median estimator

(6.2.25)

$$\hat{\boldsymbol{\theta}}_{n(M)} = (\hat{\theta}_{n1(M)}, \ldots, \hat{\theta}_{np(M)}), \qquad \hat{\theta}_{nj(M)} = \text{median}\,\{X_{j\alpha}, \alpha = 1, \ldots, n\}$$

and the least squares estimator

$$(6.2.26)\quad \bar{\mathbf{X}}_n = (\bar{X}_{n1}, \ldots, \bar{X}_{np}), \qquad \bar{X}_{nj} = \frac{1}{n}\sum_{\alpha=1}^{n} X_{j\alpha}; \qquad j = 1, \ldots, p.$$

Using the results of section 4.5 on the asymptotic normality of the (vector) sign statistics, it follows readily (Exercise 6.2.7) that

$$(6.2.27)\qquad n^{1/2}(\hat{\boldsymbol{\theta}}_{n(M)} - \boldsymbol{\theta}) \rightarrow N(\mathbf{0}, \boldsymbol{\tau}^*_{(M)}) \quad \text{as} \quad n \rightarrow \infty$$

where $\boldsymbol{\tau}^*_{(M)} = ((\tau^*_{jk(M)}))$ is given by

$$(6.2.28)\qquad \tau^*_{jk(M)} = \begin{cases} \dfrac{1}{4f^{*2}_{[j]}(0)} & \text{if } j = k, \\ [F^*_{[j,k]}(0, 0) - \frac{1}{4}]/f^*_{[j]}(0)f^*_{[k]}(0) & \text{if } j \neq k, \end{cases}$$

where $f^*_{[j]}(0)$ is the density of the marginal distribution $F^*_{[j]}(x)$ of the jth variate (corresponding to the c.d.f. $F^*(\mathbf{x})$) at the population median $\mathbf{0}$, and $F^*_{[j,k]}(x, y)$ is the bivariate c.d.f. of the (j, k)th variates corresponding to $F^*(\mathbf{x})$.

Finally, if $\boldsymbol{\Sigma}^* = ((\sigma^*_{jk}))$ is the population dispersion matrix of $F^*(\mathbf{x})$, then

$$(6.2.29)\qquad n^{1/2}(\bar{\mathbf{X}}_n - \boldsymbol{\theta}) \rightarrow N(\mathbf{0}, \boldsymbol{\Sigma}^*) \quad \text{as} \quad n \rightarrow \infty.$$

6.3. POINT ESTIMATION OF LOCATION IN THE MULTIVARIATE TWO-SAMPLE PROBLEM

As in section 6.2, we first indicate briefly the method of estimating the location in the univariate two-sample problem. Suppose that $X_1, \ldots, X_{n_1}$ and $Y_1, \ldots, Y_{n_2}$, $N = n_1 + n_2$ are independent samples from the absolutely continuous c.d.f.'s $F(x)$ and $F(x - \Delta)$, respectively. Consider a test statistic

$$h(\mathbf{X}_{n_1}; \mathbf{Y}_{n_2}) = h(X_1, \ldots, X_{n_1}; Y_1, \ldots, Y_{n_2})$$

for the hypothesis $\Delta = 0$ against the alternatives $\Delta > 0$. Assume that

$$(6.3.1)\qquad h(\mathbf{X}_{n_1}, \mathbf{Y}_{n_2} + a\mathbf{1}_{n_2}) \text{ where } \mathbf{1}_{n_2} \text{ is the } n_2 \text{ vector, } (1, 1, \ldots, 1)$$

is a nondecreasing function of a for each $\mathbf{X}_{n_1}$ and $\mathbf{Y}_{n_2}$,

$$(6.3.2)\qquad \text{when } \Delta = 0,\ E(h(\mathbf{X}_{n_1}, \mathbf{Y}_{n_2})) = \mu_n \quad \text{independent of } F.$$

Set

$$
(6.3.3)\qquad \begin{cases} \Delta_N^* = \sup\{a\colon h(\mathbf{X}_{n_1}, \mathbf{Y}_{n_2} - a\mathbf{1}_{n_2}) > \mu_n\} \\ \Delta_N^{**} = \inf\{a\colon h(\mathbf{X}_{n_1}, \mathbf{Y}_{n_2} - a\mathbf{1}_{n_2}) < \mu_n\} \end{cases}
$$

and

$$
(6.3.4)\qquad \hat{\Delta}_N = (\Delta_N^* + \Delta_N^{**})/2.
$$

Then, for a suitable function h, we propose $\hat{\Delta}_N$ as estimator of Δ. It is easy to note that the estimator $\hat{\Delta}_N$ of Δ is based on the same alignment consideration which lead to the estimator $\hat{\theta}_n$ of θ in the one-sample problem.

Here also the estimator $\hat{\Delta}_N$ forms a general class of estimators of Δ. (i) Let $h(\mathbf{X}_{n_1}, \mathbf{Y}_{n_2})$ be the Wilcoxon statistic. Then the resulting estimator $\hat{\Delta}_{N(R)}$ turns out to be the median of the set of mn differences $Y_j - X_i, i = 1, \ldots, n_1$; $j = 1, \ldots, n_2$. (cf. Hodges and Lehmann, 1963; and Sen, 1963b). (ii) Let $h(\mathbf{X}_{n_1}, \mathbf{Y}_{n_2})$ be the median test (cf. Exercise 3.3.6); then the resulting estimator $\hat{\Delta}_{N(M)}$ becomes median $(Y_1, \ldots, Y_{n_2})$ − median $(X_1, \ldots, X_{n_1})$. (iii) Let $h(\mathbf{X}_{n_1}, Y_{n_2})$ be the normal scores statistic. In this case the resulting estimator $\hat{\Delta}_{N(\Phi)}$ has to be computed by the trial-and-error method as in the one-sample problem.

We now consider the multivariate two-sample problem. Let $\mathbf{X}_\alpha = (X_{1\alpha}, \ldots, X_{p\alpha}), \alpha = 1, \ldots, n_1$ and $Y_\beta = (Y_{1\beta}, \ldots, Y_{p\beta}), \beta = 1, \ldots, n_2$ be independent vector-valued random variables drawn from p-variate continuous c.d.f.'s $F(\mathbf{x})$ and $G(\mathbf{x}) = F(\mathbf{x} - \mathbf{\Delta})$, respectively, where $\mathbf{\Delta} = (\Delta_1, \ldots, \Delta_p)$ is the shift parameter (vector). By analogy with section 6.2, we shall consider estimators based on rank tests studied in chapter 5.

As in chapter 5, consider the two-sample problem (by letting $c = 2$), and define the rank order statistics $T_{Nj} = T_{Nj}^{(1)}, j = 1, \ldots, p$, as in (5.4.4). Rewrite T_{Nj} as $h_j(\mathbf{X}_{n_1}^{(j)}, \mathbf{Y}_{n_2}^{(j)})$, where $\mathbf{X}_{n_1}^{(j)} = (X_{1j}, \ldots, X_{n_1 j})$ and $\mathbf{Y}_{n_2}^{(j)} = (Y_{1j}, \ldots, Y_{n_2 j})$ for $j = 1, \ldots, p$, and let $\mathbf{X}_{n_1}^* = (\mathbf{X}_{n_1}^{(1)\prime}, \ldots, \mathbf{X}_{n_1}^{(p)\prime})$ and $\mathbf{Y}_{n_2}^* = (\mathbf{Y}_{n_2}^{(1)\prime}, \ldots, \mathbf{Y}_{n_2}^{(p)\prime})$. Define

$$
(6.3.5)\qquad \begin{cases} \Delta_j^* = \sup\{a\colon h_j(\mathbf{X}_{n_1}^{(j)}, \mathbf{Y}_{n_2}^{(j)} - a\mathbf{1}_{n_2}) > \bar{E}_N^{(j)}\} \\ \Delta_j^{**} = \inf\{a\colon h_j(\mathbf{X}_{n_1}^{(j)}, \mathbf{Y}_{n_2}^{(j)} - a\mathbf{1}_{n_2}) < \bar{E}_N^{(j)}\}, \qquad j = 1, \ldots, p, \end{cases}
$$

where $\bar{E}_N^{(j)}$ is defined by (5.4.8). The proposed estimator of Δ is

$$
(6.3.6)\qquad \hat{\mathbf{\Delta}}_N = (\hat{\Delta}_{N1}, \ldots, \hat{\Delta}_{Np}), \quad \hat{\Delta}_{Nj} = \tfrac{1}{2}(\Delta_j^* + \Delta_j^{**}), \qquad j = 1, \ldots, p.
$$

Here also the estimator $\hat{\mathbf{\Delta}}_N$ forms a general class. Important members are (i) the Wilcoxon scores estimator $\hat{\mathbf{\Delta}}_{N(R)}$ and (ii) the normal scores estimator $\hat{\mathbf{\Delta}}_{N(\Phi)}$. The median scores estimator is $\hat{\mathbf{\Delta}}_{N(M)} = (\hat{\Delta}_{N1(M)}, \ldots, \hat{\Delta}_{Np(M)})$ where $\hat{\Delta}_{Ni(M)}$ = median $(y_{i1}, \ldots, y_{in_2})$ − median $(X_{i1}, \ldots, X_{in_1}), i = 1, \ldots, p$. However, for normal scores or for general $E_{N\alpha}^{(j)}$'s, the estimator $\hat{\mathbf{\Delta}}_N$ has to be computed by the trial-and-error method.

Let us now write $\hat{\Delta}_N$ as $\hat{\Delta}_N(X^*_{n1}, Y^*_{n2})$. Recalling that the rank order statistics in (5.4.4) satisfy the conditions that

$$h_j(\mathbf{X}^{(j)}_{n1} + a\mathbf{1}_{n_1}, \mathbf{Y}^{(j)}_{n2} + a\mathbf{1}_{n_2}) = h_j(\mathbf{X}^{(j)}_{n1}, \mathbf{Y}^{(j)}_{n2}), \quad j = 1, \ldots, p, \tag{6.3.7}$$

for all real a, we obtain from (6.3.5), (6.3.6), and (6.3.7) that

$$\begin{aligned}\hat{\boldsymbol{\Delta}}_N(\mathbf{X}_1, \ldots, \mathbf{X}_{n_1}, \mathbf{Y}_1 + \mathbf{a}, \ldots, \mathbf{Y}_{n_2} + \mathbf{a}) \\ = \hat{\boldsymbol{\Delta}}_N(\mathbf{X}_1, \ldots, \mathbf{X}_{n_1}, \mathbf{Y}_1, \ldots, \mathbf{Y}_{n_2}) + \mathbf{a}\end{aligned} \tag{6.3.8}$$

for all $\mathbf{a} \in R^p$, that is, $\hat{\boldsymbol{\Delta}}_N$ is translation invariant.

Theorem 6.3.1. *The conditions that $h_j(\mathbf{X}^{(j)}_{n1} + a\mathbf{1}_{n_1}, \mathbf{Y}_{n_2})$ is nondecreasing in a for all $j = 1, \ldots, p$ and $F(\mathbf{x})$ is (absolutely) continuous imply that the distribution of $\boldsymbol{\Delta}_N$ is also (absolutely) continuous.*

The proof follows as in Theorem 6.2.2 and hence is left as Exercise 6.3.4. It may be noted that as the scores $E^{(j)}_{N\alpha}$'s are assumed to be monotonic (for location problem, cf. chapter 5), $h_j(\mathbf{X}^{(j)}_{n1} + a\mathbf{1}_{n_1}, \mathbf{Y}_{n_2})$ will also be monotonic in a. It may also be noted that by virtue of (5.4.4) and $N = n_1 + n_2$,

$$h_j(\mathbf{X}^{(j)}_{n}, \mathbf{Y}^{(j)}_{n2}) + h_j(-\mathbf{X}^{(j)}_{n1}, -\mathbf{Y}^{(j)}_{n2}) = 2\bar{E}^{(j)}_N, \quad j = 1, \ldots, p; \tag{6.3.9}$$

$$n_1 h_j(\mathbf{X}^{(j)}_{n}, \mathbf{Y}^{(j)}_{n2}) + n_2 h_j(\mathbf{Y}^{(j)}_{n2}, \mathbf{X}^{(j)}) = N\bar{E}^{(j)}_N, \quad j = 1, \ldots, p. \tag{6.3.10}$$

Hence, proceeding essentially along the same lines as in Theorem 6.2.1, we obtain the following.

Theorem 6.3.2. *The distribution of $\boldsymbol{\Delta}_N$ is diagonally symmetric about $\boldsymbol{\Delta}$, if either (6.3.9) holds and $F(\mathbf{x})$ is diagonally symmetric about its median (vector), or $n_1 = n_2$ and (6.3.10) holds.*

For the study of the asymptotic properties of $\hat{\boldsymbol{\Delta}}_N$, we shall assume that $0 < \lim_{N\to\infty} n_1/N = \lambda < 1$. We also define $\boldsymbol{\nu}(F) = ((\nu_{jk}(F)))$ as in (5.4.29) and $B_{(j)}(F), j = 1, \ldots, p$ as in (5.6.4). Then, we have the following theorem.

Theorem 6.3.3. *If the rank scores $\{E^{(j)}_{N\alpha}\}$'s satisfy the conditions of Theorem 5.5.1, then $(n_1n_2/N)^{1/2}(\hat{\boldsymbol{\Delta}}_N - \boldsymbol{\Delta})$ has asymptotically a multinormal distribution with null mean vector and dispersion matrix $\nu^*(F) = ((\nu^*_{jk}(F)))$, where $\nu^*_{jk}(F) = \nu^*_{jk}(F)/B_{(j)}(F)B_{(k)}(F), j, k = 1, \ldots, p$.*

Proof: Denote

$$\mathbf{h}(\mathbf{X}^*_{n1}, \mathbf{Y}^*_{n2}) = (h_1(\mathbf{X}^{(1)}_{n1}, \mathbf{Y}^{(1)}_{n2}), \ldots, h_p(\mathbf{X}^{(p)}_{n1}, \mathbf{Y}^{(p)}_{n2})),$$

$\mathbf{a} = (a_1, \ldots, a_p)$ and $\bar{\mathbf{E}}_N = (\bar{E}^{(1)}_N, \ldots, \bar{E}^{(p)}_N)$. Then, proceeding as in the proof of Theorem 6.2.3, we obtain

$$\begin{aligned}P\{\mathbf{h}(\mathbf{X}_1, \ldots, \mathbf{X}_{n_1}; \mathbf{Y}_1 - \mathbf{a}, \ldots, \mathbf{Y}_{n_2} - \mathbf{a}) < \bar{\mathbf{E}}_N\} \le P\{\hat{\boldsymbol{\Delta}}_N < \mathbf{a}\} \\ \le P\{\mathbf{h}(\mathbf{X}_1, \ldots, \mathbf{X}_{n_1}; \mathbf{Y}_1 - \mathbf{a}, \ldots, \mathbf{Y}_{n_2} - \mathbf{a}) \le \bar{\mathbf{E}}_N\}.\end{aligned} \tag{6.3.11}$$

Consequently,

$$\lim_{N\to\infty} P_{\Delta}\{(n_1n_2/N)^{1/2}(\boldsymbol{\Delta}_N - \hat{\boldsymbol{\Delta}}) \le \mathbf{a}\}$$

$$(6.3.12)\qquad = \lim_{N\to\infty} P_0\left\{\mathbf{h}\left(\mathbf{X}_1, \ldots, \mathbf{X}_{n_1}, \mathbf{Y}_1 - \left(\frac{N}{n_1n_2}\right)^{1/2}\mathbf{a}, \ldots, \mathbf{Y}_{n_2} - \left(\frac{N}{n_1n_2}\right)^{1/2}\mathbf{a}\right) \le E_N\right\},$$

where P_0 denotes the probability when $\boldsymbol{\Delta} = \mathbf{0}$. Now, from Theorem 5.6.1 it follows that when $\boldsymbol{\Delta} = (N/n_1n_2)^{-1/2}\mathbf{a}$, $N^{1/2}[\mathbf{h}(\mathbf{X}^*_{n_1}, \mathbf{Y}^*_{n_2}) - \bar{\mathbf{E}}_N]$ has asymptotically a multinormal distribution with mean vector $(n_1/n_2)^{1/2}(a_1B_{(1)}, \ldots, a_pB_{(p)})$ and dispersion matrix $(n_1/n_2)\boldsymbol{\nu}(F)$, where $\boldsymbol{\nu}(F)$ is defined by (5.4.29) and the B_j's by (5.6.4). Now, the right-hand side of (6.3.12) can be written as $\lim_{N\to\infty} P_{\Delta_N}\{\mathbf{h}(\mathbf{X}^*_{n_1}, \mathbf{Y}^*_{n_2}) \le \bar{\mathbf{E}}_N\}$, which is equal to $\Phi(a_1B_{(1)}, \ldots, a_pB_{(p)}; \mathbf{0}, \boldsymbol{\nu}(F))$, for all $\mathbf{a}$. The rest of the proof follows as in Theorem 6.2.3. ◁

Here also we note that for the median estimator, instead of using Theorem 5.6.1, we have to use Theorem 5.3.2 for the asymptotic multinormality of the median statistics (vector).

For the estimation of $\boldsymbol{\Delta}$, the generalized least squares estimator is $\bar{\mathbf{Z}}_N = (\bar{\mathbf{Z}}_{N1}, \ldots, \bar{\mathbf{Z}}_{Np})$, where

$$(6.3.13)\qquad Z_{Nj} = \frac{1}{n_2}\sum_{\beta=1}^{n} Y_{j\beta} - \frac{1}{n_1}\sum_{\alpha=1}^{n} X_{j\alpha}, \qquad j = 1, \ldots, p.$$

The asymptotic distribution of $(n_1n_2/N)^{1/2}(\mathbf{Z}_N - \boldsymbol{\Delta})$ is normal with mean vector $\mathbf{0}$ and covariance matrix $\boldsymbol{\Sigma} = ((\sigma_{jk}))$ (which is the covariance matrix of $F(\mathbf{X})$), as follows from the multivariate central limit theorem.

In the next section we study the asymptotic relative efficiency of the proposed estimators.

6.4. ASYMPTOTIC RELATIVE EFFICIENCY OF THE ESTIMATORS

To obtain an idea of the relative performance of two estimators, we employ the notion of "the generalized variance," a concept introduced by Wilks (1962, p. 166). The "generalized variance" of a p-variate random vector $(X_1, \ldots, X_p)$ with nonsingular covariance matrix $\boldsymbol{\Sigma} = (\rho_{jk}\sigma_j\sigma_k)$ is defined to be the var $\mathbf{X} = \sigma_1^2, \ldots, \sigma_p^2\,|\rho_{jk}| = |\boldsymbol{\Sigma}|$, where $|\ \ |$ denote the determinant. Now suppose that the two asymptotically unbiased estimators T and T^* of $\boldsymbol{\theta}$ with asymptotically nonsingular matrices $\boldsymbol{\Sigma}$ and $\boldsymbol{\Sigma}^*$ require n and n^* observations to achieve equal asymptotic generalized variances. Then the asymptotic relative efficiency of T with respect to T^* is defined as

(6.4.1) $$e_{T,T^*} = \lim_{n\to\infty} \frac{n^*}{n} = \{|\mathbf{\Sigma}^*|/|\mathbf{\Sigma}|\}^{1/p}.$$

Consider now the one-sample location problem discussed in section 6.2. From Theorem 6.2.3, we obtain

(6.4.2) $$e_{\hat{\boldsymbol{\theta}}_n, \bar{\mathbf{X}}_n} = \{|\mathbf{\Sigma}^*|/|\mathbf{T}^*|\}^{1/p},$$

where $\mathbf{\Sigma}^*$ is the population covariance matrix of F^*, and $\mathbf{T}^* = (\tau^*_{jk})$ is as defined in Theorem 6.2.3.

The efficiency (6.4.2) depends only on the parent c.d.f. F^* and the score function $\mathbf{J}^* = (J^*_1, \ldots, J^*_p)$. We now assume for the simplicity of presentation that $F_1 = \cdots = F_n = F$ so that $F^* = F$. The results hold even when not all F_i, $i = 1, \ldots, n$ are identical provided we replace F by F^*. With this specialization, we notice that

(6.4.3) $$\operatorname{var}(n^{1/2}\hat{\boldsymbol{\theta}}_n) = \left(\prod_{j=1}^{p} \zeta_{jj}/B^2_{(j)}(F)\right) \cdot |\rho^{\theta}_{jk}|,$$

where

(6.4.4) $$\zeta_{jk} = \begin{cases} \displaystyle\int_0^1 J_j^2(x)\,dx & \text{if } j = k = 1, \ldots, p; \quad J_j = \psi_j^{-1} \\ \displaystyle\int_{-\infty}^{\infty}\int_{-\infty}^{\infty} J_j^*[F_j(x)]J_k^*[F_k(y)]\,dF_{j,k}(x, y) & \\ & \text{if } j \neq k = 1, \ldots, p; \quad J_j^* = \psi_j^{*-1}, \end{cases}$$

$B_{(j)}(F)$ is given by (5.6.4), and

(6.4.5) $$\rho^{\theta}_{jk} = \zeta_{jk}\zeta_{jj}^{-1/2}\zeta_{kk}^{-1/2}; \qquad j, k = 1, \ldots, p.$$

(6.4.6) $$\operatorname{var}(n^{1/2}\hat{\boldsymbol{\theta}}_{n(M)}) = \left(\prod_{j=1}^{p} 4^{-1}f_j^{-2}(0)\right) \cdot |\rho^M_{jk}|$$

where

(6.4.7) $$\rho^M_{jk} = \begin{cases} 1 & \text{if } j = k = 1, \ldots, p \\ 4[F_{j,k}(0, 0) - \frac{1}{4}] & \text{if } j \neq k = 1, \ldots, p. \end{cases}$$

We shall study in detail the relative asymptotic performances of $\hat{\boldsymbol{\theta}}_{n(\Phi)}$, $\hat{\boldsymbol{\theta}}_{n(R)}$, $\hat{\boldsymbol{\theta}}_{n(M)}$, and $\bar{\mathbf{X}}_n$. Thus using (6.4.1), (6.4.3), (6.4.6), and (6.2.29), we obtain

(6.4.8) $$e_{\hat{\boldsymbol{\theta}}_{n(\Phi)}, \bar{\mathbf{X}}_n} = \left[\prod_{j=1}^{p} \sigma_j^2 B_j^2(F_j, \Phi)\right]^{p^{-1}} \cdot [|\rho_{jk}|/|\rho^{\Phi}_{jk}|]^{p^{-1}},$$

where ρ^{Φ}_{jk} is given by (4.6.22) and $B_j(F_j, \Phi)$ is the same as $B_j(F)$ with $J_j^* = \Phi^{-1}$.

Now since

(6.4.9) $$\sigma_j^2 B_j^2(F_j, \Phi) \geq 1 \quad \text{for all } j = 1, \ldots, p,$$

it follows that

$$e_{\hat{\boldsymbol{\theta}}_{n(\Phi)}, \bar{\mathbf{X}}_n} \geq [|\rho_{jk}|/|\rho_{jk}^{\Phi}|]^{p^{-1}}. \tag{6.4.10}$$

For arbitrary $F(\mathbf{x})$, the bounds for the right-hand side of (6.4.9) cannot be obtained. Hence, we may consider a few special cases.

(i) Let $F(\mathbf{x})$ be the p-variate normal distribution with mean vector $\mathbf{0}$ and nonsingular covariance matrix $\boldsymbol{\Sigma} = (\rho_{jk}\sigma_j\sigma_k)$. Then it is easy to check that the efficiency (6.4.8) is 1. *Thus when the parent c.d.f. is nonsingular normal, the normal scores and the least squares estimators are asymptotically equally efficient.*

(ii) Let $F(\mathbf{x}) = \prod_{j=1}^{p} F_j(x_j)$ for all $\mathbf{x} \in R^p$. In this case, the covariance matrices (ρ_{jk}) and (ρ_{jk}^{Φ}) reduce to diagonal matrices and hence (6.4.8) ≥ 1. *Thus if $F(\mathbf{x})$ has pairwise independent coordinates, the normal scores estimator is at least as efficient as the least squares estimator.* In fact, in this case a comparatively stronger statement may be made as follows: Let $\mathcal{F}_p^*$ be a class of p-variate c.d.f.'s $\{F(\mathbf{x})\}$ for which the total correlation coefficients are equal to zero (that is, $\rho_{jk} = 0$ for all $j \neq k = 1, \ldots, p$). Then $|\rho_{jk}| = 1$ and

$$|\rho_{jk}^{\Phi}| \leq 1 \tag{6.4.11}$$

where the equality holds only when $\rho_{jk}^{\Phi} = 0$ for all $j \neq k = 1, \ldots, p$. Hence

$$\inf_{F \in \mathcal{F}_p^*} e_{\hat{\boldsymbol{\theta}}_{n(\Phi)}, \bar{\mathbf{X}}_n} \geq 1, \tag{6.4.12}$$

and the equality holds only when $F(\mathbf{x})$ is normal.

Let us now consider the Wilcoxon estimator $\boldsymbol{\theta}_{n(R)}$. In this case

$$e_{\hat{\boldsymbol{\theta}}_{n(R)}, \bar{\mathbf{X}}_n} = \left[\prod_{j=1}^{p} 12\sigma_j^2 B_j^2(F_j, R)\right]^{p^{-1}} [|\rho_{jk}|/|\rho_{jk}^{R}|]^{p^{-1}} \tag{6.4.13}$$

where

$$\rho_{jk}^{R} = \begin{cases} 12\displaystyle\int_{-\infty}^{\infty}\int_{-\infty}^{\infty} F_j(x)F_k(y)\, dF_{jk}(x, y) - 3 & \text{if } j \neq k = 1, \ldots, p, \\ 1 \quad \text{if } j = k = 1, \ldots, p, \end{cases} \tag{6.4.14}$$

and $B_j(F_j, R)$ is the same as $B_j(F)$ with $J_j^*(x) = 2x - 1$.

The first factor on the right-hand side of (6.4.13) is the geometric mean of the efficiencies of the (univariate) one-sample Wilcoxon test with respect to the t test, and is always greater than or equal to 0.864. Hence whatever the continuous c.d.f. $F(\mathbf{x})$ be

$$e_{\hat{\boldsymbol{\theta}}_{n(R)}, \bar{\mathbf{X}}_n} \geq 0.864[|\rho_{jk}|/|\rho_{jk}^{R}|]^{p^{-1}}. \tag{6.4.15}$$

Now since the bounds for $|\rho_{jk}|/|\rho_{jk}^{R}|$ depend upon the unknown association pattern of $F(\mathbf{x})$, the exact numerical values or bounds for (6.4.13) do not exist. When $F(x)$ is nonsingular multinormal with mean vector $\mathbf{0}$ and covariance matrix $\boldsymbol{\Sigma} = (\rho_{jk}\sigma_j\sigma_k)$, then $\rho_{jk}^R = (6/\pi)\sin^{-1}(\rho_{jk}/2)$ and so

$$e_{\hat{\boldsymbol{\theta}}_{n(R)},\bar{\mathbf{X}}_n} = \frac{3}{\pi}\cdot\left[|\rho_{jk}| \Big/ \left|\frac{6}{\pi}\sin^{-1}(\rho_{jk}/2)\right|\right]^{p^{-1}} = g(\rho_{jk},\ 1 \le j < k \le p). \tag{6.4.16}$$

In the case of $p = 2$, (6.4.16) reduces to

$$\left\{(1-\rho^2)\Big/\left[1 - \frac{36}{\pi^2}\left(\sin^{-1}\frac{\rho}{2}\right)^2\right]\right\}^{1/2}. \tag{6.4.17}$$

Hence maximizing and minimizing (6.4.17) with respect to $\rho(-1 \le \rho \le 1)$, we obtain (for the case when the underlying distribution $F(\mathbf{x})$ is bivariate normal)

$$0.91 \le e_{\hat{\boldsymbol{\theta}}_{n(R)},\bar{\mathbf{X}}_n} \le 0.95, \tag{6.4.18}$$

where the lower bound is obtained when $|\rho| \to 1$.

For $p > 2$, it has been shown by Bickel (1964) that $e_{\hat{\boldsymbol{\theta}}_{n(R)},\bar{\mathbf{X}}_n}$ can be arbitrarily close to zero.

Since, when the underlying distribution is normal, $e_{\hat{\boldsymbol{\theta}}_n(R),\hat{\boldsymbol{\theta}}_n(\Phi)} = e_{\hat{\boldsymbol{\theta}}_n(R),\bar{\mathbf{X}}_n}$, (6.4.18) also holds for $e_{\hat{\boldsymbol{\theta}}_{n(R)},\hat{\boldsymbol{\theta}}_n(\Phi)}$.

Here also we note that when $F(\mathbf{x})$ has pairwise independent coordinates, that is, $\rho_{jk} = 0$ for all $j \ne k = 1, \ldots, p$, then (6.4.13) is bounded below by 0.864 and (6.4.16) by $3/\pi$.

Finally, for the median estimator $\hat{\boldsymbol{\theta}}_{n(M)}$,

$$e_{\hat{\boldsymbol{\theta}}_{n(M)},\bar{\mathbf{X}}_n} = \left[\prod_{j=1}^{p}\{4\sigma_j^2 f_j^2(0)\}\right]^{1/p}[|\rho_{jk}|/|\rho_{jk}^M|], \tag{6.4.19}$$

where ρ_{jk}^M is given by (6.4.7).

For the univariate c.d.f.'s, the first factor on the right-hand side of (6.4.19) is bounded below by $\frac{1}{3}$, is equal to $2/\pi$ for normal c.d.f.'s, and has known values for various other c.d.f.'s. However, the second factor on the right-hand side of (6.4.19) again depends on the unknown association pattern of $F(\mathbf{x})$, and no bounds may exist. For parent normal c.d.f.'s

$$e_{\hat{\boldsymbol{\theta}}_{n(M)},\bar{\mathbf{X}}_n} = \frac{2}{\pi}\left\{|\rho_{jk}| \Big/ \left|\frac{2}{\pi}\sin^{-1}\rho_{jk}\right|\right\}^{1/p} \tag{6.4.20}$$

again depends solely on the ρ_{jk}'s. For coordinatewise independent c.d.f.'s, it is easy to check (by applying the results of the univariate theory) that $(6.4.19) \ge 0.33$ for all unimodal F_j. Analogous results hold for the corresponding two-sample problem and are given as exercises.

6.5. AN ALTERNATIVE APPROACH TO THE TWO-SAMPLE CASE

In the two-sample case, the generalized least square estimator $\bar{\mathbf{X}}_n$ is the difference of the two individual sample mean vectors. Analogously, it may be enquired whether the difference of the two individual sample estimators in (6.2.10) can be used as an alternative estimator of the shift vector $\boldsymbol{\Delta}$. The answer to this question depends upon the c.d.f.'s $F(x)$ and $G(x)$ of the two populations sampled as well as on $J_j(u)$, $j = 1, \ldots, p$, on which the rank order statistics are based. It may be noted that the estimator of $\boldsymbol{\Delta}$ in (6.3.6) is based upon the assumption that $G(\mathbf{x}) = F(\mathbf{x} - \boldsymbol{\Delta})$ and it does not require the additional assumption that $F(\mathbf{x})$ is diagonally symmetric about its median (vector). On the other hand, the estimators in (6.2.10), in general, demand the diagonal symmetry of $F(\mathbf{x})$ around its median (an exception being the median scores estimator). So two main questions arise: (i) is it necessary to assume that $F(\mathbf{x})$ and $G(\mathbf{x})$ are both diagonally symmetric in order that the difference of the individual sample estimators in (6.2.10) be an estimator of $\boldsymbol{\Delta}$, and (ii) how does this estimator compare with that of (6.2.10)? An important case where F and G may differ from each other (even when $\boldsymbol{\Delta} = \mathbf{0}$) is the generalized Behrens-Fisher situation where $G(\mathbf{x})$ and $F(\mathbf{x})$, apart from a shift in locations, also differ in their dispersion matrices, though they have the same functional form. It is also of some interest to study the relative performances of the two estimators in such a case.

Now suppose that $F(\mathbf{x})$ and $G(\mathbf{x})$ are both diagonally symmetric about $\boldsymbol{\mu}$ and $\boldsymbol{\mu} + \boldsymbol{\Delta}$, respectively, but otherwise, may be quite arbitrary. For $\boldsymbol{\mu}$ and $\boldsymbol{\mu} + \boldsymbol{\Delta}$, we consider the individual sample estimators in (6.2.10), and the estimator of $\boldsymbol{\Delta}$ obtained from their difference is denoted by $\boldsymbol{\Delta}_N^*$. Since the estimators in (6.2.10) are shown to have distributions symmetric about their population counterparts, it readily follows that $\boldsymbol{\Delta}_N^*$ is also distributed symmetrically about $\boldsymbol{\Delta}$. Theorem 6.2.2 also holds for $\boldsymbol{\Delta}_N^*$, and using Theorem 6.2.3, we obtain that if $\lim_{N\to\infty} m/N \to \lambda$: $0 < \lambda < 1$, then $N^{1/2}(\boldsymbol{\Delta}_N^* - \boldsymbol{\Delta})$ converges in law to a multivariate normal distribution with null mean vector and dispersion matrix $T^{**} = ((\tau_{jk}^{**}))$, where

$$\tau_{jk}^{**} = (1/\lambda)\tau_{jk}^*(F) + (1/(1-\lambda))\tau_{jk}^*(G),$$

where the τ_{jk}^*'s are defined by (4.4.40), with F^* replaced by F and G, respectively. If $F(\mathbf{x})$ and $G(\mathbf{x})$ are both diagonally symmetric about their respective medians, we have the following.

Lemma 6.5.1. *If $F(\mathbf{x})$ and $G(\mathbf{x})$ are both diagonally symmetric about a common median vector and if $J_j(u)$, $j = 1, \ldots, p$ are all skew-symmetric* (*i.e.*,

$J_j(u) + J_j(1-u) = 2J_j(0)$ *for all* $0 < u \leq \frac{1}{2}$ *and* $j = 1, \ldots, p$), *then*

$$\int_{-\infty}^{\infty} J_j[\lambda F_j(x) + (1-\lambda)G_j(x)]dF_j(x) = \int_0^1 J_j(u)du \quad \textit{for all} \quad j = 1, \ldots, p.$$

Proof: Without loss of generality, assume that $J_j(0) = 0$. Then under the conditions stated in the lemma, $\int_0^1 J_j(u)\,du = \int_0^{1/2} [J_j(u) + J_j(1-u)]\,du = 0$. Since $F_j(x)$ and $G_j(x)$ are both symmetric about a common median (say, μ_j), it follows that $\lambda F_j(x) + (1-\lambda)G_j(x)$ is also symmetric about the same point. Also, without any loss of generality, we may assume that $\mu_j = 0$. Then, $J[\lambda F_j(x) + (1-\lambda)G_j(x)] + J[\lambda F_j(-x) + (1-\lambda)G_j(-x)] = 2J_j(0) = 0$. Consequently,

$$\int_{-\infty}^{\infty} J_j[\lambda F_j(x) + (1-\lambda)G_j(x)]\,dF_j(x) = \int_{-\infty}^{0} [J_j[\lambda F_j(x) + (1-\lambda)G_j(x)] + J_j[\lambda F_j(-x)) + (1-\lambda)G_j(-x)]\,dF_j(x) = 0,$$

as $F_j(x)$ is symmetric about μ_j $(=0)$. Q.E.D. ◅

Lemma 6.5.1 shows that the estimator $\hat{\mathbf{\Delta}}_N$, defined by (6.3.4), also remains valid when $G(x) \neq F(\mathbf{x} - \mathbf{\Delta})$, but both are diagonally symmetric about their respective median vectors which differ by $\mathbf{\Delta}$. Theorems 6.3.1 and 6.3.2 remain valid even if F and G are diagonally symmetric but otherwise arbitrary. Moreover, if we define

$$\text{(6.5.1)} \quad B_j^* = \int_{-\infty}^{\infty} \frac{d}{dx} J_j[\lambda F_j(x) + (1-\lambda)G_j(x)]\,dF_j(x), \qquad j = 1, \ldots, p,$$

and $\sigma_{N[j,j']}$ as in Theorem 5.5.1 (putting $c = 2$ and dropping the superscripts k, q as both q then are equal to 1), for $j, j' = 1, \ldots, p$, then using Lemma 6.5.1 and proceeding as in the proof of Theorem 6.3.1, we see that under the conditions of Theorem 6.3.1, $N^{1/2}(\hat{\mathbf{\Delta}}_N - \mathbf{\Delta})$ has asymptotically a multinormal distribution with null mean vector and dispersion matrix having elements

$$\text{(6.5.2)} \qquad \nu_{jk}^{**} = \sigma_{[j,k]}/B_j^* B_k^* \quad \text{for} \quad j, k = 1, \ldots, p,$$

where $\sigma_{[j,k]} = \lim_{N\to\infty} \sigma_{N[j,k]}$. Let us denote by

$$\text{(6.5.3)} \qquad \boldsymbol{\nu}^{**} = ((\nu_{jk}^{**})).$$

Then to compare $\mathbf{\Delta}_N^*$ and $\hat{\mathbf{\Delta}}_N$, according to the criterion of generalized variance in Section 6.4, we have to find the value of

$$\text{(6.5.4)} \qquad \{|\boldsymbol{\nu}^{**}|/|\mathbf{T}^{**}|\}^{1/p},$$

which is the asymptotic efficiency of $\mathbf{\Delta}_N^*$ with respect to $\hat{\mathbf{\Delta}}_N$.

As in Section 6.4, for arbitrary F and G, it may be quite difficult to obtain bounds for (6.5.4) in the general case of p (≥ 1) variates. However, in the particular case of $p = 1$, and some specific situations, (6.5.4) can be shown to be bounded below by 1, in the Behrens-Fisher situation (cf. Exercise 6.5.16).

It may be noted that if F and G are both multinormal c.d.f.'s, differing possibly in dispersion matrices, and if we use $\boldsymbol{\Delta}_N^*$ with normal scores, T^{**} will be asymptotically equal to the covariance matrix of $\bar{\mathbf{X}}_N$, the two-sample least squares estimator of $\boldsymbol{\Delta}$. Consequently, $\boldsymbol{\Delta}_N^*$ will be asymptotically most efficient in the sense that it has asymptotically the minimum generalized variance. On the other hand, the dispersion matrix of $N^{1/2}(\hat{\boldsymbol{\Delta}}_N - \boldsymbol{\Delta})$ will be different from that of $N^{1/2}(\bar{\mathbf{X}}_N - \boldsymbol{\Delta})$, and by numerous examples, it can be shown that $\hat{\boldsymbol{\Delta}}_N$ cannot be asymptotically more efficient than $\boldsymbol{\Delta}_N^*$. This adds some justification for using $\boldsymbol{\Delta}_N^*$ in the generalized Behrens-Fisher situation.

Referred to the one-sample case, the estimator $\hat{\boldsymbol{\theta}}_n$ (of $\boldsymbol{\theta}$) estimates the true median vector when $F(\mathbf{x})$ is diagonally symmetric about $\boldsymbol{\theta}$. A slightly weaker condition on F may be as follows. Let

$$\xi_j(F) = \int_0^\infty J_j[F_j(x) - F_j(-x)]\, dF_j(x), \quad j = 1, \ldots, p, \tag{6.5.5}$$

where $J_j(x) = \Psi_j^{*-1}(u)$, $0 < u < 1$, Ψ_j^* being defined by (6.2.7). Let $\mathcal{F}_p^*$ be the class of p-variate c.d.f.'s,

$$\mathcal{F}_p^* = \left\{F: \xi_j(F) = \frac{1}{2}\int_0^1 J_j(u)\, du, \quad j = 1, \ldots, p\right\}. \tag{6.5.6}$$

Then, for all $F(\mathbf{x} - \boldsymbol{\theta})$, $F \in \mathcal{F}_p^*$, $\hat{\boldsymbol{\theta}}_n$ estimates θ, and its normality can be proved along the same line as in Theorem 6.2.3. Thus, if both F and G belong to $\mathcal{F}_p^*$, $\boldsymbol{\Delta}_N^*$ will be an estimator of $\boldsymbol{\Delta}$. For arbitrary J_j's, it is difficult to specify $\mathcal{F}_p^*$. On the other hand, let $\mathcal{F}_p^{**}$ be the class of p-variate c.d.f.'s, such that if both F and G belong to $\mathcal{F}_p^{**}$,

$$\int_{-\infty}^{\infty} J_j[\lambda F_j(x) + (1 - \lambda)G_j(x)]\, dF_j(x) = \int_0^1 J_j(u)\, du, \quad j = 1, \ldots, p. \tag{6.5.7}$$

Then, for F, G belonging to $\mathcal{F}_p^{**}$ (when $\boldsymbol{\Delta} = \mathbf{0}$), or $F(\mathbf{x})$, $G(\mathbf{x} - \boldsymbol{\Delta})$ belong to $\mathcal{F}_p^{**}$, $\hat{\boldsymbol{\Delta}}_N$, defined by (6.3.4), estimates $\boldsymbol{\Delta}$. Hence $\boldsymbol{\Delta}_N^*$ and $\hat{\boldsymbol{\Delta}}_N$ estimate the common parameter $\boldsymbol{\Delta}$, when $F, G \in \mathcal{F}_p^* \cap \mathcal{F}_p^{**}$, which of course includes the class of all diagonally symmetric c.d.f.'s. For general J_j's, it is difficult to specify $\mathcal{F}_p^* \cap \mathcal{F}_p^{**}$ in more simple forms. However, for specific J_j's, and the corresponding efficiencies, see Exercises 6.5.8 to 6.5.16.

6.6. POINT ESTIMATION OF CONTRASTS IN ONE-WAY LAYOUT

Consider $c\ (\geq 2)$ independent sets of random variables. In the ith set, let $\mathbf{X}_{i1}, \ldots, \mathbf{X}_{in_i}$ be n_i independent and identically distributed random variables having a common continuous c.d.f. $F_i(\mathbf{x})$, $i = 1, \ldots, c$. Write

$$F_i(\mathbf{x}) = F(\mathbf{x} - \boldsymbol{\theta}_i), \qquad i = 1, \ldots, c, \tag{6.6.1}$$

and define a contrast in $\boldsymbol{\theta}$'s as

$$\boldsymbol{\phi} = \sum_{i=1}^{c} l_i \boldsymbol{\theta}_i, \qquad \sum_{i=1}^{c} l_i = 0. \tag{6.6.2}$$

According to the classical linear estimation theory, ϕ is estimable and its best estimator is

$$\hat{\boldsymbol{\phi}}_N = \sum_{i=1}^{c} l_i (1/n_i) \sum_{\alpha=1}^{n_i} \mathbf{X}_{i\alpha} \qquad \left(\text{where } N = \sum_{1}^{c} n_i\right). \tag{6.6.3}$$

We shall consider some robust estimators of $\boldsymbol{\phi}$ based on the rank order statistics studied in chapter 5. For this, consider the pair (i, i') of samples, and define

$$\Delta_{ii'(j)} = \theta_{ij} - \theta_{i'j} \quad \text{for} \quad i, i' = 1, \ldots, c; \qquad j = 1, \ldots, p. \tag{6.6.4}$$

Based on the n_i observations of the ith sample and $n_{i'}$ observations from the i'th sample ($i \neq i'$), we define the estimator $\hat{\Delta}_{ii'(j)}$ in the same manner as in (6.3.3), for $i \neq i' = 1, \ldots, c$ and $j = 1, \ldots, p$. Also, conventionally, we let $\hat{\Delta}_{ii(j)} = 0$ for $i = 1, \ldots, c$ and $j = 1, \ldots, p$.

Our proposed estimators of $\boldsymbol{\phi}$'s are based on the set of raw estimators $\{\hat{\Delta}_{ii'(j)}, j = 1, \ldots, p; i, i' = 1, \ldots, c\}$. The estimators $\Delta_{ii'(j)}$ have the same properties as of $\hat{\boldsymbol{\Delta}}$ in section 6.3. However, they are incompatible in the following sense. Suppose we want to estimate $\boldsymbol{\theta}_2 - \boldsymbol{\theta}_1$. A different answer may be obtained when we estimate this difference directly from the first and second samples than when we take the sum of the estimators of $\boldsymbol{\theta}_3 - \boldsymbol{\theta}_1$ and $\boldsymbol{\theta}_2 - \boldsymbol{\theta}_3$. Thus, we may not obtain a single natural estimate of $\boldsymbol{\phi}$. To remove this difficulty, one may consider the following compatible or adjusted estimators. Let

$$\hat{\Delta}_{i\cdot(j)} = (1/c) \sum_{i'=1}^{c} \hat{\Delta}_{ii'(j)}, \qquad i = 1, \ldots, c, \quad j = 1, \ldots, p. \tag{6.6.5}$$

Then by minimizing $\sum_{i \neq i'} [\hat{\Delta}_{ii'(j)} - (\theta_{ij} - \theta_{i'j})]^2$, for each ($j = 1, \ldots, p$) one obtains the compatible estimators of $\Delta_{ii'(j)}$ as

$$\Delta^*_{ii'(j)} = \hat{\Delta}_{i\cdot(j)} - \hat{\Delta}_{i'\cdot(j)} \quad \text{for} \quad i, i' = 1, \ldots, c, \quad j = 1, \ldots, p. \tag{6.6.6}$$

Then, on writing $\phi_j = \sum_{i=1}^{c} l_i \theta_{ij}$ equivalently as $\sum_{i=1}^{c} l_i \Delta_{ii'(j)}$, (for some $i' = 1, \ldots, p$), we obtain from (6.6.6) its compatible estimator as

$$\phi^*_j = \sum_{i=1}^{c} l_i \hat{\Delta}_{i\cdot(j)} \quad \text{for} \quad j = 1, \ldots, p. \tag{6.6.7}$$

The property of translation invariance of ϕ^*_j follows from (6.3.3)–(6.6.7), and its compatibility is evident. However, the estimator ϕ_j^* depends not only on the set of samples for which $l_i \neq 0$, but also on those which are quite unrelated to ϕ_j^*. For example, for the estimation of $\boldsymbol{\theta}_2 - \boldsymbol{\theta}_1$, the estimator $\boldsymbol{\phi}^*$ in (6.6.7) depends on all the c sample observations, whereas the least squares estimator depends only on the first and second sample observations.

Now the set $\boldsymbol{\Phi}$ of all possible contrasts may be spanned by any set of $p(c-1)$ linearly independent contrasts, say,

$$\text{(6.6.8)} \quad \phi_{j,l} = \sum_{i=1}^{c} c_{j,i}^{(l)}\theta_{ij}, \qquad \sum_{i=1}^{c} c_{j,i}^{(l)} = 0 \text{ for } \quad l = 1, \ldots, c-1; \quad j = 1, \ldots, p.$$

The corresponding rank order estimators are

$$\text{(6.6.9)} \quad \phi_{j,l}^* = \sum_{i=1}^{c} c_{j,i}^{(l)} \hat{\Delta}_{i\cdot(j)}, \quad j = 1, \ldots, p; \qquad l = 1, \ldots, c-1.$$

For the asymptotic theory, it will be assumed that $N \to \infty$ with

$$\text{(6.6.10)} \qquad n_i/N = \lambda_i\colon 0 < \lambda_i < 1 \quad \text{for all} \quad i = 1, \ldots, c.$$

We define $J_j(u)$ $(j = 1, \ldots, p)$ as in chapter 5 (section 5.4), $\mu_j = \int_0^1 J_j(u)\,du$, $j = 1, \ldots, p$ and $\boldsymbol{\nu}(F)$ as in (5.4.29). Also, $B_j(F), j = 1, \ldots, p$ are defined as in (5.6.4). Let then

$$\text{(6.6.11)} \qquad \gamma_{j,j'}^{l,l'} = \nu_{jj'}\left(\sum_{i=1}^{c} c_{j,i}^{(l)} c_{j',i}^{(l')}/\lambda_i\right)\Big/B_j(F)B_{j'}(F)$$

for $j, j' = 1, \ldots, p$; $l, l' = 1, \ldots, c-1$.

Theorem 6.6.1. *If the rank scores satisfy the conditions* 5.5.1,

$$[N^{1/2}\{\phi_{j,l}^* - \phi_{j,l}\}, 1 \leq l \leq c-1, 1 \leq j \leq p]$$

has asymptotically a $p(c-1)$*-variate normal distribution with null mean vector and dispersion matrix* $\boldsymbol{\Gamma}$ *having elements* $\gamma_{j,j'}^{l,l'}$, *defined by* (6.6.11).

The proof of this theorem rests on the following.

Theorem 6.6.2. *Consider a sequence of c-tuplets of c.d.f.'s* $\{F_{i,N}(\mathbf{x}) = F(\mathbf{x} + N^{-1/2}\mathbf{a}_i), i = 1, \ldots, c\}$ *where* $\mathbf{a}_1, \ldots, \mathbf{a}_c$ *are p-vectors having real and finite elements. Assume that the conditions of Theorem* 5.5.1 *hold and define* $h_j(\mathbf{X}_i^{(j)}, \mathbf{X}_{i'}^{(j)})$ *as in section* 6.3. *Then the stochastic vector*

$$[N^{1/2}\{(\lambda_i + \lambda_c)/\lambda_c\}\{h_j(\mathbf{X}_i^{(i)}, \mathbf{X}_c^{(j)}) - \mu_i\}, \quad 1 \leq i \leq c-1, \quad j = 1, \ldots, p]$$

has asymptotically a $p(c-1)$*-variate normal distribution with a mean vector* $[B_j(F)(a_{ij} - a_{cj}), \ i = 1, \ldots, c-1; \ j = 1, \ldots, p]$ *and dispersion matrix with elements*

$$\text{(6.6.12)} \quad \nu_{jj'}\{1/\lambda_c + \delta_{ii'}/\lambda_i\}, \quad j, j' = 1, \ldots, p; \qquad i, i' = 1, \ldots, c-1.$$

The proof of the theorem follows directly from Theorem 5.6.1 with straightforward modifications and hence is omitted.

Proceeding now as in the proof of Theorem 6.3.3, we see that it follows from Theorem 6.6.2 that $[N^{1/2}(\hat{\Delta}_{ic(j)} - \Delta_{ic(j)}),\ i = 1, \ldots, c-1,\ j = 1, \ldots, p]$ has asymptotically a multinormal distribution with a null mean vector and dispersion matrix with elements

$$\text{Cov}\,\{N^{1/2}(\hat{\Delta}_{ic(j)} - \Delta_{ic(j)}), N^{1/2}(\hat{\Delta}_{i'c(j')} - \Delta_{i'c(j')})\} = \nu_{jj'}(1/\lambda_c + \delta_{ii'}/\lambda_i)/B(F_j)B_{j'}(F), \tag{6.6.13}$$

for $j, j' = 1, \ldots, p$; $i, i' = 1, \ldots, c-1$.

Now, using (6.6.13) and some routine computations, it follows that $|N^{1/2}(\Delta^*_{ii'(j)} - \hat{\Delta}_{ii'(j')}| \xrightarrow{p} 0$ for all $i, i' = 1, \ldots, c$ and $j = 1, \ldots, p$, (actually the convergence is in the quadratic mean). Hence

$$N^{1/2}\left(\phi^*_{j,l} - \sum_{i=1}^{c} c^{(l)}_{j,i}\hat{\Delta}_{ic(j)}\right) \xrightarrow{p} 0 \quad \text{for all} \quad l, j. \tag{6.6.14}$$

The rest of the proof follows directly from the asymptotic multinormality of $[N^{1/2}(\Delta_{ic(j)} - \hat{\Delta}_{ic(j)}),\ 1 \le i \le c-1,\ 1 \le j \le p]$, and this completes the proof of Theorem 6.6.1.

We shall now consider the A.R.E. of the set of rank order estimators in (6.6.9) with respect to the least squares estimators. As in section 6.4, we compare the reciprocals of their generalized variances. Here also, we let $\sigma_{jj'}$ be the covariance of the (j, j')th variates following the joint c.d.f. $F(\mathbf{x})$ for $j, j' = 1, \ldots, p$. Then, the covariance of the least squares estimator of $\phi_{j,l}$ and $\phi_{j',l'}$, (as defined by (6.6.3)) may be shown to be equal to

$$(1/N)\sigma_{jj'}\left(\sum_{i=1}^{c} c^{(l)}_{j,i}c^{(l')}_{j',i}/\lambda_i\right), \quad \text{for} \quad j, j' = 1, \ldots, p; \quad l, l' = 1, \ldots, c-1. \tag{6.6.15}$$

Consequently, on denoting by $\mathbf{T}/N$ the $p(c-1) \times p(c-1)$ matrix whose elements are given by (6.6.15), and $\mathbf{\Gamma}$ by those in (6.6.11), the A.R.E. of the rank order estimates with respect to the least squares estimator is

$$\{|\mathbf{T}|/|\mathbf{\Gamma}|\}^{1/p}. \tag{6.6.16}$$

For $p = 1$, (6.6.16) will be independent of $c^{(l)}_{j,l}$'s, and will be equal to $\sigma_{11}B_1^2(F)/\nu_{11}$, and the bounds for it studied in section 6.4 will reveal the bounds for the desired A.R.E. But for $p > 1$ (6.6.16) depends on $c^{(l)}_{j,i}$'s, unless we are dealing with coordinatewise independent c.d.f.'s. As in section 6.4, (6.6.16) can be factored into marginal and joint factors. The marginal factor is independent of the $c^{(l)}_{j,l}$'s, while the joint factor is not. Thus, bounds for the joint factor (over possible variation of the $c_{j,i}^{(i)}$'s) provide the bounds for the desired A.R.E. Now, these bounds are nothing but the largest and smallest characteristic roots of $(B_1^2(F)\cdots B_p^2(F))\boldsymbol{\Sigma}\boldsymbol{\nu}^{-1}$, where $\boldsymbol{\Sigma} = ((\sigma_{jj'}))$ and $\boldsymbol{\nu}$ is defined by (5.1.2). The problem thus reduces to the one-sample or two-sample situation and hence, the results of section 6.4 are applicable.

6.7. INTERVAL ESTIMATION IN UNIVARIATE ONE-SAMPLE AND TWO-SAMPLE LOCATION PROBLEMS

We first consider the one-sample problem. Suppose that $X_1, \ldots, X_n$ are independently distributed with distributions $F_1(x-\theta), \ldots, F_n(x-\theta)$, respectively, where $F_\alpha(x)$, $\alpha = 1, \ldots, n$ are all continuous and symmetric about 0. The problem is to construct a robust confidence interval for θ. We define $h(\mathbf{X}_n)$ as in (6.2.6) for $p = 1$ and note that when $\theta = 0$, the distribution of $h(\mathbf{X}_n)$ is symmetric about $\frac{1}{2}\bar{E}_n$ when $\bar{E}_n = \sum_{\alpha=1}^n E_{n,\alpha}/n$. Furthermore, in this case each $h(\mathbf{X}_n)$ can have 2^n equally likely realizations. Thus it is possible to select a constant $h^0_{n,\alpha}$ and an α_n close to α such that

$$P_{\theta=0}\{|h(\mathbf{X}_n) - \tfrac{1}{2}\bar{E}_n| \le h^0_{n,\alpha}\} = 1 - \alpha_n. \tag{6.7.1}$$

For small values of n and standard $h(\mathbf{X}_n)$ such as the Wilcoxon signed rank statistic, exact values of $h^0_{n,\alpha}$ may be obtained from the existing tables. However, if n is large, then (cf. Theorem 4.4.2 for $p = 1$) $h(\mathbf{X}_n)$ has a normal distribution in the limit, and so asymptotically

$$n^{1/2}\,|h^0_{n,\alpha} - \tfrac{1}{2}\tau_{\alpha/2}A| \to 0 \quad \text{as} \quad n \to \infty \tag{6.7.2}$$

where τ_α is in $100\alpha\%$ point of the standard normal distribution, and $A^2 = \int_0^1 J^2(u)\,du$.

By definition, $h(\mathbf{X}_n - a\mathbf{1}_n)$ is nondecreasing in a. Hence defining

$$\theta^*_{L,n} = \inf\{a\colon h(\mathbf{x}_n - a\mathbf{1}_n) \le \tfrac{1}{2}\bar{E}_n + h^0_{n,\alpha}\} \tag{6.7.3}$$

$$\theta^*_{U,n} = \sup\{a\colon h(\mathbf{x}_n - a\mathbf{1}_n) \ge \tfrac{1}{2}\bar{E}_n - h^0_{n,\alpha}\} \tag{6.7.4}$$

and using

$$P_\theta\{|h(\mathbf{X}_n - \theta\mathbf{1}_n) - \tfrac{1}{2}\bar{E}_n| \le h^0_{n,\alpha}\} = 1 - \alpha_n \tag{6.7.5}$$

we obtain

$$P_\theta\{\theta^*_{L,n} \le \theta \le \theta^*_{U,n}\} = 1 - \alpha_n \tag{6.7.6}$$

which gives the desired confidence interval for θ with confidence coefficient $1 - \alpha_n$. It may be noted that (6.7.6) is a translation invariant confidence interval for θ, since $\theta^*_{L,n}$ and $\theta^*_{U,n}$ are both translation invariant. The computation of $\theta^*_{L,n}$ and $\theta^*_{U,n}$ requires a trial-and-error method (cf. Puri and Sen,

1967a). However, for the special case of Wilcoxon scores we have the following direct method. Consider the $n(n+1)/2$ ordered values of $\frac{1}{2}(X_i + X_j)$, $1 \le i < j \le n$ and denote them by $W_{(1)} < \cdots < W_{n^*}$ where $n^* = n(n+1)/2$. Now the event (for Wilcoxon scores) $h(\mathbf{x}_n - a\mathbf{1}_n) = c$ is equivalent to saying that among the n^* values of $W_{(j)} - a$, c are positive and the rest negative. Hence on writing $M_i = n(n+1)[\frac{1}{2}\bar{E}_n + (-1)^i h^0_{n,\alpha}]$, $i = 1, 2$, we have $\theta^*_{L,n} = W_{(M_1)}$ and $\theta^*_{U,n} = W_{(M_2)}$.

Let us now consider the univariate two-sample location problem. Dropping the subscript j, we define $h(\mathbf{X}_{n_1}, \mathbf{Y}_{n_2})$ as in section 6.3 and also borrow the other notations used there. Then, under the hypothesis H: $G(x) = F(x - \Delta)$, it follows that

$$(6.7.7) \qquad Z_N = \{N\lambda_N/A^2(1-\lambda_N)\}^{1/2} \cdot \{h_N(\mathbf{X}_{n_1} + \Delta\mathbf{1}_{n_1}, \mathbf{Y}_{n_2}) - \mu_n\}$$

is a strictly distribution-free statistic, and it follows from Theorem 3.6.1 that asymptotically Z_N has a normal distribution with zero mean and unit variance. Let $Z_N^{(1)}$ and $Z_N^{(2)}$ be such that

$$(6.7.8) \qquad P_\Delta\{Z_N \le Z_N^{(1)}\} = \alpha_1, \qquad P_\Delta\{Z_N \ge Z_N^{(2)}\} = \alpha_2,$$

(ideally α_1 and α_2 are equal). The values of $Z_N^{(1)}$ and $Z_N^{(2)}$ can be computed from the known distribution of Z_N (independently of F). For large N, we take $\alpha_1 = \alpha_2 = \alpha/2$, and get the following:

$$(6.7.9) \qquad Z_N^{(1)} + Z_N^{(2)} \to 0, \qquad Z_N^{(1)} \to \tau_{\alpha/2},$$

where $\tau_{\alpha/2}$ is the lower $100\alpha/2\%$ point of the standard normal distribution. Let now

$$(6.7.10) \qquad \hat{\Delta}_{U,N} = \sup\{\theta: Z_N < Z_N^{(2)}\}, \quad \hat{\Delta}_{L,N} = \inf\{\theta: Z_N > Z_N^{(1)}\},$$

and

$$(6.7.11) \qquad I_N = \{\Delta: \hat{\Delta}_{L,N} \le \Delta \le \hat{\Delta}_{U,N}\}.$$

The width of I_N is

$$(6.7.12) \qquad \delta_N = \hat{\Delta}_{U,N} - \hat{\Delta}_{L,N}.$$

Then $\{\hat{\Delta}_{L,N} \le \Delta \le \hat{\Delta}_{U,N}\}$ provides a $100(1-\alpha)\%$ confidence interval for θ, whose width is δ_N.

It may be noted that $\hat{\Delta}_{U,N}$ and $\hat{\Delta}_{L,N}$ are both translation invariant. Hence, δ_N and I_N are also translation invariant. Furthermore, Theorem 6.3.1 also holds for I_N and δ_N. Analogous to Theorem 6.3.3, we have the following.

Theorem 6.7.1. *If $F(x)$ satisfies the regularity conditions of Exercise 3.8.5, then the joint distribution of $N_0^{1/2}((\hat{\Delta}_{U,N} - \Delta), (\hat{\Delta}_{L,N} - \Delta))$ converges in law to* *a normal distribution on the plane concentrating on a line*

$$N_0^{1/2}(\hat{\Delta}_{U,N} - \Delta) = N_0^{1/2}(\hat{\Delta}_{L,N} - \Delta) + 2A\,|\tau_{\alpha/2}|/B(F), \tag{6.7.13}$$

where $N_0 = N\lambda_N(1 - \lambda_N)$ *and* $B(F)$ *and* A *are defined in* (5.6.4) *for* $p = 1$ *and* (3.6.20), *respectively*.

Outline of proof: Let a be any real and finite quantity, and define

$$H_{(a)}(x) = \lambda_N F(x) + (1 - \lambda_N)F(x + N_0^{-1/2}a). \tag{6.7.14}$$

Then, under the stated regularity conditions, it can be shown precisely along the same lines as in the proof of Theorem 3.6.1 that for any real and finite a

$$\begin{aligned}[N_0^{1/2}/A(1 - \lambda_N)]\{h_N(\mathbf{X}_{n_1} + (\Delta + N_0^{-1/2}a)\mathbf{1}_{n_1}, \mathbf{Y}_{n_2} - \mu\}\\ = aB(F)/A + (N_0^{1/2}/A)[B_{1,N}^{(a)} - B_{2,N}^{(a)}] + o_p(1),\end{aligned} \tag{6.7.15}$$

where $\mu = \int_0^1 J(u)\,du$ and

$$B_{1,N}^{(a)} = \int_{-\infty}^{\infty} [G_{n_2}(x + N_0^{-1/2}a) - F(x + N_0^{-1/2}a)]J'(H_{(a)}(x))\,dF(x), \tag{6.7.16}$$

$$B_{2,N}^{(a)} = \int_{-\infty}^{\infty} [F_{n_1}(X) - F(x)]J'(H_{(a)}(x))\,dF(x + N_0^{-1/2}a), \tag{6.7.17}$$

$$F_{n_1}(x) = n_1^{-1}\,[\text{number of } X_i \le x,\ i = 1, \ldots, n_1] \tag{6.7.18}$$

$$G_{n_2}(x) = n_2^{-1}\,[\text{number of } (Y_i - \theta) \le x,\ i = 1, \ldots, n_2]. \tag{6.7.19}$$

Further, the proof of Theorem 3.6.1 implies that for any real and finite a, $(N_0^{1/2}/A)[B_{1,N}^{(a)} + B_{2,N}^{(a)}]$ converges in law to a normal distribution with zero mean and unit variance. Hence, using the same technique as in Theorem 6.3.3 we get from (6.7.10) that

$$\lim_{N\to\infty} P_\Delta\{N_0^{1/2}(\hat{\Delta}_{U,N} - \Delta) \le a\} = \Phi(\tau_{\alpha/2} + aB(F)/A); \qquad \Phi(\tau_{\alpha/2}) = \alpha/2, \tag{6.7.20}$$

where Φ is the c.d.f. of the standard normal variate. (6.7.20) implies that

$$|N_0^{1/2}(\hat{\Delta}_{U,N} - \Delta) + \tau_{\alpha/2}A/B(F)| \quad \text{is bounded in probability.} \tag{6.7.21}$$

Similarly, it can be shown that

$$|N_0^{1/2}(\hat{\Delta}_{L,N} - \Delta) - \tau_{\alpha/2}A/B(F)| \quad \text{is bounded in probability.} \tag{6.7.22}$$

Now, if (a, b) be any two real and finite quantities, then it can be shown using (6.7.16) and some simple algebraic manipulations that

$$E(B_{1,N}^{(a)} - B_{1,N}^{(b)}) = 0, \tag{6.7.23}$$

$$n_2E\{[B_{1,N}^{(a)}]^2\} = A^2 + o(1), \qquad n_2E\{[B_{1,N}^{(b)}]^2\} = A^2 + o(1); \tag{6.7.24}$$

$$n_2E\{B_{1,N}^{(a)} \cdot B_{1,N}^{(b)}\} = A^2 + o(1). \tag{6.7.25}$$

Thus, from (6.6.23), (6.6.24), and (6.6.25), we get

$$n_2 E\{[B_{1,N}^{(a)} - B_{1,N}^{(b)}]^2\} = o(1), \tag{6.7.26}$$

and hence by Chebychev's lemma, we get

$$|n_2^{1/2}[B_{1,N}^{(a)} - B_{1,N}^{(b)}]| = o_p(1). \tag{6.7.27}$$

Similarly, we have

$$|n_1^{1/2}[B_{2,N}^{(a)} - B_{2,N}^{(b)}]| = o_p(1). \tag{6.7.28}$$

Consequently, from (6.6.15)–(6.6.17), (6.6.27), and (6.6.28), we note that for any two real and finite (u, b)

$$\begin{aligned}[N_0^{1/2}/A(1-\lambda_N)]\{h_N(\mathbf{X}_{n_1} + (\Delta + N_0^{-1/2}a)\mathbf{1}_{n_1}, \mathbf{Y}_{n_2}) \\ - h_N(\mathbf{X}_{n_1} + (\Delta + N_0^{-1/2}b)\mathbf{1}_{n_1}, \mathbf{Y}_{n_2})\} = (a-b)B(F)/A + o_p(1).\end{aligned} \tag{6.7.29}$$

Thus, from (6.7.10), (6.7.21), (6.7.22), and (6.7.29) we conclude that

$$N_0^{1/2}(\hat{\theta}_{U,N} - \hat{\theta}_{L,N})B(F)/A = 2\,|\tau_{\alpha/2}| + o_p(1), \tag{6.7.30}$$

which asserts the truth of (6.7.13). The rest of the theorem follows directly from (6.7.15), (6.7.29), (6.7.30), and the asymptotic normality of $(N_0^{1/2}/A)[(B_{1,N}^{(a)} - B_{2,N}^{(a)}), (B_{1,N}^{(b)} - B_{2,N}^{(b)})]$. Hence the theorem.

Under the conditions of Theorem 6.7.1, we obtain

$$\begin{aligned}\hat{B}(F) &= \{(Z_N^{(2)} - Z_N^{(1)})/\hat{\delta}_N[N\lambda_N(1-\lambda_N)]^{1/2}\} \\ &\underset{p}{\sim} \{2A\,|\tau_{\alpha/2}|/\hat{\delta}_N[N\lambda_N(1-\lambda_N)]^{1/2}\}.\end{aligned} \tag{6.7.31}$$

Since $\hat{\delta}_N$ has been shown to be translation invariant, we get from (6.6.17) that $\hat{B}(F)$ is also so. This property of $\hat{B}(F)$ ensures its consistency even when the null hypothesis H_0: $F \equiv G$ does not hold.

Often the reciprocal of the expected squared length of a confidence interval is used as a measure of its efficiency. From Theorem 6.7.1, we readily arrive at the following theorem.

Theorem 6.7.2. *If corresponding to a given confidence coefficient* $1 - \alpha$, $\{I_N\}$ *and* $\{I_N^*\}$ *be two sequences of confidence intervals for the shift parameter* Δ *based on two sequences of rank order statistics* $\{T_N\}$ *and* $\{T_N^*\}$ *both satisfying the conditions of Theorem* 6.7.1, *then the reciprocal of the asymptotic ratio of squared lengths of the confidence intervals is equal to the relative efficiency of the test which generates them.*

The above results can be easily extended to the one-sample location problem (cf. Exercise 6.7.1).

6.8. INTERVAL ESTIMATION OF CONTRASTS IN ANOVA (ONE-WAY LAYOUT)

Two problems arise in practice, namely (i) to attach a confidence interval to any contrast $\boldsymbol{\phi}$, defined by (6.6.2), and (ii) to attach a simultaneous confidence region to all possible contrasts.

The monotonicity of the rank statistics in sections 6.2, 6.3, and 6.7, does not necessarily apply to contrasts in such statistics, and hence, the method of inversion, implicit in the derivation of the results of section 6.7, is no longer tenable for an arbitrary contrast. This presents some difficulty in obtaining an exact confidence interval to a contrast ϕ, based on rank tests (unless Bonferroni type inequality is used). For large samples, however, Theorem 6.6.1 can be used and from each pair (i, i'), $i < i' = 1, \ldots, c$, of samples (6.7.31) can be used to estimate $B(F)$, and these estimates may be pooled together. Then we obtain

$$P\{N^{1/2}\,|\phi_{j,l}^{*} - \phi_{j,l}| \le \tau_{\alpha/2}(\hat{\gamma}_{jj}^{ll})^{1/2}\} = 1 - \alpha \tag{6.8.1}$$

where $\hat{\gamma}_{jj}^{ll}$ is obtained from (6.6.11) upon substituting the estimate of $B_j(F)$ (note that ν_{jj} is a known constant). (6.8.1) provides a large-sample confidence interval to ϕ. On the other hand, for small sample sizes, we may write ϕ in (6.6.2) as $\sum_{i=1}^{c-1} l_i(\theta_i - \theta_c)$, and by the method discussed in section 6.7, attach a $[1 - \alpha/(c-1)]$ confidence interval to $\theta_i - \theta_c$ for all $i = 1, \ldots, c-1$. These may be used to derive the lower and upper bounds to $\sum_{i=1}^{c-1} l_i(\theta_i - \theta_c)$ having a confidence coefficient $\ge 1 - \alpha$. However, this usually results in a less efficient procedure, and moreover, it depends on the choice of the cth sample too.

For the simultaneous confidence region problem, both small-sample and large-sample solutions exist, and will be considered here in detail. These may be regarded as nonparametric generalization of the multiple-comparison procedures due to Tukey and Scheffé.

Nonparametric T Method of Multiple Comparisons. By analogy with the homoscedasticity condition implicit in the use of Tukey's T method of multiple comparisons, we require here that

$$n_1 = \cdots = n_c = n. \tag{6.8.2}$$

We let $\mathbf{X}_i = (X_{i1}, \ldots, X_{in})$, $i = 1, \ldots, c$, and define $h(\mathbf{X}_i, \mathbf{X}_j)$, $i \ne j = 1, \ldots, c$, as in section 6.3 (i.e., the two-sample case). Here, also, we assume that

(6.8.3)

$$h(\mathbf{X}_i + a\mathbf{1}_n, \mathbf{X}_j) \text{ is nondecreasing in } a \text{ for all } \mathbf{X}_i, \mathbf{X}_j, \quad i \ne j = 1, \ldots, c.$$

The procedure to be considered here is based on the following statistic:

$$W_n = \max_{1 \le i < j \le c} [2n^{1/2} A_n^{-1} |h_n(X_i, X_j) - \bar{E}_n|], \tag{6.8.4}$$

where

$$\bar{E}_n = \frac{1}{2n} \sum_{\alpha=1}^{2n} E_{n,\alpha}, \qquad A_n^2 = \frac{1}{2n-1} \sum_{\alpha=1}^{2n} \{E_{n,\alpha} - \bar{E}_n\}^2, \tag{6.8.5}$$

and $E_{n,\alpha}$, $\alpha = 1, \ldots, 2n$, are the rank scores satisfying the conditions of Theorem 3.6.1.

Theorem 6.8.1. *Under (6.6.1) with $\theta_1 = \cdots = \theta_c$,*

$$\lim_{n \to \infty} P\{W_n \le t\} = \chi_c(t),$$

where $\chi_c(t)$ is the c.d.f. of the sample range in a sample of size c drawn from the standard normal distribution.

Proof: Let $G(x)$ be the c.d.f. of $X_{i\alpha} - \theta_i$, and let us define

$$Y = B(X) = \int_{x_0}^{X} J'[G(x)]\, dG(x), \qquad G(x_0) = \tfrac{1}{2}. \tag{6.8.6}$$

Then $Y_{ij} = B(X_{ij})$, $j = 1, \ldots, n$; $i = 1, \ldots, c$ are N $(= nc)$ i.i.d.r.v.'s. From Theorem 3.6.1, it is known Y has finite absolute moment of order $2 + \eta$, $\eta > 0$, and the variance of Y is A^2 (defined in (3.8.20)). Denote the mean of Y by ξ and define

$$Z_{ni} = n^{-1/2} A^{-1} \sum_{j=1}^{n} [Y_{ij} - \xi], \qquad i = 1, \ldots, c. \tag{6.8.7}$$

It then follows that $Z_{n,1}, \ldots, Z_{n,c}$ are i.i.d.r.v.'s distributed (asymptotically) according to the normal distribution with zero mean and unit variance. Consequently, for any given c,

$$\lim_{n \to \infty} P\left\{ \max_{i \le i, j \le c} |Z_{n,i} - Z_{n,j}| \le t \right\} = \chi_c(t), \tag{6.8.8}$$

where $\chi_c(t)$ is defined in the statement of the theorem. Now, proceeding precisely on the same line as in the proof of Theorem 3.6.1, and then using Poincaré's theorem on total probability, it is easily seen that

$$[n^{1/2} A^{-1} \{h_n(X_i, X_j) - \mu\} - \tfrac{1}{2}(Z_{n,i} - Z_{n,j})] = o_p(1), \tag{6.8.9}$$

simultaneously for all $i \ne j = 1, \ldots, c$ where $\mu = \int_0^1 J(u)\, du$. From (6.8.4) and (6.8.9), we obtain

$$\left\{ W_n - \max_{1 \le i < j \le c} |Z_{ni} - Z_{nj}| \right\} = o_p(1), \tag{6.8.10}$$

The proof of the theorem now follows from (6.8.8).

In small samples, the exact distribution of W_n (when $\theta_1 = \cdots = \theta_c$) is quite involved. However, we can use a permutation procedure of evaluating the exact null distribution of W_n. Let us write $X_N = (\mathbf{X}_1, \ldots, \mathbf{X}_c)$. Under $\theta_1 = \cdots = \theta_c$, X_N is composed of N i.i.d.r.v.'s and hence conditioned on the given X_N, all possible ($N!$) permutations of the variates among themselves are equally likely. Thus, conditioned on the given X_N, all possible $(N!/(n!)^c)$ partitionings of these N variables into c subsets of equal size are equally likely, each having the (conditional) probability $[N!/(n!)^c]^{-1}$. Thus, if we consider the set of all these partitionings and for each one of them compute the value of W_n, we will arrive at the permutation distribution function of W_n. Since G is assumed to be continuous (so that the possibility of ties may be ignored, in probability), and as $h(\mathbf{X}_i, \mathbf{X}_j)$ $(i \neq j = 1, \ldots, c)$ are all rank order statistics, it follows that the permutation distribution of W_n derived in this manner will agree with the exact null distribution of W_n. This procedure may be quite useful for small or moderately large values of n (particularly if some modern computing facilities are available), while for large samples, we may use Theorem 6.8.1 to approximate the true null distribution of W_n by $\chi_c(t)$, tables for which are available in the Biometrika volume (pp. 165–171).

Let now α: $0 < \alpha < 1$ be our preassigned level of significance. We denote by $W_{n,\alpha}$ and $R_{c,\alpha}$, the upper $100\alpha\%$ point of the exact null distribution of W_n and of $\chi_c(t)$, respectively, so that

$$P\{W_n < W_{n,\alpha} \mid H_0\} = 1 - \alpha, \qquad \chi_c(R_{c,\alpha}) = 1 - \alpha, \tag{6.8.11}$$

and by Theorem 6.8.1, $W_{n,\alpha} \to R_{c,\alpha}$ as $n \to \infty$.

The simplest type of *paired comparison test* may now be formulated as follows:

1. For all $1 \leq i < j \leq c$, compute the values of $h(\mathbf{X}_i, \mathbf{X}_j)$; the values of the remaining set will be obtained from the relation that

$$h_n(X_i, X_j) + h_n(X_j, X_i) = 2\bar{E}_n \quad \text{for all} \quad i, j = 1, \ldots, c. \tag{6.8.12}$$

2. Compute the value of $W_{n,\alpha}$ corresponding to the preassigned α.
3. Referring to (6.6.1), regard those $(\theta_i - \theta_j)$ to be significantly different from zero for which

$$2n^{1/2}A_n^{-1}\,|h_n(X_i, X_j) - \bar{E}_n| \geq W_{n,\alpha}. \tag{6.8.13}$$

It is easily seen that the test is an exact size $\alpha(0 < \alpha < 1)$ multiple-comparison (similar) test.

Now, often we are not merely satisfied with the detection of those pairs (θ_i, θ_j) for which $\theta_i \neq \theta_j$, but also want to attach a simultaneous confidence region to all possible $\theta_i - \theta_j$, $i \neq j = 1, \ldots, c$. For this, we write

$$\Delta_{ij} = \theta_i - \theta_j \quad \text{for} \quad i, j = 1, \ldots, c. \tag{6.8.14}$$

From (6.6.1), we then have

$$F_i(x) = F(x - \theta_i) = F_j(x - \Delta_{ij}), \quad \text{for} \quad i, j = 1, \ldots, c. \tag{6.8.15}$$

Define

$$\begin{aligned} \mu_n^{(1)} &= \bar{E}_n - \tfrac{1}{2} n^{-1/2} A_n W_{n,\alpha} \\ \mu_n^{(2)} &= \bar{E}_n + \tfrac{1}{2} n^{-1/2} A_n W_{n,\alpha}. \end{aligned} \tag{6.8.16}$$

Now, by (6.8.3), $h_n(\mathbf{X}_i + a\mathbf{1}_n, \mathbf{X}_j)$ is monotonic in a. Hence, by the sliding principle, we arrive at the following two values:

$$\begin{aligned} \hat{\Delta}_{ij,L} &= \inf \{a\colon h_n(\mathbf{X}_i + a\mathbf{I}_n, \mathbf{X}_j) > \mu_n^{(1)}\}, \\ \hat{\Delta}_{ij,U} &= \sup \{a\colon h_n(\mathbf{X}_i + a\mathbf{I}_n, \mathbf{X}_j) < \mu_n^{(2)}\}; \end{aligned} \tag{6.8.17}$$

which defines an interval

$$I_{ij} = \{\Delta_{ij}\colon \hat{\Delta}_{ij,L} \leq \Delta_{ij} \leq \hat{\Delta}_{ij,U}\}, \qquad i \neq j = 1, \ldots, c. \tag{6.8.18}$$

Then, it follows from (6.8.13) through (6.8.18) that *the probability is* $1 - \alpha$ *that the inequalities* $\hat{\Delta}_{ij,L} \leq \Delta_{ij} \leq \hat{\Delta}_{ij,U}$ *hold simultaneously for all* $i \neq j = 1, \ldots, c$.

We shall now consider certain asymptotic properties of the proposed paired-comparison procedure. To justify the approach theoretically and to avoid the limiting degeneracy of (6.8.18), we conceive of a sequence of c tuplets of c.d.f.'s $\{F_{n,i}(x),\ i = 1, \ldots, c\}$, for which (6.6.1) holds and

$$n^{1/2}\boldsymbol{\theta} \to \boldsymbol{\lambda} = (\lambda_1, \ldots, \lambda_c) \quad \text{as} \quad n \to \infty, \tag{6.8.19}$$

where $\boldsymbol{\theta} = (\theta_1, \ldots, \theta_c)$ and λ_i, $i = 1, \ldots, c$ are real and finite. We also define

$$\lambda_{ij} = \lambda_i - \lambda_j \quad \text{for} \quad i, j = 1, \ldots, c. \tag{6.8.20}$$

We will be then interested in paired comparisons in λ_i's instead of θ_i's. It may be noted that as for the simultaneous confidence region $\{I_{ij}\colon 1 \leq i, j \leq c\}$, we may consider a somewhat more general formulation, viz.,

$$n^{1/2}(\Delta_{ij} - \Delta_{ij}^0) \to \lambda_{ij}, \quad \text{as} \quad n \to \infty, \tag{6.8.21}$$

Δ_{ij}^0 being some (fixed) real quantity, not necessarily equal to zero. Since the confidence intervals of the type (6.8.18) are all translation invariant for the study of the asymptotic properties, it is immaterial whether we take $\Delta_{ij}^0 = 0$ or not, for all $i, j = 1, \ldots, c$. The above formulation is analogous to Pitman's translation alternatives usually adopted to study the efficiency aspects of the nonparametric analysis of variance test. For this, we consider the estimate $\hat{\Delta}_{ij}$ of Δ_{ij} $(= \theta_i - \theta_j)$, defined by (6.3.3), on the basis of $\mathbf{X}_i = (X_{i1}, \ldots, X_{in})$ and $\mathbf{X}_j = (X_{j1}, \ldots, X_{jn})$. Then, by a straightforward extension of Theorem 6.7.1, it follows that asymptotically I_{ij} in (6.8.18) reduces to

$$I_{ij} = \{\lambda_{ij}: |\lambda_{ij} - \hat{\lambda}_{ij}| \leq AR_{c,\alpha}/B(F)\}, \tag{6.8.22}$$

where

$$\hat{\lambda}_{ij} = n^{1/2}(\hat{\Delta}_{ij} - \Delta^0_{ij}), \quad A^2 = \int_0^1 J^2(u)\, du - \left(\int_0^1 J(u)\, du\right)^2,$$

$$B(F) = \int_{-\infty}^{\infty} \frac{d}{dx} J[F(x)]\, dF(x),$$

and $R_{c,\alpha}$ is given by (6.8.11).

To compare (6.8.22) with the corresponding confidence interval obtained by the parametric T method, we define $\bar{\Delta}_{ij} = \bar{X}_i - \bar{X}_j$, as the difference of the ith and jth sample means, and if $R_{c,c(n-1),\alpha}$ is the upper $100\alpha\%$ point of the Studentized range $R_{c,c(n-1)}$, then we have the probability $1 - \alpha$ that the inequalities

$$\bar{\Delta}_{ij} - n^{-1/2} s R_{c,c(n-1),\alpha} \leq \Delta_{ij} \leq \bar{\Delta}_{ij} + n^{-1/2} s R_{c,c(n-1),\alpha} \tag{6.8.23}$$

hold simultaneously for all $i, j = 1, \ldots, c$, where s^2 is the unbiased estimate of σ^2 having $c(n-1)$ degrees of freedom (d.f.). As it is well known that

$$s^2 \xrightarrow{p} \sigma^2, \quad \text{and} \quad R_{c,c(n-1),\alpha} \longrightarrow R_{c,\alpha} \quad \text{as} \quad n \to \infty, \tag{6.8.24}$$

we get from (6.8.23) that asymptotically the confidence interval in (6.8.24) reduces to

$$I^0_{ij} = \{\lambda_{ij}: |\lambda_{ij} - \bar{\lambda}_{ij}| \leq \sigma R_{c,\alpha}\}, \tag{6.8.25}$$

where $\bar{\lambda}_{ij} = n^{1/2}(\bar{\Delta}_{ij} - \Delta^0_{ij})$. Thus, if we take the ratio of the square of the width of the confidence intervals as a measure of the asymptotic efficiency, the asymptotic relative efficiency of our proposed method with respect to the T method reduces to

$$e(J, F) = \sigma^2 B^2(F)/A^2. \tag{6.8.26}$$

Thus, we arrive at the following theorem.

Theorem 6.8.2. *The A.R.E of the proposed nonparametric generalization of the T method of paired comparisons with respect to the T method itself is equal to the A.R.E of the two-sample rank order test (on which the proposed method is based) with respect to the Student's t-test.*

If we estimate $B(F)$, separately for each $i \neq j (= 1, \ldots, c)$ by (6.7.30) and combine these estimators together by an unweighted average (as the sample sizes are all equal) $\hat{B}(F)$, then it follows from (6.8.22) that asymptotically the confidence interval for $\lambda_i - \lambda_j$ may be written as

$$\hat{\lambda}_{ij} - AR_{c,\alpha}/\hat{B}(F) \leq \lambda_i - \lambda_j \leq \hat{\lambda}_{ij} + AR_{c,\alpha}/\hat{B}(F). \tag{6.8.27}$$

We may remark that the estimators $\hat{\Delta}_{ij}$ are incompatible in the sense that they may not satisfy the transitive relations, viz. $\hat{\Delta}_{ij} + \hat{\Delta}_{jk} = \hat{\Delta}_{ik}$ which is true for the corresponding parametric estimators. To eliminate this drawback we consider the adjusted estimators. Let

$$\hat{\Delta}_{i\cdot} = \frac{1}{c}\sum_{j=1}^{c}\hat{\Delta}_{ij} \quad \text{for} \quad i = 1, \ldots, c \tag{6.8.28}$$

$$Z_{ij} = \hat{\Delta}_{i\cdot} - \hat{\Delta}_{j\cdot} \quad \text{for} \quad i \neq j = 1, \ldots, c. \tag{6.8.29}$$

It is easy to note that the Z_{ij}'s satisfy the aforesaid transitive relations. It is easy to check (Exercise 6.8.2) that

$$n^{1/2}(Z_{ij} - \hat{\Delta}_{ij}) = o_p(1) \quad \text{for all} \quad i, j = 1, \ldots, c. \tag{6.8.30}$$

Lemma 6.8.3.

$$\operatorname*{range}_i \{Z_{ij} - \Delta_{ij}\} = \operatorname*{range}_j \{Z_{ij} - \Delta_{ij}\} = \max_{1\leq k,l\leq c} |Z_{kl} - \Delta_{kl}|$$

Proof: Suppose for any fixed i, the range of $(Z_{ij} - \Delta_{ij})$ is attained by the pair of paired suffixes (i, k) and (i, l). Then, using (6.8.29) we get

$$\begin{aligned}\operatorname*{range}_j (Z_{ij} - \Delta_{ij}) &= (Z_{ik} - \Delta_{ik}) - (Z_{il} - \Delta_{il}) \\ &= (Z_{lk} - \Delta_{lk}).\end{aligned} \tag{6.8.31}$$

Since the right-hand side of (6.8.31) is independent of i, it holds for all $i = 1, \ldots, c$. Hence

$$\operatorname*{range}_i (Z_{ij} - \Delta_{ij}) = \max_{1\leq k,l\leq c} (Z_{kl} - \Delta_{kl}).$$

The other relation also holds similarly.

Hence we have the following

Lemma 6.8.4. *If $\phi = \sum_1^c l_i\theta_i$ be any contrast in θ and if for some $m (= 1, \ldots, c)$, we define $\hat{\phi}_m = \sum_1^c l_i Z_{im}$, then*

$$\sup_m |\hat{\phi}_m - \phi| \leq \frac{1}{2}\sum_1^c |l_i| \cdot \max_{1\leq j,k\leq c} |Z_{jk} - \Delta_{jk}| .$$

Proof: We can rewrite ϕ as $\phi_m = \sum_1^c l_i \Delta_{im}$, and hence,

$$|\hat{\phi}_m - \phi| = \left|\sum_1^c l_i [Z_{jm} - \Delta_{im}]\right|$$

$$\text{(6.8.32)} \qquad \leq \frac{1}{2}\sum_1^c |l_i| \operatorname{range}_i [Z_{im} - \Delta_{im}]$$

$$= \frac{1}{2}\sum_1^c |l_i| \max_{1 \leq j,k \leq c} |Z_{kj} - \Delta_{kj}| \quad \text{(by Lemma 6.8.3).}$$

Since the right-hand side of (6.8.32) is independent of m, and the inequality holds for all $m = 1, \ldots, c$, the lemma follows directly from (6.8.32).

If we now let

$$\text{(6.8.33)} \qquad l_{ij} = \frac{1}{c} l_i \quad \text{for } j = 1, \ldots, c, \quad i = 1, \ldots, c,$$

then the contrast ϕ may also be expressed as $\sum_{i=1}^c \sum_{j=1}^c l_{ij}\Delta_{ij}$. Consequently, from Lemma 6.8.4 we get that

$$\text{(6.8.34)} \qquad \left|\phi - \sum_{i=1}^c \sum_{j=1}^c l_{ij} Z_{ij}\right| \leq \left(\frac{1}{2}\sum_{i=1}^c |l_i|\right) \max_{1 \leq j,k \leq c} |Z_{jk} - \Delta_{jk}|\,.$$

Now corresponding to the $c(c-1)$ estimators $\hat{\Delta}_{ij}$, we compute the values of Z_{ij} for $i \neq j = 1, \ldots, c$. Further, from the $c(c-1)$ simultaneous confidence intervals I_{ij} in (6.8.18), we compute the value of

$$\text{(6.8.35)} \qquad \max_{1 \leq j,k \leq c} |Z_{kj} - \Delta_{kj}| \quad \text{subject to} \quad \Delta_{ij} \in I_{ij} \quad \text{for all } i \neq j = 1, \ldots, c.$$

We denote this maximum by $H_{n,\alpha}$. So that from (6.8.18) and the probability statement made just after it, we get

$$\text{(6.8.36)} \qquad P\left\{\max_{1 \leq j,k \leq c} |Z_{jk} - \Delta_{jk}| \leq H_{n,\alpha}\right\} \geq 1 - \alpha.$$

Consequently, from (6.8.34) and (6.8.36), we conclude that the *probability is at least* $1 - \alpha$ *that the inequalities*

$$\text{(6.8.37)} \qquad \begin{aligned} \sum_{i=1}^c \sum_{j=1}^c l_{ij} Z_{ij} - \tfrac{1}{2} H_{n,\alpha} \sum_1^c |l_i| \leq \phi = \sum_1^c l_i \theta_i \\ \leq \sum_{i=1}^c \sum_{j=1}^c l_{ij} Z_{ij} + \tfrac{1}{2} H_{n,\alpha} \sum_1^c |l_i| \end{aligned}$$

hold simultaneously for all ϕ.

This may be regarded as a nonparametric generalization of the well-known T method of multiple comparisons. (6.8.37) may be used to attach a simultaneous confidence interval to any number of contrasts in θ or to test the significance of them.

Now using (6.8.18), (6.8.22), (6.8.35), and (6.8.36), we see that it readily follows that asymptotically

$$n^{1/2}H_{n,\alpha} \to AR_{c,\alpha}/B(F). \tag{6.8.38}$$

Thus, if we define the derived estimates

$$\hat{\lambda}^0_{ij} = n^{1/2}(Z_{ij} - \Delta^0_{ij}), \qquad i, j = 1, \ldots, c \tag{6.8.39}$$

(Δ^0_{ij} being defined in (6.8.21)), then from (6.8.37)–(6.8.39), we find that (6.8.37) asymptotically reduces to

$$\begin{aligned} -\frac{1}{2}\sum_1^c |l_i|\, AR_{c,\alpha}/B(F) &\le \sum_{i=1}^c \sum_{j=1}^c l_{ij}(\lambda_{ij} - \hat{\lambda}^0_{ij}) \\ &\le \frac{1}{2}\sum_1^c |l_i|\, AR_{c,\alpha}/B(F). \end{aligned} \tag{6.8.40}$$

If we now compare (6.8.40) with Tukey's results, as adopted in the case of one-way analysis of variance with equal number of observations, we again get the same $A.R.E$ as obtained in (6.8.26). Hence, we have the following.

Theorem 6.8.3. *The conclusions of Theorem 6.8.2 also hold for the multiple-comparison tests considered above.*

The procedure discussed above is applicable only when $n_1 = \cdots = n_c = n$. The method to be considered now overcomes this drawback, and it may be regarded as a nonparametric generalization of Scheffé's s method of multiple comparisons. This method is essentially a confidence region procedure based on the construction of a simultaneous confidence region for the set of all possible contrasts. The procedure will be very appropriate if one of the c populations may be regarded as control and the rest as treatment group. On the other hand, if there is no natural control population, there remains some arbitrariness in the choice of the control population. However, for large sample sizes, the procedure is insensitive to the choice of any such control population.

In this situation, the sample sizes $n_1, \ldots, n_c$ are not necessarily equal. Let us denote by

$$N_{ij} = n_i + n_j \quad \text{for} \quad i, j = 1, \ldots, c \quad \text{and} \quad N = n_1 + \cdots + n_c. \tag{6.8.41}$$

Then, for the (i, j)th samples, we define $h(\mathbf{X}_i, \mathbf{X}_j)$ precisely in the same manner as in section 6.3, with the only change that here we have N_{ij} instead of $2n$

rank scores, the mean of which is denoted by $\bar{E}_{N_{ij}}$. Let us also define

$$V_{N\cdot ij} = \frac{N_{ij}}{n_i n_j} \sum_{\alpha=1}^{N_{ij}} [E_{N_{ij},\alpha} - \bar{E}_{N_{ij}}] Z_{N_{ij},\alpha} \tag{6.8.42}$$
$$= \frac{N_{ij}}{n_j} [h_N(X_i, X_j) - \bar{E}_{N_{ij}}], \quad \text{for} \quad i \neq j = 1, \ldots, c.$$

Conventionally, we let $V_{N\cdot ii} = 0$ for $i = 1, \ldots, c$ and we regard the first sample to constitute the control group. Then, we define

$$S_N = \frac{1}{A^2} \sum_{i=2}^{c} \sum_{j=2}^{c} [n_i(\delta_{ij} N - n_j)/N] V_{N\cdot 1i} V_{N\cdot 1j}, \tag{6.8.43}$$

where δ_{ij} is the usual Kronecker delta. If we write

$$\bar{V}_{N\cdot 1\cdot} = \frac{1}{N} \sum_{i=1}^{c} n_i V_{N\cdot 1i} = \frac{1}{N} \sum_{i=2}^{c} n_i V_{N\cdot 1i}, \tag{6.8.44}$$

then (6.8.43) may also be written as

$$S_N = A^{-2} \sum_{i=1}^{c} n_i [V_{N\cdot 1i} - \bar{V}_{N\cdot 1\cdot}]^2. \tag{6.8.45}$$

We may adopt a similar permutation approach (as earlier in this section) to find out the exact null distribution of S_N. On the other hand, if N is large subject to

$$n_i/N \to \rho_i \quad \text{as} \quad N \to \infty: 0 < \rho_i < 1 \quad \text{for all} \quad i = 1, \ldots, c, \tag{6.8.46}$$

then by an adaptation of the same proof as in Bhuchongkul and Puri (1965), we readily arrive at the conclusion that under $\theta_1 = \cdots = \theta_c$, S_N has asymptotically a chi-square distribution with $c - 1$ d.f.

Now from (6.8.15) we have

$$F_1(x) = F_i(x - \Delta_{1i}) \quad \text{for} \quad i = 1, \ldots, c. \tag{6.8.47}$$

So, if we define

$$V_{N\cdot ij}(a) = \frac{N_{ij}}{n_j} [h(\mathbf{X}_i + a\mathbf{1}_{n_i}, \mathbf{X}_j) - \bar{E}_{N_{ij}}] \quad \text{for} \quad i, j = 1, \ldots, c, \tag{6.8.48}$$

then it follows from (6.8.44) and (6.8.58) that for all $\boldsymbol{\theta}$

$$S_N(\boldsymbol{\theta}) = A^{-2} \sum_{i=1}^{c} n_i [V_{N\cdot 1i}(\Delta_{1i}) - \bar{V}_{N\cdot 1\cdot}(\boldsymbol{\Delta})]^2 \tag{6.8.49}$$

(where $\boldsymbol{\Delta}$ stands for the vector $(\Delta_{12}, \ldots, \Delta_{1c})$, and $\bar{V}_{N\cdot 1\cdot}(\boldsymbol{\Delta})$ is defined by (6.8.37) and (6.8.41) with a's replaced by Δ_{1i}, $i = 2, \ldots, c$) has the same distribution as that of S_N when $\theta_1 = \cdots = \theta_c$ hold. Further, it is easily seen that $V_{N \cdot ij}(a)$ is nondecreasing in a for all $i \neq j = 1, \ldots, c$. So if we denote by $S_{N,\alpha}$ the upper $100\alpha\%$ point of the null distribution of S_N (so that

(6.8.50) $$S_{N,\alpha} \xrightarrow{p} \chi^2_{c-1,\alpha} \quad \text{where} \quad P\{\chi^2 \geq \chi^2_{c-1,\alpha}\} = \alpha),$$

then from (6.8.49) we get

(6.8.51) $$P\{S_N(\boldsymbol{\theta}) < S_{N,\alpha} \mid \boldsymbol{\theta}\} = 1 - \alpha.$$

Let us now denote by

(6.8.52) $$V = (V_2, \ldots, V_c)$$

the running coordinate of the points $V_N(\boldsymbol{\Delta}) = (V_{N \cdot 12}(\Delta_{12}), \ldots, V_{N \cdot 1c}(\Delta_{1c}))$ on the boundary of the ellipsoid in (6.8.51). For any particular V, if V_i is positive, we find out a value of Δ_{1i}, say Δ^*_{1i}, such that

(6.8.53) $$\Delta^*_{1i} = \inf\{\Delta_{1i} : V_{N \cdot 1i}(\Delta_{1i}) \geq V_i\};$$

on the other hand, if V_i is negative, we define

(6.8.54) $$\Delta^*_{1i} = \sup\{\Delta_{1i} : V_{N \cdot 1i}(\Delta_{1i}) \leq V_i\},$$

for each $i = 2, \ldots, c$. In this manner, any point V on the interior neighborhood of the boundary of the ellipsoid in (6.8.51) is mapped into a point

(6.8.55) $$\boldsymbol{\Delta}^*_{1\cdot} = (\Delta^*_{12}, \ldots, \Delta^*_{1c}).$$

Further, (6.8.55) is the equation of a closed convex set of points $\{S_N(\boldsymbol{\theta})\}$, and as each $V_{N \cdot 1i}(a)$ is nondecreasing in a, $i = 2, \ldots, c$, it follows that the set of points

(6.8.56) $$C(\boldsymbol{\Delta}_{1\cdot}) = \{(\Delta_{12}, \ldots, \Delta_{1c}) : S_N(\boldsymbol{\theta}) \leq S_{N,\alpha}\}$$

will also be a closed convex set in $(\Delta_{12}, \ldots, \Delta_{1c})$, having the property that

(6.8.57) $$P\{(\Delta_{12}, \ldots, \Delta_{1c}) \in C(\boldsymbol{\Delta}_{1\cdot}) : \boldsymbol{\theta}\} = 1 - \alpha.$$

For small samples, S_N will have essentially a finite number of discrete mass points and hence on the interior boundary of the ellipsoid $S_N < S_{N,\alpha}$, there will be only a finite number of points $\{V\}$. So, if for each of these points, we find out (by the process in (6.8.53) and (6.8.54)) the corresponding (finite number of) points $\{\boldsymbol{\Delta}^*_{1\cdot}\}$ in (6.8.55), then the convex hull of these set of points will be our desired simultaneous confidence region for $(\Delta_{12}, \ldots, \Delta_{1c})$. We shall now use (6.8.56) and (6.8.57) to derive a simultaneous confidence region for any number of contrasts. Note that any contrast $\boldsymbol{\phi} = \sum_{i=1}^{c} l_i \theta_i$, $\Sigma l_i = 0$ may also be written as $\sum_{i=2}^{c} l_i \Delta_{i1}$. Since $c(\boldsymbol{\Delta}_{1\cdot})$ in (6.8.56) is a closed convex set, its convex hull is contained within the intersection of all the supporting hyperplanes of $c(\boldsymbol{\Delta}_{1\cdot})$. Thus, for any $\sum_2^c l_i \Delta_{i1}$ (which represents the equation of a $(c-2)$-dimensional hyperplane), we can always find two parallel $(c-2)$-dimensional hyperplanes having the equations $\sum_{i=2}^{c} l_i \Delta_{i1} = c_j$, $j = 1, 2$, such that the convex hull of $c(\boldsymbol{\Delta}_{1\cdot})$ is contained fully within the $(c-1)$-dimensional strip between these two hyperplanes. Thus, if we take $c_1 < c_2$, we obtain

$$P\left\{c_1 \leq \phi = \sum_{1}^{c} l_i \theta_i \leq c_2, \text{ for all } \phi\right\} \geq 1 - \alpha. \tag{6.8.58}$$

The procedure simplifies considerably for large samples. For this, we define the estimators $\hat{\Delta}_{i\cdot}$ $(i = 1, \ldots, c)$ as in (6.8.28) (the subscript j is dropped, as we are dealing with the univariate model). Also, let

$$\delta^2 = A^2 \chi^2_{c-1,\alpha} / B^2(F) \quad \text{and} \quad \hat{\delta}^2 = A^2 \chi^2_{c-1,\alpha} / \hat{B}^2(F), \tag{6.8.59}$$

where $\hat{B}(F)$ is the average of all the $\binom{c}{2}$ estimates. Then, proceeding as in Theorem 6.6.1, we obtain

$$\left[\sum_{1}^{c} n_i [\hat{\Delta}_{i\cdot} - \theta_i]^2 - \frac{1}{N}\left(\sum_{i=1}^{c} n_i [\Delta_{i\cdot} - \theta_i]\right)^2\right] \delta^2 = S_N \quad \text{(say)}, \tag{6.8.60}$$

has asymptotically a χ^2 distribution with $c - 1$ degrees of freedom, where of course, we put (without any loss of generality) that $\sum_1^c \theta_i = 0$. Using (6.8.56), it is easily seen that the equations of the two parallel $(c-2)$-dimensional hyperplanes $\sum_1^c l_i \theta_i = c_j$, $j = 1, 2$, reduce to

$$\left| \sqrt{N}\left[\sum_{i=1}^{c} l_i \hat{\Delta}_{i\cdot} - \sum_{i=1}^{c} l_i \theta_i\right] \right| = \pm \hat{\delta} \sqrt{\sum_{1}^{c} l_i^2 / \rho_i}. \tag{6.8.61}$$

Consequently, we obtain that *asymptotically the probability is* $1 - \alpha$ *that the inequalities*

$$\sum_{i=1}^{c} l_i \hat{\Delta}_{i\cdot} - \frac{\hat{\delta}}{\sqrt{N}}\left(\sum_{1}^{c} \frac{l_i^2}{\rho_i}\right)^{1/2} \leq \sum_{i=1}^{c} l_i \theta_i \leq \sum_{i=1}^{c} l_i \hat{\Delta}_{i\cdot} + \frac{\hat{\delta}}{\sqrt{N}}\left(\sum_{1}^{c} \frac{l_i^2}{\rho}\right)^{1/2}, \tag{6.8.62}$$

hold simultaneously for all ϕ.

Again comparing (6.8.61) with the parametric S method of multiple comparisons, we conclude that the asymptotic relative efficiency of (6.8.61) with respect to the parametric procedure is given by (6.8.26).

6.9. INTERVAL ESTIMATION OF CONTRASTS IN ONE-WAY MANOVA

In the multivariate one-sample, two-sample, or several-sample location problems, the problem of attaching simultaneous confidence regions for various estimable parameters becomes comparatively difficult (because of the interrelations of the variates). Since the parent distribution is of unspecified form, the covariance matrix of the rank statistics is also so. Nevertheless, it can be estimated from the samples. Among various possibilities, we shall briefly sketch the following two procedures.

(a) (i) Use of Bonferroni-inequality. Consider first the one-sample problem. Let $\mathbf{X}_\alpha = (X_{1\alpha}, \ldots, X_{p\alpha})'$, $\alpha = 1, \ldots, n$ be an independent sample from an absolutely continuous cumulative distribution function (c.d.f.) $F(\mathbf{x} - \boldsymbol{\theta})$, where $\mathbf{x} = (x_1, \ldots, x_p)'$ and $\boldsymbol{\theta} = (\theta_1, \ldots, \theta_p)'$. $F(\mathbf{x})$ is assumed to be diagonally symmetric about $\mathbf{0}$. The problem is to obtain a confidence region for $\boldsymbol{\theta}$.

For every univariate sample $\mathbf{X}^{(i)} = (X_{i1}, \ldots, X_{in})$, consider the rank order statistic

$$T_{n,i}(\mathbf{X}^{(i)}) = (1/n) \sum_{\alpha=1}^{n} E_{n,\alpha}^{(i)} Z_{n,\alpha}^{(i)}, \qquad i = 1, \ldots, p \tag{6.9.1}$$

where $Z_{n,\alpha}^{(i)}$ is one or zero according as the αth smallest observation among $|X_{i1}|, \ldots, |X_{in}|$ is from a positive X or not, and $E_{n,\cdot}^{(i)}$ is the expected value of the αth-order statistic of a sample of size n from a distribution $\Psi_i^*(x)$, given by

$$\Psi_i^*(x) = \Psi_i(x) - \Psi_i(-x), \quad \text{if} \quad x \geq 0, \quad \text{and is 0 otherwise} \tag{6.9.2}$$

where $\Psi_i(x)$ is symmetric about 0; $i = 1, \ldots, p$. Now when $\boldsymbol{\theta} = \mathbf{0}$, the distribution of each $T_{n,i}(\mathbf{X}^{(i)})$ is symmetric about $\bar{E}_n^{(i)}/2$ where $\bar{E}_n^{(i)} = \sum_{1=\alpha}^{N} E_{n,\alpha}^{(i)}/n$. Furthermore, in this case, each $T_{n,i}(\mathbf{X}^{(i)})$ can have 2^n equally likely realizations. Thus for each $T_{n,i}(\mathbf{X}^{(i)})$, it is possible to select a constant $a_{n,\alpha^*}^{(i)}$ such that

$$P\{|T_{n,i}(\mathbf{X}^{(i)}) - \tfrac{1}{2}\bar{E}_n^{(i)}| \leq a_{n,\alpha^*}^{(i)} \mid \boldsymbol{\theta} = 0\} \geq 1 - \alpha^*, \qquad i = 1, \ldots, p. \tag{6.9.3}$$

Thus, on setting $\alpha^* = \alpha/p$, where $1 - \alpha$ is the desired confidence coefficient for $\boldsymbol{\theta}$, we obtain by using the Bonferroni inequality,

$$P\{|T_{n,i}(\mathbf{X}^{(i)}) - \tfrac{1}{2}\bar{E}_n^{(i)})| \leq a_{n,\alpha^*}^{(i)}, \quad i = 1, \ldots, p \mid \boldsymbol{\theta} = 0\} \geq 1 - \alpha. \tag{6.9.4}$$

For small values of n and standard $T_{n,i}$'s (such as the Wilcoxon signed rank statistic), exact values of $a_{n,\alpha^*}^{(i)}$'s may be obtained from the existing tables. However, if n is large, then since each $T_{n,i}$ has a normal distribution in the limit, we find that asymptotically as $n \to \infty$,

$$|n^{1/2} a_{n,\alpha^*}^{(i)} - (\tfrac{1}{2})\tau_{\alpha/2p} A_{n,i}| \to 0 \tag{6.9.5}$$

where τ_ϵ is the upper $100\epsilon\%$ point of the standard normal distribution, and

$$A_{n,i}^2 = \frac{1}{n-1} \sum_{\alpha=1}^{n} (E_{n,\alpha}^{(i)})^2. \tag{6.9.6}$$

By definition, each $T_{n,i}(\mathbf{X}^{(i)} - t_i\mathbf{I}_n)$ is a unit vector of n elements, is non-increasing in t_i, $i = 1, \ldots, p$. Hence defining

$$\hat{\theta}_U^{(i)} = \sup\{t_i : T_{n,i}(\mathbf{X}^{(i)} - t_i\mathbf{I}_n) \geq \tfrac{1}{2}\bar{E}_n^{(i)} - a_n^{(i)}\} \tag{6.9.7}$$

$$\hat{\theta}_L^{(i)} = \inf\{t_i : T_{n,i}(\mathbf{X}^{(i)} - t_i\mathbf{I}_n) \leq \tfrac{1}{2}\bar{E}_n^{(i)} + a_n^{(i)}\} \tag{6.9.8}$$

we obtain

$$P\{\hat{\theta}_L^{(i)} \leq \theta_i \leq \hat{\theta}_U^{(i)},\quad i = 1, \ldots, p\} \geq 1 - \alpha \tag{6.9.9}$$

which gives the desired confidence region for $\boldsymbol{\theta}$ with confidence coefficient greater than or equal to $1 - \alpha$. It may be noted that (6.9.9) is a translation invariant confidence region for $\boldsymbol{\theta}$, since $\hat{\theta}_u^{(i)}$ and $\hat{\theta}_L^{(i)}$ are both translation invariant $i = 1, \ldots, p$.

***(a) (ii) Asymptotically Scheffé's Bounds for* $\boldsymbol{\theta}$.** Let us denote by $F_{[i]}(x)$ the marginal c.d.f. of $X_{i\alpha}$ and by $J_{(i)}^*(u) = \Psi_i^{*-1}(u)$: $0 < u < 1, i = 1, \ldots, p$. Also let

$$A_i^2 = \lim_{n\to\infty} A_{n,i}^2 = \int_0^1 J_{(i)}^{2*}(u)\,du, \qquad i = 1, \ldots, p, \tag{6.9.10}$$

$$B_i = \int_{-\infty}^{\infty} \frac{d}{dx} J_{(i)}[F_{[i]}(x)]\,dF_{[i]}(x); \quad J_i(u) = \Psi_i^{-1}(u), \quad i = 1, \ldots, p. \tag{6.9.11}$$

Then (Exercise 6.9.1)

$$B_i = (2\tau_{\alpha^*/2}A_i)/n^{\frac{1}{2}}(\hat{\theta}_U^{(i)} - \hat{\theta}_L^{(i)}), \qquad i = 1, \ldots, p \tag{6.9.12}$$

is a consistent estimator of B_i for each $i = 1, \ldots, p$, no matter whether $\boldsymbol{\theta}$ is $\mathbf{0}$ or not. Next, let $R_{i\alpha}^*$ be the rank of $X_{i\alpha}$ among $(X_{i1}, \ldots, X_{in})$. Then (Exercise 6.9.2)

$$\hat{\boldsymbol{\nu}} = ((\hat{\nu}_{ij})), \quad \text{where} \quad \hat{\nu}_{ij} = \sum_{\alpha=1}^{n} E_{n,R_{i\alpha}^*}^{(i)} E_{n,R_{j\alpha}^*}^{(j)} \tag{6.9.13}$$

is a consistent and translation-invariant estimator of the convariance matrix of $(n^{1/2}\mathbf{T}_n)/2$. Here $E_{n,\alpha}^{(i)}$ is the expected value of the αth statistic of a sample of size n from a distribution $\Psi_i^*(x)$. Hence, from the results of chapter 4 it follows that

$$\mathfrak{L}_n^* = 4n\sum_{i=1}^{p}\sum_{j=1}^{p}\hat{\nu}^{ij}[T_{n,i} - \tfrac{1}{2}\bar{E}_n^{(i)}][T_{n,j} - \tfrac{1}{2}\bar{E}_n^{(j)}], \tag{6.9.14}$$

where $((\hat{\nu}_{ij}))^{-1} = ((\hat{\nu}^{ij}))$, has asymptotically the chi-square distribution with p degrees of freedom. Now let us denote

$$\hat{\theta}_i^{(1)} = \sup\{t_i\colon T_{n,i}(\mathbf{X}^{(i)} - t_i\mathbf{I}_n) > \tfrac{1}{2}\bar{E}_n^{(i)}\}, \tag{6.9.15}$$

$$\hat{\theta}_i^{(2)} = \inf\{t_i\colon T_{n,i}(\mathbf{X}^{(i)} - t_i\mathbf{I}_n) < \tfrac{1}{2}\bar{E}^{(i)}\}, \qquad i = 1, \ldots, p \tag{6.9.16}$$

and

$$(6.9.17) \qquad \hat{\theta}_i = (\hat{\theta}_i^{(1)} + \theta_i^{(2)})/2, \qquad i = 1, \ldots, p.$$

Then, it is easy to verify (Exercise 6.9.3) that

$$(6.9.18) \qquad n \sum_{i=1}^{p} \sum_{j=1}^{p} \hat{\nu}^{ij}(\hat{\theta}_i - \theta_i)(\hat{\theta}_j - \theta_j)\hat{B}_i\hat{B}_j$$

has asymptotically the chi-square distribution with p degrees of freedom. Thus

$$(6.9.19) \qquad P\left\{n \sum_{i=1}^{p} \sum_{j=1}^{p} \hat{\nu}^{ij}(\hat{\theta}_i - \theta_i)(\hat{\theta}_j - \theta_j)\hat{B}_i\hat{B}_j \leq \chi^2_{p,\alpha}\right\} \simeq 1 - \alpha.$$

This gives a Scheffé-type confidence region for $\boldsymbol{\theta}$.

The asymptotic relative efficiency of the Scheffé-type confidence region defined in (6.9.19) with respect to the one based on Hotelling's T^2 statistic as measured by the inverse ratio of the volumes of the two confidence regions raised to the power $1/p$ (cf. Wilks, 1962, p. 385) is the same as the asymptotic relative efficiency of the point estimator $\hat{\boldsymbol{\theta}} = (\hat{\boldsymbol{\theta}}_1, \ldots, \hat{\boldsymbol{\theta}}_p)$ with respect to the sample mean vector $\bar{\mathbf{X}} = (\bar{\mathbf{X}}_1, \ldots, \bar{\mathbf{X}}_p)$ raised to the power $1/p$. Since the latter is studied in detail in section 6.4, we omit the details of the asymptotic relative efficiencies of the confidence regions. However, it is worth mentioning that the asymptotic relative efficiency of the Scheffé-type confidence region based on the normal scores statistic with respect to the one based on Hotelling's T^2 statistic is 1 when the underlying distribution is nonsingular p-variate normal.

We now consider the two-sample problem.

Let $\mathbf{X}_\alpha = (X_{1\alpha}, \ldots, X_{p\alpha})$, $\alpha = 1, \ldots, n_1$ and $\mathbf{Y}_\beta = (Y_{1\beta}, \ldots, Y_{p\beta})$, $\beta = 1, \ldots, n_2$, $n = n_1 + n_2$ be two independent samples from the p-variate absolutely continuous c.d.f. $F(\mathbf{x})$ and $F(\mathbf{x} - \boldsymbol{\Delta})$ respectively. The problem is to obtain the confidence region for $\boldsymbol{\Delta} = (\Delta_1, \ldots, \Delta_p)$. As before we work with the sequences of rank functions. Let $R_{i\alpha}^{(1)}$ and $R_{i\beta}^{(2)}$ be the ranks of $X_{i\alpha}$ and $Y_{i\beta}$ respectively when the observations corresponding to the ith variate of both the samples, that is, $(X_{i1}, \ldots, X_{in_1}, Y_{i1}, \ldots, Y_{in_2})$ are arranged in ascending order of magnitude. Consider now the following rank order statistic based on $\mathbf{X}^{(i)} = (X_{i1}, \ldots, X_{in_1})$ and $\mathbf{Y}^{(i)} = (Y_{i1}, \ldots, Y_{in_2})$:

$$(6.9.20) \qquad T_{n,i}(\mathbf{X}^{(i)}, \mathbf{Y}^{(i)}) = (1/n_1)\sum_{\alpha=1}^{n_1} E^{(i)}_{n,R_{i\alpha}^{(1)}}, \qquad i = 1, \ldots, p,$$

and write

$$(6.9.21) \qquad \bar{E}_n^{(i)} = \sum_{\alpha=1}^{n} E_{n,\alpha}^{(i)}/n; \qquad A_{n,i}^2 = \frac{1}{n-1}\sum_{\alpha=1}^{n}[E_{n,\alpha}^{(i)} - \bar{E}_n^{(i)}]^2, \quad i = 1, \ldots, p,$$

where we assume that $E_{n,\alpha}^{(i)}$ is the expected value of the αth statistic of a sample of size n from some distribution $\Psi_i(x)$.

(b) (i) Bonferroni Confidence Bounds for $\boldsymbol{\Delta}$. Under the null hypothesis $\boldsymbol{\Delta} = 0$, each $T_{n,i}$ can have $n!$ equally likely realizations obtained by all possible permutations of the ranks of $(X_{i1}, \ldots, X_{in_1}, Y_{i1}, \ldots, Y_{in_2})$ over $1, \ldots, n$. Thus, it is possible to select two values of $T_{n,i}$, say, $a_n^{(i)}$ and $b_n^{(i)}$ such that for each $i = 1, \ldots, p$

$$P\{a_n^{(i)} \leq T_{n,i}(\mathbf{X}^{(i)}, \mathbf{Y}^{(i)}) \leq b_n^{(i)} \mid \boldsymbol{\Delta} = \mathbf{0}\} = 1 - \alpha^*. \tag{6.9.22}$$

Thus, on setting $\alpha^* = \alpha/p$, we obtain by using the Bonferroni inequality that corresponds to the confidence coefficient $1 - \alpha$,

$$P\{a_n^{(i)} \leq T_{n,i}(\mathbf{X}^{(i)}, \mathbf{Y}^{(i)}) \leq b_n^{(i)}, i = 1, \ldots, p \mid \boldsymbol{\Delta} = 0\} \geq 1 - \alpha. \tag{6.9.23}$$

For small values of n, and specific $T_{n,i}$'s such as the Wilcoxon two-sample statistic, $a_n^{(i)}$ and $b_n^{(i)}$ may be obtained from the existing tables. However, if n is large, then since $\mathbf{T}_n = (T_{n1}, \ldots, T_{np})$ has a p-variate normal distribution in the limit, we have asymptotically

$$\left| a_n^{(i)} - \bar{E}_n^{(i)} + \tau_{\alpha/2p} A_{n,i} \left(\frac{n}{n_1 n_2}\right)^{1/2} \right| = o_p(1) \tag{6.9.24}$$

and

$$\left| b_n^{(i)} - \bar{E}_n^{(i)} + \tau_{\alpha/2p} A_{n,i} \left(\frac{n}{n_1 n_2}\right)^{1/2} \right| = o_p(1). \tag{6.9.25}$$

Now, by definition, for each $i = 1, \ldots, p$, $T_{n,i}(\mathbf{X}^{(i)} - t_i \mathbf{I}_{n_1}, \mathbf{Y}^{(i)})$ is nonincreasing in t_i. Hence, defining

$$\begin{aligned} \hat{\Delta}_{iU} &= \sup\{t_i\colon T_{n,i}(\mathbf{X}^{(i)} - t_i \mathbf{I}_{n_1}, \mathbf{Y}^{(i)}) \geq a_n^{(i)}\} \\ \hat{\Delta}_{iL} &= \inf\{t_i\colon T_{n,i}(\mathbf{X}^{(i)} - t_i \mathbf{I}_{n_1}, \mathbf{Y}^{(i)}) \leq b_n^{(i)}\} \end{aligned} \tag{6.9.26}$$

it can be shown (Exercise 6.9.5) that

$$P\{\hat{\Delta}_{iL} \leq \Delta_i \leq \hat{\Delta}_{iU}, i = 1, \ldots, p\} \geq 1 - \alpha, \tag{6.9.27}$$

which gives the desired confidence region for $\boldsymbol{\Delta}$. It may be noted that $\hat{\Delta}_{iL}$ and $\hat{\Delta}_{iU}$ are both translation invariant estimates of Δ_i.

(b) (ii) Asymptotically Scheffé Bounds for $\boldsymbol{\Delta}$. Let us denote by $F_{[i]}(x)$ the marginal c.d.f. of the ith variate of the c.d.f. $F(x)$, and by $J_{(i)}(u) = \Psi_i^{-1}(u)$, $i = 1, \ldots, p$. Also let

$$A_i^2 = \lim_{n\to\infty} A_{n,i}^2 = \int_0^1 J_{(i)}^2(u)\, du - \left(\int_0^1 J_{(i)}(u)\, du\right)^2, \qquad i = 1, \ldots, p \tag{6.9.28}$$

$$B_i = \int_{-\infty}^{+\infty} \frac{d}{dx} J_{(i)}[F_{[i]}(x)]\, dF_{[i]}(x), \qquad i = 1, \ldots, p. \tag{6.9.29}$$

Then, irrespective of $\mathbf{\Delta} = \mathbf{0}$ or not,

$$\hat{B}_i = (n/n_1n_2)^{1/2}(2A_i\tau_{\alpha/2p})/(\hat{\Delta}_{iU} - \hat{\Delta}_{iL}) \tag{6.9.30}$$

is a consistent estimator of B_i for all $i = 1, \ldots, p$ (Exercise 6.9.6).

Let now $R^{*(1)}_{1\alpha}$ be the rank of $X_{i\alpha}$ among $(X_{i1}, \ldots, X_{in_1})$, $\alpha = 1, \ldots, n_1$ and $R^{*(2)}_{i\beta}$ be the rank of $Y_{i\beta}$ among $(Y_{i1}, \ldots, Y_{in_2})$ $\beta = 1, \ldots, n_2$ for each $i = 1, \ldots, p$. Denote

$$\hat{\nu}^{(1)}_{ij} = \frac{1}{n_1}\sum_{\alpha=1}^{n_1} E^{(i)}_{n_1,R^{*(1)}_{i\alpha}} E^{(j)}_{n_1,R^{*(1)}_{j\alpha}} - \left(\frac{1}{n_1}\sum_{\alpha=1}^{n_1} E^{(i)}_{n_1\alpha}\right)\left(\frac{1}{n_1}\sum_{\alpha=1}^{n_1} E^{(j)}_{n_1,\alpha}\right) \tag{6.9.31}$$

$$\hat{\nu}^{(2)}_{ij} = \frac{1}{n_2}\sum_{\alpha=1}^{n_2} E^{(i)}_{n_2,R^{*(2)}_{i\alpha}} E^{(j)}_{n_2,R^{*(2)}_{j\alpha}} - \left(\frac{1}{n_2}\sum_{\alpha=1}^{n_2} E^{(i)}_{n_2,\alpha}\right)\left(\frac{1}{n_2}\sum_{\alpha=1}^{n_2} E^{(j)}_{n_2,\alpha}\right) \tag{6.9.32}$$

$$\hat{\nu}_{ij} = (n_1\nu^{(1)}_{ij} + n_2\hat{\nu}^{(2)}_{ij})/n, \qquad i, j = 1, \ldots, p, \tag{6.9.33}$$

$$\hat{\nu} = ((\hat{\nu}_{ij})), \qquad \hat{\nu}^{-1} = ((\hat{\nu}^{ij}))^{-1}. \tag{6.9.34}$$

Then, since (cf. Chapter 5), $\sqrt{n}(\mathbf{T}_n - \bar{E}_n)$ (where $\bar{E}_n = (\bar{E}^{(1)}_n, \ldots, \bar{E}^{(p)}_n)$) has a multinormal distribution in the limit, it follows that under $\mathbf{\Delta} = \mathbf{0}$,

$$\frac{n_2n}{n_1}\sum_{i=1}^{p}\sum_{j=1}^{p}\hat{\nu}^{ij}(T_{n,i}(\mathbf{X}^{(i)}, \mathbf{Y}^{(i)}) - \bar{E}^{(i)}_n)(T_{n,j}(\mathbf{X}^{(j)}, \mathbf{Y}^{(j)}) - \bar{E}^{(j)}_n) \tag{6.9.35}$$

has asymptotically the chi-square distribution with p degrees of freedom. Let us now define

$$\begin{aligned} \hat{\Delta}^{(1)}_i &= \sup\{t_i\colon\ T_{n,i}(\mathbf{X}^{(i)} - t_i\mathbf{I}_m, \mathbf{Y}^{(i)}) > \bar{E}^{(i)}_n\} \\ \hat{\Delta}^{(2)}_i &= \inf\{t_i\colon\ T_{n,i}(\mathbf{X}^{(i)} - t_i\mathbf{I}_m, \mathbf{Y}^{(i)}) < \bar{E}^{(i)}\}. \end{aligned} \tag{6.9.36}$$

Then (Exercise 6.9.7) letting $\hat{\Delta}_i = \frac{1}{2}(\hat{\Delta}^{(1)}_i + \hat{\Delta}^{(2)}_i)$,

$$\frac{n_2n}{n_1}\sum_{i=1}^{p}\sum_{j=1}^{p}\hat{\nu}^{ij}(\hat{\Delta}_i - \Delta_i)(\hat{\Delta}_j - \Delta_j)\hat{B}_i\hat{B}_j \tag{6.9.37}$$

has asymptotically the chi-square distribution with p degrees of freedom. Consequently

$$P\left\{\sum_{i=1}^{p}\sum_{j=1}^{p}\hat{\nu}^{ij}\hat{B}_i\hat{B}_j(\hat{\Delta}_i - \Delta_i)(\Delta_j - \Delta_j) \le \chi^2_{p,\alpha}\right\} \simeq 1 - \alpha \tag{6.9.38}$$

which is the desired confidence bound for $\mathbf{\Delta}$.

The asymptotic efficiency of the Scheffé type of confidence region defined in (6.9.38) with respect to the one based on Hotelling's T^2 statistic is the same as that of the corresponding point estimate $\mathbf{\Delta} = (\hat{\Delta}_1, \ldots, \hat{\Delta}_p)$ with respect to the sample mean vector $\bar{\mathbf{Y}} - \bar{\mathbf{X}}$; the details of which are discussed in section 6.4.

EXERCISES

Section 6.2

6.2.1. Let $h(\mathbf{X}_n)$ be the Wilcoxon signed rank statistic. Then prove that $\hat{\theta}_n$ defined in (6.2.5) is the median of the $n(n+1)/2$ values of $(X_\alpha + X_\beta)/2$ for $1 \le \alpha \le \beta \le n$.

6.2.2. Let $h(\mathbf{X}_n)$ be the sign statistic. Then prove that $\hat{\theta}_n$ defined in (6.2.5) is the median of $X_1, \ldots, X_n$.

6.2.3. Prove the assertion (6.2.8).

6.2.4. Prove that $\hat{\theta}_n$ defined in (6.2.10) is translation invariant.

6.2.5. Using the results of the Corollary 4.4.3.1, prove that $n^{1/2}\{\mathbf{h}(\mathbf{X}_1 - n^{-1/2}\mathbf{a}, \ldots, \mathbf{X}_n - n^{-1/2}\mathbf{a}) - \boldsymbol{\mu}_0^*\}$ has as $n \to \infty$, the limiting normal distribution with mean vector $(a_1c_1, \ldots, a_pc_p)$ and covariance matrix $\frac{1}{4}\hat{\nu}_n^*$ where $c_j = \frac{1}{2}B(F_j^*)$ is given by (4.4.88) with F replaced by F^*, and $\hat{\nu}_n^* = ((\nu_{njk}^*))$ is defined in (4.4.30).

6.2.6. Verify that for the Wilcoxon and the normal scores estimators the dispersion matrix T^* defined in Theorem 6.2.3 reduces to (4.6.18) and (4.6.22) respectively.

6.2.7. Prove that $n^{1/2}(\hat{\theta}_{n(M)} - \theta)$ has, as $n \to \infty$, the limiting normal distribution with mean vector $\mathbf{0}$, and dispersion matrix $\boldsymbol{\tau}_{(M)}^*$ defined in (6.2.28).

Section 6.3

6.3.1. Let $h(\mathbf{X}_{n_1}, \mathbf{Y}_{n_2})$ be the Wilcoxon statistic. Then prove that the estimator $\hat{\Delta}_n$ defined in (6.3.4) is the median of the n_1n_2 differences $Y_j - X_i$, $i = 1, \ldots, n_1, j = 1, \ldots, n_2$.

6.3.2. Let $h(\mathbf{X}_{n_1}, \mathbf{Y}_{n_2})$ be the median statistic. Then prove that $\tilde{\Delta}_n$ defined in (6.3.4) is median $(Y_1, \ldots, Y_{n_2})$—median $(X_1, \ldots, X_{n_1})$.

6.3.3. Prove that $\hat{\boldsymbol{\Delta}}_{n(\Phi)}$, $\hat{\boldsymbol{\Delta}}_{n(R)}$, and $\hat{\boldsymbol{\Delta}}_{n(M)}$ are all translation-invariant estimators.

6.3.4. Prove Theorems 6.3.1 and 6.3.2.

Section 6.4

6.4.1. Verify the assertions (6.4.3), (6.4.6), and (6.4.8).

6.4.2. Verify the assertions (6.4.13), (6.4.18), and (6.4.20).

Section 6.5

6.5.1. Let $X_1, \ldots, X_{n_1}$ and $Y_1, \ldots, Y_{n_2}$, $N = n_1 + n_2$ be independent samples from the absolutely continuous c.d.f.'s $F(x)$ and $G(x - \Delta)$ respectively. Consider the statistic $h(\mathbf{X}_{n_1}, \mathbf{Y}_{n_2}) = \sum_{j=1}^{n} E_\Psi[V^{(S_j)}]$ where $S_i, \ldots, S_n$ are the ranks of $Y_1, \ldots, Y_{n_2}$ in the combined sample and $V^{(1)} < \cdots < V^{(N)}$ is an ordered statistic from a c.d.f. $\Psi(x)$. Suppose that $E\{h(X_{n_1}, Y_{n_2})\} = \mu$ when $F = G$ and $\Delta = 0$. Prove that if $F(x)$ and $G(x)$ are symmetric about 0, then $\hat{\Delta}_N$ defines in (6.3.4) has a distribution symmetric about Δ. (Ramachandramurty, 1965.)

6.5.2. (*Continued.*) Prove that

$$\text{(i)}\ \hat{\Delta}_N(\mathbf{X}_{n_1}, \mathbf{Y}_{n_2} + a\mathbf{l}_{n_2}) = \hat{\Delta}_N(\mathbf{X}_{n_1}, \mathbf{Y}_{n_2}) + a \quad \text{for all } \mathbf{X}_{n_1} \text{ and } \mathbf{Y}_{n_2},$$

$$\text{(ii)}\ P_\Delta(\hat{\Delta}_N - \Delta \le u) = P_0(\hat{\Delta}_N \le u),$$

$$\text{(iii)}\ P\{h(\mathbf{X}_{n_1}, \mathbf{Y}_{n_2} - a\mathbf{l}_{n_2}) < \mu\} \le P\{\hat{\Delta}_N \le a\} \le P\{h(\mathbf{X}_{n_1}, \mathbf{Y}_{n_2} - a\mathbf{l}_{n_2}) \le \mu\}$$

where P_Δ indicates that the probability is taken when the true distributions are $F(x)$ and $G(x - \Delta)$. (Ramachandramurty, 1965.)

6.5.3. (*Continued.*) Let $a, c_1, c_2, \ldots$ be real constants and suppose $\Delta_N = a/c_N$. Furthermore, suppose that

$$\lim_{N\to\infty} P_N\{c_N[h(\mathbf{X}_{n_1},\mathbf{Y}_{n_2}) - \mu_N] \le u\} = \Phi\left(\frac{u + aB}{A}\right)$$

where $\Phi(x)$ is the c.d.f. of the standard normal distribution function, and P_N denotes the probability when the true distributions are $F(x)$ and $G(x + \Delta_N)$. Then for fixed Δ,

$$\lim_{N\to\infty} P_\Delta\{c_N(\hat{\Delta}_N - \Delta) \le a\} = \Phi\left(\frac{aB}{A}\right).$$

(Ramachandramurty, 1965.)

6.5.4. (*Continued.*) Let $h(\mathbf{X}_{n_1}, \mathbf{Y}_{n_2})$ be as defined in Exercise 6.5.1. Define $G_N(x) = (X + a_N^{-1/2})$; $H_N(x) = \lambda_N F(x) + (1 - \lambda_N)G_N(x)$, $H_0(x) = \lambda_0 F(x) + (1 - \lambda_0)G(x)$ where $0 < \lambda_0 = \lim_{N\to\infty} \lambda_N < 1$, $\lambda_N = m/N$,

$$a_N = \int_{-\infty}^{\infty} J[H_N(x)]\, dG_N(x)$$

where $J(x) = \Psi^{-1}(x)$, $a_0 = \int_{-\infty}^{\infty} J[H(x)]\, dG(x)$. Assume that (i) Ψ has a bounded density ψ, (ii) $aB = -\lim_{N\to\infty} N^{1/2}(a_N - a_0)$ exists and is finite, and (iii) the statistic h satisfies the assumptions of Theorem 5.5.1. Then prove that

$$\lim_{N\to\infty} P_N\{N^{1/2}[h(\mathbf{X}_{n_1}, \mathbf{Y}_{n_2}) - a_0] \le \mu\} = \Phi\left(\frac{u + aB}{A}\right).$$

where

$$A^2 = 2\lambda_0\left[\sigma_1^2 + \frac{\lambda_0}{1 - \lambda_0}\sigma_2^2\right],$$

$$\sigma_1^2 = \iint_{-\infty<x<y<\infty} F(x)[1 - F(y)]J'[H_0(x)]J'[H_0(y)]\, dG(x)\, dG(y),$$

and

$$\sigma_2^2 = \iint_{-\infty<x<y<\infty} G(x)[1 - G(y)]J'[H_0(x)]J'[H_0(y)]\, dF(x)\, dF(y).$$

6.5.5. (*Continued.*) Under the assumptions of Exercise 6.5.4,

$$\lim_{N\to\infty} P_\Delta\{N^{1/2}(\hat{\Delta}_{(NR)} - \Delta) \leq a\} = \Phi(aB/A)$$

where

$$A^2 = \lambda_0^2\left[\frac{1}{\lambda_0}\int G^2\,dF + \frac{1}{1-\lambda_0}\int F^2\,dG - \frac{1}{4\lambda_0(1-\lambda_0)}\right]$$

and

$$B = \lambda_0\int_{-\infty}^{\infty} g(x)f(x)\,dx, \qquad g(x) = G'(x), \qquad f(x) = F'(x).$$

6.5.6. (*Continued.*) Suppose (*a*) $F(x)$ has a bounded density $f(x)$, (*b*) the statistic $h(\mathbf{X}_{n_1}, \mathbf{Y}_{n_2})$ satisfies the conditions (i) and (iii) of problem 6.5.6, and (*c*) $f(x)J'[F(x)]$ is bounded. Then

$$\lim_{N\to\infty} P[N^{1/2}(\hat{\Delta}_N - \Delta) \leq a] = \Phi(aB/A)$$

where

$$A^2 = \frac{\lambda_0}{1-\lambda_0}\left[\int_0^1 J^2(u)\,du - \left\{\int_0^1 J(u)\,du\right\}^2\right]$$

$$B = \lambda_0\int J'[F(x)]f^2(x)\,dx.$$

6.5.7. (*Continued.*) Let (i) $\Psi(x)$ be symmetric, and $J = \Psi^{-1}$ satisfy the conditions of Theorem 5.5.1, (ii) $F(x)$ and $G(x)$ be symmetric about the same point, and (iii) $J'[\lambda F(x) + (1-\lambda)G(x)][\lambda f(x) + (1-\lambda)g(x)]$ be bounded uniformly in λ in the neighborhood of λ_0. Then

$$\lim_{N\to\infty} P[N^{1/2}(\hat{\Delta}_N - \Delta) \leq a] = \Phi(aB/A)$$

where A^2 is the same as in Exercise 6.5.4, and

$$B = \lambda_0\int_{-\infty}^{0} J'[H_0(x)]f(x)g(x)\,dx.$$

6.5.8. Let $X_1, \ldots, X_n$ be an independent sample from an absolutely continuous c.d.f. $F(x-\theta)$. Let $S_1, \ldots, S_m$ be the ranks of the positive X's among $|X_1|, \ldots, |X_n|$. Let $V^{(1)} < \cdots < V^{(n)}$ be an order statistic from the distribution Ψ. Consider the statistic

$$h(\mathbf{X}_n) = h(X_1, \ldots, X_n) = \frac{1}{n}\sum_{j=1}^{m} E_\Psi[V^{(S_j)}].$$

Let $E\{h(\mathbf{X}_n)\} = \mu$. Then prove that

(i) $\hat{\theta}_n(\mathbf{X}_n + a\mathbf{1}_n) = \hat{\theta}(\mathbf{X}_n) + a$ for all $\mathbf{X}_n$,

(ii) $P_\theta(\hat{\theta}_n - \theta \leq u) = P_0(\hat{\theta} \leq u)$,

(iii) $P[h(\mathbf{X}_n - a\mathbf{1}_n) < \mu] \leq P_0(\hat{\theta}_n \leq a) \leq P[h(\mathbf{X}_n - a\mathbf{1}_n) \leq \mu]$

where $\hat{\theta}_n$ is defined by (6.2.5), and P_θ means that the probability is computed when the true distribution is $F(x-\theta)$.

6.5.9. (*Continued.*) Let $\Psi_0(x) = \Psi(x) - \Psi(-x)$ if $x \geq 0$ and $\Psi_0(x) = 0$ otherwise. Assume that (i) $F(x)$ is symmetric about 0, (ii) $J_0 = \Psi_0^{-1}$ satisfies the conditions of Theorem 4.4.3, (iii) $\Psi(x)$ is symmetric and unimodal with density ψ, (iv) $f(x)$ is bounded in the neighborhood of the origin, and (v) $J_0'[F(x)]f(x)$ is bounded. Then prove that

$$\lim_{N\to\infty} P_\theta[n^{1/2}(\hat{\theta}_n - \theta) \leq a] = \Phi(aB/\beta)$$

where

$$B = \frac{1}{2}\int_0^\infty J_0'[Q(x)]q^2(x)\,dx,$$

$Q(x) = F(x) - F(-x)$ if $x \geq 0$, and $Q(x) = 0$ otherwise, $q(x) = Q'(x)$, and $\beta^2 = \frac{1}{4}$ (second moment of Ψ_0).

6.5.10. (*Continued.*) Let $\Psi(x)$ be uniform over (0, 1). Under the assumptions (ii), (iv), and (v) of Exercise 6.5.9, prove that

$$\lim_{n\to\infty} P_\theta\{n^{1/2}(\theta_{n(R)} - \theta) \leq a\} = \Phi(aB/\beta)$$

where

$$B = \int f(-x)f(x)\,dx \quad \text{and} \quad \beta^2 = \left[\int f^2(-x)f(x)\,dx - \tfrac{1}{4}\right].$$

Comparison of the Estimators. Denote the estimator $\hat{\Delta}_n$ based on the distribution Ψ in Exercise 6.5.1 by $\hat{\Delta}_{n(\Psi)}$. Denote the one-sample estimator for the location of Y's based on the distribution Ψ in Exercise 6.5.8 by $\hat{\theta}^{(2)}_{n(\Psi)}$ and the corresponding estimator for the location of X's by $\hat{\theta}^{(1)}_{n(\Psi)}$. Write $\hat{\hat{\Delta}}_{n(\Psi)} = \hat{\theta}^{(2)}_{n(\Psi)} - \hat{\theta}^{(1)}_{n(\Psi)}$. Let U be the c.d.f. of the uniform distribution over (0, 1). Denote

$$\hat{\hat{\Delta}}_{n(U)} = \underset{1\leq i\leq j\leq n_2}{\text{median}}\,[(Y_i + Y_j)/2] - \underset{1\leq i\leq j\leq n_1}{\text{median}}\,[(X_i + X_j)/2]$$

and

$$\hat{\Delta}_{n(U)} = \text{median}\,[(Y_j - X_i), \quad i = 1, \ldots, n_1; \quad j = 1, \ldots, n_2].$$

6.5.11. Suppose that (i) $J = \Psi^{-1}$ satisfy the regularity conditions of Theorem 4.4.3, and $\Psi(x)$ is symmetric about 0, (ii) F has a bounded symmetric density f, (iii) $G(x) = F(x)$, and (iv) $f(x)J'[F(x)]$ is bounded. Then

$$e(\hat{\Delta}_{n(\Psi)}, \hat{\hat{\Delta}}_{n(\Psi)}) = 1 \quad \text{for all} \quad F.$$

6.5.12. (*Continued.*) Suppose (i) F has a bounded symmetric density f, (ii) Ψ is symmetric and unimodal with density ψ, and $J_0 = \Psi_0^{-1}$ satisfies the conditions of Theorem 4.4.3, (iii) $f(x)J_0'[F(x)]$ is bounded, and (iv) $G(x) = F(cx)$ with $c \neq 1$. Then

$$e(\hat{\Delta}_{n(\Psi)}, \overset{*}{\Delta}_n) = \sigma^2\left[\int J'[F(x)]f^2(x)\,dx\right]^2$$

where $\Delta_n^* = \bar{Y}_{n_2} - \bar{X}_{n_1}$ and σ^2 is the variance of F.

6.5.13. (*Continued.*) Under the conditions of Exercise 6.5.12,

$$e_{(\hat{\hat{\Delta}}_{n(\Phi)},\ \hat{\hat{\Delta}}_{n(U)})} = 12\left(\int f^2(x)\,dx\right)^2\left[\int J_0'[F(x)]f^2(x)\,dx\right]^{-2}.$$

6.5.14. (*Continued.*) Let (i) $\Psi^{-1}(x) = J(x)$ satisfy the regularity conditions of Theorem 5.5.1, (ii) $J'\{\lambda_0 F(x) + (1-\lambda_0)F(cx)\}f(cx)$ is bounded for λ in a neighborhood of λ_0. Then

$$e_{(\hat{\Delta}_n(\Psi),\Delta_n^*)} = \sigma^2\left(\frac{\lambda_0}{c^2} + (1-\lambda_0)\right)(B^2/2A^2)$$

where

$$B = \int J'[\lambda_0 F(x) + (1-\lambda_0)F(cx)]cf(x)f(cx)\,dx,$$

$$A^2 = \alpha_1^2 + \alpha_2^2,$$

$\alpha_1^2 = \lambda_0\sigma_2^2$ and $\alpha_2^2 = (1-\lambda_0)\sigma_1^2$ where σ_1^2 and σ_2^2 are defined in Exercise 6.5.4 with $G(x) = F(cx)$.

6.5.15. (*Continued.*) Show that when $G(x) = F(cx)$

$$\lim_{c\to 0} e_{(\hat{\Delta}_n(\Psi),\Delta_n^*)} = 4\sigma^2 f^2(0),$$

$$\lim_{c\to\infty} e_{(\hat{\Delta}_n(\Psi),\Delta_n^*)} = 4\sigma^2 f^2(0).$$

6.5.16. (*Continued.*) Prove that if $F(x)$ is normal and $G(x) = F(cx)$

$$e_{(\hat{\Delta}_n(\Phi),\Delta_n^*)} \le e_{(\hat{\hat{\Delta}}_n(\Phi),\Delta_n^*)}; \qquad e_{(\hat{\hat{\Delta}}_n(\Phi),\hat{\Delta}_n(\Phi))} \ge 1$$

6.5.17. Prove that $N^{1/2}(\hat{\boldsymbol{\Delta}}_N - \boldsymbol{\Delta})$ has asymptotically, as $N \to \infty$, a multivariate normal distribution with null mean vector and dispersion matrix $T^{**} = ((\tau_{jk}^{**}))$, where $\tau_{jk}^{**} = (1/\lambda)\tau_{jk}^*(F) + (1/(1-\lambda))\tau_{jk}^*(G)$, where τ_{jk}^*'s are defined by (4.6.9) with F^* replaced by F and G respectively, and $\lambda = \lim_{N\to\infty} m/N$.

Section 6.6

6.6.1. Prove that the estimator φ_j^* defined in (6.6.7) is translation invariant.
6.6.2. Complete the proof of Theorem 6.6.2.
6.6.3. Prove that $\Delta_{ii'(j)}^* - \hat{\Delta}_{ii'(j)} = o_p(N^{-1/2})$.

Section 6.7

6.7.1. Develop the one-sample analogues of Theorems 6.7.1 and 6.7.2.

Section 6.8

6.8.1. Verify the assertion (6.8.9).

6.8.2. Verify the assertion (6.8.30).

6.8.3. Prove that under $\theta_1 = \cdots = \theta_c$, $S_{\mathcal{N}}$ defined in (6.8.43) has asymptotically, as $N \to \infty$, the chi-square distribution with $c - 1$ d.f.

Section 6.9

6.9.1. Prove that $\hat{B}_i$ defined in (6.9.12) is a consistent estimator of B_i.

6.9.2. Prove that $\hat{\mathbf{v}}$ defined in (6.9.13) is a consistent and translation invariant estimator of the covariance matrix of the limiting distribution of $n^{1/2}T_n/2$.

6.9.3. Prove that $n \sum_{i=1}^{p} \sum_{j=1}^{p} \hat{v}_{ij}(\hat{\theta}_i - \theta_i)(\hat{\theta}_j - \theta_j)\hat{B}_i\hat{B}_j$ has asymptotically, as $n \to \infty$, the chi-square distribution with p degrees of freedom.

6.9.4. Verify the assertions (6.9.24) and (6.9.25).

6.9.5. Verify the assertion (6.9.27).

6.9.6. Prove that $\hat{B}_i$ defined in (6.9.30) is a consistent estimator of B_i defined in (6.9.29).

6.9.7. Prove that

$$\frac{n_2 n}{n_1} \sum_{i=1}^{p} \sum_{j=1}^{p} \hat{\nu}^{ij}(\hat{\Delta}_i - \Delta_i)(\hat{\Delta}_j - \Delta_j)\hat{B}_i\hat{B}_j$$

has asymptotically, as $n \to \infty$, the chi-square distribution with p degrees of freedom.

PRINCIPAL REFERENCES

Bhuchongkul and Puri (1965), Bickel (1964), Gastwirth (1966), Hodges and Lehmann (1963), Hoyland (1964, 1968), Huber (1964), Lehmann (1963a, b, c), Moses (1965), Puri and Sen (1967b, 1968b, c) Ramachandramurty (1965), Sarhan and Greenberg (1962), Sen (1963b, 1966a, 1968c, f), Scheffé (1953, 1959), Sen and Puri (1969), Tukey (1960).

CHAPTER 7

Rank Procedures In Factorial Experiments

7.1. INTRODUCTION

In factorial experiments, two or more factors each at two or more levels are considered simultaneously. Some of these factors are of real interest while the others are introduced to reduce the variability of the experimental material, and thereby, to increase the accuracy of the statistical analysis of the experiment. The simplest situation is the randomized block layout where p (≥ 2) treatments (or the levels of a factor) are applied to p different plots of a block, and the experiment is replicated in n (≥ 2) blocks. More complicated layouts may arise when (i) different treatments are replicated different number of times (i.e., nonorthogonal layouts), (ii) not all treatments occur in all blocks (i.e., incomplete layouts), or (iii) the p treatments actually represent the combinations of two or more factors where interactions are not negligible. The reader is referred to Cochran and Cox (1957) and Kempthorne (1952) for details.

The classical least squares method for the point estimation of the parameters of interest in various linear models rest on the assumptions that (i) the block and treatment effects are additive, (ii) errors are independent and homoscedastic, and (iii) there is no interaction between blocks and treatments (unless each cell contains two or more observations). The theory of confidence intervals or significance tests for these parameters requires in addition that the errors are normally distributed.

We shall see that the assumption of normality of errors is completely relaxed here, while the other assumptions are relaxed to varying degrees in the different procedures to be considered. We classify the developments on rank procedures under the following categories:

(*a*) the method of n ranking (i.e., intrablock ranking procedures),
(*b*) ranking after alignment, and
(*c*) asymptotically distribution-free procedures based on robust rank estimates of the parameters of interest.

Also, for convenience of presentation, we shall start with the univariate ANOVA models, and later generalize them to the MANOVA models. Finally, for simplicity of presentation, the case of complete block designs will be considered in detail, and appropriate references would be made to possible incomplete or nonorthogonal designs.

7.2. THE METHOD OF n RANKINGS

Consider first the simple model where $p\ (\geq 2)$ different treatments are applied once each to $n\ (\geq 2)$ different blocks, so that each treatment is replicated n times and each block contains p treatments. Let X_{ij} be the chance variable associated with the yield of the jth treatment of the ith block, and let $F_{ij}(x) = P\{X_{ij} \leq x\}$, for $j = 1, \ldots, p$, $i = 1, \ldots, n$. In the parametric case, it is assumed that

$$F_{ij}(x) = F(x - \beta_i - \theta_j), \qquad \sum \theta_j = 0, \tag{7.2.1}$$

where $\beta_1, \ldots, \beta_n$ are the block effects (nuisance parameters) and $\theta_1, \ldots, \theta_p$ are the treatment effects (parameters of interest). For confidence regions for $\boldsymbol{\theta} = (\theta_1, \ldots, \theta_p)'$ or any test of significance on $\boldsymbol{\theta}$, we require further that $F(x)$ is a normal c.d.f. with zero mean and (unknown) variance σ^2. The null hypothesis of no treatment effects, viz., $\boldsymbol{\theta} = \mathbf{0}$, implies under the additivity in (7.2.1) that $F_{i1}(x) = \cdots = F_{ip}(x) = F(x - \beta_i)$ for $i = 1, \ldots, n$.

The method of n rankings neither assumes the additivity of the treatment effects nor that $F_1, \ldots, F_n$ differ only in locations. Let $\mathcal{F}$ be the class of all (univariate) continuous c.d.f.'s. We frame the null hypothesis of no treatment effects as

$$H_0\colon\ F_{i1}(x) = \cdots = F_{ip}(x) = F_i(x) \in \mathcal{F} \quad \text{for} \quad i = 1, \ldots, n, \tag{7.2.2}$$

where the F_i's can be arbitrarily different from each other. Thus, the normality of the F_i's, the additivity of the β_i's, and the homoscedasticity of the errors (over the different blocks) are not required. The random variable X_{ij} is said to be stochastically larger (smaller) than $X_{ij'}$ if $P\{X_{ij} > X_{ij'}\}$ is $> (<) \frac{1}{2}$. The treatment j is said to be stochastically better (worse) than the treatment j', if $P\{X_{ij} > X_{ij'}\} \geq (\leq) \frac{1}{2}$ for all $i = 1, \ldots, n$, with the strict inequality holding for at least one i. Then, the alternative hypothesis may be stated as follows:

(7.2.3) K: At least one of the treatment effects is stochastically larger (smaller) than the rest.

The specification of the alternative hypothesis becomes considerably simpler under the additivity of the treatment effects, viz.,

$$(7.2.4)\quad F_{ij}(x) = F_i(x - \theta_j), \quad j = 1, \ldots, p; \qquad F_i \in \mathcal{F}, \quad i = 1, \ldots, n.$$

Under (7.2.4), we may set (7.2.3) equivalently as

$$(7.2.5)\qquad K: \ \boldsymbol{\theta} = (\theta_1, \ldots, \theta_p)' \neq \mathbf{0},$$

where of course the null hypothesis in (7.2.2) states that $\boldsymbol{\theta} = \mathbf{0}$.

The method of n rankings rests on the set of n intrablock rank p-tuples $\mathbf{r}_i = (r_{i1}, \ldots, r_{ip})'$, $i = 1, \ldots, n$, where r_{ij} stands for the rank of X_{ij} among $X_{i1}, \ldots, X_{ip}$, $j = 1, \ldots, p$, $i = 1, \ldots, n$. Since $F_{ij} \in \mathcal{F}$, ties among the the observations may be ignored, in probability. Now, under (7.2.2), $X_{i1}, \ldots, X_{ip}$ are independent and identically distributed random variables, and hence, the distribution of $\mathbf{r}_i$ over the $p!$ possible realizations (which are the permutations of the numbers $1, \ldots, p$) is uniform, whatever be the c.d.f. $F_i \in \mathcal{F}$. Moreover, by definition, $\mathbf{r}_1, \ldots, \mathbf{r}_n$ are independent stochastic vectors. Consequently, under H_0 in (7.2.2), the rank collection matrix $\mathbf{R}_n = (\mathbf{r}_1, \ldots, \mathbf{r}_n)$ can have $(p!)^n$ equally likely realizations (over the permutations of the elements of each of its column), whatever be $F_i \in \mathcal{F}$, $i = 1, \ldots, n$. Let us denote by $\mathcal{T}_n$ the probability law over these $(p!)^n$ equally likely rank permutations. Since $\mathcal{T}_n$ is completely specified, we can always select a test function ϕ: $0 \leq \phi \leq 1$, such that $E_{\mathcal{T}_n}(\phi) = \epsilon, 0 < \epsilon < 1$, where ϵ is the preassigned level of significance of the test and ϕ depends on the X_{ij}'s only through $\mathbf{R}_n$. This characterizes the existence of distribution-free tests for H_0 in (7.2.2) valid for all $F_i \in \mathcal{F}$, $i = 1, \ldots, n$. Incidentally, the method of n ranking does not require that the X_{ij}'s should be all observable, it is sufficient that the rank tuplets $\mathbf{r}_i$, $i = 1, \ldots, n$ be observable; a case that may arise in various situations. Kendall (1955) gives a detailed account of this specialized aspect of the method of n rankings.

Two well-known tests by the method of n rankings have been proposed by Friedman (1937) and Brown and Mood (1951), and these are respectively based on the following test statistics:

$$(7.2.6)\quad \chi_r^2 = \frac{12}{np(p+1)} \sum_{j=1}^{p} \left(R_j - \frac{n(p+1)}{2}\right)^2; \quad R_j = \sum_{i=1}^{n} r_{ij}, \quad j = 1, \ldots, p,$$

$$(7.2.7)\quad B_r = \frac{p(p-1)}{na(p-a)} \sum_{j=1}^{p} (M_j - na/p)^2; \quad M_j = \sum_{i=1}^{n} m_{ij}, \quad j = 1, \ldots, p,$$

where $m_{ij} = 1$ or 0 according as $r_{ij} \leq a$ or not, and $1 \leq a < p$; a is the largest integer contained in $(p + 1)/2$. Instead of studying these tests separately, we shall study a general class of tests which includes both these tests as special cases.

7.2.1. A Class of Tests Based on the Method of n Ranking

Let J_r, $r = 1, \ldots, p$ be p (real) known scores, and define

$$\bar{J} = p^{-1}\sum_{j=1}^{p} J_j, \qquad A^2(J) = (p-1)^{-1}\sum_{j=1}^{p}(J_j - \bar{J})^2. \tag{7.2.8}$$

We assume that J_r is not constant for all r $(= 1, \ldots, p)$, and this implies that $A^2(J) > 0$. We then define

$$T_{n,j} = n^{-1}\sum_{i=1}^{n} J_{r_{ij}}, \qquad j = 1, \ldots, p; \qquad \mathbf{T}_n = (T_{n,1}, \ldots, T_{n,p})'. \tag{7.2.9}$$

Thus, $T_{n,j}$ stands for the average value of the scores for the jth treatment, for $j = 1, \ldots, p$. Under H_0 in (7.2.2), we have

$$E(\mathbf{T}_n) = \bar{J}\cdot \ell_p \quad \text{and} \quad nE[(T_n - \bar{J}\cdot\ell_p)(\mathbf{T}_n - \bar{J}\cdot\ell_p)'] = A^2(J)\left[\mathbf{I}_p - \frac{1}{p}\ell_p\ell_p'\right], \tag{7.2.10}$$

where $\ell_p' = (1, \ldots, 1)$ and $\mathbf{I}_p$ is the identity matrix of order p. Thus, considering the generalized inverse of the covariance matrix in (7.2.10) (where we note that $\mathbf{T}_n$ has rank $p - 1$, as $\mathbf{T}_n'\ell_p = p\bar{J}$), and constructing a quadratic form in $\mathbf{T}_n$ with this matrix as its discriminant, we obtain the following statistic

$$S_n = nA^{-2}(J)\sum_{j=1}^{p}(T_{n,j} - \bar{J})^2. \tag{7.2.11}$$

Naturally, S_n will be larger when $T_{n,j}$'s are stochastically different from each other, and the stochastic differences among the $T_{n,j}$ is expected when H_0 in (7.2.2) does not hold. Hence, it seems logical to base a test on the statistic S_n in (7.2.11), rejecting the null hypothesis (7.2.2) only when S_n exceeds a preassigned value $S_{n,\epsilon}$, where $P_{H_0}\{S_n \geq S_{n,\epsilon}\} \leq \epsilon$, the level of significance. That (7.2.6) and (7.2.7) are particular cases of (7.2.11) follows readily by letting $J_{r_{ij}} = r_{ij}$ and m_{ij} respectively. Also, instead of having the level of significance $\leq \epsilon$, we can have it equal to ϵ, by constructing a randomized test instead of the nonrandomized test considered above.

The null distribution of S_n has to be obtained by considering the probability distribution generated by the $(p!)^n$ equally likely realizations of $\mathbf{R}_n$. The task becomes prohibitively laborious as n increases. For this purpose, we consider the large-sample distribution of S_n when H_0 in (7.2.2) holds.

Theorem 7.2.1. *Under H_0 in (7.2.2), S_n converges in law to a chi-square distribution with $p - 1$ degrees of freedom, uniformly in $F_i \in \mathcal{F}$, $i = 1, \ldots, n$.*

The proof follows as a special case of the next theorem and hence is omitted. Thus, for large n, the test procedure reduces to that of rejecting H_0 when S_n exceeds $\chi^2_{p-1,\epsilon}$, the $100\epsilon\%$ point of the chi-square distribution with $p - 1$ degrees of freedom.

7.2.2. Asymptotic Distribution of S_n (General Case)

Let us define

$$(7.2.12)\quad p_{i,r}^{(j)} = P\{r_{ij} = r\},\quad p_{i,rs}^{(j,j')} = P\{r_{ij} = r, r_{ij'} = s\},\quad j \neq j',\quad r \neq s = 1, \ldots, p,$$

for $i = 1, \ldots, n$. We also let conventionally

$$(7.2.13)\quad p_{i,rr}^{(j,j')} = \delta_{jj'}p_{i,r}^{(j)} \text{ and } p_{i,rs}^{(j,j)} = \delta_{rs}p_{i,r}^{(j)},\quad j, j', r, s = 1, \ldots, p,\quad i = 1, \ldots, n,$$

where $\delta_{jj'}$ and δ_{rs} are the usual Kronecker deltas. Further, let

$$(7.2.14)\quad \boldsymbol{\mu}_n = (\mu_{n,1}, \ldots, \mu_{n,n})';\qquad \mu_{n,j} = n^{-1}\sum_{i=1}^{n} p_{i,r}^{(j)}J_r,\quad j = 1, \ldots, p;$$

$$(7.2.15)\quad \boldsymbol{\Sigma}_n = ((\sigma_{n,jj'}));\qquad \sigma_{n,jj'} = n^{-1}\sum_{i=1}^{n}\sum_{r=1}^{p}\sum_{s=1}^{p} J_rJ_sp_{i,rs}^{(j,j')} - \mu_{n,j}\mu_{n,j'},$$

for $j, j' = 1, \ldots, p$. Finally, let $\mathbf{F}_i = (F_{i1}, \ldots, F_{ip})'$, $i = 1, \ldots, n$, $\mathbf{F}_n^* = (\mathbf{F}_1, \ldots, \mathbf{F}_n)$, and let $\mathcal{F}_n^* = \{\mathbf{F}_n^*\colon \text{rank } [\boldsymbol{\Sigma}_n] \geq 1\}$, $A_n^* = \{\mathbf{a}\colon \mathbf{a}'\boldsymbol{\Sigma}_n\boldsymbol{\delta} > 0\}$. Then we have the following.

Theorem 7.2.2. *For all $\{\mathbf{F}_n^*\} \in \{\mathcal{F}_n^*\}$ and $\{\mathbf{a}\} \in \{A_n^*\}$, the sequence of distributions of $\{n^{1/2}(\mathbf{T}_n - \mathbf{u}_n)'\mathbf{a}/[\mathbf{a}'\boldsymbol{\Sigma}_n\mathbf{a}]^{1/2}\}$ converges (as $n \to \infty$) to the standard normal distribution.*

Outline of the Proof: By definition, $\sum_{j=1}^{p} (T_{n,j} - \mu_{n,j}) = 0$. Hence, we can always take $\mathbf{a} \perp \boldsymbol{\ell}_p$. Let then

$$(7.2.16)\quad Z_n = n^{1/2}\mathbf{a}'(\mathbf{T}_n - \boldsymbol{\mu}_n) = n^{-1/2}\sum_{i=1}^{n} U_i;\quad U_i = \sum_{j=1}^{p} a_j\left[J_{r_{ij}} - \sum_{r=1}^{p} J_r p_{i,r}^{(j)}\right],$$

$i = 1, \ldots, n$. Now, by definition $U_1, \ldots, U_n$ are independent random variables with zero means, and $1/n \sum_{i=1}^{n} E[U_i^2] = \mathbf{a}\boldsymbol{\Sigma}_n\mathbf{a} > 0$ (by the hypothesis that $\mathbf{a} \in A_n^*$). Also $1/n \sum_{i=1}^{n} E\,|U_i|^3 \leq \{\max_j |a_j|^3\}\{\text{range}_r J_r\}^3$, where $\text{range}_r\, J_r = J_{\max} - J_{\min}$ is finite. The proof now follows by using the Berry-Essèen theorem (cf. section 2.7). Q.E.D. ◁

Under H_0 in (7.2.2), $\boldsymbol{\Sigma}_n = A^2(J)[\mathbf{I}_p - (1/p)\boldsymbol{\ell}_p\boldsymbol{\ell}_p']$, so that $\mathbf{a}'\boldsymbol{\Sigma}_n\mathbf{a} = A^2(J)\mathbf{a}'\mathbf{a} > 0$ for all non-null $\mathbf{a} \perp \boldsymbol{\ell}_p$. Also the rank of $\boldsymbol{\Sigma}_n$ is $p - 1$ for all $F_i \in \mathcal{F}$, $i = 1, \ldots, n$. Hence the distribution of $n^{1/2}(\mathbf{T}_n - \bar{J}\cdot\boldsymbol{\ell}_p)$ converges to a normal distribution with mean $\mathbf{0}$ and dispersion matrix $A^2(J)[\mathbf{I}_p - (1/p)\boldsymbol{\ell}_p\boldsymbol{\ell}_p']$, uniformly in $\{\mathbf{F}_n^*\}$. The proof of Theorem 7.2.1 follows then from Theorem 2.8.3.

The asymptotic distribution of S_n has to be deduced from that of $\mathbf{T}_n$, and, in general, it is quite involved, unless we restrict ourself to local alternatives.

7.2.3. Asymptotic Power Properties of the S_n Test for Local Translation Alternatives

We consider the following sequence $\{K_n\}$ of alternative hypotheses:

$$(7.2.17)\quad K_n: F_{ij}(x) = F_{ij,n}(x) = F_i(x - n^{-1/2}\tau_j),\qquad j = 1, \ldots, p, \quad i = 1, \ldots, n,$$

where $\boldsymbol{\tau}_p = (\tau_1, \ldots, \tau_p)'$ is a real p vector orthogonal to ℓ_p, $F_i \in \mathcal{F}_0$, $i = 1, \ldots, n$, and $\mathcal{F}_0$ is the class of all absolutely continuous distributions having continuous density functions satisfying

$$(7.2.18)\qquad \int_{-\infty}^{\infty} \{(d/dx)F_i(x)\}^2\, dx < \infty \quad \text{for all} \quad F_i \in \mathcal{F}_0.$$

Let $f_i(x) = (d/dx)F_i(x)$, and

$$(7.2.19)\quad \beta_{s,p-2}^{(i)} = \binom{p-2}{s} \int_{-\infty}^{\infty} [F_i(x)]^s [1 - F_i(x)]^{p-2-s} f_i^2(x)\, dx,\qquad s = 0, 1, \ldots, p-2,$$

and conventionally let $\beta_{-1,p-2}^{(i)} = \beta_{p-1,p-2}^{(i)} = 0$, for $i = 1, \ldots, n$. Also, let

$$(7.2.20)\qquad \lambda_{r,n} = n^{-1} \sum_{i=1}^{n} \{\beta_{r-1,p-2}^{(i)} - \beta_{r-2,p-2}^{(i)}\}, \qquad r = 1, \ldots, p.$$

Finally, let $G_n(x) = P\{S_n \leq x \mid K_n\}$ and let $\chi_n(x; \Delta_n(\mathbf{J}))$ be the cumulative probability distribution of a random variable following a noncentral chi-square distribution with $p - 1$ degrees of freedom and the noncentrality parameter

$$(7.2.21)\quad \Delta_n(\mathbf{J}) = [p^2/A^2(J)]\left[\sum_{j=1}^{p} \theta_j^2\right]\left[\sum_{r=1}^{p} \lambda_{r,n} J_r\right]^2; \qquad \mathbf{J} = (J_1, \ldots, J_p)'.$$

Then, we have the following.

Theorem 7.2.3. *Under* (7.2.17) *and* (7.2.18)

$$(7.2.22)\quad \lim_{n\to\infty} |G_n(x) - \chi_n(x, \Delta_n(\mathbf{J}))| = 0 \quad \textit{for all} \quad -\infty < x < \infty.$$

Proof: By virtue of Theorems 7.2.2 and 2.8.2, it suffices to show that under (7.2.17) and (7.2.18), as $n \to \infty$

$$(7.2.23)\qquad n^{1/2}[\boldsymbol{\mu}_n - \bar{J} \cdot \ell_p] + p\left(\sum_{r=1}^{p} \lambda_{r,n} J_r\right)\boldsymbol{\tau} = o(1),$$

$$(7.2.24)\qquad \left(\boldsymbol{\Sigma}_n - A^2(J)\left\{\mathbf{I}_p - \frac{1}{p}\ell_p \ell_p'\right\}\right) \to \mathbf{0}^{p\times p}.$$

Now, by definition in (7.2.12)

$$p_{i,r}^{(j)} = \sum_{S_j} \int_{-\infty}^{\infty} \prod_{j'=1}^{r-1} F_{is_{j'}}(x) \prod_{j'=r+1}^{p} [1 - F_{is_{j'}}(x)]\, dF_{ij}(x) \tag{7.2.25}$$

where the summation S_j extends over all possible choices of $(s_1, \ldots, s_{r-1})$ from $(1, \ldots, j-1, j+1, \ldots, p)$ and $(s_{r+1}, \ldots, s_p)$ is the complementary set. Also by (7.2.17), we have

(7.2.26)

$$F_{il}(x) = F_{ij}(x - n^{-1/2}[\tau_l - \tau_j]) \quad \text{for} \quad j, l = 1, \ldots, p, \quad i = 1, \ldots, n.$$

From (7.2.25) and (7.2.26), we obtain after some simplifications that

$$p_{i,r}^{(j)} = p^{-1} + n^{-1/2} p \tau_j (\beta_{r-2,p-2}^{(i)} - \beta_{r-1,p-2}^{(i)}) + o(n^{-1/2}), \tag{7.2.27}$$

$$j, r = 1, \ldots, p, \quad i = 1, \ldots, n.$$

(7.2.23) follows readily from (7.2.14) and (7.2.27). Proceeding in a similar manner, it can be shown that under (7.2.17) and (7.2.18)

$$p_{i,rs}^{(j,j')} = 1/p(p-1) + o(1) \quad \text{for all} \quad j \neq j', \quad r \neq s = 1, \ldots, p, \tag{7.2.28}$$

$$i = 1, \ldots, n.$$

Thus, (7.2.24) follows readily from (7.2.15), (7.2.23), (7.2.27), and (7.2.28.) Q.E.D. ◁

Now, by definition $\sum_{r=1}^{p} \lambda_{r,n} = 0$. Hence, we may rewrite $\Delta_n(\mathbf{J})$ in (7.2.21) as

$$\Delta_n(\mathbf{J}) = \left\{ p^2(p-1) \sum_{j=1}^{p} \tau_j^2 \right\} \left\{ \sum_{r=1}^{p} \lambda_{r,n}^2 \right\} \rho^2(\mathbf{J}, \boldsymbol{\lambda}_n), \tag{7.2.29}$$

where

$$\rho(\mathbf{J}, \boldsymbol{\lambda}_n) = \frac{\sum_{r=1}^{p} (J_r - \bar{J}) \lambda_{r,n}}{\{[\sum_{r=1}^{p} (J_r - \bar{J})^2][\sum_{r=1}^{p} \lambda_{r,n}^2]\}^{1/2}}. \tag{7.2.30}$$

Thus, $\rho^2(\mathbf{J}, \boldsymbol{\lambda}_n)$ is bounded above by 1, and hence

$$\sup_{\{\mathbf{J}\}} \Delta_n(\mathbf{J}) = \left(p^2(p-1) \sum_{j=1}^{p} \tau_j^2 \right) \left(\sum_{r=1}^{p} \lambda_{r,n}^2 \right) = \Delta_n^0 \quad \text{(say)}. \tag{7.2.31}$$

Now, Δ_n^0 represents the maximum noncentrality parameter which can be achieved by a proper choice of $\mathbf{J}$. Since for any $\mathbf{J}$, S_n has asymptotically (undeı (7.2.17)) a noncentral chi-square distribution, maximizing $\Delta_n(\mathbf{J})$ amounts to maximizing the asymptotic power of the test based on S_n. Thus, $\Delta_n(\mathbf{J})$ can be interpreted as a measure of the efficiency of the test based on S_n. Hence, in this sense, Δ_n^0 in (7.2.31) relates to the efficiency of the test based on S_n which has the maximum local power. We shall define Δ_n^0 as the *intrinsic efficiency of the method of n ranking*, as Δ_n^0 does not depend on $\mathbf{J}$ and relates to the maximum value of $\Delta_n(\mathbf{J})$. The relative efficiency of any particular S_n test based on scores $\mathbf{J}$, is thus equal to $\rho^2(\mathbf{J}, \boldsymbol{\lambda}_n)$, defined by (7.2.30). Thus, $\rho^2(\mathbf{J}, \boldsymbol{\lambda}_n)$ measures the amount of information utilized by the particular $\mathbf{J}$.

Now, $\rho^2(\mathbf{J}, \boldsymbol{\lambda}_n)$ will be equal to 1, iff

$$J_r - \bar{J} = k\lambda_{r,n}, \quad r = 1, \ldots, p, \quad \text{where} \quad k \neq 0. \tag{7.2.32}$$

For $p = 2$, by (7.2.19) and (7.2.20), $\lambda_{1,n} = -\lambda_{2,n}$, and also by definition $J_1 - \bar{J} = -[J_2 - \bar{J}]$. Hence, (7.2.32) holds for all (J_1, J_2). Thus, any S_n test will achieve the intrinsic accuracy of the method of n ranking. In fact, this is evident from the fact that all these tests (for different $\mathbf{J}$) are equivalent to the simple sign test based on the differences $X_{i1} - X_{i2}$, $i = 1, \ldots, n$. For $p = 3$, if F_i is symmetric about μ_i, then $\beta_{0,1}^{(i)} = \beta_{1,1}^{(i)}$, for $i = 1, \ldots, n$. Hence, in this case, $\lambda_{1,n} = -\lambda_{3,n}$ and $\lambda_{2,n} = 0$. So by (7.2.32), the optimum S_n is based on χ_r^2, defined by (7.2.6). However, if $F_1, \ldots, F_n$ are not all symmetric, the optimality of χ_r^2 cannot be established. For $p \geq 4$, the optimum S_n depends explicitly on $F_1, \ldots, F_n$, and in general, it is quite difficult to obtain the optimum $\mathbf{J}$ in a simple manner. However, if the *block effects are additive*, that is, if the distributions $F_1, \ldots, F_n$ differ only in locations from the distribution $F(x)$, we have from (7.2.15), $\beta_{s,p-2}^{(i)} = \beta_{s,p-2}$, for all $i = 1, \ldots, n$, where

$$\beta_{s,p-2} = \frac{1}{p-1} E\{f(X_{s,p-1})\}, \qquad s = 1, \ldots, p-1, \tag{7.2.33}$$

where $X_{s,p-1}$ is the sth smallest observation of a sample of size $p - 1$ from the distribution $F(x)$ whose density function is $f(x)$. In many cases, (7.2.33) and hence the optimum $\mathbf{J}$ can be evaluated very easily. The solutions for uniform, logistic, normal, double exponential, and exponential distributions are left as Exercises 7.2.1 to 7.2.5.

7.2.4. Comparison with the Standard Variance Ratio Test

We shall now compare the S_n test with the parametric ANOVA test based on the variance ratio $(\mathcal{F}_{p-1,(n-1)(p-1)})$ criterion

$$\mathcal{F}_{p-1,(n-1)(p-1)} = \frac{n \sum_{j=1}^{p} (X_{\cdot j} - X_{\cdot\cdot})^2}{\sum_{i=1}^{n} \sum_{j=1}^{p} (X_{ij} - X_{i\cdot} - X_{\cdot j} + X_{\cdot\cdot})^2/(n-1)(p-1)} \tag{7.2.34}$$

(where a $\cdot$ in the subscript stands for the average over the subscript replaced by the $\cdot$). When the X_{ij} are distributed independently and normally with a common (unknown) variance σ^2 and the means $\mu + \beta_i + \theta_j$, then under H_0: $\theta_1 = \cdots = \theta_p = 0$, $\mathcal{F}_{p-1,(n-1)(p-1)}$ has the $\mathcal{F}$ distribution with $p - 1$,

$(n-1)(p-1)$ d.f. Thus, under H_0, $(p-1)\mathcal{F}_{p-1,(n-1)(p-1)}$ has asymptotically a chi-square distribution with $p-1$ D.F.

Now, consider the situation where the normality or the homoscedasticity may not hold. Suppose that $X_{ij}-\theta_j$, $j=1,\ldots,p$ are independent and identically distributed according to a c.d.f. F_i, such that $\int |x|^{2+\delta}\,dF_i(x)<\infty$ for some $\delta>0$ and all $i=1,\ldots,n$. Let then

$$(\bar{\sigma}_n)^2 = n^{-1}\sum_{i=1}^{n}\sigma_i^2. \tag{7.2.35}$$

It is then easy to show (cf. Exercise 7.2.6) that whatever be $\boldsymbol{\theta}$,

$$\left|\frac{1}{(n-1)(p-1)}\sum_{i=1}^{n}\sum_{j=1}^{p}(X_{ij}-X_{i\cdot}-X_{\cdot j}+X_{\cdot\cdot})^2-(\bar{\sigma}_n)^2\right|\xrightarrow{p}0. \tag{7.2.36}$$

Also, using the classical central limit theorem (under the Liapounoff condition; cf. Section 2.7), we obtain that $n^{1/2}(X_{\cdot j}-EX_{\cdot j})/\bar{\sigma}_n$, $j=1,\ldots,p$ are distributed independently and asymptotically normally with zero means and unit variance. Hence, under H_0: $\boldsymbol{\theta}=\mathbf{0}$,

$$n\sum_{j=1}^{p}(X_{\cdot j}-X_{\cdot\cdot})^2/\bar{\sigma}_n^2 \tag{7.2.37}$$

converges in law to a chi-square distribution with $p-1$ D.F. Thus, from (7.2.34), (7.2.36) and (7.2.37), we obtain that under H_0: $\boldsymbol{\theta}=\mathbf{0}$, $(p-1)\mathcal{F}_{p-1,(n-1)(p-1)}$ has asymptotically a chi-square distribution with $p-1$ D.F.

Consider now the sequence of alternative hypotheses $\{K_n\}$ in (7.2.17), and define

$$\Delta_n(\mathcal{F}) = \left[\sum_{j=1}^{p}\tau_j^2\right]\Big/\bar{\sigma}_n^2. \tag{7.2.38}$$

Finally, let $\chi(x,\Delta_n(\mathcal{F}))$ be the c.d.f. of a noncentral chi-square variable with $p-1$ D.F. and the noncentrality parameter $\Delta_n(\mathcal{F})$, defined by (7.2.38), and let $H_n(x)$ be the actual c.d.f. of $(p-1)\mathcal{F}_{p-1,(n-1)(p-1)}$ under K_n in (7.2.17). Then, proceeding as in (7.2.35) through (7.2.37) and noting that under K_n, $n^{1/2}E(X_{\cdot j}-X_{\cdot\cdot})=\tau_j$, $j=1,\ldots,p$, we obtain that

$$\lim_{n\to\infty}|H_n(x)-\chi(x,\Delta_n(\mathcal{F}))|=0 \quad \text{for all} \quad 0\le x<\infty. \tag{7.2.39}$$

Thus, a comparison of $\chi(x,\Delta_n(\mathbf{J}))$ and $\chi(x,\Delta_n(\mathcal{F}))$ reveals the relative performances of the tests based on S_n and $\mathcal{F}_{p-1,(n-1)(p-1)}$. However, the computation of the A.R.E. requires the existence of

$$\lim_{n\to\infty}\Delta_n(\mathbf{J})=\Delta(\mathbf{J}) \quad \text{and} \quad \lim_{n\to\infty}\Delta_n(\mathcal{F})=\Delta(\mathcal{F}). \tag{7.2.40}$$

Under (7.2.40), the A.R.E. of the S_n test with respect to the variance ratio test is equal to

$$e_{S,\mathcal{F}} = \Delta(\mathbf{J})/\Delta(\mathcal{F}). \tag{7.2.41}$$

Now, $\lim_{n\to\infty} \Delta_n(\mathcal{F}) = \Delta(\mathcal{F})$ exists whenever $\bar{\sigma}_n^2$ converges to a limit σ^2 as $n \to \infty$, while $\lim_{n\to\infty} \Delta_n(\mathbf{J}) = \Delta(\mathbf{J})$ exists when $\lambda_{r,n} \to \lambda_r$, $r = 1, \ldots, p$. Conditions for these are considered below.

(I) The Classical* ANOVA *Model. Here, block effects are additive and the errors are homoscedastic. Here, $F_1, \ldots, F_n$ differ only in shifts (as the block effects are additive), so that $\sigma_i^2 = \sigma^2$ for all i and $\lambda_{r,n} = \lambda_r$ for all $n \geq 1$. Consequently,

$$e_{S,\mathcal{F}} = (p^2(p-1)\sigma^2)\left(\sum_{r-1}^{p} \lambda_r^2\right)\rho^2(\mathbf{J}, \boldsymbol{\lambda}), \tag{7.2.42}$$

where $\rho(\mathbf{J}, \boldsymbol{\lambda})$ is defined by (7.2.30) with $\lambda_{r,n}$ replaced by λ_r. In the particular case of χ_r^2, defined by (7.2.6), we have

$$e_{\chi_r^2,\mathcal{F}} = \frac{12p\sigma^2}{(p+1)}\left[\int_{-\infty}^{\infty} f^2(x)\,dx\right]^2, \tag{7.2.43}$$

where $f(x) = F'(x)$. Thus, (7.2.43) is equal to $p/(p+1)$ times the A.R.E. of the Wilcoxon test with respect to the Student's t test. In particular when F is normal, then (7.2.43) equals $3p/\pi(p+1)$, which for specific values of p assumes the values:

p	2	3	4	5	10	15	20	∞
$\dfrac{3p}{\pi(p+1)}$	0.637	0.716	0.764	0.796	0.868	0.895	0.910	0.955

Thus, the loss of efficiency is more than 20% for $p < 5$ and more than 10% for $p \leq 15$. Since in actual practice p is usually small, this indicates that the use of χ_r^2 or S_n in general may lead to some loss of efficiency. We shall discuss more about it in section 7.3.

(II) Replication Model. Suppose that n_i blocks corresponds to a common c.d.f. F_i (apart from additive block and treatment effects), $i = 1, \ldots, b$ (≥ 1), where $n = \Sigma\, n_i$, and as $n \to \infty$ $n_i/n \to \nu_i$: $0 < \nu_i < 1$, $\Sigma\, \nu_i = 1$. This means that conceptually we replicate the set of b initial blocks a large number of times keeping their proportion fixed. In this case, $\bar{\sigma}_n^2$ converges to $\Sigma\, \nu_i \sigma_i^2$ where σ_i^2 is the variance of F_i. Also, defining $\beta_{s,p-2}^{(i)}$ as in (7.2.19), we let $\bar{\beta}_{s,p-2} = \Sigma\, \nu_i \beta_{s,p-2}^{(i)}$, $s = 0, 1, \ldots, p-2$, and let $\bar{\beta}_{-1,p-2} = \bar{\beta}_{p-1,p-2} = 0$. Let then $\bar{\lambda}_r = \bar{\beta}_{r-1,p-2} - \bar{\beta}_{r-2,p-2}$, $r = 1, \ldots, p$.

In this model, (7.2.41) reduces to

$$p^2(p-1)(\Sigma\nu_i\sigma_i^2)\left(\sum_{r=1}^{p}\bar{\lambda}_r^2\right)\rho^2(\mathbf{J},\bar{\boldsymbol{\lambda}}), \tag{7.2.44}$$

where $\rho(\mathbf{J},\bar{\boldsymbol{\lambda}})$ is defined by (7.2.30) with $\lambda_{r,n}$ replaced by $\bar{\lambda}_r$, $r = 1, \ldots, p$. In particular for the χ_r^2 test, (7.2.44) simplifies to

$$\frac{12p}{(p+1)}(\Sigma\nu_i\sigma_i^2)\left[\Sigma\nu_i\int_{-\infty}^{\infty} f_i^2(x)\,dx\right]^2, \quad \text{where} \quad f_i = F_i'. \tag{7.2.45}$$

A special case is the heteroscedastic model, where the c.d.f.'s $F_1, \ldots, F_b$ differ possibly in the variances only. Thus, we let

$$F_i(x) = F((x-\mu)/\sigma_i) = f_i(x) = \frac{1}{\sigma_i}f\left(\frac{x-\mu}{\sigma_i}\right), \qquad i = 1, \ldots, b, \tag{7.2.46}$$

where $\sigma_i > 0$, $i = 1, \ldots, b$ are all real and finite. Then (7.2.45) reduces to

$$\frac{12p}{(p+1)}(\Sigma\nu_i\sigma_i^2)\left[\Sigma(\nu_i/\sigma_i^2)\int_{-\infty}^{\infty} f^2\left(\frac{x-\mu}{\sigma_i}\right)dx\right]^2. \tag{7.2.47}$$

Now, by definition,

$$[12p/(p+1)]\sigma_i^2\left[\frac{1}{\sigma_i^2}\int_{-\infty}^{\infty} f^2\left(\frac{x-\mu}{\sigma_i}\right)dx\right]^2 = e(F), \qquad i = 1, \ldots, b. \tag{7.2.48}$$

Hence, from (7.2.46) and (7.2.48), we obtain that (7.2.47) is equal to

$$(\Sigma\nu_i\sigma_i^2)[\Sigma(\nu_i/\sigma_i)\sqrt{e(F)}]^2 = e(F)\cdot(\Sigma\nu_i/\sigma_i^2)(\Sigma\nu_i/\sigma_i)^2. \tag{7.2.49}$$

Now, by elementary moment inequalities,

$$\Sigma\nu_i\sigma_i^2 \geq (\Sigma\nu_i\sigma_i)^2 \geq (\Sigma\nu_i/\sigma_i)^2, \tag{7.2.50}$$

where the equality sign holds only when $\sigma_1 = \cdots = \sigma_b$. Thus, (7.2.49) is bounded below by $e(F)$ and the lower bound is attained only in the homoscedastic case. This explains the robustness of χ_r^2 for heteroscedastic errors. A similar property is possessed by S_n (cf. Exercise 7.2.7).

(III) Outlier Model. Suppose the block effects are additive and the errors are homoscedastic in a set of n_1 blocks, while in the remaining $n_2 = n - n_1$, at least one of the two assumptions does not hold. If $n_2/n \to 0$ as $n \to \infty$, it clearly follows that the A.R.E. will still be given by (7.2.42). On the other hand, if $n_2/n \to \nu$: $0 < \nu < 1$, we may proceed as in (II).

(IV) Convergence Model. If the c.d.f.'s $F_1, \ldots, F_n$ satisfy the conditions that $F_n \to F$ as $n \to \infty$, then of course $\bar{F}_n = n^{-1}\Sigma F_i \to F$ as $n \to \infty$. The implication is that $\bar{\sigma}_n^2$, defined by (7.2.35), will converge to σ^2, the variance of the c.d.f. F, as $n \to \infty$. Similarly, $\lambda_{r,n}$ will converge to λ_r, the scores for the c.d.f. F. Consequently, in this case, the A.R.E. will again be given by (7.2.42).

Coming back to the basic structure of the method of n rankings, we observe that the independence of the observations within the same block can easily be replaced by symmetric dependence or interchangeability of the errors. That is, we may assume that $\mathbf{X}_i = (X_{i1}, \ldots, X_{ip})'$ has a p-variate continuous cumulative distribution function $G_i(\mathbf{x} - \boldsymbol{\theta})$, $\boldsymbol{\theta} = (\theta_1, \ldots, \theta_p)'$, where G_i is symmetric in its p arguments. Thus, under H_0: $\boldsymbol{\theta} = \mathbf{0}$, $X_{i1}, \ldots, X_{ip}$ are interchangeable random variables, whose joint distribution remains invariant under any permutation of these variables. This implies that under H_0: $\boldsymbol{\theta} = \mathbf{0}$, $\mathbf{r}_i = (r_{i1}, \ldots, r_{ip})'$ can have $p!$ equally likely realizations over the permutations of $(1, \ldots, p)$. Consequently, the null distribution theory of S_n remains valid under this interchangeable error model. This covers the situations which may arise in random effects or mixed effects models. Exercises 7.2.8 and 7.2.9 extends the non-null distribution theory of χ_r^2 and of S_n to such exchangeable error models.

So far, we have considered the case of a complete two-factor layout. A very simple extension is the balanced incomplete block design. Here each of the b blocks contains k treatments $(k \leq p)$ and each treatment is replicated r times $(r \leq b)$ in such a way that no treatment occurs more than once in any block and each treatment is compared with any other treatment in λ blocks, where $b \geq p \geq r \geq k \geq \lambda \geq 1$. If the jth treatment occurs in the ith block, we denote by r_{ij} the rank of the corresponding response among the k responses in the ith block; otherwise, we let $r_{ij} = 0$. Let then $R_j = \sum_i r_{ij}$, $j = 1, \ldots, p$. Analogous to (7.2.6) we define here

$$\chi_r^2 = \frac{12}{p\lambda(k+1)} \sum_{j=1}^{p} [R_j - \tfrac{1}{2}r(k+1)]^2. \tag{7.2.51}$$

Exercise 7.2.10 shows that under the hypothesis of no treatment effects, χ_r^2 has asymptotically a chi-square distribution with $p - 1$ degrees of freedom. Also problem 7.2.11 shows that the A.R.E. of χ_r^2 with respect to the parametric ANOVA test is given by (7.2.43) and (7.2.45) with p replaced by k. Problem 7.2.12 shows that χ_r^2 provides a valid test even for interchangeable errors. Finally, Exercise 7.2.13 extends the S_n statistic to such an incomplete layout model. We also refer to the interesting papers by Bernard and Elteren (1953), Durbin (1951), and Bradley (1955) for further results concerning the method of n rankings in the ANOVA models. The results considered here have been studied by Elteren and Noether (1959) and Sen (1967g, 1968b, 1971).

7.2.5. Tests for MANOVA in Two-Way Layouts Based on the Method of *n* Ranking

We shall now consider multivariate generalizations of the method of n rankings. For simplicity of presentation, we shall explicitly consider the generalization of the χ_r^2 test, though the generalization of the S_n test follows on the same line.

Suppose we are given q-variate data which form a complete two-way layout. Let $\mathbf{X}_{ij} = (X_{ij}^{(1)}, \ldots, X_{ij}^{(q)})'$ be the response from the plot in the ith block that received the jth treatment; $i = 1, \ldots, n$; $j = 1, \ldots, p\ (\geq 2)$. Assume that $\mathbf{X}_{ij}$ has the continuous cumulative distribution function $F_{ij}(\mathbf{x})$, $\mathbf{x} \in R^q$. We wish to test the null hypothesis.

$$H_0\colon F_{i1} = \cdots = F_{ip} = F_i, \quad \text{for} \quad i = 1, \ldots, n, \tag{7.2.52}$$

against the set of alternatives of the form

$$K\colon F_{ij}(\mathbf{x}) = F_i(\mathbf{x} - \boldsymbol{\theta}_j), \qquad j = 1, \ldots, p, \quad i = 1, \ldots, n, \tag{7.2.53}$$

where $\boldsymbol{\theta}_j = (\theta_j^{(i)}, \ldots, \theta_j^{(q)})'$, $j = 1, \ldots, p$. Under this type of alternative, we may write H_0: $\boldsymbol{\theta}_1 = \cdots = \boldsymbol{\theta}_p = \mathbf{0}$, while K states that at least one $\boldsymbol{\theta}_j$ is different from the rest. The parametric test for H_0 in (7.2.52) is based upon the assumption that $F_i(\mathbf{x}) = F(\mathbf{x} - \beta_i)$, where F is a multinormal c.d.f. with a positive-definite dispersion matrix $\boldsymbol{\Lambda}$. Thus, the homoscedasticity of the errors and the additivity of the block effects are presumed.

As in the univariate case, the assumptions of homoscedasticity of the errors (over the different blocks) and the multinormality of the F_i are not needed here. The ranking is made within each block and for each variate separately. Let then $r_{ij}^{(s)}$ be the rank of $X_{ij}^{(s)}$ among the observations $X_{it}^{(s)}$, $t = 1, \ldots, p$, for $j = 1, \ldots, p$, $i = 1, \ldots, n$, $s = 1, \ldots, q$. This leads to n matrices

$$\mathbf{R}_i = \begin{pmatrix} r_{i1}^{(1)} & \cdots & r_{ip}^{(1)} \\ \cdot & \cdot & \cdot \\ \cdot & \cdot & \cdot \\ \cdot & \cdot & \cdot \\ r_{i1}^{(q)} & \cdots & r_{ip}^{(q)} \end{pmatrix}, \qquad i = 1, \ldots, n.$$

Each row of $\boldsymbol{R}_i$ is some permutation of the numbers $1, \ldots, p$. Define $\mathbf{R}_1^*$ to be the matrix derived from $\mathbf{R}_i$ by permuting the columns in such a way that the numbers $1, \ldots, p$ appear in sequence in the first row. We say that two matrices (of the same order) $\mathbf{A}$ and $\mathbf{B}$ are permutationally equivalent if $\mathbf{A}$ can be obtained from $\mathbf{B}$ by a finite number of permutations of the columns of $\mathbf{B}$. Let $S(\mathbf{R}_i^*)$ be the set of matrices which are permutationally equivalent to $\mathbf{R}_i^*$, so that $S(\mathbf{R}_i^*)$ contains $p!$ elements.

The distribution of $\mathbf{R}_i^*$ over all its $(p!)^{q-1}$ realizations will depend upon the parent distributions, even under H_0 (unless the q variates are stochastically independent). However, given a particular realization of $\mathbf{R}_i^*$, the distribution of $\mathbf{R}_i$ will be uniform over $S(\mathbf{R}_i^*)$, when H_0 holds. That is, for any $\mathbf{R}_i^0 \in S(\mathbf{R}_i^*)$,

$$P\{\mathbf{R}_i = \mathbf{R}_i^0 \mid S(\mathbf{R}_i^*), H_0\} = 1/p! \quad \text{for all} \quad F_i. \tag{7.2.54}$$

Also, $\mathbf{R}_1, \ldots, \mathbf{R}_n$ are stochastically independent. Hence, if $(\mathbf{R}_1^0, \ldots, \mathbf{R}_n^0) \in [S(\mathbf{R}_1^*, \ldots, S(\mathbf{R}_n^*)]$ then

$$(7.2.55) \quad P\{\mathbf{R}_i = \mathbf{R}_i^0,\ i = 1, \ldots, n \mid S(\mathbf{R}_1^*), \ldots, S(\mathbf{R}_n^*), H_0\} = (p!)^{-n},$$

whatever be $F_1, \ldots, F_n$. We denote the conditional probability law in (7.2.55) by $\mathcal{T}_n$. Since $\mathcal{T}_n$ is completely known, the existence of (conditionally) distribution-free tests for H_0 follows from section 3.3. Note that for $q = 1$, $S(R_i^*)$ is the set of all $p!$ permutations of $(r_{i1}, \ldots, r_{ip})$, and hence, the conditional and the unconditional null distributions do agree with each other. Thus, tests based on the rank matrices $\boldsymbol{R}_1, \ldots, \boldsymbol{R}_n$ are unconditionally distribution-free.

Define $T_{n,j}^{(s)} = n^{-1} \sum_{i=1}^n r_{ij}^{(s)}$, $j = 1, \ldots, p$, $s = 1, \ldots, q$. It is then easily seen that

$$(7.2.56) \quad E_{\mathcal{T}_n}\{T_{n,j}^{(s)}\} = n^{-1} \sum_{i=1}^{n} E_{\mathcal{T}_n}\{r_{ij}^{(s)}\} = (p+1)/2, \qquad j = 1, \ldots, p, \quad s = 1, \ldots, q.$$

Also, it is easy to verify that

$$(7.2.57) \quad \mathrm{cov}_{\mathcal{T}_n}\{r_{ij}^{(s)}, r_{i'j'}^{(s')}\} = \frac{\delta_{ii'}(\delta_{jj'}p - 1)}{p(p-1)}\left[\sum_{t=1}^{p} r_{it}^{(s)} r_{it}^{(s')} - \frac{p(p+1)^2}{4}\right],$$

for $i, i' = 1, \ldots, n; j, j' = 1, \ldots, p$ and $s, s' = 1, \ldots, q$, where $\delta_{ii'}$ and $\delta_{jj'}$ are the usual Kronecker delta. This yields that

$$(7.2.58) \quad n\,\mathrm{cov}_{\mathcal{T}_n}\{T_{n,j}^{(s)}, T_{n,j'}^{(s')}\} = (\delta_{jj'} - p^{-1})\sigma_{ss'}(\mathcal{T}_n), \qquad j, j' = 1, \ldots, p, \quad s, s' = 1, \ldots, q,$$

where $\boldsymbol{\Sigma}(\mathcal{T}_n) = ((\sigma_{ss'}(\mathcal{T}_n)))$ is defined by

$$(7.2.59) \quad \sigma_{ss'}(\mathcal{T}_n) = \frac{1}{n(p-1)} \sum_{i=1}^{n} \sum_{t=1}^{p} r_{it}^{(s)} r_{it}^{(s')} - \frac{p(p+1)^2}{4(p-1)}, \quad s, s' = 1, \ldots, q.$$

Thus, if we write $\mathbf{T}_n = (T_{n,1}^{(1)}, \ldots, T_{n,1}^{(q)}, \ldots, T_{n,p-1}^{(1)}, \ldots, T_{n,p-1}^{(q)})$, we have

$$(7.2.60) \quad n V_{\mathcal{T}_n}(\mathbf{T}_n) = [\mathbf{I}_{p-1} - p^{-1}\ell_{p-1}\ell_{p-1}'] \otimes \boldsymbol{\Sigma}(\mathcal{T}_n); \qquad \ell_{p-1} = (1, \ldots, 1)'.$$

Then, considering a quadratic form in $\mathbf{T}_n - \frac{1}{2}(p+1)\ell_{q(p-1)}$, using the inverse of $V_{\mathcal{T}_n}(\mathbf{T}_n)$ as its discriminant and finally using the relation that $\sum_{j=1}^{p-1} T_{n,j}^{(s)} = -T_{n,p}^{(s)}$, $s = 1, \ldots, q$, we obtain the test statistic

$$(7.2.61) \quad \begin{aligned} \mathcal{L}_n &= n(\mathbf{T}_n - \tfrac{1}{2}(p+1)\ell_{q(p-1)})' V_n^{-1}(\mathcal{T}_n)(\mathbf{T}_n - \tfrac{1}{2}(p+1)\ell_{q(p-1)}) \\ &= n \sum_{s=1}^{q} \sum_{s'=1}^{q} \sigma^{ss'}(\mathcal{T}_n) \sum_{j=1}^{p} \left(T_{n,j}^{(s)} - \frac{p+1}{2}\right)\left(T_{n,j}^{(s')} - \frac{p+1}{2}\right), \end{aligned}$$

where $((\sigma^{ss'}(\mathcal{T}_n))) = \mathbf{\Sigma}^{-1}(\mathcal{T}_n)$ is assumed to be positive definite. (It will be seen later on that for large n and with some mild restrictions on F_i, $i = 1, \ldots, n$, $\mathbf{\Sigma}(\mathcal{T}_n)$ is positive definite, in probability. If, in practice, it is found to be singular, then elimination of the proper variables will remedy the problem.)

The permutation distribution of $\mathcal{L}_n$ can be computed to construct a conditionally distribution-free test for H_0. However, as in other permutation tests (cf. chapters 4–5), the labor increases prohibitively as the sample size increases. For this reason, we study the asymptotic permutation distribution of $\mathcal{L}_n$ with a view to simplifying the large-sample test procedure.

7.2.6. Asymptotic Permutation Distribution of $\mathcal{L}_n$

Define $F_{ij[s]}(x)$ and $F_{ij[s,s']}(x, y)$ to be the marginal c.d.f.'s of $X_{ij}^{(s)}$ and $(X_{ij}^{(s)}, X_{ij}^{(s')})$ for $s \neq s' = 1, \ldots, q, j = 1, \ldots, p, i = 1, \ldots, n$. Let then

$$\sigma_{ss'}(\mathbf{F}_i) = p(p+1)/12, \quad \text{if} \quad s = s' \quad (= 1, \ldots, q);$$

$$(7.2.62)\quad = \frac{1}{p-1}\left\{\sum_{j \neq j' \neq t=1}^{p} \int_{-\infty}^{\infty}\int_{-\infty}^{\infty} F_{it[s]}(x)F_{it'[s']}(y)\, dF_{ij[s,s']}(x, y) + \sum_{j \neq t=1}^{p} \int_{-\infty}^{\infty}\int_{-\infty}^{\infty} F_{it[s,s']}(x, y)\, dF_{ij[s,s']}(x, y) - \frac{p(p-1)^2}{4}\right\},$$

$$s \neq s' = 1, \ldots, q; \quad i = 1, \ldots, n;$$

$$(7.2.63)\qquad \mathbf{\Sigma}(\mathbf{F}_i) = ((\sigma_{ss'}(\mathbf{F}_i))), \quad i = 1, \ldots, n.$$

Theorem 7.2.4. *If the $F_{ij}(\mathbf{x})$ are all continuous, then*

$$\mathbf{\Sigma}(\mathcal{T}_n) - n^{-1}\sum_{i=1}^{n}\{\mathbf{\Sigma}(\mathbf{F}_i)\} \xrightarrow{p} \mathbf{0} \quad \text{as} \quad n \to \infty.$$

Outline of the Proof: For $s = s'$, it follows from (7.2.59) that

$$\sigma_{ss}(p_n) = \frac{p(p+1)}{12}$$

Hence, we shall only consider the case $s \neq s'$. Let $u(x)$ be 1 or 0 according as $x > 0$ or not. Then, $r_{ij}^{(s)} = 1 + \sum_{t=1}^{p} u(X_{ij}^{(s)} - X_{it}^{(s)})$. Hence, from (7.2.59), we have

$$(7.2.64)\qquad \sigma_{ss'}(\mathcal{T}_n) = n^{-1}\sum_{i=1}^{n}\sigma_{ss'}(\mathbf{R}_i^*),$$

where for $s \neq s'$,

$$(7.2.65)\quad \sigma_{ss'}(\mathbf{R}_i^*) = \frac{1}{p-1} E\left[\sum_{t=1}^{p}\{u(X_{ij}^{(s)} - X_{it}^{(s)}) + u(X_{ij}^{(s')} - X_{it}^{(s')})\} + \sum_{t \neq t'=1}^{p} u(X_{ij}^{(s)} - X_{it}^{(s)})u(X_{ij}^{(s')} - X_{it'}^{(s')}) - \frac{p(p+1)^2}{4} + p\right],$$

$i = 1, \ldots, n$. Now, by definition

$$E\{u(X_{ij}^{(s)} - X_{it}^{(s)})\} = 0, \qquad j = t = 1, \ldots, p$$

(7.2.66)
$$= \int_{-\infty}^{\infty} F_{it[s]}(x)\, dF_{ij[s]}(x), \qquad j \neq t = 1, \ldots, p;$$

(7.2.67)
$$\sum_{j \neq t = 1}^{p} \int_{-\infty}^{\infty} F_{it[s]}(x)\, dF_{ij[s]}(x) = \frac{p(p-1)}{2}$$

for $s = 1, \ldots, q$; $i = 1, \ldots, n$.

Also, for $s \neq s'$, we have

$$E\{u(X_{ij}^{(s)} - X_{it}^{(s)})u(X_{ij}^{(s')} - X_{it'}^{(s')})\}$$

(7.2.68)
$$= \int_{-\infty}^{\infty}\int_{-\infty}^{\infty} F_{it[s]}(x)F_{it'[s']}(y)\, dF_{ij[s,s']}(x, y), \qquad t \neq t' \neq j,$$

$$= \int_{-\infty}^{\infty}\int_{-\infty}^{\infty} F_{it[s,s']}(x, y)\, dF_{ij[s,s']}(x, y), \qquad t = t', \quad t \neq j,$$

$$= 0, \qquad \text{otherwise}.$$

Since the random variables $u(X_{ij}^{(s)} - X_{it}^{(s)})$ are all bounded, the proof follows by using the law of large numbers, after showing that (by virtue of (7.2.65)–(7.2.68)) the right-hand side of (7.2.64) has the expectation $n^{-1}\sum_{i=1}^{n} \mathbf{\Sigma}(\mathbf{F}_i)$. The details are left as Exercise 7.2.14.

Now, under H_0 in (7.2.51), $F_{ij} = F_i$ for $j = 1, \ldots, p$; $i = 1, \ldots, n$. We define $F_{ij[s]} = F_{i[s]}$ and $F_{ij[s,s']} = F_{i[s,s']}$ and let

(7.2.69)
$$\tau_{i,ss'} = 4\int_{-\infty}^{\infty}\int_{-\infty}^{\infty} F_{i[s,s']}(x, y)\, dF_{i[s,s']}(x, y) - 1$$

(7.2.70)
$$\rho_{gi,ss'} = 12\left\{\int_{-\infty}^{\infty}\int_{-\infty}^{\infty} F_{i[s]}(x)F_{i[s']}(y)\, dF_{i[s,s']}(x, y) - \tfrac{1}{4}\right\},$$

for $s, s' = 1, \ldots, q$, $i = 1, \ldots, n$, (where we note that for $s = s'$, both τ and ρ_g are equal to 1). Hence, from Theorem 7.2.4, we obtain the following.

Corollary 7.2.4. *Under H_0, $n^{-1}\sum_{i=1}^{n} \mathbf{\Sigma}(\mathbf{F}_i) = \mathbf{\Sigma}_n = ((\sigma_{n,ss'}))$, where*

(7.2.71)
$$\sigma_{n,ss'} = \begin{cases} p(p+1)/12, & s = s' = 1, \ldots, p, \\ n^{-1}\displaystyle\sum_{i=1}^{n}\left[\frac{p}{4}\tau_{i,ss'} + \frac{p(p-2)}{12}\rho_{gi,ss'}\right], & s \neq s' = 1, \ldots, q. \end{cases}$$

Thus, if at least one of the matrices $\overline{\mathbf{T}}_n = n^{-1}\sum_{i=1}^{n}((\tau_{i,ss'}))$ and $\overline{\mathbf{G}}_n = n^{-1}\sum_{i=1}^{n}((\rho_{gi,ss'}))$ is positive definite, so will be $\mathbf{\Sigma}_n$. Now, if the distributions

$F_i(\mathbf{x})$ $(i = 1, \ldots, n)$ are nonsingular in the sense that there is no $q - 1$ (or lower)-dimensional space containing the scatter of the points, both $\bar{T}_n$ and $\bar{G}_n$ will be positive definite, and hence $\mathbf{\Sigma}_n$ will also be so. In practice, this is no severe restriction on the F_i. Of course, we do not require all the $\mathbf{T}_i$'s or $\mathbf{G}_i$'s to be positive definite; a sufficient condition may be their positive definiteness for a subset of m i's where $m/n \to c > 0$ as $n \to \infty$. When the null hypothesis is not necessarily true, the situation can be studied under the condition that the distribution $F_{ij}(\mathbf{x})$ is nonsingular in the sense that no $q - 1$ (or lower) dimensional space contains the scatter of the points, for all $j = 1, \ldots, p$ and $i = 1, \ldots, n$. Then from (7.2.62) and some simple algebraic manipulations, it follows that $\mathbf{\Sigma}(\mathbf{F}_i)$ can be expressed as the sum of two matrices where the first one is positive definite and the other is positive semidefinite. Hence, $n^{-1} \sum_{i=1}^{n} \mathbf{\Sigma}(\mathbf{F}_i)$ will also be positive definite. Thus, if we assume that

$$(7.2.72) \qquad n^{-1} \sum_{i=1}^{n} \mathbf{\Sigma}(\mathbf{F}_i) \text{ is positive definite for all } n \geq n_0,$$

then from Theorem 7.2.4 it follows that

$$(7.2.73) \qquad \text{under (7.2.72)}, \quad \mathbf{\Sigma}(\mathcal{T}_n) \quad \text{is positive definite, in probability as } n \to \infty.$$

Theorem 7.2.5. *If the F_{ij} are all continuous and (7.2.72) holds, the joint permutation distribution of $\{n^{1/2}(T_{n,j}^{(s)} - (p + 1)/2),\ s = 1, \ldots, q,\ j = 1, \ldots, p - 1\}$ is asymptotically, in probability, a $q(p - 1)$-variate normal with null mean vector and dispersion matrix $(\mathbf{I}_{p-1} - p^{-1}\ell_{p-1}\ell'_{p-1}) \otimes \mathbf{\Sigma}(\mathcal{T}_n)$.*

Proof: Since $\sum_{j=1}^{p} T_{n,j}^{(s)} = p(p + 1)/2$ for $s = 1, \ldots, q$, it suffices to show that for arbitrary constants a_{sj}, $s = 1, \ldots, q$, $j = 1, \ldots, p$, such that $\sum_{j=1}^{p} a_{sj} = 0$, $s = 1, \ldots, q$, the permutation distribution of

$$(7.2.74) \qquad Y_n = n^{1/2} \sum_{s=1}^{q} \sum_{j=1}^{p} a_{sj} T_{n,j}^{(s)}$$

is asymptotically (in probability) normal. We may write

$$(7.2.75) \quad Y_n = n^{-1/2} \sum_{i=1}^{n} U_i; \qquad U_i = \sum_{s=1}^{q} \sum_{j=1}^{p} a_{sj} r_{ij}^{(s)}, \qquad i = 1, \ldots, n.$$

Using (7.2.56) and (7.2.57), we obtain after some simplifications that $E_{\mathcal{T}_n}(Y_n) = 0$ and

$$(7.2.76) \qquad n^{-1} \sum_{i=1}^{n} E_{\mathcal{T}_n}(U_i^2) = \sum_{s=1}^{q} \sum_{s'=1}^{q} \left\{ \sum_{j=1}^{p} a_{sj} a_{s'j} \right\} \sigma_{ss'}(\mathcal{T}_n).$$

Hence, by (7.2.72) (and (7.2.73)), $\sum_{i=1}^{n} E_{\mathcal{T}_n}\{U_i^2\} = O_p(n)$. Since all the $r_{ij}^{(s)}$ are bounded above by p, we have

$$(7.2.77) \qquad n^{-1} \sum_{i=1}^{n} E_{\mathcal{T}_n}\{|U_i|^3\} \leq n \left(p \sum_{s=1}^{q} \sum_{j=1}^{p} |a_{sj}| \right)^3.$$

Thus, for the independent random variables $\{U_1, \ldots, U_n\}$,

$$\lim_{n\to\infty} \frac{\sum_{i=1}^n E_{\mathcal{S}_n}[|U_i|^3]}{[\sum_{i=1}^n E_{\mathcal{S}_n}[U_i^2]]^{3/2}} = 0, \quad \text{in probability.} \tag{7.2.78}$$

Hence, the theorem follows from the classical central limit theorem under Liapounoff's condition. Q.E.D.

From Theorem 7.2.5, we readily arrive at the following:

Theorem 7.2.6. *Under* (7.2.72), *the permutation distribution of* $\mathfrak{L}_n$ *is asymptotically, in probability, chi-square with* $q(p-1)$ D.F.

Thus, for large values of n, the permutation test based on $\mathfrak{L}_n$ reduces as follows:

$$\begin{aligned} &\text{Reject } H_0 \text{ in (7.2.51)} \quad \text{if} \quad \mathfrak{L}_n \geq \chi^2_{q(p-1),\alpha}, \\ &\text{accept } H_0, \qquad\qquad\quad \text{if} \quad \mathfrak{L}_n < \chi^2_{q(p-1),\alpha}. \end{aligned} \tag{7.2.79}$$

The consistency of the test follows readily from the following:

(i)

$$\begin{aligned} \mathfrak{L}_n &\geq \max_{s,j} \left[n\left(T_{n,j}^{(s)} - \frac{p+1}{2}\right)^2 \Big/ \sigma_{ss}(\mathcal{S}_n) \right] \\ &= \frac{12n}{p(p+1)} \left\{ \max_{s,j} \left(T_{n,j}^{(s)} - \frac{p+1}{2}\right)^2 \right\}, \end{aligned} \tag{7.2.80}$$

(ii) by the law of large numbers,

$$T_{n,j}^{(s)} - \frac{p+1}{2} \xrightarrow{p} n^{-1} \sum_{i=1}^n \left\{ \int_{-\infty}^{\infty} F_i(x)\, dF_{ij}(x) - \frac{1}{2} \right\}, \tag{7.2.81}$$

where $F_i = p^{-1} \sum_{j=1}^p F_{ij}(x)$ (and hence the maximum (over s, j) of (7.2.81) is strictly positive for shift alternatives), and (iii) the right-hand side of (7.2.80) is $O_p(n)$, as compared to $\chi^2_{q(p-1),\alpha}$ which is $O(1)$.

7.2.7. Asymptotic Non-null Distribution of $[T_{n,j}^{(s)}, j = 1, \ldots, p, s = 1, \ldots, q]$

Let

$$\mu_{n,j}^{(s)} = n^{-1} \sum_{i=1}^n \left\{ \frac{1}{2} + \sum_{t=1}^p \int_{-\infty}^{\infty} F_{it[s]}(x)\, dF_{ij[s]}(x) \right\}, \tag{7.2.82}$$

$$j = 1, \ldots, p, \quad s = 1, \ldots, q,$$

$$\begin{aligned} \pi_{i,jt,j't'}^{(s,s')} &= P\{X_{ij}^{(s)} > X_{it}^{(s)}, X_{ij'}^{(s')} > X_{it'}^{(s')}\} \\ &\quad - P\{X_{ij}^{(s)} > X_{it}^{(s)}\} P\{X_{ij'}^{(s')} > X_{it'}^{(s')}\}, \end{aligned} \tag{7.2.83}$$

for $j \neq t,\ j' \neq t' = 1, \ldots, p,\ s, s' = 1, \ldots, q,\ i = 1, \ldots, n$ (where of course

$$\pi_{i,jt,jt}^{(s,s)} = P\{X_{ij}^{(s)} > X_{it}^{(s)}\}[1 - P\{X_{ij}^{(s)} > X_{it}^{(s)}\}], \qquad j \neq t = 1, \ldots, p, \quad s = 1, \ldots, q). \tag{7.2.84}$$

Also, let

$$\sigma_{n,js,j's'}^{*} = n^{-1} \sum_{i=1}^{n} \left\{ \sum_{t=1(\neq j)}^{p} \sum_{t'=1(\neq j')}^{p} \pi_{i,jtj't'} \right\}, \tag{7.2.85}$$

and let $\mathbf{\Sigma}_n^*$ be the $pq \times pq$ matrix whose elements are the $\sigma_{n,js,j's'}^*$. Let now $\mathbf{F}_i = (F_{i1}, \ldots, F_{ip})'$, $i = 1, \ldots, n$, $\mathbf{F}_n^* = (\mathbf{F}_1, \ldots, \mathbf{F}_n)$, $\mathcal{F}_n^* = \{\mathbf{F}_n^*$: rank of $\mathbf{\Sigma}_n^* \geq 1\}$ and let for any set $\mathbf{A} = ((a_{sj}))_{s=1,\ldots,q,\ j=1,\ldots,p}$ (of real constants), $H_n(x; \mathbf{A})$ be the actual distribution of $n^{1/2} \sum_{j=1}^{p} \sum_{s=1}^{q} a_{sj}(T_{n,j}^{(s)} - \mu_{n,j}^{(s)})/\gamma_n(\mathbf{A})$, where

$$\gamma_n^2(\mathbf{A}) = \sum_{j=1}^{p} \sum_{j'=1}^{p} \sum_{s=1}^{q} \sum_{s'=1}^{q} a_{sj} a_{s'j'} \sigma_{n\ js,j's'}^{*}.$$

Finally, let $\Phi(x)$ be the standard normal cumulative distribution function.

Theorem 7.2.7. *For all* $\mathbf{A}$ *for which* $\inf_n \gamma_n(A) > 0$, $\lim_{n\to\infty} H_n(x; \mathbf{A}) = \Phi(x)$, *uniformly in* x: $-\infty < x < \infty$.

Proof: As we have

$$E\{r_{ij}^{(s)}\} = \frac{1}{2} + \sum_{t=1}^{p} \int_{-\infty}^{\infty} F_{it[s]}(x)\, dF_{ij[s]}(x),$$

we may write

$$\begin{aligned} Y_n &= n^{1/2} \sum_{j=1}^{p} \sum_{s=1}^{q} a_{sj}(T_{n,j}^{(s)} - \mu_{n,j}^{(s)}) = n^{-1/2} \sum_{i=1}^{n} \sum_{s=1}^{p} \sum_{j=1}^{q} a_{sj}(r_{ij}^{(s)} - Er_{ij}^{(s)}) \\ &= n^{-1/2} \sum_{i=1}^{n} U_i; \qquad U_i = \sum_{j=1}^{p} \sum_{s=1}^{q} a_{sj}(r_{ij}^{(s)} - Er_{ij}^{(x)}), \qquad i = 1, \ldots, n. \end{aligned} \tag{7.2.86}$$

Since $U_1, \ldots, U_n$ are independent, finite-valued (as $|U_i| \leq p \sum_{j=1}^{p} \sum_{s=1}^{q} |a_{sj}|$) random variables with $EU_i = 0$, the asymptotic normality follows from the Berry-Esséen theorem (cf. section 2.7), provided $n^{-1} \sum_{i=1}^{n} EU_i^2 > 0$. Now, by definition, $n^{-1} \sum_{i=1}^{n} EU_i^2 = \gamma_n^2(\mathbf{A}) > 0$, by the hypothesis of the theorem. Hence the theorem.

It is useful to note that under H_0 in (7.2.51) $\mathbf{\Sigma}_n^* = [\mathbf{I}_p - (1/p)\ell_p \ell_p'] \otimes \mathbf{\Sigma}_n$, where $\mathbf{\Sigma}_n$ is defined by (7.2.71) and is assumed (in (7.2.72)) to be positive definite. Hence the rank of $\mathbf{\Sigma}_n^*$ is equal to $q(p - 1)$. Consequently, for any $\mathbf{A}$ (for which $\sum_{j=1}^{p} a_{sj} = 0, s = 1, \ldots, q$), $\gamma_n^2(\mathbf{A})$ is positive. Thus, the asymptotic distribution of the $T_{n,j}^{(k)}$ will be singular and of rank $q(p - 1)$.

7.2.8. Asymptotic Distribution of $\mathfrak{L}_n$ under Local Translation Alternatives

We consider now the sequence $\{K_n\}$ of translation alternatives

$$(7.2.87) \qquad K_n\colon F_{ij}(\mathbf{x}) = F_{n,ij}(\mathbf{x}) = F_i(\mathbf{x} + n^{-1/2}\boldsymbol{\alpha}_j), \qquad j = 1, \ldots, p, \quad i = 1, \ldots, n,$$

where $\boldsymbol{\alpha}_j = (\alpha_j^{(1)}, \ldots, \alpha_j^{(q)})'$, $j = 1, \ldots, p$ are q vectors of real constants. We set $\sum_{j=1}^{p} \boldsymbol{\alpha}_j = \mathbf{0}$ (without any loss of generality) and assume that under the alternative hypotheses $\boldsymbol{\alpha}_1, \ldots, \boldsymbol{\alpha}_n$ are not all equal (i.e., equal to $\mathbf{0}$). We denote by $\mathcal{F}_n^0$ the class of all absolutely continuous $\mathbf{F}_n = (F_1, \ldots, F_n)$ for which $\boldsymbol{\Sigma}_n$ (defined by (7.2.71)) is positive definite. Let us also define $\mathbf{a}_n = (a_n^{(1)}, \ldots, a_n^{(q)})$, where

$$(7.2.88) \qquad a_n^{(s)} = (p/n) \sum_{i=1}^{n} \int_{-\infty}^{\infty} f_{i[s]}(x)\, dF_{i[s]}(x), \qquad s = 1, \ldots, q.$$

Then, from (7.2.82) and (7.2.87), we obtain that

$$(7.2.89) \qquad \left| E\left\{ n^{1/2}\left(T_{n,j}^{(k)} - \frac{p+1}{2} \right) \middle| K_n \right\} - \alpha_j^{(s)} a_n^{(s)} \right| \to 0 \quad \text{as} \quad n \to \infty,$$

for $j = 1, \ldots, p$, $s = 1, \ldots, q$. Now, under the four different models considered in section 7.2.4, we have $\lim_{n\to\infty} a_n^{(s)} = a^{(s)}$, $s = 1, \ldots, q$. Also, from (7.2.83), (7.2.85), and (7.2.87), it follows that under $\{K_n\}$

$$(7.2.90) \qquad \boldsymbol{\Sigma}_n^* - \left(\mathbf{I}_p - \frac{1}{p} \ell_p \ell_p' \right) \otimes \boldsymbol{\Sigma}_n \xrightarrow[p]{} \mathbf{0}^{pq \times pq}, \quad \text{as} \quad n \to \infty,$$

where again under the four models in section 7.2.4, $\boldsymbol{\Sigma}_n \to \boldsymbol{\Sigma}$ as $n \to \infty$. Finally, from Theorem 7.2.4, we have $\boldsymbol{\Sigma}(\mathcal{T}_n) \xrightarrow{p} \boldsymbol{\Sigma}$. Hence, from Theorem 7.2.7, we obtain the following.

Theorem 7.2.8. *If* $\mathbf{F}_n \in \mathcal{F}_n^0$ $(n \geq 1)$, *then under* $\{K_n\}$ *in* (7.2.87), $\mathfrak{L}_n$, *defined by* (7.2.61), *has asymptotically a noncentral chi-square distribution with* $q(p-1)$ D.F. *and noncentrality parameter*

$$(7.2.91) \qquad \Delta_{\mathfrak{L}} = \sum_{s=1}^{q} \sum_{s'=1}^{q} \sigma^{ss'} a^{(s)} a^{(s')} \sum_{p=1}^{p} \alpha_j^{(s)} \alpha_j^{(s')},$$

where $((\sigma^{ss'})) = \boldsymbol{\Sigma}^{-1}$.

7.2.9. Asymptotic Power Efficiency of $\mathfrak{L}_n$

When the $\mathbf{X}_{ij}$ are distributed according to multinormal distributions (with a common positive definite dispersion matrix), the mostly used test is based on the likelihood ratio criterion. This test has been studied in detail (although in a more general setup of an arbitrary linear hypothesis) in Anderson (1958, pp. 187–210). Since the statistic involves the normal theory maximum likelihood estimators (which are also the least squares estimators), and these estimators are linear in the variables $\mathbf{X}_{ij}$, the asymptotic distribution theory of the statistic can also be studied when the assumption of normality or homogeneity of dispersion matrices may not hold. Assume that $\mathbf{X}_{ij}$ satisfies (7.2.87) and has finite moments of order $2 + \delta$, $\delta > 0$, for all i and j. Then, leaving the details (cf. Exercise 7.2.15), *we say that under* $\{K_n\}$ *in* (7.2.87) *and the models of section* 7.2.4, *the likelihood ratio statistic (computed under the assumed normality and homogeneity of the dispersion matrices of* $F_1, \ldots, F_n$) *(actually* $-2 \log$ *of the statistic* U_n) *has asymptotically the noncentral chi-square distribution with* $q(p-1)$ D. F. *and noncentrality parameter*

$$\Delta_U = \sum_{j=1}^{p} \boldsymbol{\alpha}_j' \bar{\boldsymbol{\Lambda}}^{-1} \boldsymbol{\alpha}_j, \tag{7.2.92}$$

where the $\boldsymbol{\alpha}_j$*'s are defined by* (7.2.87), $\bar{\boldsymbol{\Lambda}} = \lim_{n\to\infty} \bar{\boldsymbol{\Lambda}}_n$, $\bar{\boldsymbol{\Lambda}}_n = n^{-1} \sum_{i=1}^{n} \boldsymbol{\Lambda}_i$, *and* $\boldsymbol{\Lambda}_i$ *is the dispersion matrix of* F_i*; the existence of the limit of* $\bar{\boldsymbol{\Lambda}}_n$ *follows under the models in section* 7.2.4.

The A.R.E. of $\mathcal{L}_n$ with respect to U_n, as measured by $\Delta_{\mathcal{L}}/\Delta_U$, depends not only on $\bar{\boldsymbol{\Lambda}}$ and $\boldsymbol{\Sigma}$ but also on $\boldsymbol{\alpha}_j$, $j = 1, \ldots, p$. Some results for various special models are given in the form of Exercises 7.2.16–7.2.20. The results in sections 7.2.5–7.2.8 are due to Gerig (1969), and can be extended to (i) incomplete layouts and (ii) to general scores as in section 7.2.3. For brevity we do not discuss these complicated situations and refer to Sen (1969d).

7.3. RANKING AFTER ALIGNMENT

The method of n rankings does not require the additivity of the block effects or the homoscedasticity of the errors over the different blocks. Instead, it is based on the intrablock rankings, and hence it does not utilize the possible information contained in the interblock comparisons. This accounts for the usual low efficiency of the method of n rankings, particularly for small values of p (the number of plots per block). To illustrate this, let us consider the following. Suppose (X_{i1}, X_{i2}) is the response of the first and the second treatment in the ith block, $i = 1, \ldots, n$. The test by the method of n rankings is equivalent to the sign test based on $Z_i = X_{i1} - X_{i2}$, $i = 1, \ldots, n$ (cf. section 7.2). When the block effects are additive and the errors are homoscedastic, the A.R.E. of the sign test with respect to the parametric ANOVA test is only $2/\pi$ ($\doteqdot 0.63\%$) for normal parent c.d.f. On the other hand, if we use the Wilcoxon signed rank test based on $Z_1, \ldots, Z_n$, the A.R.E. moves up to $3/\pi$ (= 95%). This gain in A.R.E. can be attributed to the fact that whereas the signed rank statistic utilizes the information on interblock comparisons (i.e., comparisons of $Z_1, \ldots, Z_n$), the sign test does not do so. This suggests that in the model in (7.2.1) if β_i's can be eliminated by some intrablock transformations of the variables X_{ij}, and if a ranking procedure be based on these transformed variables (disregarding blocks), then the interblock comparisons of the observations are not overlooked and the method should improve the efficiency of the allied tests. A discouraging feature of this ranking procedure is that the transformed variables within each block are usually dependent and this introduces complications in the distribution theory of suitable rank tests. Nevertheless, it will be shown that under certain permutational invariance structure of the transformed variables, tests based on suitable rank statistics are actually conditionally distribution-free. In large samples, the procedure simplifies considerably and yields optimum tests too.

We start with the randomized block layout again. The response of the plot receiving the jth treatment and placed in the ith block is denoted by X_{ij} and is expressed as

$$X_{ij} = \mu + \alpha_i + \theta_j + \epsilon_{ij}, \qquad j = 1, \ldots, p, \quad i = 1, \ldots, n, \quad \Sigma\theta_j = 0, \tag{7.3.1}$$

where μ stands for the main effect, $\alpha_1, \ldots, \alpha_n$ for the block effects (may or may not be stochastic), $\theta_1, \ldots, \theta_p$ for the treatment effects (assumed to be nonstochastic) and ϵ_{ij}'s are the residual error components. It is assumed that $\boldsymbol{\epsilon}_i = (\epsilon_{i1}, \ldots, \epsilon_{ip})$, $i = 1, \ldots, n$ are independent random variables (vectors) having continuous (p-variate) cumulative distribution functions $G_i(x_1, \ldots, x_p)$, $i = 1, \ldots, n$, each of which is symmetric in its p arguments. This includes the conventional situation of independence and identity of distributions of all the np error components as a special case.

The alignment procedure essentially consists in eliminating the block effects α_i from X_{ij} by subtracting from each observation in a block, a translation invariant symmetric function of the block observations. Thus, let $\tilde{X}_i$ be a symmetric function of $X_{i1}, \ldots, X_{ip}$, such that $\tilde{X}_i + a$ is the same function of $X_{i1} + a, \ldots, X_{ip} + a$ for all $\infty < a < \infty$. Typical $\tilde{X}_i$ are the block average ($X_{i\cdot}$), the median of $X_{i1}, \ldots, X_{ip}$, the Winsorized or trimmed mean, etc. We then define the aligned observations as

$$Y_{ij} = X_{ij} - \tilde{X}_i, \qquad j = 1, \ldots, p, \quad i = 1, \ldots, n. \tag{7.3.2}$$

Since $\tilde{X}_i$ is symmetric in $X_{i1}, \ldots, X_{ip}$ and under H_0: $\boldsymbol{\theta} = \mathbf{0}$, $X_{i1}, \ldots, X_{ip}$ are symmetric dependent random variables, it follows that under H_0: $\boldsymbol{\theta} = \mathbf{0}$, $(Y_{i1}, \ldots, Y_{ip})$ are also interchangeable random variables for each $i = 1, \ldots, n$. On the other hand, if $\boldsymbol{\theta} \neq \mathbf{0}$, X_{ij}'s are interchangeable only after proper adjustments of locations. Hence, Y_{ij}'s are susceptible to shifts. Thus, the test for the null hypothesis H_0: $\boldsymbol{\theta} = \mathbf{0}$, reduces to that testing the interchangeability of $Y_{i1}, \ldots, Y_{ip}$ (for each $i = 1, \ldots, n$), against shift alternatives.

Let us now arrange the np $(= N)$ observations $Y_{11}, \ldots, Y_{np}$ in order of magnitude, and let R_{ij} be the rank of Y_{ij} for $j = 1, \ldots, p$, $i = 1, \ldots, n$. Since, the errors ϵ_{ij}'s are assumed to be interchangeable random variables, it follows that if the joint distribution of $(\epsilon_{i1}, \ldots, \epsilon_{ip})$ has the rank >1 or is absolutely continuous, then ties among Y_{ij}'s may be neglected, in probability. In the sequel, it will be assumed that $p > 2$ and the rank of the joint c.d.f. of $Y_{i1}, \ldots, Y_{ip}$ is >1. For $p = 2$, we can apply the one-sample location test (based on $X_{i1} - X_{i2}$, $i = 1, \ldots, n$) studied in detail in chapter 4. Since here the ranking is made on the aligned observations Y_{ij}'s, we term the procedure as the ranking after alignment.

For each N, we define a sequence of rank scores

$$\text{(7.2.3)} \quad \mathbf{E}_N = (E_{N,1}, \ldots, E_{N,N}); \qquad E_{N,j} = J_N(j/(N+1)), \qquad 1 \leq j \leq N,$$

where the function J_N is defined in accordance with the Chernoff-Savage convention, i.e., we assume that J_N satisfies the conditions (a), (b), and (c) of section 3.6.1. We define $Z_N^{(j)} = 1$, if the αth smallest observation among the N values of Y_{ij}'s is from the jth treatment and let $Z_{N\alpha}^{(j)} = 0$, otherwise, for $\alpha = 1, \ldots, N$, $j = 1, \ldots, p$. Then, the tests to be considered are based on the statistic

$$\text{(7.3.4)} \quad \mathbf{T}_N = (T_{N,1}, \ldots, T_{N,p})'; \qquad T_{N,j} = n^{-1}\sum_{\alpha=1}^{N} Z_{N\alpha}^{(j)} E_{N,\alpha}, \quad j = 1, \ldots, p.$$

7.3.1. The Test Statistic and Its Rationality

Under the null hypothesis H_0: $\boldsymbol{\theta} = \mathbf{0}$, the joint c.d.f. of $(Y_{i1}, \ldots, Y_{ip})$ remains invariant under the $p!$ permutations of the coordinates among themselves, for each $i = 1, \ldots, n$. Thus, there exists a group $\mathcal{G}_n$ of $(p!)^n$ intrablock permutations which maps the sample space onto itself and leaves the distribution of the sample point invariant (under H_0). Hence, conditioned on the n sets of ordered observations $(Y_{i(1)}, \ldots, Y_{i(p)})$, $i = 1, \ldots, n$ (where $Y_{i(\alpha)}$ is the αth smallest observations among $Y_{i1}, \ldots, Y_{ip}$), the conditional distribution of $(Y_{i1}, \ldots, Y_{ip})$, $i = 1, \ldots, n$, over the $(p!)^n$ intrablock permutations will be equally likely, each realization having the common conditional probability $(p!)^{-n}$. Let us denote this permutational (conditional) probability measure by $\mathcal{T}_n$. Then, by simple arguments, it follows that

$$\text{(7.3.5)} \qquad E_{\mathcal{T}_n}(T_N) = \bar{E}_N \ell_p; \qquad \bar{E}_N = N^{-1}\sum_{\alpha=1}^{N} E_{N,\alpha}.$$

Let us also define

$$E_{N,R_{i\cdot}} = p^{-1}\sum_{j=1}^{p} E_{N,R_{ij}}, \qquad i = 1, \ldots, n;$$

(7.3.6) $$\sigma^2(\mathcal{S}_n) = [n(p-1)]^{-1}\sum_{i=1}^{n}\sum_{j=1}^{p}(E_{N,R_{ij}} - E_{N,R_{i\cdot}})^2.$$

Then, by routine computations, we have

(7.3.7) $$nE_{\mathcal{S}_n}\{[\mathbf{T}_N - \bar{E}_N\ell_p][\mathbf{T}_N - \bar{E}_N\ell_p]'\} = [\mathbf{I}_p - p^{-1}\ell_p\ell_p']\sigma^2(\mathcal{S}_n).$$

Thus, considering a quadratic form in $\mathbf{T}_N - \bar{E}_N\ell_p$ with the generalized inverse of (7.3.7) as its discriminant, we obtain the statistic

(7.3.8) $$S_N = n\sum_{j=1}^{p}(T_{N,j} - \bar{E}_N)^2/\sigma^2(\mathcal{S}_n),$$

which we consider as a test statistic.

For small values of n, the (exact) permutation distribution of S_N (under $\mathcal{G}_n$) can be computed and a similar (conditional) test of size ϵ $(0 < \epsilon < 1)$ can be constructed. The labor increases prohibitively with the increase in n, and for this reason, we consider the following large-sample approach.

7.3.2. Large-Sample Permutation Distribution of S_N

We define the marginal c.d.f. of Y_{ij} by $F_{i[j]}(x)$ and the joint c.d.f. of $(Y_{ij}, Y_{ij'})$ by $F_{i[j,j']}(x, y)$, for $j \neq j' = 1, \ldots, p$, $i = 1, \ldots, n$. Also, let

(7.3.9) $$H_i(x) = p^{-1}\sum_{j=1}^{p} F_{i[j]}(x), \quad i = 1, \ldots, n; \qquad \bar{H}_N(x) = n^{-1}\sum_{i=1}^{n} H_i(x);$$

(7.3.10)
$$H_i^*(x, y) = \frac{2}{p(p-1)}\sum_{j<j'} F_{i[j,j']}(x, y), \quad i = 1, \ldots, n;$$
$$\bar{H}_N^*(x, y) = n^{-1}\sum_{i=1}^{n} H_i^*(x, y).$$

Further, let

(7.3.11)
$$F_{[j]n}(x) = n^{-1}\,[\text{number of } Y_{ij} \le x,\ i = 1, \ldots, n], \qquad j = 1, \ldots, p;$$

(7.3.12) $$H_N(x) = p^{-1}\sum_{j=1}^{p} F_{[j]n}(x);$$

(7.3.13) $$F_{[j,j']n}(x, y) = n^{-1}\,[\text{number of } (Y_{ij}, Y_{ij'}) \le (x, y),\ i = 1, \ldots, n], \quad j \neq j' = 1, \ldots, p,$$

(7.3.14) $$H_N^*(x, y) = \binom{p}{2}^{-1}\sum_{j<j'} F_{[j,j']n}(x, y).$$

We define (as in section 3.6.1) $J(u) = \lim_{N\to\infty} J_N(u)$: $0 < u < 1$ and assume that

$$\lim_{N\to\infty}\left[\int_0^1 \{J_N(u) - J(u)\}^2\,du\right] = 0; \tag{7.3.15}$$

in addition, assumptions (a), (b), and (c) of section 3.6.1 are imposed. (7.3.15) implies that

$$\begin{aligned}\int_{-\infty}^{\infty}\int_{-\infty}^{\infty}&\left\{J_N\left(\frac{N}{N+1}H_N(x)\right)J_N\left(\frac{N}{N+1}H_N(y)\right)\right.\\ &\left.-J\left(\frac{N}{N+1}H_N(x)\right)J\left(\frac{N}{N+1}H_N(y)\right)\right\}dF_{[j,j']n}(x,y)\\ &= o_p(1) \quad \text{for all} \quad j \neq j' = 1,\ldots,p\end{aligned} \tag{7.3.16}$$

Finally, we define $\bar{F}^*_{n[j,j']}(x, y) = n^{-1}\sum_{i=1}^n F_{i[j,j']}(x, y)$, and let

$$\nu_{jj',n} = \int_{-\infty}^{\infty}\int_{-\infty}^{\infty} J(\bar{H}_N(x))J(\bar{H}_N(y))\,d\bar{F}^*_{n[j,j']}(x, y), \quad j, j' = 1,\ldots,p \tag{7.3.17}$$

Note that $\nu_{jj,n} = \int_{-\infty}^{\infty} J^2(\bar{H}_N(x))\,d\bar{F}_{n[j]}(x)$, $j = 1,\ldots,p$, where $\bar{F}_{n[j]}(x) = n^{-1}\sum_{i=1}^n F_{i[j]}(x)$. We let then $\boldsymbol{\nu}_n = ((\nu_{n[jj]}))$, and assume that

$$\textit{rank of } \boldsymbol{\nu}_n \geq 2 \textit{ for all } n \geq n_0. \tag{7.3.18}$$

Lemma 7.3.1. *Let* $A^2 = \int_0^1 J^2(u)\,du$ *and* $\bar{\nu}_n = (p/2)^{-1}\sum_{j'>j}\nu_{jj',n}$. *Then, under* (7.3.18)

$$A^2 - \bar{\nu}_n > 0 \quad \textit{for all} \quad n \geq n_0.$$

Proof: By definition

$$\begin{aligned}\sum_{j=1}^p \nu_{jj,n} &= \sum_{j=1}^p \int_{-\infty}^{\infty} J^2(\bar{H}_N(x))\,d\bar{F}_{n[j]}(x)\\ &= p\int_{-\infty}^{\infty} J^2(\bar{H}_N(x))\,d\bar{H}_N(x) = pA^2.\end{aligned} \tag{7.3.19}$$

Hence we have

$$\begin{aligned}A^2 - \bar{\nu}_n &= \frac{1}{p}\sum_{j=1}^p \nu_{jj,n} - \frac{1}{p(p-1)}\sum_{j\neq j'=1}^p \nu_{jj',n}\\ &= \frac{p}{p-1}\left\{\frac{1}{p}\sum_{j=1}^p \nu_{jj,n} - \frac{1}{p^2}\sum_{j=1}^p\sum_{j'=1}^p \nu_{jj',n}\right\}.\end{aligned} \tag{7.3.20}$$

Also by the Cauchy-Schwarz inequality

$$\left|\frac{1}{p^2}\sum_{j=1}^p\sum_{j'=1}^p \nu_{jj',n}\right| \leq \left\{\frac{1}{p}\sum_{j=1}^p [\nu_{jj,n}]^{1/2}\right\}^2, \tag{7.3.21}$$

where the equality sign holds iff $\nu_{jj,n} = \{\nu_{jj,n}\nu_{j'j',n}\}^{1/2}$ for all $j \neq j' = 1, \ldots, p$. Now, by (7.3.18), there exists at least one pair (j,j'), $j \neq j'$, such that $\nu_{jj,n}\nu_{j'j',n} - \{\nu_{jj',n}\}^2 > 0$, and hence, the strict inequality sign in (7.3.21) holds. Also,

$$\left[\frac{1}{p}\sum_{j=1}^{p}\{\nu_{jj,n}\}^{1/2}\right]^2 \leq \frac{1}{p}\sum_{j=1}^{p}\nu_{jj,n} = A^2, \tag{7.3.22}$$

where the equality sign holds only when $\nu_{jj,n}$, $j = 1, \ldots, p$, are all equal. Hence, the lemma follows from (7.3.20), (7.3.21), and (7.3.22), with the strict inequality in (7.3.21) (by virtue of (7.3.18)). Q.E.D.

Lemma 7.3.2. *Under the conditions* (*a*), (*b*), *and* (*c*) *of section* 3.6.1, *and* (7.3.15) *and* (7.3.18), $\sigma^2(\mathcal{T}_n)$, *defined by* (7.3.6), *is stochastically equivalent to* $A^2 - \bar{\nu}_n$, *and hence is positive in probability as* $n \to \infty$.

Proof: We may rewrite (7.3.6) as

$$N^{-1}\sum_{\alpha=1}^{N} E_{N,\alpha}^2 - \frac{1}{np(p-1)}\sum_{i=1}^{n}\sum_{j\neq j'=1}^{p} E_{N,R_{ij}}E_{N,R_{ij'}}. \tag{7.3.23}$$

Now, by (7.3.15) it is easy to show that

$$N^{-1}\sum_{\alpha=1}^{N} E_{N,\alpha}^2 = \int_0^1 J_N^2(u)\,du \to \int_0^1 J^2(u)\,du = A^2. \tag{7.3.24}$$

Again, the second term of (7.2.23) can be written (after using (7.3.16)) as

$$\int_{-\infty}^{\infty}\int_{-\infty}^{\infty} J\left(\frac{N}{N+1}H_N(x)\right)J\left(\frac{N}{N+1}H_N(y)\right)dF_{[j,j']n}(x, y) + o_p(1), \tag{7.3.25}$$

for $j \neq j' = 1, \ldots, p$. Since $H_N(x)$ is the average of the p empirical c.d.f.'s $F_{[j]n}$, $j = 1, \ldots, p$, and as on each of these c.d.f.'s, (2.11.70) holds, it follows along the same line as in Theorems 4.4.1 and 5.4.2 that (7.3.25) is stochastically equivalent to $\nu_{jj,n}$, for all $j \neq j' = 1, \ldots, p$. (The details are left as Exercise 7.3.1.) Hence, the first part of the lemma follows from (7.3.23) and (7.3.24) while the second part is a direct consequence of Lemma 7.3.1. Q.E.D.

Theorem 7.3.3. *Under the conditions of Lemma* 7.3.2, *the permutation distribution of* S_N *asymptotically, in probability, reduces to a chi-square distribution with* $p - 1$ D.F.

Proof: Let us first prove that under $\mathcal{T}_n$, $\{n^{1/2}(T_{N,j} - \bar{E}_N), j = 1, \ldots, p\}$ has asymptotically, in probability, a multinormal distribution of rank $p - 1$. As we have $\sum_{j=1}^{p}(T_{N,j} - \bar{E}_N) = 0$, it follows that the rank of $\mathbf{T}_N$ may be at most equal to $p - 1$. So, if we can show that for any non-null

$\boldsymbol{\delta} = (\delta_1, \ldots, \delta_{p-1})$, $\sum_{j=1}^{p-1} \delta_j n^{1/2}(T_{N,j} - \bar{E}_N)$ has a nondegenerate and asymptotically normal (permutation) distribution, the desired result will follow. Now, we can also write for $\boldsymbol{\delta} \neq \mathbf{0}$,

$$n^{1/2} \sum_{j=1}^{p-1} \delta_j (T_{N,j} - \bar{E}_N) = n^{1/2} \sum_{j=1}^{p} \delta'_j T_{N,j}, \tag{7.3.26}$$

where $\sum_{j=1}^{p} \delta'_j = 0$ and $(\delta'_1, \ldots, \delta'_p) \neq \mathbf{0}$. Thus, it is sufficient to show that $n^{1/2}$ times any arbitrary contrast in $\mathbf{T}_N$ has asymptotically a normal (permutation) distribution. Now, using condition (*b*) of section 3.6.1 we may write (7.3.26) as

$$n^{-1/2} \sum_{i=1}^{n} \sum_{j=1}^{p} \delta'_j J\left(\frac{R_{ij}}{N+1}\right) + o_p(1). \tag{7.3.27}$$

Let us then define

$$Y_{N,i}(\mathbf{R}_N) = \sum_{j=1}^{p} \delta'_j J\left(\frac{R_{ij}}{N+1}\right), \qquad i = 1, \ldots, n. \tag{7.3.28}$$

It thus follows from (7.3.27) and (7.3.28) that we are only to show that $n^{-1/2} \sum_{i=1}^{n} Y_{N,i}(\mathbf{R}_N)$ has asymptotically (under $\mathcal{T}_n$) a nondegenerate normal distribution, in probability. Now, under $\mathcal{T}_n$, there are $p!$ equally likely permutations of $(R_{i1}, \ldots, R_{ip})$ among themselves, and hence $Y_{N,i}(\mathbf{R}_N)$ can have only $p!$ equally likely permuted values, each with probability $1/p!$. Thus,

$$E_{\mathcal{T}_n}\{Y_{N,i}(\mathbf{R}_N)\} = \sum_{j=1}^{p} \delta'_j \left(p^{-1} \sum_{l=1}^{p} J\left(\frac{R_{il}}{N+1}\right)\right) = 0, \tag{7.3.29}$$

$$\begin{aligned} &E_{\mathcal{T}_n}\{Y^2_{N,i}(\mathbf{R}_N)\} \\ &= \sum_{j=1}^{p} (\delta'_j)^2 \left\{\frac{1}{p-1} \sum_{l=1}^{p} \left[J\left(\frac{R_{il}}{N+1}\right) - \frac{1}{p} \sum_{s=1}^{p} J\left(\frac{R_{is}}{N+1}\right)\right]^2\right\}, \end{aligned} \tag{7.3.30}$$

for $i = 1, \ldots, n$. Since the within-block permutations are independent for different blocks, $Y_{N,1}, \ldots, Y_{N,n}$ are all mutually independent (under $\mathcal{T}_n$). Hence, to prove the asymptotic normality, it suffices to show that

$$\frac{\sum_{i=1}^{n} E_{\mathcal{T}_n}\{|Y_{N,i}(\mathbf{R}_N)|^3\}}{[\sum_{i=1}^{n} E_{\mathcal{T}_n}\{Y^2_{N,i}(\mathbf{R}_N)\}]^{3/2}} = o_p(1), \tag{7.3.31}$$

For this, we note that by (7.3.30),

$$\begin{aligned} &\frac{1}{n} \sum_{i=1}^{n} E_{\mathcal{T}_n}\{Y^2_{N,i}(\mathbf{R}_N)\} \\ &= \sum_{j=1}^{p} (\delta'_j)^2 \frac{p}{p-1} \left\{\frac{1}{N} \sum_{i=1}^{n} \sum_{j=1}^{p} \left[J\left(\frac{R_{ij}}{N+1}\right) - \frac{1}{p} \sum_{l=1}^{p} J\left(\frac{R_{il}}{N+1}\right)\right]^2\right\} \\ &\underset{p}{\sim} \left[\sum_{j=1}^{p} (\delta'_j)^2\right](A^2 - \bar{\nu}_n) > 0, \qquad \text{by Lemmas 7.3.1 and 7.3.2.} \end{aligned} \tag{7.3.32}$$

Also, using condition (c) of section 3.6.1, we have

$$n^{-1}\sum_{i=1}^{n} E_{\mathcal{S}_n}\{|Y_{N,i}(\mathbf{R}_N)|^3\}$$

(7.3.33)
$$\le \sum_{j=1}^{p}|\delta_j'|\left[\max_{1<\alpha<N}\left|J\left(\frac{\alpha}{N+1}\right)\right|\right]\left[\frac{1}{n}\sum_{i=1}^{n} E_{\mathcal{S}_n}\{Y_{N,i}^2(\mathbf{R}_N)\}\right]$$
$$\le KN^{-\frac{1}{2}-\delta'}\sum_{j=1}^{p}|\delta_j'|\cdot\left[\frac{1}{n}\sum_{i=1}^{n} E_{\mathcal{S}_n}\{Y_{N,i}^2(\mathbf{R}_N)\}\right].$$

Now, (7.3.31) follows directly from (7.3.32) and (7.3.33). The rest of the proof of the theorem follows from Theorem 2.8.2. Q.E.D.

By virtue of Theorem 7.3.3, the (exact) permutation test based on S_N can be approximated asymptotically by the following test function:

(7.3.34)
$$\phi(S_N) = \begin{cases} 1, & S_N \ge \chi^2_{p-1,\epsilon}, \\ 0, & S_N < \chi^2_{p-1,\epsilon}. \end{cases}$$

For the study of the large-sample power properties of the test based on S_N, we need to study the asymptotic distribution theory of $\mathbf{T}_N$, when H_0 may not hold. This we consider below.

7.3.3. Asymptotic Non-null Distribution of $\mathbf{T}_N$

Let us define

(7.3.35)
$$\mu_{N,j} = \int_{-\infty}^{\infty} J[\bar{H}_N(x)]\,d\bar{F}_{n[j]}(x), \quad j = 1, \ldots, p;$$
$$\boldsymbol{\mu}_N = (\mu_{N,1}, \ldots, \mu_{N,p})',$$

(7.3.36)
$$\beta_{jj,kk'}^{(i)} = \iint_{-\infty<x<y<\infty} F_{i[j]}(x)[1 - F_{i[j]}(y)]J'[\bar{H}_N(x)]J'[\bar{H}_N(y)]\,d\bar{F}_{n[k]}(x)$$
$$\times\, d\bar{F}_{n[k']}(y) + \iint_{-\infty<x<y<\infty} F_{i[j]}(x)[1 - F_{i[j]}(y)]$$
$$\times\, J'[\bar{H}_N(x)]J'[\bar{H}_N(y)]\,d\bar{F}_{n[k']}(x)\,d\bar{F}_{n[k]}(y);$$

(7.3.37)
$$\beta_{jj',kk'}^{(i)} = \int_{-\infty}^{\infty}\int_{-\infty}^{\infty} [F_{i[j,j']}(x, y) - F_{i[j]}(x)F_{i[j']}(y)]$$
$$\times\, J'[\bar{H}_N(x)]J'[\bar{H}_N(y)]\,d\bar{F}_{n[k]}(x)\,d\bar{F}_{n[k']}(y),$$

for $j \ne j' = 1, \ldots, p$, $k, k' = 1, \ldots, p$, $i = 1, \ldots, n$, and let

(7.3.38)
$$\beta_{jj',n}^{*} = \frac{1}{np}\sum_{i=1}^{n}\left\{\sum_{k=1}^{p}\sum_{k'=1}^{p}[\beta_{jj',kk'}^{(i)} + \beta_{kk',jj'}^{(i)} - \beta_{kj',jk'}^{(i)} - \beta_{jk',kj'}^{(i)}]\right\},$$
$$j, j' = 1, \ldots, p,$$

(7.3.39)
$$\boldsymbol{\beta}_n^{*} = ((\beta_{jj',n}^{*})).$$

Theorem 7.3.4. *Under the conditions (a), (b), and (c) of section 3.6.1, the (random) vector* $N^{1/2}[(T_{N,j} - \mu_{N,j}),\ j = 1, \ldots, p]$ *has asymptotically a multinormal distribution with null mean vector and dispersion matrix* $\boldsymbol{\beta}_N^*$.

Proof: Proceeding as in the proof of Theorems 4.4.3 and 5.5.1, we write

$$T_{N,j} = \mu_{N,j} + B_{1,N}^{(j)} + B_{2,N}^{(j)} + \sum_{r=1}^{4} C_{r,N}^{(j)}, \qquad j = 1, \ldots, p, \tag{7.3.40}$$

where

$$B_{1,N}^{(j)} = \int_{-\infty}^{\infty} J[\bar{H}_N(x)]\, d[F_{[j]n}(x) - \bar{F}_{n[j]}(x)], \tag{7.3.41}$$

$$B_{2,N}^{(j)} = \int_{-\infty}^{\infty} [H_N(x) - \bar{H}_N(x)]J'[\bar{H}_N(x)]\, d\bar{F}_{n[j]}(x), \tag{7.3.42}$$

$$C_{1,N}^{(j)} = -\frac{1}{N+1}\int_{-\infty}^{\infty} H_N(x)J'[\bar{H}_N(x)]\, dF_{[j]n}(x), \tag{7.3.43}$$

$$C_{2,N}^{(j)} = \int_{-\infty}^{\infty} [H_N(x) - \bar{H}_N(x)]J'[\bar{H}_N(x)]\, d[F_{[j]n}(x) - \bar{F}_{n[j]}(x)], \tag{7.3.44}$$

$$\begin{aligned} C_{3,N}^{(j)} = \int_{-\infty}^{\infty} \Big[& J\Big(\frac{N}{N+1} H_N(x)\Big) - J[\bar{H}_N(x)] \\ & - \Big(\frac{N}{N+1} H_N(x) - \bar{H}_N(x)\Big) J'[\bar{H}_N(x)] \Big]\, dF_{[j]n}(x), \end{aligned} \tag{7.3.45}$$

and

$$C_{4,N}^{(j)} = \int_{-\infty}^{\infty} \Big[J_N\Big(\frac{N}{N+1} H_N(x)\Big) - J\Big(\frac{N}{N+1} H_N(x)\Big)\Big]\, dF_{[j]n}(x), \qquad j = 1, \ldots, p. \tag{7.3.46}$$

Now, by condition (*b*) of section 3.6.1, $C_{4,N}^{(j)} = o_p(N^{-1/2})$, uniformly in the class of all distributions, and $j = 1, \ldots, p$, while in the appendix (cf. section 10.4) it is shown that $C_{1,N}^{(j)}$, $C_{2,N}^{(j)}$, and $C_{3,N}^{(j)}$ are also $o_p(N^{-1/2})$ for all $j = 1, \ldots, p$. Thus,

$$|\sqrt{N}\,(T_{N,j} - \mu_{N,j}) - \sqrt{N}\,(B_{1,N}^{(j)} + B_{2,N}^{(j)})| \xrightarrow{p} 0, \qquad j = 1, \ldots, p. \tag{7.3.47}$$

Hence, it suffices to show that $[\sqrt{N}\,(B_{1,N}^{(j)} + B_{2,N}^{(j)}), j = 1, \ldots, p]$ satisfies the conditions of the central limit theorem. Now, integrating (7.3.41) by parts and adding it to (7.3.42), we obtain that

$$B_{1,N}^{(j)} + B_{2,N}^{(j)} = \frac{1}{n}\sum_{i=1}^{n} \left\{\frac{1}{p}\sum_{k=1}^{p} [B_{j:k}^{(i)}(Y_{ij}) - B_{k:j}^{(i)}(Y_{ik})]\right\}, \tag{7.3.48}$$

where

$$(7.3.49) \quad B_{k:q}^{(i)}(X_{ik}) = \int_{-\infty}^{\infty} [c(x - X_{ik}) - F_{i[k]}(x)] J'[\bar{H}_N(x)]\, d\bar{F}_{n[q]}(x),$$

for $k, q = 1, \ldots, p$, $i = 1, \ldots, n$. Now, for any $\boldsymbol{\delta} \neq \mathbf{0}$, the random variable $N^{1/2} \sum_{j=1}^{p} \delta_j [B_{1,N}^{(j)} + B_{2,N}^{(j)}]$ can be written as

$$(7.3.50) \quad N^{1/2} \sum_{k=1}^{p} \sum_{k'=1}^{q} \delta_{kk'}^{*} \left\{ \frac{1}{n} \sum_{i=1}^{n} B_{k:k'}^{(i)}(Y_{ij}) \right\},$$

where $\delta^* = (\delta_{11}^*, \ldots, \delta_{pp}^*)$ is also non-null. Hence, if we write

$$(7.3.51) \quad B(\mathbf{Y}_i, \boldsymbol{\delta}^*) = \sum_{k=1}^{p} \sum_{k'=1}^{p} \delta_{k:k'}^{*} B_{k:k'}^{(i)}(Y_{ik}), \qquad i = 1, \ldots, n,$$

(7.3.50) can be expressed as an average over n independent random variables. Using condition (c) of section 3.6.1 and following a few simple steps (cf. Exercise 7.3.2), it can be shown that

$$\sup_n n^{-1} \sum_{i=1}^{n} E\{|B(\mathbf{Y}_i, \boldsymbol{\delta}^*)|^{2+\delta}\} < \infty, \qquad \delta > 0.$$

Hence, the asymptotic normality of (7.3.50) will follow from the Liapounoff version of the central limit theorem, whenever $n^{-1} \sum_{i=1}^{n} E\{B^2(\mathbf{Y}_i, \boldsymbol{\delta}^*)\}$ is positive in the limit. The computation of the covariance terms in (7.3.38) follows from (7.3.48), (7.3.49), and by an application of the Fubini's theorem (cf. Theorem 2.10.6); the details are left as Exercise 7.3.3.

Under the null hypothesis, $F_{i[1]} = \cdots = F_{i[p]} = F_i$, $i = 1, \ldots, n$, and hence, we obtain from Theorem 7.3.4 that $\mu_{N,j} = \int_0^1 J(u)\, du$, $j = 1, \ldots, p$ and

$$(7.3.52) \quad n\beta_{jj'}^* = [(\delta_{jj'}p - 1)/p](A^2 - \bar{\nu}_n), \qquad j, j' = 1, \ldots, p;$$

where $A^2 = \int_0^1 J^2(u)\, du$ and $\bar{\nu}_n = \int_{-\infty}^{\infty} \int_{-\infty}^{\infty} J[\bar{H}_N(x)] J[\bar{H}_N(y)]\, d\bar{H}_N^*(x, y)$. This along with Theorem 7.3.2 leads to the following corollary. ◁

Corollary 7.3.4.1. *Under H_0: $\boldsymbol{\theta} = \mathbf{0}$ and the conditions of Theorem 7.3.4, S_N has asymptotically (unconditionally) the chi-square distribution with $p - 1$ D.F.*

Thus a large-sample (unconditional) test can also be based on S_N, using the rejection rule in (7.3.34).

7.3.4. Asymptotic Efficiency of S_N

We shall now compare the A.R.E. of the test based on S_N with respect to the tests in section 7.2 and the standard parametric test. For this, we conceive the following sequence $\{K_N\}$ of alternative hypotheses

$$(7.3.53) \quad K_N: \boldsymbol{\theta} = \boldsymbol{\theta}_N = N^{-1/2}\boldsymbol{\tau}, \qquad \boldsymbol{\tau} \perp \boldsymbol{\ell}_p.$$

Thus, under K_N,

(7.3.54) $$F_{i[j]}(x) = F_{i[j],N}(x) = F_i(x - N^{-1/2}\tau_j), \qquad j = 1, \ldots, p,$$

(7.3.55) $$F_{i[j,j']}(x, y) = F_{i[j,j'],N}(x, y) = F_i^*(x - N^{-1/2}\tau_j, y - N^{-1/2}\tau_{j'}), \qquad j \neq j' = 1, \ldots, p,$$

where $F_i^*(u, v) \equiv F_i^*(v, u)$. Now, under the models of section 7.2.4, (7.3.54), and (7.3.55), (*a*) $\bar{H}_N(x)$ is asymptotically equal to $F(x)$, which is the limit of $\bar{F}_n(x) = (1/n)\sum_{i=1}^n F_i(x)$, and (*b*) $\bar{H}_N^*(x, y)$ is asymptotically equal to $F^*(x, y)$ which is the limit of $\bar{F}_n^*(x, y) = n^{-1}\sum_{i=1}^n F_i(x, y)$. Hence, by Theorem 7.3.2, we obtain that under $\{K_N\}$

(7.3.56) $$\sigma^2(\mathcal{T}_N) \xrightarrow{p} A^2 - \bar{\nu} = \gamma_{00}(1 - \rho_J),$$

where

(7.3.57) $$\gamma_{00} = \int_0^1 J^2(u)\, du - \left(\int_0^1 J(u)\, du\right)^2,$$

(7.3.58) $$\rho_J\gamma_{00} = \int_{-\infty}^{\infty}\int_{-\infty}^{\infty} J[F(x)]J[F(y)]\, dF^*(x, y) - \left[\int_0^1 J(u)\, du\right]^2.$$

Then, if $(d/dx)J[F(x)]$ is bounded as $x \to \pm\infty$ (where of course $F(x)$ is assumed to be absolutely continuous in x), we obtain from (7.3.35), (7.3.54), and some simple steps that under $\{K_N\}$

(7.3.59) $$N^{1/2}\left[\mu_{N,j} - \int_0^1 J(u)\, du\right] \to \tau_j B(F), \qquad (j = 1, \ldots, p) \text{ as } N \to \infty,$$

where

(7.3.60) $$B(F) = \int_{-\infty}^{\infty} (d/dx)J[F(x)]\, dF(x).$$

Further, from Theorem 7.3.4, we obtain that under $\{K_N\}$, the covariance matrix $\boldsymbol{\beta}_n^*$ converges (as $N \to \infty$) to $(\mathbf{I}_p p - \ell_p\ell_p')\gamma_{00}(1 - \rho_J)$. Hence, from (7.3.56), (7.3.59), (7.3.60), and Theorem 7.3.4, we arrive at the following.

Theorem 7.3.5. *Under the conditions stated above, and for the sequence of alternative hypotheses $\{K_N\}$ in (7.3.53), S_N has asymptotically a noncentral chi-square distribution with $p - 1$ D.F. and noncentrality parameter*

$$\Delta_S = \left[p^{-1}\sum_{j=1}^{p}\tau_j^2\right][B^2(F)/\gamma_{00}(1 - \rho_J)].$$

In section 7.2.4, we have considered the standard parametric test when the errors are independent. It can be shown (cf. Exercise 7.3.4) that even when the errors within each block are symmetrically dependent, (7.2.39) holds. Moreover, under the models of section 7.2.4, $\Delta_n(\mathcal{F})$ converges to $\Delta(\mathcal{F})$ as

$n \to \infty$. The only change will be to replace each σ_i^2 by $\sigma_i^2(1-\rho_i)$, where $\sigma_i^2 = \text{var}(X_{ij})$ and $\rho_i = \text{corr}(X_{ij}, X_{ij'})$, $i = 1, \ldots, n$. Thus, under the models of section 7.2.4,

$$\lim_{n\to\infty} \frac{1}{n}\sum_{i=1}^{n} \sigma_i^2(1-\rho_i) = \sigma^2(1-\rho_\epsilon), \tag{7.3.61}$$

where

$$\sigma^2 = \lim_{n\to\infty} n^{-1}\sum_{i=1}^{n}\sigma_i^2 \quad \text{and} \quad \rho_\epsilon = 1 - \left\{\lim_{n\to\infty} \frac{\sum_{i=1}^{n}\sigma_i^2(1-\rho_i)}{\sum_{i=1}^{n}\sigma_i^2}\right\}.$$

Hence, from Theorem 7.3.5, we arrive at the following

Theorem 7.3.6. *For the sequence of alternative hypotheses $\{K_N\}$ in (7.3.53), the* A.R.E. *of the S_N test with respect to the variance ratio test is*

$$e_{S,\mathcal{F}} = \sigma^2(1-\rho_\epsilon)B^2(F)/\gamma_{00}(1-\rho_J). \tag{7.3.62}$$

Now, the right-hand side of (7.3.62) depends on σ^2, ρ_ϵ, ρ_J, and $B(F)$. It can be conveniently bounded below by a quantity independent of ρ_ϵ and ρ_J when in the alignment procedure $\tilde{X}_i$ is taken as $X_{i\cdot} = p^{-1}\sum_{j=1}^{p} X_{ij}$, $i = 1, \ldots, n$. Then the variance of $e_{ij} = \epsilon_{ij} - p^{-1}\sum_{j=1}^{p}\epsilon_{ij}$ is equal to $\sigma_i^2(e) = \sigma_i^2(p-1)(1-\rho_i)/p$, $i = 1, \ldots, n$, and as F is the limiting form of $n^{-1}\sum F_i(x)$, where $F_i(x)$ is the marginal c.d.f. of e_{ij}, the variance of the c.d.f. F is nothing but

$$\sigma^2(F) = [(p-1)/p]\sigma^2(1-\rho_\epsilon). \tag{7.3.63}$$

Hence, (7.3.62) reduces to

$$e_{S,\mathcal{F}} = \left\{\frac{p}{(p-1)(1-\rho_J)}\right\}\{\sigma^2(F)B^2(F)/\gamma_{00}\}, \tag{7.3.64}$$

which is independent of ρ_ϵ. Let us then consider the following two lemmas.

Lemma 7.3.7. *If the c.d.f. $G(x_1, \ldots, x_p)$ of $\mathbf{X} = (X_1, \ldots, X_p)'$ is symmetric in its p arguments and the univariate marginals of G are nondegenerate, then ρ_J, defined by* (7.3.58), *(for G replacing F) is bounded below by $(p-1)^{-1}$.*

Proof: Consider the random variable $Z = \sum_{j=1}^{p} J[G_0(X_j)]$ where G_0 is the univariate marginal of G. Then $\text{var}(Z) = p\gamma_{00} + p(p-1)\rho_J\gamma_{00} - p\gamma_{00}[1+(p-1)\rho_J] \geq 0$, where $\gamma_{00} > 0$ as $J(u)$ $(0 < u < 1)$ is not constant and the univariate marginals of G are nondegenerate. Hence, $\rho_J \geq -1/(p-1)$. Q.E.D. ◁

Lemma 7.3.8. *If $G(x_1, \ldots, x_p)$ is a totally symmetric continuous p-variate $(p > 2)$ singular distribution on the $(p-1)$-flat $\sum_{j=1}^{p} x_j = 0$, while there is no lower-dimensional space containing the scatter of the points of G, then $\rho_J = -1/(p-1)$ iff $J[G_0(x)]$ is a linear function of x, with probability one.*

Proof: Define Z as in the preceding lemma. Then var $(Z) = 0$ *iff* $Z = \sum_{j=1}^{p} J[G_0(X_j)] =$ constant, with probability one. Also, by hypothesis, $\sum_{j=1}^{p} X_j = 0$, with probability one. Hence, the *if* part follows trivially. To prove the *only if* part, suppose that on a set of points (of measure nonzero), $J[G_0(x)]$ is not linear in x. Then, for that set of points, the restriction $\sum_{j=1}^{p} X_j = 0$ along with $\sum_{j=1}^{p} J[G_0(X_j)] =$ constant implies that there is a $p - 2$ (or lower)-dimensional space with a positive probability. This contradicts the hypothesis that $G(\mathbf{x})$ is of rank $p - 1$. Hence the lemma. ◁

(Note that for $p = 2$, Y_{i1} and Y_{i2} are exactly negatively correlated and hence $\rho_J = -1$, no matter whether $J[G_0(x)]$ is linear in x or not.)

By virtue of the preceding two lemmas, we obtain that

$$e_{S,\mathcal{F}} \geq \sigma^2(F)B^2(F)/\gamma_{00}, \tag{7.3.65}$$

where the equality sign holds iff $J[G_0(x)]$ is a linear function of x, with probability one.

We shall now study specifically the normal scores and the rank sum tests based on the aligned observations. For the normal scores ($\mathcal{L}_N$) test, we let $E_{N,\alpha}$ be the expected value of αth smallest observation of a sample of size N from the standard normal distribution $\Phi(x)$, whose density is $\phi(x)$. Then (7.3.64) reduces to

$$e_{\mathcal{L},\mathcal{F}} = \frac{p}{(p-1)(1-\rho_\phi)}\left\{\sigma^2(F)\left[\int_{-\infty}^{\infty} \frac{f^2(x)}{\phi(\Phi^{-1}(F(x)))}\,dx\right]^2\right\}, \tag{7.3.66}$$

where ρ_ϕ stands for the value of ρ_J when $J(u) = \Phi^{-1}(u)$ and $f = F'$. Consequently, from (7.3.65), we obtain that

$$e_{\mathcal{L},\mathcal{F}} \geq \sigma^2(F)\left[\int_{-\infty}^{\infty} \{f^2(x)/\phi(\Phi^{-1}(F(x)))\}\,dx\right]^2, \tag{7.3.67}$$

where the equality sign holds *iff* $J[F(x)]$ is a linear function of x, with probability one, i.e., $F(x) = \Phi(a + bx)$ is also a normal c.d.f. Now, the right-hand side of (7.3.67) is bounded below by 1 (cf. Exercise 3.3.9), and the lower bound is attained only when $F(x)$ is a normal c.d.f. Hence, (7.3.66) is bounded below by 1 for all $F(x)$, where the lower bound is attained only for normal F. Now, e_{ij} is a linear function of $\epsilon_{i1}, \ldots, \epsilon_{ip}$, for all $i = 1, \ldots, p$. Hence, the marginal distribution of $n^{-1/2}\sum_{i=1}^{n} e_{ij}$ will converge to a normal one only when the corresponding joint distribution of $n^{-1/2}\sum_{i=1}^{n} \boldsymbol{\epsilon}_i$, $[\boldsymbol{\epsilon}_i = (\epsilon_{i1}, \ldots, \epsilon_{ip})']$ converges to a multinormal one. In fact, when the distributions of $\boldsymbol{\epsilon}_i$'s are the same, say, $G(\boldsymbol{\epsilon})$, then F cannot be normal unless G itself is multinormal. This clearly explains the robustness and efficiency of the aligned normal scores test.

For the rank-sum ($\mathcal{H}_N$) test, we have $E_{N,\alpha} = \alpha/(N + 1)$, $1 \leq \alpha \leq N$. In this case, (7.3.64) reduces to

$$e_{\mathcal{H},\mathcal{F}} = \frac{p}{(p-1)(1-\rho_g)}\left\{12\sigma^2(F)\left[\int_{-\infty}^{\infty} f^2(x)\,dx\right]^2\right\}, \tag{7.3.68}$$

where

$$\rho_g = 12\int_{-\infty}^{\infty}\int_{-\infty}^{\infty}[F(x)-\tfrac{1}{2}][F(y)-\tfrac{1}{2}]\,dF^*(x,y). \tag{7.3.69}$$

Thus, from (7.3.65), we obtain that

$$e_{\mathcal{H},\mathcal{F}} \geq 12\sigma^2(F)\left[\int_{\infty}^{\infty} f^2(x)\,dx\right]^2, \tag{7.3.70}$$

where the equality sign holds only when $F(x) = a + bx$ except on a set of measure 0.

In Chapter 3, we have seen that (cf. Exercise 3.8.7)

$$\inf_{F\in\mathcal{F}_0}\left\{12\sigma^2(F)\left[\int_{-\infty}^{\infty} f^2(x)\,dx\right]^2\right\} = 0.864, \tag{7.3.71}$$

where $\mathcal{F}_0$ is the family of all univariate continuous c.d.f.'s, and where the lower bound 0.864 is attained for the density function $f(x) = a(x^2 - b^2)$, for $|x| \leq b$ and $f(x) = 0$, otherwise, where a and b are real and finite. On the other hand, the equality sign in (7.3.70) holds only when $F(x)$ is a rectangular distribution. Thus, the lower bound in (7.3.71) is not attainable. If the limiting distribution $G(\mathbf{x})$ is multinormal, then we have

$$\rho_g = (6/\pi)\sin^{-1}\left(\frac{-1}{2(p-1)}\right), \tag{7.3.72}$$

and hence, (7.3.68) reduces to

$$\frac{3}{\pi}\cdot\frac{p}{(p-1)[1+(6/\pi)\sin^{-1}(\frac{1}{2}(p-1))]} = e_p \quad \text{(say)}. \tag{7.3.73}$$

It may be noted that for $p = 2$, e_p is equal to $3/\pi$, while for $p > 2$, e_p is strictly greater than $3/\pi$. In section 7.2, we obtained that the A.R.E. of the χ_r^2 test with respect to the variance ratio test for normal alternatives is equal to $3p/\pi(p+1)$ ($= e_p^*$, say), which is strictly less than $3/\pi$. Table 7.1 illustrates the gain in efficiency due to the alignment procedure.

Table 7.1

p	2	3	4	5	10	15	20	∞
e_p	0.955	0.966	0.965	0.963	0.960	0.958	0.957	0.955
e_p^*	0.637	0.716	0.764	0.896	0.860	0.895	0.910	0.955
e_p/e_p^*	1.500	1.349	1.263	1.210	1.116	1.070	1.052	1.000

(Note that the A.R.E. $3p/\pi(p+1)$ of the χ_r^2 test computed in section 7.2 for independent errors has been shown to be unaffected by the symmetric dependence of the errors in Sen (1971).)

7.3.5. Rank Order Tests Based on Aligned Observations for Several Observations per Cell

So far we have considered the case of a single observation per cell. We may consider the more general case where the jth treatment is applied to m_j (≥ 1) plots within each block, for $j = 1, \ldots, p$ (so that the number of plots per block is equal to $M = \sum_{j=1}^{p} m_j$ ($\geq p$) and the total number of observations is equal to $N' = nM$). Here also, we may work with some translation-invariant symmetric function $\tilde{X}_{i\cdot}$ of the within-block observations, and by subtraction, get the aligned observations. Thus, after alignment (ignoring treatment) to the N' observations, N' rank scores $\{E_{N',\alpha} = J_{N'}(\alpha/(N'+1)),\ 1 \leq \alpha \leq N'\}$ will correspond. The average rank scores of the nm_j aligned observations corresponding to the jth treatment is denoted by $T_{N',j}$, $j = 1, \ldots, p$. The (pooled) within-block mean square of the rank scores (i.e., $\sigma^2(\mathcal{T}_n)$) is defined as in (7.3.6), with p replaced by M. This yields that

$$E_{\mathcal{T}_n}(T_{N',j}) = (N')^{-1}\sum_{\alpha=1}^{N'} E_{N',\alpha} = \bar{E}_{N'}, \qquad j = 1, \ldots, p; \tag{7.3.74}$$

$$n \operatorname{cov}_{\mathcal{T}_n}(T_{N',j}, T_{N',j'}) = \sigma^2(\mathcal{T}_n)(\delta_{jj'}M - m_{j'})/(Mm_{j'}) \quad \text{for} \quad j, j' = 1, \ldots, p, \tag{7.3.75}$$

where $\delta_{jj'}$ is the Kronecker delta. This leads to the test statistic

$$S_{N'} = [n/\sigma^2(\mathcal{T}_n)]\sum_{j=1}^{p} m_j[T_{N',j} - \bar{E}_{N'}]^2. \tag{7.3.76}$$

It can be shown similarly (cf. problem 7.3.55) that under H_0: $\boldsymbol{\theta} = \mathbf{0}$, S_N has asymptotically, in probability (under $\mathcal{T}_n$) a chi-square distribution with $p-1$ D.F. Also, by direct generalization of Theorems 7.3.4 and 7.4.5, it follows that under $\{K_{N'}\}$ in (7.3.53) (with N replaced by N'), $S_{N'}$ has asymptotically a noncentral chi-square distribution with $p-1$ D.F. and the noncentrality parameter

$$\Delta_S^* = \left\{M^{-1}\sum_{j=1}^{p} m_j(\theta_j - \bar{\theta})^2\right\}\{B^2(F)/\gamma_{00}(1-\rho_J)\}, \tag{7.3.77}$$

where $\bar{\theta} = M^{-1}\sum_{j=1}^{p} m_j\theta_j$. Comparison with the parametric test based on the variance ratio criterion again leads to the A.R.E. considered in (7.3.64) and (7.3.65), with p replaced by M. Thus, the bounds considered earlier remain valid in this case too.

7.3.6. Nonparametric Tests for MANOVA in Two-Way Layouts

Let us consider a complete two-way layout comprising n blocks of p plots each where p different treatments are applied. The response of the plot in the ith block receiving the jth treatment is a q (≥ 1) vector $\mathbf{X}_{ij} = (X_{ij}^{(1)}, \ldots, X_{ij}^{(q)})'$, $j = 1, \ldots, p$, $i = 1, \ldots, n$. We consider the model

$$\mathbf{X}_{ij} = \boldsymbol{\mu} + \boldsymbol{\alpha}_i + \boldsymbol{\tau}_j + \boldsymbol{\epsilon}_{ij}, \qquad j = 1, \ldots, p, \quad i = 1, \ldots, n, \tag{7.3.78}$$

where $\boldsymbol{\mu}$ (mean effect), $\boldsymbol{\alpha}_1, \ldots, \boldsymbol{\alpha}_n$ (block effects), $\boldsymbol{\tau}_1, \ldots, \boldsymbol{\tau}_p$ (treatment effects) and $\boldsymbol{\epsilon}_{11}, \ldots, \boldsymbol{\epsilon}_{np}$ are all q vectors. It is assumed that the joint distribution function $G_i(\boldsymbol{\epsilon}_{i1}, \ldots, \boldsymbol{\epsilon}_{ip})$ is symmetric in the p vectors $\boldsymbol{\epsilon}_{i1}, \ldots, \boldsymbol{\epsilon}_{ip}$ and may be different for $i = 1, \ldots, n$. (Tests without assuming the additivity of the block effects were considered in section 7.2.) Let $\tilde{\mathbf{X}}_{i\cdot}$ be some translation invariant symmetric function of $\mathbf{X}_{i1}, \ldots, \mathbf{X}_{ip}$ (such as $n^{-1}\sum_{j=1}^{p} \mathbf{X}_{ij}$, etc.). We define the aligned observations as

$$\mathbf{Y}_{ij} = \mathbf{X}_{ij} - \tilde{\mathbf{X}}_{i\cdot}, \qquad j = 1, \ldots, p, \quad i = 1, \ldots, n. \tag{7.3.79}$$

For the kth variate, we rank the observations $Y_{11}^{(k)}, \ldots, Y_{np}^{(k)}$ in ascending order of magnitude and denote by $R_{ij}^{(k)}$ the rank of $Y_{ij}^{(k)}$ in this set, for $i = 1, \ldots, n$, $j = 1, \ldots, p$; $k = 1, \ldots, q$. Thus, corresponding to the aligned observation Y_{ij}, we have a rank vector $\mathbf{R}_{ij} = (R_{ij}^{(1)}, \ldots, R_{ij}^{(q)})'$, $j = 1, \ldots, p$, $i = 1, \ldots, n$. For each positive integer N ($= np$) and each k ($= 1, \ldots, q$), we define the rank scores $\mathbf{E}_N^{(k)} = (E_{N,1}^{(k)}, \ldots, E_{N,N}^{(k)})$ as in (7.3.3), and as in (7.3.4), we define the rank order statistic $T_{N,j}^{(k)}$ for the kth variate and jth treatment, for $j = 1, \ldots, p$, $k = 1, \ldots, q$. Let then

$$\mathbf{T}_N = ((T_{N,j}^{(k)}))_{j=1,\ldots,p;\,k=1,\ldots,q}. \tag{7.3.80}$$

Also, we define $\bar{E}_N^{(k)}$ as in (7.3.5). The tests to be considered are based on $\{T_{N,j}^{(k)} - \bar{E}_N^{(k)}$ for $k = 1, \ldots, q, j = 1, \ldots, p\}$.

We define the rank collection matrix (of order $q \times N$) by

$$\mathbf{R}_N^0 = (\mathbf{R}_{11}, \ldots, \mathbf{R}_{np}). \tag{7.3.81}$$

$\mathbf{R}_N$ can now be partitioned into n submatrices of order $q \times p$ each; i.e.,

$$\mathbf{R}_N^0 = (\mathbf{R}_1^{q\times p}, \ldots, \mathbf{R}_n^{q\times p}); \qquad \mathbf{R}_i = (\mathbf{R}_{i1}, \ldots, \mathbf{R}_{ip}), \quad i = 1, \ldots, n. \tag{7.3.82}$$

Under the null hypotheses

$$H_0\colon \boldsymbol{\tau}_1 = \cdots = \boldsymbol{\tau}_p = \mathbf{0}, \tag{7.3.83}$$

the distribution of $\mathbf{Y}_{i1}, \ldots, \mathbf{Y}_{ip}$ is symmetric in the p vectors and thereby remains invariant under any permutation of these p vectors. Thus, the joint distribution of

$$\mathbf{Y}_N = (\mathbf{Y}_{11}, \ldots, \mathbf{Y}_{1p}, \ldots, \mathbf{Y}_{n1}, \ldots, \mathbf{Y}_{np}) \tag{7.3.84}$$

remains invariant under the finite group $\mathcal{G}_n$ (of transformations $\{g_n\}$) which maps the sample space onto itself. The number of elements of $\mathcal{G}_n$ is equal to $(p!)^n$ and typically a g_n is such that

$$g_n\mathbf{Y}_N = (\mathbf{Y}_{11}^*, \ldots, \mathbf{Y}_{1p}^*, \ldots, \mathbf{Y}_{n1}^*, \ldots, \mathbf{Y}_{np}^*) = \mathbf{Y}_N^*, \tag{7.3.85}$$

where $(\mathbf{Y}_{i1}^*, \ldots, \mathbf{Y}_{ip}^*)$ is a permutation of $\mathbf{Y}_{i1}, \ldots, \mathbf{Y}_{ip}$, $i = 1, \ldots, n$. We denote by $\mathbf{R}_N^*$ the rank collection matrix corresponding to $\mathbf{Y}_N^*$. It follows that $\mathbf{R}_N^*$ is also given by $g_n\mathbf{R}_N^0$. Thus, for every $g_n \in \mathcal{G}_n$, there exists a $\mathbf{R}_N^* = g_n\mathbf{R}_N^0$ which is permutationally equivalent to $\mathbf{R}_N^0$. The distribution of $\mathbf{R}_N^0$ over its $(N!)^q$ possible realizations will depend on the unknown c.d.f. G, even when (7.3.83) holds. However, when (7.3.83) holds, $\mathbf{Y}_N^*$ has the same distribution as $\mathbf{Y}_N$ for all $g_n \in \mathcal{G}_n$, and hence, the conditional distribution of $\mathbf{Y}_N$ over $\{\mathbf{Y}_N^* = g_n\mathbf{Y}_N: g_n \in \mathcal{G}_n\}$ will be uniform, each realization having the common conditional probability $(p!)^{-n}$. This leads to the following probability law:

Under H_0 in (7.3.83), the conditional distribution of $\mathbf{R}_N^0$ over the $(p!)^n$ realizations $\{\mathbf{R}_N^* = g_n\mathbf{R}_N^0: g_n \in \mathcal{G}_n\}$ is uniform, each realization having the conditional probability $(p!)^{-n}$.

Let us denote this conditional probability measure by $\mathcal{T}_n$. Since $\mathcal{T}_n$ is completely specified, the existence of conditionally distribution-free tests for H_0 in (7.3.83) is thus established.

It readily follows that

$$E_{\mathcal{T}_n}(T_{N,j}^{(k)}) = \bar{E}_N^{(k)} \quad \text{for} \quad j = 1, \ldots, p,\ k = 1, \ldots, q. \tag{7.3.86}$$

Let us define then

$$\bar{E}_{N R_{i.}^{(k)}}^{(k)} = \frac{1}{p}\sum_{j=1}^{p} E_{N,R_{ij}^{(k)}}^{(k)} \qquad i = 1, \ldots, n, \quad k = 1, \ldots, q,$$

as the intrablock averages. Also let

$$v_{kk'}(\mathbf{R}_N^0) = \frac{1}{n(p-1)}\sum_{i=1}^{n}\sum_{j=1}^{p}\{E_{N,R_{ij}^{(k)}}^{(k)} - \bar{E}_{N,R_{i.}^{(k)}}^{(k)}\} \times \{E_{N,R_{ij}^{(k')}}^{(k')} - \bar{E}_{N,R_{i.}^{(k')}}^{(k')}\}, \tag{7.3.87}$$

for $k, k' = 1, \ldots, q$;

$$\mathbf{V}_N(\mathbf{R}_N^0) = ((V_{kk'}(\mathbf{R}_N^0)))_{k,k'=1,\ldots,q}. \tag{7.3.88}$$

It is then easy to verify that

$$\text{cov}_{\mathcal{T}_n}\{T_{N,j}^{(k)}, T_{N,j'}^{(k')}\} = \frac{1}{np}(\delta_{jj'}p - 1)v_{kk'}(\mathbf{R}_N^0), \tag{7.3.89}$$

for $k, k' = 1, \ldots, q$, $j, j' = 1, \ldots, p$, where $\delta_{jj'}$ is the usual Kronecker delta. For the time being, let us assume that $\mathbf{V}_N(\mathbf{R}_N^0)$ is positive definite, and denote its reciprocal matrix by

$$\mathbf{V}_N^{-1}(\mathbf{R}_N^0) = ((v^{kk'}(\mathbf{R}_N^0)))_{k,k'=1,\ldots,q}.$$

Our proposed test statistic S_N can then be expressed as

$$(7.3.90)\qquad S_N = n\sum_{k=1}^{q}\sum_{k'=1}^{q} v^{kk'}(\mathbf{R}_N^0)\sum_{j=1}^{p}[T_{N,j}^{(k)} - \bar{E}_N^{(k)}][T_{N,j}^{(k')} - \bar{E}_N^{(k')}],$$

and it may be noted that S_N is essentially a non-negative stochastic variable. We shall see later on that under certain regularity conditions on $G_i(\mathbf{Y}_{i1}, \ldots, \mathbf{Y}_{ir})$, $i = 1, \ldots, n$, $\mathbf{V}_N(\mathbf{R}_N^0)$ is positive definite with a very high probability (precisely, in probability as $n \to \infty$). However, if $\mathbf{V}_N(\mathbf{R}_N^0)$ fails to be nonsingular, we may work with the highest-order principal minor of it which is positive definite and work with the quadratic form involving only these variables.

For small values of n, we may evaluate the exact permutation distribution of S_N and obtain the critical values of it. For large n, we proceed as follows:

7.3.7. Asymptotic Permutation Distribution of S_N

Let us define

$$(7.3.91)\quad F_{N[j]}^{(k)}(x) = \frac{1}{n}\,[\text{number of } Y_{i,j}^{(k)} \le x], \quad k = 1, \ldots, q, \quad j = 1, \ldots, p;$$

$$(7.3.92)\qquad H_N^{(k)}(x) = \frac{1}{p}\sum_{j=1}^{p} F_{N[j]}^{(k)}(x), \quad k = 1, \ldots, q;$$

$$(7.3.93)\qquad F_{N[j,l]}^{(k,k')}(x, y) = \frac{1}{n}\,[\text{number of } (Y_{ij}^{(k)}, Y_{il}^{(k')}) \le (x, y)],$$

for $k, k' = 1, \ldots, q$, $j, l = 1, \ldots, p$ with either $j \ne l$ or $k \ne k'$ or both. Now, corresponding to the joint c.d.f. G_i, let us denote the marginal c.d.f. of $Y_{ij}^{(k)}$ and of $(Y_{ij}^{(k)}, Y_{il}^{(k')})$ by $F_{i[j]}^{(k)}(x)$ and $F_{i[j,l]}^{(k,k')}(x, y)$ respectively, for $j, l = 1, \ldots, p$, $k, k' = 1, \ldots, q$, with at least one of $j \ne l$, $k \ne k'$ being true, and let

$$(7.3.94)\qquad \bar{H}_N^{(k)}(x) = \frac{1}{np}\sum_{i=1}^{n}\sum_{j=1}^{p} F_{i[j]}^{(k)}(x), \quad \text{for} \quad k = 1, \ldots, q.$$

Further, we assume that for each $k = 1, \ldots, q$, $E_{N,\alpha}^{(k)} = J_N^{(k)}(\alpha/(N+1))$ satisfies the following conditions, in addition to the conditions (*a*), (*b*), and (*c*) of section 3.6.1: (*a*)

$$(7.3.95)\qquad \int_0^1 [J_N^{(k)}(u) - J^{(k)}(u)]^2\, du \to 0 \quad \text{as} \quad N \to \infty,$$

and (*b*)

(7.3.96) $\qquad \boldsymbol{\nu}_N$ as defined below is positive definite.

For the definition of $\mathbf{v}_N$, let

$$(7.3.97) \quad Z_{ij}^{(k)} = J^{(k)}(\bar{H}_N^{(k)}(Y_{ij}^{(k)})), \qquad k = 1, \ldots, q, \quad j = 1, \ldots, p;$$

$$(7.3.98) \quad \mathbf{Z}'_{ij} = (Z_{ij}^{(1)}, \ldots, Z_{ij}^{(p)})', \qquad j = 1, \ldots, p, \quad i = 1, \ldots, n;$$

$$(7.3.99) \quad a_{kk'\cdot jl,i} = E\{Z_{ij}^{(k)} \cdot Z_{il}^{(k')}\}, \qquad k, k' = 1, \ldots, q, \quad j, l = 1, \ldots, p;$$

$$(7.3.100) \quad a_{kk',jl}^{(n)} = n^{-1} \sum_{i=1}^{n} {}_k a_{k',jl}, i \qquad k, k' = 1, \ldots, q, \quad j, l = 1, \ldots, p;$$

$$(7.3.101) \quad A_{jl}^{(n)} = ((a_{kk'\cdot jl}^{(n)}))_{k,k'=1,\ldots,q}, \quad \text{for} \quad j, l = 1, \ldots, p;$$

$$(7.3.102) \quad \nu_{kk',N} = \frac{1}{p} \sum_{j=1}^{p} a_{kk'\cdot jj}^{(n)} - \frac{1}{p^2} \sum_{j=1}^{p} \sum_{l=1}^{p} a_{kk'\cdot jl}^{(n)}, \quad \text{for} \quad k, k' = 1, \ldots, q;$$

$$(7.3.103) \qquad \mathbf{v}_N = ((\nu_{kk',N}))_{k,k'=1,\ldots,q}.$$

Let us study first some conditions under which (7.3.96) holds. For this, we define

$$(7.3.104) \quad \mathbf{A}_{(j,l)}^{(n)} = [\mathbf{A}_{jj}^{(n)} + \mathbf{A}_{ll}^{(n)} - 2\mathbf{A}_{jl}^{(n)}], \qquad j \neq l = 1, \ldots, p.$$

Lemma 7.3.9. (7.3.96) *holds if*

$$\max_{j \neq l = 1, \ldots, p} [\text{rank } \mathbf{A}_{(j,l)}^{(n)}] = p.$$

Proof: For any non-null $\mathbf{a} = (a_1, \ldots, a_q)'$, let $t_j = a'(n^{-1/2} \sum_{i=1}^n \mathbf{Z}_{ij})$. It is then easily verified that

$$(7.3.105) \qquad \mathbf{a}'\mathbf{v}_N\mathbf{a} = p^{-1} \sum_{j=1}^{p} E(t_j^2) - E(t_.)^2; \qquad t_. = p^{-1} \sum_{j=1}^{p} t_j.$$

Now, (7.3.105) is strictly positive when $t_1, \ldots, t_p$ are not all equal. Also, by definition, $E(t_j - t_l)^2 = \mathbf{a}' A_{(j,l)}^{(n)} \mathbf{a} > 0$ for at least one $j \neq l = 1, \ldots, p$, by (7.3.104). Hence, the lemma. ◁

Lemma 7.3.10. $\mathbf{V}_N(\mathbf{R}) \underset{p}{\sim} \nu_N$, *and hence,* $\mathbf{V}_N(\mathbf{R}_N)$ *is positive definite, in probability, when* (7.3.96) *holds.*

The proof is a straightforward extension of Lemma 7.3.2 and is left as Exercise 7.3.6.

Theorem 7.3.11. *Under the conditions stated above, the permutation distribution of* S_N *converges, in probability, to a chi-square distribution with* $q(p - 1)$ D.F.

The proof follows along the same line as in Theorem 7.3.3 and is left as Exercise 7.3.7.

7.3.8. Asymptotic Non-null Distribution of S_N

We consider first the asymptotic distribution theory of $\mathbf{T}_N$. For this, we define $\bar{F}_{n[j]}^{(k)} = n^{-1}\sum_{i=1}^{n} F_{i[j]}^{(k)}$, $1 \leq j \leq p$, $1 \leq k \leq q$, and let

$$\mu_{N,j}^{(k)} = \int_{-\infty}^{\infty} J^{(k)}[\bar{H}_N^{(k)}(x)]\, d\bar{F}_{n[j]}^{(k)}(x), \qquad j = 1, \ldots, p, \quad k = 1, \ldots, q; \tag{7.3.106}$$

$$\beta_{jj',ll',i}^{(k,k')} = \int_{-\infty}^{\infty}\int_{-\infty}^{\infty} [F_{i[j,j']}^{(k,k')}(x, y) - F_{i[j]}^{(k)}(x)F_{i[j']}^{(k')}(y)]J^{(k)\prime}[\bar{H}_N^{(k)}(x)] \times J^{(k')\prime}[\bar{H}_N^{(k')}(y)]\, d\bar{F}_{n[l]}^{(k)}(x)\, d\bar{F}_{n[l']}^{(k')}(y), \tag{7.3.107}$$

for $j, j', l, l' = 1, \ldots, p$, $k, k' = 1, \ldots, q$ with either $j \neq j'$ or $k \neq k'$ or both, while

$$\beta_{jj,ll',i}^{(k,k)} = \iint\limits_{-\infty<x<y<\infty} F_{i[j]}^{(k)}(x)[1 - F_{i[j]}^{(k)}(y)]J^{(k)\prime}[\bar{H}_N^{(k)}(x)]J^{(k)\prime}[\bar{H}_N^{(k)}(y)] \times \{d\bar{F}_{n[l]}^{(k)}(x)\, d\bar{F}_{n[l']}^{(k)}(y) + d\bar{F}_{n[l']}^{(k)}(x)\, d\bar{F}_{n[l]}^{(k)}(y)\}, \tag{7.3.108}$$

for $j, l, l' = 1, \ldots, p$, $k = 1, \ldots, q$, $i = 1, \ldots, n$. Finally, let

$$\beta_{jj',n}^{(k,k')} = \frac{1}{np}\sum_{i=1}^{n}\left\{\sum_{l=1}^{p}\sum_{l'=1}^{p}[\beta_{jj',ll',i}^{(k,k')} + \beta_{ll',jj',i}^{(k,k')} - \beta_{jl',lj',i}^{(k,k')} - \beta_{lj',jl',i}^{(k,k')}]\right\}, \tag{7.3.109}$$

for $k, k' = 1, \ldots, q, j, j' = 1, \ldots, p$.

Theorem 7.3.12. *Under the conditions (a), (b), and (c) of section* 3.6.1, *the vector* $[N^{1/2}(T_{N,j}^{(k)} - \mu_{N,j}^{(k)}), j = 1, \ldots, p, k = 1, \ldots, q]$ *has asymptotically a multinormal distribution with null means and dispersion matrix with elements* $\beta_{jj,n}^{(k,k')}$, *defined by* (7.3.109).

The proof is a straightforward generalization of the proof of Theorem 7.3.4 and hence is left as Exercise 7.3.8.

We have already noted that the asymptotic normal distribution of Theorem 7.3.12 is singular and of rank at most equal to $p(r - 1)$. If the null hypothesis H_0 is true, $G_i(\mathbf{Y}_{i1}, \ldots, \mathbf{Y}_{ir})$ will be a symmetric function of the r vectors, and hence it is easily seen that (i) the marginal c.d.f. of $Y_{ij}^{(k)}$ will be the same for all $j = 1, \ldots, p$, and is denoted by $H_i^{(k)}(x)$ for $k = 1, \ldots, q$; (ii) the marginal c.d.f. of $(Y_{ij}^{(k)}, Y_{ij}^{(k')})$ $(k \neq q)$ will not depend on j, and is denoted by $H_{1,i}^{(k,k')}(x, y)$ for $k \neq k' = 1, \ldots, q$, and (iii) the marginal c.d.f. of $(Y_{ij}^{(k)}, Y_{il}^{(k')})$ $(j \neq l)$ will not depend on $(j \neq l)$, and is denoted by $H_{2,i}^{(k,k')}(x, y)$ for $j \neq l = 1, \ldots, p$, $k, k' = 1, \ldots, q$, $i = 1, \ldots, n$. Thus, it follows from (7.3.107) and (7.3.108) that

$$\begin{aligned} \beta_{jj':ll',i}^{(k,k')} &= a_{kk',jj',i} = a_{kk',i}^{(1)}, \quad \text{if} \quad j = j' = 1, \ldots, p, \\ &= a_{kk',i}^{(2)}, \quad \text{if} \quad j \neq j' = 1, \ldots, p, \end{aligned} \tag{7.3.110}$$

where $a^{(1)}_{kk',i}$ depends only on $H^{(k,k')}_{1,i}(x, y)$ and $a^{(2)}_{kk',i}$ on $H^{(k,k')}_{2,i}(x, y)$, respectively, $i = 1, \ldots, n$. Thus, from (7.3.102) and (7.3.110), we get that in this case $\nu_{kk',N}$ defined by (7.3.102) reduces to

$$\nu_{kk',N} = [(p-1)/p](\bar{a}^{(1)}_{kk',n} - \bar{a}^{(2)}_{kk',n}), \qquad k, k' = 1, \ldots, q, \tag{7.3.111}$$

and

$$\beta^{(k,k')}_{jl,n} = (\delta_{jl}p - 1)\nu_{kk',N}, \qquad j, l = 1, \ldots, p, \quad k, k' = 1, \ldots, q, \tag{7.3.112}$$

where δ_{jl} is the usual Kronecker delta and $\bar{a}^{(j)}_{kk',n} = n^{-1}\sum_{i=1}^{n} a^{(j)}_{kk',i}, j = 1, 2$. Consequently, it is easily seen that under H_0,

$$S^*_N = n\sum_{k=1}^{q}\sum_{k'=1}^{q} \nu_N^{kk'} \sum_{j=1}^{p}(T^{(k)}_{N,j} - \mu^{(k)}_{.})(T^{(k')}_{N,j} - \mu^{(k')}_{.}) \tag{7.3.113}$$

(where $((\nu_N^{kk'}))$ is the reciprocal of $((\nu_{kk',N}))$, and $\mu^{(k)}_{.} = \int_0^1 J^{(k)}(u)\,du$, $k = 1, \ldots, q$) has asymptotically a chi-square distribution with $q(p-1)$ D.F. Also, by condition (a) of section 3.6.1,

$$|N^{1/2}(\bar{E}^{(k)}_N - \mu^{(k)}_{.})| = o(1), \quad \text{for} \quad k = 1, \ldots, q, \tag{7.3.114}$$

and by Lemma 7.3.10 we have under (7.3.96) that

$$\mathbf{V}_N(\mathbf{R}^0_N) \overset{p}{\sim} \boldsymbol{\nu}_N; \quad \text{i.e.,} \quad \mathbf{V}_N^{-1}(\mathbf{R}^0_N) \overset{p}{\sim} \boldsymbol{\nu}_N^{-1}. \tag{7.3.115}$$

Hence, under H_0: $S_N \overset{p}{\sim} S^*_N$. This leads to the following.

Theorem 7.3.13. *Under H_0 and granted the conditions* (7.3.45), (7.3.96), *and* (a), (b), *and* (c) *of section* 3.6.1, *S_N has asymptotically a chi-square distribution with $q(p-1)$ degrees of freedom.*

Let now $\hat{\boldsymbol{\nu}}_N$ be any consistent estimator of $\boldsymbol{\nu}_N$, defined by (7.3.103) and (7.3.112). If $\hat{\boldsymbol{\nu}}_N$ is positive definite and we denote its reciprocal by $\hat{\boldsymbol{\nu}}_N^{-1} = ((\hat{\boldsymbol{\nu}}_N^{kk'}))$, then we can have an *asymptotically distribution-free test based on*

$$\hat{S}_N = n\sum_{k=1}^{q}\sum_{k'=1}^{q} \hat{\nu}_N^{kk'} \sum_{j=1}^{p}(T^{(k)}_{N,j} - \bar{E}^{(k)}_N)(T^{(k')}_{N,j} - \bar{E}^{(k')}_N). \tag{7.3.116}$$

Since $\mathbf{V}_N(\mathbf{R}_N)$ has already been shown to be a consistent estimate of $\boldsymbol{\nu}_N$, it follows that a large-sample (unconditional) test can also be based on S_N, using the rejection rule (7.3.39), with $p-1$ replaced by $q(p-1)$.

As in section 7.2.8, we now consider a sequence of local translation alternatives and study the distribution of S_N. Thus, we consider a sequence $\{K_N\}$ of alternative hypotheses where K_N specifies that the random variables $\mathbf{X}_{ij} - N^{-1/2}\boldsymbol{\tau}_j, j = 1, \ldots, p$ are interchangeable, for each $i = 1, \ldots, n$ and

$\boldsymbol{\tau}_1, \ldots, \boldsymbol{\tau}_p$ are p real q vectors with $\sum_{j=1}^{p} \boldsymbol{\tau}_j = \mathbf{0}$. Thus, under $\{K_N\}$

$$F_{i[j]}^{(k)}(x) = F_i^{(k)}(x - N^{-\frac{1}{2}}\tau_j^{(k)}), \qquad j = 1, \ldots, p, \quad k = 1, \ldots, q; \tag{7.3.117}$$

$$F_{i[j,j']}^{(k,k')}(x, y) = F_{i,\delta_{jj'}+1}^{(k,k')}(x - N^{-\frac{1}{2}}\tau_j^{(k)}, Y - N^{-\frac{1}{2}}\tau_{j'}^{(k')}), \qquad j, j' = 1, \ldots, p, \quad k, k' = 1, \ldots, q, \tag{7.3.118}$$

where $\delta_{jj'}$ is the usual Kronecker delta. We assume that $F_{i[j]}^{(k)}(x), j = 1, \ldots, p$, $k = 1, \ldots, q$, $i = 1, \ldots, n$ are all absolutely continuous. Also, under the models of section 7.2.4, we have

$$\lim_{n \to \infty} n^{-1} \sum_{i=1}^{n} F_i^{(k)}(x) = F^{(k)}(x), \tag{7.3.119}$$

$$\lim_{n \to \infty} \frac{1}{n} \sum_{i=1}^{n} F_{i,\delta_{jj'}+1}^{(k,k')}(x, y) = F_{\delta_{jj'}+1}^{(k,k')}(x, y), \qquad \delta_{jj'} = 0, 1, \tag{7.3.120}$$

for all $k \neq k' = 1, \ldots, q$. Hence, by Lemma 7.3.10, $\mathbf{V}_N(\mathbf{R}_N^0) \sim \boldsymbol{\nu}$, where $\boldsymbol{\nu}$ is defined as in (7.3.103) with $a_{kk',jj'}$ all defined for the limiting distribution. We also define

$$B(F^{(k)}) = \int_{-\infty}^{\infty} [(d/dx)J^{(k)}[F^{(k)}(x)]]\, dF^{(k)}(x), \qquad k = 1, \ldots, q. \tag{7.3.121}$$

Then from Theorem 7.3.12, we obtain the following.

Theorem 7.3.14. *Under the sequence of alternatives* $\{K_N\}$ *and granted the conditions of Theorem* 7.3.13 *and the existence of the limits in* (7.3.119) *and* (7.3.120), S_N *has asymptotically a noncentral chi-square distribution with* $q(p-1)$ D.F. *and the noncentrality parameter*

$$\Delta_S = \sum_{k=1}^{q} \sum_{k'=1}^{q} \nu^{kk'} B(F^{(k)}) B(F^{(k')}) \left\{ \frac{1}{p} \sum_{j=1}^{p} \tau_j^{(k)} \tau_j^{(k')} \right\}. \tag{7.3.122}$$

In section 7.2.8, we have considered the standard parametric test and studied its asymptotic distribution theory. Let us define

$$\sigma_{kk',i}^{(1)} = \operatorname{cov}(X_{ij}^{(k)}, X_{ij}^{(k')}), \qquad \sigma_{kk',i}^{(2)} = \operatorname{cov}(X_{ij}^{(k)}, X_{ij'}^{(k')}), \tag{7.3.123}$$

for $j \neq j' = 1, \ldots, p$, $k, k' = 1, \ldots, q$, $i = 1, \ldots, n$, and let

$$\bar{\boldsymbol{\Lambda}}_n = \frac{1}{n} \sum_{i=1}^{n} ((\sigma_{kk',i}^{(1)} - \sigma_{kk',i}^{(2)}))_{k,k'=1,\ldots,q} = ((\bar{\lambda}_{n,kk'})). \tag{7.3.124}$$

Then, under the models of section 7.2.4, $\lim_{n \to \infty} \bar{\lambda}_{n,kk'} = \lambda_{kk'}$ exists for all $k, k' = 1, \ldots, q$, and we denote

$$\boldsymbol{\Lambda} = ((\lambda_{kk'})); \qquad \boldsymbol{\Lambda}^{-1} = ((\lambda^{kk'})). \tag{7.3.125}$$

Then, along the same line as in section 7.2.8, it can be shown that under $\{K_N\}$ in (7.3.117) and (7.3.118), the parametric (U_N-) test has asymptotically a noncentral chi-square distribution with $q(p-1)$ D.F. and the noncentrality parameter

$$\Delta_U = \frac{p}{p-1} \sum_{k=1}^{q} \sum_{k'=1}^{q} \lambda^{kk'} \frac{1}{p} \sum_{j=1}^{p} \tau_j^{(k)} \tau_j^{(k')}. \tag{7.3.126}$$

The A.R.E. of the S_N test with respect to the U_N test is given by the ratio of Δ_S and Δ_0 in (7.3.122) and (7.3.126) respectively. This depends not only on $\mathbf{\Lambda}$ and $\mathbf{v}$, but also on $\boldsymbol{\tau}_j$, $j = 1, \ldots, p$. The maximum and the minimum values of Δ_S/Δ_U are, however, the largest and smallest characteristic roots of $[p/(p-1)]\mathbf{\Lambda}\mathbf{v}^{-1}$, and therefore the results studied in chapters 4 and 5 (see sections 4.6 and 5.8) are applicable. Some extensions of these results to a general class of incomplete block designs are considered in Sen (1969d).

7.4. ASYMPTOTICALLY DISTRIBUTION-FREE PROCEDURES BASED ON ROBUST ESTIMATORS OF THE PARAMETERS

In this section, we shall first extend the estimation theory considered in chapter 6 to two-way layouts and then employ these estimates to construct suitable (but only asymptotically distribution-free) rank procedures for testing and estimating parameters of interest.

For simplicity, we consider first the ANOVA problem, and subsequently extend it to the MANOVA problem.

We start with the model in (7.3.1) with the interpretation of α_i, θ_j, and ϵ_{ij} as there. Suppose now we want to estimate a contrast $\xi = \sum_{j=1}^{p} l_j\theta_j$ (where $\sum_{j=1}^{p} l_j = 0$). The intrablock estimates of ξ are

$$V_i = \sum_{j=1}^{p} l_j X_{ij}, \qquad i = 1, \ldots, n. \tag{7.4.1}$$

Denote the c.d.f. of V_i by $F_i(x)$ for $i = 1, \ldots, n$. The variables $\sum_{j=1}^{p} l_j\epsilon_{ij}$, $i = 1, \ldots, n$ are all independent, and if they are symmetrically distributed about zero, then it follows from (7.3.1) and (7.4.1) that $V_1, \ldots, V_n$ are also independent and distributed symmetrically about ξ. Thus, we may proceed as in section 6.2 (namely (6.2.6) through (6.2.10)) and consider the estimator $\hat{\xi}_n$ of ξ, where $\hat{\xi}_n$ is defined as in (6.2.9) and (6.2.10). The various properties (such as invariance, symmetry, asymptotic normality, and its relative efficiency with respect to the estimator $\bar{V}_n = n^{-1}\sum_{i=1}^{n} V_i$) follow from the results of sections 6.2 and 6.4, and hence are not reproduced. However, we may note the following disadvantages of $\hat{\xi}_N$ as an estimate of ξ. First, for an arbitrary contrast, the symmetry of the distribution of V_i (about ξ) may not hold under the same generality as in (7.3.1). For example, if $V_i = X_{i1} - X_{i2}$

(or $X_{i1} + X_{i2} - X_{i3} - X_{i4}$, $p \geq 4$), the symmetry of the distribution of V_i will follow for any continuous $G_i(x_1, \ldots, x_p)$ when G_i is symmetric in its p arguments (as assumed under (7.3.1)). On the other hand, if $V_i = X_{i1} + X_{i2} - 2X_{i3}$, the symmetry of its distribution requires further restrictions on G_i. Second, unlike the least squares estimator, $\hat{\xi}_n$ may lead to ambiguities. For example, in estimating $\theta_2 - \theta_1$ a different answer may be obtained when we estimate this difference directly than we take the sum of the estimators of $\theta_3 - \theta_1$ and $\theta_2 - \theta_3$. Thus, we may not obtain a single natural estimate. As an illustration, consider Table 7.2. (The data from Anderson and Bancroft, 1952, p. 238.)

Table 7.2.* *Numerical Examples Showing the Incompatibility of the Estimators $\hat{\xi}_n$

Block	Treatments					
	1	2	3	(ii) − (i)	(iii) − (i)	(ii) + (iii) − 2(i)
	(i)	(ii)	(iii)	(iv)	(v)	(vi)
1	164	172	177	8	13	21
2	771	197	184	20	7	27
3	168	167	187	−1	19	18
4	156	161	169	5	13	18
5	172	180	179	8	7	15
6	195	190	197	−5	2	−3
Least squares estimates (sample averages)				5.83	10.17	16.00
Wilcoxon scores estimates				6.5	10.0	18.0

The Wilcoxon scores estimates are the medians of the midranges of the observations in columns (iv), (v), and (vi), and it is easy to verify that they do not satisfy the compatibility condition that the sum of the estimates in (iv) and (v) should be the estimate in (vi). The normal scores estimator encounters the same difficulty. Moreover, an iterative procedure is required for the computation of this estimator. This is illustrated below. The expected values $E_{6,\alpha}$ ($\alpha = 1, \ldots, 6$) of the order statistics of a sample of size 6 from a chi distribution with one degree of freedom are as follows:

α	1	2	3	4	5	6
$E_{6,\alpha}$	0.183	0.377	0.589	0.835	1.149	1.654.

Consider the estimate of $\xi = \theta_2 - \theta_1$. Start with column (iv) of Table 7.2 and the trial value $\xi = 6.0$. The values of $V_\alpha - \xi$ are then

$$2, \quad 14, \quad -7, \quad -1, \quad 2, \quad -11. \tag{7.4.2}$$

To handle tied observations, we adopt the convention that the total scores for the tied ranks should be equally divided among them. Thus, the value of h_n defined in (4.4.38) for $\xi = 6.0$ is 0.437, which is greater than $\mu_n = 0.399$. Observe that the ranks of the absolute values of the residuals (7.4.2) remain unaltered as long as $\xi < 6.5$. For $\xi = 6.5$, the residuals are 1.5, 13.5, -7.5, -1.5, 1.5, -10.5, and as a result h_n equals 0.403, which is still greater than 0.399. But if we slightly increase ξ (i.e., $\xi = 6.5^+$), the value of h_n equals 0.369, which is less than 0.399. Consequently, the normal score estimate is 6.5. In a similar way, the normal scores estimates of $\theta_3 - \theta_1$ and $\theta_2 + \theta_3 - 2\theta_1$ can be obtained as 10.0 and 18.0 respectively. Hence their incompatibility follows, as in the case of Wilcoxon scores estimates. Incidentally, in this example, the Wilcoxon scores estimates and the normal scores estimates coincide.

Thus it is desirable to replace the incompatible estimates considered so far by a mutually compatible system. With this end in view, we consider first the paired differences $\theta_j - \theta_{j'}$, $j \neq j' = 1, \ldots, p$. Incidentally, for these differences, the distribution of $V_{ij} - V_{ij'}$ is also symmetric about $\theta_j - \theta_{j'}$.

7.4.1. Compatible Estimates of Contrasts

Let us define

$$X^*_{i,jj'} = X_{ij} - X_{ij'}, \qquad \Delta_{jj'} = \theta_j - \theta_{j'} \quad \text{and} \quad \epsilon_{i,jj'} = \epsilon_{ij} - \epsilon_{ij'}, \tag{7.4.3}$$

for $j \neq j' = 1, \ldots, p$, $i = 1, \ldots, n$. Since the joint c.d.f. $G_i(\epsilon_1, \ldots, \epsilon_p)$ of $(\epsilon_{i1}, \ldots, \epsilon_{ip})$ is symmetric in its p arguments, the distribution of $\epsilon_{i,jj'}$ is necessarily symmetric about 0. We denote by $F_i(x)$ the c.d.f. of $\epsilon_{i,jj'}$, so that the c.d.f. of $X^*_{i,jj'}$ is $F_i(x - \Delta_{jj'})$ and is symmetric about $\Delta_{jj'}$. Hence, we may proceed as in (6.2.6) through (6.2.10) and obtain the rank order estimator of $\Delta_{jj'}$, which we denote by $\hat{\Delta}_{n,jj'}$, for $j \neq j' = 1, \ldots, p$. As has been noted earlier, these estimators are incompatible. To derive compatible estimators, we define

$$\hat{\Delta}_{n,j\cdot} = p^{-1} \sum_{j'=1}^{p} \hat{\Delta}_{n,jj'}; \qquad \hat{\Delta}_{n,jj} = 0, \quad j = 1, \ldots, p. \tag{7.4.4}$$

Then, by minimizing $\Sigma_{j \neq j'} (\hat{\Delta}_{n,jj'} - \{\theta_j - \theta_{j'}\})^2$ with respect to the θ's, we obtain the compatible or adjusted estimator of $\Delta_{jj'}$ as

$$Z_{n,jj'} = \hat{\Delta}_{n,j\cdot} - \hat{\Delta}_{n,j'\cdot}, \qquad j \neq j' = 1, \ldots, p. \tag{7.4.5}$$

These estimators satisfy the linear relations which are satisfied by the $\Delta_{jj'}$'s. The contrast $\xi = \sum_{j=1}^{p} l_j \theta_j$ with $\sum_{j=1}^{p} l_j = 0$ can also be written as

$\sum_{j=1}^{p}\sum_{j'=1}^{p} d_{jj'}\Delta_{jj'}$ where $d_{jj'} = l_j/p$ for $j' = 1, \ldots, p$. Thus, we may propose the compatible estimator of ξ as

$$\xi_{n(c)} = \sum_{j=1}^{p}\sum_{j'=1}^{p} d_{jj'}Z_{n,jj'} = \sum_{j=1}^{p} l_j\hat{\Delta}_{n,j\cdot}. \tag{7.4.6}$$

For illustration, consider again the data in Table 7.2. The values of $\hat{\Delta}_{n,1\cdot}$, $\hat{\Delta}_{n,2\cdot}$, and $\hat{\Delta}_{n,3\cdot}$ are -5.5, 0.5, and 5.0, respectively. Hence, the compatible estimators of $\theta_2 - \theta_1$, $\theta_3 - \theta_1$, and $\theta_2 + \theta_3 - 2\theta_1$ are 16.5, 6.0, and 10.5 respectively (as compared to the incompatible estimates 18.0, 6.5, and 10.0).

Note that the compatible estimator $Z_{n,jj'}$ of $\Delta_{jj'}$ depends not only on the jth and j'th treatment observations but also on the remaining $p - 2$ treatment observations which are unrelated to $\Delta_{jj'}$. This pecularity may be considered a disadvantage. On the other hand, the compatible estimators (like the least squares estimators) possess the merit of providing a single natural estimator of any contrast, and they do not require any further restriction on the distribution of the errors to insure the symmetry of the distributions of the adjusted errors.

7.4.2. Large-Sample Properties of the Estimators

Here we study the asymptotic properties of the compatible, incompatible, and (conventional) least squares estimators of contrasts. In this context, it is shown that for all parent distributions, asymptotically, (i) the compatible estimator is at least as efficient as the incompatible estimator for all score functions $\Psi(x)$ in (6.2.7), and (ii) the compatible normal scores estimator is at least as efficient as the least squares estimator. As basis for this study, we consider the following notations first.

We denote by $F_i(x)$ the c.d.f. of $\epsilon_{i,jj'}$ ($j \neq j' = 1, \ldots, p$) and by $F_1^*(x, y)$ the joint c.d.f. of $(\epsilon_{ijj'}, \epsilon_{i,jl'})$ where $j \neq j' \neq l'$, for $j = 1, \ldots, n$. Let then

$$\bar{F}_n(x) = n^{-1}\sum_{i=1}^{n} F_i(x) \quad \text{and} \quad \bar{F}_n^*(x, y) = n^{-1}\sum_{i=1}^{n} F_i^*(x, y). \tag{7.4.7}$$

Now, under the models of section 7.2.4, $\bar{F}_n(x)$ and $\bar{F}_n^*(x, y)$ both approach limits $F(x)$ and $F^*(x, y)$, respectively, as $n \to \infty$. Also, let $J(u) = \Psi^{*-1}(u)$: $0 < u < 1$ be the score function, defined by (6.2.7), and

$$A^2 = \int_0^1 J^2(u)\,du, \qquad B = \int_{-\infty}^{\infty} (d/dx)J[F(x)]\,dF(x); \tag{7.4.8}$$

$$\lambda_J(F^*) = \int_{-\infty}^{\infty}\int_{-\infty}^{\infty} J[F(x)]J[F(y)]\,dF^*(x, y). \tag{7.4.9}$$

Now, by definition in (6.2.7), $\Psi^*(x)$ is symmetric about 0. Hence, we have

$J(\frac{1}{2}) = 0$ and

$$J(u) + J(1 - u) = 0 \quad \text{for all} \quad 0 < u \leq \tfrac{1}{2}. \tag{7.4.10}$$

For any skew-symmetric function $h(x)$ (i.e., $h(x) + h(-x) = 0$), let $\zeta_{i,0} = E\{h(\epsilon_{i,jj'})h(\epsilon_{i,ll'})\}$ where $j \neq j' \neq l \neq l'$, $\zeta_{i,1} = E\{h(\epsilon_{i,jj'})h(\epsilon_{i,jl'})$, where $j \neq j' \neq l'$ and $\zeta_{i,2} = E\{h^2(\epsilon_{i,jj'})\}, j \neq j'$. Then we have the following:

Lemma 7.4.1. *If* $\epsilon_{i1}, \ldots, \epsilon_{ip}$ *are interchangeable random variables and* $h(x)$ *is skew-symmetric, then* (i) $E\{h(\epsilon_{i,jj'})\} = 0$ *and* (ii) $\zeta_{i,0} = 0$.

Proof: (i) follows trivially from the fact that the distribution of $\epsilon^*_{i,jj'} = \epsilon_{ij} - \epsilon_{ij'}$ is symmetric about zero and $h(x)$ is skew-symmetric. To prove (ii), let $\mathbf{t} = \{t_1 \leq t_2 \leq t_3 \leq t_4\}$ be the order statistic corresponding to ϵ_{ij}, $\epsilon_{ij'}$, ϵ_{il}, $\epsilon_{il'}$. Since the ϵ_{ij} are interchangeable random variables, the conditional distribution of ϵ_{ij}, $\epsilon_{ij'}$, ϵ_{il}, $\epsilon_{il'}$, given $\mathbf{t}$, is uniform over the 24 equally likely permutations of t_1, t_2, t_3 and t_4. Hence,

$$E\{h(\epsilon_{i,jj'})h(\epsilon_{i,ll'}) \mid t\} = \tfrac{1}{24}\textstyle\sum^* h(t_{i_1} - t_{i_2})h(t_{i_3} - t_{i_4}), \tag{7.4.11}$$

where the summation $\sum^*$ extends over the 24 permutations of (i_1, i_2, i_3, i_4) over $(1, 2, 3, 4)$. Since $h(x)$ is skew-symmetric, $h(t_{i_2} - t_{i_4}) + h(t_{i_4} - t_{i_2}) = 0$. Hence, the right-hand side of (7.4.11) can easily be shown to be equal to zero. Thus, writing $\zeta_{i,0}$ equivalently as $E_{\mathbf{t}}\{E[h(\epsilon_{i,jj'})h(\epsilon_{i,ll'}) \mid \mathbf{t}]\}$, the desired result follows. Q.E.D. ◁

Lemma 7.4.2. *Under the conditions of Lemma* 7.4.1, $\zeta_{i,1} \leq \frac{1}{2}\zeta_{i,2}$, *where the equality sign holds only when* $h(x) = bx$, *with probability one. If, in addition,* $h(x)$ *is monotonic*, $0 \leq \zeta_{i,1} \leq \frac{1}{2}\zeta_{i,2}$.

Proof: Define $Z_i = h(\epsilon_{i,12}) + h(\epsilon_{i,23}) + h(\epsilon_{i,31})$. Then, using Lemma 7.4.1, we obtain that

$$V(Z_i) = 3\zeta_{i,2}[1 - 2\zeta_{i,1}/\zeta_{i,2}] \geq 0, \tag{7.4.12}$$

where the equality sign holds only when $Z_i \equiv 0$, with probability one. Now, (7.4.12) implies that $\zeta_{i,1} \leq \frac{1}{2}\zeta_{i,2}$. Also, by definition $\epsilon_{i,12} + \epsilon_{i,23} + \epsilon_{i,31} = 0$. Hence $Z_i \equiv 0$, with probability one, along with the skew symmetry of $h(x)$, implies that

$$\begin{aligned} h(\epsilon_{i,12}) + h(\epsilon_{i,23}) &= -h(\epsilon_{i,31}) = h(-\epsilon_{i,31}) \\ &= h(\epsilon_{i,12} + \epsilon_{i,23}), \qquad \text{with probability one,} \end{aligned} \tag{7.4.13}$$

and this implies that $h(x) = bx$, with probability one. This completes the proof of the first part. Let now $\mathbf{t} = \{t_1 \leq t_2 \leq t_3\}$ be the order statistics corresponding to $(\epsilon_{i1}, \epsilon_{i2}, \epsilon_{i3})$. Then as in Lemma 7.4.1, we have

$$\begin{aligned} E\{h(\epsilon_{i,12})h(\epsilon_{i,13}) \mid \mathbf{t}\} &= E\{h(\epsilon_{i1} - \epsilon_{i2})h(\epsilon_{i1} - \epsilon_{i3}) \mid \mathbf{t}\} \\ &= \tfrac{1}{3}[h(t_1 - t_2)h(t_1 - t_3) + h(t_3 - t_1)h(t_3 - t_2) - h(t_2 - t_1)h(t_3 - t_2)], \end{aligned} \tag{7.4.14}$$

as $h(-x) = -h(x)$. Assume that $h(x)$ is $\uparrow$ in x (as otherwise, work with $-h(x)$). Then

$$(7.4.15)\quad 0 \le h(t_2 - t_1) \le h(t_3 - t_1), \qquad 0 \le h(t_3 - t_2) \le h(t_3 - t_1).$$

Consequently, by (7.4.15), the right-hand side of (7.4.14) is necessarily non-negative. Hence, integrating over the distribution of $\mathbf{t}$, it follows that $\zeta_{i,1} \ge 0$. Q.E.D. ◁

For $\Delta_{jj'}$, defined by (7.4.3), consider the following sequence of alternative hypotheses $\{K_n\}$, where

$$(7.4.16)\quad K_n\colon\ \Delta_{jj'} = n^{-1/2}a_{jj'}, \qquad a_{jj} = 0, \quad \text{for } j, j' = 1, \ldots, p,$$

where the $a_{jj'}$ are real constants. Also, define $T^*_{n,jj'}$ as in (4.4.38), based on the values of $X^*_{i,jj'}$, $i = 1, \ldots, n$ and the scores $E^*_{n,\alpha} = J^*_n(\alpha/(n+1))$, where $J^*(u) = J((1+u)/2)$, and $J(u)$ satisfies (7.4.10). Finally, let $\mu = \int_0^1 J^*(u)\,du$.

Theorem 7.4.3. *Under the sequence of alternatives in* (7.4.16) *and the models of section* 7.2.4, $\{2n^{1/2}(T^*_{n,jj'} - \frac{1}{2}\mu),\ 1 \le j < j' \le p\}$ *has asymptotically a* $\binom{p}{2}$*-variate (singular) multinormal distribution with mean* $\{a_{jj'}B,\ 1 \le j < j' \le p\}$ *and dispersion matrix* $\mathbf{T} = ((\tau_{jj',ll'}))$, *where*

$$(7.4.17)\quad \tau_{jj',ll'} = \begin{cases} A^2, & j = l,\ j' = l', \\ \lambda_J(F^*), & j = l,\ j' \ne l',\ \text{or}\ j \ne l,\ j' = l' \\ -\lambda_J(F^*), & j = l',\ j' \ne l,\ \text{or}\ j \ne l',\ j' = l, \\ 0, & j \ne j' \ne l \ne l', \end{cases}$$

and A^2, B, *and* $\lambda_J(F^*)$ *are defined by* (7.4.8) *and* (7.4.9).

The proof of this theorem follows directly from Theorem 4.4.3 and Lemmas 7.4.1 and 7.4.2 (along the line of Corollary 4.4.3.1), and hence is left as Exercise 7.4.2.

Theorem 7.4.4. *The joint limiting distribution of the random variables* $n^{1/2}(\hat{\Delta}_{n,jj'} - \Delta_{jj'})$, $1 \le j < j' \le p$, *is a* $\binom{p}{2}$*-variate normal distribution with zero means and a covariance matrix* $\mathbf{\Gamma} = ((\gamma_{jj',ll'}))$, *where*

$$(7.4.18)\quad \tau_{jj',ll'} = \begin{cases} A^2/B^2, & j = l,\ j' = l', \\ \lambda_J(F^*)/B^2, & j = l,\ j' \ne l'\ \text{or}\ j \ne l,\ j' = l', \\ -\lambda_J(F^*)/B^2, & j = l',\ j' \ne l\ \text{or}\ j \ne l',\ j' = l, \\ 0, & j \ne j' \ne l \ne l', \end{cases}$$

where A^2, B, *and* $\lambda_J(F^*)$ *are defined by* (7.4.8) *and* (7.4.9).

The proof of the theorem follows directly from Theorems 6.2.3 and 7.4.3, and hence is left as Exercise 7.4.3.

Since the compatible estimators $Z_{n,jj'}$, defined by (7.4.4) and (7.4.5), are linear functions of $\{\hat{\Delta}_{n,jj'}, 1 \leq j < j' \leq p\}$, we arrive at the following theorem from Theorem 7.4.4 through a few simple steps.

Theorem 7.4.5. *The joint distribution of* $[n^{1/2}(Z_{n,jp} - \Delta_{jp}), 1 \leq j \leq p-1]$ *is asymptotically* $(p-1)$*-variate normal with null mean vector and a covariance matrix* $\boldsymbol{\beta} = ((\beta_{jj'}))$, *where*

$$\beta_{jj} = 2\{A^2 + (p-2)\lambda_J(F^*)\}/pB^2, \tag{7.4.19}$$

$$\beta_{jj'} = \{A^2 + (p-2)\lambda_J(F^*)\}/pB^2, \quad \textit{for} \quad j \neq j' = 1, \ldots, p-1, \tag{7.4.20}$$

and A^2, B, *and* $\lambda_J(F^*)$ *are defined by* (7.4.8) *and* (7.4.9).

Now, by Theorem 6.2.3, the random variable $n^{1/2}(\hat{\Delta}_{n,jj'} - \Delta_{jj'})$ has asymptotically a normal distribution with zero mean and variance A^2/B^2, where A^2 and B are defined by (7.4.8). Hence, the A.R.E. of the compatible estimator $Z_{n,jj'}$ with respect to the raw (incompatible) estimator $\hat{\Delta}_{n,jj'}$ is equal to

$$e_{Z_n,\hat{\Delta}_n} = pA^2/\{2[A^2 + (p-2)\lambda_J(G)]\}. \tag{7.4.21}$$

Now, since $\bar{F}_n(x)$ is symmetric about zero, so also is $F(x)$. Also $J(u)$ satisfies (7.4.10). Hence $J[F(x)]$ is a skew-symmetric function of x. Thus, from Lemma 7.4.2 and (7.4.9) we obtain that

$$\lambda_J(F^*) \leq \frac{1}{2}\int_0^1 J^2(u)\,du = \tfrac{1}{2}A^2, \tag{7.4.22}$$

where the equality sign holds only when $J[F(x)]$ = constant with probability one, i.e., $J(u) = F^{-1}(u)$, up to a multiplicative constant. Hence, from (7.4.21) and (7.4.22), we have

$$e_{Z_n,\hat{\Delta}_n} \geq 1, \tag{7.4.23}$$

where the equality sign holds *iff* $J[F(x)] = bx$, with probability one.

In general, if $\xi = \sum_{j=1}^{p} l_j\theta_{n,j}$ is any constant, we can write (as in (7.4.6)) the compatible estimator of ξ as $\hat{\xi}_{n(i)} = \sum_{j=1}^{p} l_j\hat{\Delta}_{n,j\cdot}$. On the other hand, the incompatible estimator is not unique. For example, for any given $j\,(= 1, \ldots, p)$ $\xi_{n(u)}^{(j)} = \sum_{j'=1}^{p} l_{j'}\hat{\Delta}_{n,j'j}$ is an incompatible estimator of ξ. However, with the aid of Theorems 7.4.4 and 7.4.5, we arrive at the following theorem which justifies the use of compatible estimators.

Theorem 7.4.6. *The A.R.E. of* $\hat{\xi}_{n(c)}$ *with respect to* $\xi_{n(u)}^{(j)}$ *is*

$$e_{c,2}^{(j)} = \frac{p(A^2 - \lambda_J(F^*))}{A^2 + (p-2)\lambda_J(F^*)} \geq \frac{pA^2}{2[A^2 + (p-2)\lambda_J(F^*)]} \geq 1, \tag{7.4.24}$$

where the equality sign holds only when $J[F(x)] = bx$, *with probability one.*

Let us next compare the compatible estimator $\hat{\xi}_{n(c)}$ and the conventional least squares estimator $\xi_n^* = \sum_{j=1}^{p} l_j X_{\cdot j}$, where $X_{\cdot j} = n^{-1} \sum_{i=1}^{n} X_{ij}$, $j = 1, \ldots, p$. The asymptotic variance of $\hat{\xi}_{n(c)}$ is obtained from Theorem 7.4.5 as

$$\left(\sum_{j=1}^{p} l_j^2\right)([A^2 + (p-2)\lambda_J(F^*)]/pB^2).$$

Also, if the variance of $\epsilon_{i,jj'}$ be denoted by σ_i^2, then $n^{-1}\sum_{i=1}^{n} \sigma_i^2$ converges (as $n \to \infty$) to σ^2, the variance of the c.d.f. $F(x)$. Consequently, it can be shown by the classical central limit theorem that if the X_{ij}'s have finite moments up to the order $2 + \delta$ for some $\delta > 0$, then $\sqrt{n}\,(\xi_n^* - \xi)$ has asymptotically a normal distribution with zero mean and variance $\frac{1}{2}(\sum_{j=1}^{p} l_j^2)\sigma^2$. Thus,

$$\text{(7.4.25)} \qquad e_{\hat{\xi}_{n(c)}, \xi_n^*} = \frac{p\sigma^2 B^2}{2[A^2 + (p-2)\lambda_J(F^*)]} = \frac{\sigma^2 B^2}{A^2} \cdot \frac{p}{2[1 + (p-2)\lambda_J(F^*)]} \geq \frac{\sigma^2 B^2}{A^2},$$

by Lemma 7.4.2 and (7.4.22). Now $\sigma^2 B^2/A^2$ is the A.R.E. of the allied one-sample rank order test with respect to the Student's t test. Hence, for normal scores, it follows from the results of chapter 3 that $\sigma^2 B^2/A^2 \geq 1$, where the equality sign holds iff $F(x)$ is normal. Hence the compatible normal scores estimator $\hat{\xi}_{n(c)}$ (i.e., with $J(u)$ as the inverse of the standard normal distribution) is asymptotically at least as efficient as the least squares estimator ξ_n^* for arbitrary $\ell \perp (1, \ldots, 1)$.

Finally, we desire to compare the compatible estimator $\hat{\xi}_{n(c)}$ with the rank order estimator based on the variables $V_1, \ldots, V_n$, defined in (7.4.1). Here, we assume that the distribution of $V_i - \xi$ is symmetric about 0 for $i = 1, \ldots, n$, and that the average of these distributions converges (as $n \to \infty$) to the c.d.f. $H(x)$. Then, from Theorem 6.3.2 we obtain that on denoting by $\hat{\xi}_{n(v)}$, the rank order estimator based on $V_1, \ldots, V_n$,

$$\text{(7.4.26)} \quad \mathcal{L}(n^{1/2}[\hat{\xi}_{n(v)} - \xi]) \to \mathcal{N}\left(0, A^2 \Big/ \left[\int_{-\infty}^{\infty} (d/dx)J[H(x)]\, dH(x)\right]^2\right).$$

Thus, the A.R.E. of $\hat{\xi}_{n(c)}$ with respect to $\hat{\xi}_{n(v)}$ is

$$\text{(7.4.27)} \qquad e_{\hat{\xi}_{n(c)}, \hat{\xi}_{n(v)}} = \frac{\sigma^2(F)\left[\int_{-\infty}^{\infty} (d/dx)J[F(x)]\, dF(x)\right]^2}{\sigma^2(H)\left[\int_{-\infty}^{\infty} (d/dx)J[H(x)]\, dH(x)\right]^2} \times \frac{pA^2}{2[A^2 + (p-2)\lambda_J(F^*)]},$$

where $\sigma^2(H)$ is the variance of the c.d.f. $H(x)$. Now, $\sigma^2(F)B^2/A^2$ is the A.R.E. of the rank test based on $T^*_{n,j}$ with respect to the Student's t test when the parent c.d.f. is $F(x)$, and we denote this by $e(J, F)$. Hence, we have

$$(7.4.28)\quad e_{\hat{\xi}_{n(c)},\hat{\xi}_{n(v)}} = \frac{e(J, F)}{e(J, H)} \frac{pA^2}{2[A^2 + (p-2)\lambda_J(G)]} \geq \frac{e(J, F)}{e(J, H)},$$

by Lemma 7.4.2. Now, the right-hand side of (7.2.28) depends, in general, on both the c.d.f.'s H and F, unless they differ by location and scale only (in which case, the ratio becomes unity). Thus, if both F and H are normal c.d.f.'s, (7.2.28) is bounded below by 1, and the lower bound is attained from the normal scores estimators, as then $\lambda_J(G) = \frac{1}{2}$.

7.4.3. Confidence Intervals and Tests for Contrasts

Consider now the problem of providing a confidence limit for a contrast $\xi = \sum_{j=1}^{p} l_j\theta_j$. We define $V_1, \ldots, V_n$ as in (7.4.1). If we have the reasons to accept the hypothesis that $V_i - \xi$ has a distribution symmetric about zero, then proceeding as in section 6.7 (namely as in (6.7.1)–(6.7.6)), we may attach an exact confidence interval to ξ based on suitable rank order statistics. Similarly, we may test for any H_0: $\xi = \xi_0$ (known), by using the rank statistics of section 3.6.2 based on the variables $V_1 - \xi_0, \ldots, V_n - \xi_0$.

The test and confidence intervals given above are subject to the same criticisms as the unadjusted rank order estimator based on $V_1, \ldots, V_n$, considered in the beginning of section 7.4. In large samples, we can overcome these difficulties as follows. By Theorem 7.4.5, we have for the compatible estimator $\hat{\xi}_{n(c)}$ in (7.4.6),

$$(7.4.29)\quad \mathcal{L}(n^{1/2}(\hat{\xi}_{n(c)} - \xi)) \to \mathcal{N}\left(0, \left(\sum_{j=1}^{p} l_j^2\right)([A^2 + (p-2)\lambda_J(F^*)]/pB^2)\right),$$

where A^2, B and $\lambda_J(F^*)$ are defined by (7.4.8) and (7.4.9). Thus, the unknown quantities entering into the expression for the asymptotic variance are $\lambda_J(F^*)$ and B, and these we estimate as follows.

As in section 7.4.2, we define $J(u) = \Psi^{*-1}(u)$, $0 < u < 1$, and let $a_{n,\alpha} = J_n(\alpha/(n+1))$ be the expected value of the αth-order statistic of a sample of size n from the distribution $\Psi^*(x)$, $\alpha = 1, \ldots, n$. Also, we define $X^*_{i,jj'}$ as in (7.4.3), and let $R_{i,jj'}$ be the rank of $X^*_{i,jj'}$ among $(X^*_{1,jj'}, \ldots, X^*_{n,jj'})$ for $i = 1, \ldots, n$ and $j' > j = 1, \ldots, p$. Finally, let

$$(7.4.30)\quad L_{j:j',l,n} = \frac{1}{n}\sum_{\alpha=1}^{n} a_{n,R_{\alpha,jj'}} a_{n,R_{\alpha,jl}}, \qquad j \neq j' \neq l = 1, \ldots, p.$$

Theorem 7.4.7.

$$(7.4.31)\quad L_n = \frac{2}{p(p-1)(p-2)} \sum_{j=1}^{p} \sum_{j'>l=1}^{p} L_{j:j'l,n}, \qquad j \neq j' \neq l,$$

is a translation invariant consistent estimator of $\lambda_J(F^*)$.

Proof: Since the ranks are invariant under the change of origin, if we define $Y_{ijj'} = X^*_{i,jj'} - \Delta_{jj'}$, $i = 1, \ldots, n$, the rank of $Y_{ijj'}$ among $Y_{1jj'}, \ldots, Y_{njj'}$ will also be $R_{ijj'}$, $i = 1, \ldots, n$. Define then

$$F_{n[j,j']}(x) = \frac{1}{n}\,[\text{number of } Y_{i,jj'} \le x,\ i = 1, \ldots, n], \tag{7.4.32}$$

$$F^*_{n[j:j'l]}(x, y) = \frac{1}{n}\,[\text{number of } (Y_{i,jj'}, Y_{i,jl}) \le (x, y),\ i = 1, \ldots, n]. \tag{7.4.33}$$

Then (7.4.30) can be rewritten as

$$\int_{-\infty}^{\infty}\int_{-\infty}^{\infty} J_n[F_{n[j,j']}(x)]J_n[F_{n[j,l]}(y)]\,dF^*_{n[j:j',l]}(x, y). \tag{7.4.34}$$

Now, by definition, $E[F_{n[j,j']}(x)] = F(x)$ for all $j \neq j' = 1, \ldots, p$ and $E[F^*_{n[j:j',l]}(x, y)] = F^*(x, y)$ for all $j \neq j' \neq l = 1, \ldots, p$. Further, $(Y_{ijj'}, Y_{ijl})$, $i = 1, \ldots, n$ are independent random vectors. Hence, proceeding as in the proof of Theorem 4.1, it follows that (7.4.34) converges, in probability (as $n \to \infty$), to $\lambda_J(F^*)$, for all $j \neq j' \neq l = 1, \ldots, p$. Since L_n is the arithmetic mean of the quantities in (7.4.34), the rest of the proof follows trivially. Q.E.D. ◁

Now, by definition, $X^*_{i,jj'}$, $i = 1, \ldots, n$ are independent random variables whose distributions are symmetric about the common median $\Delta_{jj'}$. Hence, proceeding as in (6.7.1) to (6.7.6), we obtain the result

$$P\{\Delta^*_{jj',L,n} \le \Delta_{jj'} \le \Delta^*_{jj',U,n} \mid \Delta_{jj'}\} = 1 - \alpha, \tag{7.4.35}$$

where $\Delta^*_{jj',L,n}$ and $\Delta^*_{jj',U,n}$ are defined as in (6.7.3) and (6.7.4). Then, define

$$\hat{B}_{jj',n} = (2A\tau_{\alpha/2})/[n^{1/2}(\Delta^*_{jj',U,n} - \Delta^*_{jj',L,n})], \tag{7.4.36}$$

for $j' > j = 1, \ldots, p$;

$$\hat{B}_n = \binom{p}{2}^{-1} \sum_{j<j'=1}^{p} \hat{B}_{jj',n}. \tag{7.4.37}$$

Theorem 7.4.8. *$\hat{B}_n$ is a translation invariant consistent estimator of B, defined by* (7.4.8).

Proof: Proceeding as in the proof of Theorem 6.7.1, it follows that $\hat{B}_{jj',n}$ is a translation invariant consistent estimator of B, defined by (7.4.8), for all $1 \le j < j' \le p$. Hence, $\hat{B}_n$, being the arithmetic mean of several consistent and translation invariant estimators, is also so. Q.E.D. ◁

Once B and $\lambda_J(F^*)$ are estimated by $\hat{B}_n$ and L_n, respectively, we have no difficulty in providing a confidence interval for $\xi = \sum_{j=1}^{p} l_j\theta_j$ (or in testing any hypothetical value of ξ), based on the asymptotic normality of

$$n^{1/2}(\hat{\xi}_{n(c)} - \xi)\Big/\left\{\left(\sum_{j=1}^{p} l_j^2\right)[A^2 + (p - 2)L_n]\Big/ p\hat{B}_n^2\right\}^{1/2}. \tag{7.4.38}$$

Now, suppose we have $r(1 \le r \le p - 1)$ linearly independent contrasts

$$(7.4.39) \qquad \xi_s = \sum_{j=1}^{p} l_{sj}\theta_j; \qquad \sum_{j=1}^{p} l_{sj} = 0, \qquad s = 1, \ldots, r.$$

We desire to have a simultaneous confidence region for $\xi_1, \ldots, \xi_r$ or to test the null hypothesis that $\xi_s = \xi_s^0$ (known), $s = 1, \ldots, r$. Denote by

$$(7.4.40) \quad \mathbf{T} = ((\tau_{ss'})), \quad \tau_{ss'} = \sum_{j=1}^{p} l_{sj} l_{s'j}, \qquad s, s' = 1, \ldots, r_s, \quad \mathbf{T}^{-1} = ((\tau^{ss'})).$$

Then, by a straightforward generalization of Theorem 7.4.5 and using Theorems 7.4.7 and 7.4.8, we obtain the following.

Theorem 7.4.9. *Let $\tilde{\xi}_{n(c),s}$ be the compatible estimator of ξ_s, $s = 1, \ldots, r$. Then, the statistic*

$$(7.4.41) \qquad S_n = \left[\sum_{s=1}^{r} \sum_{s'=1}^{r} \tau^{ss'}(\tilde{\xi}_{n(c),s} - \xi_s)(\tilde{\xi}_{n(c),s'} - \xi_{s'})\right] \times \left[\frac{n}{p\hat{B}_n^2}(A^2 + (p-2)L_n)\right]$$

has asymptotically a chi-square distribution with r D.F.

The statistic S_n may be used to provide a large-sample test for H_0: $\xi_s = \xi_s^0$, $s = 1, \ldots, r$ or to construct a confidence ellipsoid for $(\xi_1, \ldots, \xi_r)$.

7.4.4. Estimation in Incomplete Block Designs

We consider an incomplete block design D consisting of I blocks of (constant) size b to which p ($\ge b$) treatments are applied, there being n_i replication of the ith block, for $i = 1, \ldots, I$. Let $n = \sum_{i \in I} n_i$, and let S_i consist of the numbers of the b treatments applied in the ith block, $i = 1, \ldots, I$. The observable random variables are

(7.4.42)

$$X_{ij\alpha} = \mu_i + \beta_{i\alpha} + \xi_j + U_{ij\alpha}, \qquad \alpha = 1, \ldots, n_i, \qquad j \in S_i, \quad i = 1, \ldots, I.$$

where μ_i is the ith replication effect, $\beta_{i\alpha}$ is the effect of the αth block in the ith replication set, ξ_j is the jth treatment effect, and the error variables $\{U_{ij\alpha}, j \in S_i\}$ have a joint c.d.f. $G(x_1, \ldots, s_b)$ (for all $\alpha = 1, \ldots, n_i$, $i = 1, \ldots, I$), where G is symmetric in its arguments. [Since not all treatments occur in all blocks, the assumption of homogeneity of the distribution of the errors simplifies the asymptotic theory considerably. If we allow G to depend on i and α, the average (for each i) over α, will, in general, be different from each other, and make the results complicated.] We denote by F the c.d.f. of $U_{ij\alpha} - U_{ij\alpha}$, $j, j' \in S_i$, and by $F^*(x, y)$ the joint c.d.f. of $U_{ij\alpha} - U_{ij'\alpha}$,

$U_{ij\alpha} - U_{il\alpha}$ where $j, j', l \in S_i$. We may set, without any loss of generality,

$$\sum_{j=1}^{p} \xi_j = 0 \quad \text{and} \quad \sum_{\alpha=1}^{n_i} \beta_{i\alpha} = 0 \quad \text{for} \quad i = 1, \ldots, I. \tag{7.4.43}$$

Let us now define

(7.4.44)

$$Y_{i,jj'.\alpha} = X_{ij\alpha} - X_{ij'\alpha}, \qquad \Delta_{jj'} = \xi_j - \xi_{j'}, \qquad U_{i,jj'.\alpha} = U_{ij\alpha} - U_{ij'\alpha},$$

for all $\alpha = 1, \ldots, n_i$, $j, j' \in S_i$, $i = 1, \ldots, I$. Based on the n_i observations $Y_{i,jj'.\alpha}$, $\alpha = 1, \ldots, n_i$, we consider the rank order estimates $\hat{\Delta}^{(i)}_{jj'}$, defined as in (6.2.9) and (6.2.10). Let us denote

$$\hat{\Delta}^{(i)}_{j\cdot} = b^{-1} \sum_{j' \in S_i} \hat{\Delta}^{(i)}_{jj'}, \qquad j \in S_i, \quad \text{where} \quad \hat{\Delta}^{(i)}_{jj} = 0. \tag{7.4.45}$$

Then, the adjusted or the compatible estimators of Δ_{jj} are defined by

$$Z^{(i)}_{jj'} = \hat{\Delta}^{(i)}_{j\cdot} - \hat{\Delta}^{(i)}_{j'\cdot}, \qquad j, j' \in S_i, \quad i = 1, \ldots, I. \tag{7.4.46}$$

For the study of the asymptotic distribution theory of the estimates in (7.4.46), we assume that

$$\lim_{n \to \infty} n_i/n = \rho_i: \quad 0 < \rho_i < 1, \quad i = 1, \ldots, I. \tag{7.4.47}$$

Then we have the following.

Theorem 7.4.10. *The random variables* $n^{1/2}(\hat{\Delta}^{(i)}_{jj'} - \Delta_{jj'})$, $j, j' \in S_i$, $i = 1, \ldots, I$ *have (jointly) asymptotically a multinormal distribution with means zero and a covariance matrix having the elements*

(7.4.48) $\sigma_{ijj',i'll'}$

$$= \begin{cases} 0, & i \neq i' \text{ or } i = i', \ j \neq l \neq j' \neq l'; \\ A^2/B^2\rho_i, & i = i', \ j = l, \ j' = l'; \\ S^2/2\rho_i, & i = i', \ j = l, \ j' \neq l', \ j \neq l', \text{ or } j \neq l, \ j' = l', \ j \neq l'; \\ -S^2/2\rho_i, & i = i', \ j = l', \ j' \neq l \neq j, \text{ or } j \neq l' \neq j', \ j' = l, \end{cases}$$

or $j, j' \in S_i$, $l, l' \in S_{i'}$, $i, i' = 1, \ldots, I$, *where*

$$S^2 = (2/b)[A^2 + (b-2)\lambda_J(F^*)]/B^2, \tag{7.4.49}$$

and A^2, B, *and* $\lambda_J(F^*)$ *are defined by* (7.4.8) *and* (7.4.9) *respectively. Hence,* $n^{1/2}(Z^{(i)}_{jj'} - \Delta_{jj'})$, $j, j' \in S_i$, $i = 1, \ldots, I$ *have asymptotically a (singular) normal distribution with mean* 0 *and a covariance matrix with elements* $\sigma^*_{ijj',i'll'}$, *where*

$$\begin{aligned} \sigma^*_{ijj',i'll'} &= \sigma_{ijj',i'll'} \quad \text{if} \quad (j, j') \neq (l, l'), \\ &= S^2/\rho_i \quad \text{if} \quad i = i', \ j = l, \ j' = l'. \end{aligned} \tag{7.4.50}$$

The proof of the first part of the theorem follows along the lines of Theorems 7.4.3 and 7.4.4, while the second part follows from the first part and (7.4.45)–(7.4.46). The details are left as Exercise 7.4.4.

Let us now define

$$C_{jj'}^{(i)} = n_i^{-1}\sum_{\alpha=1}^{n_i}(X_{ij\alpha} - X_{ij'\alpha}), \quad j, j' \in S_i, \quad i = 1, \ldots, I. \tag{7.4.51}$$

Then, the classical least squares estimator of $\Delta_{jj'}$ has the form

$$C_{jj'}^{*} = \sum_{i=1}^{I}\sum_{l,l' \in S_i} A_{ll',i}^{(j,j')} C_{ll'}^{(i)}, \tag{7.4.52}$$

where $A_{ll',i}^{(j,j')}$ are constants. It is easy to verify (cf. Exercise 7.4.4) that the covariance matrix of the random variables $n^{1/2}(C_{jj'}^{(i)} - \Delta_{jj'})$, $j, j' \in S_i$, ($i \doteq 1, \ldots,$) has the elements

(7.4.53)

$$\gamma_{ijj',i'll'} = \begin{cases} 0, & i \neq i' \ \text{ or } \ i = i' \ \text{ but } \ j \neq j' \neq l \neq l', \\ \sigma^2(F)/\rho_i, & i = i', \ j = l, \ j' = l', \\ \sigma^2(F)/2\rho_i, & i = i', \ j = l, \ j' \neq l', \ \text{ or } \ j \neq l, \ j' = l; \\ -\sigma^2(F)/2\rho_i, & i = i', \ j = l', \ j' \neq l', \ \text{ or } \ j \neq l', \ j' = l, \end{cases}$$

where $\sigma^2(F)$ is the variance of the c.d.f. F. Thus,

$$\sigma_{ijj',i'll'}^{*}/\gamma_{ijj',i'll'} = S^2/\sigma^2(F), \quad \text{when} \quad \gamma_{ijj',i'll'} \neq 0, \tag{7.4.54}$$

and otherwise, both of them are equal to zero. Since the two covariance matrices are proportional to each other, the same linear function of the $Z_{jj'}^{(i)}$ as the classical least squares estimate is of the $C_{jj}^{(i)}$ (i.e., (7.4.52)) has minimum asymptotic variance among unbiased estimates which are linear functions of the $Z_{jj'}^{(i)}$. This leads to the asymptotically optimum estimator

$$Z_{jj'}^{*} = \sum_{i=1}^{I}\sum_{l,l' \in S_i} A_{ll',i}^{(j,j')} Z_{ll'}^{(i)}, \tag{7.4.55}$$

where the constants $A_{ll',i}^{(j,j')}$ are the same as in (7.4.52). It is easy to verify that $Z_{jj'}^{*} + Z_{j'l}^{*} = Z_{jl}^{*}$ for all $j \neq j' \neq l$ (cf. Exercise 7.4.5). Thus, for any contrast $\sum_{j=1}^{p} l_j \xi_j$, the corresponding estimator $\sum_{j=1}^{T} l_j Z_{jj}^{*}$ (for any $j' = 1, \ldots, p$) is a compatible estimator.

Let us now compare the estimates in (7.4.52) and (7.4.55). Since the dispersion matrices in (7.4.48) and (7.4.53) are proportional, and the coefficients in (7.4.52) and (7.4.55) are the same, it follows that $[n^{1/2}(C_{jj'}^{*} - \Delta_{jj'}),$ $\neq j = 1, \ldots, p]$ and $[(\sigma(F)/S)n^{1/2}(Z_{jj'}^{*} - \Delta_{jj'}), \ j \neq j' = 1, \ldots, p]$ both have the same limiting normal distribution. Hence, the A.R.E. of $Z_{jj'}^{*}$ with respect to $C^*_{jj'}$ is

(7.4.56) $$e^*_{Z,C} = \frac{\sigma^2(F)}{S^2} = \left\{\frac{\sigma^2(F)B^2}{A^2}\right\}\left\{\frac{bA^2}{2[A^2 + (b-2)\lambda_J(F^*)]}\right\}$$
$$= e(F)\left\{\frac{bA^2}{2[A^2 + (b-2)\lambda_J(F^*)]}\right\},$$

where $e(F) = \sigma^2(F)B^2/A^2$ represents the A.R.E. of the two-sample rank order statistic with respect to the Student's t statistic. By virtue of (7.4.22), the second factor on the right-hand side of (7.4.56) is (a) bounded below by 1 and (b) an increasing function of b: $2 \leq b \leq p$. Thus,

(7.4.57) $$e(F) \leq e^*_{Z,C} \leq e(F)pA^2/\{2[A^2 + (p-2)\lambda_J(F^*)]\}.$$

By virtue of the compatibility of the $Z^*_{jj'}$ and of (7.4.48), (7.4.52), (7.4.53), and (7.4.55), it follows that (7.4.56) and (7.4.57) hold also for the corresponding estimators of contrasts among $\xi_1, \ldots, \xi_p$. Since the bounds for the left-hand side of (7.4.57) for specific scores are already studied in chapter 3, and for the right-hand side of (7.4.57) in section 7.4.2, we have no difficulty in studying similar bounds for $e^*_{Z,C}$.

7.4.5. Multivariate Generalizations

Consider now the model (7.3.78), where $\mathbf{X}_{ij} = (X^{(1)}_{ij}, \ldots, X^{(q)}_{ij})'$, $j = 1, \ldots, p$, $i = 1, \ldots, n$. As in (7.4.3), define $\mathbf{X}^*_{i,jj'}$, $\mathbf{\Delta}_{jj'}$, and $\boldsymbol{\epsilon}_{i,jj'}$, each of which is now a q vector. For each of the q variates, consider the coordinate-wise raw and adjusted estimators as in section 7.4.1, and let the vector of adjusted estimators so obtained be denoted by

(7.4.58) $$\mathbf{Z}_{n,jj'} = (Z^{(1)}_{n,jj'}, \ldots, Z^{(q)}_{n,jj'})', \qquad j \neq j' = 1, \ldots, p.$$

For any contrast (vector) $\boldsymbol{\xi} = \sum_{j=1}^p l_j \boldsymbol{\theta}_j$, the corresponding adjusted estimator is obtained as in (7.4.6) as

(7.4.59) $$\boldsymbol{\xi}_{n(c)} = \sum_{j=1}^p \sum_{j'=1}^p (l_j/p)\mathbf{Z}_{n,jj'}.$$

The marginal c.d.f. of $\epsilon^{(k)}_{i,jj'}$ is denoted by $F_{i[k]}(x)$, and the joint c.d.f. of $\epsilon^{(k)}_{i,jj'}$, $\epsilon^{(k')}_{i,jl}$ is denoted by

(7.4.60) $$\begin{cases} F^{(1)}_{i[k,k']}(x, y) & \text{if } j' = l, \quad k \neq k', \\ F^{(2)}_{i[k,k']}(x, y) & \text{if } j' \neq l, \quad k \neq k', \\ F^{(2)}_{i[k,k]}(x, y) & \text{if } j \neq l, \quad k = k'. \end{cases}$$

The corresponding average c.d.f.'s $F_{n[k]}$, $F^{(1)}_{n[k,k']}$, $F^{(2)}_{n[k,k']}$ and $F^{(2)}_{n[k,k]}$ are defined as in (7.4.7), and under the models of section 7.2.4, it is assumed that these c.d.f.'s converge (as $n \to \infty$) to $F_{[k]}$, $F^{(1)}_{[k,k']}$, $F^{(2)}_{[k,k']}$, and $F^{(2)}_{[k,k]}$, respectively. As in (6.2.7), the limiting score function for the kth variate is denoted by $J_k(u)$: $0 < u < 1$, $k = 1, \ldots, q$. Let then

$$B_k = B(F_{[k]}, J_k) = \int_{-\infty}^{\infty} (d/dx) J_k[F_{[k]}(x)]\, dF_{[k]}(x), \quad k = 1, \ldots, q; \tag{7.4.61}$$

$$\nu^{(1)}_{kk} = \int_0^1 J_k^2(u)\, du; \tag{7.4.62}$$

$$\nu^{(1)}_{kk'} = \int_{-\infty}^{\infty}\int_{-\infty}^{\infty} J_k[F_{[k]}(x)] J_{k'}[F_{[k']}(y)]\, dF^{(1)}_{[k,k']}(x, y),$$

for $k \neq k' = 1, \ldots, q$; $\boldsymbol{\nu}^{(1)} = ((\nu^{(1)}_{kk'}))$, and

$$\nu^{(2)}_{kk'} = \int_{-\infty}^{\infty}\int_{-\infty}^{\infty} J_k[F_{[k]}(x)] J_{k'}[F_{[k']}(y)]\, dF^{(2)}_{[k,k']}(x, y), \tag{7.4.63}$$

$$k, k' = 1, \ldots, q;$$

$$\boldsymbol{\nu}^{(2)} = ((\nu^{(2)}_{kk'})). \tag{7.4.64}$$

As in (7.4.10), $J_k(u)$, $k = 1, \ldots, q$ are all assumed to be skew-symmetric.

Theorem 7.4.11. *The joint limiting distribution of the $q\binom{p}{2}$ random variables $n^{1/2}(\hat{\Delta}^{(k)}_{n,jj'} - \Delta^{(k)}_{jj'})$, $j < j' = 1, \ldots, p$, $k = 1, \ldots, q$ is multinormal with means zero and covariances*

$$\gamma^{(k,k')}_{jj',ll'} = \begin{cases} \nu^{(1)}_{kk'}/B_k B_{k'}, & \text{if } j = l,\ j' = l', \\ \nu^{(2)}_{kk'}/B_k B_{k'}, & \text{if } j = l,\ j' \neq l' \text{ or } j \neq l,\ j' = l', \\ -\nu^{(2)}_{kk'}/B_k B_{k'}, & \text{if } j \neq l',\ j' = l \text{ or } j \neq l',\ j' = l, \\ 0, & \text{otherwise}, \end{cases} \tag{7.4.65}$$

where $\hat{\Delta}^{(k)}_{n,jj'}$ is the raw (unadjusted) estimator of $\Delta^{(k)}_{jj'}$.

The proof is a direct extension of that of Theorem 7.4.4, and hence is left as Exercise 7.4.6.

Now, the adjusted estimators $\mathbf{Z}_{n,jj'}$ are defined by $\hat{\boldsymbol{\Delta}}_{n,j\cdot} - \hat{\boldsymbol{\Delta}}_{n,j'\cdot}$, where $\hat{\boldsymbol{\Delta}}_{n,j\cdot} = p^{-1}\sum_{j'=1}^{p} \hat{\boldsymbol{\Delta}}_{n,jj'}$, $\hat{\boldsymbol{\Delta}}_{n,jj} = \hat{\boldsymbol{\Delta}}_{n,jj} = \mathbf{0}$. Hence, from the above theorem, we obtain the following.

Theorem 7.4.12. *The joint distribution of $[n^{1/2}(\mathbf{Z}_{n,jp} - \boldsymbol{\Delta}_{jp})$, $j = 1, \ldots, p-1]$ is asymptotically $q(p-1)$-variate normal with zero means and covariances*

$$\beta^{(k,k')}_{jj'} = (1 + \delta_{jj'})(\nu^{(1)}_{kk'} + (p-2)\nu^{(2)}_{kk'})/(pB_k B_{k'}), \tag{7.4.66}$$

for $k, k' = 1, \ldots, q$, $j, k' = 1, \ldots, p-1$, where $\delta_{jj'}$ is the Kronecker delta.

It may be noted that $\nu_{kk}^{(2)} \leq \frac{1}{2}\nu_{kk}^{(1)}$ (by Lemma 7.4.2), but such an inequality is not necessarily true for $\nu_{kk'}^{(2)}$ and $\nu_{kk'}^{(1)}$. Thus, for any (fixed) k $(= 1, \ldots, q)$, the efficiency results of section 7.4.2 hold, but these are not necessarily true for the vectors $\mathbf{Z}_{n,jj'}$ and $\hat{\mathbf{\Delta}}_{n,jj'}$ as a whole. The A.R.E. of $\hat{\mathbf{\Delta}}_{n,jj'}$ with respect to $\mathbf{Z}_{n,jj'}$, as measured by the generalized variances of the estimates (cf. (6.4.2)), comes out as

$$\left\{ \begin{vmatrix} \nu_{11}^{*} & \cdots & \nu_{1q}^{*} \\ \cdot & & \\ \cdot & & \\ \cdot & & \\ \nu_{q1}^{*} & \cdots & \nu_{qq}^{*} \end{vmatrix} \div \begin{vmatrix} \nu_{11}^{(1)} & \nu_{12}^{(1)} & \cdots & \nu_{1q}^{(1)} \\ & \cdot & & \\ & \cdot & & \\ & \cdot & & \\ \nu_{q1}^{(1)} & \nu_{q2}^{(1)} & \cdots & \nu_{qq}^{(1)} \end{vmatrix} \right\}^{1/q}, \tag{7.4.67}$$

where $\nu_{kk'}^{*} = 2\{\nu_{kk'}^{(1)} + (p-2)\nu_{kk'}^{(2)}\}/p$ for $k, k' = 1, \ldots, q$. The actual value is bounded below and above by the characteristic roots of

$$\left[\frac{2}{p}\boldsymbol{\nu}^{(1)} + \frac{2(p-2)}{p}\boldsymbol{\nu}^{(2)}\right][\boldsymbol{\nu}^{(1)}]^{-1}$$

i.e., of

$$\frac{2}{p}\mathbf{I}_q + \left[\frac{2(p-2)}{p}\right]\boldsymbol{\nu}^{(2)}(\boldsymbol{\nu}^{(1)})^{-1}$$

(For specific scores, see Exercises 7.4.7 and 7.4.8 for these bounds).

Now, Theorem 7.4.7 can be used to estimate each of the quantities $\nu_{kk'}^{(1)}$, $k \neq k' = 1, \ldots, q$ and $\nu_{kk'}^{(2)}$, $k, k' = 1, \ldots, q$, and Theorem 7.4.8 extends for each $k = 1, \ldots, q$. Hence, we have no problem in extending the results of Section 7.4.3 to the multivariate case.

7.5. RANK ORDER TESTS FOR ORDERED ALTERNATIVES IN RANDOMIZED BLOCKS

Often, in testing the null hypothesis H_0: $\boldsymbol{\theta} = \mathbf{0}$ (referred to (7.3.1)), we have the set of alternative hypotheses specified by

$$K^{*}: \theta_1 \leq \cdots \leq \theta_p, \quad \text{with at least one strict inequality.} \tag{7.5.1}$$

The rank procedures considered in sections 7.2–7.4 can be used to develop suitable tests for this problem.

(a) Procedures Based on Intrablock Rankings. Let q_i be any measure of agreement of the ranks $r_{i1}, \ldots, r_{ip}$ (of $X_{i1}, \ldots, X_{ip}$ among themselves) with the natural integers $1, \ldots, p$ for $i = 1, \ldots, n$. Then, a test procedure may be based on the statistic

(7.5.2) $$Q_n = \sum_{i=1}^{n} q_i.$$

Two particular tests are notable and are due to Page (1963) and Jonckheere (1954b). Jonckheere's test is based on

(7.5.3) $$J_n = \sum_{i=1}^{n} t_i,$$

where t_i is the Kendall τ between $(r_{i1}, \ldots, r_{ip})$ and $(1, \ldots, p)$, i.e.,

(7.5.4) $$t_i = \binom{p}{2}^{-1} \sum_{\alpha<\beta} c(\beta - \alpha)c(r_{i\beta} - r_{i\alpha}), \qquad i = 1, \ldots, n,$$

where $c(u)$ is 1, 0, or -1 according as $u >, =,$ or <0. The test by Page is based on

(7.5.5) $$P_n = \sum_{i=1}^{n} S_i,$$

where S_i is the Spearman rank correlation between $(r_{i1}, \ldots, r_{ip})$ and $(1, \ldots, p)$, i.e.,

(7.5.6) $$S_i = \frac{12}{p(p^2-1)} \sum_{\alpha=1}^{p} \left(\alpha - \frac{p+1}{2}\right)\left(r_{i\alpha} - \frac{p+1}{2}\right), \qquad i = 1, \ldots, n.$$

Since under the null hypothesis $\boldsymbol{\theta} = \mathbf{0}$, the distributions of t_i (or S_i), $i = 1, \ldots, n$ are identical and completely known, the distribution of J_n (or P_n) can be computed by convolution. Thus, the tests based on J_n or P_n are distribution-free. Since t_i (or S_i), $i = 1, \ldots, n$ are all finite-valued random variables, the asymptotic normality of $n^{1/2}(J_n - EJ_n)$ or $n^{1/2}(P_n - EP_n)$ readily follows from the classical central limit theorems. It follows by routine computations (cf. Exercise 7.5.1) that under H_0: $\boldsymbol{\theta} = \mathbf{0}$,

(7.5.7) $$\mathcal{L}(n^{-1/2}J_n/\sigma_J) \to \mathcal{N}(0, 1),$$

(7.5.8) $$\mathcal{L}(n^{-1/2}P_n/\sigma_P) \to \mathcal{N}(0, 1),$$

where

(7.5.9) $$\sigma_J^2 = 2(2p+5)/9p(p-1) \quad \text{and} \quad \sigma_P^2 = 1/(p-1).$$

For the study of the asymptotic power properties of these tests, we conceive of the following sequence of alternative hypotheses

(7.5.10) $$\{K_n^*\}, \quad \text{where} \quad K_n^*\colon \boldsymbol{\theta} = n^{-1/2}\boldsymbol{\lambda},$$

where $\boldsymbol{\lambda} = (\lambda_1, \ldots, \lambda_p)$ and $\lambda_1 \leq \cdots \leq \lambda_p$. Let us denote by

(7.5.11) $$\rho(\boldsymbol{\lambda}) = [12/p(p^2-1)] \sum_{\alpha=1}^{p} \left(\alpha - \frac{p+1}{2}\right)\lambda_\alpha.$$

As in section 7.2, we denote the c.d.f. of X_{ij} by $F_{ij} = F_i(x - n^{-1/2}\lambda_j)$, $j = 1, \ldots, p$, $i = 1, \ldots, n$, and under the models of section 7.2.4, we assume that $\bar{F}_n = n^{-1/2} \sum_{i=1}^n F_i \to F$ as $n \to \infty$. Let us then assume that $F(x)$ has a continuous density function $f(x)$. Then, using (7.5.2), (7.5.3), and (7.5.9), we obtain after some simplifications that

$$(7.5.12)\quad E\{n^{-1/2}J_n \mid K_n^*\} = (\tfrac{2}{3})(p + 1)\rho(\boldsymbol{\lambda})\left[\int_{-\infty}^{\infty} f^2(x)\,dx\right]^2 + o(1),$$

$$(7.5.13)\quad \text{var}\,\{n^{-1/2}J_n \mid K_n^*\} = \sigma_J^2 + o(1) = 2(2p + 5)/9p(p - 1) + o(1).$$

Thus, under $\{K_n^*\}$

$$(7.5.14)\qquad \mathcal{L}(n^{-1/2}J_n/\sigma_J) \to \mathcal{N}(\tau, 1),$$

where

$$(7.5.15)\quad \tau^2 = [p(p^2 - 1)2(p + 1)/(2p + 5)]\rho^2(\boldsymbol{\lambda})\left[\int_{-\infty}^{\infty} f^2(x)\,dx\right]^2.$$

Similarly, from (7.5.4), (7.5.5), and (7.5.9), we obtain that

$$(7.5.16)\qquad E\{n^{-1/2}P_n \mid K_n^*\} = \rho(\boldsymbol{\lambda})p\left(\int_{-\infty}^{\infty} f^2(x)\,dx\right) + o(1),$$

Thus, under $\{K_n^*\}$,

$$(7.5.17)\qquad \mathcal{L}(n^{-1/2}P_n/\sigma_p) \to \mathcal{N}(\xi, 1),$$

where

$$(7.5.18)\qquad \xi^2 = p^2(p - 1)\rho^2(\boldsymbol{\lambda})\left[\int_{-\infty}^{\infty} f^2(x)\,dx\right]^2.$$

Thus, from (7.5.13), (7.5.14), (7.5.17), and (7.5.18), we obtain that the A.R.E. of the test based on J_n with respect to the one based on P_n is

$$(7.5.19)\qquad e_{J,P} = 2(p + 1)^2/p(2p + 5),$$

which is independent of the parent c.d.f. $F(x)$: Actual computations show that $e_{J,P}$ is decreasing in p for $2 \le p \le 5$ and then increasing in p (≥ 6). Also,

$$(7.5.20)\qquad 0.96 = \tfrac{24}{25} \le e_{J,P} \le 1.$$

It is quite clear that the optimum (i.e., the likelihood ratio) test for H_0: $\boldsymbol{\theta} = \mathbf{0}$ (vs. K_n^*) not only requires the knowledge of the parent c.d.f.'s, but also of $\boldsymbol{\lambda}$ in (7.5.9). The test mostly used in practice is based on the assumption of normal parent distributions with common variance (σ^2), and for which $\lambda_i = \lambda_1 + (i - 1)c$, $c > 0$, $i = 1, \ldots, p$. The test statistic is

$$(7.5.21)\qquad Q_n = \left[\frac{12}{p(p^2 - 1)}\right]^{1/2} \sum_{j=1}^{p}\left(j - \frac{p + 1}{2}\right)\left(\frac{1}{n}\sum_{i=1}^{n} X_{ij}\right)\Big/\hat{\sigma},$$

where $\hat{\sigma}^2$ is a suitable estimate of the common variance σ^2. It is easy to verify that the test based on Q_n is consistent for any ordered alternative. If X_{ij}'s have all finite moments up to the order $2 + \delta$, $\delta > 0$, it follows by standard techniques using the central limit theorem (cf. Exercise 7.5.3) that

$$\mathcal{L}(n^{1/2}Q_n \mid H_0) \to \mathcal{N}(0, 1). \tag{7.5.22}$$

Also, under $\{K_n^*\}$

$$\mathcal{L}(n^{1/2}Q_n) \to \mathcal{N}(Q, 1), \tag{7.5.23}$$

where

$$Q^2 = [p(p^2 - 1)/12][\rho^2(\boldsymbol{\lambda})/\sigma^2], \tag{7.5.24}$$

where σ^2 is the limit of the average variance $\bar{\sigma}_n^2 = (1/n)\sum_{i=1}^n \sigma_i^2$. Thus, for $\{K_n^*\}$, we obtain the following results:

$$e_{J,Q} = [2(p + 1)/(2p + 5)]\left[12\sigma^2\left(\int_{-\infty}^{\infty} f^2(x)\,dx\right)^2\right], \tag{7.5.25}$$

where the second factor on the right-hand side of (7.4.25) is the A.R.E. of the Wilcoxon test with respect to the Student test, while the first factor is $\uparrow$ in p: $2 \leq p < \infty$ and is bounded below and above by $\frac{2}{3}$ and 1, respectively;

$$e_{P,Q} = [p/(p + 1)]\left[12\sigma^2\left(\int_{-\infty}^{\infty} f^2(x)\,dx\right)^2\right], \tag{7.5.26}$$

which agrees with the A.R.E. of the Friedman test with respect to the classical ANOVA test (cf. (7.2.43)).

For small values of p and nearly normal c.d.f.'s, the values of (7.5.25) and (7.5.26) are usually quite low. We shall now consider some aligned rank order tests which are of higher efficiency.

Define the aligned rank order statistics $T_{N,j}$, $j = 1, \ldots, p$ as in (7.3.4) and $\sigma^2(\mathcal{T}_n)$ as in (7.3.6). Then, consider the statistic

$$T_N^* = \{12n/p(p^2 - 1)\sigma^2(\mathcal{T}_n)\}^{1/2}\sum_{j=1}^{p}\left(j - \frac{p + 1}{2}\right)(T_{N,j} - \bar{E}_N). \tag{7.5.27}$$

Under $\mathcal{T}_n$, the distribution of T_N^* (over the $(p!)^n$ conditionally equally likely realizations of the collection rank matrices) can be used (for small values of n) to determine the critical values of T_N^*. For large n, we may proceed as in Theorem 7.3.3 and show that

$$\mathcal{L}_{\mathcal{T}_n}(T_n^*) \to \mathcal{N}(0, 1), \qquad \text{in probability}. \tag{7.5.28}$$

(Note that the permutation distribution of T_N^* is a conditional distribution.) Again, using Theorem 7.3.4, it follows that under $\{K_n^*\}$ in (7.5.9),

$$\mathcal{L}(T_n^* \mid K_n^*) \to \mathcal{N}(\delta, 1), \tag{7.5.29}$$

where

$$\delta^2 = \left[\frac{12}{p(p^2-1)}\,\rho^2(\boldsymbol{\lambda})\right][B^2(F)/\gamma_{00}(1-\rho_J)], \tag{7.5.30}$$

and $\rho(\boldsymbol{\lambda})$, $B(F)$, γ_{00}, and ρ_J are defined by (7.5.10), (7.3.60), (7.3.57), and (7.3.58), respectively. Thus, the A.R.E. of T_N^* with respect to Q_n is

$$e_{T^*,Q} = \sigma^2 B^2(F)/\gamma_{00}(1-\rho_J) \tag{7.5.31}$$

which agrees with the A.R.E. of the aligned rank order test (when we put $\rho_\epsilon = 0$) with respect to the classical ANOVA test (cf. (7.7.62)). Thus, for normal scores $T_{n,j}$, $j = 1, \ldots, p$, the A.R.E. in (7.5.31) is bounded below by 1. If we use the rank sums for $T_{N,j}$, $j = 1, \ldots, p$, it follows from (7.3.70) that (7.5.31) is bounded below by $12\sigma^2(\int_{-\infty}^{\infty} f^2(x)\,dx)^2$, which is always greater than (7.5.25) and (7.5.26), and is not deflated by small values of p.

Another type of asymptotically distribution-free tests may be posed as follows. We define, as in (7.4.3), $X^*_{ijj'}$, for $j' \neq j = 1, \ldots, p$ and $i = 1, \ldots, n$. Let then $T_{n,jj'}$ be a suitable one-sample rank order statistic based on the values of $X^*_{1,jj'}, \ldots, X^*_{n,jj'}$, for $j \neq j' = 1, \ldots, p$, and let $T_{n,jj}$ be conventionally defined as 0, $j = 1, \ldots, p$. Let then

$$T_{n,j\cdot} = p^{-1}\sum_{j'=1}^{p} T_{n,jj'}, \qquad j = 1, \ldots, p, \tag{7.5.32}$$

$$T_n^0 = \frac{12\sqrt{n}}{p(p-1)}\left[\sum_{j=1}^{p}\left(j - \frac{p+1}{2}\right)T_{n,j\cdot}\right]\Big/\hat{\sigma}^0, \tag{7.5.33}$$

where

$$(\hat{\sigma}^0)^2 = [3/(p^2-1)][A^2 + (p-2)L_n], \tag{7.5.34}$$

where A^2 and L_n are defined by (7.4.8) and (7.4.31), respectively. It follows from Theorem 7.4.1 that $\hat{\sigma}_0^2$ consistently estimates

$$(\sigma^0)^2 = [3/(p^2-1)][A^2 + (p-2)\lambda_J(F^*)], \tag{7.5.35}$$

where $\lambda_J(F^*)$ is defined by (7.4.9). From Theorem 4.5.1, it follows that under H_0: $\boldsymbol{\theta} = \mathbf{0}$, T_n^0 has asymptotically a normal distribution with zero mean and unit variance, while under $\{K_n^*\}$ in (7.5.9), it has asymptotically a normal distribution with mean δ^* and unit variance, where

$$(\delta^*)^2 = \left[\frac{12}{p(p^2-1)}\,\rho^2(\boldsymbol{\lambda})\right][pB^2(F)/\{2A^2 + 2(p-2)\lambda_J(F^*)\}]. \tag{7.5.36}$$

Thus, the A.R.E. of T_n^0 with respect to Q_n is

$$e_{T^0,Q} = [\sigma^2 B^2(F)/A^2][pA^2/2\{A^2 + (p-2)\lambda_J(F^*)\}], \tag{7.5.37}$$

which agrees with (7.4.25). Hence the details are omitted. Also, from (7.5.30) and (7.5.36), we have

$$e_{T^*,T^0} = \frac{2[A^2 + (p-2)\lambda_J(F^*)]}{p\gamma_{00}(1-\rho_J)}, \qquad \gamma_{00} = A^2, \tag{7.5.38}$$

which depends on the unknown ρ_J and $\lambda_J(F^*)$. Usually, (7.5.38) is very close to one (cf. Exercise 7.5.4). The reader is referred to Hollander (1967), Doksum (1967), Puri and Sen (1968a) and Sen (1968d), for details.

7.6. MULTIPLE COMPARISONS OF TREATMENTS

As in section 6.5, we may be interested in simultaneous tests and confidence bounds for various contrasts among $\theta_1, \ldots, \theta_p$. A simultaneous test for all possible pairs $(\theta_j, \theta_{j'})$, $j \neq j' = 1, \ldots, p$ can be based either on the intrablock rankings or on the aligned rank order statistics.

Let us define $T_{n,j}$ as in (7.2.9) and $A^2(J)$ as in (7.2.8). Consider then the statistic

$$W_n = \max_{1 \le j < j' \le p} [n^{1/2}\sqrt{(p-1)/p}\, |T_{n,j} - T_{n,j'}|/A(J)]. \tag{7.6.1}$$

Under H_0: $\theta_j = \theta_{j'}$, $\forall j \neq j'$, the exact distribution of W_n can be traced by reference to the $(p!)^n$ equally likely permutations of the intrablock ranks, and thus, the critical values of W_n may be evaluated. For large n, it follows from Theorem 7.2.2 that under H_0, W_n has asymptotically the distribution of the sample range of a sample of size p from a standard normal distribution. Let $R_{p,\epsilon}$ denote the upper $100\epsilon\%$ point of the latter distribution. Then, we have the following simultaneous test procedure:

in the model (7.2.4), regard those pairs $(\theta_j, \theta_{j'})$, $j \neq j'$, to be significantly different from each other for which

$$[n(p-1)/pA^2(J)]^{1/2}\, |T_{n,j} - T_{n,j'}| \ge R_{p,\epsilon}, \tag{7.6.2}$$

for $j \neq j' = 1, \ldots, p$.

The analogous parametric test is based on Tukey's T method of multiple comparisons, considered in detail in Scheffé (1959, chapter 3) and Miller (1966, chapters 1 and 2). By virtue of the results in section 7.2, the test in (7.6.2) is consistent against the set of alternatives that $\theta_j \neq \theta_{j'}$ for at least one $j \neq j' = 1, \ldots, p$. As such, we consider the sequence of alternatives that $\boldsymbol{\theta} = \boldsymbol{\theta}_n = n^{-1/2}\boldsymbol{\tau}$, where the elements of $\boldsymbol{\tau}$ are not all equal. For such local alternatives, we can again use Theorem 7.2.2 to study the non-null distribution of W_n, and that of the parametric statistic

$$\max_{1 \le j < j' \le p} [n^{1/2}\sqrt{(p-1)/p}\, |\bar{X}_j - \bar{X}_{j'}|/S] = Z_n \quad \text{(say)} \tag{7.6.3}$$

(where $\bar{X}_j = n^{-1} \sum_{i=1}^{n} X_{ij}, j = 1, \ldots, p$ and S^2 is an unbiased estimate of σ^2, the common variance of X_{ij}) follows from the central limit theorem. A comparison of these two asymptotic distribution reveals that the A.R.E. of W_n with respect to Z_n is given by (7.2.41), which is independent of $\boldsymbol{\tau}$. The details are left as Exercise 7.6.1. The results of section 7.2 indicate its low A.R.E. particularly for small values of p. This drawback can be removed to a great extent by the aligned rank statistics, which we consider below.

Define the aligned rank statistics $T_{N,j}, j = 1, \ldots, p$ as in (7.3.4) and $\sigma^2(\mathfrak{I}_n)$ as in (7.3.6), and let then

$$W_n^* = \max_{1 \leq j < j' \leq p} [n^{\frac{1}{2}} |T_{N,j} - T_{N,j'}|/\sigma(\mathfrak{I}_n)]. \tag{7.6.4}$$

For small values of n, the conditional (permutational) distribution of W_n^* can be evaluated by reference to the $(p!)^n$ conditionally equally likely permutations of the rank collection matrices, and this can be used to construct a conditionally distribution-free test for H_0: $\theta_j = \theta_{j'}$, $\forall j \neq j'$. For large n, following along the lines of Theorem 7.3.3, it follows that the permutation distribution of W_n^* converges (as $n \to \infty$), in probability, to the distribution of the sample range of a sample of size p from a standard normal distribution. Thus, the same test procedure as in (7.6.2), with $T_{n,j}$ and $A^2(J)p/(p - 1)$ replaced by $T_{N,j}$ and $\sigma^2(\mathfrak{I}_n)$ respectively, can be used to test for the simultaneous hypothesis H_0: $\theta_j = \theta_{j'}$, $\forall j \neq j'$. The asymptotic distribution of W_n^* under the sequence of alternatives $\boldsymbol{\theta} = \boldsymbol{\theta}_n = n^{-\frac{1}{2}}\boldsymbol{\tau}$ can be studied with the help of Theorem 7.3.4, and hence, comparing the same with that of the distribution of Z_n, we obtain that the A.R.E. of W_n^* with respect to Z_n is given by (7.3.62) (cf. Exercise 7.6.2). This explains the increased efficiency of W_n^* (as compared to that of W_n).

The procedures considered above have certain difficulties in providing simultaneous confidence bounds or tests for contrasts among $\theta_1, \ldots, \theta_p$, which are not paired differences. For this, we may proceed as follows. By definition,

$$Y_{ij} = X_{ij} - X_{i\cdot} = \tau_j + e_{ij}, \qquad j = 1, \ldots, p, \quad i = 1, \ldots, n, \tag{7.6.5}$$

where the (joint) c.d.f. $F_i(\mathbf{e})$ of $\mathbf{e}_i = (e_{i1}, \ldots, e_{ip})$ is symmetric in its p arguments, singular and of rank $p - 1$. Let then

$$\mathbf{Y}_n^{(j)} = (Y_{1j}, \ldots, Y_{nj}), \qquad j = 1, \ldots, p; \tag{7.6.6}$$

$$\mathbf{E}_{2n} = (E_{2n,1}, \ldots, E_{2n,2n}) \quad \text{where} \quad E_{2n,\alpha} = J_{2n}\left(\frac{\alpha}{2n + 1}\right), \quad 1 \leq \alpha \leq 2n \tag{7.6.7}$$

where $J_{2n}(u)$ satisfies the conditions (*a*), (*b*), and (*c*) of section 3.6.1. Let then

$$(7.6.8)\qquad A_{2n}^2 = \frac{1}{2n-1}\sum_{\alpha=1}^{2n}(E_{2n,\alpha} - \bar{E}_{2n})^2; \qquad \bar{E}_{2n} = \frac{1}{2n}\sum_{\alpha=1}^{2n} E_{2n,\alpha}.$$

Based on the two matched samples $\mathbf{Y}_n^{(j)}$, $\mathbf{Y}_n^{(j')}$, we consider the usual two-sample statistic (cf. section 3.6.1) viz.,

$$(7.6.9)\qquad h(\mathbf{Y}_n^{(j)}, \mathbf{Y}_n^{(j')}) = n^{-1}\sum_{\alpha=1}^{2n} E_{2n,\alpha} Z_{2n,\alpha}^{(j,j')},$$

where $Z_{2n,\alpha}^{(j,j')}$ is 1 or 0 according as the αth smallest observation in the combined (j, j')th set is from the jth set or not, $\alpha = 1, \ldots, 2n$. Consider then the statistic

$$(7.6.10)\qquad \tilde{W}_n = \max_{1 \le j < j' \le p} [2\{n(p-1)/p\}^{1/2}\, |h(\mathbf{Y}_n^{(j)}, \mathbf{Y}_n^{(j')}) - \bar{E}_{2n}|/A_{2n}].$$

Theorem 7.6.1. *Under* H_0: $\boldsymbol{\theta} = \mathbf{0}$, $\lim_{n\to\infty} P\{\tilde{W}_n \le R_{p,\epsilon}\} \ge 1 - \epsilon$, *where the equality sign holds only when* ρ_J, *defined by* (7.3.58) *is equal to* $-1/(p-1)$.

Proof: Let us define $F_{[i]}(x)$ as the univariate marginal c.d.f. of $F_i(\mathbf{x})$, and let

$$(7.6.11)\qquad Z_{ij}^{(n)} = \int_{-\infty}^{\infty} [c(x - Y_{ij}) - F_{[i]}(x)]J'[\bar{F}_n(x)]\, d\bar{F}_n(x),$$
$$j = 1, \ldots, p, \quad i = 1, \ldots, n,$$

where $\bar{F}_n = n^{-1}\sum_{i=1}^n F_i$. Then, from Theorem 7.3.4, it follows that under H_0: $\boldsymbol{\theta} = \mathbf{0}$,

$$(7.6.12)\qquad n^{1/2}\, |\{h(\mathbf{Y}_n^{(j)}, \mathbf{Y}_n^{(j')}) < \bar{E}_{2n}\} - \tfrac{1}{2}(Z_{\cdot j'}^{(n)} - Z_{\cdot j}^{(n)})| = o_p(1),$$

where

$$(7.6.13)\qquad Z_{\cdot j}^{(n)} = \frac{1}{n}\sum_{i=1}^{n} Z_{ij}^{(n)}, \qquad j = 1, \ldots, p.$$

Let now, $F_i^*(x, y)$ be the bivariate c.d.f. of $(e_{ij}, e_{ij'})$, and let $\bar{F}_n^* = (1/n)\sum F_i^*$. Also, let

(7.6.14)

$$\rho_J^{(n)} A^2 = \int_{-\infty}^{\infty}\int_{-\infty}^{\infty} [\bar{F}_n^*(x, y) - \bar{F}_n(x)\bar{F}_n(y)]J'[\bar{F}_n(x)]J'[\bar{F}_n(y)\, d\bar{F}_n(x)\, d\bar{F}_n(y)]$$

where $A^2 = \int_0^1 J^2(u)\, du - [\int_0^1 J(u)\, du]^2$. Now, under the models of section 7.2.4, $\bar{F}_n$ and $\bar{F}_n^*$ converge respectively to F and F^* (as $n \to \infty$), and as a result $\rho_J^{(n)}$ converges (as $n \to \infty$) to ρ_J, defined by (7.3.58). Also, by definition and theorem 7.3.4, under H_0: $\boldsymbol{\theta} = \mathbf{0}$, $n^{1/2}(Z_{\cdot 1}, \ldots, Z_{\cdot p})$ converges in law to a multinormal distribution with null mean vector and dispersion matrix

$A^2((\delta_{jj'}(1 - \rho_J) + \rho_J))$. Hence, under H_0: $\boldsymbol{\theta} = \mathbf{0}$, we have (cf. Exercise 7.6.3)

$$\text{(7.6.15)} \quad \lim_{n\to\infty} P\{ \max_{1\le j<j'\le p} |n^{1/2}(Z_{\cdot j} - Z_{\cdot j'})|/A(1 - \rho_J)^{1/2} \le R_{p,\epsilon} \mid H_0\} = 1 - \epsilon.$$

Now, under H_0: $\boldsymbol{\theta} = \mathbf{0}$, $Z_{\cdot 1}, \ldots, Z_{\cdot p}$ are interchangeable random variables, and hence, by Lemma 7.3.7, $\rho_J \ge -1/(p-1)$, where the equality sign holds only when $J[F(x)]$ is linear in x, with probability one. The rest of the proof follows from (7.6.15) by noting that $A(1-\rho_J)^{1/2} \le A(p/(p-1))^{1/2}$. Q.E.D.

Now, when $\boldsymbol{\theta} \ne \mathbf{0}$, by definition $\mathbf{Y}_n^{(j)} - \Delta_{jj'}\mathbf{1}_n$ and $\mathbf{Y}_n^{(j')}$ (where $\Delta_{jj'} = \theta_j - \theta_{j'}$) have the same distribution, for $j \ne j' = 1, \ldots, p$. Thus, we have

$$\text{(7.6.16)} \quad \lim_{n\to\infty} P\left\{ \max_{1\le j<j'\le p} |n^{1/2}\{h(\mathbf{Y}_n^{(j)} - \Delta_{jj'}\ell_n, \mathbf{Y}_n^{(j')}) - \bar{E}_{2n}| \right.$$
$$\left. \le \tfrac{1}{2}A_{2n}\left(\frac{p}{p-1}\right)^{1/2} R_{p,\epsilon} \quad \text{for all} \quad j \ne j' = 1, \ldots, p \right\} \ge 1 - \epsilon.$$

Hence, proceeding as in section 6.8 (namely, (6.8.16)–(6.8.18)), we obtain that

$$\text{(7.6.17)} \qquad \lim_{n\to\infty} P\{\hat{\Delta}_{jj',L} \le \Delta_{jj'} \le \hat{\Delta}_{jj',U}, \quad j \ne j'\} \ge 1 - \epsilon,$$

where $\hat{\Delta}_{jj',L}$ and $\hat{\Delta}_{jj',U}$ are defined by (6.8.17), with $\mu_n^{(j)}$, $j = 1, 2$ being replaced by $\bar{E}_{2n} \pm \frac{1}{2}n^{-1/2}A_{2n}(p/(p-1))^{1/2}R_{p,\epsilon}$. (6.8.18) supplies the desired simultaneous confidence bounds to $\Delta_{jj'}$.

The general case of multiple comparisons for contrasts other than paired differences then follows along the same line as in section 6.8. The details are therefore omitted; the reader is referred to Sen (1969c). The expressions for the A.R.E. are bounded below by the corresponding expressions in Theorems 6.8.2 and 6.8.3, as $\rho_J \ge -1/(p-1)$.

7.7. RANK ORDER TESTS FOR INTERACTIONS IN FACTORIAL EXPERIMENTS

For simplicity of presentation, we shall consider in detail only the case of replicated two-factor experiments with one observation per cell. The procedures and arguments below, being perfectly general, can be extended to cover the case of several factors and/or observations per cell. Let X_{ijk} be the response on the plot in the ith block receiving the jth level of the first factor and the kth level of the second factor. Consider the model

$$X_{ijk} = \mu_i + \nu_j + \tau_k + \gamma_{jk} + \epsilon_{ijk}, \qquad 1 \le i \le n, \quad 1 \le j \le p, \quad 1 \le k \le q,$$

where the replication effects μ_i's and the main effects ν_j's and τ_k's, and the interactions γ_{jk}'s satisfy the side conditions: $\mu. = \nu. = \tau. = \gamma_{j\cdot} = \gamma_{\cdot k} = 0$, and $\cdot$ stands for the average over the subscript replaced by $\cdot$ It is assumed that $(\epsilon_{i11}, \ldots, \epsilon_{ipq})$, $i = 1, 2, \ldots, n$ are independently distributed with (jointly) continuous distribution function (c.d.f.) $G_i(x_{11}, \ldots, x_{pq})$ which is symmetric in the arguments and satisfies the property that $P[\epsilon_{ijk} = \epsilon_{jk'k'}] = 0$ for any two distinct pairs (j, k) and (j', k'). This includes the conventional assumptions of independence, continuity, and identity of the distributions of all U_{ijk}'s. We want to test

$$H_0: \mathbf{\Gamma} = (\gamma_{jk}) = \mathbf{O}^{p\times q} \tag{7.7.1}$$

against alternatives that $\mathbf{\Gamma}$ is non-null. By using the following intrablock transformations,

$$\mathbf{Z}_i = \left(\mathbf{I}_p - \frac{1}{p}\ell_p'\ell_p\right)\mathbf{X}_i\left(\mathbf{I}_q - \frac{1}{q}\ell_q'\ell_q\right), \qquad i = 1, \ldots, n, \tag{7.7.2}$$

$$\mathbf{E}_i = \left(\mathbf{I}_p - \frac{1}{p}\ell_p'\ell_p\right)\boldsymbol{\epsilon}_i\left(\mathbf{I}_q - \frac{1}{q}\ell_q'\ell_q\right), \qquad i = 1, \ldots, n, \tag{7.7.3}$$

where $\mathbf{X}_i = (X_{ijk})$, $\boldsymbol{\epsilon}_i = (\epsilon_{ijk})$, $\mathbf{Z}_i = (Z_{ijk})$, $\mathbf{E}_i = (E_{ijk})$, $1 \leq i \leq n$ are $p \times q$ matrices, $\mathbf{I}_t$ is the identity matrix of order t, and ℓ_t is the (row) t vector with each element equal to unity, we obtain

$$\mathbf{Z}_i = \mathbf{\Gamma} + \mathbf{E}_i, \qquad i = 1, 2, \ldots, n. \tag{7.7.4}$$

In the sequel we shall work with model (7.7.4), which is free of the nuisance parameters μ_i's, ν_j's, and τ_k's. Also we will consider only the case where p, $q \geq 3$. For $p = q = 2$, we may use the one-sample location tests (in chapter 4) based on Z_{i11}, $k = 1, \ldots, n$. (as $Z_{i11} = -Z_{i12} = -Z_{i21} = Z_{i22}$, $i = 1, \ldots, n$). Also, for $p > 2$, $q = 2$, $Z_{ij1} = -Z_{ij2}$ for $j = 1, \ldots, p$, $i = 1, \ldots, n$. Hence, based on the vectors $(Z_{i11}, \ldots, Z_{ip1})$, $i = 1, \ldots, n$, we may use the test considered earlier in section 7.3.

Now, by definition, the c.d.f. $G_i(x_{11}, \ldots, x_{pq})$ of $(\epsilon_{i11}, \ldots, \epsilon_{ipq})$ is symmetric in its pq arguments. Hence, from (7.3.1) and some straightforward computations, it follows that the joint distribution of $(E_{i11}, \ldots, E_{ipq})$ (defined by (7.7.3)) remains invariant under any permutations of the labels $(1, \ldots, p)$ of the first factor and the labels $(1, \ldots, q)$ of the second factor. This yields a finite group $\mathcal{G}$ of $p!q!$ transformations on the variables $(E_{i11}, \ldots, E_{ipq})$ which maps the sample space onto itself and leaves the distribution invariant. Thus, considering the n independent matrices $\mathbf{E}_i$, $i = 1, \ldots, n$, and using the same group of transformations for each of them, we get a compound group $\mathcal{G}_n^*$ which contains $(p!\,q!)^n$ transformations, and each of these transformations maps the sample space (or npq dimension) onto itself and leaves the distribution of the sample point invariant. Hence, when H_0 in (7.7.1) holds, so that $\mathbf{Z}_i = \mathbf{E}_i$ for all $i = 1, \ldots, n$, the same group of transformations $\mathcal{G}_n^*$ also works on $\mathbf{Z}_1, \ldots, \mathbf{Z}_n$. Considering then the set $\mathfrak{Z}_n$ of all possible $(p!\,q!)^n$ realizations of $(\mathbf{Z}_1, \ldots, \mathbf{Z}_n)$ obtained by permuting the rows and columns of each $\mathbf{Z}_i$, we conclude from the above discussions that under H_0 in (7.7.1), the conditional distribution of $(Z_1, \ldots, Z_n)$ over $\mathfrak{Z}_n$ is uniform, each realization having the (conditional) probability $(p!\,q!)^{-n}$. Let us denote this conditional (permutational) probability measure by $\mathcal{P}_n$. Since $\mathcal{P}_n$ is completely specified, proceeding as in section 7.3, the existence of similar size α tests for H_0 in (7.7.1) follows.

We shall specifically consider a class of conditional tests based on the following rank order statistics.

Let r_{ijk} denote the rank of Z_{ijk} in a combined ranking of the totality of $N = pqn$ aligned observations Z_{ijk}'s, where on account of the assumptions made earlier, ties among Z_{ijk}'s may be ignored with probability one. Let $\{J_{N,1}, \ldots, J_{N,N}\}$ be a sequence of real numbers such that the function $J_N(u)$, $0 < u < 1$, satisfies the conditions of (*a*), (*b*), and (*c*) of section 3.6.1. Let $J_{N,r_{ijk}} = \eta_{ijk}$ and

$$\mathbf{T}_N = (T_{N,jk}); \qquad T_{N,jk} = \frac{1}{n}\sum \eta_{ijk} = \eta_{\cdot jk} \tag{7.7.5}$$

$$\mathbf{T}_N^* = \left(\mathbf{I}_p - \frac{1}{p}\,\ell_p'\ell_p\right)\mathbf{T}_N\left(\mathbf{I}_q - \frac{1}{q}\,\ell_q'\ell_q\right) = T_{N,jk}^*. \tag{7.7.6}$$

Then it is easily seen that

$$T_{N,jk}^* = \frac{1}{n}\sum_{i=1}^{n}(\eta_{ijk} - \eta_{i\cdot k} - \eta_{ij\cdot} + \eta_{i\cdot\cdot}). \tag{7.7.7}$$

Let $\mathbf{r}_i^{(1)}(\mathbf{r}_i^{(2)})$ stand for the (observed) partition of the ranks in the ith replication into the $p(q)$ sets of $q(p)$ ordered elements and let the configuration $\{\mathbf{r}_2^{(1)}, \mathbf{r}_i^{(2)}: 1 \le i \le n\}$ be denoted by $\mathcal{E}$. We observe that the $\mathcal{P}_n$ defined earlier is simply the conditional probability measure given $\mathcal{E}$, and

$$E(\mathbf{T}_N^* \mid \mathcal{P}_N) = \mathbf{0} \tag{7.7.8}$$

$$nE(\mathbf{T}_{N^*}^0 \mathbf{T}_{N^*}^0) = \left(\mathbf{I}_p - \frac{1}{p}\,\ell_p'\ell_p\right) \otimes \left(\mathbf{I}_q - \frac{1}{q}\,\ell_q'\ell_q\right)\sigma^2(\mathcal{P}_n), \tag{7.7.9}$$

where $\mathbf{T}_N^{0*} = (T_{N11}^*, \ldots, T_{Npq}^*)$, and

$$\sigma^2(\mathcal{P}_n) = \frac{1}{n(p-1)(q-1)}\sum_{i=1}^{n}\sum_{j=1}^{p}\sum_{k=1}^{q}(\eta_{ijk} - \eta_{i\cdot k} - \eta_{ij\cdot} + \eta_{i\cdot\cdot})^2. \tag{7.7.10}$$

Thus, considering the generalized inverse of the $pq \times pq$ matrix in (7.7.9) and employing it to construct a quadratic form in the elements of $\mathbf{T}_N^*$, we derive the following test statistic

$$\mathfrak{L}_N = [n/\sigma^2(\mathfrak{T}_N)] \sum_{j=1}^{p} \sum_{k=1}^{q} \{T^*_{N,jk}\}^2. \tag{7.7.11}$$

For small n, p, and q the exact permutation distribution of $\mathfrak{L}_N$ can be obtained by considering the $(p!\,q!)^n$ (conditionally) equally likely row and column permutations of the matrices $H_i = (\eta_{ijk})$, $i = 1, \ldots, n$. This procedure becomes prohibitively laborious for large values of n, p, or q. For this reason we consider the following large-sample approach.

Let $J(u) = \lim_{N\to\infty} J_N(u)$, $0 < u < 1$, $\gamma_{00} = \int_0^1 J^2(u)\,du - [\int_0^1 J(u)\,du]^2$, and

$$\delta^2 = \int_{-\infty}^{\infty} J[H(x)]\,dH(x) - \int_{-\infty}^{\infty} J[H(x)]J[H(y)]\,d[H_{10}(x, y) + H_{01}(x, y) - H_{11}(x, y)], \tag{7.7.12}$$

where H_{01}, H_{10}, H_{11}, and H are defined as follows. Let $H_{i,01}$, $H_{i,10}$, $H_{i,11}$, and H_i be respectively, the bivariate c.d.f.'s of $(Z_{ijk}, Z_{ijk'})$, $k \neq k'$, $(Z_{ijk}, Z_{ij'k})$, $j \neq j'$, $(Z_{ijk}, Z_{ij'k'})$, $j \neq j'$, $k \neq k'$ and the common univariate c.d.f. of Z_{ijk}, under H_0 in (7.7.1). Then, $H_{rs} = \lim_{n\to\infty} n^{-1} \sum_{i=1}^{n} H_{i,rs}$, for $r, s = 0, 1$ (where $H_{00} = H$ and $H_{i,00} = H_i$); the justification of these limits again follows under the models of section 7.2.4. Now, if we assume that

$$P[\{J[H(Z_{jk})] - J(H(Z_{j'k})] - J[H(Z_{jk'})] + J[H(Z_{j'k'})]\} = \text{constant}] < 1 \tag{7.7.13}$$

for at least one pair $j \neq j'$ and $k \neq k'$, then as in Lemma 7.3.1, it can be shown that (7.7.12) is strictly positive. Then, we have the following.

Theorem 7.7.1. *Under the conditions* (*a*), (*b*), *and* (*c*) *of section* 3.6.1 *and* (7.7.13), *the conditional* (*permutational*) *distribution of* $\mathfrak{L}_N$ (*under* H_0: $\mathbf{\Gamma} = \mathbf{0}$), *converges, in probability, to a chi-square distribution with* $(p - 1)(q - 1)$ D.F.

Proof: By virtue of (7.7.8) to (7.7.9), it suffices to show that for any non-null $\mathbf{A} = (a_{jk})$, $Y_n = n^{1/2} \sum_{j=1}^{p} \sum_{k=1}^{q} a_{jk} T^*_{N,jk}$ converges in law (under $\mathfrak{T}_N$) to a normal distribution as $n \to \infty$. By using the conditions (*a*) and (*b*) of section 3.6.1, (7.7.5), and (7.7.6), we write

$$Y_n = n^{-1/2} \sum_{i=1}^{n} Y_{ni} + o_p(1); \qquad Y_{ni} = \sum_{j=1}^{p} \sum_{k=1}^{q} a^*_{jk} J\left(\frac{R_{ijk}}{N + 1}\right), \quad i = 1, \ldots, n, \tag{7.7.14}$$

where the a_{jk}^*'s are linear functions of the a_{jk}'s and satisfy the constraints $a_{j\cdot}^* = a_{\cdot k}^* = 0$ for all j and k. By observing that $Y_{n1}, \ldots, Y_{nn}$ are stochastically independent under $\mathcal{F}_N$, we obtain $E[Y_{ni} \mid \mathcal{F}_N] = 0$ and

$$\begin{aligned}\frac{1}{n}\sum_{i=1}^{n} E(Y_{ni}^2 \mid \mathcal{F}_N) &= \frac{1}{n(p-1)(q-1)}\sum_{i=1}^{n}\Bigg\{\sum_{j=1}^{p}\sum_{k=1}^{q} J^2\left(\frac{R_{ijk}}{N+1}\right) \\ &\quad - \frac{1}{p}\sum_{k=1}^{q}\left[\sum_{j=1}^{p} J\left(\frac{R_{ijk}}{N+1}\right)\right]^2 \\ &\quad - \frac{1}{q}\sum_{j=1}^{p}\left[\sum_{k=1}^{q} J\left(\frac{R_{ijk}}{N+1}\right)\right]^2 \qquad (7.7.15) \\ &\quad + \frac{1}{pq}\left[\sum_{j=1}^{p}\sum_{k=1}^{q} J\left(\frac{R_{ijk}}{N+1}\right)\right]^2\Bigg\} \sim_p \delta^2 \sum_{j=1}^{p}\sum_{k=1}^{q}(a_{jk}^*)^2 \\ &> 0,\end{aligned}$$

where δ^2 is given by (7.7.12) and in proving (7.7.15) we use the arguments of Lemma 7.3.2 and some routine analysis. Similarly, when we use the condition (*c*) of section 3.6.1, it follows that

$$\frac{1}{n}\sum_{i=1}^{n} E[|Y_{ni}|^{2+\eta} \mid \mathcal{F}_N] < \infty \quad \text{for some} \quad \eta > 0. \qquad (7.7.16)$$

By the Berry-Esséen theorem (cf. section 2.7) the asymptotic normality of Y_n follows from (7.7.15) and (7.7.16); hence the theorem. ◁

Theorem 7.7.1 simplifies the large-sample test based on $\mathcal{L}_N$ and suggests that the chi-square distribution with $(p-1)(q-1)$ degrees of freedom can be used to compute the critical values of $\mathcal{L}_N$. For studying the asymptotic efficiency of the test based on $\mathcal{L}_N$, we consider the sequence of Pitman alternatives $\{K_N\}$ where

$$K_N\colon \mathbf{\Gamma} = \mathbf{\Gamma}_n = n^{-1/2}\mathbf{\Lambda}, \qquad \mathbf{\Lambda} = (\lambda_{jk}), \qquad (7.7.17)$$

where the λ_{jk}'s are real and finite and satisfy $\lambda_{j\cdot} = \lambda_{\cdot k} = 0$ for each $1 \le j < p$, $1 \le k \le q$. Further, set

$$B(H) = \int_{-\infty}^{\infty} \{(d/dx)J[H(x)]\}\, dH(x) \qquad (7.7.18)$$

$$A^2 = \int_0^1 J^2(u) - \left[\int_0^1 J(u)\, du\right]^2 \qquad (7.7.19)$$

$$\rho_{ij} = \frac{1}{A^2}\left[\int_{-\infty}^{\infty}\int_{-\infty}^{\infty} J[H(x)]J[H(y)]\, dH_{ij}(x, y) - \left(\int_0^1 J^2(u)\, du\right)^2\right] \qquad (7.7.20)$$

for $(i, j) = (0, 1), (1, 0), (1, 1)$.

The proof of the following theorem can be accomplished by directly extending the proof of Theorems 7.3.4 and 7.3.5 (cf. Exercise 7.7.1).

Theorem 7.7.2. *If* (i) $\{K_N\}$ *in* (7.7.17) *and* (ii) *the conditions of Theorem* 7.3.5 *hold, then* $\mathcal{L}_N$, *defined by* (7.7.11), *has asymptotically a noncentral chi-square distribution with* $(p-1)(q-1)$ D.F. *and noncentrality parameter*

$$\Delta = \left[\left(\frac{1}{pq}\right)\sum_{j=1}^{p}\sum_{k=1}^{q}\lambda_{jk}^2\right]\cdot\left[\frac{B^2(H)}{A^2(1-\rho_{10}-\rho_{01}+\rho_{11})}\right]. \tag{7.7.21}$$

Proof: By using the same technique as in the proof of Theorem 7.3.5, it follows that, under K_N,

$$n^{1/2}E\{\mathbf{T}_N^* \mid H_N\} = [(pq)^{-1/2}B(H)]\boldsymbol{\Lambda} + o(1), \tag{7.7.22}$$

$$nE\{\mathbf{T}_{N^*}^0,\mathbf{T}_{N^*}^0 \mid H_N\} = \left(\mathbf{I}_p - \frac{1}{p}\,\ell_p'\ell_p\right)\otimes\left(\mathbf{I}_q - \frac{1}{q}\,\ell_q'\ell_q\right) \cdot A^2(1-\rho_{10}-\rho_{01}+\rho_{11}) + o(1), \tag{7.7.23}$$

and, similarly as in (7.7.15), that under H_N,

$$\sigma^2(\mathcal{T}_N) \xrightarrow{p} \gamma_{00}(1-\rho_{10}-\rho_{01}+\rho_{11})$$

where $\sigma^2(\mathcal{T}_N)$, $B(H)$, and the ρ_{ij}'s are defined, respectively, by (7.7.10) and (7.7.18)–(7.7.20). The asymptotic normality of $(n^{1/2}T_{N,jk})$, (7.7.22), and (7.7.23) complete the proof of the theorem. Q.E.D.

Let σ_e^2 denote the limiting value of the average of the variances of e_{ijk}. Then as can easily be seen (cf. Exercise 7.7.2) the classical ANOVA test statistic for H_0 has asymptotically, under H_N, a noncentral chi-square distribution with $(p-1)(q-1)$ degrees of freedom and the noncentrality parameter

$$\Delta_Q = \left[\left(\frac{1}{pq}\right)\sum_{j=1}^{p}\sum_{k=1}^{q}\lambda_{jk}^2\right]\left[\frac{pq\sigma_e^2}{\{(p-1)(q-1)\}}\right]. \tag{7.7.24}$$

We obtain from (7.7.24) and Theorem 7.7.2 the following:

Theorem 7.7.3. *When the conditions of Theorem* 7.7.2 *hold, the asymptotic relative efficiency of the* $\mathcal{L}_N$ *test with respect to the classical analysis of variance test is given by*

$$e_{(\{\mathcal{L}_N\},\{Q_N\})} = \left[\frac{pq}{\{(p-1)(q-1)(1-\rho_{10}-\rho_{01}+\rho_{11})\}}\right] \times \left[\frac{\sigma_e^2B^2(H)}{A^2}\right]. \tag{7.7.25}$$

From Lemmas 7.3.7 and 7.3.8 it follows that $\rho_{10} \geq -1/(p-1)$ and $\rho_{01} \geq -1/(q-1)$ with equality if and only if $J = H^{-1}$ (apart from an additive constant), so that from (7.7.24) we obtain that

$$e_{(\{\mathfrak{L}_N\},\{Q_N\})} \geq \left[1 - \frac{1}{pq}(1 - (p-1)(q-1)\rho_{11})\right]^{-1}\left[\frac{\sigma_e^2 B^2(H)}{A^2}\right] \tag{7.7.26}$$

which leads to the following:

Corollary 7.7.3. *A sufficient condition for* $e_{(\{\mathfrak{L}_N\},\{Q_n\})}$ *to be at least as large as* $[\sigma_e^2 B^2(H)/A^2]$ *is that* $\rho_{11} \leq 1/[(p-1)(q-1)]$.

For the normal scores version of $\mathfrak{L}_N$, say $\mathfrak{L}_N^*$, it follows, using the lower bound (1) for $\sigma_e^2 B^2(H)/\gamma_{00}$, that

$$\begin{aligned} e_{(\{\mathfrak{L}_N^*\},Q\{N\})} &\geq \left[1 - \frac{1}{pq}(1 - (p-1)(q-1)\rho_{11})\right]^{-1} \\ &\geq 1 \quad \text{if} \quad \rho_{11} \leq \frac{1}{(p-1)(q-1)} \\ &\geq \tfrac{1}{2} \quad \text{in general,} \end{aligned} \tag{7.7.27}$$

(using the fact that $\rho_{11} \leq 1$). When the parent c.d.f., F, is normal, the l.h.s. of (7.3.27) takes the value 1, so $\{\mathfrak{L}_N^*\}$ and $\{Q_N\}$ are asymptotically equally efficient. Similarly for the Wilcoxon version, $\{W_N\}$ of $\{\mathfrak{L}_N\}$,

$$\begin{aligned} e_{(\{W_N\},\{Q_N\})} &\geq (0.864)\left[1 - \frac{1}{pq}\{1 - (p-1)(q-1)\rho_{11}\}\right]^{-1} \\ &\geq 0.432, \qquad \text{in general.} \end{aligned} \tag{7.7.28}$$

For normal F, it is known that

$$\rho_{10} = \left(\frac{6}{\pi}\right)\sin^{-1}\left[\frac{-1}{2(p-1)}\right],$$

$$\rho_{01} = \left(\frac{6}{\pi}\right)\sin^{-1}\left[\frac{-1}{2(q-1)}\right] \quad \text{and} \quad \rho_{11} = \left(\frac{6}{\pi}\right)\sin^{-1}\left[\frac{1}{2(p-1)(q-1)}\right].$$

Hence, from (7.7.25), we find that $e_{(\{W_N\},\{Q_N\})}$ is equal to

$$\begin{aligned} 3pq\Bigg\{\pi(p-1)(q-1)\Bigg[1 + \frac{6}{\pi}\Bigg\{\sin^{-1}\frac{1}{2(p-1)} \\ + \sin^{-1}\frac{1}{2(q-1)} + \sin^{-1}\frac{1}{2(p-1)(q-1)}\Bigg\}\Bigg]\Bigg\}^{-1}. \end{aligned} \tag{7.7.29}$$

Actual computations show that (7.7.29) is bounded below by 0.955 and above by 0.975. The above theory is based on the results of Mehra and Sen (1969).

7.8. ANALYSIS OF COVARIANCE BASED ON GENERAL RANK SCORES

In section 5.9, the multivariate rank order tests for the one-way MANOVA problem have been employed to construct suitable tests for the multivariate analysis of covariance (MANOCA) problem. In the same way, the rank statistics of sections 7.2.5–7.2.9 and 7.3.1–7.3.6 can be used to provide suitable tests for the MANOCA problem. Because of the close similarity of the approaches, we shall present only a brief description of MANOCA. Also, we shall consider only the ranking after alignment procedures; the procedures based on the intrablock rankings will follow along the same line, with direct adaptations from sections 7.2.5–7.2.9.

Let $\mathbf{Z}_{ij} = (\mathbf{X}_{ij}, \mathbf{Y}_{ij})'$ (where $\mathbf{X}$ and $\mathbf{Y}$ are q- and r-vectors), $j = 1, \ldots, p$, $i = 1, \ldots, n$ be $N (= np)$ stochastic $(q + r)$-vectors. It is assumed that $\mathbf{Z}_i = (\mathbf{Z}_{i1}, \ldots, \mathbf{Z}_{ip})$ has a continuous $p(q + r)$-variate c.d.f. $G_i(\mathbf{Z})$, $\mathbf{Z} \in R^{p(q+r)}$, $i = 1, \ldots, n$. The basic assumption (parallel to (5.9.1)) is that the joint c.d.f. of $\mathbf{Y}_i = (\mathbf{Y}_{i1}, \ldots, \mathbf{Y}_{ip})$ is symmetric in the p r-vectors. The null hypothesis states that the conditional c.d.f. of $\mathbf{X}_i = (\mathbf{X}_{i1}, \ldots, \mathbf{X}_{ip})$ given $\mathbf{Y}_i = \mathbf{y}_i$, is symmetric in its p q-vectors, for all $i = 1, \ldots, n$, while the set of admissible alternatives relates to possible shifts (due to treatment effects). (Note that under the null hypothesis, $G_i(\mathbf{Z})$ is symmetric in its p $(q + r)$-vectors, for all $i = 1, \ldots, n$.)

We consider the aligned observations $\mathbf{Z}_{ij}^* = \mathbf{Z}_{ij} - p^{-1} \sum_{l=1}^{p} \mathbf{Z}_{il}$, $j = 1, \ldots, p$; $\mathbf{Z}_i^* = (\mathbf{Z}_{i1}^*, \ldots, \mathbf{Z}_{ip}^*)$, $i = 1, \ldots, n$. Based on these aligned observations, we define for the jth treatment the mean rank scores corresponding to the q primary and r concomitant variates

$$\mathbf{T}_{N,j} = (T_{N,j}^{(1)}, \ldots, T_{N,j}^{(q)})', \qquad S_{N,j} = (S_{N,j}^{(1)}, \ldots, S_{N,j}^{(r)}), \tag{7.8.1}$$

where the $T_{N,j}^{(k)}$ and the $S_{N,j}^{(k)}$ are defined as in (7.3.4). Also, we define the $(q + r) \times (q + r)$ permutation covariance matrix as in (7.3.87) and (7.3.88) (with q replaced by $q + r$), and partition it as

$$\mathbf{V}_N(\mathbf{R}_N) = \begin{pmatrix} \mathbf{V}_{N,11} & \mathbf{V}_{N,12} \\ \mathbf{V}_{N,21} & \mathbf{V}_{N,22} \end{pmatrix}, \tag{7.8.2}$$

where $\mathbf{V}_{N,11}$, $\mathbf{V}_{N,22}$ and $\mathbf{V}_{N,12} = \mathbf{V}_{N,21}'$ stand respectively for the dispersion matrices of the primary, the concomitant variate scores, and the covariance matrix of the two sets of scores. Let us then define

$$\mathbf{T}_{N,j}^* = \mathbf{T}_{N,j} - \overline{\mathbf{E}}_N + \mathbf{V}_{N,12}\mathbf{V}_{N,22}^{-1}[\mathbf{S}_{N,j} - \overline{\mathbf{E}}_N^*], \quad j = 1, \ldots, p, \tag{7.8.3}$$

where $\overline{\mathbf{E}}_N = (\bar{E}_N^{(1)}, \ldots, \bar{E}_N^{(q)})'$ and $\overline{\mathbf{E}}_N^* = (\bar{E}_N^{*(1)}, \ldots, \bar{E}_N^{*(r)})$ stand for the mean scores of the (combined treatment) primary and the concomitant variates. $\mathbf{V}_{N,22}^{-1}$ may be replaced by the generalized inverse of $\mathbf{V}_{N,22}$. Let us then define

(7.8.4) $$\mathbf{V}_{N,11\cdot 2} = \mathbf{V}_{N,11} - \mathbf{V}_{N,12}\mathbf{V}_{N,22}^{-1}\mathbf{V}_{N,21}.$$

Then, the test statistic may be written as

(7.8.5) $$\mathcal{L}_N^* = n\sum_{j=1}^{p}\mathbf{T}_{N,j}^{*\prime}\mathbf{V}_{N,11\cdot 2}^{-1}\mathbf{T}_{N,j}^*.$$

It may be noted that under the conditional measure $\mathcal{T}_n$ in section 7.3.6, $\mathbf{V}_{N,11\cdot 2}$ is invariant, while $(\mathbf{T}_{N,1}^*, \ldots, \mathbf{T}_{N,p}^*)$ can have $(p!)^n$ conditionally equally likely realizations (not all distinct). This may be used to construct a conditionally distribution-free test for the MANOCA problem based on $\mathcal{L}_N^*$. For large n, it follows from Theorem 7.3.11 and some routine computations that under H_0, $\mathcal{L}_N^*$ has asymptotically (in probability, under $\mathcal{T}_n$) a chi-square distribution with $q(p-1)$ D.F.

We now write $\mathbf{Z}_i^* = (\mathbf{X}_i^*, \mathbf{Y}_i^*)$, and note that by assumption the distribution of $\mathbf{Y}_i^*$ is symmetric in its p r-vectors. For the study of the asymptotic non-null distribution of $\mathcal{L}_N^*$, we write $G_i(\mathbf{Z}) = G_i(\mathbf{x}, \mathbf{y})$, $\mathbf{x} \in R^{pq}$, $\mathbf{y} \in R^{pr}$, and consider the following sequence of alternative hypotheses:

(7.8.6) $$H_N\colon\ G_i(\mathbf{z}) = G_{i,N}(\mathbf{Z}) = G_i(\mathbf{x} - N^{-1/2}\boldsymbol{\tau}, y), \qquad i = 1, \ldots, n,$$

where $\boldsymbol{\tau} = ((\tau_{jk}))$ stands for a $p \times q$ finite matrix of treatment effects. The marginal c.d.f. of the kth primary variate corresponding to the c.d.f. G_i is denoted by $G_{i[k]}$, $k = 1, \ldots, q$, $i = 1, \ldots, n$. Then, under the models of section 7.2.4, $n^{-1}\sum_{i=1}^{n} G_{i[k]} \to G_{[k]}$, for $k = 1, \ldots, q$. The density function corresponding to $G_{[k]}$ is assumed to be continuous and is denoted by $g_{[k]}(x)$. We define as in (7.3.121), $B(G_{[k]}, J^{(k)}) = \int_{-\infty}^{\infty} (d/dx)J^{(k)}[G_{[k]}(x)]\, dG_{[k]}(x)$, $k = 1, \ldots, q$, and let $\eta_{jk} = \tau_{jk}B(G_{[k]}, J^{(k)})$, $j = 1, \ldots, p$, $k = 1, \ldots, q$. Further, we define a $p(q+r) \times p(q+r)$ matrix $\boldsymbol{\nu}_N$, with elements defined by (7.3.102) (with q replaced by $q+r$). Then, under the models of section 7.2.4, $\boldsymbol{\nu}_N \to \boldsymbol{\nu}$ as $n \to \infty$, where $\boldsymbol{\nu}$ is defined in the same way by replacing the average c.d.f.'s by their limiting forms. We now partition $\boldsymbol{\nu}$ as

(7.8.7) $$\boldsymbol{\nu} = \begin{pmatrix} \boldsymbol{\nu}_{11} & \boldsymbol{\nu}_{12} \\ \boldsymbol{\nu}_{21} & \boldsymbol{\nu}_{22} \end{pmatrix}; \qquad \begin{array}{ll} \boldsymbol{\nu}_{11} \text{ is } q \times q, & \boldsymbol{\nu}_{12} \text{ is } q \times r, \\ \boldsymbol{\nu}_{21} \text{ is } r \times q, & \boldsymbol{\nu}_{22} \text{ is } r \times r. \end{array}$$

Let then

(7.8.8) $$\boldsymbol{\nu}_{11\cdot 2} = \boldsymbol{\nu}_{11} - \boldsymbol{\nu}_{12}\boldsymbol{\nu}_{22}^{-1}\boldsymbol{\nu}_{21} \quad \text{and} \quad \boldsymbol{\nu}_{11\cdot 2}^{-1} = ((\nu_{11\cdot 2}^{kk'})).$$

Then, from Theorem 7.3.12 and some standard analysis we obtain the following result (on using the convergence of $\boldsymbol{\nu}_{N,11\cdot 2}$ to $\boldsymbol{\nu}_{11\cdot 2}$, consequent on

Lemma 7.3.10):

Under $\{H_N\}$ in (7.8.6), $\mathfrak{L}_N^*$ has asymptotically a noncentral chi-square distribution with $q(p-1)$ D.F. and the noncentrality parameter

$$\Delta_{\mathfrak{L}^*} = \sum_{k=1}^{q} \sum_{k'=1}^{q} \nu_{11\cdot 2}^{kk'} \sum_{j=1}^{p} \eta_{jk}\eta_{jk'}. \tag{7.8.9}$$

By definition in (7.8.8), the characteristic roots of the matrix $\mathbf{\nu}_{11}^{-1}\mathbf{\nu}_{11\cdot 2}$ are all bounded above by 1, and hence, comparing (7.5.122) and (7.6.9), it can be shown (cf. Exercise 7.8.1) that

$$\Delta_{\mathfrak{L}^*}/\Delta_{\mathfrak{L}} \geq 1, \qquad \text{uniformly in } \tau_{jk}\text{'s and } G(z). \tag{7.8.10}$$

Thus, the A.R.E. of the MANOCA relative to the ANOCA test based on the same type of rank statistics is bounded below by 1, as is expected. If $p = 1$, then the efficiency results of section 5.9 also go through in the two-way layout case. But for $p > 1$, such simple results are difficult to obtain.

EXERCISES

Section 7.2

7.2.1. Show that the optimum $\{J_r, r = 1, \ldots, p\}$ in (7.2.32) when the F_i's are all uniform distributions over (0, 1) is

$$J_r = \begin{cases} 1, & r = 1 \\ 0, & 2 \leq r \leq p-1 \\ -1, & r = p. \end{cases}$$

(Sen, 1968b.)

7.2.2. For exponential distribution, show that the optimum scores are

$$J_r = \begin{cases} 1/p, & \\ -1/p(p-1), & r = 2, \ldots, p. \end{cases}$$

(Sen, 1968b.)

7.2.3. For logistic distribution, show that the optimal scores are

$$J_r = r \quad \text{for} \quad 1 \leq r \leq p.$$

(Sen, 1968b.)

7.2.4. Obtain the optimal scores for double exponential distribution. (Sen, 1968b.)

7.2.5. Obtain the optimal scores for normal distribution and approximate these by suitable functions of the expected values and variances of the normal order statistics. (Sen, 1968b.)

7.2.6. Prove the convergence in (7.2.36). (*Hints:* Express the denominator on the right-hand side of (7.2.34) as an average of n independent and identically distributed random variables plus another random variable which converges to zero as $n \to \infty$. Apply the law of large numbers to the average.)

7.2.7. Using (7.2.19), (7.2.20), (7.2.30), and (7.2.44) show that for the heteroscedastic model (7.2.46), the A.R.E. in (7.2.44) is minimum when the σ_i, $i = 1, \ldots, b$ are all equal.

7.2.8. Obtain the non-null distribution of χ_r^2 for exchangeable errors when the sequence of alternative hypotheses is K_n: $\boldsymbol{\theta} = \boldsymbol{\theta}_n = n^{-1/2}\boldsymbol{\tau}_n$. In particular for parent normal distribution, show that the A.R.E. of the χ_r^2 with respect to the parametric ANOVA test is also equal to $3p/\pi(p + 1)$. (Sen, 1971.)

7.2.9. Generalize the results of Exercise 7.2.8 to any arbitrary S_N test based on scores $\{J_1, \ldots, J_p\}$.

7.2.10. Show that χ_r^2 defined by (7.2.50), has (under the null hypothesis) asymptotically a chi-square distribution with $p - 1$ D.F. (Elteren and Noether 1959.)

7.2.11. Obtain the non-null distribution theory of χ_r^2 in (7.2.50) under the sequence of alternatives in (7.2.17), and derive its A.R.E. with respect to the parametric ANOVA test for the BIB design. (Elteren and Noether, 1959.)

7.2.12. Show that χ_r^2 in (7.2.50) provides a valid test for the hypothesis of no treatment effects even when the errors are interchangeable. (Sen, 1971.)

7.2.13. Extend the S_N test in (7.2.11) to the incomplete block layouts when the errors are exchangeable.

7.2.14. Show that the right-hand side of (7.2.64) has the expectation

$$n^{-1} \sum_{i=1}^{n} \sigma_{ss'}(F_i),$$

for all $s, s' = 1, \ldots, q$. (Gerig, 1969.)

7.2.15. Obtain the asymptotic distribution of the normal theory likelihood ratio statistic, as has been stated above (7.2.92).

Section 7.3

7.3.1. Show that (7.3.25) is stochastically equivalent to $\nu_{jj',n}$, defined by (7.3.17).

7.3.2. Prove that $\sup_n n^{-1} \sum_{i=1}^{n} E\{|B(\mathbf{Y}_i, \boldsymbol{\delta}^*)|^{2+\delta}\} < \infty$, where $B(\mathbf{Y}_i, \boldsymbol{\delta}^*)$ is defined by (7.3.51).

7.3.3. Using (7.3.48) and (7.3.49), compute the covariance terms in (7.3.38).

7.3.4. Prove the relation in (7.2.39) when the errors are not independent, but are within block symmetric dependent. (Sen, 1968e.)

7.3.5. Obtain the large-sample permutation distribution of S_N', defined by (7.3.76).

7.3.6. Supply the proof of Lemma 7.3.10. (Sen, 1969a.)

7.3.7. Obtain a formal proof of Theorem 7.3.11. (Sen, 1969a.)

7.3.8. Prove Theorem 7.3.12. (Sen, 1969a.)

Section 7.4

7.4.1. If $\mathbf{X} = (X_1, \ldots, X_p)'$ has the c.d.f. $G(x)$, symmetric in its p arguments, show that

(i) the distribution of $X_i - X_j$ is symmetric for all $i \neq j = 1, \ldots, p$,

(ii) the distribution of $\sum_{i=1}^{p} l_i X_i$, where $\sum_{i=1}^{p} l_i = 0$ and the l_i's are either $+1$, 0, or -1, is also symmetric about 0,

(iii) the distribution of $X_i + X_j - 2X_k$ is not necessarily symmetric (consider a counterexample).

7.4.2. Prove the Theorem 7.4.3. (Puri and Sen, 1967b.)

7.4.3. Prove Theorem 7.4.4. (Puri and Sen, 1967b.)

7.4.4. Prove Theorem 7.4.10 and (7.4.53). (Puri and Sen, 1967a.)

7.4.5. Show that the estimates $Z^*_{jj'}$, defined by (7.4.55) are compatible in the sense that $Z^*_{jj'} + Z^*_{j'l} = Z^*_{jl}$.

7.4.6. Supply the proof of Theorem 7.4.11. (Sen, 1969a.)

7.4.7. For $\mathbf{v}^{(1)}$ and $\mathbf{v}^{(2)}$, defined by (7.4.62) and (7.4.64), obtain the extreme characteristic roots of $\mathbf{v}^{(2)}\mathbf{v}^{(1)-1}$, when the underlying distribution is normal, and (i) the Wilcoxon scores and (ii) the normal scores statistics are used.

7.4.8. In Exercise 7.4.7, supply the bounds for (7.4.66).

Section 7.5

7.5.1. Prove (7.5.6) and (7.5.7).

7.5.2. Prove (7.5.11), (7.5.12), (7.5.15), and (7.5.16).

7.5.3. Prove (7.5.22) and (7.5.23).

7.5.4. Obtain the value of (7.5.37) for (i) normal, (ii) logistic, (iii) uniform, and (iv) double exponential distributions.

Section 7.6

7.6.1. Obtain the asymptotic non-null distribution of the statistic in (7.6.3) for the sequence of alternatives: $\boldsymbol{\theta} = \boldsymbol{\theta}_n = n^{-\frac{1}{2}}\boldsymbol{\tau}$. Show that this distribution has the same form as the limiting distribution of W_n in (7.6.1), but they differ in the noncentrality parameters. Hence, or otherwise, obtain the allied A.R.E. results.

7.6.2. Obtain similar results for W^*_n, defined by (7.6.4).

7.6.3. Prove the identity in (7.6.15).

Section 7.7

7.7.1. Supply the proof of Theorem 7.7.2.

7.7.2. Obtain the limiting distribution of the classical ANOVA test statistics, as stated just before (7.7.24).

Section 7.8

7.8.1. Obtain the proof of the inequality in (7.8.10).

7.8.2. Obtain the expressions for $\Delta_{\mathfrak{L}^*}/\Delta_{\mathfrak{L}}$ when (i) $p = q = 1$ and (ii) $p = 2$, $q = 1$, and in both cases the Wilcoxon scores are used.

PRINCIPAL REFERENCES

Anderson and Bancroft (1952), Anderson (1958), Benard and Elteren (1953), Bradley (1955), Brown and Mood (1951), Cochran and Cox (1952), Doksum (1967), Durbin (1951), Friedman (1937), Gerig (1969), Hodges and Lehmann (1962), Hollander (1967), Jonckhree (1954b), Kempthorne (1952), Kendall (1955), Lehmann (1964), Mehra and Sarangi (1967), Mehra and Sen (1969), Page (1963), Puri and Sen (1967a, 1967b, 1968a), Sen (1967g, 1967h, 1967i, 1968b, 1968c, 1968d, 1968e, 1969a, 1969c, 1969d, 1970e, 1971).

CHAPTER 8

Rank Tests for Independence

8.1. INTRODUCTION

There are three basic problems connected with the study of the interrelationships of the different variates of a stochastic p-vector ($p \geq 2$). First, we may desire to test the hypotheses relating to the stochastic independence of two or more of these variates. As we shall see, these hypotheses of independence are characterized by appropriate hypotheses of invariance (under suitable groups of transformations), and so distribution-free tests for those hypotheses exist. These will be studied in detail in this chapter. The second problem is to study the degree of association of the different variates in a stochastic vector. For this purpose, suitable measures of association, analogous to the measures based on the classical product moment correlations, are studied in section 8.2. The tests for independence are in fact based on the sample counterparts of these measures of association. The third problem—the nonparametric analogue of the classical regression problem—requires a different approach, and will not be considered here. However, at the end of the chapter, some appropriate references are provided for the convenience of the reader.

Let $\mathbf{X}_\alpha = (X_{1\alpha}, \ldots, X_{p\alpha})$, $\alpha = 1, \ldots, n$ be n i.i.d.r.v.'s having a p-variate continuous c.d.f. $F(\mathbf{x})$, where $\mathbf{x} \in R^p$, the real p-space. Let $\mathbf{X}_\alpha$ be partitioned into q (≥ 2) subvectors; that is,

$$\mathbf{X}_\alpha = (\mathbf{X}_\alpha^{(1)}, \ldots, \mathbf{X}_\alpha^{(q)}), \qquad \alpha = 1, \ldots, n, \tag{8.1.1}$$

where $\mathbf{X}_\alpha^{(j)}$ is of i_j (≥ 1) components, $j = 1, \ldots, q$; $\sum_{j=1}^q i_j = p$. Let the marginal c.d.f. of $\mathbf{X}_\alpha^{(j)}$ be $F^{(j)}(\mathbf{x}^{(j)})$, where $\mathbf{x}^{(j)} \in R^{i_j}$, $j = 1, \ldots, q$. Our first problem is to test the null hypothesis that $\mathbf{X}_\alpha^{(1)}, \ldots, \mathbf{X}_\alpha^{(q)}$ are stochastically independent, i.e.,

$$H_0^{(q)}\colon F(\mathbf{x}) = \prod_{j=1}^{p} F^{(j)}(\mathbf{x}^{(j)}), \quad \text{for all} \quad \mathbf{x} \in R^p. \tag{8.1.2}$$

Let us denote the marginal c.d.f. of $X_{j\alpha}$ by $F_{[j]}(x)$, $j = 1, \ldots, p$, and the marginal joint c.d.f. of $(X_{j\alpha}, X_{j'\alpha})$ by $F_{[j,j']}(x, y)$, for $j \neq j' = 1, \ldots, p$. $X_{1\alpha}, \ldots, X_{p\alpha}$ are said to be totally independent if

$$(8.1.3) \qquad H_0^{(p)}: F(\mathbf{x}) = \prod_{j=1}^{p} F_{[j]}(x_j), \quad \text{for all} \quad \mathbf{x} \in R^p.$$

Thus, $H_0^{(p)}$ is a particular case of $H_0^{(q)}$ when $p = q$, that is, when $i_1 = \cdots = i_p = 1$. Also, $X_{1\alpha}, \ldots, X_{p\alpha}$ are said to be pairwise independent if

$$(8.1.4) \qquad H_0^{(*)}: F_{[j,j']}(x, y) = F_{[j]}(x)F_{[j']}(y),$$

for all $(x, y) \in R^2$, and for all $j \neq j' = 1, \ldots, p$. It may be noted that (8.1.3) implies (8.1.4) but not conversely.

In the particular case of $F(\mathbf{x})$ being a multinormal c.d.f., pairwise independence implies total independence and vice versa. Moreover, in this case uncorrelation implies independence and vice versa. Also, the multinormal c.d.f. is completely specified by its mean vector and covariance matrix. Thus, all the three problems stated above reduce to certain specific structures for the covariance matrix. If the covariance matrix is partitioned into q^2 submatrices of orders $i_j \times i_l$; $j, l = 1, \ldots, q$, then the first hypothesis $H_0^{(q)}$ is equivalent to the hypothesis that all off-diagonal partitioned matrices are null matrices, while the hypotheses $H_0^{(p)}$ and H_0^* are both equivalent to the hypothesis that the covariance matrix itself is a diagonal matrix. Tests for these hypotheses are considered in Anderson (1958, chapter 9), a good source for details of background and motivation.

In this chapter we generalize the above problems in a completely nonparametric setup. For arbitrary (continuous) multivariate distributions the covariance matrix may not exist, and even if it exists it may not play the fundamental role that it does in the case of the multinormal distributions. For this reason, we shall formulate a class of *association parameters* which are regular functionals of the c.d.f. $F(\mathbf{x})$ and which are defined for a wider class of c.d.f.'s. This will provide us with suitable nonparametric competitors of the classical covariance matrix and also increase the scope of applications. Further, for multinormal c.d.f.'s zero correlation and independence are equivalent, but the same is not necessarily true for non-normal c.d.f.'s. Thus, to be precise about the class of alternative hypotheses, we also define some *dependence function* with some emphasis on association alternatives. The tests to be considered will be shown to be well behaved for such alternatives. Finally, for arbitrary c.d.f.'s, since pairwise independence does not imply total independence, we prefer to consider the hypothesis (8.1.4) in some detail. Throughout the chapter we consider the general case of $p \geq 2$; the particular case of bivariate independence follows by substituting $p = 2$. However, for clarification of ideas, we have treated the bivariate case in detail in the exercises.

8.2. FORMULATION OF A CLASS OF ASSOCIATION PARAMETERS

Apart from the necessity of the existence of the second-order moments, the estimator of the usual covariance matrix is quite sensitive to outlying observations. Here we shall consider some alternative measures of association which are really functionals of the parent c.d.f.'s. For this, we consider the marginal c.d.f.'s $F_{[i]}$ and $F_{[i,j]}$ $(i \neq j = 1, \ldots, p)$ as in section 8.1. Also, let $J_{(i)}(u)$, $0 < u < 1$ $(i = 1, \ldots, p)$ be some absolutely continuous function defined on the open interval (0, 1) and suppose that $J_{[i]}(u)$ is normalized in the following manner:

$$(8.2.1) \quad \int_0^1 J_{(i)}(u)\,du = 0 \quad \text{and} \quad \int_0^1 J_{(i)}^2(u)\,du = 1 \quad \text{for} \quad i = 1, \ldots, p.$$

Later, we shall impose certain regularity conditions on $J_{(i)}(u)$. Let us now consider the transformed variables

$$(8.2.2) \qquad Y_i = J_{(i)}(F_{[i]}(X_i)) \quad \text{for} \quad i = 1, \ldots, p$$

and write

$$(8.2.3) \qquad \mathbf{Y} = (Y_1, \ldots, Y_p).$$

Y_i will be called the *grade functional* of X_i, $i = 1, \ldots, p$. The joint distribution of $\mathbf{Y}$ can be obtained from $F(\mathbf{x})$, but it will naturally depend upon the association pattern of $F(\mathbf{x})$. Let $J_{[i]}(u)$ be strictly increasing for all $0 < u < 1$ $(i = 1, \ldots, p)$, then it is easy to verify that independence of (X_i, X_j) implies the independence of (Y_i, Y_j) and vice versa. Further, this property is preserved under any monotone transformation of X_i; $i = 1, \ldots, p$. So when we are interested in the problems of stochastic independence we may work with $\mathbf{Y}_\alpha$, $\alpha = 1, \ldots, n$, and suitably chosen $J_{(i)}$, $i = 1, \ldots, p$, often lead to some nice properties of such alternative measures. Let us now define

$$(8.2.4) \quad \theta_{ij} = \int_{-\infty}^{\infty}\int_{-\infty}^{\infty} J_{(i)}(F_{[i]}(x))J_{(j)}(F_{[j]}(y))\,dF_{[i,j]}(x, y); \quad i, j = 1, \ldots, p.$$

It may be noted that by (8.2.1), $\theta_{ii} = 1$ for all $i = 1, \ldots, p$ and $|\theta_{ij}| \leq 1$ for all $i \neq j = 1, \ldots, p$. Further, if $X_{i\alpha}$ and $X_{j\alpha}$ are independent, then from (8.2.1) and (8.2.4), we see that $\theta_{ij} = 0$. In fact θ_{ij} is the product moment correlation of $Y_{i\alpha}$ and $Y_{j\alpha}$. Thus, we define a *grade functional association matrix* by

$$(8.2.5) \qquad \mathbf{\Theta} = ((\theta_{ij}))_{i,j=1,\ldots,p}.$$

Naturally, the choice of the transformations $J_{(i)}$, $i = 1, \ldots, p$ plays an important role in the procedure to be considered. We shall consider a general class of such functionals and subsequently we shall briefly consider the problem of selecting some particular measures.

Now, usually the c.d.f.'s $F_{[i]}$, $i = 1, \ldots p$ are all unknown, and hence $\mathbf{Y}$ is also unknown. However, (8.2.4) has a close analogy with the type of functionals considered by Von Mises (1947), and following his line of approach we shall frame suitable estimators of $\mathbf{\Theta}$. With this end in view, let us define

$$F_{n[i]}(x) = (\text{number of } X_{i\alpha} \leq x,\ \alpha = 1, \ldots, n)/n; \qquad i = 1, \ldots, p, \tag{8.2.6}$$

$$F_{n[i,j]}(x, y) = (\text{number of } (X_{i\alpha}, X_{j\alpha}) \leq (x, y),\ \alpha = 1, \ldots, n)/n, \qquad i \neq j' = 1, \ldots, p. \tag{8.2.7}$$

Since the $F_{n[i]}$ are all step functions, and are unbiased estimators of their population counterparts, we consider some score functions $J_{n(i)}(u)$, $i = 1, \ldots, p$, as in the preceding chapters, and define the scores as

$$J_{n(i)}(F^*_{n[i]}(x)) \quad \text{where} \quad F^*_n = nF_n/(n + 1). \tag{8.2.8}$$

Consequently, from (8.2.4), (8.2.6), (8.2.7), and (8.2.8), we may frame the estimator of θ_{ij} as $T_{n,ij}$ where

$$\begin{aligned} T_{n,ij} &= \int_{-\infty}^{\infty}\int_{-\infty}^{\infty} J_{n(i)}(F^*_{n[i]}(x))J_{n(j)}(F^*_{n[j]}(y))\, dF_{n[i,j]}(x, y) \\ &= \frac{1}{n}\sum_{\alpha=1}^{n} J_{n(i)}(F^*_{n[i]}(X_{i\alpha}))J_{n(j)}(F^*_{n[j]}(X_{j\alpha})); \qquad i, j = 1, \ldots, p. \end{aligned} \tag{8.2.9}$$

It may be noted that $T_{n,ii}$ are nonstochastic (if the $J_{n(i)}$'s are so) and by (8.2.1), they converge to unity as $n \to \infty$. Our proposed tests are based on suitable functions of the stochastic matrix $\mathbf{T}_n$, where

$$\mathbf{T}_n = ((T_{n,ij})); \qquad i, j = 1, \ldots, p. \tag{8.2.10}$$

In the sequel, $\mathbf{T}_n$ and $\mathbf{\Theta}$ will be termed respectively the sample and population dispersion matrices. It may be noted that many well-known nonparametric measures of association belong to the class considered above. For example, if we let

$$J_{n(i)}\left(\frac{\alpha}{n + 1}\right) = \left(\frac{12}{n^2 - 1}\right)^{1/2}\left\{\alpha - \frac{n + 1}{2}\right\}; \qquad \alpha = 1, \ldots, n; \quad i = 1, \ldots, p, \tag{8.2.11}$$

then $T_{n,ij}$ reduces to the Spearman rank correlation, while for $J_{(i)}(u) = \sqrt{12}\,(u - \frac{1}{2})$, $0 < u < 1$, $i = 1, \ldots, p$, θ_{ij} reduces to the grade correlation coefficient; these measures were studied in chapter 3. Some other important measures will be considered in later sections of this chapter.

Now, in multinormal distributions, any stochastic association is reflected in the correlation matrix. But, in general, deviation from independence may not be properly reflected by the correlations. Thus, to consider the class of alternative hypotheses, we formulate some *dependence functions*, apparently first studied by Sibuya (1959). Denote by

$$\Omega(\mathbf{x}) = F(\mathbf{x}) \Big/ \prod_{i=1}^{p} F_{[i]}(x_i), \tag{8.2.12}$$

$$\Omega_{ij}(x, y) = F_{[i,j]}(x, y)/F_{[i]}(x_i)F_{[j]}(y), \qquad i \neq j = 1, \ldots, p. \tag{8.2.13}$$

Both Ω and Ω_{ij} are functions of the marginal distributions as well as the joint distributions. So we shall prefer to write them as $\Omega(F_{[1]}, \ldots, F_{[p]})$ and $\Omega_{ij}(F_{[i]}, F_{[j]})$ respectively. Hereafter we shall term Ω and Ω_{ij} dependence functions. (It may be noted that if $X_1, \ldots, X_p$ are totally independent (or if X_i and X_j are independent), then Ω (or Ω_{ij}) will be equal to unity of all $\mathbf{x}$ (or all x_i, x_j) and so, the divergence from unity on a set of points of measure nonzero relates to stochastic dependence). In subsequent sections, we shall make repeated use of this type of dependence function, along with some others to be developed in section 8.7.

8.3. PERMUTATIONALLY DISTRIBUTION-FREE TESTS FOR H_0^q

It may be noted that if X_i and X_j belong to two different subsets, then under $H_0^{(q)}$: $\theta_{ij} = 0$. Let us denote by $\mathcal{F}_r$ the class of all r-variate continuous c.d.f.'s for $r = 1, \ldots, p$. Furthermore define

$$\mathcal{F}_p^0 = \left\{ F(\mathbf{x}):\ F(\mathbf{x}) = \prod_{j=1}^{q} F^{(j)}(\mathbf{x}^{(j)}), \qquad F^{(j)} \in \mathcal{F}_{i_j},\ \ j = 1, \ldots, q \right\}. \tag{8.3.1}$$

Thus $\mathcal{F}_p^0$ is a subclass of all p-variate continuous c.d.f.'s for which the q specified subsets of variates are mutually stochastically independent. Thus, one may write equivalently

$$H_0^{(q)}:\ F(\mathbf{x}) \in \mathcal{F}_p^{(0)} \tag{8.3.2}$$

and may be interested in the set of alternatives that $F(\mathbf{x}) \subset \mathcal{F}_p - \mathcal{F}_p^0$. However, such a formulation will give rise to considerable amount of mathematical complications since $F(\mathbf{x})$ is otherwise of completely unknown form. In this chapter, we shall restrict ourselves to the class of alternatives which relates to the lack of independence through the values of the functional matrix $\boldsymbol{\Theta}$

defined in (8.2.5). Let us now partition $\mathbf{\Theta}$ as

$$\mathbf{\Theta} = \begin{pmatrix} \mathbf{\Theta}_{11} & \mathbf{\Theta}_{12} & \cdots & \mathbf{\Theta}_{1q} \\ \mathbf{\Theta}_{21} & \mathbf{\Theta}_{22} & \cdots & \mathbf{\Theta}_{2q} \\ \cdot & \cdot & & \cdot \\ \cdot & \cdot & & \cdot \\ \cdot & \cdot & & \cdot \\ \mathbf{\Theta}_{q1} & \mathbf{\Theta}_{q2} & & \mathbf{\Theta}_{qq} \end{pmatrix} \tag{8.3.3}$$

where $\mathbf{\Theta}_{kl}$ is of order $i_k \times i_l$; $k, l = 1, \ldots, q$. Now from (8.3.2), we have

$$\mathbf{\Theta}_{kl} = 0 \quad \text{for} \quad k \neq l = 1, \ldots, q \quad \text{under} \quad H_0^{(q)}. \tag{8.3.4}$$

We frame the class of alternatives as the set of all p-variate c.d.f.'s for which the equality sign in (8.3.4) does not hold for at least one pair (k, l), $k \neq l = 1, \ldots, q$. Now recalling the definition of θ_{ij} in (8.2.4) and Ω_{ij} in (8.2.13), we may write

$$\begin{aligned} \theta_{ij} &= \int_{-\infty}^{\infty}\int_{-\infty}^{\infty} [F_{[i,j]}(x, y) - F_{[i]}(x)F_{[j]}(y)]J'_{(i)}[F_{[i]}(x)] \\ &\qquad \times J'_{(j)}[F_{[j]}(y)]\, dF_{[i]}(x)\, dF_{[j]}(y) \\ &= \int_{-\infty}^{\infty}\int_{-\infty}^{\infty} F_{[i]}(x)F_{[j]}(y)[\Omega_{ij}(F_{[i]}(x), F_{[j]}(y)) - 1] \\ &\qquad \times J'_{(i)}[F_{[i]}(x)]J'_{(j)}[F_{[j]}(y)]\, dF_{[i]}(x)\, dF_{[j]}(y) \end{aligned} \tag{8.3.5}$$

(Here J' denotes the first derivative of J.)

In most cases (as we shall later see) $J_{(i)}(u)$ is monotone in u, $0 < u < 1$ for $i = 1, \ldots, p$, so that if $\Omega_{ij}(x, y)$ is uniformly (in x, y) greater than (less than) or equal to 1 with the strict inequality on at least a set of points of nonzero measure, then $\theta_{ij} \neq 0$. In many types of factorial dependence (cf. Konijn, 1956; Bhuchongkul, 1964; for the bivariate case), it is easy to verify that Ω_{ij} is greater than or less than one, depending upon the coefficients of the stochastic factors, and hence such an alternative may be quite suitable. Though we have explained the model in greater generality, we shall for simplicity of presentation consider in detail the tests for $H_0^{(q)}$ when $q = 2$, and then indicate briefly how the theory can be extended to the case of more than two subsets. With this end in mind, we define

$$\mathbf{X}_\alpha = (\mathbf{X}_\alpha^{(1)}, \mathbf{X}_\alpha^{(2)}); \qquad \begin{aligned} \mathbf{X}_\alpha^{(1)} &= (X_{1\alpha}, \ldots, X_{l\alpha}); \\ \mathbf{X}_\alpha^{(2)} &= (X_{l+1,\alpha}, \ldots, X_{p\alpha}), \quad \alpha = 1, \ldots, n, \end{aligned} \tag{8.3.6}$$

where $1 \leq l$, $m = p - l < p$. We denote the sample point $\mathbf{E}_n$ by

$$\mathbf{E}_n^{p\times n} = (\mathbf{X}_1', \ldots, \mathbf{X}_n') = \begin{pmatrix} \mathbf{X}_1^{(1)\prime}, \ldots, \mathbf{X}_n^{(1)\prime} \\ \mathbf{X}_1^{(2)\prime}, \ldots, \mathbf{X}_n^{(2)\prime} \end{pmatrix} \tag{8.3.7}$$

The joint distribution function of $\mathbf{E}_n^{p\times n}$ is given by

$$G(\mathbf{E}_n) = \prod_{\alpha=1}^{n} F(\mathbf{X}_\alpha) \tag{8.3.8}$$

and under $H_0^{(2)}$

$$G(\mathbf{E}_n) = \prod_{\alpha=1}^{n} \{F^{(1)}(\mathbf{X}_\alpha^{(1)})F^{(2)}(\mathbf{X}_\alpha^{(2)})\}. \tag{8.3.9}$$

Let $\mathbf{R}_n = (R_1, R_2, \ldots, R_n)$ be any permutation of $(1, 2, \ldots, n)$ and denote by $\mathcal{R}_n$, the set of all the $n!$ permutations of $(1, 2, \ldots, n)$. Furthermore, denote

$$\mathbf{E}(\mathbf{R}_n) = \begin{pmatrix} \mathbf{X}_1^{(1)\prime}, \ldots, \mathbf{X}_n^{(1)\prime} \\ \mathbf{X}_{R_1}^{(2)\prime}, \ldots, \mathbf{X}_{R_n}^{(2)\prime} \end{pmatrix}, \qquad \mathbf{R}_n \in \mathcal{R}_n \tag{8.3.10}$$

$$S(\mathbf{E}_n) = \{\mathbf{E}(\mathbf{R}_n)\colon\ \mathbf{R}_n \in \mathcal{R}_n\}. \tag{8.3.11}$$

Then it follows from (8.3.9) and (8.3.10) that the joint distribution of $\mathbf{E}(\mathbf{R}_n)$ remains invariant under the group of transformations $\mathbf{R}_n \in \mathcal{R}_n$ if $H_0^{(2)}$ holds. We shall call $S(\mathbf{E}_n)$, the permutation invariant set of $\mathbf{E}_n$. Let now $\mathcal{E}_n$ be the np-dimensional Euclidean space, so that $\mathbf{E}_n \in \mathcal{E}_n$. Under $H_0^{(2)}$, the probability distribution of $\mathbf{E}_n$ over the set of $n!$ points in $S(\mathbf{E}_n)$ is uniform, and hence, if $\phi(\mathbf{E}_n)$ is any test function chosen in such a way that

$$\sum_{\mathbf{E}_n^* \in S(\mathbf{E}_n)} \phi(\mathbf{E}_n^*) = n!\,\epsilon; \qquad 0 < \epsilon < 1 \tag{8.3.12}$$

for all $\mathbf{E}_n \in \mathcal{E}_n$, then

$$E_{H_0^{(2)}}\{\phi(\mathbf{E}_n)\} = \epsilon, \qquad \text{the level of significance.} \tag{8.3.13}$$

($E_{H_0^{(q)}}\{\cdot\}$ denotes the expectation of $\{\cdot\}$ under $H_0^{(q)}$).

Now to formulate $\phi(\mathbf{E}_n)$ in a suitable manner and to make it invariant under monotone transformation of the variables, we proceed as follows. For each integer n, let

$$\mathbf{L}_n^{(i)} = (L_{n,1}^{(i)}, \ldots, L_{n,n}^{(i)}); \qquad i = 1, \ldots, p. \tag{8.3.14}$$

be the p sets of real constants, where

$$L_{n,\alpha}^{(i)} = J_n^{(i)}\left(\frac{\alpha}{n+1}\right); \qquad \alpha = 1, \ldots, n; \quad i = 1, \ldots, p. \tag{8.3.15}$$

and where the functions $J_n^{(i)}$ have the same interpretation as in section 8.2. With this notation we can rewrite $T_{n,ij}$ (cf. (8.2.9)) as

$$T_{n,ij} = \frac{1}{n}\sum_{\alpha=1}^{n} L_{n,R_{i\alpha}}^{(i)} L_{n,R_{j\alpha}}^{(j)}; \qquad i, j = 1, \ldots, p, \tag{8.3.16}$$

where $R_{i\alpha}$ is the rank of $X_{i\alpha}$ among $(X_{i1}, \ldots, X_{in})$ for $i = 1, \ldots, p$. Under $H_0^{(2)}$, $X_{i\alpha}$ and $X_{j\alpha}$ are stochastically independent for all $i = 1, \ldots, l$ and $j = l + 1, \ldots, p$.

Let us now partition $\mathbf{T}_n = ((T_{n,ij}))$, $i, j = 1, \ldots, p$ as we did $\boldsymbol{\Theta}$ in (8.3.3) for $q = 2$. We obtain

$$\mathbf{T}_n = \begin{pmatrix} ((T_{n,ij}^{(1,1)})) & ((T_{n,ij}^{(1,2)})) \\ ((T_{n,ij}^{(2,1)})) & ((T_{n,ij}^{(2,2)})) \end{pmatrix} \tag{8.3.17}$$

Note that

$$\begin{aligned} ((T_{n,ij}^{(1,1)})) &= ((T_{n,ij})), \qquad i, j = 1, \ldots, l; \\ ((T_{n,ij}^{(1,2)}))' = ((T_{n,ij}^{(2,1)})) &= ((T_{n,ij})), \qquad i = l + 1, \ldots, p; \\ & \qquad\qquad j = 1, \ldots, l \\ ((T_{n,ij}^{(2,2)})) &= ((T_{n,ij})), \qquad i = l + 1, \ldots, p, \quad j = l + 1, \ldots, p. \end{aligned} \tag{8.3.18}$$

Let us now denote by $\mathcal{T}_n$ the permutational probability law generated by the $n!$ equally likely (conditional) realizations of $\mathbf{E}_n$ over $S(\mathbf{E}_n)$. It is seen that

$$E(T_{n,ij} \mid \mathcal{T}_n) = 0 \quad \text{for all} \quad i = 1, \ldots, l \quad \text{and} \quad j = l + 1, \ldots, p. \tag{8.3.19}$$

Furthermore, since under $\mathcal{T}_n$, the $n!$ realizations of $\mathbf{R}_n$ over $\mathcal{R}_n$ are conditionally equally likely, we obtain

$$\begin{aligned} \operatorname{cov}\{T_{n,ij}, T_{n,i'j'} \mid \mathcal{T}_n\} &= E\{T_{n,ij}T_{n,i'j'} \mid \mathcal{T}_n\} \\ &= \frac{1}{n^2}\sum_{\alpha=1}^{n}\sum_{\beta=1}^{n} E\{L_{n,R_{i\alpha}}^{(i)} L_{n,R_{j\alpha}}^{(j)} L_{n,R_{i'\beta}}^{(i')} L_{n,R_{j'\beta}}^{(j')} \mid \mathcal{T}_n\}. \end{aligned} \tag{8.3.20}$$

Since i, i' range over $1, \ldots, l$ and j, j' over $l + 1, \ldots, p$, we get for $\alpha = \beta$

$$\begin{aligned} & E\{L_{n,R_{i\alpha}}^{(i)} L_{n,R_{j\alpha}}^{(j)} L_{n,R_{i'\alpha}}^{(i')} L_{n,R_{j'\alpha}}^{(j')} \mid \mathcal{T}_n\} \\ &= \frac{1}{n^2}\sum_{\alpha=1}^{n}\sum_{\beta=1}^{n} L_{n,R_{i\alpha}}^{(i)} L_{n,R_{j'\alpha}}^{(i')} L_{n,R_{j\beta}}^{(j)} L_{n,R_{j'\beta}}^{(j')} = T_{n,ii'}T_{n,jj'}. \end{aligned} \tag{8.3.21}$$

For $\alpha \neq \beta$

$$\begin{aligned} & E\{L_{n,R_{i\alpha}}^{(i)} L_{n,R_{j\alpha}}^{(j)} L_{n,R_{i'\beta}}^{(i')} L_{n,R_{j'\beta}}^{(j')} \mid \mathcal{T}_n\} \\ &= \frac{1}{n^2(n-1)^2}\left\{\sum_{\alpha \neq \alpha'=1}^{n} L_{n,R_{i\alpha}}^{(i)} L_{n,R_{i'\alpha'}}^{(i')} \cdot \sum_{\beta \neq \beta'=1}^{n} L_{n,R_{j\beta}}^{(j)} L_{n,R_{j'\beta'}}^{(j')}\right\} \\ &= \frac{1}{(n-1)^2} T_{n,ii'}T_{n,jj'}, \text{ as } \sum_{\alpha=1}^{n} E_{n,R_{j\alpha}}^{(i)} = 0 \quad \text{for all } i = 1, \ldots, p. \end{aligned} \tag{8.3.22}$$

Using (8.3.21) and (8.3.22) we obtain

$$\text{(8.3.23)} \quad \operatorname{cov}(T_{n,ij}, T_{n,i'j'} \mid \mathfrak{T}_n) = \frac{1}{n-1} T_{n,ii'} T_{n,jj'}$$

for all $i, i' = 1, \ldots, l$ and $j, j' = l+1, \ldots, p$.

Since we are interested in a comprehensive test for $H_0^{(2)}$ (i.e., when $q = 2$), following some simple steps we may consider the test statistic

$$\text{(8.3.24)} \quad V_{n(2)}^{(J)} = (n-1) \sum_{i=1}^{l} \sum_{i'=1}^{l} \sum_{j=l+1}^{p} \sum_{j'=l+1}^{p} T_{n,ij} T_{n,i'j'}, T_{n(1)}^{ii'} T_{n(2)}^{jj'}$$

where $T_{n(1)}^{ii'}$ is the (i, i')th term in $((T_{n,ij}^{(1,1)}))^{-1}$ and $T_{n(2)}^{jj'}$ is the (j, j')th term in $((T_{n,ij}^{(2,2)}))^{-1}$. However, we shall find it more convenient to work with the test statistic $S_{n(2)}^{(J)}$ defined as

$$\text{(8.3.25)} \quad S_{n(2)}^{(J)} = \frac{\|T_{nij}\|}{\|T_{n,ij}^{(1,1)}\| \, \|T_{n,ij}^{(2,2)}\|}$$

where $\|\cdot\|$ is the determinant of the matrix $((\cdot))$. S_n is analogous to the usual likelihood ratio test (cf. Anderson, 1958, p. 233) where instead of $T_{n,ij}$'s the sample product moment correlations are used. We shall first consider the relation between $S_{n(2)}^{(J)}$ and $V_{n(2)}^{(J)}$.

Theorem 8.3.1. *If* $\mathbf{X}$ *has a nonsingular distribution in the sense that* $\mathbf{\Theta}$, *defined by* (3.4) *and* (3.5), *is positive definite, then* $\|T_n\|$ *converges in probability to* $\|\mathbf{\Theta}\|$ *as* $n \to \infty$, *provided* $E\,|T_{n,ij}|^{(2+\delta)/2} < \infty$ *for some* $\delta > 0$.

The proof of this theorem is similar to that of Theorem 5.4.2 and is therefore left as an exercise.

Theorem 8.3.2. *Under* $H_0^{(2)}$ *defined in* (8.1.2),

$$\text{(8.3.26)} \quad |T_{n,ij}| = O_p(n^{-1/2}) \quad \text{for all} \quad i = 1, \ldots, l; j = l+1, \ldots, p.$$

The same result also holds under the permutational law $\mathfrak{T}_n$.

Proof: First note that under $H_0^{(2)}$, $E(T_{n,ij}) = 0$, $i = 1, \ldots, l$; $j = l + 1, \ldots, p$. Furthermore, since

$$\frac{1}{n} \sum_{\alpha=1}^{n} J_{n(i)}^2 \left(\frac{\alpha}{n+1} \right)$$

is finite and bounded (in the limit as $n \to \infty$) for all $i = 1, \ldots, p$, we find after a few simple steps that

$$\text{(8.3.27)} \quad \operatorname{var}_{H_0}(T_{n,ij}) = \frac{c}{n-1}, \qquad c < \infty \quad \text{for all} \quad i = 1, \ldots, p;$$

$$j = l+1, \ldots, p.$$

Using (8.3.22), (8.3.23), and (8.3.27) and Chebychev's inequality we obtain the desired result. ◁

Theorem 8.3.2. *If* $|T_{n,ij}| = O_p(n^{-1/2})$ *for all* $i = 1, \ldots, l; j = l + 1, \ldots, p$ *then*

$$|V_{n(2)}^{(J)} + n \log S_{n(2)}^{(J)}| = O_p(n^{-1}), \tag{8.3.28}$$

provided $\boldsymbol{\theta}$ *is positive definite.*

Proof: By Laplace's expansion of the determinants we have

(8.3.29)

$$\|\mathbf{T}_n\| = \sum_{1 \le q_1 < q_2 < \cdots < q_l < p} \left\| T\begin{pmatrix} 1 & 2 & \cdots & l \\ q_1 & q_2 & \cdots & q_l \end{pmatrix} \right\| \cdot \left\| T^*\begin{pmatrix} 1 & 2 & \cdots & l \\ q_1 & q_2 & \cdots & q_l \end{pmatrix} \right\|$$

where $\left\| T\begin{pmatrix} 1 & 2 & \cdots & l \\ q_1 & q_2 & \cdots & q_l \end{pmatrix} \right\|$ is the determinant consisting of the rows 1, $2, \ldots, l$ and columns $q_1, q_2, \ldots, q_l$, and $\left\| T^*\begin{pmatrix} 1 & 2 & \cdots & l \\ q_1 & q_2 & \cdots & q_l \end{pmatrix} \right\|$ is the complimentary determinant. By virtue of Theorem 8.3.1, we have the following results.

If $r(\geq 1)$ of $q_1, q_2, \ldots, q_l$ is different from $1, 2, \ldots, l$, then

(8.3.30)

$$\left\| T\begin{pmatrix} 1 & 2 & \cdots & l \\ q_1 & q_2 & \cdots & q_l \end{pmatrix} \right\| = O_p(n^{-r/2}); \qquad \left\| T^*\begin{pmatrix} 1 & 2 & \cdots & l \\ q_1 & q_2 & \cdots & q_l \end{pmatrix} \right\| = O_p(n^{-r/2})$$

Consequently, from (8.3.29) and (8.3.30), we obtain

$$\begin{aligned} \|T_{n,ij}\| &= \|T_{n,ij}^{(1,1)}\| \, \|T_{n,ij}^{(2,2)}\| \\ &\quad + \sum_{i=1}^{l} \sum_{i'=1}^{l} \sum_{j=l+1}^{p} \sum_{j'=l+1}^{p} T_{n,ij} T_{n,i'j'} \|T_{n,ii'}^{(1)}\| \cdot \|T_{n,jj'}^{(2)}\| + O_p(n^{-2}), \end{aligned} \tag{8.3.31}$$

where

$$\|T_{n,ii'}^{(1)}\| = \text{cofactor of } T_{n,ii} \text{ in } ((T_{n,ij}^{(1,1)})) \qquad \text{(cf. (8.3.18))} \tag{8.3.32}$$

and

$$\|T_{n,jj'}^{(2)}\| = \text{cofactor of } T_{n,jj} \text{ in } ((T_{n,ij}^{(2,2)})) \qquad \text{(cf. (8.3.18))}.$$

Now by definition

$$T_{n(1)}^{ii'} = \|T_{n,ii'}^{(1)}\| / \|T_{n,ij}^{(1,1)}\| \quad \text{and} \quad T_{n(2)}^{jj'} = \|T_{n,jj'}^{(2)}\| / \|T_{n,ij}^{(2,2)}\|. \tag{8.3.33}$$

Hence, from (8.3.25), (8.3.31), (8.3.33), and Theorem 8.3.1 we obtain

$$-n \log S_{n(2)}^{(J)} = n \sum_{i=1}^{l} \sum_{i'=1}^{l} \sum_{j=l+1}^{p} \sum_{j'=l+1}^{p} T_{n,ij} T_{n,i'j'} \cdot T_{n(1)}^{ii'} T_{n(2)}^{jj'} + O_p(n^{-1}) \tag{8.3.34}$$

which proves the desired result. ◁

Theorem 8.3.3. *Under the assumption of Theorem* 8.3.2, *the permutation distribution of* $V_{n(2)}^{(J)}$ (*or* $-n \log S_{n(2)}^{(J)}$) *converges asymptotically, in probability, to a chi-square distribution with* $l(p - l)$ *degrees of freedom.*

Proof: It suffices to show that any linear combination

$$U_n = n^{-1/2} \sum_{i=1}^{l} \sum_{j=l+1}^{p} d_{ij} T_{n,ij} \tag{8.3.35}$$

has asymptotically (under $\mathcal{T}_n$) a normal distribution in the limit with mean zero and finite variance. That the mean is zero and variance is finite follows by using (8.3.19) and (8.3.20). To prove the asymptotic normality of U_n, let us denote

$$Z_\alpha(\mathbf{R}_n) = \sum_{i=1}^{l} \sum_{j=l+1}^{p} d_{ij} L_{n,R_{i\alpha}}^{(i)} L_{n,R_{j\alpha}}^{(j)}, \qquad \alpha = 1, \ldots, n \tag{8.3.36}$$

so that

$$U_n = n^{-1/2} \sum_{\alpha=1}^{n} Z_\alpha(\mathbf{R}_n). \tag{8.3.37}$$

Now for any $\mathbf{E}_n \in S(\mathbf{E}_n)$, defined in (8.3.11), $Z_\alpha(\mathbf{R}_n)$ can assume $n!$ equally likely values obtained by letting $\mathbf{R}_n \in \mathcal{R}_n$. By following the method of proof as in the Wald-Wolfowitz-Noether theorem, it is seen that

$$E_{\mathcal{T}_n}\{U_n^r\}/\{E_{\mathcal{T}_n}(V_n^2)\}^{r/2} = \begin{cases} \dfrac{2k!}{k!\,2^k} + o_p(1) & \text{if} \quad r = 2k \\ o_p(1) & \text{if} \quad r = 2k + 1 \end{cases} \tag{8.3.38}$$

for $k = 0, 1, 2, \ldots$. Hence the theorem. ◁

Thus for large samples, the permutation test based on $V_{n(2)}^{(J)}$ (or $-n \log S_{n(2)}^{(J)}$) reduces to the following rule:

$$\text{If } V_{n(2)}^{(J)} \text{ or } -n \log S_{n(2)}^{(J)} \begin{cases} > \chi^2_{l(p-l),\alpha}, & \text{reject } H_0^{(1)}, \\ < \chi^2_{l(p-l),\alpha}, & \text{accept } H_0^{(1)}, \end{cases} \tag{8.3.39}$$

where $\chi^2_{r,\alpha}$ is the $100(1 - \alpha)\%$ point of a chi-square distribution with r degrees of freedom.

We will now discuss briefly the case of $q(q > 2)$ subsets. Here we partition

$\mathbf{X}_\alpha$ as in (8.1.1), $\mathbf{\Theta}$ as in (8.3.3), and $\mathbf{T}_n$ accordingly. Then

$$\mathbf{T}_n = \begin{pmatrix} ((T_{n,ij}^{(1,1)})), \ldots, ((T_{n,ij}^{(1,q)})) \\ \cdots\cdots\cdots \\ ((T_{n,ij}^{(q,1)})), \ldots, ((T_{n,ij}^{(q,q)})) \end{pmatrix} \tag{8.3.40}$$

where $((T_{n,ij}^{(k,l)}))$ is of the order $i_k \times i_l$; $k, l = 1, \ldots, q$. Then, by analogy with the parametric likelihood ratio test statistic, for the problem of testing

$$H_0^{(q)}\colon \mathbf{\theta}_{k,l} = \mathbf{\theta}'_{l,k} = \mathbf{0}, \qquad k \neq l = 1, \ldots, p, \tag{8.3.41}$$

we consider the test statistic

$$S_{n(q)}^{(J)} = \frac{\|T_{n,ij}\|}{\prod_{k=1}^{q} \|T_{n,ij}^{(k,k)}\|}. \tag{8.3.42}$$

Proceeding as in Theorem 8.3.2, we can show that $-n \log S_{n(q)}^{(J)}$ is asymptotically equivalent to a positive definite quadratic form in $\sum_{k<l=1}^{p} i_k i_l$ statistics $T_{n,kl}$, where $X_{k\alpha}$ and $X_{l\alpha}$ ($k < l = 1, \ldots, p$; $\alpha = 1, \ldots, n$) belong to two different subsets of variates. The permutation argument considered earlier readily extends to the present case, where we shall have $(n!)^{q-1}$ equally likely realizations. Further, as a straightforward generalization of Theorem 8.3.3, we shall arrive at the conclusion that under the permutation model, $-n \log S_{n(q)}^{(J)}$ has asymptotically, in probability, a chi-square distribution with $\sum_{k<l} i_k i_l$ degrees of freedom. Thus we have a test criterion very similar to (8.3.39) with the only difference that the degrees of freedom will be equal to $\sum_{k<l=1}^{q} i_k i_l$.

As a special case, we consider $q = p$. Here $i_1 = \cdots = i_p = 1$, and the problem reduces to that of testing the hypothesis of total independence of all the p variates. Here we may work with the test criterion

$$V_{n(p)}^{(J)} = (n-1) \sum_{i<j=1}^{p} T_{n,ij}^2 / T_{n,ii} T_{n,jj} \tag{8.3.43}$$

(which is asymptotically equivalent to $-\log S_{n(p)}^{(J)}$ where $S_{n(p)}^{(J)}$ is defined by (8.3.42) by putting $q = p$). In this case permutation distribution of $V_{n(p)}^{(J)}$ coincides with the unconditional null distribution, and following the lines of Theorem 8.3.3, it can be shown that under $H_0^{(p)}$ defined in (8.1.3), $V_{n(p)}^{(J)}$ has asymptotically a chi-square distribution with $p(p-1)/2$ degrees of freedom.

8.4. ASYMPTOTIC NORMALITY OF $\mathbf{T}_n$ FOR ARBITRARY $F(\mathbf{x})$

In this section we study the asymptotic normality of the statistics $((T_{n,ij}))$, defined by (8.2.9), for arbitrary $F(\mathbf{x})$. We make the following assumptions:

(I) $\lim_{n\to\infty} J_{n(i)}(u) = J_{(i)}(u)$ exists for $0 < u < 1$ and is not constant, $i = 1, \ldots, p$;

(II) $\int_{-\infty}^{\infty}\int_{-\infty}^{\infty} [J_{n(i)}(F^*_{n[i]}(x_i))J_{n(j)}(F^*_{n[j]}(x_j)) - J_{(i)}(F^*_{n[i]}(x_i))J_{(j)}(F^*_{n[j]}(x_j))]$
$dF_{n[ij]}(x_i, x_j) = o_p(n^{-1/2})$, $i, j = 1, \ldots, p$;

(III) $J_{(i)}(u)$ is absolutely continuous in u: $0 < u < 1$, and

$$|J_{(i)}(u)| \leq K[u(1-u)]^{-\beta}, \quad |J'_{(i)}(u)| \leq K/u(1-u), \quad i = 1, \ldots, p, \tag{8.4.1}$$

where $0 < \beta < \frac{1}{8}$ and K is a positive constant. Note that (8.4.1) is more restrictive than (c) of section 3.6.1.

Finally, we write

$$\mu_{n(i,j)} = \int_{-\infty}^{\infty}\int_{-\infty}^{\infty} J_{(i)}[F_{(i)}(x)]J_{(j)}[F_{(j)}(x_j)]\, dF_{(i,j)}(x_i, x_j) \tag{8.4.2}$$

$$n\sigma^2_{n(i,j)} = n\sigma_{n\{(i,j),(i,j)\}} = \operatorname{var}\left\{\sum_{l=1}^{3} U^{(\alpha)}_{(i,j);l}\right\} \tag{8.4.3}$$

$$n\sigma_{n\{(i,j),(r,s)\}} = \sum_{l=1}^{3}\sum_{l'=1}^{3} \operatorname{cov}(U^{(\alpha)}_{(i,j);l}, U^{(\alpha)}_{(r,s);l'}) \tag{8.4.4}$$

where

$$\begin{aligned} U^{(\alpha)}_{(r,s);l} &= J_{(r)}[F_{[r]}(X_{r\alpha})]J_{(s)}[F_{[s]}(X_{s\alpha})] \quad \text{if} \quad l = 1, \\ &= \int_{-\infty}^{+\infty}\int_{-\infty}^{+\infty} c(x_r - X_{r\alpha}) - F_{[r]}(x_r)]J_{(s)}[F_{[s]}(x_s)] \\ &\quad \times J'_{(r)}[F_{[r]}(x_r)]\, dF_{[r,s]}(x_r, x_s), \quad \text{if} \quad l = 2 \\ &= \int_{-\infty}^{\infty}\int_{-\infty}^{\infty} c(x_s - X_{s\alpha}) - F_{[s]}(x_s)]J_{(r)}[F_{[r]}(x_r)] \\ &\quad \times J'_{(s)}[F_{[s]}(x_s)]\, dF_{[r,s]}(x_r, x_s) \quad \text{if} \quad l = 3. \end{aligned} \tag{8.4.5}$$

and where

$$c(u) \text{ is 1 or 0 according as } u \text{ is} \geq \text{or} < 0. \tag{8.4.6}$$

We now prove the following theorem.

Theorem 8.4.1. *Under the assumptions* I, II, *and* III, *the random vector with elements* $n^{1/2}[(T_{n(i,j)} - \mu_{n(i,j)}), 1 \leq i < j \leq p]$, *where* $\mu_{n(i,j)}$ *is given by* (8.4.2), *has a limiting normal distribution with mean vector zero, and covariance matrix* $\mathbf{\Sigma} = n((\sigma_{n\{(i,j),(r,s)\}}))$ *given by* (8.4.3) *and* (8.4.4) *respectively.*

Proof: We have

$$\begin{aligned} J_{n(i)}[F^*_{n[i]}(x_i)]J_{n(j)}[F^*_{n[j]}(x_j)] = \{&J_{n(i)}[F^*_{n[i]}(x_i)]J_{n(j)}[F^*_{n[j]}(x_j)] \\ &- J_{(i)}[F^*_{n[i]}(x_i)]J_{(j)}[F^*_{n[j]}(x_j)]\} \\ &+ J_{(i)}[F^*_{n[i]}(x_i)]J_{(j)}[F^*_{n[j]}(x_j)]\} \end{aligned} \tag{8.4.7}$$

and, (by Taylor's theorem)

$$
\begin{aligned}
J_{(i)}[F^*_{n[i]}(x_i)]J_{(j)}[F^*_{n[j]}(x_j)]
&= J_{(i)}[F_{[i]}(x_i)]J_{(j)}[F_{[j]}(x_j)] + [F^*_{n[i]}(x_i) - F_{[i]}(x_i)]J'_{(i)}[F_{[i]}(x_i)] \\
&\quad \times J_{(j)}[F_{[j]}(x_j)] + [F^*_{n[j]}(x_j) - F_{[j]}(x_j)]J'_{(j)}[F_{[j]}(x_j)] \\
&\quad \times J_{(i)}[F_{[i]}(x_i)] + [J_{(i)}(F^*_{n[i]}(x_i)) - J_{(i)}(F_{[i]}(x_i)) \\
&\quad - \{F^*_{n[i]}(x_i) - F_{[i]}(x_i)\}J'_{(i)}(F_{[i]}(x_i))]J_{(j)}(F^*_{n[j]}(x_j)) \\
&\quad + [J_{(j)}(F^*_{n[j]}(x_j)) - J_{(j)}(F_{[j]}(x_j)) \\
&\quad - \{F^*_{n[j]}(x_j) - F_{[j]}(x_j)\}J'_{(j)}(F_{[j]}(x_j))]J_{(i)}(F_{[i]}(x_i)) \\
&\quad + [F^*_{n[i]}(x_i) - F_{[i]}(x_i)][J_{(j)}(F^*_{n[j]}(x_j)) - J_{(j)}(F_{[j]}(x_j))] \\
&\quad \times J'_{(i)}(F_{[i]}(x_i)),
\end{aligned} \tag{8.4.8}
$$

and

$$dF_{n[i,j]}(x_i, x_j) = dF_{[i,j]}(x_i, x_j) + d[F_{n[i,j]}(x_i, x_j) - F_{[i,j]}(x_i, x_j)]. \tag{8.4.9}$$

Then we have by (II), (8.4.8), and (8.4.9) that

$$T_{n,ij} = A^{(i,j)}_{1n} + A^{(i,j)}_{2n} + A^{(i,j)}_{3n} + \sum_{r=1}^{5} B^{(i,j)}_{rn} \tag{8.4.10}$$

where

$$A^{(i,j)}_{1n} = \int_{-\infty}^{\infty}\int_{-\infty}^{\infty} J_{(i)}[F_{[i]}(x_i)]J_{(j)}[F_{[j]}(x_j)]\, dF_{n[i,j]}(x_i, x_j) \tag{8.4.11}$$

$$
\begin{aligned}
A^{(i,j)}_{2n} &= \int_{-\infty}^{\infty}\int_{-\infty}^{\infty} [F^*_{n[i]}(x_i) - F_{[i]}(x_i)]J_{(j)}[F_{[j]}(x_j)] \\
&\quad \times J'_{(i)}[F_{[i]}(x_i)]\, dF_{[i,j]}(x_i, x_j)
\end{aligned} \tag{8.4.12}
$$

$$
\begin{aligned}
A^{(i,j)}_{3n} &= \int_{-\infty}^{\infty}\int_{-\infty}^{\infty} [F^*_{n[j]}(x_j) - F_{[j]}(x_j)]J_{(i)}[F_{[i]}(x_i)] \\
&\quad \times J'_{(j)}[F_{[j]}(x_j)]\, dF_{[i,j]}(x_i, x_j),
\end{aligned} \tag{8.4.13}
$$

$$
\begin{aligned}
B^{(i,j)}_{1n} &= \int_{-\infty}^{\infty}\int_{-\infty}^{\infty} [F^*_{n[i]}(x_i) - F_{[i]}(x_i)]J'_{(i)}[F_{[i]}(x_i)] \\
&\quad \times J_{(j)}[F_{[j]}(x_j)]\, d[F_{n[i,j]}(x_i, x_j) - F_{[i,j]}(x_i, x_j)],
\end{aligned} \tag{8.4.14}
$$

$$
\begin{aligned}
B^{(i,j)}_{2n} &= \int_{-\infty}^{\infty}\int_{-\infty}^{\infty} [F^*_{n[j]}(x_j) - F_{[j]}(x_j)]J'_{(j)}[F_{[j]}(x_j)] \\
&\quad \times J_{(i)}[F_{[i]}(x_i)]\, d[F_{n[i,j]}(x_i, x_j) - F_{[i,j]}(x_i, x_j)],
\end{aligned} \tag{8.4.15}
$$

(8.4.16)
$$B_{3n}^{(i,j)} = \int_{-\infty}^{\infty}\int_{-\infty}^{\infty} [J_{(i)}(F^*_{n[i]}(x_i)) - J_{(i)}(F_{[i]}(x_i)) - \{F^*_{n[i]}(x_i) - F_{[i]}(x_i)\} J'_{(i)}(F_{[i]}(x_i))J_{(j)}(F^*_{n[j]}(x_j))]\, dF_{n[i,j]}(x_i, x_j),$$

(8.4.17)
$$B_{4n}^{(i,j)} = \int_{-\infty}^{\infty}\int_{-\infty}^{\infty} [J_{(j)}(F^*_{n[j]}(x_j)) - J_{(j)}(F_{[j]}(x_j)) - \{F^*_{n[j]}(x_j) - F_{[j]}(x_j)\} \cdot J'_{(j)}(F_{[j]}(x_j))J_{(i)}(F_{[i]}(x_i))]\, dF_{n[i,j]}(x_i, x_j),$$

(8.4.18)
$$B_{5n}^{(i,j)} = \int_{-\infty}^{\infty}\int_{-\infty}^{\infty} [F^*_{n[i]}(x_i) - F_{[i]}(x_i)]J'_{(i)}(F_{[i]}(x_i)) \cdot [J_{(j)}(F^*_{n[j]}(x_j)) - J_{(j)}(F_{[j]}(x_j))]\, dF_{n[i,j]}(x_i, x_j).$$

Exercise 8.4.1 asks you to show that the B terms are all $o_p(n^{-1/2})$. Thus the difference $n^{1/2}[T_{n,ij} - \sum_{r=1}^{3} A_{rn}^{(i,j)}]$ tends to zero in probability as n tends to infinity, and so the vectors $n^{1/2}[T_{n,ij}, 1 \le i < j \le p]$ and $n^{1/2}[\sum_{r=1}^{3} A_{rn}^{(i,j)}, i \le i < j \le p]$ have the same limiting distributions, if they have one at all. Thus, to prove the theorem, it suffices to show that for any real $\lambda_{ij}(1 \le i < j \le p)$, not all zero, $\sum_{i=1}^{p}\sum_{j=1}^{p} \lambda_{ij}(A_{1n}^{(i,j)} + A_{2n}^{(i,j)} + A_{3n}^{(i,j)})$ has normal distribution in the limit. Now $\sum_{r=1}^{3} A_{rn}^{(i,j)}$ can be expressed as

(8.4.19)
$$\sum_{r=1}^{3} A_{rn}^{(i,j)} = \frac{1}{n}\sum_{r=1}^{n} [U_{(i,j):1}^{(r)} + U_{(i,j):2}^{(r)} + U_{(i,j):3}^{(r)}]$$

where the $U_{(i,j)}$'s are given by (8.4.5). Hence

(8.4.20)
$$\sum_{\substack{i=1\\i<j}}^{p}\sum_{j=1}^{p} \lambda_{ij}\left\{\sum_{r=1}^{3} A_{rn}^{(i,j)}\right\} = \frac{1}{n}\sum_{r=1}^{n}\left[\sum_{\substack{i=1\\i<j}}^{p}\sum_{j=1}^{p} \lambda_{ij}\{U_{(i,j);1}^{(r)} + U_{(i,j):2}^{(r)} + U_{(i,j):3}^{(r)}\}\right].$$

The right-hand side of (8.4.20) is the average of n independent and identically distributed random variables, each having mean $\sum_{i=1}^{p}\sum_{j=1,i<j}^{p} \lambda_{ij}\mu_{n(i,j)}$ and finite third moments. The asymptotic normality follows. Furthermore using (8.4.19) it is easy to check that the variance-covariance matrix $\mathbf{\Sigma} = n((\sigma_{n\{(i,j),(r,s)\}}))$ is given by (8.4.3) and (8.4.4). The theorem follows. ◁

The following theorem gives a simple sufficient condition under which the assumptions 8.4.1, 8.4.2, and 8.4.3 hold; the proof is left as an exercise.

Theorem 8.4.2. *If $J_{n(i)}(\alpha/(n+1)(i = 1, \ldots, p)$ is the expected value of the αth order statistic of a sample size from a population whose cumulative distribution is the inverse function of $J_{(i)}$, $i = 1, \ldots, p$, and if the assumption* III *of Theorem* 8.4.1 *is satisfied, then the assumptions* I *and* II *are satisfied.*

With the use of this theorem it is easy to verify that if $J_{n(i)}(\alpha/(n+1))$ is the expected value of the αth order statistic of a sample of size n from (i) the standard normal distribution, (ii) the logistic distribution, (iii) the double exponential distributions, (iv) the exponential distribution, and (v) the uniform distribution, then the vector $n^{1/2}[(T_{n,ij} - \mu_{n(i,j)}), 1 \leq i < j \leq p]$ has a limiting normal distribution.

We shall use the results of Theorem 8.4.1 in deriving large-sample power properties of the statistics associated with (i) tests of independence of $q \geq 2$ sets of variates, and (ii) tests for pairwise independence. First, we consider the problem of testing independence of two sets of variates.

8.5. TESTING INDEPENDENCE OF TWO SETS OF VARIATES

Let the p-component vector $\mathbf{X}_\alpha$ be partitioned into two subvectors $\mathbf{X}_\alpha^{(1)}$ and $\mathbf{X}_\alpha^{(2)}$ as defined in (8.3.6).

Let the population dispersion matrix $\mathbf{\Theta}$ defined in (8.2.4) and (8.2.5) be partitioned as we did the sample dispersion matrix $\mathbf{T}_n$ defined in (8.2.10). Then we have

$$\mathbf{\Theta} = \begin{pmatrix} \mathbf{\Theta}_{11} & \mathbf{\Theta}_{12} \\ \mathbf{\Theta}_{21} & \mathbf{\Theta}_{22} \end{pmatrix} \tag{8.5.1}$$

Note that $\mathbf{\Theta}_{11}$ is of order $l \times l$, $\mathbf{\Theta}_{12}$ is of order $l \times m$. Then, for the problem of testing $H_0^{(2)}$

$$H_0^{(2)}\colon\ \mathbf{\Theta}_{12} = \mathbf{\Theta}_{21}' = \mathbf{0}, \tag{8.5.2}$$

we propose to consider the statistic $S_{n(2)}^{(J)}$ defined by (8.3.25). The test consists in rejecting $H_0^{(2)}$ if $S_{n(2)}^{(J)}$ exceeds some predetermined number h_ϵ. We shall prove below that if $H_0^{(2)}$ is true, the statistic $-n \log S_{n(2)}^{(J)}$ has a limiting chi-square distribution with lm degrees of freedom as $n \to \infty$. This provides the user of this $S_{n(2)}^{(J)}$ test with a large-sample approximation of the value of h_ϵ for any $0 < \epsilon < 1$.

We shall now study the asymptotic distribution of $S_{n(2)}^{(J)}$ assuming a sequence of alternative hypotheses $\{H_n^{(2)}, n = 1, 2, \ldots\}$ which specifies that

$$\Omega_{ij}(F_{[i]}, F_{[j]}) = 1 + n^{-1/2}\omega_{ij}(F_{[i]}, F_{[j]}); \qquad i = 1, \ldots, l, \quad j = l+1, \ldots, p, \tag{8.5.3}$$

where Ω_{ij} is defined by (3.13), ω_{ij} is some function of $(F_{[i]}, F_{[j]})$, and $\omega_{ij} \neq 0$ for all (i,j); $i = 1, \ldots, l$; $j = l+1, \ldots, p$. It may be pointed out that the sequence of the alternative hypotheses $\{H_n^{(2)}, n = 1, 2, \ldots\}$ defined in (8.5.3) implies that

$$\mathbf{\Theta}_{12} = \mathbf{\Theta}_{21}' = n^{-1/2}\mathcal{B}_{12} \tag{8.5.4}$$

where $\mathcal{B}_{12} = ((\beta_{ij}))$ is an $l \times m$ matrix where elements are

$$\beta_{ij} = \int_{-\infty}^{\infty}\int_{-\infty}^{\infty} F_{[i]}(x)F_{[j]}(y)\omega_{ij}(F_{[i]}, F_{[j]})J'_{(i)}[F_{[i]}(x)]$$

$$\times J'_{(j)}[F_{[j]}(y)]\, dF_{[i]}(x)\, dF_{[j]}(y), \tag{8.5.5}$$

$$i = 1, \ldots, l; \quad j = l+1, \ldots, p.$$

Exercise 8.5.1 is to prove the following lemmas.

Lemma 8.5.1. *Under the assumptions of Theorem* 8.4.1.

(*a*) $\mathbf{T}_n \to \mathbf{\Theta}$ *in probability as* $n \to \infty$ *for all* $F \in \mathcal{F}_p$.
(*b*) $T^{ii'}_{n(1)} \to \theta^{ii'}_{(1)}$ *in probability as* $n \to \infty$ *for all* $F \in \mathcal{F}_p$, $i, i' = 1, \ldots, l$.
(*c*) $T^{jj'}_{n(2)} \to \theta^{jj'}_{(2)}$ *in probability as* $n \to \infty$ *for all* $F \in \mathcal{F}_p$, $j, j' = l+1, \ldots, p$.

where

(8.5.6) $\theta^{ii'}_{(1)}$ *and* $\theta^{jj'}_{(2)}$ *are the* (i, i')th *and* (j, j')th *elements of* $\mathbf{\Theta}^{-1}_{11}$ *and* $\mathbf{\Theta}^{-1}_{22}$ *respectively*.

Lemma 8.5.2. *If, for each* $i < j = 1, \ldots, p$ (i) *the conditions of Theorem* 8.4.1 *are satisfied, and* (ii) *the hypothesis* $\{H^{(2)}_n\}$ *defined by* (8.5.3) *is valid, then the matrix with elements* $n^{1/2}[(T_{n,ij} - \mu_{n(i,j)}), i = 1, \ldots, l;\ j = l+1, \ldots, p]$ *has a limiting normal distribution with means zero and variance-covariance matrix* $\Sigma^* = ((\sigma^*_{\{(i,j),(r,s)\}}))$

$$\sigma^{2*}_{(i,j)} = \sigma^*_{\{(i,j),(i,j)\}} = 1 \quad \text{if} \quad i = 1, \ldots, l; \quad j = l+1, \ldots, p, \tag{8.5.7}$$

$$\sigma^*_{\{(i,j),(r,s)\}} = \theta_{ir}\theta_{js}, \quad i, r = 1, \ldots, l; \quad j, s = l+1, \ldots, p \tag{8.5.8}$$

(*note that* $\theta_{ii} = 1$).

Theorem 8.5.1. *Under the assumptions of Lemma* 8.5.2, *the limiting distribution of* $-n \log S_{n(2)}$ *is noncentral chi-square with lm degrees of freedom and noncentrality parameter*

$$\Delta_{S^{(J)}_{n(2)}} = \sum_{i=1}^{l}\sum_{i'=1}^{l}\sum_{j=l+1}^{p}\sum_{j'=l+1}^{p} \beta_{ij}\beta_{i'j'}\theta^{ii'}_{(1)}\theta^{jj'}_{(2)}. \tag{8.5.9}$$

Proof: Under the assumed conditions $n^{1/2}T_{n,ij}$ is bounded in probability for all $i = 1, \ldots, l; j = l+1, \ldots, p$: Hence applying Laplace expansion for the determinant $\|\mathbf{T}_n\|$ and proceeding as in Theorem 8.3.3, we obtain

$$-n \log S^{(J)}_{n(2)} = n \sum_{i=1}^{l}\sum_{i'=1}^{l}\sum_{j=l+1}^{p}\sum_{j'=l+1}^{p} T_{n,ij}T_{n,i'j'}T^{ii'}_{n(1)}T^{jj'}_{n(2)} + O_p(n^{-1}). \tag{8.5.10}$$

Using Lemma 8.5.1, we notice that $-n \log S_{n(2)}^{(J)}$ is asymptotically equivalent to

$$n \sum_{i=1}^{l} \sum_{i'=1}^{l} \sum_{j=l+1}^{p} \sum_{j'=l+1}^{p} T_{n,ij} T_{n,i'j'} \theta_{(1)}^{ii'} \theta_{(2)}^{jj'} \tag{8.5.11}$$

where $\theta_{(1)}^{ii'}$ and $\theta_{(2)}^{jj'}$ are defined in (8.5.6). The result now follows as an application of Lemma 8.5.2. Q.E.D. ◁

We shall consider the special forms of the function $\Delta_{S_{n(2)}^{(J)}}$ for suitable choices of the function J in section 8.8, where we consider the asymptotic properties of the $S_{n(2)}^{(J)}$ tests in relation to their parametric competitors based on the likelihood ratio test (cf. Anderson, 1958; p. 233).

In passing we may remark that for the problem of testing the independence of q sets of variates against the sequence of alternatives $\{H_n^{(q)}, n = 1, 2, \ldots\}$ which specify that $\Omega_{ij}^{(n)}$ is of the form (8.5.3) for all i, j belonging to different sets, it can be shown (Exercise 8.5.2) by the use of Theorem 8.4.1 that $-n \log S_{n(q)}^{(J)}$ (cf. (8.3.42)) has asymptotically a noncentral chi-square distribution with $\sum_{k<l=1}^{p} i_k i_l$ degrees of freedom and some appropriate noncentrality parameter.

In a relatively simple case, when $q = p$, the statistic (8.3.43) has under the sequence of alternatives (8.5.3), asymptotically, a noncentral chi-square distribution with $p(p-1)/2$ degrees of freedom and noncentrality parameter $\Delta_{S_{n(p)}^{(J)}}$ given by

$$\Delta_{S_{n(p)}^{(J)}} = \sum_{i<j=1}^{p}{}' \beta_{ij}^2 \tag{8.5.12}$$

where β_{ij} is defined in (8.5.5).

8.6. PAIRWISE INDEPENDENCE

In view of the fact that the pairwise independence does not in general imply total independence, we shall consider this problem separately in this section. To be precise we consider the problem of testing

$$H_0^*: F_{[i,j]}(x_i, x_j) = F_{[i]}(x_i) \cdot F_{[j]}(x_j) \quad \text{for all pairs} \quad (i, j) \tag{8.6.1}$$

against the sequence of alternatives $\{H_n^*, n = 1, 2, \ldots\}$ which specifies that

$$F_{ij}(x_i, x_j) = F_i(x_i) \cdot F_j(x_j)\{1 + n^{-1/2}\omega_{ij}(F_i, F_j)\} \tag{8.6.2}$$

where $\omega_{ij}(F_i, F_j)$ is not identically equal to zero (a.e.), for at least one pair (i, j); $1 \leq i < j \leq p$. The test statistic proposed for this problem is based on the statistic

$$\mathfrak{L}_{n(J)} = n \mathbf{T}_n^* \hat{\Gamma}^{-1} \mathbf{T}_n^{*\prime} \tag{8.6.3}$$

where

$$\mathbf{T}_n^* = (T_{n,12}, T_{n,13}, \ldots, T_{n,(p-1),p}) \tag{8.6.4}$$

and $\hat{\Gamma}^{-1}$ is a consistent estimator of Γ^{-1}, which is the inverse of the covariance matrix of $n^{1/2}T_n^*$ defined by (8.6.7).

Theorem 8.6.1. *If, for each* $i < j = 1, \ldots, p$, (i) *the conditions of Theorem* 8.4.1 *are satisfied, and* (ii) *the hypothesis* H_n^* *defined by* (8.6.2) *is valid, then the statistic* $\mathfrak{L}_{n(j)}$ *defined by* (8.6.3) *has asymptotically a noncentral chi-square distribution with* $p(p-1)/2$ *degrees of freedom, and noncentrality parameter* $\Delta_{\mathfrak{L}_{n(J)}}$ *defined as*

$$\Delta_{\mathfrak{L}_{n(J)}} = \mathbf{\Delta\Gamma}^{-1}\mathbf{\Delta} \tag{8.6.5}$$

where

$$\mathbf{\Delta} = (\beta_{12}, \beta_{13}, \ldots, \beta_{p-1,p}). \tag{8.6.6}$$

Proof: The proof of this theorem is an immediate consequence of the fact that under the given assumptions, the random vector $\sqrt{n}\,\mathbf{T}_n^*$ has the limiting normal distribution with mean vector $\mathbf{\Delta}$ and the covariance matrix $\mathbf{\Gamma} = ((\tau_{\{(i,j),(r,s)\}}))$ where

$$\begin{aligned}
\tau_{\{(i,j),(r,s)\}} &= 1, \quad \text{if} \quad i = r, \quad j = s \\
&= E_0[J_{(i)}(F_{[i]}(X_i))J_{(j)}(F_{[j]}(X_j))J_{(r)}(F_{[r]}(X_r))J_{(s)}(F_{[s]}(X_s))] \\
&\qquad\qquad \text{if} \quad i \neq r, \quad j \neq s, \\
&= E_0[J_{(i)}^2(F_{[i]}(X_i))J_{(j)}(F_{[j]}(X_j))J_{(s)}(F_{[s]}(X_s))] \\
&\qquad\qquad \text{if} \quad i = r, \quad j \neq s, \\
&= E_0[J_{(i)}(F_{[i]}(X_i))J_{(j)}^2(F_{[j]}(_jX))J_{(r)}(F_{[r]}(X_r))] \\
&\qquad\qquad \text{if} \quad i \neq r, \quad j = s, \\
&= E_0[J_{(i)}^2(F_{[i]}(X_i))J_{(j)}(F_{[j]}(X_j))J_{(r)}(F_{[r]}(X_r))] \\
&\qquad\qquad \text{if} \quad i = s, \quad j \neq r,
\end{aligned} \tag{8.6.7}$$

and the fact that $\hat{\mathbf{\Gamma}}^{-1}$ is a consistent estimator of $\mathbf{\Gamma}^{-1}$. ◁

From Theorem 8.6.1, it is clear that any consistent estimator of $\mathbf{\Gamma}^{-1}$ will preserve the asymptotic distribution of the statistic $\mathfrak{L}_{n(J)}$. A natural consistent estimator of (8.6.7) may be obtained readily by replacing $J_{(i)}(F_i(X_i))$ by $J_{n(i)}(F_{n(i)}^*(X_{i\alpha}))$ for all i, and the expectation by sample averages.

8.7. ANOTHER MODEL FOR DEPENDENCE BETWEEN TWO SETS OF VARIABLES

Let $(X_1, \ldots, X_l)$, $(X_{l+1}, \ldots, X_p)$ be two sets of random variables. As a class of alternatives to the independence of $(X_1, \ldots, X_l)$ and $(X_{l+1}, \ldots, X_p)$, consider

$$X_i = U_i + \Delta Z_i, \qquad X_j = V_j + \Delta Z_j \tag{8.7.1}$$

$i = 1, \ldots, q$ and $j = q + 1, \ldots, p$ where $U = (U_1, \ldots, U_l)$, $V = (V_{l+1}, \ldots, V_p)$ and $Z = (Z_1, \ldots, Z_l, Z_{l+1}, \ldots, Z_p)$ are independent vector random variables with l-variate, m-variate, and p-variate distributions F, G, and M respectively. Further, M is such that

$$\text{cov}\,(Z_i, Z_j) \neq 0 \tag{8.7.2}$$

for $i = 1, 2, \ldots, l$ and $j = l + 1, \ldots, p - 1, p$. Notice that F and G may have any dependence pattern. The hypothesis of independence of $(X_1, \ldots, X_l)$ and $(X_{l+1}, \ldots, X_p)$ corresponds to

$$H_0\colon \Delta = 0 \quad \text{for} \quad i = 1, \ldots, l, \quad j = l + 1, \ldots, p. \tag{8.7.3}$$

Such a model may prove useful, for example, in analyzing group tests in psychology. For example, the outcomes (scores) of two reading tests and two mathematical tests can be described by a (linear) combination of individual group factors pertaining to the reading or mathematical abilities and common factors corresponding to intelligence or comprehension. The same tests considered in the preceding sections can be used for this model.

8.8. PARAMETRIC THEORY

In the parametric theory a commonly used test for the hypothesis $H_0^{(2)}$ is based on the statistic (see Anderson, 1958, p. 242)

$$V^* = |A|/|A_{11}| \cdot |A_{22}|, \tag{8.8.1}$$

$$A = \sum_{\alpha=1}^{n} (\mathbf{X}_\alpha - \overline{\mathbf{X}})'(\mathbf{X}_\alpha - \mathbf{X}) = \begin{bmatrix} A_{11} & A_{12} \\ A_{21} & A_{22} \end{bmatrix}, \tag{8.8.2}$$

$$A_{ij} = \sum_{\alpha=1}^{n} (\mathbf{X}_\alpha^{(i)} - \overline{\mathbf{X}}^{(i)})'(\mathbf{X}_\alpha^{(j)} - \overline{\mathbf{X}}^{(j)}); \qquad i = 1, 2, \tag{8.8.3}$$

$$\overline{\mathbf{X}}^{(i)} = \sum_{\alpha=1}^{n} (\mathbf{X}_\alpha^{(i)}/n). \tag{8.8.4}$$

Under the null hypothesis (8.1.2) (with $q = 2$), the elements of $(n - 1)^{-1}A_{12}$ all have expectation zero. Moreover, as these are all U statistics, it follows that if H_0 in (8.1.2) holds and $E\,|X_{i\alpha}|^4 < \infty$ for all $i = 1, \ldots, p$, then the elements of $(n - 1)^{-1/2}A_{12}$ all have finite variances. Consequently, under (8.1.2), the elements of $(1/n)A_{12}$ are all $O_p(n^{-1/2})$. Hence, using (8.8.1)–(8.8.3), and Theorem 8.3.3, and proceeding as in Theorem 8.3.4, it follows

(Exercise 8.8.1) that the permutation distribution of $-n \log V^*$ is asymptotically, in probability, a chi-square distribution with lm degrees of freedom.

The following theorem gives the limiting distributions of $-n \log V^*$ under the sequence of alternatives $H_0^{(2)}$ defined in (8.5.3).

Theorem 8.8.1. *Under the sequence of alternatives defined in* (8.5.3), $-n \log V^*$ *has asymptotically a noncentral chi-square distribution with* lm *degrees of freedom and noncentrality parameter* Δ_{V^*} *given by*

$$\Delta_{V^*} = \sum_{i=1}^{l} \sum_{i'=1}^{l} \sum_{j=l+1}^{p} \sum_{j'=l+1}^{p} \lambda_{ij}\lambda_{i'j'}\rho_{(1)}^{ii'}\rho_{(2)}^{jj'} \tag{8.8.5}$$

where $((\rho_{(1)}^{ii'}))$ *and* $((\rho_{(2)}^{jj'}))$ *are the reciprocals of* $\mathcal{R}_{ii}$, $i = 1, 2$, *defined by* (8.8.6).

Proof: It follows from (8.8.2) that under the sequence of alternative hypotheses in (8.6.2), $E\{n^{-1/2}A_{12}\} \to A_{12}^*$ as $n \to \infty$, where A_{12}^* has the elements $\sigma_i\sigma_j\lambda_{ij}$, $i = 1, \ldots, l$, $j = l + 1, \ldots, p$, which are all real and finite. Let us denote by R the sample correlation matrix, which we partition into R_{ij}'s, $i, j = 1, 2$. Then, it follows that

$$E(R_{ii}) = \mathcal{R}_{ii}, \quad i = 1, 2 \quad \text{and} \quad n^{1/2}E\{R_{12} \mid H_n^*\} \to \Lambda_{12} = ((\lambda_{ij})), \tag{8.8.6}$$

where $\mathcal{R}_{11}$, $\mathcal{R}_{22}$, and A_{12} have all finite elements. It is easy to show that

$$\begin{aligned}
\operatorname{cov}(r_{ij}, r_{kl}) = {} & \frac{\rho_{ij}\rho_{kl}}{n}\left(\frac{\mu_{0202}(i, j, k, l)}{\mu_{02}(i, j)\mu_{02}(k, l)} + \frac{\mu_{0220}(i, j, k, l)}{\mu_{02}(i, j)\mu_{20}(k, l)}\right. \\
& \left. + \frac{\mu_{2002}(i, j, k, l)}{\mu_{20}(i, j)\mu_{02}(k, l)} + \frac{\mu_{2020}(i, j, k, l)}{\mu_{20}(i, j)\mu_{20}(k, l)}\right) \\
& - \frac{\rho_{kl}}{2n}\left(\frac{\mu_{1102}(i, j, k, l)}{\mu_{02}(k, l)(\mu_{02}(i, j)\mu_{20}(i, j))^{1/2}}\right. \\
& \left. + \frac{\mu_{1120}(i, j, k, l)}{\mu_{20}(k, l)(\mu_{02}(i, j)\mu_{20}(i, j))^{1/2}}\right) \\
& - \frac{\rho_{ij}}{2n}\left(\frac{\mu_{0211}(i, j, k, l)}{\mu_{02}(i, j)(\mu_{02}(k, l)\mu_{20}(k, l))^{1/2}}\right. \\
& \left. + \frac{\mu_{2011}(i, j, k, l)}{\mu_{20}(i, j)(\mu_{02}(k, l)\mu_{20}(k, l))^{1/2}}\right) \\
& + \frac{1}{n}\frac{\mu_{1111}(i, j, k, l)}{(\mu_{20}(i, j)\mu_{02}(i, j))^{1/2}(\mu_{20}(k, l)\mu_{02}(k, l))^{1/2}} + O(n^{-3/2}),
\end{aligned} \tag{8.8.7}$$

where

$$\mu_{\alpha\beta\gamma\cdots}(i, j, k, \ldots) = \int \cdots \int [(x_i - E(x_i))^\alpha (x_j - E(x_j))^\beta (x_k - E(x_k))^\gamma \cdots] \times dF_{ijk} \cdots (x_i, x_j, x_k \cdots)$$

where $F_{ijk\ldots}$ is the corresponding marginal of F. Thus, under H_n^* in (8.5.3), $n^{1/2}R_{12}$ has the covariance matrix $\mathfrak{R}_{11} \otimes \mathfrak{R}_{22}$. Hence, using (8.8.6) and applying the Chebychev inequality we obtain

$$r_{ij} = O_p(n^{-1/2}) \quad \text{for all} \quad i = 1, \ldots, l; j = 1, \ldots, p, \tag{8.8.8}$$

while R_{ii} $(i = 1, 2)$ are consistent estimates of $\mathfrak{R}_{ii}$, $i = 1, 2$. Thus, writing (8.8.1) equivalently as $|R|/\{|R_{11}| \cdot |R_{22}|\}$, using (8.8.8) and then proceeding as in Theorem 8.3.3, we see that under $\{H_n^*\}$

$$\left| -n \log V^* - n \sum_{i=1}^{l} \sum_{i'=1}^{l} \sum_{j=l+1}^{p} \sum_{j'=l+1}^{p} r_{ij} r_{i'j'} r_{(1)}^{ii'} r_{(2)}^{jj'} \right| = O_p(n^{-1}), \tag{8.8.9}$$

where $((r_{(1)}^{ii'})) = R^{11}$ and $((r_{(2)}^{jj'})) = R^{22}$. Since $R^{ii} \xrightarrow{p} \mathfrak{R}^{ii}$, $i = 1, 2$, the second term on the left-hand side of (8.8.9) is again equivalent in probability to

$$n \sum_{i=1}^{l} \sum_{i'=1}^{l} \sum_{j=l+1}^{p} \sum_{j'=l+1}^{p} r_{ij} r_{i'j'} \rho_{(1)}^{ii'} \rho_{(2)}^{jj'}, \tag{8.8.10}$$

where $\rho_{(1)}^{ii'}$ and $\rho_{(2)}^{jj'}$ are defined by (8.8.6). Since the denominator of r_{ij} converges in probability to $\sigma_i \sigma_j$, while the numerator is an average over n independent random variables, by the classical (vector-valued) central limit theorem, it follows that if $E|X_{i\alpha}|^4 < \infty$, $i = 1, \ldots, p$, then under $\{H_n^*\}$, $n^{1/2}R_{12} - \Lambda_{12}$ has asymptotically a multinormal distribution with null mean vector and dispersion matrix $\mathfrak{R}_{11} \otimes \mathfrak{R}_{22}$. Consequently, (8.8.10) will have asymptotically the noncentral chi-square distribution with lm degrees of freedom and the noncentrality parameter (8.8.5).

Furthermore, since under the normal theory model, pairwise independence is equivalent to total independence, therefore the tests for $H_0^{(p)}$ and H_0^* are based on the same statistic, viz.

$$U^* = \frac{|A|}{\prod_{i=l}^{p} a_{ii}} \tag{8.8.11}$$

where $A = ((a_{ij}))_{i,j=1,\ldots,p}$ is defined in (8.8.2), and

$$a_{ij} = \sum_{\alpha=1}^{n} (X_{i\alpha} - \bar{X}_i)(X_{j\alpha} - \bar{X}_j); \qquad \bar{X}_i = \sum_{\alpha=1}^{n} X_{i\alpha}/n. \tag{8.8.12}$$

It turns out by the same logic as in the case of V^* that (problem 8.8.2) $-n \log U^*$ has under the sequence of alternatives (8.6.2) (which in the case of the normal theory model imply $\rho_{ij}^{(n)} = (1/\sqrt{n})\lambda_{ij}$ for all $i, j = 1, \ldots, p$) asymptotically the noncentral chi-square distribution with $p(p-1)/2$ degrees of freedom and noncentrality parameter

$$\Delta_{U^*} = \boldsymbol{\lambda\lambda}' \tag{8.8.13}$$

where

$$\boldsymbol{\lambda} = (\lambda_{12}, \lambda_{13}, \ldots, \lambda_{1p}). \tag{8.8.14}$$

8.9. ASYMPTOTIC RELATIVE EFFICIENCY

In this section we shall make large-sample power comparisons between the $S_{n(2)}^{(J)}$ tests, $\mathfrak{L}_{n(2)}^{(J)}$ tests, and their parametric competitors based on the likelihood ratio test. First we consider the interesting particular cases of the $S_{n(2)}^{(J)}$ and $\mathfrak{L}_{n}^{(J)}$ tests.

(a) Special Cases of the $S_{n(2)}^{(J)}$ Tests. ***a(1).*** Let $J_{(i)}(u) = \Phi^{-1}(u)$ where Φ is the standard normal distribution function. The $-n \log S_{n(2)}^{(J)}$ test then reduces to the normal scores $-n \log S_{n(2)}^{(\Phi)}$ test. The noncentrality parameter in this case is $\Delta_{S_{n(2)}^{(\Phi)}}$ where

$$\Delta_{S_{n(2)}^{(\Phi)}} = \sum_{i=1}^{l} \sum_{i'=1}^{l} \sum_{j=l+1}^{p} \sum_{j'=l+1}^{p} \beta_{ij}\beta_{i'j'}\theta_{(1)}^{ii'}\theta_{(2)}^{jj'} \tag{8.9.1}$$

where

$$\begin{aligned} \beta_{ij} = \int_{-\infty}^{\infty}\int_{-\infty}^{\infty} & F_{[i]}(x)F_{[j]}(y)\omega_{ij}(F_{[i]}, F_{[j]}) \\ & \times \frac{1}{\phi[\Phi^{-1}(F_{[i]}(x)]} \cdot \frac{1}{\phi[\Phi^{-1}[F_{[j]}(x)]} \, dF_{[i]}(x)\, dF_{[j]}(y) \end{aligned} \tag{8.9.2}$$

and

(8.9.3) $\theta^{ii'}, \theta^{jj'}$ are the (i, i')th and (j, j')th terms of

$$\boldsymbol{\Theta}_{11}^{-1} = ((\theta_{ij}))_{l\times l}^{-1} \quad \text{and} \quad \boldsymbol{\Theta}_{22}^{-1} = ((\theta_{ij}))_{m\times m}^{-1} \text{ respectively,}$$

and where

$$\theta_{ij} = \int_{-\infty}^{\infty}\int_{-\infty}^{\infty} \Phi^{-1}[F_{[i]}(x)]\Phi^{-1}[F_{[j]}(y)]\, dF_{[i,j]}(x, y). \tag{8.9.4}$$

In an important case, where $F(\mathbf{x})$ is $N(\boldsymbol{\mu}, \boldsymbol{\Sigma})$ where

$$\boldsymbol{\Sigma} = ((\rho_{ij}^{(n)}))_{p\times p}; \quad \rho_{ij}^{(n)} = \frac{1}{\sqrt{n}}\lambda_{ij}, \quad i = 1, \ldots, l; \; j = l+1, \ldots, p \tag{8.9.5}$$

$$\text{or} \quad i = l+1, \ldots, p; \; j = 1, \ldots, l$$

and $\rho_{ij}^{(n)} = \rho_{ij}$ otherwise. We have

(8.9.6) $$\omega_{ij}(F_{[i]}, F_{[j]}) = \sqrt{n}\left\{\frac{\Phi_{ij}(x, y, \rho_{ij}^{(n)})}{\Phi_i(x)\Phi_j(y)} - 1\right\}$$

where $\Phi_{ij}(x, y, \rho_{ij}^{(n)})$ is the normal c.d.f. of (X_i, X_j) where X_i is $N(0, 1)$, X_j is $N(0, 1)$ and $\rho_{ij}^{(n)}$ is the correlation coefficient between X_i and X_j. Expanding $\Phi_{ij}(x, y, \rho_{ij}^{(n)})$ about $\rho_{ij}^{(n)} = 0$, and neglecting the terms of order $o(n^{-1/2})$, we obtain (also see Sibuya, 1959)

(8.9.7) $$\omega_{ij}(F_{[i]}, F_{[j]}) = \lambda_{ij}\frac{\phi_i(x)\phi_j(y)}{\Phi_i(x)\Phi_j(y)}$$

where ϕ_i is the density of Φ_i; $i = 1, \ldots, p$. Hence, when $F(x)$ is normal $N(\boldsymbol{\mu}, \boldsymbol{\Sigma})$

(8.9.8) $$\beta_{ij} = \lambda_{ij}; \qquad i = 1, \ldots, l; \quad j = l + 1, \ldots, p.$$

(8.9.9) $$\theta_{ij} = \rho_{ij};\ i = 1, \ldots, l; j = 1, \ldots, l;\ \text{or } i = l + 1, \ldots, p; \quad j = l + 1, \ldots, p.$$

Hence from (8.9.1), in case $F(\mathbf{x})$ is normal

(8.9.10) $$\Delta_{S_{n(2)}}^{(\Phi)} = \sum_{i=1}^{l}\sum_{i'=1}^{l}\sum_{j=l+1}^{p}\sum_{j'=l+1}^{p}\lambda_{ij}\lambda_{i'j'}\rho_{(1)}^{ii'}\rho_{(2)}^{jj'},$$

where

(8.9.11) $\rho_{(1)}^{ii'}$ and $\rho_{(2)}^{jj'}$ are the (i, i')th and (j, j')th terms of

$$\boldsymbol{\rho}_{11}^{-1} = ((\rho_{ij}))_{l\times l}^{-1} \quad \text{and} \quad \boldsymbol{\rho}_{22}^{-1} = ((\rho_{ij}))_{m\times m}^{-1},$$

a(2). Let $J_{(i)}(u) = \sqrt{12(u - \frac{1}{2})}$; then the $-n \log S_{n(2)}^{(J)}$ test reduces to the rank statistic $-n \log S_{n(2)}^{(R)}$ which is a multivariate analog of the Spearman's rank correlation statistic (cf. chapter 3). The noncentrality parameter in this case reduces to $\Delta_{S_{n(2)}^{(R)}}$, where

(8.9.12) $$\Delta_{S_{n(2)}^{(R)}} = \sum_{i=1}^{l}\sum_{i'=1}^{l}\sum_{j=l+1}^{p}\sum_{j'=l+1}^{p}\beta_{ij}\beta_{i'j'}\theta_{(1)}^{ii'}\theta_{(2)}^{jj'}$$

where

(8.9.13) $$\beta_{ij} = 12\int_{-\infty}^{\infty}\int_{-\infty}^{\infty}F_{[i]}(x)F_{[j]}(y)\omega_{ij}(F_{[i]}, F_{[j]})\,dF_{[i]}(x)\,dF_{[j]}(y)$$

and $\theta_{(1)}^{ii'}$, $\theta_{(2)}^{jj'}$ are the (i, i')th and (j, j')th terms of $\boldsymbol{\Theta}_{11}^{-1} = ((\theta_{ij}))_{i,j=1,\ldots,l}$ and $\boldsymbol{\Theta}_{22}^{-1} = ((\theta_{ij}))_{i,j=l+1,\ldots,p}$, respectively, and where

(8.9.14) $$\theta_{ij} = 12\int_{-\infty}^{\infty}\int_{-\infty}^{\infty}(F_{[i]}(x) - \tfrac{1}{2})(F_{[j]}(y) - \tfrac{1}{2})\,dF_{[i,j]}(x, y).$$

In case $F(\mathbf{x})$ is $N(\boldsymbol{\mu}, \boldsymbol{\Sigma})$ where $\boldsymbol{\Sigma} = ((\rho_{ij}^{(n)}))$, $\rho_{ij}^{(n)} = \rho_{ij}$ for $i, j = 1, \ldots, l$ or $i, j = l+1, \ldots, p$ and $\rho_{ij}^{(n)} = (1/\sqrt{n})\lambda_{ij}$ for $i = 1, \ldots, l; j = l+1, \ldots, p$ or $i = l+1, \ldots, p;\ j = 1, \ldots, l$, the noncentrality parameter $\Delta_{S_{n(2)}^{(R)}}$ reduces to

$$\Delta_{S_{n(2)}^{(R)}} = \frac{9}{\pi^2} \sum_{i=1}^{l} \sum_{i'=1}^{l} \sum_{j=l+1}^{p} \sum_{j'=l+1}^{p} \lambda_{ij}\lambda_{i'j'}\rho_{(1)}^{*ii'}\rho_{(2)}^{*jj'} \tag{8.9.15}$$

where

(8.9.16) $\rho_{(1)}^{*ii'}$ and $\rho_{(2)}^{*jj'}$ are the (i, i')th and (j, j')th terms of

$$\boldsymbol{\rho}_{11}^{-1*} = \left(\left(\frac{6}{\pi}\sin^{-1}\frac{\rho_{ij}}{2}\right)\right)^{-1}_{i,j=1,\ldots,l} \quad \text{and} \quad \boldsymbol{\rho}_{22}^{-1*} = \left(\left(\frac{6}{\pi}\sin^{-1}\frac{\rho_{ij}}{2}\right)\right)^{-1}_{i,j=l+1,\ldots,p}$$

respectively.

(b) Special Cases of $\mathfrak{L}_n^{(J)}$ Tests. ***b(1)*** Let $J_{(i)}(u) = \Phi^{-1}(u)$. Then the $\mathfrak{L}_n(J)$ reduces to the normal scores $\mathfrak{L}_n(\Phi)$ test. The noncentrality parameter is then

$$\Delta_{\mathfrak{L}_n}(\Phi) = \boldsymbol{\lambda}\mathbf{T}^{-1}\boldsymbol{\lambda}' \tag{8.9.17}$$

where $\boldsymbol{\lambda} = (\lambda_{12}, \lambda_{13}, \ldots, \lambda_{p-1,p})$ and $\mathbf{T}^{-1} = ((\tau_{\{(i,j),(r,s)\}}))$ are given by (8.6.7).

In a special case when $F(x)$ is $N(\boldsymbol{\mu}, \boldsymbol{\Sigma})$, where $\boldsymbol{\Sigma} = ((\rho_{ij}^{(n)}))$ pairwise independence is equivalent to total independence and, hence from (8.6.7)

$$\Delta_{\mathfrak{L}_n}(\Phi) = \boldsymbol{\lambda}\boldsymbol{\lambda}'. \tag{8.9.18}$$

b(2). Let $J_{(i)}(u) = \sqrt{12}(u - \frac{1}{2})$; then the $\mathfrak{L}_n(J)$ test becomes the rank sum $\mathfrak{L}_n(R)$ test. The noncentrality parameter, for the case when $F(\mathbf{x})$ is $N(\boldsymbol{\mu}, \boldsymbol{\Sigma})$, is

$$\Delta_{\mathfrak{L}_n}(R) = \frac{9}{\pi^2}\boldsymbol{\lambda}\boldsymbol{\lambda}'. \tag{8.9.19}$$

Hence, denoting $e_{[T,T^*]}$, as the asymptotic efficiency of a test T relative to T^*, we have

$$e_{[S_{n(2)}^{(\Phi)}, V^*]} = \Delta_{S_{n(2)}^{(\Phi)}}/\Delta_{V^*} \tag{8.9.20}$$

where $\Delta_{S_{n(2)}^{(\Phi)}}$ and Δ_{V^*} are defined in (8.9.1) and (8.8.5) respectively, and

$$e_{[S_{n(2)}^{(R)}, V^*]} = \Delta_{S_{n(2)}^{(R)}}/\Delta_{V^*} \tag{8.9.21}$$

where $\Delta_{S_{n(2)}^{(R)}}$, and Δ_{V^*} are defined in (8.9.12) and (8.8.5), respectively.

In particular, if $F(x)$ is $N(\boldsymbol{\mu}, ((\rho_{ij}^{(n)})))$, it is easy to notice from (8.9.10),

(8.9.15), and (8.9.21) that

$$e_{[S_{n(2)}^{(\Phi)}, V^*]} = 1 \tag{8.9.22}$$

$$e_{[S_{n(2)}^{(R)}, V^*]} = \frac{9}{\pi^2} \frac{\sum_{i=1}^{l} \sum_{i'=1}^{l} \sum_{j=l+1}^{p} \sum_{j'=l+1}^{p} \lambda_{ij}\lambda_{i'j'}\rho_{(1)}^{*ii'}\rho_{(2)}^{*jj'}}{\sum_{i=1}^{l} \sum_{i'=1}^{l} \sum_{j=l+1}^{p} \sum_{j'=l+1}^{p} \lambda_{ij}\lambda_{i'j'}\rho_{(1)}^{ii'}\rho_{(2)}^{jj'}} \tag{8.9.23}$$

where $\rho_{(1)}^{*ii'}$, $\rho_{(2)}^{*jj'}$, $\rho_{(1)}^{ii'}$, and $\rho_{(2)}^{jj'}$ are defined in (8.9.11) and (8.9.16) respectively.

From (8.9.22) and (8.9.23) we note that whereas in the case of normal distributions, the optimality, of the bivariate normal scores test of independence (Exercise 8.10.8) relative to the likelihood ratio test is preserved in the multivariate case, the same is not true in the case of multivariate rank sum $S_{n(2)}^{(R)}$ test (see also Exercise 8.10.11). However, for some special cases one may consider finding the bounds of (8.9.23), which are considered below.

I. $p = 3, q = 2(l = 1, m = 2)$. In this case (8.9.23) reduces to

$$\left(\frac{9}{\pi^2}\right)\left\{\sum_{j=2}^{3} \sum_{j'=2}^{3} \lambda_{1j}\lambda_{1j'}\rho_{(2)}^{*jj'} \Big/ \sum_{j=2}^{3} \sum_{j'=2}^{3} \lambda_{1j}\lambda_{1j'}\rho_{(2)}^{jj'}\right\}. \tag{8.9.24}$$

Since this is a ratio of two quadratic forms, using Courant's theorem we obtain the maximum and minimum values of (8.9.24) as $(9/\pi^2)$ (maximum or minimum characteristic root of $\mathcal{R}_{22}^{*-1}\mathcal{R}_{22}$). Since each matrix is of order 2×2, straightforward computation yields that the maximum (and minimum) possible values of (8.9.24) are

$$\max \text{ (or min) } \left\{\frac{1 + \rho_{23}}{1 + \rho_{23}^*}, \frac{1 - \rho_{23}}{1 - \rho_{23}^*}\right\} \qquad (\text{over } -1 \le \rho_{23} \le 1). \tag{8.9.25}$$

Since $\rho_{23}^* = (6/\pi)\sin^{-1}(\rho_{23}/2)$, from (8.9.25) we obtain (Exercise 8.9.1)

$$0.827 \le e_{[S_{n(2)}^{(R)}, V^*]} \le 0.93. \tag{8.9.26}$$

II. $p = 4, q = 2(l = m = 2)$. In this case, it can be shown (Exercise 8.9.1) by using (8.9.23) and Courant's theorem that the bounds are

$$(3/\pi)\ [\max \text{ (min) characteristic roots of } (\mathcal{R}_{11} \otimes \mathcal{R}_{22})(\mathcal{R}_{11}^* \otimes \mathcal{R}_{22}^*)^{-1}]. \tag{8.9.27}$$

Since the characteristic roots of a Kronecker product are products of the characteristic roots of the individual matrices, it follows that the maximum (minimum) value of (8.9.28) is $(3/\pi)$ times the products of the largest (smallest) characteristic roots of $\mathcal{R}_{11}\mathcal{R}_{11}^{*-1}$ and $\mathcal{R}_{22}\mathcal{R}_{22}^{*-1}$. Since $\mathcal{R}_{ii}\mathcal{R}_{ii}^{*-1}$ is a 2×2 matrix, $i = 1, 2$, again by straightforward computations, we get (Exercise 8.9.1)

$$0.75 \le e_{[S_{n(2)}^{(R)}, V^*]} \le 0.95. \tag{8.9.28}$$

III. *It is of some interest to show that* (8.9.23) *cannot exceed* 1. This will follow simply by showing that $(\pi^2/9)(\mathfrak{R}_{11}^* \otimes \mathfrak{R}_{22}^*) - (\mathfrak{R}_{11} \otimes \mathfrak{R}_{22})$ is positive semidefinite. For this purpose, it is again sufficient to show that $(\pi/3)\mathfrak{R}_{ii}^* - \mathfrak{R}_{ii}$ is positive semidefinite, for $i = 1, 2$. For $p = 2$, a proof of this is given by Chatterjee and Sen (1964), while for $p > 2$, see Bellman (1960, pp. 91–92).

IV. (8.9.23) *can be arbitrarily close to zero if l or $m \geq 3$*. To prove this, let $p = 4$, $q = 2(l = 1, m = 3)$ and consider the intraclass correlation model, viz., $\rho_{ij} = \rho(|\rho| < \frac{1}{2})$, $i \neq j = 2, 3, 4$. In this case, (8.9.23) reduces to

$$\frac{(9/\pi^2)(1 + 2\rho)}{(1 + (12/\pi)\sin^{-1}\frac{1}{2}\rho)}. \tag{8.9.29}$$

As (8.9.29) $\to 0$ as $\rho \to -\frac{1}{2}$, the proof follows.

V. We now consider the efficiencies of the V^* test relative to the $S_{n(2)}^{(J)}$ tests for the model (8.7.1). To this end consider a sequence of alternatives $\Delta_n = (\Delta_n(1, q + 1), \ldots, \Delta_n(q, p))$ such that

$$\Delta_n(i, j) = \frac{\Delta_0}{n^{1/4}}, \qquad i = 1, \ldots, q, \quad j = q + 1, \ldots, p \tag{8.9.30}$$

where $\Delta_0 \neq 0$. Then we have

$$\rho_{nij} = \frac{\Delta_n^2}{1 + \Delta_n^2} \operatorname{cov}(Z_i, Z_j), \tag{8.9.31}$$

provided we assume that X_i, Y_j and Z_k have unit variances.

The joint distribution of $X_i \cdot X_j \cdot X_k \cdot X_q$ with i, j, k and q all distinct is given by

$$\begin{aligned} H_{ijkq}^{(n)}(x_i, x_j, x_k, x_q) = \iiiint & F_{ik}(x_i - \Delta_n z_i, x_k - \Delta_n z_k) \\ & \times G_{jq}(x_j - \Delta_n z_j, x_q - \Delta_n z_k) \\ & \times dM_{ijkq}(z_i, z_j, z_k, z_q). \end{aligned} \tag{8.9.32}$$

The marginal distribution of X_i is given by

$$H_i^{(n)}(x_i) = \int F_i(x_i - \Delta_n z_i)\, dM_i(z_i) \tag{8.9.33}$$

and similarly the one for X_j.

Now suppose that the fourth moments of $H_i^{(n)}$, $i = 1, \ldots, p$, are uniformly bounded in n. It then easily follows that the covariance matrix of the r_{ij} for $i = 1, \ldots, l, j = l + 1, \ldots, p$, has the same limit under the alternatives Δ_n as under the hypothesis (8.7.3). Also, then the statistic $-n \log V^*$ has asymptotically a noncentral chi-square distribution with lm degrees of freedom and noncentrality parameter

$$Q_n(V^*) = n\boldsymbol{\rho}'_{nv}\boldsymbol{\Pi}^{-1}\boldsymbol{\rho}_{nv} \tag{8.9.34}$$

with

$$\boldsymbol{\rho}'_{nv} = \Delta_n^2[\text{cov}(Z_1, Z_{l+1}), \ldots, \text{cov}(Z_l, Z_p)]', \tag{8.9.35}$$

$$\boldsymbol{\Pi} = \mathcal{K}_{11} \otimes \mathcal{K}_{22} \tag{8.9.36}$$

$\mathcal{K}_{11}$ being the correlation matrix of F and $\mathcal{K}_{22}$ the correlation matrix of G.

Consider, analogously, the covariances of T_{nij} and T_{nkj}. Under the alternatives Δ_n these covariances are given by (8.6.3) but now all the underlying distributions depend on n as derived from the expression (8.9.32) for $H_{ijkj'}^{(n)}$.

Now in the course of the proof for the joint asymptotic normality of the T_{nij} (cf. section 8.4) it is shown that the third moments of linear combinations of (8.4.5) are uniformly bounded in n. This is sufficient to ensure that as $n \to \infty$ i.e., a $\Delta_n \to 0$ the covariances of T_{nij} and T_{nkj} have the same limit as under the hypothesis.

Thus letting $\mu_{ij}(\Delta_n)$ denote the asymptotic mean of T_{nij} under the alternatives Δ_n, the statistics $-n \log S_n^{(J)}$ have asymptotically the noncentral chi-square distribution with lm degrees of freedom and noncentrality parameter

$$Q_n(S^{(J)}) = n\boldsymbol{\mu}'_{nv}\boldsymbol{\Lambda}^{-1}\boldsymbol{\mu}_{nv} \tag{8.9.37}$$

with

$$\boldsymbol{\mu}'_{nv} = (\mu_{1l+1}(\Delta_n), \ldots, \mu_{lp}(\Delta_n))' \tag{8.9.38}$$

$$\boldsymbol{\Lambda} = \boldsymbol{\Lambda}_{11} \otimes \boldsymbol{\Lambda}_{22} \tag{8.9.39}$$

$$\boldsymbol{\Lambda}_{11} = ((\mu_{ik}))_{l\times l}\ \boldsymbol{\Lambda}_{22} = ((\mu_{jj'}))_{m\times m} \tag{8.9.40}$$

$$\mu_{ik} = \iint J(F_i(u_i))J(F_k(u_k))\, dF_{ik}(u_i, u_k) \tag{8.9.41}$$

$$\mu_{jj'} = \iint J(G_j(v_j))(G_{j'}(v_{j'}))\, dG_{jj'}(v_j, v_{j'}).$$

The asymptotic relative efficiency of the test $S_{n(2)}^{(J)}$ relative to the test V^* based on correlation coefficients is thus obtained from the following theorem.

Theorem 8.9.1. *Under the above assumptions and notations, the asymptotic relative efficiency of the $S_{n(2)}^{(J)}$ test relative to the V^* test is*

$$e_{S^{(J)},V}(F, G, M) = \lim_{n\to\infty} \frac{\boldsymbol{\mu}'_{nv}\boldsymbol{\Lambda}^{-1}\boldsymbol{\mu}_{nv}}{\boldsymbol{\rho}'_{nv}\boldsymbol{\Pi}^{-1}\boldsymbol{\rho}_{nv}}. \tag{8.9.42}$$

Observe that this efficiency, besides depending upon F and G, depends also upon M through cov (Z_i, Z_j) except in a special case where $Z_1 = Z_2 = \cdots = Z_p = Z$, say. An extremely special case of the above theorem is the following corollary.

Corollary 8.9.1. *Let $Z_1 = Z_2 = \cdots = Z$, let $F_1 = F_1 = \cdots = F_q = \bar{F}$, $G_{q+1} = \cdots = G_p = \bar{G}$ (that is F and G each have identical marginals), and let the J_n functions be the same for all the variables belonging to each set. Then*

$$e_{S,V}(F, G, M) = e_{S,V}(F, G) = \frac{S(\mathbf{\Lambda}^{-1})}{S(\mathbf{\Pi}^{-1})} \lim_{n\to\infty} \frac{\mu(\Delta_n)^2}{\Delta_n^2} \tag{8.9.43}$$

where

$$\mu(\Delta_n) = \iint J(\bar{F}_{\Delta_n}(x)) J(\bar{G}_{\Delta_n}(y))\, dH_{\Delta_n}(x, y) \tag{8.9.44}$$

and $S(A)$ denotes the sum of the elements of the matrix A.

The proof is straightforward and is given as Exercise 8.9.3. As a final specialization, if F and G have pairwise independent coordinates, we have the obvious

Corollary 8.9.2. *If in addition to the assumptions of Corollary 8.9.1, F and G have pairwise independent coordinates, then*

$$e_{S,V}(F, G) = \lim_{n\to\infty} \frac{\mu(\Delta_n)^2}{\Delta_n^2}. \tag{8.9.45}$$

For the studies of the bounds of (8.9.42) (which may be very close to 0) in various limiting degenerate cases, see Exercises 8.10.8–8.10.11.

EXERCISES

Section 8.3

8.3.1. Prove Theorem 8.3.1.

8.3.2. Prove that under the assumptions of Theorem 8.3.3, $-n \log S_{n(q)}^{(J)}$ has asymptotically a chi-square distribution with $\sum_{k<l} i_k i_l$ degrees of freedom.

8.3.3. Prove that under the assumptions of Theorem 8.3.3, $V_{n(p)}^{(J)}$ has asymptotically a chi-square distribution with $p(p-1)/2$ degrees of freedom.

Section 8.4

8.4.1. Prove that the B terms defined in (8.4.14)–(8.4.21) are all $o_p(n^{-1/2})$.

(*Hints:* For $F^*_{n[i]}(x) - F_{[i]}(x)$, $i = 1, \ldots, p$, use the bound in (2.11.70), while for $J'_{(i)}(u)$, $i = 1, \ldots, p$, use (8.4.1). In (8.4.14) and (8.4.15) use the dominated convergence theorem, while in (8.4.16)–(8.4.18), first use the

Schwartz inequality and then use (2.11.70) and (8.4.1). Under the assumption of an additional bound on the second derivative of $J(u)$, a detailed treatment of the higher-order terms is due to Bhuchongkul (1964).)

8.4.2. Complete the proof of Theorem 8.4.2.

Section 8.5

8.5.1. Complete the proofs of Lemmas 8.5.1 and 8.5.2.

8.5.2. Using Theorem 8.4.1 and the assumptions of Lemma 8.5.2 show that $-n \log S_{n(q)}^{(J)}$ has asymptotically the noncentral chi-square distribution with $\sum_{k<l=1}^{p} i_k i_l$ degrees of freedom. Obtain the noncentrality parameter.

Section 8.6

8.6.1. Complete the proof of Theorem 8.6.1.

Section 8.8

8.8.1. Show that the permutation distribution of $-n \log V^*$ is asymptotically that of a chi-square variable with lm degrees of freedom.

8.8.2. Show that under (8.6.2), $-n \log U^*$ has asymptotically the noncentral chi-square distribution with $p(p-1)/2$ degrees of freedom and noncentrality parameter Δ_{U^*} defined in (8.8.13).

Section 8.9

8.9.1. Prove the assertions (8.9.25)–(8.9.28).

8.9.2. Prove that $e_{[S_{n(2)}^{(R)}, V^*]}$ defined in (8.9.23) cannot exceed 1.

8.9.3. Prove Corollaries 8.9.1 and 8.9.2.

8.9.4. Take $p = 3$, $q = 1$, and assume that the conditions of Corollary 6.9.1 are satisfied. Denote by $\rho(G)$ the correlation coefficient corresponding to G, and by $\rho^*(G)$ the correlation coefficient between $J(G_2(V_2))$ and $J(G_3(V_3))$; then prove that

$$e_{S,V}(F, G, M) = \lim_{n\to\infty} \left[\frac{\mu(\Delta_n)}{\Delta_n^2}\right]^2 \frac{1 + \rho(G)}{1 + \rho^*(G)}$$

$$\leq \lim_{n\to\infty} \left[\frac{\mu(\Delta_n)}{\Delta_n^2}\right]^2 \left[\frac{1 - \rho^2(G)}{1 - \rho^{*2}(G)}\right]^{1/2}.$$

8.9.5. (*Continued.*) Let the distribution G be defined as $G(x, y) = G_{\epsilon,b}(x, y) = (1 - \epsilon)\Lambda(x, y) + \epsilon b\psi(x, y) + \epsilon(1 - b)\psi_1(x)\psi_2(y)$ and assume that Λ is standardized uniform $\psi_2(y)$ and ψ is standardized double exponential on the line $x = y$. Then prove that

$$\lim_{b\to 1} \lim_{\epsilon\to 0} \frac{1 - \rho^2(G)}{1 - \rho_R^{*2}(G)} = 0$$

where

$$\rho_R^*(G) = 12 \iint G_2(x) G_3(y)\, dG(x, y) - 3.$$

8.9.6. Using the results of (8.9.8), show that if F is a standard normal distribution, then $e_{R,V}(F, G)$ can be made arbitrarily low.

8.10. TESTS FOR INDEPENDENCE IN BIVARIATE POPULATIONS

8.10.1. Let $(X_1, Y_1), \ldots, (X_n, Y_n)$ be n mutually independent pairs of random variables from absolutely continuous c.d.f. $H(x, y)$ with marginal distributions $F(x)$ and $G(y)$ respectively. We wish to test the null hypothesis H^* that X_i and Y_i are independent for all i, that is $H(x, y) = F(x) \cdot G(y)$ against various alternatives under which X_i and Y_i are dependent. If H is a bivariate normal distribution with correlation coefficient ρ, then the problem is reduced to that of testing $\rho = 0$ against $\rho \neq 0$.

Let us consider first the general model of dependence introduced by Hájek and Šidák (1967). Let

$$X_i = X_i^* + \Delta Z_i, \qquad Y_i = Y_i^* + \Delta Z_i, \qquad i = 1, \ldots, n.$$

Assume that X_i^*, Y_i^* and Z_i are mutually independent, their distributions do not depend on i, and Δ is a non-negative parameter. In this set up, the hypothesis of independence H^* reduces to $\Delta = 0$.

(*a*) *Sample Correlation Coefficient Test.* This test is based on the statistic

$$R = \sum_{i=1}^{n} (X_i - \bar{X})(Y_i - \bar{Y})/S_1 S_2$$

where

$$S_1^2 = \sum_{i=1}^{n} (X_i - \bar{X})^2, \qquad S_2^2 = \sum_{i=1}^{n} (Y_i - \bar{Y})^2,$$

$$\bar{X} = \sum_{i=1}^{n} X_i/n, \qquad \bar{Y} = \sum_{i=1}^{n} Y_i/n.$$

Prove that this test is locally most powerful against the alternative that $H(x, y)$ is bivariate normal with $\Delta > 0$.

(*b*) *The Spearman Rank Correlation Coefficient Test.* This test is based on the statistic

$$S = \frac{\sum_{i=1}^{n} (R_i - \bar{R})(S_i - \bar{S})}{\sqrt{\sum_{i=1}^{n} (R_i - \bar{R})^2 \sum_{i=1}^{n} (S_i - \bar{S})^2}}$$

$$= \frac{12}{N^3 - N} \sum_{i=1}^{n} \left(R_i - \frac{N+1}{2}\right)\left(S_i - \frac{N+1}{2}\right),$$

where R_i is the rank of X_i among the X's, S_i is the rank of Y_i among the Y's, $\bar{R} = \sum_{i=1}^{n} R_i/n$, and $\bar{S} = \sum_{i=1}^{n} S_i/n$.

Let the sample $(X_1, Y_1), \ldots, (X_n, Y_n)$ be arranged according to the increasing value of the X's and let the result be $(X_{(1)}, Z_1), \ldots, (X_{(n)}, Z_n)$

where $X_{(1)} < \cdots < X_{(n)}$ and the Z's are a permutation of the Y's. Let R_i^0 be the rank of Z_i among the Z's. Then an alternate expression for S is

$$S = 1 - \frac{6}{n^3 - n} \sum_{i=1}^{n} (R_i^0 - i)^2 = 1 - \frac{6}{n^3 - n} \sum_{i=1}^{n} (S_i - R_i)^2.$$

Write

$$s(u) = \begin{cases} -1 & \text{if } u < 0, \\ 0 & \text{if } u = 0, \\ +1 & \text{if } u > 0. \end{cases}$$

Then S can also be written as

$$S = \frac{3}{n^3 - n} \sum_{\alpha=1}^{n} \sum_{\beta=1}^{n} \sum_{\gamma=1}^{n} s(x_\alpha - x_\beta) s(y_\alpha - y_\gamma)$$

$$= [(n-2)k + 3r]/(n+1),$$

where

$$r = \frac{1}{n(n-1)} \sum_{\alpha \neq \beta} s(x_\alpha - x_\beta) s(y_\alpha - y_\beta)$$

and

$$k = \frac{3}{n(n-1)(n-2)} \sum s(x_\alpha - x_\beta) s(y_\alpha - y_\beta)$$

where the summation $\sum$ is over all α, β, γ which are different. r is called the *difference-sign covariance.*

The test based on S is locally most powerful against $\Delta > 0$ for the case where both F and G are logistics.

(*c*) *The Fisher-Yates Normal Scores Test.* This test is based on the statistic

$$M = \sum_{i=1}^{n} E_{n,R_i} E_{n,S_i},$$

where $E_{n,i}$ is the expected value of this ith-order statistic in a sample of size n from the standard normal distribution. Prove that the test based on M is locally most powerful against $\Delta > 0$ for the case when F and G are both normal.

(*d*) *The Van der Waerden Test.* This test is based on the statistic

$$M^* = \sum_{i=1}^{n} \Phi^{-1}\left(\frac{R_i}{n+1}\right) \Phi^{-1}\left(\frac{S_i}{n+1}\right),$$

where Φ is the standard normal distribution. Prove that the statistics M and M^* are asymptotically equivalent.

8.10.2. (*Continued.*) Prove that under H^*

$$ES = 0 \qquad \text{var } S = 1/(n-1)$$

$$EM = 0, \qquad \text{var } M = \left\{\sum_{i=1}^{n} [E_{n,i}]^2\right\} \Big/ n(n-1)$$

$$EM^* = 0, \qquad \text{var } M^* = \left\{\sum_{i=1}^{n} [\Phi^{-1}(i/n+1)]^2\right\}^2 \Big/ n(n-1).$$

8.10.3. (*Continued.*) Let $(X_1, Y_1), \ldots, (X_n, Y_n)$ be as defined in Exercise 8.10.1. Let $F_n(x)$, $G_n(y)$, and $H_n(x, y)$ be the sample c.d.f.'s of the X's, the Y's, and the (X, Y)'s respectively. Consider the statistic

$$T_n = \int_{-\infty}^{\infty} \int_{-\infty}^{\infty} J_n[F_n(x)]L_n[F_n(y)]\, dH_n(x, y)$$

where the functions J_n and L_n satisfy the assumptions (8.4.1) to (8.4.4). Then prove that

$$\lim_{n\to\infty} p\left[\frac{T_n - \mu_n}{\sigma_n} \le t\right] = (2\pi)^{-1/2} \int_{-\infty}^{t} e^{-x^2/2}\, dx,$$

provided $\sigma_n \ne 0$ where

$$\mu_n = \int_{-\infty}^{\infty} \int_{-\infty}^{\infty} J[F(x)]L[G(y)]\, dH(x, y)$$

and

$$n\sigma_n^2 = \text{var}\,\{J[F(x)]L[G(y)]$$
$$+ \int_{-\infty}^{\infty} \int_{-\infty}^{\infty} [\varphi_X(u) - F(u)]J'[F(u)]L[G]\, dH(u, v)$$
$$+ \int_{-\infty}^{\infty} \int_{-\infty}^{\infty} [\varphi_Y(v) - G(y)]J[F(u)]L'[G]\, dH(u, v)\}$$

where $\varphi_Z(u) = 1$ or 0 according as $u \le Z$ or not. (Bhuchongkul, 1964.)*

8.10.4. (*Continued.*) Prove that under the null hypothesis H^* defined in Exercise 8.10.1

$$n\,\text{var}_0\,(T_n) = \text{var}_0\,\{J[F_0(X)]L[G_0(Y)]\} - E_0^2\{L[G_0(Y)]\}\,\text{var}_0\,\{J[F_0(X)]\}$$
$$- E_0^2\{J[F_0(X)]\}\,\text{var}_0\,\{L[G_0(Y)]\}$$

where the functions with subscript 0 are computed under H^*.

(Bhuchongkul, 1964.)

8.10.5. (*Continued.*) Show that if $J = L = \Phi^{-1}$, the statistic T_n is equivalent to the normal scores statistic M defined in Exercise 8.10.1. Show that in this case $\sigma_n^2 = 1/n$.

* For other nonparametric tests for bivariate independence, see Hájek and Šidák (1967) and Kendall and Stuart (1961).

8.10.6. (*Continued.*) Show that if $J_n(\alpha/n) = L_n(\alpha/n) = \alpha/n$, then the statistic T_n of Exercise 8.10.3 is equivalent to the Spearman rank correlation statistic S defined in Exercise 8.10.1. Show that in this case $\sigma_n^2 = 1/144n$.

8.10.7. Consider the dependence model under which

$$X_i = (1 - \Delta)U_i + \Delta Z_i, \qquad Y_i = (1 - \Delta)V_i + Z_i$$

where $0 \leq \Delta \leq 1$ and U_i, V_i, Z_i are independent random variables having distributions which do not depend on i.

Let f_0, g_0, and f_0^* be the density functions of U_i, V_i and Z_i respectively. Let f_Δ and g_Δ denote the marginal density functions of X and Y, and h_Δ their joint density functions. Furthermore, assume that (i) U_i, V_i, Z_i have finite fourth moments; (ii) f_0, g_0, and f_0^* are positive and continuous on $(-\infty, \infty)$; (iii) f_0 and g_0 are twice differentiable on $(-\infty, \infty)$; and

(iv)

$$\frac{d^i}{d\Delta^i}\mu_n(\Delta)\bigg|_{\Delta=0} = \iint \frac{d^i}{d\Delta^i}\{J_0[F_0(n)]L_0[G_\Delta(y)]h_\Delta(x, y)\}\,|_{\Delta=0}\,dx\,dy$$

$$\text{for} \quad i = 1, 2, \quad \text{and} \quad J_0 = L_0 = \Phi^{-1}.$$

Then the A.R.E. (of the normal scores M test with respect to the sample correlation coefficient R test is

$$e_{M,R} = \left\{\int J_0'[f_0(x)]\,f_0^2(x)dx \cdot \int J_0'[G_0(y)]g_0^2(y)\,dy\right\}^2$$

(Bhuchongkul, 1964.)

8.10.8. (*Continued.*) Prove that $e_{M,R} \geq 1$ for all F_0 and G_0 and the equality holds only if $F_0 = G_0 = \Phi$. (Bhuchongkul, 1964.)

8.10.9. (*Continued.*) Prove that the results of Exercise 8.10.8 hold for the more general model of dependence given in Exercise 8.10.1.

8.10.10. Prove that

$$e_{S,M} = \left\{\frac{12\int f_0^2(x)\,dx \cdot \int g_0^2(y)\,dy}{\int J_0'[F_0(x)]f_0^2(u)\,dx \cdot \int J_0'[G_0(y)]g_0^2(y)\,dy}\right\}^2.$$

8.10.11. (*Continued.*) Show that (i) $e_{S,M} \leq 6/\pi$, and (ii) $e_{S,R} \geq 0.864$.

PRINCIPAL REFERENCES

Bellman (1960), Bhuchongkul (1964), Blomquist (1950), Blum, Kiefer and Rosenblatt (1961), Hájek (1962, 1963, 1968a, 1968b), Hoeffding (1947, 1948a, 1948b), Hotelling and Pabst (1936), Kendall (1938, 1955), Konijn (1956), Kruskal (1958), Lehmann (1966), Puri and Sen (1969a,d), Puri, Sen and Gokhale (1970), Sen (1967a, 1967e, 1968a, 1969b), and Sibuya (1959).

CHAPTER 9

Rank Tests for Homogeneity of Dispersion Matrices

9.1. INTRODUCTION

In this chapter we study the problems of testing the homogeneity of several dispersion matrices through the use of rank order statistics. In the parametric theory the problem of testing homogeneity of dispersion matrices has nothing to do with the homogeneity of location parameters. However, in the non-parametric case the formulation of the hypothesis of invariance demands the identity of distributions, and this, in turn, requires the homogeneity of location parameters. Thus we consider the following problems:

(i) Tests for the homogeneity of dispersion matrices assuming the identity of location parameters. Some permutationally distribution-free tests for this problem are considered here and their large-sample properties are studied.

(ii) Tests for the homogeneity of dispersion matrices without assuming the identity of location parameters. Here the basic permutation argument is not tenable, and only large-sample properties are considered.

(iii) The problem of testing simultaneously the homogeneity of location parameters as well as the dispersion matrices is briefly discussed.

In the parametric case, tests for the homogeneity of dispersion matrices are based on the sample covariance matrices (see Anderson, 1958, Chapter 10). However, in the nonparametric setup, the sample covariance matrix is not usually looked upon as a suitable statistic (matrix) because it is sensitive to outlying observations; it depends on the existence of the moments of the parent distributions, and it may not retain its sufficiency or optimality for non-normal distributions. For these reasons, we shall formulate a class of regular functionals which seems desirable for nonparametric procedures. The corresponding rank order estimates are then used to formulate the proposed tests.

9.2. PRELIMINARY NOTIONS

Let $\mathbf{X}_{\alpha}^{(k)} = (X_{1\alpha}^{(k)}, \ldots, X_{p\alpha}^{(k)})$, $\alpha = 1, \ldots, n_k$ be n_k independent and identically distributed (vector-valued) random variables having a p-variate absolutely continuous cumulative distribution function (c.d.f.) $F^{(k)}(\mathbf{x})$, for $k = 1, \ldots, c\ (\geqq 2)$; all these c samples being assumed to be mutually independent. Write $N = \sum_{k=1}^{c} n_k$ and assume that for all N, the inequalities

$$(9.2.1)\quad 0 < \lambda_0 \leqq \lambda_N^{(k)} = n_k/N \leqq 1 - \lambda_0 < 1, \quad \text{for} \quad k = 1, \ldots, c$$

hold for some fixed $\lambda_0 \leqq 1/c$.

Let $F_{(i)}^{(k)}(x)$ and $F_{(i,j)}^{(k)}(x, y)$ be the marginal c.d.f.'s of $X_{i\alpha}^{(k)}$ and $(X_{i\alpha}^{(k)}, X_{j\alpha}^{(k)})$ respectively for $i < j = 1, \ldots, p$ and $k = 1, \ldots, c$.

Write

$$(9.2.2)\qquad H_{(i)}(x) = \sum_{k=1}^{c} \lambda_N^{(k)} F_{(i)}^{(k)}(x).$$

$H_{(i)}(x)$ is the combined population c.d.f. of the ith variate for $i = 1, \ldots, p$. Let $J_{(i)}(u)$ for $i = 1, \ldots, p$ be some absolutely continuous function of u, defined in the open interval $(0, 1)$, and suppose that $J_{(i)}(u)$ is normalized in the following manner:

$$(9.2.3)\quad \int_0^1 J_{(i)}(u)\, du = 0, \qquad \int_0^1 J_{(i)}^2(u)\, du = 1; \qquad i = 1, \ldots, p.$$

Later on we shall impose certain regularity conditions on $J_{(i)}(u)$. Let us define

$$(9.2.4)\quad Y_{i\alpha}^{(k)} = J_{(i)}[H_{(i)}(X_{i\alpha}^{(k)})]; \qquad i = 1, \ldots, p; \qquad k = 1, \ldots, c.$$

$Y_{i\alpha}^{(k)}$ will be called the grade functional of $X_{i\alpha}^{(k)}$.

We shall employ these grade functionals to formulate the desired class of tests. Write

$$(9.2.5)\quad \begin{aligned} \theta_{ij}(H) &= \operatorname{cov}\,[(J_{(i)}[H_{(i)}(X)], J_{(j)}[H_{(j)}(Y)]) \mid H(\mathbf{x})] \\ &= \int_{-\infty}^{\infty}\int_{-\infty}^{\infty} J_{(i)}[H_{(i)}(x)] J_{(j)}[H_{(j)}(y)]\, dH_{(i,j)}(x,y) \end{aligned}$$

where

$$(9.2.6)\qquad H_{(i,j)}(x, y) = \sum_{k=1}^{c} \lambda_N^{(k)} F_{(i,j)}^{(k)}(x, y)$$

for $i \neq j = 1, \ldots, p$ and $H_{(i)}(x)$ is as defined in (9.2.2). Also write

$$(9.2.7)\quad \begin{aligned} \theta_{ij}^{(k)} &= \operatorname{cov}\,[(J_{(i)}[H_{(i)}(X)], J_{(j)}[H_{(j)}(Y)]) \mid F^{(k)}] \\ &= \int_{-\infty}^{\infty}\int_{-\infty}^{\infty} J_{(i)}[H_{(i)}(x)] J_{(j)}[H_{(j)}(y)]\, dF_{(i,j)}^{(k)}(x, y) - \mu_{N,i}^{(k)} \mu_{N,j}^{(k)} \end{aligned}$$

where

$$(9.2.8)\quad \mu_{N,i}^{(k)} = \int_{-\infty}^{\infty} J_{(i)}[H_{(i)}(x)]\, dF_{(i)}^{(k)}(x); \qquad 1, \ldots, p; \qquad k = 1, \ldots, c.$$

Finally, let

$$(9.2.9)\qquad \mathbf{\Theta}^* = ((\theta_{ij}(H)))_{i,j=1,\ldots,p};$$

$$\mathbf{\Theta}^{(k)} = ((\theta_{ij}^{(k)}))_{i,j=1,\ldots,p}; \quad k = 1.\ldots.c.$$

(Note that all these matrices depend on N through $\lambda_N^{(1)}, \ldots, \lambda_N^{(c)}$). Thus, $\mathbf{\Theta}^{(k)}$ is the dispersion matrix of the grade functionals of the kth sample, $k = 1, \ldots, c$. The purpose of introducing the dispersion matrices in (9.2.9) is to work with a class of dispersion matrices which are invariant under certain transformations of the variables, are less sensitive to outlying observations, and at the same time, are reasonably informative measures. In particular, if $J_{(i)}(u)$ is monotone in $u(0 < u < 1)$ for all $i = 1, \ldots, p$, then $\mathbf{\Theta}^{(k)}$, $k = 1, \ldots, c$ defined in (9.2.9) will be invariant under monotone transformations of the coordinate variables.

Now since the c.d.f.'s $F^{(1)}, \ldots, F^{(c)}$ are all unknown, so are also the $c + 1$ matrices defined in (9.2.9). To estimate these we follow Von Mises' (1947) approach, and define for any $q \leqq p$

$$(9.2.10)\quad F_{N(i_1,\ldots,i_q)}^{(k)}(x_1, \ldots, x_q) = \frac{1}{n_k}\,[\text{number of } (X_{i_1\alpha}^{(k)}, \ldots, X_{i_q\alpha}^{(k)}) \leqq (x_1, \ldots, x_q),\ \ \alpha = 1, \ldots, n_k]$$

$$(9.2.11)\quad H_{N(i_1,\ldots,i_q)}(x_1, \ldots, x_q) = \sum_{k=1}^{c} \lambda_N^{(k)} F_{N(i_1,\ldots,i_q)}^{(k)}(x_1, \ldots, x_q)$$

for all $i_1 \neq \cdots \neq i_q = 1, \ldots, p$; $q = 1, \ldots, p$. Again, let

$$(9.2.12)\qquad F_{(i_1,\ldots,i_q)}^{(k)}(x_1, \ldots, x_q) = P\{X_{ij\alpha}^{(k)} \leq x_j;\ j = 1, \ldots, q\}$$

$$(9.2.13)\qquad H_{(i_1,\ldots,i_q)}(x_1, \ldots, x_q) = \sum_{k=1}^{c} \lambda_N^{(k)} F_{(i_1,\ldots,i_q)}^{(k)}(x_1, \ldots, x_q)$$

for all $i_1 \neq \cdots \neq i_q = 1, \ldots, p$; $q = 1, \ldots, p$.

Now in actual practice, we have a sequence of functions $J_{N(i)}(u)$ which converges to $J_{N(i)}(u)$ as $N \to \infty$ for all $0 < u < 1$, and $i = 1, \ldots, p$. Thus we may estimate $\theta_{ij}(H)$ defined in (9.2.5) by

$$(9.2.14)\qquad \begin{aligned} T_{N,ij}^* &= \int_{-\infty}^{\infty}\int_{-\infty}^{\infty} J_{N(i)}[H_{N(i)}(x)]J_{N(j)}[H_{N(j)}(y)]\, dH_{N(i,j)}(x, y) \\ &= \frac{1}{N}\sum_{k=1}^{c}\sum_{\alpha=1}^{n_k} J_{N(i)}[H_{N(i)}(X_{i\alpha}^{(k)})]J_{N(j)}[H_{N(j)}(X_{j\alpha}^{(k)})] \end{aligned}$$

for all $i, j = 1, \ldots, p$. Since $T^*_{N,ij} = T^*_{N,ji}$, the number of statistics $T^*_{N,ij}$ is actually $p(p+1)/2$. Now let us write

$$\mathbf{T}^*_N = \{T^*_{N,ij},\ i \leq j = 1, \ldots, p\} \tag{9.2.15}$$

$$T^{(k)}_{N,i} = \int_{-\infty}^{\infty} J_{N(i)}[H_{N(i)}(x)]\, dF^{(k)}_{N(i)}(x); \qquad i = 1, \ldots, p; \quad k = 1, \ldots, c, \tag{9.2.16}$$

$$\begin{aligned} T^{(k)}_{N,ij} &= \int_{-\infty}^{\infty}\int_{-\infty}^{\infty} J_{N(i)}[H_{N(i)}(x)] J_{N(j)}[H_{N(j)}(y)]\, dF^{(k)}_{N(i,j)}(x, y) \\ &\quad - T^{(k)}_{N,i} T^{(k)}_{N,j} \\ &= \frac{1}{n_k}\sum_{\alpha=1}^{n_k} J_{N(i)}[H_{N(i)}(X^{(k)}_{i\alpha})] J_{N(j)}[H_{N(j)}(X^{(k)}_{j})] - T^{(k)}_{N,i} T^{(k)}_{N,j} \end{aligned} \tag{9.2.17}$$

and, finally

$$\mathbf{T}^{(k)}_N = \{T^{(k)}_{N,ij},\ i \leq j = 1, \ldots, p\}; \qquad k = 1, \ldots, c. \tag{9.2.18}$$

Then, our proposed tests are based on some functions of $\mathbf{T}^{(k)}_N$, $k = 1, \ldots, c$ and $\mathbf{T}^*_N$.

It may be noted that the class of dispersion matrices considered above contains many well-known nonparametric measures as special cases. For example, if we let $J_{N(i)}(\alpha/(N+1)) = [12/(N^2 - 1)]^{1/2}(\alpha - (N+1)/2)$, $\alpha = 1, \ldots, N$; $i = 1, \ldots, p$; then $\mathbf{T}^*_n$ reduces to the Spearman's rank correlations while for $J_{(i)}(u) = \sqrt{12}\,(u - \frac{1}{2})$, $\mathbf{\Theta}^*$ reduces to the grade correlations matrix. The nonparametric measures by Blomqvist (1950) and Bhuchongkul (1964) among others considered in Chapter 8 belong to this class.

We now consider the statistical formulations of the problem of testing the homogeneity of dispersion matrices.

First we consider equality of covariance matrices, assuming the identity of locations. In such a case our null hypothesis is

$$H_0^{(1)}:\ F^{(1)}(\mathbf{x}) = \cdots = F^{(c)}(\mathbf{x}) \tag{9.2.19}$$

(which implies that $\mathbf{\Theta}^{(1)} = \cdots = \mathbf{\Theta}^{(c)}$), against the set of alternatives that $\mathbf{\Theta}^{(1)}, \ldots, \mathbf{\Theta}^{(c)}$ are not all identical. We may note that $H^{(1)}$ will include a test for identity of scales and of association patterns of the c c.d.f.'s. Next we consider the equality of dispersion matrices without assuming the identity of locations. In such a case, writing

$$F^{(k)}(\mathbf{x}) = F_0^{(k)}(\mathbf{x} - \mathbf{\delta}^{(k)}), \qquad k = 1, \ldots, c \tag{9.2.20}$$

where $\mathbf{\delta}^{(k)} = (\delta_1^{(k)}, \ldots, \delta_p^{(k)})$, $k = 1, \ldots, c$ are c real p vectors, we test

$$H_0^{(2)}:\ F_0^{(1)}(\mathbf{x}) = \cdots = F_0^{(c)}(\mathbf{x}) \tag{9.2.21}$$

against a similar type of alternatives involving $\mathbf{\Theta}^{(1)}, \ldots, \mathbf{\Theta}^{(c)}$, where of course, in the definition of $\mathbf{\Theta}^{(k)}$'s; $F^{(1)}, \ldots, F^{(c)}$ are replaced by $F_0^{(1)}, \ldots, F_0^{(c)}$ respectively. Finally, we consider the hypothesis that both locations and dispersion matrices are the same; that is, we test

$$H_0^{(3)}: F^{(1)}(\mathbf{x}) = \cdots = F^{(c)}(\mathbf{x}) \tag{9.2.22}$$

where $F^{(k)}(\mathbf{x})$, $k = 1, \ldots, c$ is given by (9.2.20), against the alternatives that at least one of the following two equalities is not true.

$$\boldsymbol{\delta}^{(1)} = \cdots = \boldsymbol{\delta}^{(c)}; \qquad \mathbf{\Theta}^{(1)} = \cdots = \mathbf{\Theta}^{(c)}. \tag{9.2.23}$$

9.3. PERMUTATION TESTS FOR $H_0^{(1)}$

Let us pool the $N = \sum_{k=1}^{c} n_k$ observation $\mathbf{X}_\alpha^{(k)}$, $\alpha = 1, \ldots, n_k$; $k = 1, \ldots, c$ into a combined sample, and denote the sample point by the $p \times N$ matrix

$$\mathbf{Z}_N' = (\mathbf{X}_1^{(1)'}, \ldots, \mathbf{X}_{N_c}^{(c)'}) \tag{9.3.1}$$

and the sample space by $\mathfrak{Z}_N$. Under $H_0^{(1)}$, the joint distribution of Z_N remains invariant under the finite group of $N!$ permutations of the N columns of Z_N'. Hence, conditionally given $\mathbf{Z}_N$, all the $N!$ sample points generated by these $N!$ permutations of the columns of $\mathbf{Z}_N'$ are equiprobable, each having the conditional probability $1/N!$. Let us denote this conditional probability measure (defined over the $N!$ equiprobable realizations) by $\mathcal{P}_N$. Now, ranking the N elements in each row of $\mathbf{Z}_N'$ in increasing order of magnitude, we obtain a $p \times N$ matrix

$$R_N = \begin{pmatrix} R_{11}^{(1)} & \cdots & R_{1n_1}^{(1)} & \cdots & R_{1n_c}^{(c)} \\ \cdot & & \cdot & & \cdot \\ \cdot & & \cdot & & \cdot \\ \cdot & & \cdot & & \cdot \\ R_{p1}^{(1)} & \cdots & R_{pn_1}^{(1)} & \cdots & R_{pn_c}^{(c)} \end{pmatrix} \tag{9.3.2}$$

where by virtue of the continuity of the c.d.f.'s, the possibility of ties may be ignored in probability. For every $i\,(= 1, \ldots, p)$, replacing the ranks $\alpha\,(= 1, \ldots, N)$ in the ith line of R_N by a set of general scores $\{E_{N,\alpha}^{(i)} - J_{N(i)}(\alpha/(N+1), \alpha = 1, \ldots, N)\}$, we get the corresponding score matrix

$$\mathbf{E}_N = \begin{pmatrix} E_{N,R_{11}^{(1)}}^{(1)} & \cdots & E_{N,R_{1n_1}^{(1)}}^{(1)} & \cdots & E_{N,R_{1n_c}^{(c)}}^{(1)} \\ \cdot & & \cdot & & \cdot \\ \cdot & & \cdot & & \cdot \\ \cdot & & \cdot & & \cdot \\ E_{N,R_{p1}^{(1)}}^{(p)} & \cdots & E_{N,R_{pn_1}^{(1)}}^{(p)} & \cdots & E_{N,R_{pn_c}^{(c)}}^{(p)} \end{pmatrix}. \tag{9.3.3}$$

With this terminology, we can rewrite $T_{N,ij}^{(k)}$ and $T_{N,ij}^{*}$ as

$$(9.3.4)\quad T_{N,ij}^{(k)} = \frac{1}{n_k}\sum_{\alpha=1}^{n_k} E_{N,R_{i\alpha}^{(k)}}^{(i)} E_{N,R_{j\alpha}^{(k)}}^{(j)} - \left[\frac{1}{n_k}\sum_{\alpha=1}^{n_k} E_{N,R_{i\alpha}^{(k)}}^{(i)}\right]\left[\frac{1}{n_k}\sum_{\alpha=1}^{n_k} E_{N,R_{j\alpha}^{(k)}}^{(j)}\right]$$

$$i \le j = 1, \ldots, p; \qquad k = 1, \ldots, c$$

$$(9.3.5)\quad T_{N,ij}^{*} = \frac{1}{n}\sum_{k=1}^{c}\sum_{\alpha=1}^{n_k} E_{N,R_{i\alpha}^{(k)}}^{(i)} E_{N,R_{j\alpha}^{(k)}}^{(j)}; \qquad i \le j = 1, \ldots, p.$$

Operationally, it will be much simpler to work with the slightly adjusted statistics

$$(9.3.6)\quad S_{N,ij}^{(k)} = \frac{1}{n_k - 1}\left\{\sum_{\alpha=1}^{n_k} E_{N,R_{i\alpha}^{(k)}}^{(i)} E_{N,R_{j\alpha}^{(k)}}^{(j)} - \frac{1}{n_k}\left[\sum_{\alpha=1}^{n_k} E_{N,R_{i\alpha}^{(k)}}^{(i)}\right]\left[\sum_{\alpha=1}^{n} E_{N,R_{j\alpha}^{(k)}}^{(j)}\right]\right\},$$

$$i \le j = 1, \ldots, p; \quad k = 1, \ldots, c,$$

$$(9.3.7)\quad S_{N,ij}^{*} = \frac{1}{N - 1}\left\{\sum_{k=1}^{c}\sum_{\alpha=1}^{n_k} E_{N,R_{i\alpha}^{(k)}}^{(i)} E_{N,R_{j\alpha}^{(k)}}^{(j)} - N\bar{E}_N^{(i)}\bar{E}_N^{(j)}\right\}$$

where

$$(9.3.8)\quad \bar{E}_N^{(i)} = \frac{1}{N}\sum_{k=1}^{c}\sum_{\alpha=1}^{n_k} E_{N,R_{i\alpha}^{(k)}}^{(i)} = \int_0^1 J_{N(i)}[H_{N(i)}(x)]\, dH_{N(i)}(x), \; i = 1, \ldots, p.$$

Note that

$$(9.3.9)\qquad S_{N,ij}^{(k)} = [n_k/(n_k - 1)]T_{N,ij}^{(k)}$$

and

$$(9.3.10)\qquad S_{N,ij}^{*} = |(N/(N - 1))|\,|T_{N,ij}^{*} - \bar{E}_N^{(i)}\bar{E}_N^{(j)}|.$$

Now under the permutational probability measure $\mathcal{T}_N$, there are $N!$ equally likely (conditionally) permutations of the columns of $\mathbf{R}_N$ (and hence of $\mathbf{E}_N$). Hence, by an adaptation of the same argument as in section 5.4, we have

$$E(S_{N,ij}^{(k)} \mid \mathcal{T}_N) = S_{N,ij}^{*} \quad \text{for all} \quad i \le j = 1, \ldots, p.$$

Furthermore, $S_{N,ij}^{(k)}$ in (9.3.5) can also be expressed as U statistics in the scores $E_{N,R_{i\alpha}^{(k)}}^{(i)}$; $i = 1, \ldots, p$; $\alpha = 1, \ldots, n_k$ as follows:

$$(9.3.11)\quad S_{N,ij}^{(k)} = \frac{1}{2}\binom{n_k}{2}^{-1}\sum_{\alpha<\beta=1}^{n_k}(E_{N,R_{i\alpha}^{(k)}}^{(i)} - E_{N,R_{i\beta}^{(k)}}^{(i)})(E_{N,R_{j\alpha}^{(k)}}^{(j)} - E_{N,R_{j\beta}^{(k)}}^{(j)})$$

$$\text{for} \quad i, j = 1, \ldots, p; \quad k = 1, \ldots, c$$

Let us now define

$$(9.3.12)\quad 4\zeta^{(1)}_{j,i'j'}(\mathbf{R}_N) = \frac{1}{N(N-1)(N-2)} \sum_{k=1}^{c}\sum_{q=1}^{c}\sum_{r=1}^{c}\sum_{\alpha=1}^{n_k}\sum_{\beta=1}^{n_q}\sum_{\gamma=1}^{n_r} (E^{(i)}_{N,R^{(k)}_{i\alpha}} - E^{(i)}_{N,R^{(q)}_{i\beta}})(E^{(j)}_{N,R^{(k)}_{j\alpha}} - E^{(j)}_{N,R^{(q)}_{j\beta}}) \cdot (E^{(i')}_{N,R^{(k)}_{i'\alpha}} - E^{(i')}_{N,R^{(r)}_{i'\alpha}})(E^{(j')}_{N,R^{(k)}_{j'\alpha}} - E^{(j')}_{N,R^{(r)}_{j'\alpha}}) - S^*_{N,ij}\cdot S^*_{N,i'j'} \quad \text{for } i, i', j, j' = 1, \ldots, p;$$

$$(9.3.13)\quad 4\zeta^{(2)}_{ij,i'j'}(\mathbf{R}_N) = \frac{1}{N(N-1)} \sum_{k=1}^{c}\sum_{q=1}^{c}\sum_{\alpha=1}^{n_k}\sum_{\beta=1}^{n_k} (E^{(i)}_{N,R^{(k)}_{i\alpha}} - E^{(i)}_{N,R^{(k)}_{i\beta}}) \cdot (E^{(j)}_{N,R^{(k)}_{j\alpha}} - E^{(j)}_{N,R^{(q)}_{j\beta}})(E^{(i')}_{N,R^{(k)}_{i'\alpha}} - E^{(i')}_{N,R^{(q)}_{i'\beta}}) - (E^{(j')}_{N,R^{(k)}_{j'\alpha}} - E^{(j')}_{N,R^{(q)}_{j'\beta}}) - S^*_{N,ij}S^*_{N,i'j'} \text{ for } i, j, i', j' = 1, \ldots, p.$$

Then, by an adaptation of the results of Nandi and Sen (1963) and Sen (1966c), it can be shown after some algebraic manipulations that

$$(9.3.14)\quad \operatorname{cov}\{S^{(k)}_{N,ij}, S^{(q)}_{N,i'j'} \mid \mathcal{T}_N\} = \frac{\binom{n_k-2}{2}}{\binom{n_k}{2}} \sum_{d=1}^{2}\binom{2}{d}\left\{\frac{\binom{n_k-2}{2-d}}{\binom{n_k-2}{2}} - \frac{\binom{n-2}{2-d}}{\binom{n-2}{2}}\right\} \times \zeta^{(d)}_{ij,i'j'}(\mathbf{R}_N) \quad \text{if } k = q,$$

$$(9.3.15)\quad = -\binom{N-2}{2}^{-1}\sum_{d=1}^{2}\binom{2}{d}\binom{N-2}{2-d}\zeta^{(d)}_{ij,i'j'}(\mathbf{R}_N),$$

$$\text{if } k \neq q, \quad \text{for } i, i', j, j' = 1, \ldots, p; \quad k, q = 1, \ldots, c.$$

Thus, for small samples we may work with the random variables

$$(9.3.16)\quad \{S^{(k)}_{N,ij} - S^*_{N,ij};\ i \leq j = 1, \ldots, p,\ k = 1, \ldots, c\},$$

and considering the generalized inverse of their (permutation) covariance matrix, arrive at the desired quadratic form as suitable test statistics. Under $\mathcal{T}_N$, this statistic will have only $N!/\prod_1^c n_k!$ equally likely (not necessarily all distinct) values, and hence, an exact size $\epsilon(0 < \epsilon < 1)$ test for $H_0^{(1)}$ can be constructed. However, the expressions in (9.3.12) through (9.3.15) are quite complicated. If N is not small, it will be convenient to consider the following permutation test.

By virtue of (9.2.3), $\bar{E}^{(i)}_N$ defined in (9.3.8) converges to zero as $N \to \infty$. In fact, $\bar{E}^{(i)}_N = o(N^{-1/2})$ for all $i = 1, \ldots, p$. Hence from (9.3.5) and (9.3.7)

$$(9.3.17)\quad \left| T^*_{N,ij} - \frac{N-1}{N} S^*_{N,ij} \right| = o(N^{-1/2}).$$

Further, from the results of chapter 5, under $\mathcal{T}_N$

$$\left| \frac{1}{n_k} \sum_{\alpha=1}^{n_k} E_{N,R_{i\alpha}^{(k)}}^{(i)} - \bar{E}_N^{(i)} \right| = O_p(N^{-1/2}) \tag{9.3.18}$$

for all $i = 1, \ldots, p$; $k = 1, \ldots, c$. Consequently under $\mathcal{T}_N$

$$\left| S_{N,ij}^{(k)} - \frac{1}{n_k} \sum_{\alpha=1}^{n_k} E_{N,R_{i\alpha}^{(k)}}^{(i)} E_{N,R_{j\alpha}^{(k)}}^{(j)} \right| = O_p(N^{-1}) \tag{9.3.19}$$

for all $i, j = 1, \ldots, p$; $k = 1, \ldots, c$.

Now let us define

$$v_{ij,i'j'}(\mathbf{R}_N) = \frac{1}{N} \sum_{k=1}^{c} \sum_{\alpha=1}^{n_k} E_{N,R_{i\alpha}^{(k)}}^{(i)} E_{N,R_{j\alpha}^{(k)}}^{(j)} E_{N,R_{i'\alpha}^{(k)}}^{(i')} E_{N,R_{j'\alpha}^{(k)}}^{(j')} - S_{N,ij}^{*} S_{N,i'j'}^{*} \quad \text{for all} \quad i, i', j, j' = 1, \ldots, p. \tag{9.3.20}$$

Setting

$$r = [(i-1)(2p-i)/2] + j \quad \text{for} \quad i \le j = 1, \ldots, p, \tag{9.3.21}$$

we rewrite

$$\{S_{N,ij}^{(k)}, i \le j = 1, \ldots, p\} \quad \text{as} \quad \{S_{N,r}^{(k)}, r = 1, \ldots, p(p+1)/2\} = \mathbf{S}_N^{(k)} \quad \text{for} \quad k = 1, \ldots, c, \tag{9.3.22}$$

$$\{S_{N,ij}^{*}, i \le j = 1, \ldots, p\} \quad \text{as} \quad \{S_{N,r}^{*}, r = 1, \ldots, p(p+1)/2\} = \mathbf{S}_N^{*}, \tag{9.3.23}$$

and

$$\mathbf{V}_N(\mathbf{R}_N) = ((v_{rs}(\mathbf{R}_N)))_{r,s=1,\ldots,p(p+1)/2} \tag{9.3.24}$$

where r is defined as in (9.3.21) with (i, j) replaced by (i', j').

Then, as in the location problem considered in chapter 5, we consider the statistic

$$\mathcal{L}_N = \sum_{k=1}^{c} n_k [\mathbf{S}_N^{(k)} - \mathbf{S}_N^{*}] \mathbf{V}_N^{-1}(\mathbf{R}_N) [\mathbf{S}_N^{(k)} - \mathbf{S}_N^{*}]' \tag{9.3.25}$$

for the problem of testing $H_0^{(1)}$. (Here $\mathbf{V}_N^{-1}(\mathbf{R}_N)$ is the inverse of $\mathbf{V}_N(\mathbf{R}_N)$, and is assumed to be positive definite.) Thus, we have the following rule:

$$\text{if} \quad \mathcal{L}_N \begin{cases} \ge \mathcal{L}_{N,\epsilon}, & \text{reject } H_0^{(1)} \\ < \mathcal{L}_{N,\epsilon}, & \text{accept } H_0^{(1)}, \end{cases} \tag{9.3.26}$$

$\mathcal{L}_{N,\epsilon}$ is so chosen that $P\{\mathcal{L}_N \ge \mathcal{L}_{N,\epsilon} \mid H_0^{(1)}\} = \epsilon$, where $0 < \epsilon < 1$ is the preassigned level of significance of our test. In small samples, we can evaluate $\mathcal{L}_{N,\epsilon}$ by referring to the exact permutation distribution of $\mathcal{L}_N$ generated by the

$N!$ equally likely permutations of the columns of (9.3.3), while for large samples, we shall show in the next section that

$$\mathfrak{L}_{N,\epsilon} \xrightarrow{p} \chi^2_{\epsilon,(c-1)p(p+1)/2} \tag{9.3.27}$$

where $\chi^2_{\epsilon,r}$ is the $100(1-\epsilon)\%$ point of a chi distribution with r degrees of freedom.

9.4. ASYMPTOTIC PERMUTATION DISTRIBUTION OF $\mathfrak{L}_N$

As in section 9.3 we define

$$E^{(i)}_{N,\alpha} = J_{N(i)}(\alpha/(N+1)); \qquad \alpha = 1, \ldots, n; \quad i = 1, \ldots, p, \tag{9.4.1}$$

and extend the domain of $J_{n(i)}(u)$ to (0, 1) by letting $J_{N(i)}$ be constant on $(\alpha/(N+1), (\alpha+1)/(N+1))$. Furthermore, we shall make the following assumptions:

I. $J_{(i)}(u) = \lim_{n\to\infty} J_{N(i)}(u)$ exists for $0 < u < 1$ and is not a constant (for $i = 1, \ldots, p$).

II.

$$\int_{-\infty}^{\infty}\int_{-\infty}^{\infty} \{J_{N(i)}[H^*_{N(i)}(x)]J_{N(j)}[H^*_{N(j)}(y)] - J_{(i)}[H^*_{N(i)}(x)]J_{(j)}[H^*_{N(j)}(y)]\}\, dF^{(k)}_{(i,j)}(x, y) = o_p(n^{-1/2}) \tag{9.4.2}$$

$$i, j = 1, \ldots, p; \quad k = 1, \ldots, c$$

where

$$H^*_{N(i)}(x) = \frac{N}{N+1} H_{N(i)}(x). \tag{9.4.3}$$

III.

$$|J^{(r)}_{(i)}(u)| = \left|\frac{\partial^r}{\partial u^r} J_{(i)}(u)\right| \le K[u(1-u)]^{\alpha_r} \tag{9.4.4}$$

for $r = 0, 1$, where $0 < \alpha_0 < \frac{1}{8}$; $\alpha_1 = -1$, and K is some constant. Now let us write

$$\mu^{(k)}_{Nij} = \int_{-\infty}^{\infty}\int_{-\infty}^{\infty} J_{(i)}[H_{(i)}(x)]J_{(j)}[H_{(j)}(y)]\, dF^{(k)}_{(i,j)}(x, y); \quad k = 1, \ldots, c; \tag{9.4.5}$$

$$\mu^*_{ij} = \int_{-\infty}^{\infty}\int_{-\infty}^{\infty} J_{(i)}[H_{(i)}(x)]J_{(j)}[H_{(j)}(y)]\, dH_{(i,j)}(x, y) \tag{9.4.6}$$

$$\nu^*_{ij,i'j'} = \int_{-\infty}^{\infty}\int_{-\infty}^{\infty}\int_{-\infty}^{\infty}\int_{-\infty}^{\infty} J_{(i)}[H_{(i)}(x_i)]J_{(j)}[H_{(j)}(x_j)] \times J_{(i')}[H_{(i')}(x_{i'})]J_{(j')}[H_{(j')}(x_{j'})]\, dH_{(i,j,i',j')}(x_i, x_j, x_{i'}, x_{j'}) - \mu^*_{ij}\mu^*_{i'j'} \quad \text{for} \quad i, j, i', j' = 1, \ldots, p. \tag{9.4.7}$$

Adopting the suffixing system of (9.3.21), we rewrite $\nu^*_{ij,i'j'}$ as ν^*_{rs} for $r, s = 1, \ldots, p(p+1)/2$ and define

$$\text{(9.4.8)} \qquad \boldsymbol{\nu}^* = ((\nu^*_{rs}))_{r,s} = 1, \ldots, p(p+1)/2.$$

IV. $\boldsymbol{\nu}^*$ defined by (9.4.8) is positive definite. Then, we have the following theorem.

Theorem 9.4.1. *Under assumptions* I *to* V, $\mathbf{V}_N(\mathbf{R}_N)$, *defined in* (9.3.24), *is equivalent in probability to* $\boldsymbol{\nu}^*$, *defined in* (9.4.8), *as* $N \to \infty$ *uniformly in* $\lambda_N^{(1)}, \ldots, \lambda_N^{(c)}$ *and all* $F^{(1)}, \ldots, F^{(c)} \in \mathcal{F}_p$.

The proof follows precisely along the same line as in Theorem 5.4.2 with straightforward extensions, and is given as exercise 9.4.1.

Thoerem 9.4.2. *Under assumptions* I *to* V, *the permutation distribution of* $\mathfrak{L}_N$ *(defined by* (9.3.25)) *asymptotically reduces to chi-square distribution with* $(c-1)p(p+1)/2$ *degrees of freedom.*

Proof: If we define

$$Z^{(k)}_{N,ij} = \frac{1}{n_k} \sum_{\alpha=1}^{n_k} E^{(i)}_{N,R_{i\alpha}} E^{(j)}_{N,R_{j\alpha}}$$

for, $i \le j = 1, \ldots, p$; $k = 1, \ldots, c$, then it follows by elementary computations as in section 5.4 that

$$\operatorname{cov}(Z^{(k)}_{N,ij}, Z^{(q)}_{N,i'j'} \mid \mathcal{T}_N) = \frac{(\delta_{kq}N - n_k)}{n_k(N-1)} V_{ij,i'j'}(\mathbf{R}_N),$$

where δ_{kq} is the Kronecker delta. Thus, by (9.3.19) and (9.3.25), we are only to show that $\{\mathbf{S}^{(k)}_N, k = 1, \ldots, c\}$, when standardized has a multinormal distribution, under $\mathcal{T}_N$, as $N \to \infty$. Since the elements of $\mathbf{S}^{(k)}_N$ are all U-statistics in the rank vectors, the desired result directly follows from Theorem 5.7.3. Q.E.D. ◁

By virtue of Theorem 9.4.2, $\mathfrak{L}_{N,\epsilon}$ defined in (9.3.2) converges in probability to $\chi^2_{\epsilon,(c-1)p(p+1)/2}$. Hence the permutation test procedure based on $\mathfrak{L}_N$ simplifies in large samples to the following rule:

$$\text{if} \quad \mathfrak{L}_N > \chi^2_{\epsilon,(c-1)p(p+1)/2}, \qquad \text{reject } H_0^{(1)}$$
$$< \chi^2_{\epsilon,(c-1)p(p+1)/2}, \qquad \text{accept } H_0^{(1)}$$

In order to study the power properties of the test considered above, we need to study the unconditional distribution of $\mathfrak{L}_N$ under an appropriate sequence of alternative hypotheses. This, in turn, requires the study of the joint distribution of the rank order statistics $\mathbf{S}^{(k)}_N$ defined in (9.3.25).

9.5. ASYMPTOTIC NORMALITY OF $S_N^{(k)}$, $k = 1, \ldots, c$, FOR ARBITRARY $F^{(1)}, \ldots, F^{(c)}$

Theorem 9.5.1. *Under the assumptions* I *to* V *of section* 9.4, *the random variables*

$$N^{1/2}[U_{N,ij}^{(k)} - \mu_{N,ij}^{(k)} - \mu_{N,i}^{(k)}\mu_{N,j}^{(k)}, T_{N,i}^{(k)} - \mu_{N,i}^{(k)}; i \leq j = 1, \ldots, p]$$

where $T_{N,i}^{(k)}$, $\mu_{N,i}^{(k)}$ *and* $\mu_{N,ij}^{(k)}$ *are given by* (9.2.16), (9.2.8), *and* (9.4.5) *respectively, and* $U_{N,ij}^{(k)} = T_{N,ij}^{(k)} + T_{N,i}^{(k)}T_{N,j}^{(k)}$ *have asymptotically a* $cp(p+3)/2$-*variate normal distribution with null mean vector and finite covariance matrix.*

Proof: We write

$$(9.5.1) \qquad J_{N(i)}[H_{N(i)}^*]J_{N(j)}[H_{N(j)}^*] = \{J_{N(i)}[H_{N(i)}^*]J_{N(j)}[H_{N(j)}^*] - J_{(i)}[H_{N(i)}^*]J_{(j)}[H_{N(j)}^*]\} + J_{(i)}[H_{N(i)}^*]J_{(j)}[H_{N(j)}^*]$$

$$\begin{aligned}(9.5.2) \qquad J_{(i)}[H_{N[i]}^*]J_{(j)}[H_{N[j]}^*] = {} & J_{(i)}[H_{[i]}]J_{(j)}[H_{[j]}] + (H_{N[i]}^* - H_{[i]}) \\ & \times J'_{(i)}[H_{[i]}]J_{(j)}[H_{(j)}] + (H_{N[j]}^* - H_{[j]}) \\ & \times J'_{(j)}[H_{(j)}]J_{(i)}[H_{(i)}] + [J_{(i)}(H_{N[i]}^*) \\ & - J_{(i)}(H_{[i]}) - \{H_{N[i]}^* - H_{[i]}\} \\ & \times J'_{(i)}(H_{[i]})]J_{(j)}(H_{N[j]}^*) \\ & + [J_{(j)}(H_{N[j]}^*) - J_{(j)}(H_{[j]}) \\ & - \{H_{N[j]}^* - H_{[j]}\}J'_{(j)}(H_{[j]})]J_{(i)}(H_{[i]}) \\ & + [H_{N[i]}^* - H_{[i]}][J_{(j)}(H_{N[j]}^*) - J_{(j)}(H_{[j]})] \\ & \times J'_{(i)}(H_{[i]}),\end{aligned}$$

$$(9.5.3) \qquad dF_{N[i,j]}^{(k)} = d(F_{N[i,j]}^{(k)} - F_{[i,j]}^{(k)}) + dF_{[i,j]}^{(k)}.$$

Proceeding as in Theorems 8.4.1 and 5.5.1 we can express

$$(9.5.4) \qquad U_{N,ij}^{(k)} = A_{1N(i,j)}^{(k)} + A_{2N(i,j)}^{(k)} + A_{3N(i,j)}^{(k)} + \sum_{r=1}^{5} \mathcal{B}_{rN(i,j)}^{(k)}$$

and

$$(9.5.5) \qquad T_{Ni}^{(k)} = \mu_{N(i)}^{(k)} + B_{1N(i)}^{(k)} + B_{2N(i)}^{(k)} + \sum_{r=1}^{4} \mathcal{C}_{rN(i)}^{(k)}$$

where

$$(9.5.6) \qquad A_{1N(i,j)}^{(k)} = \int_{-\infty}^{\infty}\int_{-\infty}^{\infty} J_{(i)}[H_{(i)}(x_i)]\, dF_{N(i,j)}^{(k)}(x_i, x_j)$$

$$(9.5.7)\qquad A_{2N(i,j)}^{(k)} = \int_{-\infty}^{\infty}\int_{-\infty}^{\infty} [H_{N(i)}^{*}(x_i) - H_{(i)}(x_i)]J'_{(i)}[H_{(i)}(x_i)] \times J_{(j)}[H_{(j)}(x_j)]\, dF_{[i,j]}^{(k)}(x_i, x_j)$$

$$(9.5.8)\qquad A_{3N(i,j)}^{(k)} = \int_{-\infty}^{\infty}\int_{-\infty}^{\infty} [H_{N(j)}^{*}(x_j) - H_{(j)}(x_j)]J'_{(j)}[H_{(j)}(x_j)] \times J_{(i)}[H_{(i)}(x_i)]\, dF_{[i,j]}^{(k)}(x_i, x_j)$$

$$(9.5.9)\qquad B_{1N(i)}^{(k)} = \int_{-\infty}^{\infty} J_{(i)}[H_{(i)}(x_i)]\, d[F_{N(i)}^{(k)}(x_i) - F_{(i)}^{(k)}(x_i)],$$

$$(9.5.10)\qquad B_{2N(i)}^{(k)} = \int_{-\infty}^{\infty} [H_{N(i)}^{*}(x_i) - H_{(i)}(x_i)]J'_{(i)}[H_{(i)}(x_i)]\, dF_{(i)}^{(k)}(x_i),$$

$$(9.5.11)\qquad \mathcal{B}_{1N(i,j)}^{(k)} = \int_{-\infty}^{\infty}\int_{-\infty}^{\infty} [H_{N(i)}^{*} - H_{(i)}]J'_{(i)}[H_{(i)}]J_{(j)}[H_{(j)}]\, d[F_{N(i,j)}^{(k)} - F_{(i,j)}^{(k)}],$$

$$(9.5.12)\qquad \mathcal{B}_{2N(i,j)}^{(k)} = \int_{-\infty}^{\infty}\int_{-\infty}^{\infty} [H_{N(j)}^{*} - H_{(j)}]J'_{(j)}[H_{(j)}]J_{(i)}[H_{(i)}] \times d\,[F_{N(i,j)}^{(k)} - F_{(i,j)}^{(k)}],$$

$$(9.5.13)\qquad \mathcal{B}_{3N}^{(i,j)} = \int_{-\infty}^{\infty}\int_{-\infty}^{\infty} [J_{(i)}(H_{N[i]}^{*}(x_i)) - J_{(i)}(H_{[i]}(x_i)) - \{H_{N[i]}^{*}(x_i) - H_{[i]}(x_i)\}J'_{(i)}(F_{[i]}(x_i))] \cdot J_{(j)}(H_{N[j]}^{*}(x_j))\, dF_{N[i,j]}(x_i, x_j),$$

$$(9.5.14)\qquad \mathcal{B}_{4N}^{(i,j)} = \int_{-\infty}^{\infty}\int_{-\infty}^{\infty} [J_{(j)}(H_{N[j]}^{*}(x_j)) - J_{(j)}(H_{[j]}(x_j)) - \{H_{N[j]}^{*}(x_j) - H_{[j]}(x_j)\}J'_{(j)}(H_{[j]}(x_j))] \cdot J_{(i)}(H_{[i]}(x_i))\, dF_{N[i,j]}(x_i, x_j),$$

$$(9.5.15)\qquad \mathcal{B}_{5N}^{(i,j)} = \int_{-\infty}^{\infty}\int_{-\infty}^{\infty} [H_{N[i]}^{*}(x_i) - H_{[i]}(x_i)]J'_{(i)}(H_{[i]}(x_i)) \cdot [J_{(j)}(H_{N[j]}^{*}(x_j)) - J_{(j)}(H_{[j]}(x_j)]\, dF_{N[i,j]}(x_i, x_j).$$

$$(9.5.16)\qquad \mathcal{C}_{1N(i)}^{(k)} = -(N+1)^{-1}\int_{-\infty}^{\infty} H_{N(i)}J'_{(i)}[H_{(i)}]\, dF_{N(i)}^{(k)}$$

$$(9.5.17)\qquad \mathcal{C}_{2N(i)}^{(k)} = \int_{-\infty}^{\infty} (H_{N(i)} - H_{(i)})J'_{(i)}[H_{(i)}]\, d[F_{N(i)}^{(k)} - F_{(i)}^{(k)}]$$

$$(9.5.18)\qquad \mathcal{C}_{3N(i)}^{(k)} = \int_{-\infty}^{\infty} [J_{(i)}(H_{N(i)}^{*}) - J_{(i)}[H_{(i)}] - (H_{N(i)}^{*} - H_{(i)})\, J'_{(i)}[H_{(i)}]]\, dF_{N(i)}^{(k)}$$

$$(9.5.19)\qquad \mathcal{C}_{4N(i)}^{(k)} = \int_{-\infty}^{\infty} [[J_{N(i)}[H_{N(i)}^{*}] - J_{(i)}[H_{N(i)}^{*}]]\, dF_{N(i)}^{(k)}.$$

Now proceeding precisely as in Theorems 8.4.1 and 5.5.1, it can be easily shown that the $\mathfrak{B}$ and $\mathfrak{C}$ terms are all $o_p(N^{-1/2})$. Thus

$$N^{1/2}\left[U_{N,ij}^{(k)} - \sum_{h=1}^{3} A_{hN(i,j)}^{(k)}\right]$$

as well as $N^{1/2}[T_{N(i)}^{(k)} - \mu_{N(i)}^{(k)}] - N^{1/2}[B_{1N(i)}^{(k)} + B_{2N(i)}^{(k)}]$

tend to zero in probability as $n \to \infty$, for each $i, j = 1, \ldots, p$ and $k = 1, \ldots, c$. Hence it suffices to establish the joint asymptotic normality of

$$N^{1/2}\left[\sum_{h=1}^{3} A_{hN(i,j)}^{(k)};\ 1 \le i < j \le p;\ k = 1, \ldots, c\right]$$

and

$$N^{1/2}[B_{1N(i)}^{(k)} + B_{2N(i)}^{(k)};\ i = 1, \ldots, p;\ k = 1, \ldots, c].$$

Now let us denote

$$U_{1N_k,\alpha}^{(i,j)}(X_{i\alpha}^{(k)}, X_{j\alpha}^{(k)}) = J_{(i)}[H_{(i)}(X_{i\alpha}^{(k)})]J_{(j)}[H_{(j)}(X_{j\alpha}^{(k)})], \tag{9.5.20}$$

$$U_{2N_k,r,\alpha}^{(i,j)}(X_{i\alpha}^{(r)}) = \int_{-\infty}^{\infty}\int_{-\infty}^{\infty} [F_{1X_{i\alpha}}^{(r)}(x_i) - F_{[i]}^{(r)}(x_i)] \cdot J'_{(i)}[H_{(i)}]J_{(j)}[H_{(j)}]\, dF_{[i,j]}^{(k)}(x_i, x_j) \tag{9.5.21}$$

where

$$F_{1X_{i\alpha}}^{(r)}(x_i) \quad \text{if} \quad X_{i\alpha}^{(r)} \le x_i \quad \text{and is zero otherwise,} \tag{9.5.22}$$

$$U_{3N_k,r,\alpha}^{(i,j)}(X_{j\alpha}^{(r)}) = \int_{-\infty}^{\infty}\int_{-\infty}^{\infty} [F_{1X_{j\alpha}}^{(r)}(x_j) - F_{[j]}^{(r)}(x_j)] \cdot J'_{(j)}[H_{(j)}]J_{(i)}[H_{(i)}]\, dF_{[i,j]}^{(k)}. \tag{9.5.23}$$

Then, we can write

$$\sum_{h=1}^{3} A_{hn(i,j)}^{(k)} = \sum_{\substack{r=1 \\ r\neq k}}^{c}\left[\lambda_N^{(r)} \cdot \frac{1}{n_r}\sum_{\alpha=1}^{n_r}\{U_{2N_k,k,r}(X_{i\alpha}^{(r)}) + U_{3N_k,r,\alpha}^{(i,j)}(X_{j\alpha}^{(r)})\}\right] + \frac{1}{n_k}\sum_{\alpha=1}^{n_r}[U_{1N_k,\alpha}^{(i,j)}(X_{i\alpha}^{(k)}, X_{j\alpha}^{(k)}) + \lambda_N^{(k)}\{U_{2N_k,k,\alpha}(X_{j\alpha}^{(k)}) + U_{3N_k,k,\alpha}(X_{j\alpha}^{(k)})\}, \tag{9.5.24}$$

$$B_{1N(i)}^{(k)} + B_{2N(i)}^{(k)} = -\sum_{r=1}^{c}\left[\lambda_N^{(r)} \cdot \frac{1}{n_r}\sum_{\alpha=1}^{n_r}\{B_i(X_{i\alpha}^{(r)}) - EB_i(X_{i\alpha}^{(r)})\}\right] + \frac{1}{n_k}\sum_{\alpha=1}^{N_k}\{J_{(i)}[H_{(i)}(X_{i\alpha}^{(k)})] - \lambda_N^{(k)}B_i(X_{i\alpha}^{(k)}) - E[J_{(i)}[H_{(i)}(X_{i\alpha}^{(k)})] - \lambda_N^{(k)}B_i(X_{i\alpha}^{(k)})]\} \tag{9.5.25}$$

where

$$B_i(x_i^{(r)}) = \int_{x_{0i}}^{x_i^{(r)}} J'_{(i)}[H_{(i)}(y)]\, dF_{(i)}^{(r)}(y), \tag{9.5.26}$$

with x_{0i} determined somewhat arbitrarily, say by $H_{(i)}(x_{0i}) = \frac{1}{2}$; E represents the expectation, and $X_{i\alpha}^{(k)}$, $\alpha = 1, \ldots, n_k$ has the distribution $F_i^{(k)}$; $i = 1, \ldots, p$; $k = 1, \ldots, c$.

Hence with $d_{ij}^{(k)}$ and $e_i^{(k)}$, $1 \leq i \leq j \leq p$, $k = 1, \ldots, c$ any real constants, not all zero, we can write

$$\begin{aligned} (9.5.27)\quad & N^{1/2}\Bigg[\sum_{i\leq j=1}^{p}\sum_{k=1}^{c} d_{ij}^{(k)}\{A_{1N(i,j)}^{(k)} + A_{2N(i,j)}^{(k)} + A_{3N(i,j)}^{(k)}\} \\ & \qquad + \sum_{i=1}^{p}\sum_{k=1}^{c} e_i^{(k)}\{B_{1N(i)}^{(k)} + B_{2N(i)}^{(k)}\}\Bigg] \\ & = N^{1/2}\sum_{\substack{r=1\\ r\neq k}}^{c} \lambda_N^{(r)}\Bigg[\frac{1}{n_r}\sum_{\alpha=1}^{n_r}\Bigg[\sum_{i\leq j} d_{ij}^{(k)}\{U_{2N_k,r,\alpha}(X_{i\alpha}^{(e)}) + U_{3N_k,r,\alpha}(X_{j\alpha}^{(r)})\} \\ & \qquad \times \sum_{i=1}^{p} e_i^{(k)}\{B_i(X_{i\alpha}^{(r)}) - EB_i(X_{i\alpha}^{(r)})\}\Bigg]\Bigg] + N^{1/2}\frac{1}{n_k}\sum_{\alpha=1}^{n_k} \\ & \qquad \Bigg[\sum_{i\leq j} d_{ij}^{(k)}\{U_{1N_k,\alpha}^{(i,j)}(X_{i\alpha}^{(k)}, X_{j\alpha}^{(k)}) + \lambda_N^{(k)} U_{2N,k,k\alpha}(x_{i\alpha}^{(k)}) \\ & \qquad + \lambda_N^{(k)} U_{3N_k,k,\alpha}(X_{j\alpha}^{(k)})\} - \sum_{i=1}^{p} e^{(k)}\{J_{(i)}[H_{(i)}(x_{i\alpha}^{(k)})] - \lambda_N^{(k)} B_i(X_{i\alpha}^{(k)}) \\ & \qquad - E\{J_{(i)}[H_{(i)}(X_{i\alpha}^{(k)})] - \lambda_N^{(k)} B_i(X_{\alpha}^{(k)})\}\Bigg]. \end{aligned}$$

These are the c summations. They involve independent samples of identically distributed random variables, each having finite third moment. The proof follows. ◁

To compute the variance covariance terms, we note using (9.5.6) to (9.5.8) that

$$\begin{aligned} (9.5.28)\quad N \operatorname{var}\left[\sum_{h=1}^{3} A_{hN(i,j)}^{(k)}\right] &= \sum_{\substack{r=1\\ r\neq k}}^{c} \lambda_N^{(r)} \operatorname{var}\{U_{2N_k,r,\alpha}^{(i,j)}(X_{i\alpha}^{(r)}) + U_{3N_k,r,\alpha}^{(i,j)}(X_{j\alpha}^{(r)})\} \\ & \quad + \frac{1}{\lambda_N^{(k)}} \operatorname{var}\{U_{1N_k,\alpha}^{(i,j)}(X_{i\alpha}^{(k)}, X_{j\alpha}^{(k)}) + \lambda_N^{(k)} \\ & \quad U_{2N_k,k,\alpha}^{(i,j)}(X_{i\alpha}^{(k)}) + \lambda_N^{(k)} U_{3N_k,k,\alpha}^{(i,j)}(X_{j\alpha}^{(k)})\}. \end{aligned}$$

$$N \operatorname{cov}\left(\sum_{h=1}^{3} A_{hN(i,j)}^{(k)}, \sum_{h=1}^{3} A_{hN(r,s)}^{(q)}\right) = \sum_{\alpha=1}^{3}\sum_{\beta=1}^{3} \operatorname{cov}(A_{\alpha N(i,j)}^{(k)}, A_{\beta N(r,s)}^{(q)}). \tag{9.5.29}$$

The variance-covariance terms of $\{n^{1/2}(B^{(k)}_{1N(i)} + B^{(k)}_{2N(i)}),\ i = 1, \ldots, p;$ $k = 1, \ldots, c\}$ are given in the expressions (5.5.2)–(5.5.5). The other variance-covariance terms can be evaluated analogously.

We shall now obtain the limiting distributions of the statistic $\mathfrak{L}_N$ given by (9.3.25) under the null hypothesis $H_0^{(1)}$ and under a sequence of specific dispersion alternatives to be stated below. In what follows, we assume that for all $k = 1, \ldots, c$, $\lim_{N\to\infty} \lambda_N^{(k)} = \lambda^{(k)}$ exists and is bounded away from 0 and 1. Under $H_0^{(1)}$, it follows from the results of chapter 5 that $|T^{(k)}_{N,i}| = O_p(N^{-1/2})$ for all $i = 1, \ldots, p$ and $k = 1, \ldots, c$; and following routine computations (Exercise 9.5.1) one can show that under $H_0^{(1)}$, asymptotically

$$N \operatorname{cov}\{S^{(k)}_{N,ij}, S^{(q)}_{N,rs}\} \to \left\{\frac{\delta_{kq}}{\lambda^{(k)}} - 1\right\}\nu_{ij,rs} \tag{9.5.30}$$

for all $i, j, r, s = 1, \ldots, p$; $k, q = 1, \ldots, c$, where δ_{kq} is the usual Kronecker delta and ν_{ijrs} is defined as in (9.4.7) with the further simplification that $F^{(k)}_{[i]} \equiv H_{(i)}$, $F^{(k)}_{[i,j,r,s]} \equiv H_{(i,j,r,s)}$ for all $k = 1, \ldots, c;\ i, j, r, s = i, \ldots, p$. Consequently, using Theorems 9.4.1, 9.5.1, and some routine analysis, it can be shown (Exercise 9.5.2) that under $H_0^{(1)}$, $\mathfrak{L}_N$ has asymptotically (unconditionally) the chi-square distribution with $(c - 1)p(p + 1)/2$ degrees of freedom.

We now formulate a class of dispersion alternatives. For the scalar alternatives we let (as in chapter 5)

$$F^{(k)}_{(i)}(x) = F_{(i)}([x - \mu_i]\,\delta_i^{(k)}); \qquad i = 1, \ldots, p; \quad k = 1, \ldots, c, \tag{9.5.34}$$

where under $H_0^{(1)}$, all $\delta_1^{(k)}$ are equal. Next, we consider bivariate dependence functions $\Omega^{(k)}_{ij}$ defined as

$$\Omega^{(k)}_{ij} = F^{(k)}_{(i,j)}(x, y)/F^{(k)}_{(i)}(x)F^{(k)}_{(j)}(y) \tag{9.5.35}$$

for all $i \neq j = 1, \ldots, p$; $k = 1, \ldots, c$. Clearly, $\Omega^{(k)}_{ij}$ is a function of $F^{(k)}_{(i)}$ and $F^{(k)}_{(j)}$, and will be equal to unity if $X^{(k)}_{i\alpha}$ and $X^{(k)}_{j\alpha}$ are statistically independent. Various properties of such dependence functions have been studied in the previous chapter, and in nonparametric statistics they appear to have some advantages over the other measures. We denote by Ω_{ij} the same function (i.e., (5.3)) for the bivariate c.d.f. $F_{(i,j)}(x, y)$, and relate

$$\Omega^{(k)}_{ij} = \Omega_{ij} + \omega^{(k)}_{ij} \quad \text{for} \quad i \neq j = 1, \ldots, p; \quad k = 1, \ldots, c. \tag{9.5.36}$$

(9.5.34), (9.5.35), and (9.5.36) together constitute our desired class of alternatives. Now we can rewrite (9.2.7) as

$$\theta^{(k)}_{ij} - \int_{-\infty}^{\infty}\int_{-\infty}^{\infty} [F^{(k)}_{[i,j]}(x, y) - F^{(k)}_{[i]}(x)F^{(k)}_{(j)}(y)] \times J'_{(i)}[H_{(i)}(x)]J'_{(j)}[H_{(j)}(y)]\, dH_{(i)}(x)\, dH_{(j)}(y) \tag{9.5.37}$$

for all $i, j = 1, \ldots, p$; $k = 1, \ldots, c$. From (9.5.35) and (9.5.36) we note that $\theta_{ij}^{(k)}$ measures some functional of the dependence function $\Omega_{ij}^{(k)}$, for all $i, j = 1, \ldots, p$, and $k = 1, \ldots, c$.

For the study of the asymptotic non-null distribution of $\mathcal{L}_N$, we shall consider a sequence of alternatives converging to the null hypothesis in such a manner that the power of the test based on $\mathcal{L}_N$ lies in the open interval $(\epsilon, 1)$. Thus we replace $\delta_i^{(k)}$ in (9.5.34) by a sequence $\{1 + N^{-\frac{1}{2}}\beta^{(k)}\}$ where $\beta_1^{(k)}$, $i = 1, \ldots, p$; $k = 1, \ldots, c$ are real finite constants. Also, we replace (9.5.37) by

$$(9.5.38) \quad F_{[i,j]}^{(k)}(x, y) - F_{[i]}^{(k)}(x)F_{[j]}^{(k)}(y) = [H_{(i,j)}(x, y) - H_{(i)}(x)H_{(j)}(y)] + N^{-\frac{1}{2}}\xi_{ij}^{(k)}(x, y)$$

for all $i \neq j = 1, \ldots, p$; $k = 1, \ldots, c$ where $\xi_{ij}^{(k)}(x, y)$ are real and finite-valued functions. We shall denote such a sequence of alternative hypotheses by $\{H_N\}$. Thus, on defining

$$(9.5.39) \quad \zeta_{ij}^{(k)} = \int_{-\infty}^{\infty}\int_{-\infty}^{\infty} \xi_{ij}^{(k)}(x, y)J'_{(i)}[H_{(i)}(x)]J'_{(j)}[H_{(j)}(y)]\, dH_{(i)}(x)\, dH_{(j)}(y)$$

$$\text{for} \quad i, j = 1, \ldots, p; \quad k = 1, \ldots, c;$$

$$(9.5.40) \qquad \boldsymbol{\zeta} = \{\zeta_{ij}^{(k)}, i \leq j = 1, \ldots, p\}; \qquad k = 1, \ldots, c,$$

and following some routine analysis, we arrive at the following theorem.

Theorem 9.5.2. *Under $\{H_N\}$, $\mathcal{L}_N$ has asymptotically a noncentral chi-square distribution with $(c - 1)p(p + 1)/2$ degrees of freedom and the noncentrality parameter*

$$(9.5.41) \qquad \Lambda = \sum_{k=1}^{c} \lambda_k[\boldsymbol{\zeta}^{(k)}\boldsymbol{\nu}^{*-1}\boldsymbol{\zeta}^{(k)\prime}]$$

where $\boldsymbol{\nu}^{*-1}$ is the inverse of $\boldsymbol{\nu}^*$, defined in (9.4.8).

In the parametric case, the likelihood ratio (l.r.) test for this problem is given in Anderson (1958). He has only considered the distribution of the l.r. criterion under the null hypothesis. However, using Wald's (1943) methods, one can readily find out the asymptotic non-null distribution of the l.r. criterion under $\{H_N\}$. If $\boldsymbol{\Sigma}^{(k)}$, $k = 1, \ldots, c$ are the c dispersion matrices, then under $\{H_N\}$, we will have

$$(9.5.42) \qquad \boldsymbol{\Sigma}^{(k)} = \boldsymbol{\Sigma}_N^{(k)} = \boldsymbol{\Sigma} + N^{\frac{1}{2}}\mathbf{T}^{(k)}, \qquad k = 1, \ldots, c,$$

where $\mathbf{T}^{(k)}$, $k = 1, \ldots, c$ are all real and finite. Now defining

$$(9.5.43) \quad \gamma_{ij,i'j'} = \int_{-\infty}^{\infty}\int_{-\infty}^{\infty}\int_{-\infty}^{\infty}\int_{-\infty}^{\infty} x_i x_j x_{i'} x_{j'}\, dF_{(i,j,i',j')}(x_i, x_j, x_{i'}, x_{j'}) - \sigma_{ij}\sigma_{i'j}$$

and adopting the suffixing system of (9.3.21), we write

$$\Gamma = ((\gamma_{rs}))_{r,s=1,\ldots,p(p+1)/2}$$

(9.5.44)
$$\mathbf{T}_0^{(k)} = \{\tau_{ij}^{(k)}, i \leq j = 1, \ldots, p\}, \quad k = 1, \ldots, c.$$

Then, we have the following theorem, the proof of which is given as Exercise 9.5.3.

Theorem 9.5.3. *Under* (9.5.42), *the l.r. test criterion has asymptotically the noncentral chi-square distribution with* $(c - 1)p(p + 1)/2$ *degrees of freedom, and the noncentrality parameter*

(9.5.45)
$$\Delta = \sum_{k=1}^{c} \lambda_k [\mathbf{T}_0^{\prime(k)} \Gamma^{-1} \mathbf{T}_0^{\prime(k)}].$$

Comparison of (9.5.41) and (9.5.43) will yield the asymptotic relative efficiency of the $\mathfrak{L}_N$ test with respect to the l.r. test. Evidently, this depends on (i) $J_i(u)$, $i = 1, \ldots, p$; (ii) $\{H_N\}$, and (iii) the unknown c.d.f. $F(\mathbf{x})$.

Among the possible members of the $\mathfrak{L}_N$ tests, the two most appealing are (i) the Spearman type of test for which $J_{(i)}(u) = \sqrt{12}\,(u - \frac{1}{2})$, and (ii) the normal scores test for which $J_{(i)}(u)$ is the expected value of the ith order statistic of a sample of size N form a standardized normal c.d.f. For the normal scores test, it can be shown that if $F(\mathbf{x})$ is also normal, then (9.5.41) and (9.5.45) are equal, that is, the two tests are asymptotically power equivalent.

9.6. NONPARAMETRIC TESTS FOR $H_0^{(2)}$

Under $H_0^{(2)}$ defined in (9.2.20) and (9.2.21), the basic permutation argument of section 9.3 is no longer tenable, and so, it is difficult to derive the permutationally distribution-free tests. However, we shall derive a class of rank order tests by centering the observations at the respective estimates of the location parameters (as is generally done in the univariate scale problems, see for example Sukhatme (1958a), Raghavachari (1965a)), and working with these centered observations. Due to centering, the independence is vitiated, and as a result, one is faced with the question: Under what conditions is the test based on the centered observations asymptotically equivalent to the test based on the observations centered at the true locations? In this section we study this problem. For this purpose, we assume that $F^{(k)}(\mathbf{x})$ is absolutely continuous having the continuous density function $f^{(k)}(\mathbf{x})$; $k = 1, \ldots, c$, and the following conditions hold:

A. The densities $f^{(k)}(\mathbf{x})$ of $F^{(k)}(\mathbf{x})$, $k = 1, \ldots, c$ are diagonally symmetric about their location parameters.

B. $J_{(i)}(u) = \Psi^{-1}(u)$ where ψ is symmetric about zero.

C. The functions $(d/dx)J_{(i)}[F_{(i)}^{(k)}(x)]$ are bounded for each $i = 1, \ldots, p$ and $k = 1, \ldots, c$.

Let now $\hat{\delta}_i^{(k)}$ be a consistent estimator of $\delta_i^{(k)}$ defined in (9.2.20), and denote

$$X_{i\alpha}^{*(k)} = X_{i\alpha}^{(k)} - \hat{\delta}_i^{(k)}: \; i = 1, \ldots, p; \; k = 1, \ldots, c; \; \alpha = 1, \ldots, n_k. \tag{9.6.1}$$

In addition to the notations of the previous sections, we introduce the following:

$$F_{N(i)}^{*(k)}(x) = \frac{1}{n_k}\,[\text{number of } X_{i\alpha}^{(k)} \text{ such that } (X_{i\alpha}^{(k)} - \hat{\delta}_i^{(k)}) \le x; \; \alpha = 1, \ldots, n_k], \quad i = 1, \ldots, p; \; k = 1, \ldots, c, \tag{9.6.2}$$

$$H_{N(i)}^{*}(x) = \frac{N}{N+1}\sum_{k=1}^{c} \lambda_N^{(k)} F_{N(i)}^{*(k)}(x), \qquad i = 1, \ldots, p; \tag{9.6.3}$$

$$F_{N(i,j)}^{*(k)}(x, y) = \frac{1}{n_k}\,[\text{number of } (X_{i\alpha}^{(k)}, X_{j\alpha}^{(k)}) \text{ such that } (X_{i\alpha}^{(k)} - \hat{\delta}_i^{(k)}, X_{j\alpha}^{(k)} - \hat{\delta}_j^{(k)}) \le (x, y), \alpha = 1, \ldots, n_k,] \quad i, j = 1, \ldots, p. \tag{9.6.4}$$

$$H_{N(i,j)}^{*} = \sum_{k=1}^{c} \lambda_N^{(k)} F_{N(i,j)}^{*(k)}. \tag{9.6.5}$$

Now consider the statistics (where $H_N^{**} = \dfrac{N}{N+1}\, H_N^{*}$)

$$U_{N(i,j)}^{*(k)} = \int_{-\infty}^{\infty}\int_{-\infty}^{\infty} J_{N(i)}[H_{N(i)}^{**}(x_i)] J_{N(j)}[H_{N(j)}^{**}(x_j)]\, dF_{N(i,j)}^{*(k)}, \quad i \ne j = 1, \ldots, p; \; k = 1, \ldots, c, \tag{9.6.6}$$

$$T_{N,i}^{*(k)} = \int_{-\infty}^{\infty} J_{N(i)}[H_{(i)}^{**}(x_i)]\, dF_{N(i)}^{(k)}(x_i); \quad i = 1, \ldots, p; \; k = 1, \ldots, c. \tag{9.6.7}$$

Finally, concerning $\hat{\delta}_i^{(k)}$'s, we shall assume that

$$N^{1/2}\,|\hat{\delta}_i^{(k)} - \delta_i^{(k)}| = O_p(1) \quad \text{for all} \quad i = 1, \ldots, p; \quad k = 1, \ldots, c. \tag{9.6.8}$$

Then the main theorem of this section is the following.

Theorem 9.6.1. *Under the assumptions A, B, and C, the random variables* $N^{1/2}[(T_{N,ij}^{(k)} - \mu_{N,ij}^{(k)}), \; (T_{N,i}^{(k)} - \mu_{N,i}^{(k)}) i \le j = 1, \ldots, p, \; k = 1, \ldots, c]$ *and* $N^{1/2}[(U_{N,ij}^{(k)} - \mu_{N,ij}^{(k)}), (T_{N,i}^{*(k)} - \mu_{N,i}^{(k)}), i \le j = 1, \ldots, p, k = 1, \ldots, c]$ *have the same limiting distribution*

The proof of this theorem rests on the following two lemmas, of which the first is due to Raghavachari (1965a) and the second is a slight generalization of Raghavachari (1965a).

Lemma 9.6.1. (*Raghavachari.*) *If the assumption* (*C*) *holds, then* $(d/dx)J[H_{(i)}(x)]$ *is also bounded, where* $H_{(i)}(x)$ *is given by* (9.2.2).

Lemma 9.6.2. *Let* $(X_1, Y_1), \ldots, (X_N, Y_N)$ *be i.i.d.r.v.'s with continuous c.d.f.* $F(x - \xi, y - \eta)$ *where F is assumed to be symmetric about* (0, 0). *Let* $F_1(x)$ *and* $F_2(y)$ *be the marginal c.d.f.'s of* X_i *and* Y_i *respectively. Let* $\hat{\xi}(X_1, \ldots, X_N)$ *and* $\hat{\eta}(Y_1, \ldots, Y_N)$ *be consistent estimators of* ξ *and* η *respectively such that* $N^{1/2}(\hat{\xi} - \xi)$ *as well as* $N^{1/2}(\hat{\eta} - \eta)$ *is bounded in probability as* $N \to \infty$. *Assume that there exists a function* $M(x, y) = M_1(x)M_2(y)$ *such that the following conditions are satisfied.*

$$\text{(i)} \qquad E[M_1(x)]^2 < \infty, \quad E[M_2(y)]^2 < \infty,$$

$$\text{(ii)} \qquad M_1'(x) = \frac{\partial M_1(x)}{\partial x} \quad \textit{and} \quad M_2'(y) = \frac{\partial M_2(y)}{\partial y} \quad \text{exist,}$$

$$\text{(iii)} \qquad |M_1'(x - t_1)| \le K_1 T_1(x), \qquad |M_2'(y - t_2)| \le K T_2(y).$$

uniformly in t_1 *and* t_2 *for* $|t_i| \le c_i$; $i = 1, 2$; (c_i *and* K_i, $i = 1, 2$ *are constants*).

$$\text{(iv)} \qquad E[T_1(X)]^2 < \infty, \quad E[T_2(Y)]^2 < \infty$$

where X and Y have the distribution functions $F_1(x - \xi)$ *and* $F_2(y - \eta)$ *respectively.*

$$\text{(v)} \qquad E[M_1'(x - \xi)] = 0, \qquad E[M_2'(y - \eta)] = 0.$$

Then

$$Q_N = N^{-1/2} \sum_{i=1}^{N} \{M_1(x_i - \xi) \cdot M_2(y_i - \eta) - M_1(X_i - \hat{\xi}) \cdot M_2(y_i - \hat{\eta})\} \xrightarrow{p} 0$$

Proof of Lemma 9.6.2: Without loss of generality we can take $\xi = \eta = 0$. Then

$$\text{(9.6.9)} \quad Q_N = N^{-1/2} \sum_{i=1}^{N} \{M_1(X_i)M_2(Y_i) - M_1(X_i - \hat{\xi}) \cdot M_2(Y_i - \hat{\eta})\}$$

$$= -\frac{t_1}{N} \sum_{i=1}^{N} M_2(Y_i) M_1'\left(X_i - \frac{\theta_1 t_1}{\sqrt{N}}\right) - \frac{t_2}{N} \sum_{i=1}^{N} M_1(X_i)$$

$$\times M_2'\left(Y_i - \frac{\theta_2 t_1}{\sqrt{N}}\right)$$

after neglecting terms of order $O(1/N)$. Under the assumptions (i) to (iv) of the lemma, it is easy to note (by applying Chebychev's inequality) that

$$(9.6.10)\quad \frac{t_1}{N}\sum_{i=1}^{N}\left[M_2(Y_i)M_1'\left(X_i-\frac{\theta_1 t_1}{\sqrt{N}}\right)-E\left\{M_2(Y_i)M_1'\left(X_i-\frac{\theta_1 t_1}{\sqrt{N}}\right)\right\}\right]^2\to 0.$$

Furthermore, by the dominated convergence theorem,

$$(9.6.11)\qquad E\left\{M_2(Y_i)M_1'\left(X_i-\frac{\theta_1 t_1}{\sqrt{n}}\right)\right\}\to E[M_1'(X_i)M_2(Y_i)]$$

uniformly in t_1, for $|t_1|\le c_1$; and since under the assumed conditions, $E\{M_1'(x_i)M_2(y_i)\}=0$ it follows that

$$\sup_{|t_1|\le c_1}\left|\frac{t_1}{N}\sum_{i=1}^{N}M_2(Y_i)M_1'\left(X_i-\frac{\theta_1 t_1}{\sqrt{N}}\right)\right|\xrightarrow{p}0.$$

Hence the first factor on the right-hand side $\xrightarrow{p} 0$ as $N\to\infty$. Similarly the second factor $\xrightarrow{p} 0$ as $N\to\infty$. Hence $Q_N \xrightarrow{p} 0$ as $N\to\infty$.

Proof of the Theorem: Without any loss of generality we can assume that $\boldsymbol{\delta}_k=0$, $k=1,\ldots,c$. The proof of the theorem is accomplished by showing that (*a*) $N^{1/2}(U^{*(k)}_{N,ij}-U^{(k)}_{N,ij})\xrightarrow{p}0$ as $N\to\infty$ for each $i,j=1,\ldots,p$ and $k=1,\ldots,c$ and (*b*) $N^{1/2}(T^{*(k)}_{N(i)}-T^{(k)}_{N(i)})\xrightarrow{p}0$ as $N\to\infty$ for each $i=1,\ldots,p$, and $k=1,\ldots,c$.

To prove (*a*), note that proceeding as in Theorem 9.5.1 we can rewrite $U^{*(k)}_{N(i,j)}$ as

$$U^{*(k)}_{N,ij}=A^{*(k)}_{N(i,j)}+A^{*(k)}_{2N(i,j)}+A^{*(k)}_{3N(i,j)}+\sum_{r=1}^{5}\mathcal{B}^{*(k)}_{rN(i,j)},$$

$$A^{*(k)}_{1N(i,j)}=\int_{-\infty}^{\infty}\int_{-\infty}^{\infty}J_{(i)}[H_{(i)}]J_{(j)}[H_{(j)}]\,dF^{*(k)}_{N[i,j]},$$

$$A^{*(k)}_{2N(i,j)}=\int_{-\infty}^{\infty}\int_{-\infty}^{\infty}[H^*_{N(i)}-H_{(i)}]J'_{(i)}[H_{(i)}]J_{(j)}[H_{(j)}]\,dF^{(k)}_{[i,j]},$$

$$A^{*}_{3N(i,j)}=\int_{-\infty}^{\infty}\int_{-\infty}^{\infty}[H^*_{N(j)}-H_{(j)}]J'_{(j)}[H_{(j)}]J_{(i)}[H_{(i)}]\,dF^{(k)}_{[i,j]},$$

and the $\mathcal{B}^*$ terms are all $o_p(N^{-1/2})$.

Proceeding again as in Theorem 9.5.1, we obtain

$$N^{\frac{1}{2}}(A^{*(k)}_{1N(i,j)} + A^{*(k)}_{2N(i,j)} + A^{*(k)}_{3N(i,j)}) - N^{\frac{1}{2}}(A^{(k)}_{1N(i,j)} + A^{(k)}_{2N(i,j)} + A^{(k)}_{3N(i,j)})$$

$$= N^{\frac{1}{2}} \sum_{\substack{r=1 \\ r \neq k}}^{c} \Bigg[\lambda_N^{(k)} \cdot \frac{1}{n_r} \sum_{\alpha=1}^{n_r} \{[U^{(i,j)}_{2Nkr,\alpha}(X^{(r)}_{i\alpha} - \hat{\nu}^{(r)}_i) + U^{(i,j)}_{3Nkr,\alpha}(X^{(r)}_{j\alpha} - \hat{\nu}^{(r)}_j)]$$

$$- [U^{(i,j)}_{2Nk,r}(X^{(r)}_{i\alpha}) + U^{(i,j)}_{3Nkr,\alpha}(X^{(r)}_{j\alpha})]\} \Bigg] + N^{\frac{1}{2}} \frac{1}{n_k} \sum_{\alpha=1}^{n_k}$$

$$\times [U^{(i,j)}_{1Nk,\alpha}(X^{(k)}_{i\alpha} - \hat{\nu}^{(k)}_j, X^{(k)}_{j\alpha} - \hat{\nu}^{(k)}_j)$$

$$+ \lambda_N^{(k)} \{U^{(i,j)}_{2Nkk,\alpha}(X^{(k)}_{i\alpha} - \hat{\nu}^{(k)}_i) + U^{(i,j)}_{3Nkk,\alpha}(X^{(k)}_{j\alpha} - \hat{\nu}^{(k)}_j)\}$$

$$- U^{(i,j)}_{1Nk,\alpha}(X^{(k)}_{i\alpha}, X^{(k)}_{j\alpha}) - \lambda_N^{(k)} \{U^{(i,j)}_{2Nkk,\alpha}(X^{(k)}_{i\alpha}) + U^{(i,j)}_{3Nkk}(X^{(k)}_{j\alpha})\}]$$

$$= \sum_{\substack{r=1 \\ r \neq k}}^{c} [L^{(1)}_{N,r} + L^{(2)}_{N,r}] + L^{(1)}_{N,k} + L^{(2)}_{N,k} + L^{(3)}_{N,k},$$

where

$$L^{(1)}_{N,r} = N^{-\frac{1}{2}} \sum_{\gamma=1}^{n_r} \{[U^{(i,j)}_{2Nk,r,\alpha}(X^{(r)}_{i\alpha} - \hat{\nu}^{(r)}_i) - U^{(i,j)}_{2Nk,r,\alpha}(X^{(r)}_{i\alpha})]\}$$

$$r = 1, \ldots, c; \quad r \neq k,$$

$$L^{(2)}_{N,r} = N^{-\frac{1}{2}} \sum_{\alpha=1}^{n_r} \{[U^{(i,j)}_{3Nk,r,\alpha}(X^{(r)}_{j\alpha} - \hat{\nu}^{(r)}_j) - U^{(i,j)}_{3Nk,r,\alpha}(X^{(r)}_{j\alpha})]\},$$

$$r = 1, \ldots, c; \quad r \neq k.$$

$$L^{(1)}_{N,k} = N^{-\frac{1}{2}} \sum_{\alpha=1}^{n} \{[U^{(i,j)}_{1N\alpha,k}(X^{(k)}_{i\alpha} - \hat{\nu}^{(k)}_i, X^{(k)}_{j\alpha} - \hat{\nu}^{(k)}_j) - U^{(i,j)}_{1N\alpha,k}(X^{(k)}_{i\alpha}, X^{(k)}_{j\alpha})]\},$$

$$L^{(2)}_{N,k} = N^{-\frac{1}{2}} \sum_{\alpha=1}^{n_k} \{U^{(i,j)}_{2Nkk,\alpha}(X^{(k)}_{i\alpha} - \hat{\nu}^{(k)}_i) - U^{(i,j)}_{2Nkk,\alpha}(X^{(k)}_{i\alpha})\},$$

$$L^{(3)}_{N,k} = N^{-\frac{1}{2}} \sum_{\alpha=1}^{n_k} \{U^{(i,j)}_{3Nkk,\alpha}(X^{(k)}_{j\alpha} - \hat{\nu}^{(k)}_j) - U^{(i,j)}_{3Nkk,\alpha}(X^{(k)}_{j\alpha})\}.$$

Consider first

$$L^{(1)}_{N,r} = N^{-\frac{1}{2}} \sum_{\alpha=1}^{n_r} \iint [F^{(r)}_{1X_{i\alpha}}(X_i - \hat{\nu}^{(r)}_i) - F^{(r)}_{1X_{i\alpha}}(X_i)] \times J'_{(i)}[H_i(x_i)] J_{(j)}[H_{(j)}(x_j)] \, dF^{(r)}_{[i,j]}(x_i, x_j).$$

Define

$$B_{(r)}(X_{i\alpha}) = \iint F^{(r)}_{1X_{i\alpha}}(x_i) J'_{(i)}[H_i(x_i)] J_{(j)}[H_j(x_j)] f^{(r)}_{ij}(x_i, x_j)\, dx_i\, dx_j$$

$$|B'_{(r)}(x_{i\alpha} - t)| = |\, J'_{(i)}[H_i(x_i - t)]| \left| \int_{-\infty}^{\infty} J_j(H_j(x_j)) f_j(x_j)\, dx_j \right|$$

$$\leq K\, |J'_{(i)}[H_i(x_i - t)]| \leq K'$$

by virtue of Lemma 1. Furthermore $EB'_{(r)}(X_{i\alpha}) = 0$. Hence, by Lemma 2, $L^{(1)}_{N,r} \xrightarrow{p} 0$ as $n \to \infty$. Similarly $L^{(2)}_{N}$, $L^{(2)}_{N,k}$, $L^{(3)}_{N,k}$ and $L^{(1)}_{N,k}$ all $\xrightarrow{p} 0$ as $N \to \infty$. Hence $N^{1/2}[U^{*(k)}_{N(i,j)} - \mu^{(k)}_{N(i,j)};\ i, j = 1, \ldots, p;\ k = 1, \ldots, c]$ and $N^{1/2}[U^{(k)}_{N(i,j)} - \mu^{(k)}_{N(i,j)}]$ have the same limiting distributions. The proof of (*b*) can be tackled in a similar manner and is given as Exercise 9.6.1. ◁

9.7. TESTS FOR $H_0^{(3)}$

In chapter 5 the problem of testing the equality of location parameters was considered under the assumption that the covariance matrices were the same. The tests of this assumption have been considered in the previous sections. In this section we sketch very briefly the problem of testing simultaneously the identity of location parameters and covariance matrices. Here also, since the null hypothesis reduces to the identity of the c c.d.f.'s, the basic permutation arguments of section 9.3 will remain valid. Thus, we work with the set of statistics $\{T^{(k)}_{N,ij}$ and $T^{(i)}_{N,i}$ for $i \leq j = 1, \ldots, p, k = 1, \ldots, c\}$ or equivalently with the set of statistics

$$U^{(k)}_{N,ij} = \frac{1}{n_k} \sum_{\alpha=1}^{n_k} E^{(i)}_{N,R^{(k)}_{i\alpha}} E^{(j)}_{N,R^{(k)}_{j\alpha}}$$

for $k = 1, \ldots, c$, $i \leq j = 1, \ldots, p$ and

$$T^{(k)}_{N,i} = \frac{1}{n_k} \sum_{\alpha=1}^{n_k} E^{(i)}_{N,R_{i\alpha}} \quad \text{for} \quad i = 1, \ldots, p, \quad k = 1, \ldots, c.$$

Write

$$\mathbf{Q}'_k = (\mathbf{T}^{(k)'}_N, U^{(k)'}_N), \qquad k = 1, \ldots, c,$$

where $\mathbf{T}^{(k)}_N = (T^{(k)}_{N,1}, \ldots, T^{(k)}_{N,p})$ and $\mathbf{U}^{(k)}_N = (U^{(k)}_{N,ij},\ 1 \leq i \leq j \leq p)$, $k = 1, \ldots, c$; and let $\mathbf{Q}' = (\mathbf{Q}_1, \ldots, \mathbf{Q}_c)$. The permutation covariance matrix of $\mathbf{Q}'$ is then

$$\left(\left(\frac{N\delta_{kq} - n_k}{n_k(N-1)}\right)\right)_{k,q=1,\ldots,c} \otimes \mathbf{W}^*(\mathbf{R}_N)$$

where

$$\mathbf{W}^*(\mathbf{R}_N) = \begin{pmatrix} \mathbf{V}(\mathbf{R}_N) & \mathbf{Z}(\mathbf{R}_N) \\ \mathbf{Z}'(\mathbf{R}_N) & \mathbf{W}(\mathbf{R}_N) \end{pmatrix},$$

$\mathbf{V}(\mathbf{R}_N)$ is defined by (5.4.7) and (5.4.9),

$$\mathbf{Z}(\mathbf{R}_N) = ((z_{iji'}(\mathbf{R}_N)))_{i \le j, i'=1,\ldots,p}$$

$$z_{iji'}(\mathbf{R}_N) = \frac{1}{N}\sum_{k=1}^{c}\sum_{\alpha=1}^{n_k} E^{(i)}_{N,R^{(k)}_{i\alpha}} E^{(j)}_{N,R^{(k)}_{j\alpha}} E^{(i')}_{N,R^{(k)}_{i'\alpha}} - \bar{E}^{(i')}_N v_{ij}(\mathbf{R}_N),$$

$\bar{E}^{(i)}_N$ and $v_{ij}(\mathbf{R}_N)$ are defined by (5.4.7) and (5.4.8),

$$\mathbf{W}(\mathbf{R}_N) = ((w_{iji'j'}(\mathbf{R}_N)))_{i \le j, i' \le j'=1,\ldots,p},$$

and

$$w_{iji'j'}(\mathbf{R}_N) = \frac{1}{N}\sum_{k=1}^{c}\sum_{\alpha=1}^{n_k} E^{(i)}_{N,R^{(k)}_{i\alpha}} E^{(j)}_{N,R^{(k)}_{j\alpha}} E^{(i')}_{N,R^{(k)}_{i'\alpha}} E^{(j')}_{N,R^{(k)}_{j'\alpha}} - v_{ij}(\mathbf{R}_N)v_{i'j'}(\mathbf{R}_N).$$

Then, proceeding as in section 5.4, we consider the following quadratic form as a test statistic:

$$L^*_N = \sum_{k=1}^{c} n_k \mathbf{Q}'_k[\mathbf{W}^*(\mathbf{R}_N)]^{-1}\mathbf{Q}_k$$

The small sample permutation distribution theory is the same as for the hypothesis $H_0^{(1)}$. The large-sample permutation distribution is again chi-square with $(c - 1)p(p + 3)/2$ D.F. The asymptotic non-null distribution can be studied as in the previous sections. The details are omitted.

EXERCISES

Section 9.3

9.3.1. Prove the assertions (9.3.14) and (9.3.15).

Section 9.4

9.4.1. Complete the proof of Theorem 9.4.1.

Section 9.5

9.5.1. Prove the assertion (9.5.33).

9.5.2. Prove that under $H_0^{(1)}$, $\mathfrak{L}_n$ has asymptotically, as $n \to \infty$ the chi-square distribution with $(c - 1)p(p + 1)/2$ degrees of freedom.

9.5.3. Complete the proofs of Theorems 9.5.2 and 9.5.3.

Section 9.6

9.6.1. Prove that $n^{1/2}(T^{*(k)}_{n(i)} - T^{(k)}_{n(i)}) \xrightarrow{p} 0$ as $n \to \infty$ for each $i = 1, \ldots, p$ and $k - 1, \ldots, c$.

Section 9.7

9.7.1. Using the ideas of section 9.7, develop completely the test procedures for $H_0^{(3)}$.

9.7.2. Specialize the results of chapter 9 to the problems of testing $H_0^{(1)}$, $H_0^{(2)}$, and $H_0^{(3)}$ for the case of two samples.

PRINCIPAL REFERENCES

Bhuchongkul (1964), Blomquist (1950), Nandi and Sen (1963), Puri, Sen and Gokhale (1970), Raghavachari (1965a), Sen (1966c), Sen and Puri (1968).

CHAPTER 10

Appendices

10.1. INTRODUCTION

In chapters 3 (section 6), 4 (section 4), 7 (section 3), 8 (section 4) and 9 (section 5), while considering the proofs of the theorems on the asymptotic distributions of various rank order statistics, we postponed the treatment of the higher-order terms. These higher-order terms will be dealt here in a unified manner. As indicated in section 3.6, we shall mainly follow the treatment of Chernoff and Savage (1958) with some modifications to make the proofs comparatively simple. We follow the order in which these higher-order terms appeared in earlier chapters.

10.2. APPENDIX 1: THE TWO-SAMPLE CASE

We start with the univariate case of section 3.6, and supply *the proof of Lemma* 3.6.3.

We state (without proof) first the following elementary results used repeatedly in this section; the notations are the same as in section 3.6.

10.2.1. Some Elementary Results

(10.2.1) $$H \geq \lambda_N F \geq \lambda_0 F.$$

(10.2.2) $$H \geq (1 - \lambda_N)G \geq \lambda_0 G.$$

(10.2.3) $$1 - F \leq (1 - H)/\lambda_N \leq (1 - H)/\lambda_0.$$

(10.2.4) $$1 - G \leq (1 - H)/(1 - \lambda_N) \leq (1 - H)/\lambda_0.$$

(10.2.5) $$F(1 - F) \leq H(1 - H)/\lambda_N^2 \leq H(1 - H)/\lambda_0^2.$$

(10.2.6) $$G(1 - G) \leq H(1 - H)/(1 - \lambda_N)^2 \leq H(1 - H)/\lambda_0^2.$$

(10.2.7) $$dH \geq \lambda_N\, dF \geq \lambda_0\, dF.$$

(10.2.8) $$dH \geq (1 - \lambda_N)\, dG \geq \lambda_0\, dG.$$

Let (a_N, b_N) be the interval S_{N_ϵ}, where

$$S_{N_\epsilon} = \{x\colon H(x)[1 - H(x)] > \xi_\epsilon \lambda_0/N\}. \tag{10.2.9}$$

Then, there exists an $\xi_\epsilon > 0$ independent of F, G, and λ_N such that

$$P\{X_i \in S_{N_\epsilon}, Y_j \in S_{N_\epsilon}, i = 1, \ldots, m, j = 1, \ldots, n\} \geq 1 - \epsilon. \tag{10.2.10}$$

To begin with the proof of Lemma 3.6.3, let us consider C_{1N}.

$$|C_{1N}| \leq \frac{K}{N} \cdot \frac{1}{m} \left| \sum_{i=1}^{m} J'[H(X_i)] \right| \leq \frac{K}{mN} \sum_{i=1}^{m} [H(X_i)[1 - H(X_i)]]^{\delta - 3/2}. \tag{10.2.11}$$

We may assume $\delta < \frac{1}{2}$. Then using (10.2.5), we obtain

$$|C_{1N}| \leq \frac{K}{mN} \sum_{i=1}^{m} [F(X_i)\{1 - F(X_i)\}]^{\delta - 3/2}.$$

Now since $[F(X_i)\{1 - F(X_i)\}]^{\delta - 3/2}$ has a finite moment of order $2/(3 - \delta)$, we obtain by using Marcinkiewicz's theorem (cf. Loève, 1963, pp. 242–243) that

$$C_{1N} = o_p\left(\frac{1}{m^2} N^{3/2 - \delta/2}\right) = o_p(N^{-1/2}).$$

Next consider C_{2N}

$$C_{2N} = \int_{-\infty}^{\infty} [H_N(x) - H(x)] J'[H(x)]\, d(F_m(x) - F(x)].$$

Using Theorem 2.11.10 (chapter 2), it follows that for any δ': $0 < \delta' < \frac{1}{2}$, and for every $\epsilon > 0$, there exists a finite constant $c(\epsilon, \delta') > 0$ such that

$$P\left[\sup_x \frac{N^{1/2} |H_N(x) - H(x)|}{\{H(x)[1 - H(x)]\}^{\delta' - 1/2}} > C(\epsilon, \delta\,)\right] < \epsilon \tag{10.2.12}$$

for all F, G, and λ_N. Hence with probability $> 1 - \epsilon$, we have

$$\begin{aligned} &\sup_x N^{1/2} |\{H_N(x) \quad H(x)\} J'[H(x)]| \\ &\quad \leq K\{H(x)[1 - H(x)]\}^{\delta - \delta' - 1} C(\epsilon, \delta') = KC(\epsilon, \delta')\{H(x)[1 - H(x)]\}^{\delta^* - 1} \end{aligned} \tag{10.2.13}$$

where $\delta^* = \delta - \delta' > 0$.

Thus, to prove $|C_{2N}| = o_p(N^{-1/2})$, it suffices to show that

$$\int_0^1 \{H(x)[1 - H(x)]\}^{\delta^* - 1}\, dF_m(x) \xrightarrow{p} \int_0^1 \{H(x)[1 - H(x)]\}^{\delta^* - 1}\, dF(x). \tag{10.2.14}$$

The result follows by noting that

$$\int_0^1 \{H(x)[1 - H(x)]\}^{\delta^* - 1}\, dF_m(x) \leq \frac{K}{m} \sum_{i=1}^{m} \{H(X_i)[1 - H(X_i)]\}^{\delta^* - 1}$$

and applying the law of large numbers to the right-hand side. We now consider

$$C_{3N} = \int_{-\infty}^{\infty} \left\{ J\left[\frac{N}{N+1} H_N(x)\right] - J[H(x)] \right.$$

$$\left. - \left(\frac{N}{N+1} H_N(x) - H(x)\right) J'[H(x)] \right\} dF_m(x).$$

Let $Z_{N1} < \cdots < Z_{NN}$ denote the ordered observations $(X_1, \ldots, X_m, Y_1, \ldots, Y_n)$, and let

$$\lambda_{iN} = J[i/(N+1)] - J[H(Z_i)] - \left(\frac{i}{N+1} - H(X_i)\right) J'[H(Z_i)]. \tag{10.2.15}$$

Then

$$|C_{3N}| = \frac{1}{m}\sum_{i=1}^{m} J\left[\frac{N}{N+1} H_N(X_i)\right] - J[H(X_i)] \tag{10.2.16}$$

$$- \left(\frac{N}{N+1} H_N(X_i) - H(X_i)\right) J'[H(X_i)]$$

$$\leq \frac{1}{N}\sum_{i=1}^{N} \lambda_{iN}.$$

Denote $K_N = [N^{\delta''}]$ where $0 < \delta' < \delta''/2$, $\delta'' < \delta/2$, and $[x]$ denotes the largest integer $\leq x$. Then

$$|C_{3N}| \leq C_{3N}^{(1)} + C_{3N}^{(2)} + C_{3N}^{(3)} \tag{10.2.17}$$

where

$$C_{3N}^{(1)} = \frac{1}{N}\sum_{i=1}^{K_N} \lambda_{iN}, \qquad C_{3N}^{(2)} = \frac{1}{N}\sum_{i=N-K_N+1}^{N} \lambda_{iN}, \qquad C_{3N}^{(3)} = \frac{1}{N}\sum_{i=K_N+1}^{N-K_N} \lambda_{iN}. \tag{10.2.18}$$

and λ_{iN} is given by (10.2.15).

Now let us consider the random variable $C_{3N}^{(1)}$. We have

$$|C_{3N}^{(1)}| \leq N^{-1}\sum_{i=1}^{K_N} \left| J\left(\frac{i}{N+1}\right)\right| + N^{-1}\sum_{i=1}^{K_N} |J(H(Z_{Ni}))| \tag{10.2.19}$$

$$+ N^{-1}\sum_{i=1}^{K_N} \left| \frac{i}{N+1} - H(Z_{Ni})\right| \left| J'[H(Z_{Ni})]\right|.$$

By assumption (c),

$$N^{-1}\sum_{i=1}^{K_N} \left| J\left(\frac{i}{N+1}\right)\right| \leq N^{-1} \cdot K_N \cdot K\left(\frac{N}{(N+1)^2}\right)^{\frac{1}{2}-\delta} \tag{10.2.20}$$

$$= O(N^{-\frac{1}{2}-\delta+\delta''}) = o(N^{-\frac{1}{2}}).$$

Next, since $P[H(Z_{Ni}) > (1/N)\eta_\epsilon,\ i = 1, \ldots, N] > 1 - \epsilon$, therefore, with probability $>1 - \epsilon$, we have

$$\frac{1}{N}\sum_{i=1}^{K_N} |J(H(Z_{Ni}))| < \frac{K_N}{N} K(\eta_\epsilon/N)^{\delta-1/2} = o(N^{-1/2}). \tag{10.2.21}$$

Finally, using (10.2.9) and (10.2.13), we see that with probability $>1 - \epsilon$,

$$\frac{1}{N}\sum_{i=1}^{K_N} \left|\frac{i}{N+1} - H(Z_{Ni})\right| \left|J'[H(Z_{Ni})]\right| = o(N^{-1/2}). \tag{10.2.22}$$

Thus, from (10.2.19), (10.2.20), (10.2.21), and (10.2.22), we obtain

$$|C_{3N}^{(1)}| = o_p(N^{-1/2}). \tag{10.2.23}$$

The proof of $|C_{3N}^{(2)}| = o_p(N^{-1/2})$ is the same as above. We now prove that

$$|C_{3N}^{(3)}| = o_p(N^{-1/2}). \tag{10.2.24}$$

Let us define

$$S_{N,\eta}^{(1)}(\tau) = \{x: \tau \leq H(x) \leq 1 - \tau\}. \tag{10.2.25}$$

$$S_{N,\eta}^{(2)}(\tau) = \{x: Z_{NK_N} < x < H^{-1}(\tau)\}. \tag{10.2.26}$$

$$S_{N,\eta}^{(3)}(\tau) = \{x: H^{-1}(1 - \tau) < x < Z_{N,N-K_N+1}\}. \tag{10.2.27}$$

Then

$$C_{3N}^{(3)} \leq \sum_{j=1}^{3} \int_{S_{N,\eta}^{(j)}(\tau)} N^{1/2} \left| J\left(\frac{N}{N+1} H_N(x)\right) - J(H(x)) - \left(\frac{N}{N+1} H_N(x) - H(x)\right) J'(H(x)) \right| dH_N(x). \tag{10.2.28}$$

Now, by definition of the derivative of $J(u)$, and by assumption (c),

(10.2.29)

$$\sup_{|u|\leq c}\ \sup_{\tau\leq u\leq 1-c} N^{1/2} \left| J\left(u + \frac{v}{\sqrt{N}}\right) - J(u) - \left(\frac{v}{N^{1/2}}\right) J'(u) \right| \to 0, \quad \text{as } N \to \infty.$$

Hence, using (10.2.13), we have with probability $>1 - \epsilon$,

$$\int_{S_{N,\eta}^{(j)}(\tau)} N^{1/2} \left| J\left[\frac{N}{N+1} H_N(x)\right] - J[H(x)] - \left[\frac{N}{N+1} H_N(x) - H(x)\right] J'[H(x)] \right| dH_N(x) < \frac{\epsilon_1}{2} \tag{10.2.30}$$

where $\epsilon_1 > 0$ is a preassigned small number.

Also, we may write

$$(10.2.31)\quad \int_{s_{N,\eta}^{(j)}(\tau)} N^{1/2} \left| J\left[\frac{N}{N+1} H_N(x)\right] - J[(H(x)] - \left[\frac{N}{N+1} H_N(x) - H(x)\right] J'[H(x)] \right| dH_N(x)$$

$$\leq \int_{s_{N,\eta}^{(j)}(\tau)} N^{1/2} \left| \frac{N}{N+1} H_N(x) - H(x) \right| \times \left| J'\left[\phi H(x) - (1-\phi)\frac{N}{N+1} H_N(x)\right] - J'[H(x)\right| dH_N(x),$$

$$j = 2, 3, \quad 0 < \phi < 1.$$

Since

$$(10.2.32)\quad \left[\phi H(x) + (1-\phi)\frac{N}{N+1} H_N(x)\right] \Big/ H(x) = \phi + (1-\phi)\frac{N}{N+1}\frac{H_N(x)}{H(x)}$$

and

$$(10.2.33)\quad \frac{\left\{1 - \left[\phi H(x) + (1-\phi)\frac{N}{N+1} H_N(x)\right]\right\}}{[1 - H(x)]} = \phi + (1-\phi)\frac{\left(1 - \frac{N}{N+1} H_N(x)\right)}{(1 - H(x))},$$

we have with probability $>1 - 2\epsilon$,

$$(10.2.34)\quad \inf_{x \epsilon S_{N,\eta}^{(j)}(\tau)} \frac{\left[\phi H_N(x) + (1-\phi)\frac{N}{N+1} H_N(x)\right] \times \left[1 - \left\{\phi H(x) + (1-\phi)\frac{N}{N+1} H_N(x)\right\}\right]}{H(x)[1 - H(x)]} > \beta_1^2\left(\frac{N}{N+1}\right)^2, \quad j = 2, 3.$$

Thus, with probability $>1 - 2\epsilon$, the right-hand side of (10.2.31) is less than

$$(10.2.35)\quad c(\epsilon, \delta')K[1 + \beta_{1N}^{-3+2\delta}] \times \int_{S_{N,\eta}^{(j)}(\tau)} \{H(x)[1 - H(x)]\}^{\delta^* - 1}\, dH_N(x), \qquad j = 2, 3.$$

where $\beta_{1N} = N\beta_1/(N + 1)$, and $\delta^* = \delta - \delta'' > 2\delta' > 0$. Now

$$(10.2.36)\quad \sum_{j=2}^{3} \int_{S_{N,\eta}^{(j)}(\tau)} \{H(x)[1 - H(x)]\}^{\delta^*-1}\, dH_N(x)$$
$$\leq \int_0^{\tau} + \int_{1-\tau}^{1} \{H(x)[1 - H(x)]\}^{\delta^*-1}\, dH_N(x),$$

and it follows that for any $\tau < \frac{1}{2}$, the expected value of the right-hand side of (10.2.26) is

$$(10.2.37)\quad \int_0^{\tau} + \int_{1-\tau}^{1} \{H(x)[1 < H(x)]\}^{\delta^*-1}\, dH(x) \leq \left(\frac{2^{1+\delta^*}}{\delta^*}\right)\tau^{\delta^*}.$$

Now, for any fixed $\epsilon > 0$, and δ', τ may be so chosen that $[2^{1+\delta^*}/\delta^*]c(\epsilon, \delta')$ $K[1 + \beta_{1N}^{-3+2\delta}]\tau^{\delta^*} < \epsilon^{1/2}$. Then combining (10.2.35), (10.2.36), and (10.2.37), we conclude that sum of the two terms (for $j = 2$ and 3) in (10.2.31) is bounded above by $\epsilon_{1/2}$, in probability. Using this result, along with (10.2.30), we obtain (10.2.24).

Finally, $C_{4N} = o_p(N^{-1/2})$ follows from the assumption (6).

10.3. APPENDIX 2: THE ONE-SAMPLE CASE

Here we consider the higher-order terms of Theorems 3.6.4 and 4.4.3. Note that in Theorem 3.6.4 we considered the univariate problem dealing with homogeneous populations, whereas in Theorem 4.4.3 we considered the multivariate problem dealing with heterogeneous distributions. Since the higher-order terms in Theorem 4.4.3 only involve the p marginal (univariate) distributions, it seems enough to prove that the higher-order terms in (3.6.34)–(3.6.37) are all $o_p(n^{-1/2})$, even when the parent distributions are not all identical. With this end in view, we adopt the notations of section 4.4, but omit the subscript j for simplicity. Thus, we require to show only that $C_{2,n}$, $C_{3,n}$ and $C_{4,n}$ defined by (4.4.49)–(4.4.51) are $o_p(n^{-1/2})$. To avoid confusion, we denote the empirical c.d.f. by $\bar{F}_n(x)$ and the average c.d.f. by $F_n^*(x)$, $\bar{H}_n(x) = \bar{F}_n(x) - \bar{F}_n(-x-)$, and $H_n^*(x) = F_n^*(x) - F_n^*(-x)$, where the later two are defined only for $x \geq 0$.

Let (a_n^*, b_n^*) be the interval $S_{n,\eta}$ such that

$$(10.3.1)\qquad S_{n,\eta} = \{x\colon H_n^*(x)[1 - H_n^*(x)] \geq n^{-1}\eta_\epsilon\},$$

where $\epsilon > 0$ and $\eta_\epsilon(>0)$ depends on ϵ. Thus,

$$(10.3.2)\qquad \eta_\epsilon < n[H_n^*(a_n^*)], \qquad [1 - H_n^*(b_n^*)] < \eta_\epsilon[1 + 2n^{-1}\eta_\epsilon].$$

Hence, we can always select η_ϵ in such a way that

$$(10.3.3)\qquad n[H_n^*(a_n^*) + 1 - H_n^*(b_n^*)] \leq \epsilon, \qquad \epsilon \text{ arbitrarily small.}$$

Noting that $|X_i|$ has the c.d.f. H_i, $i = 1, \ldots, n$, we have

(10.3.4) $$P\{|X_i| \in S_{n,\eta}, 1 \le i \le n\} = P\{a_n^* \le |X_i| \le b_n^*, 1 \le i \le n\}$$
$$= \prod_{i=1}^{n} \{H_i(b_n^*) - H_i(a_n^*)\} = \prod_{i=1}^{n} \{1 - [H_i(a_n^*) + 1 - H_i(b_n^*)]\}$$
$$\ge 1 - \sum_{i=1}^{n} [H_i(a_n^*) + 1 - H_i(b_n^*)] = 1 - n[H_n^*(a_n^*) + 1 - H_n^*(b_n^*)]$$
$$\ge 1 - \epsilon, \qquad \text{by (10.3.3), uniformly in } \{\mathcal{F}_n\}.$$

Hence, with probability $\ge 1 - \epsilon$,

(10.3.5) $$|C_{2,n}| \le \frac{1}{n+1} \int_{S_{n,\eta}} |J'[H_n^*(x)]| \, dF_n(x) \le \frac{1}{n} \int_{S_{n,\eta}} |J'[H_n^*(x)]| \, d\bar{H}_n(x).$$

Now, from condition (c) of section 3.6.1 and (10.3.1), for all $x \in S_{n,\eta}$

(10.3.6) $$n^{-\frac{1}{2}(1-\delta)} |J'[H_n^*(x)]| \le K^*\{H_n^*(x)[1 - H_n^*(x)]\}^{-1+\delta/2}; \qquad K^* < \infty.$$

Thus, the right-hand side of (10.3.5) is bounded above by

(10.3.7) $$n^{-\frac{1}{2}(1+\delta)} K^* Q_n; \qquad Q_n = \int_{S_{n,\eta}} \{H_n^*(x)[1 - H_n^*(x)]\}^{-1+\delta/2} \, d\bar{H}_n(x).$$

Since Q_n can be written as $n^{-1} \sum_{i=1}^{n} u(|X_i|)\{H_n^*(|X_i|)[1 - H_n^*(|X_i|)]\}^{-1+\delta/2}$ (where $u(|X_i|)$ is 1 or 0 according as $|X_i| \in S_{n,\eta}$ or not), and as $E(Q_n) \le \int_0^1 [u(1-u)]^{-1+\delta/2} \, du < \infty$, by the Markov inequality, we obtain that Q_n is bounded, in probability, and hence, from (10.3.5) and (10.3.7), we have

(10.3.8) $$|C_{2,n}| = O_p(n^{-\frac{1}{2}-\delta/2}) = o_p(n^{-\frac{1}{2}}).$$

Let us now consider $C_{3,n}$. We note that by (2.11.70), for every positive ϵ' and $\delta'(0 < \delta' < \frac{1}{2})$, there exists a finite (>0) constant $c(\epsilon', \delta')$, such that for all $\mathbf{F}_n \in \mathcal{F}_n$,

(10.3.9)
$$P\left\{\sup_x \frac{n^{\frac{1}{2}} |\bar{H}_n(x) - H_n^*(x)|}{\{H_n^*(x)[1 - H_n^*(x)]\}^{\frac{1}{2}-\delta'}} > c(\epsilon', \delta')\right\} < \epsilon', \quad \text{when} \quad n \ge n_0(\epsilon', \delta').$$

(Note that (10.3.9) follows directly from (2.11.70) by noting that $\bar{H}_n$ and H_n^* are respectively the empirical and the average c.d.f.'s of the $|X_i|$, $i = 1, \ldots, n$.) Hence, with probability $\ge 1 - \epsilon$, for all $\mathbf{F}_n \in \mathcal{F}_n$ and all $0 < x < \infty$,

(10.3.10) $$n^{\frac{1}{2}} |[\bar{H}_n(x) - H_n^*(x)] J'[H_n^*(x)]| \le K^{**}[H_n^*(x)\{1 - H_n^*(x)\}]^{-1+\delta-\delta'},$$

where $K^{**} < \infty$. Hence, if we let $\delta' < \delta$, it follows from (4.4.50) and

(10.3.10) that a sufficient condition for $|C_{3,n}| = o_p(n^{-\frac{1}{2}})$, is that

$$(10.3.11)\quad \int_0^\infty \{H_n^*(x)[1 - H_n^*(x)]\}^{-1+\eta}\, d[\bar{F}_n(x) - F_n^*(x)] \xrightarrow{p} 0, \quad \text{as } n \to \infty,$$

which follows directly by noting that $d\bar{F}_n \leq d\bar{H}_n$, $dF_n^* \leq dH_n^*$, and using the dominated convergence theorem along with the law of large numbers, as in (10.3.7). Thus, $|C_{3,n}| = o_p(n^{-\frac{1}{2}})$.

The treatment of $C_{4,n}$ follows precisely on the same line as in (10.2.15) to (10.2.37), with (10.2.12) being replaced by (10.3.9). For brevity the details are omitted; the reader is referred to Sen (1970a), if necessary.

10.4. APPENDIX 3: ALIGNED RANK ORDER STATISTICS

These statistics are defined in section 7.3. Note that as in the one- or two-sample cases, the treatment of higher-order terms will be the same in both the uniresponse or the multiresponse cases. Hence, we shall consider only the ANOVA case, namely, the higher-order terms of Theorem 7.3.4.

We define the p empirical c.d.f.'s $F_{[j]n}$, $j = 1, \ldots, p$ as in (7.3.11) and the pooled empirical c.d.f. $H_N(x)$ as in (7.3.12). Also, the average c.d.f. $\bar{H}_N(x)$ is defined as in (7.3.9), and the c.d.f.'s $\bar{F}_{n[j]}$, $j = 1, \ldots, p$, as after (7.3.17). Then we consider the following lemma, which serves as a basis for the treatment of higher-order terms.

Lemma 10.4.1. *For every positive ϵ and $\delta'(0 < \delta' < \frac{1}{2})$, there exists a finite positive $c^*(\epsilon, \delta')$, such that*

$$(10.4.1)\quad \limsup_n P\left\{\sup_x \frac{n^{\frac{1}{2}}|H_N(x) - \bar{H}_N(x)|}{\{\bar{H}_N(x)[1 - \bar{H}_N(x)]\}^{\frac{1}{2}-\delta'}} > c^*(\epsilon, \delta')\right\} < \epsilon.$$

Proof: Note that by definition for all x: $-\infty < x < \infty$,

(10.4.2)

$$\bar{H}_N(x) \geq p^{-1}\bar{F}_{n[j]}(x), \qquad 1 - \bar{H}_N(x) \geq p^{-1}[1 - \bar{F}_{n[j]}(x)], \qquad j = 1, \ldots, p.$$

Therefore, for all x: $-\infty < x < \infty$,

$$(10.4.3)\quad \bar{H}_N(x)[1 - \bar{H}_N(x)] \geq \max_{1 \leq j \leq p} p^{-2}\bar{F}_{n[j]}(x)[1 - \bar{F}_{n[j]}(x)].$$

Now, by (2.11.70), for every j $(= 1, \ldots, p)$,

$$(10.4.4)\quad \limsup_n P\left\{\sup_x \frac{n^{\frac{1}{2}}|F_{[j]n}(x) - \bar{F}_{n[j]}(x)|}{\{\bar{F}_{n[j]}(x)[1 - \bar{F}_{n[j]}(x)]\}^{\frac{1}{2}-\delta'}} > c(\epsilon', \delta')\right\} < \epsilon',$$

and, by definition,

$$(10.4.5) \qquad |H_N(x) - \bar{H}_N(x)| \leq p^{-1} \sum_{j=1}^{p} |F_{[j]n}(x) - \bar{F}_{n[j]}(x)|.$$

Hence, (10.4.1) follows readily from (10.4.4) and (10.9.5) by letting $\epsilon' = \epsilon/p$ and using the Bonferroni inequality. Q.E.D. ◁

Lemma 10.4.2. *If* $K(u) \leq K[u(1-u)]^{-1+\eta}$, $0 < u < 1$, $K < \infty$, *then*

$$(10.4.6) \qquad \int_{-\infty}^{\infty} K[\bar{H}_N(x)]\, d[F_{[j]n}(x) - \bar{F}_{n[j]}(x)] \xrightarrow{p} 0, \quad \text{as} \quad n \to \infty,$$

for all $j = 1, \ldots, p$.

Proof: The left-hand side of (10.4.6) can be written as

$$(10.4.7) \qquad n^{-1} \sum_{i=1}^{n} \{K[\bar{H}_N(X_{ij})] - E(K[\bar{H}_N(X_{ij})])\},$$

where X_{ij} has the c.d.f. $F_{i,j}(x)$, $i = 1, \ldots, n$, and $\bar{F}_{n[j]} = n^{-1} \sum_{i=1}^{n} F_{i,j}$. Since the X_{ij} are independent, the result follows directly by using the Markov law of large numbers. Q.E.D. ◁

Treatment of Higher-order Terms of Theorem 7.3.4. The proof of $C_{i,N}^{(j)} = o_p(N^{-1/2})$, $j = 1, \ldots, p$, follows precisely on the same lines as in (10.2.11), with the direct use of Lemma 10.4.2. Again, the treatment of $C_{2,N}^{(j)}$, $j = 1, \ldots, p$, follows along the line of C_{2N}, in section 10.2, where we need only replace (10.2.12) by Lemma 10.4.1 and (10.2.14) by Lemma 10.4.2. Finally, the treatment of $C_{3,N}^{(j)}$, $j = 1, \ldots, p$ also follows as in $C_{3,N}$ of section 10.2, namely, as in (10.2.15)–(10.2.37), where again we need to use Lemmas 10.4.1 and 10.4.2 instead of (10.2.12) and (10.2.14). This completes the proof.

10.5. THE PROOF OF THEOREM 3.6.6

Note that by (3.6.73), $\int_{-\infty}^{\infty} |x|^{2+\delta}\, d\Psi(x) = \nu_{2+\delta} < \infty$, where $\delta > 0$. Also, $J_N(i/N+1) = E\{X_{N,i}\}$, where $X_{N,1} \leq \cdots \leq X_{N,N}$ are the ordered random variables of a sample of size N from the distribution $\Psi(x)$, where $J(u) = \Psi^{-1}(u)$: $0 < u < 1$. Thus,

$$J_N\left(\frac{i}{N+1}\right) \leq J_N(u) < J_N\left(\frac{i+1}{N+1}\right) \quad \text{for} \quad \frac{i}{N+1} \leq u < \frac{i+1}{N+1}.$$

Consequently, if $i = [(N+1)u]$, the integral part of $(N+1)u$, $J_N(i/N+1)$ is the expected value of the uth sample quantile, which by well-known results on the moments of the sample quantiles [cf. Blom (1958) and Sen

(1959)], converges to $\Psi^{-1}(u)$ as $N \to \infty$. Similarly, $J_N((i+1)/(N+1)) \to \Psi^{-1}(u) = J(u)$ as $N \to \infty$. Therefore, $\lim_{N\to\infty} J_N(u) = J(u)$: $0 < u < 1$. As such, the condition (c) of section 3.6.1 follows readily from (3.6.73). To prove (3.6.74), we note that

$$\left| \int_{-\infty}^{\infty} \left\{ J_N\left(\frac{N}{N+1} H_N(x)\right) - J\left(\frac{N}{N+1} H_N(x)\right) \right\} dF_m(x) \right| \tag{10.5.1}$$

$$\leq \frac{1}{m} \sum_{i=1}^{m} \left| J_N\left(\frac{N}{N+1} H_N(X_i)\right) - J\left(\frac{N}{N+1} H_N(X_i)\right) \right|$$

$$\leq \left(\frac{N}{m}\right)\left(\frac{1}{N}\right) \sum_{i=1}^{N} \left| J_N\left(\frac{i}{N+1}\right) - J\left(\frac{i}{N+1}\right) \right|.$$

Hence, upon noting that $m/N > \lambda_0 > 0$, it suffices to show that

$$A_N = N^{-1/2} \sum_{i=1}^{N} \left| J_N\left(\frac{i}{N+1}\right) - J\left(\frac{i}{N+1}\right) \right| = o(1) \quad \text{as} \quad N \to \infty. \tag{10.5.2}$$

Now, in (3.6.73), we let $0 < \delta \leq \frac{1}{2}$; if $\delta \geq \frac{1}{2}$, the proof becomes comparatively simple. We select a δ_1 such that

$$0 < \delta_1 < \delta/[2(2+\delta)] < \delta/4. \tag{10.5.3}$$

Also, let $N_1 = [N^{\delta_1}]$ and $N_2 = N - [N^{\delta_1}]$. Then, we have

$$A_N = N^{-1/2} \left\{ \sum_{i \leq N_1} + \sum_{N_1 < i < N_2} + \sum_{i \leq N_2} \left| J_N\left(\frac{i}{N+1}\right) - J\left(\frac{i}{N+1}\right) \right| \right\} \tag{10.5.4}$$

$$= A_N^{(1)} + A_N^{(2)} + A_N^{(3)}.$$

Note that

$$\max_{1 \leq i \leq N} \left| J_N\left(\frac{i}{N+1}\right) \right|^{1+\delta/2} = \max_{1 \leq i \leq N} |E\{X_{N,i} \mid \Psi\}|^{1+\delta/2} \tag{10.5.5}$$

$$\leq \left\{ E \max_{1 \leq i \leq N} |X_{N,i}|^{1+\delta/2} \,\middle|\, \Psi \right\}$$

$$= N \int_0^{\infty} x^{1+\delta/2} [\Psi^*(x)]^{N-1} \, d\Psi^*(x)$$

$$\text{(where } \Psi^*(x) = \Psi(x) - \Psi(-x)\text{)}$$

$$\leq N \left[\left\{ \int_0^{\infty} x^{2+\delta} \, d\Psi^*(x) \right\} \times \left\{ \int_0^{\infty} [\Psi^*(x)]^{2N-2} \, d\Psi^*(x) \right\} \right]^{1/2}$$

$$= \{N/(2N-1)^{1/2}\} \nu_{2+\delta}^{1/2} = O(N^{1/2}),$$

which implies that

$$\max_{1\le i\le N}\left|J_N\left(\frac{i}{N+1}\right)\right| = O(N^{1/(2+\delta)}). \tag{10.5.6}$$

Also, by (3.6.73),

$$\max_{1\le i\le N}\left|J\left(\frac{i}{N+1}\right)\right| \le K(N+1)^{\frac{1}{2}-\delta} = O(N^{\frac{1}{2}-\delta}).$$

Hence,

$$|A_N^{(1)}| \le N^{-\frac{1}{2}+\delta_1}\{O(N^{1/(2+\delta)}) + O(N^{\frac{1}{2}-\delta})\} = o(1), \tag{10.5.7}$$

by (10.5.3.). Similarly, $|A_N^{(3)}| = o(1)$, as $N\to\infty$. So, we need to consider only $A_N^{(2)}$. For this, we first let for $0 < t < p < 1$

$$g_p(t) = \{(p-t)/p\}^p\{(1-p+t)/(1-p)\}^{1-p}. \tag{10.5.8}$$

Then, some simple computations (cf. Hoeffding, 1963) lead to

(*a*) $\sup_{t\ge\eta>0} g_p(t) = g_p(\eta)$: $0 < g_p(\eta) < 1$ for every $0 < p < 1$,

(*b*) $\sup_{0<p<\eta} |g_p(p/2)(2/\sqrt{e})^p - 1| = o(\eta)$ for every $\eta(>0)$ close to 0.

Also, it is well known that

$$i\binom{N}{i}\int_0^{u_0} u^{i-1}(1-u)^{N-i}\,du = \sum_{r=i}^{N}\binom{N}{r}u_0^r(1-u_0)^{N-r}, \tag{10.5.9}$$

and hence, using theorem 1 of Hoeffding (1963), we have

$$\sup_{u_0\le i/N-t}\left\{i\binom{N}{i}\int_0^{u_0} u^{i-1}(1-u)^{N-i}\,du\right\} \le [g_{(i/N)}(t)]^N. \tag{10.5.10}$$

Further, by (*a*) and (*b*) following (10.5.8), we have

$$\sup_{i/N>\epsilon>0}\ \sup_{u_0\le i/N-\eta}\left\{i\binom{N}{i}\int_0^{u_0} u^{i-1}(1-u)^{N-i}\,du\right\} \le [\rho(\epsilon,\eta)]^N, \tag{10.5.11}$$

where $0 < \rho(\epsilon,\eta) < 1$, and

$$\sup_{N_1\le i\le N\epsilon}\ \sup_{u_0\le i/2N}\left\{i\binom{N}{i}\int_0^{u_0} u^{i-1}(1-u)^{N-i}\,du\right\} \le [\rho^*(\epsilon)]^{N^{\delta_1}}, \tag{10.5.12}$$

where $0 < \rho^*(\epsilon) < 1$ for every $\epsilon > 0$. Note that for N sufficiently large,

$$[\rho(\epsilon,\eta)]^N = o(N^{-k}) \quad\text{and}\quad [\rho^*(\epsilon)]^{N^{\delta_1}} = o(N^{-k}), \tag{10.5.13}$$

for any fixed $k(>0)$. Finally, by the assumed continuity of $J'(u)$ (on the open interval (0, 1)), we have for every $\epsilon(>0)$ and $\eta(>0)$, some positive $\gamma(<\epsilon)$

such that

$$\sup_{\epsilon<u<1-\epsilon}\sup_{|v|<\gamma}|J'(u+v)-J'(u)|<\eta. \tag{10.5.14}$$

We then write

$$\begin{aligned} A_N^{(2)} &= N^{-1/2}\left\{\sum_{N_1<i\le[N\epsilon]}+\sum_{[N\epsilon]<i<N-[N\epsilon]}+\sum_{N-[N\epsilon]\le i<N}\left|J_N\left(\frac{i}{N+1}\right)-J\left(\frac{i}{N+1}\right)\right|\right\} \\ &= A_{N,1}^{(2)}+A_{N,2}^{(2)}+A_{N,3}^{(2)}, \quad \text{say}. \end{aligned} \tag{10.5.15}$$

Note that for $N_1 < i \le [N\epsilon]$,

$$\begin{aligned} &\left|J_N\left(\frac{i}{N+1}\right)-J\left(\frac{i}{N+1}\right)\right| \\ &= \left|\int_0^1\left\{J(u)-J\left(\frac{i}{N+1}\right)\right\}i\binom{N}{i}u^{i-1}(1-u)^{N-i}\,du\right| \\ &\le \int_0^{u_1}+\int_{u_1}^{4u_1}+\int_{4u_1}^{1}\left|J(u)-J\left(\frac{i}{N+1}\right)\right|i\binom{N}{i}u^{i-1}(1-u)^{N-i}du \\ &= D_{i1}+D_{i2}+D_{i3}, \quad \text{say,} \quad \text{where } u_1 = i/2N. \end{aligned} \tag{10.5.16}$$

Then, by definition, and by (10.5.8) through (10.5.12),

$$\begin{aligned} |D_{i1}| &\le i\binom{N}{i}\int_0^{u_1}|J(u)|\,u^{i-1}(1-u)^{N-i}\,du \\ &\quad + \left|J\left(\frac{i}{N+1}\right)\right|\int_0^{u_1}i\binom{N}{i}u^{i-1}(1-u)^{N-i}\,du \\ &\le i\binom{N}{i}\left(\frac{i-1}{N-1}\right)^{i-1}\left(1-\frac{i-1}{N-1}\right)^{N-i}\cdot\int_0^{u_1}|J(u)| \\ &\quad \times\left[g_{(i-1)/(N-1)}\left(\frac{1}{2}\frac{i-1}{N-1}-\frac{N-1}{2N(N-1)}\right)\right]^{N-1}du \\ &\quad + \left|J\left(\frac{i}{N+1}\right)\right|\sum_{r=i}^{N}\binom{N}{r}u_1^r(1-u_1)^{N-r} \\ &\le N\left[g_{(i-1)/(N-1)}\left(\frac{i-1}{2(N-1)}-\frac{N-1}{2N(N-1)}\right)\right]^{N-1} \\ &\quad \times\int_0^1|J(u)|\,du+\left|J\left(\frac{i}{N+1}\right)\right|[\rho^*(\epsilon)]^{N^{\delta_1}} \\ &= N\cdot O(\rho^*(\epsilon))^{N^{\delta_1}}+O(N^{(1/2-\delta)(1-\delta_1)})O(\rho^*(\epsilon))^{N^{\delta_1}} \\ &= o(N^{-1/2}), \end{aligned} \tag{10.5.17}$$

by (10.5.13). Similarly, $|D_{i3}| = o(N^{-1/2})$. Further, letting $0 < \theta < 1$, we have

$$(10.5.18)\quad |D_{i2}| \leq \int_{u_1}^{4u_1} \left| u - \frac{i}{N+1} \right| \left| J'\left[\theta u + (1-\theta)\frac{i}{N+1}\right] \right|$$

$$\times\, i\binom{N}{i} u^{i-1}(1-u)^{N-i}\, du$$

$$\leq K[u_1(1-u_1)]^{-3/2+\delta} \int_{u_1}^{4u_1} \left| u - \frac{i}{N+1} \right| i\binom{N}{i} u^{i-1}$$

$$\times\, (1-u)^{N-i}\, du$$

$$= K^* u_1^{-3/2+\delta} \cdot \int_0^1 \left| u - \frac{i}{N+1} \right| i\binom{N}{i} u^{i-1}(1-u)^{N-i}\, du$$

$$\leq K^*(N/2i)^{3/2-\delta} \cdot N^{-1/2}\, \{i(N-i+1)/(N+1)^2\}^{1/2}$$

$$\leq K^{**} \cdot N^{-1/2}(i/N)^{-1+\delta}, \qquad N_1 \leq i \leq N\epsilon.$$

Hence, we have

$$(10.5.19)\quad |A_{N,1}^{(2)}| \leq N^{-1/2}\left\{N\epsilon \cdot o(N^{-1/2}) + K^* N^{-1/2} \sum_{i \leq N\epsilon} \left(\frac{i}{N}\right)^{-1+\delta}\right.$$

$$\left. + N\epsilon \cdot o(N^{-1/2})\right\}$$

$$= \epsilon \cdot o(1) + K^* N^{-1} \sum_{i \leq N\epsilon} \left(\frac{i}{N}\right)^{-1+\delta}$$

$$\leq o(\epsilon) + K^* \epsilon^{\delta}/\delta = o(1),$$

as $\delta(>0)$ is fixed and ϵ arbitrarily small. Similarly, $|A_{N,3}^{(2)}| = o(1)$. Finally, for $A_{N,2}^{(2)}$, we let, as in (10.5.14), $\eta(>0)$ sufficiently small, and have for $[N\epsilon] \leq i \leq N - [N\epsilon]$,

$$(10.5.20)\quad \left| J_N\left(\frac{i}{N+1}\right) - J\left(\frac{i}{N+1}\right) \right|$$

$$= \left| i\binom{N}{i}\left[\left[\int_0^{u_1} + \int_{u_1}^{u_2} + \int_{u_2}^1\right]\left[J(u) - J\left(\frac{i}{N+1}\right)\right]\right.\right.$$

$$\left.\left. \times\, u^{i-1}(1-u)^{N-i}\, du\right]\right|$$

$$= |D'_{i1} + D'_{i2} + D'_{i3}|,$$

where $u_1 = i/(N+1) - \eta$ and $u_2 = i/(N+1) + \eta$. Then precisely as in (10.5.17),

$$(10.5.21)\quad |D'_{i1}| \leq O(N)O[\rho(\epsilon,\eta)]^N + K^* \epsilon^{-1/2+\delta} \cdot 0[\rho(\epsilon,\eta)]^N = o(N^{-1/2}),$$

and similarly $|D'_{i3}| = o(N^{-1/2})$ for N sufficiently large. Further,

$$(10.5.22)\quad |D'_{i2}| \le \left|J'\left(\frac{i}{N+1}\right)\right| \left|\int_{u_1}^{u_2}\left(u - \frac{i}{N+1}\right) i\binom{N}{i} u^{i-1}(1-u)^{N-i}\,du\right|$$

$$+ \left[\sup_{u\in[u_1,u_2]}\left|J'(u) - J'\left(\frac{i}{N+1}\right)\right|\right]$$

$$\times \int_{u_1}^{u_2}\left|u - \frac{i}{N+1}\right| i\binom{N}{i} u^{i-1}(1-u)^{N-i}\,du$$

$$\le (K/\epsilon^{3/2-\delta})\left\{\int_0^{u_1} + \int_{u_2}^{1}\left|u - \frac{i}{N+1}\right| i\binom{N}{i} u^{i-1}(1-u)^{N-i}\,du\right\}$$

$$+ \eta\left\{\int_0^1\left(u - \frac{i}{N+1}\right)^2 i\binom{N}{i} u^{i-1}(1-u)^{N-i}\,du\right\}^{1/2},$$

as

$$\int_0^1\left(u - \frac{i}{N+1}\right) i\binom{N}{i} u^{i-1}(1-u)^{N-i}\,du = 0.$$

Now

$$(10.5.23)\quad \int_0^{u_1}\left|u - \frac{i}{N+1}\right| i\binom{N}{i} u^{i-1}(1-u)^{N-i}\,du$$

$$\le \int_0^{u_1} i\binom{N}{i} u^{i-1}(1-u)^{N-i}\,du$$

$$= \sum_{r=i}^{N}\binom{N}{r} u_1^r(1-u_1)^{N-r}$$

$$\le [\rho(\epsilon,\eta)]^N,$$

by (10.5.10), and a similar inequality holds for the other integral over $(u_2, 1)$. Finally, the last integral on the right-hand side of (10.5.22) is equal to

$$(10.5.24)\qquad \eta\left[\frac{i(N-i+1)}{(N+1)^2(N+2)}\right]^{1/2} \le \eta/2\sqrt{N}.$$

Hence, from (10.5.12), (10.5.22), (10.5.23), and (10.5.24), we have

$$(10.5.25)\qquad |D'_{i2}| \le (K/\epsilon^{3/2-\delta})\cdot(N^{-1/2}) + (\eta/2)N^{-1/2},$$

which leads to

$$(10.5.26)\quad |A_{N,2}^{(2)}| \le (1-2\epsilon)(K/\epsilon^{3/2-\delta})o(1) + \eta\left(\frac{1-2\epsilon}{2}\right) + o(1).$$

Since η is arbitrary, the proof is completed.

PRINCIPAL REFERENCES

Blom (1958), Chernoff and Savage (1958), Govindarajulu, Le Cam and Raghavachari (1966), Pyke and Shorack (1968a), Sen (1959, 1970a).

References

The set of references given here is by no means exhaustive but is closely related to the topics covered in the various chapters of the book. We use the following abbreviations for the journals most frequently referred to:

AISM	*Annals of the Institute of Statistical Mathematics*
AMS	*Annals of Mathematical Statistics*
BC	*Biometrics*
BK	*Biometrika*
CSAB	*Calcutta Statistical Association Bulletin*
JASA	*Journal of the American Statistical Association*
JRSS, B	*Journal of the Royal Statistical Society, Series B*
PCPS	*Proceedings of the Cambridge Philosophical Society*
S, A	*Sankhyā, Series A*
TAMS	*Transactions of the American Mathematical Society*
TVP	*Teoria Veroyatnostey i ee Primenyia*

Abrahamson, I. G. (1967). "Exact Bahadur efficiencies for the Kolmogorov-Smirnov and Kuiper one- and two-sample statistics." *AMS* **38,** 1775–1790.

Adichie, J. N. (1967). "Estimates of regression parameters based on rank tests." *AMS* **38,** 894–904.

Ali, M. M. and Chan, L. K. (1965). "Some bounds for expected values of order statistics." *AMS* **36,** 1055–1057.

Anderson, R. L. and Bancroft, T. A. (1952). *Statistical Theory in Research.* McGraw-Hill, New York.

Anderson, T. W. (1958). *An Introduction to Multivariate Statistical Analysis.* John Wiley, New York.

______ (1966). "Some nonparametric multivariate procedures based on statistically equivalent blocks." *Proc. 1st Internat. Symp. Mult. Analysis* (edited by P. R. Krishnaiah), 5–27.

______ and Darling, D. A. (1952). "Asymptotic theory of certain goodness of fit criteria based on stochastic processes." *AMS* **23,** 193–212.

Andrews, F. C. (1954). "Asymptotic behavior of rank tests for analysis of variance." *AMS* **25,** 724–736.

Ansari, A. R. and Bradley, R. A. (1960). "Rank-sum tests for dispersions." *AMS* **31,** 1174–1189.

Bahadur, R. R. (1960a). "Asymptotic efficiency of tests and estimates." *Sankhya,* **22,** 229–252.

______ (1960b). "Simultaneous comparison of the optimum and sign tests of a normal mean." *Contributions to Probability and Statistics—Essays in Honor of Harold Hotelling.* Stanford University Press. Stanford, Calif. 79–88.

______ (1960c). "Stochastic comparison of tests." *AMS* **31,** 276–295.

______ (1967). "Rates of convergence of estimates and test statistics." *AMS* **38,** 303–324.

______ and Ranga Rao, R. (1960). "On deviations of the sample mean." *AMS* **31,** 1015–1027.

Barton, D. E. and David, F. N. (1958). "A test for birth-order effects." *Ann. Eugenics.* **22,** 250–257.

______ and Mallows, C. L. (1965). "Some aspects of the random sequence." *AMS* **36,** 236–260.

Bell, C. B. and Doksum, K. A. (1967). "Distribution-free tests of independence." *AMS* **38,** 619–628.

______ and Haller, H. S. (1969). "Bivariate symmetry tests: parametric and nonparametric." *AMS* **40,** 259–269.

Bellman, R. (1960). *Introduction to Matrix Analysis.* McGraw-Hill, New York.

Bennett, B. M. (1962). "On Multivariate sign tests." *JRSS, B* **24,** 159–161.

______ (1964). "A bivariate signed rank test." *JRSS, B* **26,** 457–461.

Berk, R. H. (1966). "Limiting behavior of posterior distributions when the model is incorrect." *AMS* **37,** 51–58.

Benard, A., and Elteren, Ph. Van. (1953). "A generalization of the method of *m* rankings." *Indag. Math.* **15,** 358–369.

Bernstein, S. N. (1926). "Sur l'extension du théoreme limit du calcul des probabilités aux sommes des quantités dépendantes." *Math. Annalen* **97,** 1–59.

Bhapkar, V. P. (1961a). "Some nonparametric median procedures." *AMS* **32,** 846–863.

______ (1961b). "A nonparametric test for the problem of several samples." *AMS* **32,** 1108–1117.

______ (1963). "The asymptotic power and efficiency of Mood's test for two-way classification." *Jour. Indian Statist. Assoc.* **1,** 24–31.

______ (1966). "Some nonparametric tests for the multivariate several sample location problem." *Proc. 1st Internat. Symp. Mult. Analysis* (edited by P. R. Krishnaiah). 29–42.

Bhattacharyya, G. K. (1967). "Asymptotic efficiency of multivariate normal score test." *AMS* **38,** 1753–1758.

Bhattacharyya, R. N. (1968). "Berry-Esseen bounds for the Multidimensional central limit theorem." *Bull. Amer. Math. Soc.* **74,** 285–287.

Bhattacharyya, R. N. (1970). "The central limit theorem in R^k, $k \geq 1$, and normal approximation to the probabilities of Borel sets." *TVP* **15,** 69–85.

Bhuchongkul, S. (1964). "A class of nonparametric tests for independence in bivariate populations." *AMS* **35,** 138–149.

——— and Puri, M. L. (1965). "On the estimation of contrasts in linear models." *AMS* **36,** 198–202.

Bickel, P. J. (1964). "On some alternative estimates for shift in the p-variate one sample problem." *AMS* **35,** 1079–1090.

——— (1965). "On some asymptotically nonparametric competitors of Hotelling's T^2." *AMS* **36,** 160–173.

Billingsley, P. (1968). *Weak Convergence of Probability Measures.* John Wiley, New York.

Birnbaum, Z. W. (1952). "Numerical tabulation of the distribution of Kolmogorov's statistic for finite sample size." *JASA* **47,** 425.

——— (1953a). "Distribution-free tests of fit for continuous distribution functions." *AMS* **24,** 1–8.

——— (1953b). "On the power of a one-sided test of fit for continuous probability functions." *AMS* **24,** 484–489.

——— (1954). "On a use of the Mann-Whitney statistic." *Proc. Third Berkeley Symposium Math. Statist. Prob.*, vol. I, 13–17 (ed: J. Neyman).

——— and Hall, R. A. (1960). "Small sample distributions for multisample statistics of the Smirnov type." *AMS* **31,** 710–720.

——— and Klose, O. M. (1957). "Bounds for the variance of the Mann-Whitney statistics." *AMS* **28,** 933–945.

——— and Marshall, A. W. (1961). "Some multivariate Chebyshev inequalities with extensions to continuous parameter processes." *AMS* **32,** 687–703.

——— and McCarty, R. C. (1958). "A distribution-free upper confidence bound for $Pr\{Y > X\}$, based on independent samples of X and Y." *AMS* **29,** 558–562.

——— and Pyke, R. (1958). "On some distributions related to the statistic D_N^+." *AMS* **29,** 179–187.

———, Raymond, J., and Zuckerman, H. S. (1947). "A generalization of Tshebyshev's inequality to two dimensions." *AMS* **18,** 70–79.

——— and Rubin, H. (1954). "On distribution-free statistics." *AMS* **25,** 593–598.

——— and Tingey, F. H. (1951). "One-sided confidence contours for probability distribution functions." *AMS* **22,** 592–596.

Blom, G. (1958). *Statistical Estimates and Transformed Beta Variables.* John Wiley, New York and Almqvist and Wiksell, Uppsala, Sweden.

Blomqvist, N. (1950). "On a measure of dependence between two random variables." *AMS* **21,** 593–600.

Blum, J. R., Kiefer, J. and Rosenblatt, M. (1961). "Distribution free tests of independence based on the sample distribution function." *AMS* **32,** 485–498.

Blumen, I. (1958). "A new bivariate sign test." *JASA* **53,** 448–456.

Bradley, R. A. (1955). "Rank analysis of incomplete block design. III. Some large-sample results on estimation and power for a method of paired comparisons." *BK* **42,** 450–470.

Bradley, R. A. (1966). "Topics in rank order statistics." *Proc. 5th Berkeley Symp. Math. Statist. Prob.* **1,** 593–608.

Brown, G. W. and Mood, A. M. (1951). "On median tests for linear hypotheses." *Proc. Second Berkeley Symp. Math. Statist. Prob.* (J. Neyman, ed.) **1,** 159–166. Univ. of California Press.

Capon, J. (1961). "Asymptotic efficiency of certain locally most powerful rank tests." *AMS* **32,** 88–100.

Chacko, V. J. (1961). "A test of homogeneity of means under restricted alternatives." *AMS* **34,** 945–956.

Chanda, K. C. (1963). "On the efficiency of two-sample Mann-Whitney test for discrete populations." *AMS* **34,** 612–617.

Chapman, D. G. (1950). "Some two-sample tests." *AMS* **21,** 601–606.

______ (1958). "A comparative study of several one-sided goodness-of-fit tests." *AMS* **29,** 655–674.

Chatterjee, S. K. (1966a). "A multisample nonparametric scale test based on *U*-statistics." *CSAB* **15,** 109–119.

______ (1966b). "A bivariate sign test for location." *AMS* **37,** 1771–1782.

______ (1970). "Pseudo-likelihood approach for some two-decision problems under nonparametric set-ups." *Nonparametric Tech. Statist. Inf.*, Cambridge Univ. Press. (Ed: M. L. Puri.) 563–578.

______ and Sen, P. K. (1964). "Nonparametric tests for the bivariate two-sample location problem." *CSAB* **13,** 18–58.

______ and Sen, P. K. (1965). "Some nonparametric tests for the bivariate two-sample association problem." *CSAB* **14,** 14–34.

______ and Sen, P. K. (1966). "Nonparametric tests for the multivariate multi-sample location problem." *Essays in Probability and Statistics in Memory of S. N. Roy.* (Ed: by Bose et al.) Univ. of N. Carolina Press, 197–228.

Chernoff, H. (1952). "A measure of asymptotic efficiency for tests of a hypothesis based on the sum of observations." *AMS* **23,** 493–507.

______ and Savage, I. R. (1958). "Asymptotic normality and efficiency of certain nonparametric test statistics." *AMS* **29,** 972–994.

Chow, Y. S. (1960). "A martingale inequality and the law of large numbers." *Proc. Amer. Math. Soc.*, **11,** 107–111.

Chung, J. H., and Fraser, D. A. S. (1958). "Randomization tests for a multivariate two-sample problem." *JASA*, **53,** 729–735.

Cochran, W. G. (1934). "The distribution of quadratic forms in a normal system, with applications to the analysis of covariance." *PCPS Math. Phys. Sci.* **30,** 178–191.

______ and Cox, G. M. (1957). *Experimental Designs.* John Wiley, New York.

Craig, A. T. (1943). "Note on the independence of certain quadratic forms." *AMS* **14,** 195–197.

Cramér, H. (1946). *Mathematical Methods of Statistics.* Princeton Univ. Press.

Crouse, C. F. (1964). "Note on Mood's Test." *AMS* **35,** 1825–1826.

Daly, J. F. (1940). "On the unbiased character of likelihood-ratio tests for independence in normal systems." *AMS* **11,** 1–32.

Dantzig, D. van (1951). "On the consistency and power of Wilcoxon's two-sample test." *Mat. Kutato Int. Kozlemenyei* **4.** 313–319.

Darling, D. A. (1957). "The Kolmogorov-Smirnov, Cramer-von Mises tests." *AMS* **28,** 823–838.

Davidson, R. R. and Bradley, R. A. (1969). "Multivariate paired comparisons: the extension of a univariate model and associated estimation and test procedures." *BK* **56,** 81–95.

——— and Bradley, R. A. (1970). Multivariate Paired Comparisons: Some large sample results on estimation and tests of equality of preference. *Nonparametric Tech. in Statist. Inf.* Cambridge Univ. Press. (Ed: M. L. Puri), 111–125.

Deshpande, J. V. (1965). "A nonparametric test based on *U*-statistics for the problem of several samples." *Jour. Indian Statist. Assoc.* **3,** 20–29.

——— (1970). "A class of multisample distribution-free tests." *AMS* **41,** 227–236.

Doksum, K. A. (1967). "Robust procedures for some linear models with one observation per cell." *AMS* **38,** 878–883.

Doob, J. L. (1949). "Heuristic approach to the Kolmogorov-Smirnov theorems." *AMS* **20,** 393–403.

——— (1953). *Stochastic Processes*, John Wiley, New York.

Dupac, V. (1970). "A contribution to the asymptotic normality of simple linear rank statistics." *Nonparametric Tech. in Statist. Inf.* Cambridge Univ. Press. (Ed: M. L. Puri.) 75–88.

Durbin, J. (1951). "Incomplete blocks in ranking experiments." *Brit. J. Psychol.* **4,** 85–90.

——— (1970). "Some results for the bivariate goodness-of-fit problem." *Nonparametric Tech. in Statist. Inf.* Cambridge Univ Press. (Ed: M. L. Puri.) 435–449.

Dwass, M. (1953). "On the asymptotic normality of certain rank order statistics." *AMS* **24,** 303–306.

——— (1955a). "A note on simultaneous confidence intervals." *AMS* **26,** 146–147.

——— (1955b). "On the asymptotic normality of some statistics used in non-parametric tests." *AMS* **26,** 334–339.

——— (1956). "The large sample power of rank tests in the two sample problems." *AMS* **27,** 352–374.

——— (1957). "On the distribution of ranks and of certain rank order statistics." *AMS* **28,** 424–431.

——— (1960). "Some *k*-sample rank order tests." *Contributions to Prob. Statistics Essays in Honor of H. Hotelling* (ed: Olkin et al.) Stanford Univ. Press, Stanford, Calif. 198–202.

——— (1967). "Sample random walk and rank order statistics." *AMS* **38,** 1042–1053.

van Eeden, Constance (1963). "The relation between Pitman's asymptotic relative efficiency of two tests and the correlation coefficient between their test statistics." *AMS* **34,** 1442–1451.

van Elteren, Ph. and Noether, G. E. (1959). "The asymptotic efficiency of χ_r^2-test for a balanced incomplete block design." *BK* **46,** 475–477.

Feller, W. (1948). "On the Kolmogorov-Smirnov limit theorems for empirical distributions." *AMS* **19,** 177.

______ (1965). *An Introduction to Probability Theory and Its Applications.* Vol. II John Wiley, New York.

Ferrar, W. L. (1941). *Algebra.* Oxford University Press.

Fillippova, A. A. (1961). "Mises' theorem on the asymptotic behavior of functionals of empirical distribution functions and its statistical applications." *TVP.* **7,** 24–57.

Fisher, R. A. (1935). *The Design of Experiments.* Oliver & Boyd, Edinburgh.

______ and Yates, F. (1938). *Statistical Tables for Biological, Agricultural and Medical Research.* Oliver & Boyd, Edinburgh.

Fraser, D. A. S. (1951). "Sequentially determined statistically equivalent blocks." *AMS* **22,** 294.

______ (1953a). "Nonparametric tolerance regions." *AMS* **24,** 44–55.

______ (1953b). "Completeness of order statistics." *Canadian Jour. Math.*, **6,** 42–45.

______ (1956). "A vector form of the Wald-Wolfowitz-Hoeffding theorem." *AMS* **27,** 540–543.

______ (1957a). "Most powerful rank-type tests." *AMS* **28,** 1040–1043.

______ (1957b). *Nonparametric Methods in Statistics.* John Wiley, New York.

Friedman, M. (1937). "The use of ranks to avoid the assumption of normality implicit in the analysis of variance." *JASA* **32,** 675–701.

Gabriel, K. R. and Sen, P. K. (1968). "Simultaneous test procedures for one-way ANOVA and MANOVA based on rank scores." *S, A* **30,** 303–312.

Gastwirth, J. L. (1966). "On robust procedures." *JASA* **61,** 929–948.

______ and Wolff, S. S. (1968). "An elementary method for obtaining lower bounds on the asymptotic power of rank tests." *AMS* **39,** 2128–2131.

Gerig, T. M. (1969). "A multivariate extension of Friedman's χ_r^2-test." *JASA* **64,** 1595–1608.

Ghosh, M. (1969). "Asymptotically optimal nonparametric tests for miscellaneous problems of linear regression." Doctoral Dissertation, University of North Carolina, Chapel Hill.

______ and Sen, P. K. (1970). "On the almost sure convergence of von Mises' differentiable statistical functions." *CSAB* **19,** 41–44.

______ and Sen, P. K. (1971). "On a class of rank order tests for regression with partially informed stochastic predictors." *AMS* **42.** (in press).

Gleser, L. J. (1964). "On a measure of test efficiency proposed by R. R. Bahadur." *AMS* **35,** 1537–1544.

Gnedenko, B. V. (1944). "Elements of the theory of distribution functions of random vectors (in Russian)." *Uspehi. Math. Nauk* **10,** 230.

Gnedenko, B. V. and Kolmogorov, A. N. (1954). *Limit Distributions for Sums of Independent Random Variables.* Addison-Wesley, Reading, Mass.

_____ and Korolink, V. S. (1951). "On the maximum discrepancy between two empirical distributions." *Dok, Akad, Nauk SSR*, **80,** 525–528.

Govindarajulu, Z. (1960). "Central limit theorems and asymptotic efficiency for one-sample nonparametric procedures." *Technical Report no. 11*, Dept. of Statistics, University of Minnesota.

_____, Le Cam, L. and Raghavachari, M. (1966). "Generalizations of theorems of Chernoff and Savage on the asymptotic normality of test statistics." *Proc. Fifth Berkeley Symp. on Math. Stat. and Prob.* **1,** 609–638. (Ed: J. Neyman and L. Le Cam.) Univ. of California Press.

Hájek, J. (1961). "Some extensions of the Wald-Wolfowitz-Noether theorem." *AMS* **32,** 506–523.

_____ (1962). "Asymptotically most powerful rank order tests." *AMS* **33,** 1124–1147.

_____ (1963). "Extensions of the Kolmogorov-Smirnov tests to regression alternatives." *Proc. Bernoulli-Bayes-Laplace Seminar*, Berkeley (ed: L. Le Cam), Univ. of California Press. 45–60.

_____ (1968a). "Locally most powerful rank tests of independence." *Studies in Mathematical Statistics. Akadémiai Kiadó*, Edited by K. Sarkadi and I. Vincze.

_____ (1968b). "Asymptotic normality of simple linear rank statistics under alternatives." *AMS* **39,** 325–346.

_____ (1969). *A Course in Nonparametric Statistics.* Holden-Day: San Francisco.

_____ (1970). "Miscellaneous problems of rank test theory." *Nonparametric Tech. in Statist. Inf.* Cambridge Univ. Press. (Ed: M. L. Puri.) 3–17.

_____ and Rènyi, A. (1955). "Generalization of an inequality of Kolmogorov," *Acta Math, Acad. Sci. Hung.* **6,** 281.

_____ and Šidák, Z. (1967). *Theory of Rank Tests.* Academic Press, New York.

Halmos, P. R. (1946). "The theory of unbiased estimation." *AMS* **17,** 34–44.

_____ (1950). *Measure Theory.* Van Nostrand, New York.

Hannan, E. J. (1956). "The asymptotic powers of certain tests based on multiple correlations. *JRSS*, *B* **18,** 227–233.

Herr, David G. (1967). "Asymptotically optimal tests for multivariate normal distributions." *AMS* **38,** 1829–1844.

Hoadley, A. B. (1967). "On the probability of large deviations of functions of several empirical c.d.f.'s." *AMS* **38,** 360–381.

Hodges, J. L. Jr. (1955). "A bivariate sign test." *AMS* **26,** 523–527.

_____ and Lehmann, E. L. (1956). "The efficiency of some nonparametric competitors of the *t*-test." *AMS* **27,** 324–335.

_____ and Lehmann, E. L. (1961). "Comparison of the normal scores and Wilcoxon tests." *Proc. Fourth Berkeley Symp. Math. Statist. Prob.* **1,** 307–317. (Ed: by J. Neyman.) Univ. of California Press

_____ and Lehmann, E. L. (1962). "Rank methods for combination of independent experiments in analysis of variance." *AMS* **33,** 482–497.

Hodges, J. L. Jr. and Lehmann, E. L. (1963). "Estimates of location based on rank tests." *AMS* **34,** 598–611.

Hoeffding, W. (1948a). "A class of statistics with asymptotically normal distribution." *AMS* **19,** 293–325.

______ (1948b). "A nonparametric test of independence." *AMS* **19,** 546–557.

______ (1951a). "A combinatorial central limit theorem." *AMS* **22,** 558–566.

______ (1951b). "Optimum nonparametric tests." *Proc. Second Berkeley Symp. Math. Statist. Prob.*, Univ. of Calif. Press, (ed: J. Neyman), 83–92.

______ (1952). "The large-sample power of tests based on permutations of observations." *AMS* **23,** 169–192.

______ (1953). "On the distribution of the expected values of the order statistics." *AMS* **24,** 93–100.

______ (1961). "The strong law of large numbers for *U*-statistics." *Inst. Statist. Univ. North Carolina, Mimeo Ser. No. 302.*

______ (1963). "Probability inequalities for sums of bounded random variables." *JASA* **58,** 13–30.

______ (1965). "Asymptotically optimal tests for multinomial distributions." *AMS* **36,** 369–400.

______ (1966). "On probabilities of large deviations." *Proc. Fifth Berkeley Symp. Math. Statist. Prob.*, University of California Press (ed. J. Neyman and L. LeCam), **1,** 203–220.

______ (1968). "On the centering of a simple linear rank statistics." *Inst. Statist. Univ. North Carolina, Mimeo Ser. No. 585.*

______ and Rosenblatt, J. R. (1955). "The efficiency of tests." *AMS* **26,** 52–63.

Hollander, M. (1967). "Rank tests for randomized blocks when the alternatives have a priori ordering." *AMS* **38,** 867–877.

Hotelling, H., and Pabst, Margaret R. (1936). "Rank Correlation and Tests of significance involving no assumption of normality." *AMS* **7,** 29–43.

Hoyland, A. (1965). "Robustness of the Hodges-Lehmann estimates for shift." *AMS* **36,** 174–197.

______ (1968). "Robustness of the Wilcoxon estimate of location against a certain dependence." *AMS*, **39,** 1196–1201.

Huber, P. J. (1964). "Robust estimation of a location parameter." *AMS* **35,** 73–101.

Hušková, M. (1970). "Asymptotic distribution of simple linear rank statistics for testing symmetry." *Zeit. fur Wahrsch verw. Geb.* **12,** 308–322.

James, G. S. (1952). "Notes on a theorem of Cochran." *PCPS. Math. Phys. Sci.* **48,** 443–446.

Joffe, A. and Klotz, J. (1962). "Null distribution and Bahadur efficiency of the Hodges bivariate sign test." *AMS* **33,** 803–807.

Jonckheere, A. R. (1954a). "A test of significance for the relation between *m* rankings and *k* ranked categories. *Brit. J. Statist. Psych.* **7,** 93–100.

______ (1954b). "A distribution-free *k* sample test against ordered alternatives." *BK* **41,** 135–145.

Jurečková, J. (1967). "Nonparametric estimate of regression coefficients." Ph.D. Thesis. *Mathematical Institute*, Prague.

______ (1969). "Asymptotic linearity of a rank statistic in regression parameter." *AMS* **40,** 1889–1900.

Kac, M., Kiefer, J., and Wolfowitz, J. (1955). On tests of normality and other tests of goodness of fit based on distance methods." *AMS* **26,** 189–211.

Kempthorne, O. (1952). *The Design and Analysis of Experiments.* John Wiley, New York.

Kendall, M. G. (1938). "A new measure of rank correlation." *BK* **30,** 81–93.

______ (1955). *Rank Correlation Methods* (2nd Ed.). Hafner, New York.

______ and Stuart, A. (1961). *The Advanced Theory of Statistics.* Vol. 2. Griffen & Co. London.

Kiefer, J. (1959). "*K*-sample analogues of the Kolmogorov-Smirnov and Cramér-von Mises tests." *AMS* **30,** 420–447.

Klotz, J. (1962). "Nonparametric tests for scale." *AMS* **33,** 498–512.

______ (1963). "Small sample power and efficiency for the one sample Wilcoxon and normal scores tests." *AMS* **34,** 624–632.

______ (1964). "Small sample power of the bivariate sign tests of Blumen and Hodges." *AMS* **35,** 1576–1582.

______ (1965). "Alternative efficiencies for signed rank tests." *AMS* **36,** 1759–1766.

Koch, G. G. and Sen, P. K. (1968). "Some aspects of the statistical analysis of the 'mixed model'." *BC* **24,** 27–48.

Kolmogorov, A. N. (1928). "Uber die Summen durch den Zufall bestimmter unabhangiger Grosse," *Math. Annalen* **99,** 309–319.

______ (1930). "Sur la loi forte des grandes nombres." *Comp. Rend. Acad. Sc. (Paris).* **191,** 910.

______ (1933). "Sulla determinazione empirica di una legge di distribuzione." *Giorn. dell'Instituto Ital. degli Attuari* **4,** 83–91.

______ (1950). *Foundations of the Theory of Probability* (German edition, 1933). Chelsea, New York.

Konijn, H. S. (1956). "On the power of certain tests of independence in bivariate populations." *AMS* **27,** 300–323.

Kounias E. G. and Weng, T. S. (1969). "An inequality and almost sure convergence." *AMS* **40,** 1091–1093.

Krickeberg, K. (1965). *Probability Theory.* Addison-Wesley, Reading, Mass.

Kruskal, W. H. (1952). "A nonparametric test for the several sample problem." *AMS* **23,** 525–540.

______ (1958). "Ordinal measures of association." *JASA* **53,** 814–861.

______ and Wallis, W. A. (1952). "Use of ranks in one-criterion variance analysis." *JASA* **47,** 583–621.

Kuiper, N. H. (1960). "Tests concerning random points on a circle." *Proc. Konink. Ned. Akad. von Wettenschaffen Ser. A*, **63,** 38–47.

Lal, D. N. (1955). "A note on a form of Tchebycheff's inequality for two or more variables." *Sankhya* **15,** 317–32.

Lamperti, John (1966). *Probability*. W. A. Benjamin, New York.

Lehmann, E. L. (1951a). "Consistency and unbiasedness of certain nonparametric tests." *AMS* **22,** 165–179.

——— (1951b). "A general concept of unbiasedness." *AMS* **22,** 587–592.

——— (1953). "The power of rank tests." *AMS* **24,** 23–43.

——— (1959). *Testing Statistical Hypotheses*. John Wiley, New York.

——— (1963a) "Robust estimation in analysis of variance." *AMS* **34,** 957–966.

——— (1963b) "Asymptotically nonparametric inference: An alternative approach to linear models." *AMS* **34,** 1499–1506.

——— (1963c). "Nonparametric confidence interval for a shift parameter." *AMS* **34,** 1507–1512.

——— (1964). "Asymptotically non-parametric inference in some linear models with one observation per cell." *AMS* **35,** 726–734.

——— (1966). "Some concepts of dependence." *AMS* **37,** 1137–1153.

——— and Stein, C. (1949). "On the theory of some nonparametric hypotheses." *AMS* **20,** 28–45.

Loève, M. (1963). *Probability Theory*. D. Van Nostrand, Princeton. (3rd ed.)

Madow, W. G. (1948). "On the limiting distributions of estimates based on samples from finite universes." *AMS* **19,** 535–545.

Mann, H. B. and Whitney, D. R. (1947). "On a test of whether one of two random variables is stochastically larger than the other." *AMS* **18,** 50–60.

Marshall, A. W. and Olkin, I. (1960). "Multivariate Chebyshev inequalities." *AMS* **31,** 1001–1014.

Massey, F. Jr. (1950). "A note on the estimation of a distribution function by confidence limits." *AMS* **21,** 116–119.

——— (1951a). "The distribution of the maximum deviation between two sample cumulative step functions." *AMS* **22,** 125–128.

——— (1951b). "The Kolmogorov-Smirnov test for goodness of fit." *JASA* **46,** 68–78.

——— (1951c). "A note on a two sample test." *AMS*, **22,** 304–306.

Mathisen, H. C. (1943). "A method of testing the hypothesis that two samples are from the same population." *AMS* **14,** 188–194.

Mehra, K. L. and Puri, M. L. (1967). "Multi-sample analogues of some one-sample tests." *AMS* **38,** 523–549.

——— and Sarangi, J. (1967). "Asymptotic efficiency of certain rank tests for comparative experiments." *AMS* **38,** 90–107.

——— and Sen, P. K. (1969). "On a class of conditionally distribution-free tests for interactions in factorial experiments." *AMS* **40,** 658–664.

Mikulski, P. W. (1963). "On the efficiency of optimal nonparametric procedures in the two sample case." *AMS* **34,** 22–32.

Miller, R. G. (1966). *Simultaneous Statistical Inference*. McGraw-Hill, New York.

von Mises, R. (1947). "On the asymptotic distribution of differentiable statistical functions." *AMS*, **18,** 309–348.

Mood, A. M. (1950). *Introduction to the Theory of Statistics*. McGraw-Hill, New York.

Mood, A. M. (1954). 'On the asymptotic efficiency of certain nonparametric two-sample tests." *AMS* **25,** 514–522.

Moran, P. A. P. (1968). *An Introduction To Probability Theory.* Clarendon Press, Oxford.

Moses, L. E. (1965). "Confidence limits from rank tests." *Technometrics* **7,** 257–260.

Motoo, M. (1957). "On the Hoeffding's combinatorial central limit theorem." *AISM* **8,** 145–154.

Nandi, H. K. and Sen, P. K. (1963). "On the properties of *U*-statistics when the observations are not independent. Part II. Unbiased estimation of the parameters of a finite population." *CSAB* **12,** 125–143.

Neveu, Jacques, (1965). *Mathematical Foundations of the Calculus of Probability.* Holden-Day, San Francisco.

Noether, G. E. (1949a). "Confidence limits in the nonparametric case." *JASA* **44,** 89–100.

______ (1949b). "On a theorem by Wald and Wolfowitz." *AMS* **20,** 455–458.

______ (1955). "On a theorem of Pitman." *AMS* **26,** 64–68.

Olkin, I. and Pratt, J. W. (1958). "A multivariate Chebyshev inequality." *AMS* **29,** 226–234.

Owen, D. B. (1962). *Handbook of Statistical Tables.* Addison-Wesley, Reading, Mass.

Page, E. B. (1963). "Ordered Hypothesis for multiple treatments; A significance test for linear ranks." *JASA* **58,** 216–230.

Parthasarathy, R. R. (1967). *Probability Measures on Metric Spaces.* Academic Press, New York.

Pearson, K. (1900). "On a criterion that a system of deviations form the probable in the case of a correlated system of variables is such that it can be reasonably supposed to have arisen in random sampling." *Phil. Mag.*, **50,** 157–175.

Pearson, E. S. and Hartley, H. O. (1954). *Biometrika Tables for Statisticians.* Cambridge University Press.

Pitman, E. J. G. (1937a). "Significance tests which may be applied to samples from any populations. I". *Suppl. JRSS* **4,** 119–130.

______ (1937b). "Significance tests which may be applied to samples from any populations. II. The correlation coefficient test." *Suppl. JRSS* **4,** 225–232.

______ (1938). "Significance Tests which may be applied to samples from any populations. III. The analysis of variance test." *BK* **29,** 322–335.

______ (1948). "Notes on nonparametric statistical inference." Columbia University, New York (mimeographed.)

Prohorov, Yu. V. (1956). "Convergence of random processes and limit theorems in probability theory." *TVP* **1,** 157–214.

Puri, M. L. (1964). "Asymptotic efficiency of a class of *c*-sample tests." *AMS* **35,** 102–121.

______ (1965a). "On some tests of homogeneity of variances." *AISM* **17,** 323–330.

Puri, M. L. (1965b). "Some distribution-free *k* sample rank tests of homogeneity against ordered alternatives." *Communications on Pure and Applied Mathematics.* **18,** 51–63.

——— (1965c). "On the combination of independent two sample tests of a general class." *Rev. Inst. Internat. Statist.*, **33,** 229–241.

——— (1965d). "On a class of testing procedures in linear models." *Jour. Indian Statist. Assoc.*, **3,** 1–8.

——— (1967a). "Combining independent one sample tests of significance." *AISM*, **19,** 285–300.

——— (1967b). "Multisample scale problem: unknown location parameters." *AISM* **20,** 99–106.

——— (1969). "The Van Elteren W-test and non-null hypothesis." *Review of the Internat. Statist. Institute*, **37,** 166–175.

——— (editor) (1970). *Nonparametric Techniques in Statistical Inference.* Cambridge Univ. Press.

——— and Sen, P. K. (1966). "On a class of multivariate multisample rank order tests." *S, A* **28,** 353–376.

——— and Sen, P. K. (1967a). "On robust estimation in incomplete block designs." *AMS* **38,** 1587–1591.

——— and Sen, P. K. (1967b). "On some optimum nonparametric procedures in two-way layouts." *JASA* **62,** 1214–1229.

——— and Sen, P. K. (1968a). "On Chernoff-Savage tests for ordered alternatives in randomized blocks." *AMS* **39,** 967–972.

——— and Sen, P. K. (1968b). "On a class of rank order estimates of contrasts in MANOVA." *S, A* **30,** 31–36.

——— and Sen, P. K. (1968c). "Nonparametric confidence regions for some multivariate location problems." *JASA* **63,** 1373–1378.

——— and Sen, P. K. (1969a). "A class of rank order tests for a general linear hypothesis." *AMS* **40,** 1325–1343.

——— and Sen, P. K. (1969b). "On the asymptotic normality of one sample rank order test statistics." *TVP* **14,** 167–172.

——— and Sen, P. K. (1969c). "Analysis of covariance based on general rank scores." *AMS* **40,** 610–618.

——— and Sen, P. K. (1969d). "On a class of rank order tests for the identity of two multiple regression surfaces." *Z. Wahrs. verw. Geb.* **12,** 1–8.

——— and Sen, P. K. (1970). "On the estimation of location parameters in the multivariate one sample and two sample problems." *Metrika* 58–73.

———, Sen, P. K., and Gokhale, D. V. (1970). "On a class of rank order tests for independence in multivariate distributions." *S, A*, **32,** in press.

——— and Shane, H. D. (1970). "Statistical inference in incomplete block designs." *Nonparametric tech. in Statist. inf.* Camb. Univ. Press. (ed: M. L. Puri) 131–153.

Putter, J. (1955). "The treatment of ties in some nonparametric tests." *AMS* **26,** 368–386.

Pyke, R. (1968). "The weak convergence of the empirical process with random sample size." *PCPS* **64,** 155–160.

——— (1970). "Asymptotic results for rank statistics." *Nonparametric Tech. in Statist. Inf.* Cambridge Univ. Press. (ed. M. L. Puri) 21–37.

Pyke, R. and Shorack, G. R. (1968a). "Weak convergence of a two-sample empirical process and a new approach to Chernoff-Savage theorems." *AMS* **39,** 755–771.
——— and Shorack, G. R. (1968b). "Weak convergence and a Chernoff-Savage theorem for random sample sizes." *AMS* **39,** 1675–1685.

Quade, D. (1966). "On the analysis of variance for the k-sample problem." *AMS* **37,** 1747–1758.
——— (1967). "Rank analysis of covariance." *JASA* **62,** 1187–1200.

Raghavachari, M. (1965a). "The two sample scale problem when the locations are unknown." *AMS* **36,** 1236–1242.
——— (1965b). "On the efficiency of normal scores test relative to the F-test." *AMS* **36,** 1306–1307.
Ramachandramurty, P. V. (1966). "On some nonparametric estimates for shift in the Behrens-Fisher situation." *AMS* **37,** 593–610.
Rao, C. R. (1965). *Linear Statistical Inference and Its Applications.* John Wiley, New York.
Rao, J. S. (1969). "Some Contributions to the Analysis of Circular Data." Ph.D. Thesis, Indian Statistical Institute.
Rosenblatt, M. (1952). "Limit theorems associated with variants of the von Mises statistic." *AMS* **23,** 617–623.
Roy, S. N. (1957). *Some aspects of multivariate analysis.* John Wiley, New York and Asia Pub. House, Bombay.

Sanov, I. N. (1957). "On the probability of large deviations of random variables." *Sel. Transl. Math. Statist. Prob.* **1,** 213–244.
Sarhan, A. E. and Greenberg, B. G. (eds). (1962). *Contributions to Order Statistics.* John Wiley, New York.
Savage, I. R. (1956). "Contributions to the theory of rank order statistics—the two-sample case." *AMS* **27,** 590–615.
——— (1962). *Bibliography of Nonparametric Statistics*, Harvard University, Cambridge, Mass.
Scheffé, H. (1943a). "On solutions of the Fisher-Behrens problem based on the t-distribution." *AMS* **14,** 35–44.
——— (1943b). "Statistical inference in the nonparametric case." *AMS* **14,** 305–332.
——— (1959). *The Analysis of Variance.* John Wiley, New York.
Sen, P. K. (1959). "On the moments of the sample quantiles." *CSAB*, **9,** 1–19.
——— (1960). "On some convergence properties of U-statistics." *CSAB* **10,** 1–18.
——— (1962a). "On the role of a class of quantile tests in some multisample nonparametric problems." *CSAB* **11,** 125–143.
——— (1962b). "On studentized nonparametric multisample location tests." *AISM* **14,** 114–131.
——— (1963a). "On weighted rank-sum tests for dispersion." *AISM* **15,** 117–135.
——— (1963b). "On the estimation of relative potency in dilution (-direct) assays by distribution-free methods." *BC* **19,** 532–552.

Sen, P. K. (1963c). "On the properties of *U*-statistics when the observations are not independent. Part one: Estimation of the non-serial parameters of a stationary process." *CSAB* **12,** 69–92.

——— (1964). "On some properties of the rank weighted means." *J. Indian Soc. Agri. Statist.* **16,** 51–61.

——— (1965a). "On some asymptotic properties of a class of nonparametric tests based on the number of rare exceedances." *AISM* **17,** 233–256.

——— (1965b). "On some permutation tests based on *U* statistics." *CSAB* **14,** 106–126.

——— (1965c). "Some nonparametric tests for *m*-dependent time series." *JASA* **60,** 134–147.

——— (1966a). "On a distribution-free method of estimating asymptotic efficiency of a class of nonparametric tests." *AMS* **37,** 1759–1770.

——— (1966b). "On nonparametric simultaneous confidence regions and tests for one criterion analysis of variance problem." *AISM* **18,** 319–336.

——— (1966c). "On a class of bivariate two-sample nonparametric tests." *Proc. Fifth Berkeley Symp. Math. Stat. Prob.* Univ. California Press (ed: J. Neyman and L. LeCam). **1,** 638–656.

——— (1966d). "Rank methods for combination of independent experiments in MANOVA. Part one: Two treatment case." *Essays in Probability and Statistics: S. N. Roy memorial vol.* (ed: R. C. Bose et al). Univ. N. Carolina Press, 631–654.

——— (1967a). "Asymptotically most powerful rank order tests for grouped data." *AMS* **38,** 1229–1239.

——— (1967b). "*U*-statistics and combination of independent estimates of regular functionals." *CSAB* **16,** 1–14.

——— (1967c). "On pooled estimation and testing heterogeneity of shift parameters by distribution-free methods." *CSAB* **16,** 139–152.

——— (1967d). "A note on asymptotically distribution-free confidence bounds for $P\{X < Y\}$ based on two independent samples." *S, A* **29,** 95–102.

——— (1967e). "On some permutation tests for stochastic independence, I." *S, A* **29,** 157–174.

——— (1967f). "On a class of multisample permutation tests based on *U*-statistics." *JASA* **62,** 1200–1213.

——— (1967g). "A note on the asymptotic efficiency of Friedman's χ_r^2-test." *BK* **54,** 677–679.

——— (1967h). "Nonparametric tests for multivariate interchangeability Part I. The problems of location and scale in bivariate distributions." *S, A* **29,** 351–371.

——— (1967i). "On some nonparametric generalizations of Wilks' tests for H_M H_{VC} and H_{MVC}, I." *AISM* **19,** 451–471.

——— (1968a). "On some permutation tests for stochastic independence, II." *S, A* **30,** 22–31.

——— (1968b). "Asymptotically efficient tests by the method of *n*-rankings." *JRSS, B* **30,** 312–317.

——— (1968c). "On a further robustness property of the test and estimator based on Wilcoxon's signed rank statistic." *AMS* **39,** 282–285.

Sen, P. K. (1968d). "On a class of aligned rank order tests in two-way layouts." *AMS* **39,** 1115–1124.

——— (1968e). "Robustness of some nonparametric procedures in linear models." *AMS* **39,** 1913–1922.

——— (1968f). "Estimates of the regression coefficient based on Kendall's tau." *JASA* **63,** 1379–1389.

——— (1969a). "Nonparametric tests for multivariate interchangeability. Part II. The problem of MANOVA in two-way layouts." *S, A* **30,** 145–156.

——— (1969b). "On a robustness property of a class of nonparametric tests based on *U*-statistics." *CSAB* **18,** 51–60.

——— (1969c). "On nonparametric *T*-method of multiple comparisons in randomized blocks." *AISM* **21,** 329–333.

——— (1969d). "On a class of aligned rank order tests for multiresponse experiments in some incomplete block design." *Inst. Statist. Univ. North Carolina Mimeo Report No.* 607. to appear in *AMS 42* (1971) No. 3.

——— (1970a). "On the distribution of the one-sample rank order statistics." *Nonparametric Tech. Statist. Inf.* Cambridge Univ. Press. (Ed: M. L. Puri) 53–72.

——— (1970b). "The Hájek-Rènyi inequality for sampling from a finite population." *S.A.* **32,** 181–188.

——— (1970c). "Asymptotic distribution of a class of multivariate rank order statistics." *CSAB*, **19**. 22–32.

——— (1970d). "On some convergence properties of one-sample rank order statistics." *AMS* **41,** 2206–2209.

——— (1970e). "Nonparametric inference in replicated 2^m factorial experiments." *AISM* **22,** 281–294.

——— (1970f). "On the robust-efficiency of the combination of independent nonparametric tests." *AISM*, **22,** 277–280.

——— (1971). "A further note on the asymptotic efficiency of Friedman's χ_r^2-test." *Metrika*, in press.

———, Bhattacharyya, B. B. and Suh, M. W. (1969). "On the limiting behaviour of the extremum of certain sample functions." *Inst. Statist. Univ. North Carolina Mimeo Report No. 628.*

——— and David, H. A. (1968). "Paired comparison for paired characteristics." *AMS* **39,** 200–208.

——— and Ghosh, M. (1971). "On bounded length sequential confidence intervals based on one-sample rank order statistics." *AMS*, **42,** in press.

——— and Govindarajulu, Z. (1966). "On a class of *c* sample weighted rank-sum tests for location and scale." *AISM* **18,** 87–105.

——— and Puri, M. L. (1967). "On the theory of rank order tests for location in the multivariate one sample problem." *AMS* **38,** 1216–1228.

——— and Puri, M. L. (1968). "On a class of multivariate multisample rank order tests. II. Tests for the homogeneity of dispersion matrices." *S, A* **30,** 1–22.

——— and Puri, M. L. (1969). "On robust nonparametric estimation in some multivariate linear models." *Multivariate Analysis II* (edited by P. R. Krishnaiah), Academic Press, New York, 33–52.

Sen, P. K. and Puri, M. L. (1970a). "Asymptotic theory of likelihood ratio and rank order tests in some multivariate linear models." *AMS* **41,** 87–100.

——— and Puri, M. L. (1970b). "On the robustness of rank order tests and estimates in the generalized multivariate one sample location problem." (Unpublished.)

Sethuraman, J. (1964). "On the probability of large deviations of families of sample means." *AMS*, **35,** 1304–1316.

——— and Sukhatme, B. V. (1958). "Joint asymptotic distribution of *U*-statistics and order statistics." *Sankhya*, **21,** 289–298.

Shane, H. D., and Puri, M. L. (1969). "Rank order tests for multivariate paired comparisons." *AMS*, **40,** 2101–2117.

Shorack, Galen R. (1967). "Testing against order alternatives in model I analysis of variance, normal theory and nonparametric." *AMS* **38,** 1740–1752.

Sibuya, M. (1959). "Bivariate extreme statistics." *AISM*, **11,** 195–210.

Siegel, S., and Tukey, J. (1960). "A nonparametric sum of ranks procedure for relative spread in unpaired samples." *JASA*, **55,** 429–444.

Skorohod, A. V. (1956). "Limit theorems for stochastic processes." *TVP* **1,** 261–290.

Smirnov, N. (1939a). "On the estimation of the discrepancy between empirical curves of distribution for two independent samples." *Bull. Math. Univ. Moscow* **2,** 3–16.

——— (1939b). "Sur les écarts de la courve de distribution empirique." *Rec. Math. N.S.* **6,** 3–26.

Sproule, R. N. (1969). "A sequential fixed-width confidence interval for the mean of a *U*-statistic." Doctoral Dissertation, University of North Carolina, Chapel Hill.

Steck, G. P. (1969). "The Smirnov two-sample tests as rank tests." *AMS* **40,** 1449–1466.

Stieger, W. L. (1969). "A best possible Kolmogoroff-type inequality for martingales and a characteristic property." *AMS* **40,** 764–769.

Sugiura, N. (1965). "Multisample and multivariate nonparametric tests based on *U*-statistics and their asymptotic efficiencies." *Osaka Jour. Math.* **2,** 385–426.

Sukhatme, B. V. (1957). "On certain two-sample nonparametric tests for variances." *AMS* **28,** 188–194.

——— (1958a). "Testing the hypothesis that two populations differ only in location." *AMS* **29,** 60–78.

——— (1958b). "A two-sample distribution-free test for comparing variances." *BK* **45,** 544–549.

Sverdrup, E. (1952). "The limit distribution of a continuous function of random variables." *Skandinavisk Aktuarietidskrift.* **35,** 1–10.

Tamura, R. (1960). "On the nonparametric tests based on certain *U*-statistics." *Bull. Math. Statist. (Japan)* **9,** 61–68.

——— (1963). "On a modification of certain rank tests." *AMS* **34,** 1101–1103.

——— (1966). "Multivariate nonparametric several sample tests." *AMS* **37,** 611–618.

Terry, M. E. (1952). "Some rank order tests which are most powerful against specific parametric alternatives." *AMS* **23,** 346–366.

Theil, H. (1950). "A rank invariant method of linear and polynomial regression analysis." *Indag. Math. 12, Fasc.* **2,** 85–91, 173–177.

Tucker, Howard G. (1967). *A graduate course in probability*. Academic Press, New York.

Tukey, J. W. (1947). "Nonparametric estimations, II. Statistically equivalent blocks and tolerance regions—the continuous case." *AMS* **18,** 529–539.

——— (1948). "Nonparametric estimation, III. Statistically equivalent blocks and multivariate tolerance regions—the discontinuous case." *AMS* **19,** 30–39.

——— (1949). "The simplest signed rank tests." *Princeton University Stat. Res. Group, Memo. Report No. 17*, July 1949.

——— (1960). "A survey of sampling from contaminated distributions. In "Contributions to Probability and Statistics" (I. Olkin et al., eds). Stanford University Press, Stanford, Calif.

Varadarajan, V. S. (1958). "A useful convergence theorem." *Sankhya* **20,** 221–222.

van der Waerden, B. L. (1952/53). "Order tests for the two-sample problem and their power." *Proc. Koninklijke Nederlandse Akad. Van Wetenschoppen Ser. A.* **55,** p. 453.

——— (1953a). "Ein neuer test fur das Problem der zwei Stichproben." *Math. Annalen* **126,** 93–107.

——— (1953b). "Testing a distribution function." *Proc. Koninklijke Nederlandse Akad. van Wetenschoppen, Ser A.* **56,** 201–207.

Wald, A. (1943a). "An extension of Wilks' method for setting tolerance limits." *AMS* **14,** 45–55.

——— (1943b). "Tests of statistical hypotheses concerning several parameters when the number of observations is large." *TAMS* **54,** 426–482.

——— and Brookner, R. J. (1941). "On the distribution of Wilks' statistic for testing the independence of several groups of variates." *AMS* **12,** 137–152.

——— and Wolfowitz, J. (1939). "Confidence limits for continuous distribution functions." *AMS* **10,** 105–118.

——— and Wolfowitz, J. (1940). "On a test whether two samples are from the same population." *AMS* **11,** 147–162.

——— and Wolfowitz, J. (1941). "Note on confidence limits for continuous distributions functions." *AMS* **12,** 118–119.

——— and Wolfowitz, J. (1944). "Statistical tests based on permutations of the observations." *AMS* **15,** 358–372.

Walsh, J. E. (1946). "On the power function of the sign test for slippage of means." *AMS* **17,** 358–362.

——— (1949). "Some significance tests for the median which are valid under very general conditions." *AMS* **20,** 64–81.

——— (1957). "Nonparametric mean estimation of percentage points and density function values." *AISM* **8,** 167–80.

Watson, G. (1957). "The chi-square goodness-of-fit test for normal distributions." *BK* **44,** 336–348.

______ (1958). "On chi-square goodness-of-fit tests for continuous distributions." *JRSS, B* **20,** 44–61.

Welch, B. L. (1937). "The significance of the difference between two means when the population variances one unequal." *BK* **29,** 350–362.

Wilcoxon, F. (1945). "Individual comparisons by ranking methods." *BC* **1,** 80–83.

______ (1949). "Some rapid approximate statistical procedures." American Cyanamid Company, New York (pamphlet).

Wilks, S. S. (1935). "On the independence of k sets of normally distributed statistical variables." *Econometrika* **3,** 309–326.

______ (1941). "Determination of sample sizes for setting tolerance limits." *AMS* **12,** 91–96.

______ (1961). "A combinatorial test for the problem of two samples from continuous distributions." *Proc. 4th Berkeley Symp. Math. Statist. Prob.* **1,** Univ. of California Press (ed: J. Neyman). 707–718.

______ (1962). *Mathematical Statistics.* John Wiley, New York.

Wolfowitz, J. (1946). "On sequential binomial estimation·" *AMS* **17,** 489–492.

Author Index

Abrahamson, I. G., 51, 123, 143, 415
Adichie, J. N., 415
Ali, M. M., 12, 48, 194, 415
Anderson, R. L., 415
Anderson, T. W., 1, 7, 39, 48, 180, 286, 342, 344, 351, 360, 362, 377, 392, 415
Andrews, F. C., 135, 140, 143, 416
Ansari, A. R., 94, 133, 140, 143, 416

Bahadur, R. R., 3, 122, 123, 124, 142, 143, 416
Bancroft, T. A., 342, 416
Barton, D. E., 51, 133, 143, 416
Bell, C. B., 416
Bellman, R., 369, 376, 416
Benard, A., 277, 342, 416
Benett, B. M., 148, 180, 416
Berk, R. H., 60, 143, 416
Bernstein, S. N., 24, 48, 416
Berry, A. C., 24, 56, 270, 335
Bhapkar, V. P., 50, 143, 416
Bhattacharyya, B. B., 39, 48, 429
Bhattacharyya, G. K., 179, 180, 220, 426
Bhattacharyya, R. N., 25, 48, 416, 417
Bhuchongkul, S., 50, 70, 78, 91, 143, 265, 348, 372, 375, 376, 380, 399, 417
Bickel, P. J., 179, 180, 213, 233, 265, 417
Billigsley, P., 32, 48, 417
Birnbaum, Z. W., 11, 12, 13, 39, 42, 48, 417
Blom, G., 409, 417
Blomqvist, N., 376, 380, 400, 417
Blum, J. R., 376, 417
Blumen, I., 148, 180, 417
Bradley, R. A., 5, 7, 94, 133, 140, 143, 277, 342, 417, 419, 428
Brookner, R. J., 431
Brown, G. W., 3, 4, 7, 135, 187, 268, 342, 418

Capon, J., 108, 133, 138, 139, 142, 143, 418
Chacko, V. J., 418
Chan, G. K., 12, 48, 194, 415
Chanda, K. C., 128, 143, 418
Chapman, D. G., 418
Chatterjee, S. K., 130, 143, 144, 148, 178, 180, 213, 220, 369, 418
Chebyshev, P. L., 10, 11, 17, 57, 169, 243, 352, 364
Chernoff, H., 3, 7, 50, 70, 78, 86, 91, 92, 93, 95, 116, 117, 123, 131, 132, 133, 140, 142, 143, 288, 400, 414, 418
Chow, Y. S., 31, 48, 418
Chung, J. H., 420
Cochran, W. G., 26, 48, 266, 342, 418
Courant, R., 121, 122, 368
Cox, G. M., 266, 342, 418
Craig, A. T., 26, 48, 418
Cramér, H., 20, 50, 51, 143, 419
Crouse, C. F., 419

Daly, J. F., 421
Dantzig, Ph. Van, 139, 143, 419, 429
Darling, D. A., 39, 48, 51, 143, 415, 419
David, F. N., 133, 143, 416
David, H. A., 5, 7, 429
Davidson, R. R., 5, 7, 419
Deshpande, J. V., 419
Doksum, 328, 342, 416, 419
Doob, J. L., 8, 29, 31, 34, 35, 36, 48, 419

Dupac, V., 419
Durbin, J., 277, 342, 419
Dwass, M., 49, 50, 54, 70, 76, 78, 86, 90, 91, 92, 130, 143, 419

Van Eeden, Ph., 118, 142, 143, 419
Van Elteren, Ph., 277, 341, 342, 416, 419
Esseen, C. G., 24, 56, 99, 270, 335

Feller, W., 8, 23, 29, 48, 420
Ferrar, W. L., 420
Fillippova, A. A., 53, 143, 420
Fisher, R. A., 3, 50, 69, 84, 86, 94, 103, 111, 143, 374, 420
Fraser, D. A. S., 50, 51, 55, 65, 70, 133, 143
Friedman, M., 4, 7, 268, 326, 342, 420

Gabriel, K. R., 420
Gastwirth, J. L., 118, 143, 221, 265, 420
Gerig, T. M., 286, 341, 342, 420
Ghosh, M., 6, 7, 70, 75, 143, 420, 429
Gleser, L. J., 123, 143, 420
Gnedenko, B. V., 8, 25, 39, 48, 420
Gokhale, D. V., 376, 399, 426
Gorindarajulu, Z., 50, 78, 91, 93, 138, 143, 414, 421, 429
Greenberg, B. G., 221, 265, 427

Hájek, J., 2, 3, 6, 7, 12, 29, 33, 39, 48, 50, 70, 71, 72, 74, 75, 76, 93, 118, 130, 143, 373, 375, 376, 421
Hall, R. A., 39, 48, 417
Haller, H. S., 416
Halmos, P. R., 43, 48, 50, 51, 55, 143, 421
Hannan, E. J., 421
Hartley, H. O., 425
Herr, D. G., 421
Hoadley, A. B., 124, 143, 421
Hodges, J. L. Jr., 50, 117, 123, 139, 140, 143, 148, 180, 222, 228, 265, 342, 421, 422
Hoeffding, W., 2, 7, 16, 48, 50, 53, 56, 57, 60, 61, 69, 70, 72, 73, 75, 76, 77, 78, 79, 81, 86, 87, 88, 90, 91, 92, 93, 94, 110, 111, 113, 124, 125, 130, 138, 143, 376, 410, 422
Hollander, M., 328, 342, 422
Hotelling, H., 85, 143, 376, 422
Hoyland, A., 265, 422
Huber, P. J., 221, 265, 422
Husková, M., 6, 7, 93, 103, 143, 163, 180, 422

James, G. S., 26, 48, 422
Jaffe, A., 148, 180, 422
Jonekheere, A. R., 324, 342, 422
Jureckova, J., 6, 7, 93, 143, 423

Kac, M., 423
Kempthorne, O., 266, 342, 423
Kendall, M. G., 50, 55, 86, 143, 266, 324, 342, 375, 376, 423
Kiefer, 39, 48, 376, 417, 423
Klose, O. M., 417
Klotz, J., 50, 70, 78, 133, 141, 143, 148, 180, 422
Koch, G. G., 423
Kolmogorv, A. N., 8, 12, 15, 16, 18, 31, 33, 38, 48, 50, 143, 421
Konijn, H. S., 348, 376, 423
Korolink, V. S., 39, 48, 421
Kounias, E. G., 31, 48, 423
Krickeberg, K., 8, 48, 423
Kruskal, W. H., 3, 7, 50, 135, 139, 143, 213, 376, 423
Kuiper, N. H., 50, 51, 143, 423

Lal, D. N., 11, 48, 423
Lamperti, J., 424
Le Cam, L., 78, 91, 93, 143, 414, 421
Lehmann, E. L., 50, 52, 54, 63, 64, 69, 86, 91, 110, 117, 123, 128, 134, 135, 139, 140, 143, 222, 228, 265, 342, 376, 421, 422, 424
Liapounoff, A., 12, 23, 274, 295
Loève, M., 8, 9, 10, 14, 29, 31, 48, 401, 424

McCarty, R. C., 417
Madow, W. G., 70, 143, 424
Mallows, C. L., 51, 143, 416
Mann, H. B., 3, 7, 50, 55, 84, 86, 111, 112, 131, 143, 424
Marshall, A. W., 11, 12, 13, 39, 417, 424
Massey, F., Jr., 39, 48, 127, 143, 424
Mathieson, H. C., 132, 143, 424
Mehra, K. L., 50, 143, 342, 424
Milulski, P. W., 117, 143, 424
Miller, R. G., 328, 424
Von Mises, R., 50, 51, 53, 143, 346, 379, 424
Mood, A. M., 3, 4, 7, 50, 94, 127, 130, 133, 135, 136, 140, 143, 187, 268, 342, 418, 424, 425, 426, 427

Moran, P. A. P., 425
Moses, L. E., 265, 425
Motoo, M., 50, 70, 72, 76, 130, 143

Nandi, H. K., 50, 63, 79, 80, 143, 383, 399, 425
Neveu, J., 425
Noether, G. E., 3, 50, 70, 71, 72, 73, 113, 121, 143, 277, 341, 353, 420, 425, 427

Olkin, I., 11, 48, 424, 425, 427
Owen, B., 131, 143, 425

Pabst, Margaret R., 85, 86, 143, 376, 422
Page, E. B., 324, 342, 425
Parthasarathy, R., 32, 33, 36, 48, 425
Pearson, E. S., 425
Pearson, Karl, 50, 143, 425
Pitman, E. J. G., 3, 50, 70, 83, 85, 86, 90, 91, 113, 121, 123, 143, 335, 427
Pratt, J. W., 11, 425
Prohorov, Yu. V., 31, 48, 425
Puri, M. L., 5, 6, 7, 50, 70, 78, 86, 91, 93, 111, 135, 136, 137, 139, 140, 141, 143, 163, 180, 220, 222, 223, 224, 240, 265, 328, 342, 376, 399, 417, 425, 426, 429, 430
Putter, J., 84, 126, 143, 426
Pyke, R., 3, 7, 39, 42, 48, 93, 103, 143, 414, 426

Quade, D., 3, 7, 220, 427

Raghavachari, M., 78, 91, 93, 110, 134, 135, 141, 143, 393, 394, 395, 399, 400, 414, 421, 427
Ramachandramurty, P. V., 261, 265, 427
Ranga Rao, R., 124, 143, 416
Rao, C. R., 1, 7, 10, 27, 28, 48, 55, 123, 143, 427
Rao, J. S., 36, 48, 427
Raymond, J., 11, 48, 417
Rènyi, A., 12, 421
Rosenblatt, Joan, 113, 143, 422
Rosenblatt, M., 376, 417, 427
Roy, S. N., 1, 7, 427
Rubin, H., 417

Sanov, I. N., 124, 143, 427
Sarangi, J., 342, 424
Sarhan, A. E., 221, 265, 427
Savage, I. R., 3, 7, 50, 70, 78, 86, 91, 92, 93, 95, 110, 116, 117, 131, 132, 133, 140, 142, 143, 288, 400, 414, 418, 427
Scheffé, H., 22, 50, 68, 69, 130, 143, 215, 216, 220, 251, 265, 328, 427
Sen, P. K., 5, 6, 7, 39, 42, 48, 50, 59, 60, 63, 70, 71, 78, 79, 80, 81, 82, 83, 84, 86, 91, 92, 93, 103, 124, 125, 126, 127, 128, 129, 130, 131, 135, 138, 139, 140, 143, 163, 180, 213, 220, 222, 223, 224, 228, 240, 265, 277, 300, 308, 328, 341, 342, 369, 376, 383, 399, 407, 408, 409, 414, 418, 420, 424, 426, 427, 428, 429, 430, 432
Sethuraman, J., 50, 124, 143, 430
Shane, H. D., 5, 7, 426, 430, 432
Shorack, G. R., 3, 7, 42, 93, 103, 143, 414, 427, 429, 430, 432
Sibuya, M., 347, 366, 376, 430
Siegel, S., 133, 143, 430
Sidák, Z., 2, 33, 39, 48, 93, 143, 373, 375, 421
Skorohod, A. V., 31, 36, 48, 430
Smirnov, N., 31, 38, 39, 48, 430
Spearman, C., 125, 324, 346, 366, 373, 376, 380, 393
Sproule, R. N., 60, 143, 432
Steck, G. P., 432
Stein, C., 50, 69, 143, 424
Stieger, W. G., 16, 31, 48, 430
Stuart, A., 375, 423
Student, 213, 316, 321, 326
Sugiura, N., 430
Suh, M. W., 39, 48, 429
Sukhatme, B. V., 50, 91, 128, 129, 134, 141, 143, 393, 430
Sverdrup, E., 21, 430

Tamura, R., 50, 129, 134, 143, 220, 430
Terry, M. E., 50, 78, 86, 94, 110, 111, 143, 431
Theil, H., 431
Tingey, F. H., 39, 42, 417
Tucker, H. G., 8, 48, 431
Tukey, J. W., 51, 129, 132, 133, 143, 221, 265, 328, 430, 431

Varadarajan, V. S., 431
Van der Waerden, 50, 94, 132, 143, 374, 431

Wald, A., 25, 39, 48, 50, 51, 70, 71, 73, 74, 128, 130, 143, 353, 392, 431
Wallis, W. A., 3, 7, 50, 135, 139, 143, 213, 423
Walsh, J. E., 123, 129, 143, 431
Watson, G., 432
Welch, B. L., 50, 70, 90, 143, 432
Weng, T. S., 12, 18, 41, 48
Whitney, D. R., 3, 7, 50, 55, 84, 86, 111, 112, 131, 143, 424
Wilcoxon, F., 3, 7, 50, 55, 84, 85, 86, 94, 111, 112, 129, 132, 138, 139, 140, 143, 213, 326, 337, 342, 432
Wilks, S. S., 39, 48, 51, 76, 128, 143, 230, 257, 432
Wolff, S. S., 118, 143, 420
Wolfowitz, J., 25, 39, 48, 50, 70, 71, 73, 74, 128, 130, 139, 143, 353, 423, 431, 432

Yates, F., 3, 50, 94, 103, 111, 143, 374, 420

Zuckerman, H. S., 11, 48, 417

Subject Index

Additive, finitely, 44; σ, 44
ANOVA 3, estimation of contrasts in, 237; interval estimation in, 243
a.s. bounded, 14
Association (measures of), 4, 345
Asymptotically distribution, free procedures based on robust estimators, 308
Asymptotically efficient test, 142
Asymptotic distribution theory of, bivariate sign statistic, 148; Chernoff-Savage statistic, 95; multisample rank order statistic, 104; multisample multivariate order statistic, 196; multivariate one-sample statistic, 157; one-sample rank order statistic, 101; point estimates, 226, 229, 235, 238, 313, 319, 322; rank statistics for equality of dispersion matrices, 387; rank statistics for independence, 355; rank statistics in two-way layouts, 270, 282, 293, 305; U-statistics, 56, 61, 65; Von Mises' statistics, 57
Asymptotic relative efficiency (A.R.E.), 3, 4, 112, 123; A.R.E. and the correlation coefficient, 118; Bahadur efficiency, 3, 122, 123 Hodges-Lehmann A.R.E., 3, 123; non-normal case, 119; Pitman-Noether A.R.E., 113, 121, 130, 131
A.R.E. of tests: Ansari-Bradley test, 140; Bartlett's likelihood ratio test, 140, 141; Chernoff-Savage tests, 116, 123; c sample tests, 139, 140, 141; Kruskal-Wallis test, 139; Mood's tests, 140; normal scores tests, 139, 140, 141; one-way ANOVA tests, 139; sign test, 139; Sukhatme's test, 141; signed-rank (Wilcoxon) test, 139; rank-sum (Wilcoxon) test, 139; tests for independence, 365; multivariate multisample tests, 211; multivariate one-sample tests, 172; point estimates, 230, 314, 320, 323; interval estimates, 243, 251; tests in factorial layouts, 272, 275, 286, 297, 326, 336, 339
Attributes, 5

Borel Field, 8, 9
Bound, Kolmogorov-Smirnov, 39, 41
Brownian Bridge, 34, 35, 36
Brownian Movement process, 33, 35

Central limit theorem, 22; Berry-Esseen, 23, 24, 56, 99, 335; Liapounoff, 23, 156, 274, 295; Lindeberg-Feller, 23, 56; Lindeberg-Levy, 22, 24, 163; multivariate, 24, 25; permutational, 70
Chebyshev theorem, 10, 17
Concordant, 52, 124; probability of concordance, 125, 145
Confidence, bands, 3
Continuity theorem on characteristic functions, 21
Continuous, absolutely continuous distribution function, 9; from below, 44; from above, 44
Contrasts, 3; in ANOVA, 236, 244, 308, 310; in MANOVA, 254, 321
Convergence, almost sure, 13; of characteristic functions, 19; of density, 24; of distribution function, 18; of moments, 19; in probability, 13, 15; in r^{th} mean, 13, 15; of sequence of probability measures, 33; theorems, 18;

weak convergence of stochastic processes, 31
Correlation coefficient, 118; grade, 125, 380; Kendall's rank, 55; Spearman's rank, 125, 380
Courant theorem, 122, 174
Covariance: analysis of, 3, 4; matrix, 10; rank order tests for analysis of, 215, 338

Decomposition theorem, 9
Density, 9; convergence of, 25
Dependence, model for, 361
Designs, two-factor, 3, 226; incomplete block, 5, 318
Discordant, 124; probability of discordance, 125, 145
Dispersion, homogeneity of dispersion matrices, 4; matrix, 4
Distance function, 64
Distribution, diagonally symmetric, 149, 164; discrete, 10; function (cumulative), 9; empirical, 36, 52, 152, 188, 289, 346, 379; marginal, 10; multivariate normal, 27; of quadratic forms, 25; symmetric, 102
Distribution-free, 2, 3; asymptotically, 4; conditionally, 4; permutationally, 3, 4
Dominated convergence theorem, 47

Efficacy, *see* A.R.E.
Elementary events, 43
Estimation-interval, of location, 240; of difference of locations, 241; in ANOVA, 243; in MANOVA, 254; in two-way layout, 316; in incomplete block designs, 318
Estimation-point, of location, 222; of difference of locations, 227, 234; of contrasts in ANOVA, 237; in two-way layout, 308; compatible estimators, 237, 310; in incomplete block designs, 318
Estimators, m.v.u., 55, 125; minimum risk estimator, 55; rank order, *see* Estimation
Expected value, 10

Fatou-Lebesque theorem, 47
Field, σ-field, 43
Frechet-Shohat theorem, 19
Fubini's theorem, 47
Functional, regular, 51, 65, 167; degree of, 52, 206; degree vector, 64; kernel of, 52, 64, 65, 80, 206; symmetric kernel of, 52, 167; stationary of order d, 54; grade, 378; dispersion matrix of grade, 379; distribution theory of regular functionals, 49, 50, 51

General linear hypotheses, 6
Gini's mean difference, 52
Glivenko-Cantelli theorem, 22, 31
Group of (invariant) transformations, 67, 68, 149, 183, 268, 288, 332, 349, 381

Helly-Bray theorem, 20
Higher order terms, two-sample case, 401; one sample case, 405; aligned rank statistics,407
Hypothesis, of invariance, 2; of (sign) invariance, 3, 67; of (sign) invariance (multivariate), 148; of no treatment effects (in ANOVA and MANOVA), 3; of matching invariance, 67; of invariance under partitioning into subsets, 67; of interchangeability, 68; of invariance of distribution, 68

Incomplete block designs, 4, 5, 266, 318
Independence, testing, 4; sums of independent random variables, 15; tests of (two sets of variates), 358; pairwise, 360; parametric tests for, 362
Inequalities, Ali-Chan, 12; Berge, 11; Birnbaum and Marshall, 12; Cantelli, 11; Chebyshev, 11, 15; Hájek-Rènyi, 12, Holder, 11; Jensen, 12; Kolmogorov, 12, 15, 31; Liapounoff, 12; Minkowski, 11; multivariate Chebyshev, 11; Schwarz, 11
Integral (integrable), 46
Interaction, 4; rank order tests for, 331
Intra-block comparisons, 4; intra-block transformations, 4

Kolmogorov theorem, 15; (3 series) criterion, 16, 147

Laws of large numbers, 17; Bernoulli, 17; Borel Strong, 17, 22; Khinchine, 18; Kolmogorov Strong, 18; Kolmogorov sufficient condition for, 18
Lemma, Birnbaum-Marshal, 12; Borel-Cantelli, 16; Helly-Bray, 19; (extended) Helly-Bray, 19
Levy-cramer theorem, 19
Location, problem of, 3; estimation, 222, 227; identity of, 3

MANOVA, 3; interval estimation in, 254; nonparametric tests for, 301; test for (based on n-rankings), 277
Martingales, 29; reverse, 29; sub, 29; reverse sub, 29; closed martingale, 30; martingale reversed sequence, 39; submartingale reversed sequence, 31
Measure, 8, 44, 45; inner and outer, 45; probability, 8; Lebesque-Stieltjes, 9, 10; Wienir, 33; some results on, 43; measurable Lebesque, 45; function, 46; mapping, 45
Median, procedure, 4
Metric, space, 31
Models for, ANOVA, MANOVA, 3, 4; dependence, 361; linear regression, 3, 5; paired comparisons, 5
Moments, 4
Monotone, sequence, 44; class theorem, 44; convergence theorem, 46
Multiple comparisons, one-way ANOVA, 244; two-way ANOVA, 328
Multivariate analysis, a brief review of (nonparametric), 4

n-ranking, 3, 4, 267; Friedman's χ_r^2 test, 268; Brown and Mood's test, 268; a general class of tests, 269; A.R.E., 272; MANOVA, 277

Parameters, nuisance, 4; preference, 25; nonserial, 126
Permutation, basic principles, 67, 145, 151, 167, 183, 207, 268, 278, 288, 332, 347, 381; permutational limit theorems, 70, 73, 75, 76, 80, 82, 147, 152, 168, 187, 280, 289, 303, 334, 339, 353, 354, 385; tests, 1, 78, 79, 86, 172, 174, 175, 184, 215, 300, 347, 381, 398; statistic, 70, 73, 75, 79, 150, 151, 186, 279, 289, 334, 351, 384, 399
Power, asymptotic power of non-parametric tests, 86; equivalence, 88, 130
Problem, of randomness in a series of observations, 68; of sign invariance, 76; multisample scale, 136; multisample multivariate location and scale, 182; multivariate single sample location, 144; multivariate independence, 343; equality of multivariate dispersions, 378
Process, seperable, 13; realization of, 32; empirical, 36; m-dependent stationary, 63; symmetric dependent process, 63

Radon-Nikodym theorem, 47
Random variable, 9; random vector, 9; a.s. bounded random variable, 14
Ranking after alignment, 286; tests for, 289; permutation distribution theory, 291; A.R.E. 296; analysis of covariance, 338; several observations per cell, 300; MANOVA, 301; multiple comparisons, 329; interactions, 331; ordered alternatives, 326
Regression, models, 5; problems for grouped data, 6; problems with stochastic predictors, 6
Robust, estimation, 3, 5, 222, 227, 234, 237, 240, 244, 255, 308, 310, 318, 321; test, 5
Run, 128

Scheffe's theorem, 22
Simple function, 46
Space, complete separable metric space $\mathscr{Z}$, 32; existence of measure on $\mathscr{Z}$, 32; convergence of measure on $\mathscr{Z}$, 34; equivalence of two measures on $\mathscr{Z}$, 34; measure space, 44; L_r, 14; metric, 32; probability, 8, 32
Statistic, bilinear permutation, 70, 75, 79; Brown Mood, 268; Cramer-Von Mises, 51; Friedman, 268; Hotelling T_n^2, 173; Kolmogorov-Smirnov, 50; linear permutation, 70, 73, 79; rank order, 2, 3, 6, 49, 50, 70, 92, 93, 100, 118, 133, 134, 151, 184, 186, 289, 303, 333, 351, 382, 399; U-statistics, 2, 49, 53, 55, 56, 60, 63, 65, 71, 79, 81, 83, 91, 92, 124, 125, 126, 167, 207, 219; Wilcoxon-Mann-Whitney, 55
Stochastic, vector, 1; matrices, 5; predictors, 6; processes (weak convergence of), 31
S(α) structure of texts, 68
Sverdrup theorem, 21

Test, Ansari-Bradley, 94, 140; asymptotically efficient, 142; Bartlett's likelihood ratio test, 140; Brown and Mood's median, 135, 140; chi-square, 5, 6; component randomization, 76; Cramer-Von Mises, 50; Fisher-Yates', 94; Friedman, 268; function, 68; for equality of dispersion matrices, 381, 393, 398; goodness of fit, 50; Hotelling and Pabst, 86; Hotelling T^2, 3; likelihood ratio, 3; for independence of two sets of variables, 347, 358; for interaction,

331; permutation rank order, 184, 347, 358, 381; parametric (for independence), 362; Kruskal-Wallis, 50, 135; Kuiper, 51; location: one sample, 132; two sample, 131; multi-sample, 135; locally most powerful, 42, 51, 108, 110; median, 127, 132, 135, 186; Mood's two sample test, 94, 133; for MANOVA, 187, 277, 301; normal scores test: for location, 3, 50, 103, 117, 132, 133, 136, 139, 140, 176, 187; for scale, 141; for ordered alternatives, 323; Pitman's, 85, 91; permutation, 1, 49, 50, 68, 69, 70, 86; Type 1, 86, 87; Type II,86; Type III, 86; rank permutation, 78, 79; *see also* Permutation; Quantile, 132, 140; rank order, 3, 4, 6, 7, 8, 140, 215, 300, 323, 338; rank sum, 83, 84, 111, 117, 118, 131, 138, 139, 140, 187; rank permutation, 78, 79; run, 128, 139; scale: location known, 133; location unknown, 134; multisample, 136, 137; simplest rank order, 133; t, 84, 91, 116, 117, 139, 140; sign, 3, 103, 127, 132, 142, 144, 174; signed-rank, 50, 85, 103, 127, 132, 135, 139, 172, 175; Sukhatme's scale, 129; based on U-statistics, 167, 206; Wilcoxon signed rank, *see* Signed rank; Wilcoxon-Mann-Whitney, *see* Rank sum; van der Warden, 49, 94, 132

Truncation, 16

Unimodal, 12
Uniqueness theorem on moments, 19
U-shaped (distribution function), 12

Weak compactness theorem, 19